Understanding the behaviour and improving the welfare of pigs

It is widely recognised that agriculture is a significant contributor to global warming and climate change. Agriculture needs to reduce its environmental impact and adapt to current climate change whilst still feeding a growing population, i.e. become more 'climate-smart'. Burleigh Dodds Science Publishing is playing its part in achieving this by bringing together key research on making the production of the world's most important crops and livestock products more sustainable.

Based on extensive research, our publications specifically target the challenge of climate-smart agriculture. In this way we are using 'smart publishing' to help achieve climate-smart agriculture.

Burleigh Dodds Science Publishing is an independent and innovative publisher delivering high quality customer-focused agricultural science content in both print and online formats for the academic and research communities. Our aim is to build a foundation of knowledge on which researchers can build to meet the challenge of climate-smart agriculture.

For more information about Burleigh Dodds Science Publishing simply call us on +44 (0) 1223 839365, email info@bdspublishing.com or alternatively please visit our website at www.bdspublishing.com.

Related titles:

Achieving sustainable production of pig meat Volume 1: Safety, quality and sustainability
Print (ISBN 978-1-78676-088-3); Online (ISBN 978-1-78676-091-3, 978-1-78676-090-6)

Achieving sustainable production of pig meat Volume 2: Animal breeding and nutrition
Print (ISBN 978-1-78676-092-0); Online (ISBN 978-1-78676-094-4, 978-1-78676-095-1)

Achieving sustainable production of pig meat Volume 3: Animal health and welfare
Print (ISBN 978-1-78676-096-8); Online (ISBN 978-1-78676-099-9, 978-1-78676-098-2)

Improving organic animal farming
Print (ISBN 978-1-78676-180-4); Online (ISBN 978-1-78676-182-8, 978-1-78676-183-5)

Chapters are available individually from our online bookshop:
https://shop.bdspublishing.com

climate•SMART•publishing
IN AGRICULTURAL SCIENCE

BURLEIGH DODDS SERIES IN AGRICULTURAL SCIENCE
NUMBER 96

Understanding the behaviour and improving the welfare of pigs

Edited by Emerita Professor Sandra Edwards, Newcastle University, UK

burleigh dodds
SCIENCE PUBLISHING

Published by Burleigh Dodds Science Publishing Limited
82 High Street, Sawston, Cambridge CB22 3HJ, UK
www.bdspublishing.com

Burleigh Dodds Science Publishing, 1518 Walnut Street, Suite 900, Philadelphia, PA 19102-3406, USA

First published 2021 by Burleigh Dodds Science Publishing Limited

Library of Congress Control Number: 2020945603

British Library Cataloguing in Publication Data
A catalogue record for this book is available from the British Library

ISBN 978-1-78676-443-0 (Print)
ISBN 978-1-78676-446-1 (PDF)
ISBN 978-1-78676-445-4 (ePub)
ISSN 2059-6936 (print)
ISSN 2059-6944 (online)

DOI 10.19103/AS.2020.0081

Typeset by Deanta Global Publishing Services, Dublin, Ireland

Contents

Series list xi
Acknowledgements xviii
Introduction xix

Part 1 Determinants of behaviour

1 Advances in understanding the genetics of pig behaviour 3
Lotta Rydhmer, Swedish University of Agricultural Sciences, Sweden

1 Introduction 3
2 Maternal behaviour 4
3 Feeding behaviour and the consequences of selection for feed efficiency 6
4 Tail biting 8
5 Aggressive behaviour between pigs 9
6 Direct and social effects on growth rate and reproduction traits 11
7 Characteristics of pigs with high social breeding values for growth rate 14
8 Conclusion and future trends 16
9 Where to look for further information 19
10 Personal communications 20
11 References 20

2 Developmental influences on pig behaviour 27
Giuliana Miguel-Pacheco and Yolande M. Seddon, University of Saskatchewan, Canada

1 Introduction 27
2 Prenatal environmental effects on development of pig behaviour 28
3 Environmental effects during the early postnatal period on the development of behaviour 50
4 Conclusions 64
5 Future trends in research 65
6 Where to look for further information 66
7 References 67

Part 2 Management of behaviour in different production stages

3 Optimising pig welfare in breeding and gestation 87
Paul H. Hemsworth, University of Melbourne, Australia

1 Introduction 87
2 Breeding and gestation of gilts and sows 88
3 Animal management 106
4 Conclusion and future trends 108
5 Where to look for further information 110
6 References 111

4 Optimising sow and piglet welfare during farrowing and lactation 121
Emma M. Baxter, Animal Behaviour and Welfare Team, Animal and Veterinary Sciences Research Group, SRUC, UK; and Sandra Edwards, Newcastle University, UK

1 Introduction 121
2 Welfare challenges during farrowing and lactation 122
3 Mitigating welfare challenges 126
4 Conclusions 161
5 Future trends in research 161
6 Where to look for further information 162
7 References 162

5 Optimising pig welfare at the weaning and nursery stage 177
Nicole Kemper, University of Veterinary Medicine Hannover, Germany

1 Introduction 177
2 Minimising the impact of weaning 181
3 Optimised management: the human key factor 198
4 Conclusion 199
5 Future trends in research 199
6 Where to look for further information 200
7 References 200

6 Optimizing pig welfare in the growing and finishing stage 209
Arlene Garcia and John J. McGlone, Texas Tech University, USA

1 Introduction 209
2 Pig behavioral issues that impact pig welfare 212
3 Production systems 215
4 Group size 222
5 Living conditions in indoor systems 223
6 Animal health and animal care 230
7 Conclusion and future trends in research 232

8 Where to look for further information 234
9 References 235

7 Optimising pig welfare during transport, lairage and slaughter 247
Luigi Faucitano, Agriculture and Agri-Food Canada, Canada; and Antonio Velarde, Institute of Agrifood Research and Technology, Spain

1 Introduction 247
2 Welfare during transport 248
3 Welfare in lairage 256
4 Welfare during stunning and slaughter 260
5 Animal welfare audit protocols 265
6 Conclusion and future trends 267
7 Where to look for further information 268
8 References 269

Part 3 Current welfare issues

8 Evidence of pain in piglets subjected to invasive management procedures 281
Armelle Prunier and Céline Tallet, PEGASE, INRAE, Institut Agro, France; and Dale Sandercock, Scotland's Rural College (SRUC), UK

1 Introduction 281
2 The why and how of invasive management procedures in piglets 282
3 Definition and mechanisms of pain 284
4 Neural evidence of pain due to invasive management procedures in piglets 291
5 Hormonal and metabolic evidence of pain due to invasive management procedures in piglets 294
6 Behavioural evidence of pain due to invasive management procedures in piglets 297
7 Conclusion and future trends 303
8 Summary 304
9 Where to look for further information 305
10 References 305

9 Alternatives to castration of pigs 315
Emma Fàbrega, IRTA, Spain

1 Introduction 315
2 Entire male production 317
3 Immunocastration 331
4 Sex sorting of semen 342

5 Case study 342
6 Conclusion and future trends 343
7 Where to look for further information 346
8 References 347

10 Understanding and preventing tail biting in pigs 361
Sandra Edwards, Newcastle University, UK; and Anna Valros, University of Helsinki, Finland

1 Introduction 361
2 The prevalence and economic importance of the problem 363
3 Do we have a mechanistic understanding of tail biting? 366
4 Risk factors for the occurrence of tail biting 371
5 Ethical considerations associated with tail biting 381
6 Managing pigs without tail biting 383
7 Intervention measures for a tail biting outbreak 386
8 Conclusions 389
9 Future trends in research 390
10 Where to look for further information 390
11 References 391

11 The role of enrichment in optimizing pig behaviour and welfare 401
Sandra Düpjan and Liza R. Moscovice, Institute of Behavioural Physiology - Leibniz Institute for Farm Animal Biology Dummerstorf, Germany; and Birger Puppe, Institute of Behavioural Physiology - Leibniz Institute for Farm Animal Biology Dummerstorf and University of Rostock, Germany

1 Introduction 401
2 Environmental enrichment and animal welfare 404
3 The role of environmental enrichment in pig behaviour and welfare 406
4 Evaluating the biological relevance of different types of enrichment 416
5 Conclusions 419
6 Where to look for further information 420
7 References 421

Part 4 Assessment of welfare states

12 Physiological and behavioral responses to disease in pigs 431
M. D. Pairis-Garcia and B. K. Wagner, North Carolina State University, USA

1 Introduction 431
2 Physiological responses to disease 432
3 Behavioral responses to disease 435
4 Strategies for disease mitigation 439

5 Conclusion 448
6 Future trends in research 449
7 Where to look for further information 449
8 References 449

13 Assessing emotions in pigs: determining negative and positive mental states 455
Eimear Murphy, Luca Melotti and Michael Mendl, Bristol Veterinary School, University of Bristol, UK

1 Introduction 455
2 Behavioural indicators of emotion 459
3 Cognitive indicators of emotion in pigs 470
4 Neurophysiological indicators of emotion in pigs 473
5 Summary and future trends 477
6 Where to look for further information 480
7 References 481

14 Welfare assessment of pigs 497
Björn Forkman, University of Copenhagen, Denmark

1 Introduction 497
2 Welfare comparisons and frameworks 498
3 Input and output-related welfare measures 499
4 Required properties for welfare measures: validity, repeatability and feasibility 501
5 Feeding-related welfare indicators 503
6 Housing-related welfare indicators 504
7 Health- and pain-related welfare indicators 507
8 Behavioural welfare indicators 511
9 General indicators of poor and positive welfare 512
10 Future trends 515
11 Where to look for further information 515
12 References 516

15 Advances in technologies for monitoring pig welfare 525
Maciej Oczak, University of Veterinary Medicine Vienna, Austria; Kristina Maschat, FFoQSI GmbH, Austria; and Johannes Baumgartner, University of Veterinary Medicine Vienna, Austria

1 Introduction 525
2 Good feeding 528
3 Good housing 531
4 Good health 532
5 Appropriate behaviour 535

6 Legal and ethical considerations 537
7 Sensor data management and modelling 539
8 Case study 542
9 Conclusion 546
10 Future trends in research 548
11 Where to look for further information 550
12 References 551
Index 563

Series list

Title	Series number
Achieving sustainable cultivation of maize - Vol 1 From improved varieties to local applications Edited by: Dr Dave Watson, CGIAR Maize Research Program Manager, CIMMYT, Mexico	001
Achieving sustainable cultivation of maize - Vol 2 Cultivation techniques, pest and disease control Edited by: Dr Dave Watson, CGIAR Maize Research Program Manager, CIMMYT, Mexico	002
Achieving sustainable cultivation of rice - Vol 1 Breeding for higher yield and quality Edited by: Prof. Takuji Sasaki, Tokyo University of Agriculture, Japan	003
Achieving sustainable cultivation of rice - Vol 2 Cultivation, pest and disease management Edited by: Prof. Takuji Sasaki, Tokyo University of Agriculture, Japan	004
Achieving sustainable cultivation of wheat - Vol 1 Breeding, quality traits, pests and diseases Edited by: Prof. Peter Langridge, The University of Adelaide, Australia	005
Achieving sustainable cultivation of wheat - Vol 2 Cultivation techniques Edited by: Prof. Peter Langridge, The University of Adelaide, Australia	006
Achieving sustainable cultivation of tomatoes Edited by: Dr Autar Mattoo, USDA-ARS, USA & Prof. Avtar Handa, Purdue University, USA	007
Achieving sustainable production of milk - Vol 1 Milk composition, genetics and breeding Edited by: Dr Nico van Belzen, International Dairy Federation (IDF), Belgium	008
Achieving sustainable production of milk - Vol 2 Safety, quality and sustainability Edited by: Dr Nico van Belzen, International Dairy Federation (IDF), Belgium	009
Achieving sustainable production of milk - Vol 3 Dairy herd management and welfare Edited by: Prof. John Webster, University of Bristol, UK	010
Ensuring safety and quality in the production of beef - Vol 1 Safety Edited by: Prof. Gary Acuff, Texas A&M University, USA & Prof. James Dickson, Iowa State University, USA	011
Ensuring safety and quality in the production of beef - Vol 2 Quality Edited by: Prof. Michael Dikeman, Kansas State University, USA	012
Achieving sustainable production of poultry meat - Vol 1 Safety, quality and sustainability Edited by: Prof. Steven C. Ricke, University of Arkansas, USA	013
Achieving sustainable production of poultry meat - Vol 2 Breeding and nutrition Edited by: Prof. Todd Applegate, University of Georgia, USA	014

Achieving sustainable production of poultry meat - Vol 3 015
Health and welfare
Edited by: Prof. Todd Applegate, University of Georgia, USA

Achieving sustainable production of eggs - Vol 1 016
Safety and quality
Edited by: Prof. Julie Roberts, University of New England, Australia

Achieving sustainable production of eggs - Vol 2 017
Animal welfare and sustainability
Edited by: Prof. Julie Roberts, University of New England, Australia

Achieving sustainable cultivation of apples 018
Edited by: Dr Kate Evans, Washington State University, USA

Integrated disease management of wheat and barley 019
Edited by: Prof. Richard Oliver, Curtin University, Australia

Achieving sustainable cultivation of cassava - Vol 1 020
Cultivation techniques
Edited by: Dr Clair Hershey, formerly International Center for Tropical Agriculture (CIAT), Colombia

Achieving sustainable cultivation of cassava - Vol 2 021
Genetics, breeding, pests and diseases
Edited by: Dr Clair Hershey, formerly International Center for Tropical Agriculture (CIAT), Colombia

Achieving sustainable production of sheep 022
Edited by: Prof. Johan Greyling, University of the Free State, South Africa

Achieving sustainable production of pig meat - Vol 1 023
Safety, quality and sustainability
Edited by: Prof. Alan Mathew, Purdue University, USA

Achieving sustainable production of pig meat - Vol 2 024
Animal breeding and nutrition
Edited by: Prof. Julian Wiseman, University of Nottingham, UK

Achieving sustainable production of pig meat - Vol 3 025
Animal health and welfare
Edited by: Prof. Julian Wiseman, University of Nottingham, UK

Achieving sustainable cultivation of potatoes - Vol 1 026
Breeding improved varieties
Edited by: Prof. Gefu Wang-Pruski, Dalhousie University, Canada

Achieving sustainable cultivation of oil palm - Vol 1 027
Introduction, breeding and cultivation techniques
Edited by: Prof. Alain Rival, Center for International Cooperation in Agricultural Research for Development (CIRAD), France

Achieving sustainable cultivation of oil palm - Vol 2 028
Diseases, pests, quality and sustainability
Edited by: Prof. Alain Rival, Center for International Cooperation in Agricultural Research for Development (CIRAD), France

Achieving sustainable cultivation of soybeans - Vol 1 029
Breeding and cultivation techniques
Edited by: Prof. Henry T. Nguyen, University of Missouri, USA

Achieving sustainable cultivation of soybeans - Vol 2 030
Diseases, pests, food and non-food uses
Edited by: Prof. Henry T. Nguyen, University of Missouri, USA

Achieving sustainable cultivation of sorghum - Vol 1 031
Genetics, breeding and production techniques
Edited by: Prof. William Rooney, Texas A&M University, USA

Achieving sustainable cultivation of sorghum - Vol 2 032
Sorghum utilization around the world
Edited by: Prof. William Rooney, Texas A&M University, USA

Achieving sustainable cultivation of potatoes - Vol 2 033
Production, storage and crop protection
Edited by: Dr Stuart Wale, Potato Dynamics Ltd, UK

Achieving sustainable cultivation of mangoes 034
Edited by: Prof. Víctor Galán Saúco, Instituto Canario de Investigaciones Agrarias (ICIA), Spain & Dr Ping Lu, Charles Darwin University, Australia

Achieving sustainable cultivation of grain legumes - Vol 1 035
Advances in breeding and cultivation techniques
Edited by: Dr Shoba Sivasankar et al., formerly International Crops Research Institute for the Semi-Arid Tropics (ICRISAT), India

Achieving sustainable cultivation of grain legumes - Vol 2 036
Improving cultivation of particular grain legumes
Edited by: Dr Shoba Sivasankar et al., formerly International Crops Research Institute for the Semi-Arid Tropics (ICRISAT), India

Achieving sustainable cultivation of sugarcane - Vol 1 037
Cultivation techniques, quality and sustainability
Edited by: Prof. Philippe Rott, University of Florida, USA

Achieving sustainable cultivation of sugarcane - Vol 2 038
Breeding, pests and diseases
Edited by: Prof. Philippe Rott, University of Florida, USA

Achieving sustainable cultivation of coffee 039
Edited by: Dr Philippe Lashermes, Institut de Recherche pour le Développement (IRD), France

Achieving sustainable cultivation of bananas - Vol 1 040
Cultivation techniques
Edited by: Prof. Gert H. J. Kema, Wageningen University and Research, The Netherlands & Prof. André Drenth, University of Queensland, Australia

Global Tea Science 041
Current status and future needs
Edited by: Dr V. S. Sharma, formerly UPASI Tea Research Institute, India & Dr M. T. Kumudini Gunasekare, Coordinating Secretariat for Science Technology and Innovation (COSTI), Sri Lanka

Integrated weed management 042
Edited by: Emeritus Prof. Rob Zimdahl, Colorado State University, USA

Achieving sustainable cultivation of cocoa 043
Edited by: Prof. Pathmanathan Umaharan, Cocoa Research Centre – The University of the West Indies, Trinidad and Tobago

Robotics and automation for improving agriculture 044
Edited by: Prof. John Billingsley, University of Southern Queensland, Australia

Water management for sustainable agriculture 045
Edited by: Prof. Theib Oweis, ICARDA, Jordan

Improving organic animal farming 046
Edited by: Dr Mette Vaarst, Aarhus University, Denmark & Dr Stephen Roderick, Duchy College, UK

Improving organic crop cultivation 047
Edited by: Prof. Ulrich Köpke, University of Bonn, Germany

Managing soil health for sustainable agriculture - Vol 1 048
Fundamentals
Edited by: Dr Don Reicosky, Soil Scientist Emeritus USDA-ARS and University of Minnesota, USA

Managing soil health for sustainable agriculture - Vol 2 049
Monitoring and management
Edited by: Dr Don Reicosky, Soil Scientist Emeritus USDA-ARS and University of Minnesota, USA

Rice insect pests and their management 050
E. A. Heinrichs, Francis E. Nwilene, Michael J. Stout, Buyung A. R. Hadi & Thais Freitas

Improving grassland and pasture management in temperate agriculture 051
Edited by: Prof. Athole Marshall & Dr Rosemary Collins, IBERS, Aberystwyth University, UK

Precision agriculture for sustainability 052
Edited by: Dr John Stafford, Silsoe Solutions, UK

Achieving sustainable cultivation of temperate zone tree fruit and berries – Vol 1 053
Physiology, genetics and cultivation
Edited by: Prof. Gregory A. Lang, Michigan State University, USA

Achieving sustainable cultivation of temperate zone tree fruit and berries – Vol 2 054
Case studies
Edited by: Prof. Gregory A. Lang, Michigan State University, USA

Agroforestry for sustainable agriculture 055
Edited by: Prof. María Rosa Mosquera-Losada, Universidade de Santiago de Compostela, Spain & Dr Ravi Prabhu, World Agroforestry Centre (ICRAF), Kenya

Achieving sustainable cultivation of tree nuts 056
Edited by: Prof. Ümit Serdar, Ondokuz Mayis University, Turkey & Emeritus Prof. Dennis Fulbright, Michigan State University, USA

Assessing the environmental impact of agriculture 057
Edited by: Prof. Bo P. Weidema, Aalborg University, Denmark

Critical issues in plant health: 50 years of research in African agriculture 058
Edited by: Dr Peter Neuenschwander and Dr Manuele Tamò, IITA, Benin

Achieving sustainable cultivation of vegetables 059
Edited by: Emeritus Prof. George Hochmuth, University of Florida, USA

Advances in breeding techniques for cereal crops 060
Edited by: Prof. Frank Ordon, Julius Kuhn Institute (JKI), Germany & Prof. Wolfgang Friedt, Justus-Liebig University of Giessen, Germany

Advances in Conservation Agriculture – Vol 1 061
Systems and Science
Edited by: Prof. Amir Kassam, University of Reading, UK and Moderator, Global Conservation Agriculture Community of Practice (CA-CoP), FAO, Rome, Italy

Advances in Conservation Agriculture – Vol 2 062
Practice and Benefits
Edited by: Prof. Amir Kassam, University of Reading, UK and Moderator, Global Conservation Agriculture Community of Practice (CA-CoP), FAO, Rome, Italy

Achieving sustainable greenhouse cultivation 063
Edited by: Prof. Leo Marcelis & Dr Ep Heuvelink, Wageningen University, The Netherlands

Achieving carbon-negative bioenergy systems from plant materials 064
Edited by: Dr Chris Saffron, Michigan State University, USA

Achieving sustainable cultivation of tropical fruits 065
Edited by: Prof. Elhadi M. Yahia, Universidad Autónoma de Querétaro, Mexico

Advances in postharvest management of horticultural produce 066
Edited by: Prof. Chris Watkins, Cornell University, USA

Pesticides and agriculture 067
Profit, politics and policy
Dave Watson

Integrated management of diseases and insect pests of tree fruit 068
Edited by: Prof. Xiangming Xu and Dr Michelle Fountain, NIAB-EMR, UK

Integrated management of insect pests: Current and future developments 069
Edited by: Emeritus Prof. Marcos Kogan, Oregon State University, USA & Emeritus Prof. E. A. Heinrichs, University of Nebraska-Lincoln, USA

Preventing food losses and waste to achieve food security and sustainability 070
Edited by: Prof. Elhadi M. Yahia, Universidad Autónoma de Querétaro, Mexico

Achieving sustainable management of boreal and temperate forests 071
Edited by: Dr John Stanturf, Estonian University of Life Sciences , Estonia

Advances in breeding of dairy cattle 072
Edited by: Prof. Julius van der Werf, University of New England, Australia & Prof. Jennie Pryce, Agriculture Victoria and La Trobe University, Australia

Improving gut health in poultry 073
Edited by: Prof. Steven C. Ricke, University of Arkansas, USA

Achieving sustainable cultivation of barley 074
Edited by: Prof. Glen Fox, University of California-Davis, USA and The University of Queensland, Australia & Prof. Chengdao Li, Murdoch University, Australia

Advances in crop modelling for a sustainable agriculture 075
Edited by: Emeritus Prof. Kenneth Boote, University of Florida, USA

Achieving sustainable crop nutrition 076
Edited by: Prof. Zed Rengel, University of Western Australia, Australia

Achieving sustainable urban agriculture 077
Edited by: Prof. Johannes S. C. Wiskerke, Wageningen University, The Netherlands

Climate change and agriculture 078
Edited by Dr Delphine Deryng, NewClimate Institute/Integrative Research Institute on Transformations of Human-Environment Systems (IRI THESys), Humboldt-Universität zu Berlin, Germany

Advances in poultry genetics and genomics 079
Edited by: Prof. Samuel E. Aggrey, University of Georgia, USA; Prof. Huaijun Zhou, University of California-Davis, USA; Dr Michèle Tixier-Boichard, INRAE, France; and Prof. Douglas D. Rhoads, University of Arkansas, USA

Achieving sustainable management of tropical forests 080
Edited by: Prof. Jürgen Blaser, Bern University of Life Sciences, Switzerland; and Patrick D. Hardcastle, Forestry Development Specialist, UK

Improving the nutritional and nutraceutical properties of wheat and other cereals 081
Edited by: Prof. Trust Beta, University of Manitoba, Canada

Achieving sustainable cultivation of ornamental plants 082
Edited by: Emeritus Prof. Michael Reid, University of California-Davis, USA

Improving rumen function 083
Edited by: Dr C. S. McSweeney, CSIRO, Australia; and Prof. R. I. Mackie, University of Illinois, USA

Biostimulants for sustainable crop production 084
Edited by Youssef Rouphael, Patrick du Jardin, Patrick Brown, Stefania De Pascale and Giuseppe Colla

Improving data management and decision support systems in agriculture 085
Edited by: Dr Leisa Armstrong, Edith Cowan University, Australia

Achieving sustainable cultivation of bananas - Volume 2 086
Germplasm and genetic improvement
Edited by: Prof. Gert H. J. Kema, Wageningen University, The Netherlands; and Prof. Andrè Drenth, The University of Queensland, Australia

Reconciling agricultural production with biodiversity conservation 087
Edited by: Prof. Paolo Bàrberi and Dr Anna-Camilla Moonen, Institute of Life Sciences - Scuola Superiore Sant'Anna, Pisa, Italy

Advances in postharvest management of cereals and grains 088
Edited by: Prof. Dirk E. Maier, Iowa State University, USA

Biopesticides for sustainable agriculture 089
Edited by: Prof. Nick Birch, formerly The James Hutton Institute, UK; and Prof. Travis Glare, Lincoln University, New Zealand

Understanding and improving crop root function 090
Edited by: Emeritus Prof. Peter Gregory, University of Reading, UK

Understanding the behaviour and improving the welfare of chickens 091
Edited by: Prof. Christine Nicol, Royal Veterinary College - University of London, UK

Advances in measuring soil health 092
Edited by: Prof. Wilfred Otten, Cranfield University, UK

The sustainable intensification of smallholder farming systems 093
Edited by: Dr Dominik Klauser and Dr Michael Robinson, Syngenta Foundation for Sustainable Agriculture, Switzerland

Advances in horticultural soilless culture 094
Edited by: Prof. Nazim Gruda, University of Bonn, Germany

Reducing greenhouse gas emissions from livestock production 095
Edited by: Dr Richard Baines, Royal Agricultural University, UK

Understanding the behaviour and improving the welfare of pigs 096
Edited by: Emerita Prof. Sandra Edwards, Newcastle University, UK

Genome editing for precision crop breeding 097
Edited by: Dr Matthew Willman, Cornell University, USA

Understanding the behaviour and improving the welfare of dairy cattle 098
Edited by: Dr Marcia Endres, University of Minnesota, USA

Defining sustainable agriculture 099
Dave Watson

Plant genetic resources: A review of current research and future needs 100
Edited by: Dr Ehsan Dulloo, Team Leader, Effective Genetic Resources Conservation and Use Initiative - Bioversity International, Mauritius

Developing animal feed products 101
Edited by: Dr Navaratnam Partheeban, Royal Agricultural University, UK

Improving dairy herd health 102
Edited by: Prof. Emile Bouchard, University of Montreal Saint-Hyacinthe Campus, Canada

Understanding gut microbiomes as targets for improving pig gut health 103
Edited by: Prof. Mick Bailey and Emeritus Prof. Chris Stokes, University of Bristol, UK

Advances in Conservation Agriculture - Vol 3 104
Adoption and Spread
Edited by: Professor Amir Kassam, University of Reading, UK and Moderator, Global Conservation Agriculture Community of Practice (CA-CoP), FAO, Rome, Italy

Advances in Precision Livestock Farming 105
Edited by: Prof. Daniel Berckmans, Katholieke University of Leuven, Belgium

Achieving durable disease resistance in cereals 106
Edited by: Prof. Richard Oliver, Curtin University, Australia

Seaweed and microalgae as alternative sources of protein 107
Edited by: Prof. Xingen Lei, Cornell University, USA

Microbial bioprotectants for plant disease management 108
Edited by: Dr Jürgen Köhl, Wageningen University & Research, The Netherlands; and Dr Willem Ravensberg, Koppert Biological Systems, The Netherlands

Improving soil health 109
Edited by: Prof. William Horwath, University of California-Davis, USA

Improving integrated pest management (IPM) in horticulture 110
Edited by: Prof. Rosemary Collier, Warwick University, UK

Climate-smart production of coffee: Achieving sustainability and ecosystem services 111
Edited by: Prof. Reinhold Muschler, CATIE, Costa Rica

Developing smart agri-food supply chains: using technology to improve safety and quality 112
Edited by: Prof. Louise Manning, Royal Agricultural University, UK

Advances in integrated weed management 113
Edited by: Prof. Per Kudsk, Aarhus University, Denmark

Understanding and improving the functional and nutritional properties of milk 114
Edited by: Prof. Thom Huppertz, Wageningen University, The Netherlands; and Prof. Todor Vasiljevic, Victoria University, Australia

Energy-smart farming: efficiency, renewable energy and sustainability 115
Edited by: Emeritus Prof. Ralph Sims, Massey University, New Zealand

Acknowledgements

We wish to acknowledge the following for their help in reviewing a particular chapter:

- Chapter 10: Dr Hans Spoolder, Wageningen University and Research, The Netherlands; and Dr Laura Boyle, Teagasc, Ireland

Introduction

The welfare of farmed animals such as pigs is an increasing concern for consumers and regulatory agencies. This collection summarises and reviews the wealth of recent research on understanding pig behaviour and improving their welfare. Part 1 of the book provides an initial review of genetic and developmental factors influencing pig behaviour. The book then moves on to assess ways of optimising pig welfare at differing production stages in Part 2. Chapters cover breeding and gestation, farrowing and lactation, weaning, growing and finishing as well as transport, lairage and slaughter. The book then reviews our understanding of the most important current welfare issues, such as pain assessment in relation to piglet management procedures, castration and its alternatives, tail biting and docking, as well as the nature of environmental enrichment. The final part of the book addresses welfare assessment methodologies, including detection of health problems, understanding of emotional responses of pigs, practical use of welfare indicators and advances in technologies for automated monitoring of pig behaviour and welfare.

Part 1 Determinants of behaviour

The first part of the book opens with a review of advances in understanding the genetics of pig behaviour. Chapter 1 begins by discussing the maternal behaviours that are relevant for piglet survival, including savaging, crushing and nursing. It then goes on to address other behaviours that are important for pig production, focusing on feeding behaviour, tail biting and aggressive behaviour between pigs. The chapter also examines the contrast of direct and social genetic effects on growth rate and reproduction traits, with an analysis of the characteristics of pigs with high social breeding values for growth rate, and concludes with an overview of potential future research trends in the subject.

Chapter 2 focuses on developmental influences on pig behaviour. It begins by discussing prenatal effects on the development of pig behaviour, considering aspects of the pre-conception and post-conception environments. These include housing conditions and nutrition of the mother, heat stress, infectious diseases and sow management. The chapter moves on to discuss environmental effects during the early postnatal period on the development of behaviour, considering both the physical and social environment as well as human interactions.

Part 2 Management of behaviour in different production stages

Part 2 begins with Chapter 3 which considers factors affecting the welfare of pigs during breeding and gestation, covering gilts, sows and breeding boars. Particular focus is given to the factors influencing aggression and lameness which constitute important welfare issues. The chapter reviews the physical, social and nutritional aspects of various housing systems in which weaned and pregnant gilts and sows are housed and provides recommendations on optimising their welfare in commercial conditions.

Chapter 4 examines sow and piglet welfare during farrowing and lactation. Optimising welfare in this stage involves resolving the concerns regarding continued use of close confinement systems, such as the farrowing crate for the sows, and the lack of provision of environmental enrichment to provide for behavioural needs. For piglets the main welfare and health issues surround high levels of piglet mortality and its pre-disposing risk factors. The chapter concentrates on managerial and environmental interventions that attempt to reconcile the behavioural and physiological needs of both the sow and piglets to optimise their welfare whilst appreciating stockperson needs in how best to implement them. This is illustrated by a case study of the development of a novel free farrowing system.

The subject of Chapter 5 is optimising pig welfare at the weaning and nursery stage. The welfare of pigs at the nursery stage is an area of special concern because of the many factors that have the potential to seriously decrease piglet wellbeing at this time. Major challenges include separation from the mother, often at a very young weaning age in modern pig farming, the new environment they are introduced to, hitherto unknown sources of feed and water, and regrouping with non-litter mates. In this chapter, the most important sources of environmental, nutritional and social stress are explained, and strategies to increase weaner welfare are summarised. These include innovative housing systems providing greater environmental complexity and the importance of trained and skilled stockpeople.

Chapter 6 discusses optimising pig welfare at the growing and finishing stage. It begins by defining the different behavioural issues that can impact welfare during this period, such as tail biting, belly nosing and aggression. The chapter then addresses how these behavioural issues can be affected in the wide range of different production systems, which encompass differences in many aspects of the physical and social environment. It concludes by highlighting the importance of human inputs in the system to improve pig welfare and also reviews potential research trends for the future as herd size increases.

The final chapter of Part 2, Chapter 7, considers pig welfare during transport, lairage and slaughter. Preslaughter practices, such as transport and handling, as well as methods of slaughter, can significantly affect pig welfare, carcass and meat quality. Chapter 7 provides an overview of research findings on the effects of transport, handling and slaughter practices on behavioural and physiological responses of pigs, and the consequences for carcass and meat quality. It provides recommendations about best practice, illustrating how economic losses are limited when the design of trucks is improved, the recommended transport, handling, stunning and slaughter practices are used and environmental conditions are properly controlled.

Part 3 Current welfare issues

The first chapter in Part 3 analyses evidence of pain in piglets subjected to widely-used invasive management procedures. Chapter 8 begins by considering the why and how of invasive management procedures in piglets, before reviewing pain mechanisms and assessment. It examines the reactions in pigs submitted to tissue-damaging procedures, breaking these down into three groups: neural, hormonal/metabolic and behavioural consequences. The chapter focuses on the detailed evidence for acute and longer-term pain associated with the practices of surgical castration, tail docking and tooth resection that are currently contentious in the pig industry.

Chapter 9 considers alternatives to castration of pigs. Piglets have been traditionally been surgically castrated to avoid boar taint risk as well as aggression and mounting behaviour. However, castration without anaesthesia and analgesia induces pain and has become a controversial practice on welfare grounds. Alternatives to surgical pig castration that guarantee acceptable meat quality and welfare are required if castration is to be abandoned. The chapter deals with the main alternatives which are commercially available at present: raising entire males or immunocastration. It summarises the effects on welfare and meat quality, as well as dealing with other aspects such as societal and farmers' acceptance of these alternatives. The nutritional, management and breeding strategies, which can be implemented to enhance welfare and meat quality when raising entire males are described. The methodologies being used, or under development, to assess boar taint risk at the slaughter line are summarised.

The next chapter addresses understanding and preventing tail biting in pigs. As Chapter 10 shows, tail biting is a behavioural problem of pigs associated with detrimental welfare for both the perpetrator and victim. It is seen to some extent on most farms worldwide and causes considerable economic loss, leading to widespread adoption of tail docking for risk reduction. After highlighting the prevalence and economic importance of the problem of tail biting, the chapter then reviews the current understanding of the mechanisms underlying tail biting and the key risk factors. Ethical considerations associated with tail biting are also discussed. This is followed by an examination of strategies to manage pigs without tail biting, highlighting intervention measures for when a tail biting outbreak occurs, before concluding with an overview of potential future research trends.

Part 3 concludes with a discussion on the role of environmental enrichment in optimising pig behaviour and welfare. Modern pig housing environments provide animals with essential resources but, from an animal's point of view, they are often quite barren and deprive them of the opportunity to make full use of their natural behavioural repertoire. The lack of stimulation resulting from such environments compromises animal welfare. This issue can be addressed by providing enrichment across all aspects of an animal's environment that facilitate engagement in species-specific behaviours. Chapter 11 provides an overview of different types of enrichment, ranging from artificial point-source objects to social and cognitive enrichment. It discusses the existing body of evidence for welfare-enhancing effects of different enrichment provision, focussing on whether potential enrichment promotes natural behaviours, reduces abnormal behaviours, and whether effects are sustainable over the long-term. The chapter concludes with a comparative evaluation of enrichment from an animal welfare perspective, highlighting the potential of social and cognitive enrichment.

Part 4 Assessment of welfare states

The final part of the book starts with Chapter 12 which provides an overview of the physiological and behavioural responses common to pigs experiencing ill-health and describes animal-based outcomes directly impacted during an acute or chronic disease state. In addition, strategies to prevent and monitor disease at the herd level are discussed. Lastly, the chapter explores strategies for managing sick pigs as the critical next step when prevention is unsuccessful.

Chapter 13 reviews the assessment of emotions in pigs, specifically focusing on determining negative and positive mental states. It begins by examining behavioural indicators of emotion such as behavioural tests, qualitative behaviour assessment, vocalisations, play behaviour, defence cascade responses and facial expression and body posture. The chapter also assesses cognitive indicators of emotion in pigs such as judgement bias. This is followed by a discussion of neurophysiological indicators of emotion in pigs, encompassing the physiology of both negative and positive emotions. The chapter concludes by considering future trends in this rapidly evolving field.

The subject of Chapter 14 is the welfare assessment of pigs under farm conditions. The chapter begins by providing an overview of approaches to welfare comparisons and frameworks for animal welfare evaluation. It examines the measures employed in animal welfare assessments, dividing them into two groups: measures of input and measures of outcome. The required properties for welfare measures are emphasised in terms of their validity, repeatability and feasibility. Sections on practical welfare indicators relating to the dimensions of good feeding, good housing, good health and appropriate behaviour are then provided, with consideration of the welfare relevance, reproducibility and practicality of individual measures. Finally, indicators of positive welfare and the practical use of iceberg indicators are considered.

The final chapter of the book focuses on advances in technologies for automated monitoring of pig welfare. The pig sector has undergone a major transformation characterized by intensification, geographic concentration, vertical integration and increasing scales of production, which creates new challenges for animal welfare. Sensor technology offers multiple advantages for monitoring welfare in pigs under these conditions. Chapter 15 reviews recent advances in technologies for monitoring key indicators of pig welfare by considering the domains of good feeding, good housing, good health, and appropriate behaviour. As an example, it presents how an accelerometer-based nest-building detection system can contribute to improvement of pig welfare. Finally, future trends in development of monitoring technology including solution-oriented research, integration of key welfare indicators and wider adoption of image analysis are highlighted, with further possibilities for more validation studies on open datasets and advances in robotics.

Overall, the book seeks to provide practical and up-to-date information on how the welfare of farmed pigs at all production stages can be assessed and improved, and where further scientific study towards this objective can be progressed.

Part 1

Determinants of behaviour

Chapter 1

Advances in understanding the genetics of pig behaviour

Lotta Rydhmer, Swedish University of Agricultural Sciences, Sweden

1 Introduction
2 Maternal behaviour
3 Feeding behaviour and the consequences of selection for feed efficiency
4 Tail biting
5 Aggressive behaviour between pigs
6 Direct and social effects on growth rate and reproduction traits
7 Characteristics of pigs with high social breeding values for growth rate
8 Conclusion and future trends
9 Where to look for further information
10 Personal communications
11 References

1 Introduction

The pig is an explorative animal with a large ability to adapt to different environments. Its large litters (no other animal as large as the pig gives birth to so many young) demand a maternal behaviour that is very different from other livestock. A successful maternal behaviour is crucial for the efficiency of piglet production. Pigs are generally kept in groups and their social behaviour has an impact on animal welfare in positive and negative ways. Social interactions influence growth of young pigs and reproduction of sows. Common pig production routines, like tail docking and keeping sows in crates, are strongly connected to pigs' behaviour and these routines are highly problematic from an animal welfare perspective.

Apart from the joy of research in general and animal science in special, there are several reasons to study and learn more about the genetics of pig behaviour:

- Pig behaviours influence animal welfare;
- Pig behaviours influence work satisfaction of care takers;

http://dx.doi.org/10.19103/AS.2020.0081.01

- Pig behaviours influence production profit; and
- Some pig behaviours (e.g. tail biting) or their interventions (e.g. tail docking) influence consumers' acceptance of pig production.

There seems to be a genetic variation in most behavioural traits that have been studied in pigs, and many examples will be given in this chapter. The chapter starts with a review of the genetic background of various behavioural traits, with references to molecular genetic studies as well as quantitative genetic studies. The group model with its direct and social genetic effects is described and consequences of selection for social breeding values are presented. Some future perspectives on methods and breeding goals, and the room for selection for changed behaviours, are discussed.

2 Maternal behaviour

One way to express the goal for piglet production is 'a high number of healthy piglets with a low variation around the target weaning weight'. The level of this target weight depends on management system and factors like weaning age, nutrition value of piglet feed and use of nurse sows. Although several genetic studies of maternal behaviour are found in the literature, maternal behaviour traits are usually not included as selection traits in pig breeding programs. Instead, the selection traits are piglet survival and piglet growth, traits more directly reflecting the goal and also more easy to record than behavioural traits. Nevertheless, we can achieve a better understanding of opportunities and limitations of piglet production from genetic studies of nest building, savaging, crushing and nursing.

Sows obviously need straw or other construction material to build a nest, but an increased activity with a typical behavioural pattern before farrowing can be seen also in sows not provided any building material. Sows actively engaged in nest building before farrowing calms down when the nest is built and thereafter move less during farrowing, thus reducing the risk of crushing new-born piglets and increasing the possibility for all piglets to achieve colostrum (Ocepek et al., 2017). Accordingly, Ocepek and Andersen (2018) found that nest building activity is associated with a lower proportion of starved piglets and overlaid piglets. Nest building is related to oxytocin and preliminary results of Rydhmer and Jonas (2016) show an association between the oxytocin synthesis gene and nest building behaviour. An association between the oxytocin gene and stillborn piglets, but not between the gene and survival of live born piglets, was found in the same project (Jonas and Rydhmer, 2018).

Savaging of new-born piglets (infanticide) is associated with low plasma oxytocin levels at farrowing (Gilbert, 2001). Around 5–11% of primiparous sows savage their piglets (Chen et al., 2008; Gäde et al., 2008) and the repeatability is around 0.4 (Gäde et al., 2008). An ongoing threshold selection against savaging at herd level seems plausible; farmers avoid selecting gilts for replacement born by savaging sows. Old studies reported high heritability estimates for savaging, but Gäde et al. (2008) estimated the heritability at 0.02. Several QTLs for savaging have been found (Chen et al., 2009b) and for some of them genes involved in anxiety are located in homolog chromosome regions in humans (Daigle, 2018). This is an example of studies where pigs are used as model for humans in studies of psychiatric disorders (Daigle, 2018).

Bauer (2019) recently presented a doctoral thesis on genetic components of savaging. He found associations between savaging and genes involved in the regulation of dopamine, vasopressin and oxytocin levels, and also genes involved with mitochondria and energy production "suggesting that alteration of the genome impacting on the way the cells produce energy could have a behavioural impact" (Bauer, 2019). Savaging sows show a more active behaviour and they are more responsive to piglets (Jarvis et al., 2004; Chen et al., 2008). Outdoor raised sows selected for the maternal effect on piglet survival had a higher frequency of savaging (compared to a control line) when they farrowed indoors (Baxter et al., 2011). This could be interpreted as savaging being the outcome of a low ability to cope with a confined environment under high pressure. Assuming that the pressure on the sow increases with increasing litter size, this is however contradicted by a negative genetic correlation between litter size (total born) and savaging (r_g = -0.34) estimated by Gäde et al. (2008).

Farmers' judgement of sow behaviour was studied by Stratz et al., 2016. A good farrowing behaviour included remaining in lying position during farrowing, not snapping or biting piglets and having a rapid farrowing. The heritability for this farrowing behaviour was estimated at 0.07. Farrowing behaviour was correlated to piglet weight at weaning (r_g = 0.52) and to piglet vitality (r_g = 0.32). It has previously been reported that unsavaged piglets are heavier at birth than savaged piglets (Grandinson et al., 2002) and it can be hypothesised that the genotype of the piglet influences the risk of being savaged. This would motivate the use of a genetic model including two genetic effects - a direct (piglet) and a maternal (sow) - in the model when analysing savaging. No such genetic analysis has been found in the literature.

When nucleus farmers recorded how often the sow showed careless behaviour among the piglets, the heritability of carefulness was estimated at 0.1-0.2 in two different breeds (Vangen et al., 2005). Hellbrügge et al. (2007, 2008) studied maternal behaviour on thousand German Landrace sows in a nucleus herd. The main piglet mortality cause was crushing by the sow. The sow's reaction to a recorded distress call was heritable (h^2 = 0.13). A stronger reaction was genetically correlated with higher piglet survival, but the correlation was low and not significant. The sow's reaction to an unknown sound (music) was also heritable and showed a higher correlation with piglet survival (r_g = 0.26, SE ±0.18) and with number of piglets crushed during the first days (r_g = -0.28, SE ±0.19). The sow's reaction when separated from the piglets three weeks after farrowing had a low heritability but was highly correlated with the reaction to an unknown sound (r_g = 0.89, SE ±0.35). The reaction when separated from the piglets was also correlated with piglet survival (r_g = 0.44, SE± 0.31). Standard errors are given for these estimates to illustrate that although behavioural records from thousands of sows takes a lot of time to collect, it is hardly enough to estimate genetic parameters for these complex traits.

According to Stratz et al. (2016), a good nursing behaviour includes not lying on the udder, feeding piglets until satiation and having a well-developed udder with functional teats. The heritability for assessed nursing behaviour was estimated at 0.10 and the nursing behaviour was genetically correlated with piglet weight at weaning (r_g = 0.86).

Selection for increased litter sizes increase the demands on the sow; it should avoid crushing any piglet and provide enough milk to all piglets. For animal welfare reasons we want to get rid of the farrowing crates. Genetic predisposition for larger litters in combination with loose-housing are motives for selection for maternal abilities. As described above, there are several heritable maternal behaviour-traits, but recording behaviour is time-consuming and thus expensive. Selection for high piglet survival and piglet growth rate, using both direct (piglet) and maternal (sow) breeding values, may be a better alternative than introducing behavioural tests in the breeding program of dam lines.

3 Feeding behaviour and the consequences of selection for feed efficiency

Being an omnivore, the pig is an explorative animal with a high capacity to adapt to a wide range of feeds and environments (discussed by Brunberg et al., 2016). Pigs show a large variation in feeding behaviour as a consequence of the management system, such as ad libitum feeding in feeders or restricted group feeding in troughs once per day, and there is also a genetic variation (Rohrer et al., 2013). Automatic feeders are often used to record individual feed intake of growing pigs. This equipment also provides large amounts of feeding behaviour data that until now have not been of much use for selection. Kavlak and Uimari (2019) analysed data from more than 3000 pigs fed with electronic feeders at a test station. The feeding behaviour traits were number of visits per day, time spent in feeding per day, time spent feeding per visit and feed intake per visit. Heritabilities of these traits were estimated at 0.17–0.47, but none of them were genetically correlated to any production trait. The authors thus concluded that there is no reason to include feeding behaviour in the genetic evaluation.

Several quantitative trait loci (QTL) for feeding behaviour have been found. Reyer et al. (2017) found QTLs for daily feeder occupation time and number of daily feeder visits. Some candidate genes were identified; one of them (MC4R at chromosome 1) with effect on energy homeostasis influencing feed intake. Although both daily feed intake and nutrient needs change with age, strong genetic correlations have been found between feeding behaviour traits recorded at different ages (Kavlak and Uimari, 2019). In a genome wide association study, the QTLs identified for feeding behaviour recorded during two age periods on growing pigs were, however, not the same (Guo et al., 2015). When 338 Duroc boars with records on daily feed intake, number and duration of visits per day and feed intake per visit were genotyped; six SNPs associated with feeding behaviour were located in genomic regions where QTLs for feeding behaviour have been found earlier (Ding et al., 2017). Five candidate genes with biochemical and physiological roles relevant for feeding behaviour were recognized close to these markers. Several of the candidate genes were involved in the development of the hypothalamus; on organ important for the regulation of hunger. Future functional genomic studies may reveal the genetic regulation of feeding behaviour and feed intake.

The main motive for genetic studies of feeding behaviours seems to be their relation to feed efficiency; an important goal trait in most pig breeding programs. According to a French selection experiment lasting for nine generations, selection for more efficient pigs (low residual feed intake (RFI)) results in pigs with lower physical activity during both day and night, as compared to pigs selected for high RFI (Meunier-Salaün et al., 2014). The low RFI line also seemed to be less affected by tail biting, which may be a consequence of the lower activity level. When comparing the high and low RFI lines, no behavioural difference was found in a novel object test (Meunier-Salaün et al., 2014). Colpoys et al. (2014) also compared the behaviour of pigs selected for high or low RFI and concluded that low RFI pigs were more calm in the home pen as well as in novel object and human approach tests, thus less reactive to novelty.

Changes (or lack of changes) in feeding behaviour can also be used as an indicator of robustness, for example, the ability to cope with heat (Cross et al., 2018). Furthermore, feeding behaviour can be used as an indicator of behavioural traits difficult to record, such as risk of becoming a victim of tail biting (Wallenbeck and Keeling, 2013). Rohrer et al. (2013) found that pigs with a reactive coping style, recorded in a backtest, tended to eat fewer but longer meals per day. Data from feeding stations are 'for free', and with increased knowledge in bioinformatics, the use of the large amounts of data from feeding stations will maybe increase in future herd monitoring and breeding programs.

4 Tail biting

Tail biting is a multi-factorial problem, with both environmental and genetic causes. Breuer et al. (2005) found a positive genetic correlation between tail biting and lean tissue growth rate and a negative genetic correlation between tail biting and backfat thickness. Brunberg et al. (2013b) reported a genetic association between tail biting and fatness; biters and victims had a different expression of the gene PDK4 compared to pigs not involved in tail biting. The PDK4 gene has an impact on fat content in pigs (Lan et al., 2009). Selection for lean pigs may thus increase the risk of tail biting. Using metabolites from pigs as phenotypes in a genome-wide association study, Dervishi et al. (2019) showed that pigs predisposed to be bitten may have an impaired lipolysis process.

There are several challenges related to genetic analyses of tail biting and - although highly relevant for pig welfare - genetic studies of tail biting are scarce. If a pig becomes a tail biter or a victim of tail biting depends both on the pig itself and on its group mates (discussed by Brunberg et al., 2016). Data from several thousand pigs may be necessary to get accurate genetic estimates. Tail biting behaviours are often recorded as binary traits, which increases the need for large data sets even more. Bitten tails can be recorded by repeated visual monitoring, but the frequency is often low which puts high demands on the statistical model used for genetic analysis. To identify biting pigs by direct observations or video recording is very time-consuming. Furthermore, the expression of tail biting differs over time as a pig can change from being a non-biter to being a biter (Ursinus et al., 2014).

Breuer et al. (2005) estimated the heritability of performing tail biting in Landrace pigs at 0.05. More recently, Canario and Flatres-Grall (2018) estimated the heritability of being a victim at 0.06 in Tai Zumu pigs. The genetic correlation between being a biter and being a victim is not known. Brunberg et al. (2013a) found some SNPs that had the same associations to both being a biter and being a victim, but these SNPs were not associated with being a neutral pig. Neutral pigs are pigs neither being biters nor victims in pens where tail biting is ongoing. Wilson et al. (2012) found that some SNPs are associated with being neutral whereas other SNPs are associated with being a biter or a victim. A study on gene expression in the brain also suggested that biters and victims have more in common than neutral pigs (Brunberg et al., 2013b).

Neutral pigs (in pens with tail biting) performed less pig-directed behaviours, such as belly nosing and tail in mouth, compared to pigs in pens without tail biting (Brunberg et al., 2013a), and 100 transcripts were differently expressed between those two types of pigs. Several of the transcripts were also differently expressed in neutral pigs as compared to both biters and victims (Brunberg et al., 2013b). This suggests that neutral pigs' gene expression is not a consequence of not being involved in tail biting, but rather a cause for not being involved. Are neutral pigs the pigs that should be selected? We do, however, not know what would happen if all pigs in a pen were genetically predisposed to be neutral.

Camerlink et al. (2015) compared the behaviour of two groups of pigs with different social breeding values for growth rate. The pigs were studied in two environments, in barren, standard pens and in pens with lots of straw. As expected, less tail biting was performed in pens with straw. Pigs with high social breeding values in enriched pens showed least biting behaviour and pigs with low social breeding values in barren pens showed most biting behaviour.

5 Aggressive behaviour between pigs

In todays' pig production, most pigs meet unfamiliar pigs at several occasions, for example, at weaning, when moved to the finishing stable and at the abattoir. Except for a short period during the first weeks of life, such mixing of pigs leads to fighting. In addition to the intensive fighting just after mixing when a rank order is established, aggressive behaviour is also seen in stable groups of familiar pigs.

In general terms, the neuropeptide vasopressin facilitates aggression whereas serotonin inhibits aggression. Terenina et al. (2012) identified markers in young pigs indicating that aggressive behaviour is associated with genes related to vasopressin (number of attacks and number of fights won) and serotonin (number of attacks and number of fights involved in). Four markers for the dopamine receptor were also associated with aggressive behaviour (number of fights involved in, number of attacks, number of fights won). Pigs can be classified according to two coping styles: the proactive style with increased aggressiveness and the reactive style with reduced aggressiveness. A pig's reaction when placed on its back in the so-called backtest is used to test coping style. A proactive pig shows more resistance and struggling in the backtest. The result of the backtest is heritable in pigs (Rohrer et al., 2013). A proactive coping style is related to low activity in the hypothalamic-pituitary-adrenal (HPA) axis and low parasympathetic activity accompanied by high levels of sympathetic and testosterone activity. A reactive coping style is related to elevated HPA axis and parasympathetic activity and low levels of sympathetic activation and testosterone levels (Koolhaas and Boer, 2008). Amygdala and hippocampus are involved in the regulation of the HPA axis. Ponsuksili et al. (2015) identified several markers on chromosome 12 associated with pigs' reaction in a backtest. In a recent study, Ponsuksili and co-authors also demonstrated an interplay between haplotypes associated with coping behaviour in this region and transcriptome profiles in amygdala and hippocampus (Gley et al., 2019). Candidate genes were discovered which relate to the dopaminergic neurotransmitter system. Differently expressed genes also indicated immunological differences between animals with different coping styles (Gley et al., 2019). Large-scale recording of young pigs' coping style with backtest seems feasible, and it is tempting to use the test result as an indicator trait in breeding for reduced aggressive behaviour. The backtest, however, seems to tell more about the pigs' reaction to fearful situations than to the pigs' predisposition for showing aggressive behaviour (Zebunke et al., 2017).

As the development of stronger animal welfare laws deliberate sows from gestation stalls in many countries, fighting of sows becomes an issue. Eighteen percent of the pregnant sows in a nucleus herd showed aggressive behaviour at mixing, in a study by Helbrügge et al. (2007). The heritability of this binary trait was estimated at 0.32 with a threshold model. Appel et al. (2013) studied aggressive behaviour of gilts at mixing in two herds. The levels of aggression differed between herds but no genotype by environment interactions were found. The heritabilities were estimated at 0.20 for aggressive attack and 0.16 for reciprocal fighting. These traits were highly correlated (r_g = 0.95) and recording one of them would be enough for genetic evaluation. The authors conclude, "Selection for reduced aggression in group-housed animals seems to be feasible and desirable to improve the welfare of the animals" (Appel et al., 2013). In a later study, Appel et al. (2016) followed the gilts until they became sows showing maternal behaviour. The correlations between sows' response when separated from piglets one day after farrowing and aggressive behaviour (attack and fighting) were negative, that is, sows genetically predisposed to be less aggressive were also genetically predisposed to show a strong response in the separation test. This is not necessarily favourable, since the results also indicated that a strong response in separation test is genetically correlated to an increased risk of becoming a 'problem sow'. Usability (need for additional labour input) was recorded by care takers ten days after farrowing on a 3-point scale from 'no additional effort' to 'problem sow'. According to the genetic correlation, this usability was almost the same trait as sows' ability to raise a large litter of homogenous piglets according to care taker's judgement day 10 (more effort needed - more heterogenous litter, r_g = 0.91). Gilts genetically predisposed to be less aggressive were genetically predisposed to rear large, homogenous litters. The authors stress that large-scale recording of aggressive behaviour is feasible; less than two minutes per gilt was needed in their study (Appel et al., 2016). Sows' aggressive behaviour at mixing is recorded in nucleus herds by a German breeding organisation and included in the genetic evaluation of their dam line (Hubert Henne, personal communication).

Stukenborg et al. (2012) recorded the start and finish times of each aggressive interaction, the identity of the aggressor and the receiver, and whether they were the winner or loser of the fight during 48 hours, at three occasions (weaners, growing pigs and gilts). The highest heritability was estimated for the number of won fights for growing pigs (h^2=0.37). For piglets at weaning, the highest heritability was estimated for time spent in initiated fights (h^2=0.20). The genetic correlations between aggressive behaviour at weaning and later in life, when being a growing pig or a gilt, were generally low (Stukenborg et al., 2012). Within occasion, the genetic correlation between number of initiated and won fights was stronger than the correlation between number of initiated and lost fights. This relationship - pigs initiating many fights win most of the fights - was later confirmed by Sheffler et al. (2016). Performing aggressive behaviour is a more heritable trait than being a victim (Løvendahl et al., 2005). Likewise, the heritability of number of skin lesions on anterior part of the body is higher than the heritability of skin lesions on the posterior part (Wurtz et al., 2017). Based on all these studies, selection against performing aggressive behaviour at mixing seems feasible, but recording aggressive behaviour comes with additional costs for the breeding organisation. The genetic correlation between aggressive behaviour of growing pigs and aggressive behaviour of sows needs to be further investigated.

Skin lesions have been used as an indicator trait for aggressive behaviour (Turner et al., 2009). The number of lesions is a heritable trait both after mixing and in stable groups (Desire et al., 2015). The genetic correlations between number of fresh lesions after mixing at ten weeks of age and five weeks later (in the same groups) were moderate for lesions on the posterior body (r_g = 0.46) and high for lesions on the anterior body (r_g = 0.76). The position of the lesions is crucial for their use as indicators of aggressive behaviour. Posterior lesions are genetically correlated to receiving aggression and anterior lesions to reciprocal fighting and delivering aggression in nonreciprocal interactions (Turner et al., 2009). According to Desire et al. (2016), skin lesions in the front recorded shortly after mixing can be used in selection against both reciprocal and nonreciprocal aggression after mixing. Such selection may not only increase welfare, but also increase loin muscle area without negative effects on growth rate (Wurtz et al., 2017).

6 Direct and social effects on growth rate and reproduction traits

A pig's performance is governed by its genotype and influenced by the environment. Often the most important elements of this environment are the other pigs in the pen and their influence is in turn governed by their genotypes. The group mates may be genetically predisposed to influence other pigs in a favourable or an unfavourable way. In the group model (also called the social model) there are two different genetic effects; the direct effect which explains the pigs' own capacity to perform, and the social (indirect) effect which explains the effect of group members on a pig's performance; that is,the genetic ability to influence other pigs in the group (Muir, 2005; Bijma et al., 2007). In pigs, the group model including these two genetic effects has mostly been used for analysing growth rate.

The social genetic effect on growth rate is most likely a consequence of 'good' or 'bad' behaviour, but the group model does not per se describe any behaviours. It merely describes an outcome of social interactions in a group and this outcome is the growth rate of all pigs in that group. If the pigs compete over a limited resource, for example restricted feeding, the genetic correlation between the direct and the social effect will be negative and unfavourable. Such a negative correlation was found by Muir (2005) in quails. If the correlation is negative, selection for pigs with high direct breeding values will have a negative effect on the growth rate of the whole group. A zero correlation between the direct and social effects means that selection for the direct effect on growth rate will not have a negative influence on the growth rate of the group members. Even so, selection for growth rate based on a group model including both the direct and social genetic effects utilize the full genetic variation and will result in a faster genetic progress than selecting for growth rate with a model only including a direct genetic effect of the animal (Rodenburg et al., 2010). In a competitive environment where the genetic correlation between the direct and the social effect is negative, the motive for using the group model in the genetic evaluation is even stronger.

The group model assumes that an animal has the same social genetic effect on all other group members although it is known from behavioural studies that the number of social interactions differ between dyads in the group (Løvendahl et al., 2005). The number of potential interactions of course increases with group size and an effect of a variable group size can be handled by including a dilution factor in the genetic model (Bijma, 2010a). Very large data sets are needed to get accurate estimates of the genetic parameters and ideally, the groups should consist of pigs from two litters when applying the group model (Bijma, 2010b) but this is difficult to achieve in practice. Another difficulty when analysing growth rate with the social model is that some pigs are lost due to injuries or even death and they will not get records on growth rate (as discussed by Bunter et al., 2015). Assuming that the group model partly describes a consequence of receiving aggressive interactions, these pigs could have been the most informative ones if they were not lost. It would be good to include information on their growth rate until the day they were taken out of the group.

The group model has been used to analyse growth rate of pigs by for example, Bergsma et al. (2008), Chen et al. (2009a), Canario et al. (2012) and others. No behavioural records are needed in such studies, but we often assume that social genetic effects are related to competition and aggressive behaviour (Rodenburg et al., 2010). We will come back to behavioural differences between pigs with high and low social breeding values.

Nielsen et al. (2018) state that the group model "has not yet been shown to consistently improve predictive ability [of breeding values] compared to the classical animal model" and hypothesize that one reason for this is that direct and social genetic effects differ between males and females. They analysed growth data (30–94 kg) from 80 000 pigs kept in sex sorted groups with an average group size of around 11 pigs. The pigs were fed ad libitum. No pigs were castrated. Growth rate in males and females were handled as different traits in a bivariate analysis. Significant social genetic variances were found for both sexes and the total heritability estimates (including direct and social variance) were 0.32 for males and 0.27 for females. The genetic correlation between the social genetic effects of males and females was rather low (rg =0.30), indicating that growth rate is not the same trait in both sexes. The group model improved the predictive ability (i.e. the correlation between breeding values and phenotype values of growth rate) for both sexes. The genetic correlation between the direct and the social effect differed between sexes (-0.04 for males and -0.22 for females), indicating competition among females but not males (Nielsen et al., 2018). The authors admit that this is surprising, since entire males generally show more aggressive behaviour than females.

Group-housed sows in general have better welfare than sows kept in individual stalls and consequently individual stalls for pregnant sows are forbidden in the EU. For some sows the welfare in these group pens is, however, low due to aggressive interactions and competition for resources. The hypothesis of Bunter et al. (2015) was that some sows may be better adapted to group housing. They used the group model to analyse reproductive data from around 8000 sows that were group housed during gestation. The sows were mixed after pregnancy test and stayed in the groups until a few days before farrowing. The pens were designed for a maximum number of 4, 8 or 10 sows (all pens with 1.5 m^2 per animal) and a majority of the groups included the maximum number of animals. Total number of piglets born and number born alive were analysed with different versions of the group model. The direct heritabilities for litter sizes were estimated at 0.11–0.12. The social heritability estimates were much lower (0.001–0.002) and the total heritabilities were estimated at 0.14–0.16. Applying dilution factors to handle the range of group size did not improve the fit of the model. The authors propose that dilution factors are relevant only when resources are fixed per group and not per individual as in this study. The model fit was however much improved by including an effect of number of sows in relation to the maximum number for the given pen size, i.e. a combination of number of sows to interact with and area per sow. When this effect was included in the model, the correlations between the direct and the social genetic effects were close to zero for both litter sizes. Without this number-area-effect, the correlations between the direct and the social effect were estimated at 0.3–0.4 (Bunter et al., 2015). This difference between models raises questions about previous positive correlation estimates from the group model. It should, however, be remembered that growth rate, the trait typically analysed with the group model, may be directly influenced by other pigs competing more or less over feed resources as both the direct and the social genetic effects may partly reflect feed intake. For litter size, the direct genetic effect reflects ovulation rate and embryo survival whereas the social genetic effect must reflect an indirect genetic effect of, for example, aggressive behaviour causing stress (and thus embryonic loss), as discussed by Bunter et al. (2015).

7 Characteristics of pigs with high social breeding values for growth rate

Different behaviours have been compared for pigs with high and low social breeding values for growth rate. Pigs with high social breeding values seem to be calmer (more time lying down) in stable groups (Canario et al., 2012). In a small study by Hong et al. (2018a), pigs' behaviour was recorded with scan sampling during day time. One day after mixing and one month after mixing, pigs with high social breeding values spent less time on aggressive behaviour and more time on feeding than pigs with low social breeding values. They were also more often feeding together with other pigs. Two weeks after mixing, the results were less clear.

The aggressive behaviour of pigs with high social breeding values for growth rate is complicated to describe and interpret. Pigs with a high social breeding value tend to show more aggressive behaviour at mixing, but less aggressive behaviour later on (in stable groups) as compared to pigs with low social breeding values (Rodenburg et al., 2010; Canario et al. 2012). This has been interpreted as a higher ability of pigs with high social breeding values to quickly form dominance relationships in a new group. The study by Canario et al. (2012) indicates that the behaviour of pigs with high social breeding values differs with the level of competition in the group. When the environment is competitive (negative correlation between the direct and the social genetic effect), pigs with high social breeding values initiate less fights and lose more fights after mixing than pigs with low social breeding values. When there is a positive correlation, they initiate and win more fights after mixing, thereby actively participating in a quick establishment of the social hierarchy within the group.

Camerlink (2014) and Reimert (2014) both studied pigs from a one-generation selection experiment. Half of the pigs were offspring of sows and boars with extremely low social breeding values for growth rate (25-110 kg) and the other half were offspring from sows and boars with extremely high social breeding values. The direct breeding value for growth rate was equal between both the groups. The offspring (480 pigs) were reared in five batches and their behaviour was tested at several occasions. Half of them were kept in barren pens and half of them in enriched pens.

Before weaning, no difference between groups was seen in a backtest (Reimert et al., 2013). When tested in a novel object test and in a human approach test, high-social pigs were quicker to touch the object and more frequently present near humans (Reimert et al., 2013). When tested in the home pen a few weeks after weaning, high-social pigs were faster to touch a novel object but no difference was seen in a human approach test (Reimert et al., 2014a). The pigs were tested again at 13 weeks of age, this time in a novel environment. Fourteen different behaviours were recorded and the only significant difference was that high-social pigs showed less locomotion after the introduction of the novel object (Reimert et al., 2014a). No difference between groups was found in salivary cortisol measured after this test. In summary, the authors conclude that high-social pigs "might be less fearful" than low-social pigs.

Camerlink et al. (2015) found that high-social pigs showed less ear biting and chew less on jute sacks (more chewing is an indicator of pigs being predisposed to tail biting). The behavioural differences between pigs kept in barren and enriched pens were generally larger than the differences between high- and low-social pigs and no genotype by environment interactions were found. Genetic and environmental improvements are thus additive which increases the opportunities to work for improved welfare. There was no difference between high-social and low-social pigs in the number of skin lesions or number of reciprocal fights in stable groups or in confrontation with unfamiliar pigs in a re-grouping test at 9 weeks of age (Camerlink et al., 2013). Pigs from the high-social group did, however, perform less non-reciprocal fighting and showed less aggressive behaviour when meeting familiar group members after 24 h of separation during the re-grouping test (Camerlink et al., 2013). A re-grouping test is a stressful event and Reimert et al. (2014b) analysed several blood variables before and after the re-grouping test. The results indicated that high-social pigs were less affected by stress.

Dervishi et al. (2018) analysed blood metabolites on a small sample of the pigs studied by Camerlink, Reimerts and co-workers. At an age of 22 weeks, high-social pigs had lower concentration of serotonin and lower ratio of tryptophan/kynurenine (tryptophan is the precursor of serotonin) which might indicate less aggression. They also had lower concentration of epinephrine and higher concentration of phenyl-ethylamine which might indicate a lower activity of the sympathetic-adrenal-medullary system and thus less stressed animals (Dervishi et al., 2018).

In a genome-wide association study, the average growth rate of unrelated pen mates was used as the phenotype (Hong et al., 2018b). The heritability for this trait was estimated at 0.14. Around 1000 Landrace pigs were genotyped and 5 SNPs, all on chromosome 6, were detected. The authors point out that one of these markers is located upstream of the gene interferon-induced protein 44 (IF144); a gene known to influence stress-induced diseases in humans. This fits well with Reimert's (2014) conclusion that pigs with high social genetic breeding values for growth rate "may indeed be pigs that are less easily stressed". Wu et al. (2019) estimated direct and social breeding values for growth rate and performed a whole-genome re-sequencing association study for these effects in Large White pigs. They detected 151 SNPs, pointing to 42 candidate genes, for the direct breeding values and 205 SNPs, pointing to 54 candidate genes, for the social breeding values. One of the candidate genes for both direct and social effects is a protein coding gene called GABRR2. Wu et al. (2019) refer to several studies showing that this gene is involved in behavioural stress response as well as aggressiveness in animals.

Based on social breeding values for growth rate in performance test, gilts were sorted into two groups; one with high and one with low social breeding values (Hong et al., 2017). The gilts (in total 124 animals) were thereafter followed for several parities. Gilts with high social breeding values were younger at first farrowing. As sows, they gave birth to larger litters and had shorter weaning to oestrous intervals (Hong et al., 2017).

In summary, all studies on social genetic effects together show that the social genetic model has the potential to improve animal welfare, but the background of social interactions among pigs is complex. The 'best' model also seems to differ between populations (Kim et al., 2016). Some breeding companies already estimate direct and social genetic breeding values for growth rate and apply them in thegenetic evaluation, expecting benefit in reduced damaging behaviour (Egbert Knol, personal communication). There are, however, still questions regarding the optimum weight of behavioural traits in the breeding goal.

8 Conclusion and future trends

Most behavioural traits seem to be governed by many genes, each with a rather small but additive effect on the phenotype. Therefore, marker-assisted selection (MAS) seems less promising today than when the method was introduced in the 1980s. Selection for changed behaviour based on best linear unbiased prediction (BLUP) breeding values is probably a better alternative than MAS for most traits. With genomic BLUP breeding values (based on a combination of data from SNP markers from all pigs and phenotypic data from pigs in a reference population) breeding values can be estimated with higher accuracies than traditional BLUP values. Expressed in another way, less phenotype records are needed for estimating G-BLUP values with a certain accuracy as compared to ordinary BLUP values. Genomic selection is especially relevant for traits that are complicated and expensive to record, which is true for many behavioural traits. In genomic selection, knowledge about markers from genome-wide association studies like the ones previously mentioned in this chapter can improve the accuracy of genetic evaluation of behavioural traits even further. Also the accuracy of direct and social breeding values for growth rate can be improved by genomic selection (Hong et al., 2019). Genomic selection on social genetic effects however requires genotyping of almost all pigs in the pen, as discussed by Duijvesteijn (2014). This may not be done today, but could become affordable with a decreasing genotyping price.

New methods for gene editing are rapidly developing and opportunities for animal breeding are discussed (Van Eenennaam, 2017). Given the complex genetic and physiological background of behavioural traits, it seems unlikely that gene editing will be used to change the behaviour of pigs in future breeding programs. Nevertheless, gene editing may have indirect consequences on behavioural traits. Assume, for example, that boar taint can be reduced by gene editing. Then all male pigs could be reared as entire males instead of castrates. Assume also that gene editing changes the metabolism of androstenone (a pheromone causing off-flavour in pork) without influencing the level of testosterone and other hormones. That means the gene-edited pigs would show as much aggressive behaviour as today's entire males. In such a scenario, the motives for selection against aggressive behaviour (based on breeding values) would be even stronger than today.

In this chapter, genetic studies of allele effects have been reviewed from a molecular and from a quantitative genetic angle. A possible future of animal production may be to use epigenetic effects to influence gene expression. Epigenetic effects influence the phenotype by regulation of gene expression rather than alterations in the DNA sequence. Assume epigenetic effects on genes involved in aggressive behaviour can be transmitted from generation to generation as reviewed by David et al. (2019). Further assume that a good treatment of parents early in life influences DNA methylation of their offspring in such a way that alleles associated with aggressive behaviour are turned off. Then providing young AI-boar candidates in sire lines with the very best environment (enriched environment, tailor-made feed, etc.) could make a huge difference for the welfare of all pigs raised for slaughter. The number of these young boars is low (since they can be selected with a high accuracy based on genomic breeding values early in life) and thus it would be affordable to give them an enriched, more expensive environment.

It is known from several studies that gut microbiota influences social behaviour (reviewed by Sherwin et al., 2019) and the genotype of the pig influences the establishment of the microbiota (Camarinha-Silva et al., 2017). The relationships between microbiota, immune system and stress reactivity may play a role in the development of tail biting. In the synthesis article of

Brunberg et al. (2016) we discussed the possibility that tail biting pigs have a different composition of gut microbiota than pigs that do not show tail biting. If yes, is the microbiota simply transferred from the sow to its piglets? Or is being a good host for good microbiota a heritable trait? That would open up for a future possibility to select pigs with favourable microbiota, in order to decrease tail biting.

The development of the group model is still ongoing. Canario et al. (2017) showed that non-genetic social factors on growth rate are larger than genetic social factors. Piglets sharing the same environment early in life develop common social skills that generate early-life social effects (ELSEs). These ELSEs can have a strong influence on the penmates later in life. Canario et al. (2017) estimated both genetic social effects and ELSEs and showed that including ELSEs reduces bias when estimating social genetic breeding values.

Until now, the group model has mostly been used for analysing performance traits. It can, however, also be used for behavioural traits such as tail-biting (Canario and Flatres-Grall, 2018) or traits that are indicators of aggressive behaviour, such as skin lesions. Angarita et al. (2019) analysed skin lesions of 792 growing pigs in 59 pens with the group model. Aggressive behaviour within each dyad was recorded at the day of mixing. The skin lesions were counted on the following day. An ordinary group model assuming uniform interactions between dyads was compared to an intensity-based group model. Both models included the total time in aggressive interactions per pig. The intensity-based group model including unilateral attack behaviour resulted in higher heritability estimates for lesion scores. The correlations were positive for all lesion scores. Angarita et al. (2019) conclude that if behavioural observations are available, selection incorporating social genetic effects "may reduce the consequences [i.e. the skin lesions] of aggressive behaviours after mixing pigs".

The number of phenotype records is still a severe limitation when performing genetic studies of behavioural traits. Rodenburg et al. (2019) argue that a combined sensor and genomics approach can facilitate genetic studies of behavioural traits. Image-based recording methods, and sensors keeping track of individuals in a group, pave the way for new selection traits. The breeding goal can be either the behaviour itself, or the behaviour can be used as an indicator for a goal trait like robustness, for example, drinking behaviour during heat waves.

It would be interesting to study the neutral pigs, not involved in tail biting or aggressive interactions, more in depth. Can breeding values be estimated for the ability to stay neutral? How is this ability correlated to other traits, like feeding behaviour and calmness? Rauw et al. (2017) state, "Although it is generally proposed that animal welfare improves with selection for calmer animals that are less fearful and reactive to novelty, animals bred to be less sensitive with fewer desires may be undesirable from an ethical point of view."

As for other ethical questions, different persons may have different views on selection for less responsive pigs.

Selection against tail biting and aggressive behaviour seems possible and yet it is not common. More studies can always be asked for, for example, further studies of potential side effects of selection against aggressive behaviour based on behavioural observations or skin lesions. What would, for example, be the consequences for the care takers? König von Borstel et al. (2018) reported that skin lesions after mixing has a negative genetic correlation with agitation at weighing (less lesions - more agitated pigs) whereas D'Eath et al. (2009) reported that aggressive behaviour at mixing has a positive genetic correlation with activity during weighing (more aggressive - more activity). Both very agitated pigs and very non-active pigs may lead to more work for the care takers.

Peden et al. (2018) ask why "in spite of the amount of research on reducing aggression at mixing the problem has not reduced in intensive farming systems". They conclude that the industry seems to give low priority to the problem and recommend "a better alignment between research questions and industry interests to increase the success of research efforts to improve animal welfare in practice." Research results may also influence the industry indirectly, as a base for animal welfare laws. Changes in legislation can, potentially, change the economic weights given to different traits in the breeding goal. Tail docking has become forbidden in EU, and that should increase breeding organisations' interest in selection against tail biting. So far, legal consequences of tail docking are, however, negligible in many countries and neither breeding goals nor management routines seem to change. Value shifts in society, leading to consumers not accepting certain management routines, may become a future driver for changed breeding goals. Although many questions are unsolved, there is knowledge already available that can be implemented by breeding organisations ready for change.

9 Where to look for further information

Many doctoral theses include a general discussion that is easy to read and full of valuable information and interesting ideas. These texts are missed when only looking for articles published in scientific journals. I especially recommend the theses by Camerlink, Reimert, Duijvesteijn and Ursinus, all from Wageningen University, 2014. Some key researchers in genetics of pig behaviour are Laurianne Canario (INRA, France), Pieter Bijma (Wageningen University, The Netherlands) and Simon Turner (SRUC, UK) who have a long history of fruitful cooperation. The yearly conference of the European Federation for Animal Science (EAAP) gives good opportunities for meeting geneticists interested in animal behaviour.

10 Personal communications

Dr. Egbert Knol, Topigs Norsvin Research Center B.V., Beuningen, The Netherlands; 2019-11-11.

Dr. Hubert Henne, BHZP GmbH, Dahlenburg – Ellringen, Germany. 2019-11-14.

11 References

Angarita, B. K., Cantet, R. J. C., Wurtz, K. E., O'Malley, C. I. O., Siegford, J. M., Ernst, C. W., Turner, S. P. and Steibel, J. P. (2019). Estimation of indirect social genetic effects for skin lesion count in group-housed pigs by quantifying behavioral interactions. *J. Anim. Sci.* 97(9), 3658–3668.

Appel, A. K., Voß, B., Tönepöhl, B., König von Borstel, U. and Gauly, M. (2013). Variance components of aggressive behavior in genetically highly connected Pietrain populations kept under two different housing conditions. *J. Anim. Sci.* 91(12), 5557–5564.

Appel, A. K., Voß, B., Tönepöhl, B., König von Borstel, U., Gauly, M. (2016). Genetic associations between maternal traits and aggressive behaviour in Large White sows. *Animal* 10(7), 1234–1242.

Bauer, J. (2019). Investigation of the genetic components of maternal infanticide in Sus Scrofa. Doctoral thesis, University of Cambridge, UK. https://doi.org/10.17863/CAM.40257.

Baxter, E. M., Jarvis, S., Sherwood, L., Farish, M., Roehe, R., Lawrence, A. B. and Edwards, S. A. (2011). Genetic and environmental effects on piglet survival and maternal behaviour of the farrowing sow. *Appl. Anim. Behav. Sci.* 130(1–2), 28–41.

Bergsma, R., Kanis, E., Knol, E. F. and Bijma, P. (2008). The contribution of social effects to heritable variation in finishing traits of domestic pigs (Sus scrofa). *Genetics* 178(3), 1559–1570.

Bijma, P., Muir, W. M. and Van Arendonk, J. A. M. (2007). Multilevel selection 1: quantitative genetics of inheritance and response to selection. *Genetics* 175(1), 277–288.

Bijma, P. (2010a). Multilevel selection 4: modeling the relationship of indirect genetic effects and group size. *Genetics* 186(3), 1029–1031.

Bijma, P. (2010b). Estimating indirect genetic effects: precision of estimates and optimum designs. *Genetics* 186(3), 1013–1028.

Breuer, K., Sutcliffe, M. E. M., Mercer, J. T., Rance, K. A., Beattie, V. E., Sneddon, I. A. and Edwards, S. E. (2005). The effect of breed on the development of adverse social behaviours in pigs. *Appl. Anim. Beh. Sci.* 84, 59–74.

Brunberg, E., Jensen, P., Isaksson, A. and Keeling, L. J. (2013a). Behavioural and brain gene expression profiling in pigs during tail biting outbreaks - Evidence of a tail biting resistant phenotype. *PLoS ONE* 8(6), e66513.

Brunberg, E., Jensen, P., Isaksson, A. and Keeling, L. J. (2013b). Brain gene expression differences are associated with abnormal tail biting behavior in pigs. *Genes Brain Behav.* 12(2), 275–281.

Brunberg, E. I., Rodenburg, T. B., Rydhmer, L., Kjaer, J. B., Jensen, P. and Keeling, L. J. (2016). Omnivores going astray: a review and new synthesis of abnormal behavior in pigs and laying hens. *Front. Vet. Sci.* 3, 57.

Bunter, K. L., Lewis, C. R. and Newman, S. (2015). Social genetic effects influence reproductive performance of group-housed sows. *J. Anim. Sci.* 93(8), 3783-3793.

Camarinha-Silva, A., Maushammer, M., Wellmann, R., Vital, M., Preuss, S. and Bennewitz, J. (2017). Host genome influence on gut microbial composition and microbial prediction of complex traits in pigs. *Genetics* 206(3), 1637-1644.

Camerlink, I., Turner, S. P., Bijma, P. and Bolhuis, J. E. (2013). Indirect genetic effects and housing conditions in relation to aggressive behaviour in pigs. *PLOS ONE* 8(6), e65136.

Camerlink, I. (2014). Sociable swine: indirect genetic effects on growth rate and their effect on behaviour and production of pigs in different environments. PhD thesis, Wageningen University, the Netherlands.

Camerlink, I., Ursinus, W. W., Bijma, P., Kemp, B. and Bolhuis, J. E. (2015). Indirect genetic effects for growth rate in domestic pigs alter aggressive and manipulative biting behaviour. *Behav. Genet.* 45(1), 117-126.

Canario, L., Turner, S. P., Roehe, R., Lundeheim, N., D'Eath, R. B., Lawrence, A. B., Knol, E., Bergsma, R. and Rydhmer, L. (2012). Genetic associations between behavioral traits and direct-social effects of growth rate in pigs. *J. Anim. Sci.* 90(13), 4706-4715.

Canario, L., Lundeheim, N. and Bijma, P. (2017). The early-life environment of a pig shapes the phenotypes of its social partners in adulthood. *Heredity* 118(6), 534-541.

Canario, L. and Flatres-Grall, L. (2018). Genetics of tail-biting receipt in gilts from the Tai Zumu line. *Proc. World Congress Genetics Applied to Livest. Prod. (WCGALP)* 11, 572.

Chen, C., Gilbert, C. L., Yang, G., Guo, Y., Segonds-Pichon, A., Ma, J., Evans, G., Brenig, B., Sargent, C., Affara, N. and Huang, L. (2008). Maternal infanticide in sows: incidence and behavioural comparisons between savaging and non-savaging sows at parturition. *Appl. Anim. Beh. Sci.* 109(2-4), 238-248.

Chen, C. Y., Johnson, R. K., Newman, S., Kachman, S. D. and van Vleck, L. D. (2009a). Effects of social interactions on empirical responses to selection for average daily gain of boars. *J. Anim. Sci.* 87(3), 844-849.

Chen, C., Guo, Y., Yang, G., Yang, Z., Zhang, Z., Yang, B., Yan, X., Perez-Enciso, M., Ma, J., Duan, Y., Brenig, B. and Huang, L. (2009b). A genome wide detection of quantitative trait loci on pig maternal infanticide behavior in a large scale White Duroc x Erhualian resource population. *Behav. Genet.* 39(2), 213-219.

Colpoys, J. D., Abell, C. E., Young, J. M., Keating, A. F., Gabler, N. K., Millman, S. T., Siegford, J. M. and Johnson, A. K. (2014). Effects of genetic selection for residual feed intake on behavioral reactivity of castrated male pigs to novel stimuli tests. *Appl. Anim. Beh. Sci.* 159, 34-40.

Cross, A. J., Keel, B. N., Brown-Brandl, T. M., Cassady, J. P. and Rohrer, G. A. (2018). Genome-wide association of changes in swine feeding behaviour due to heat stress. *Genet. Sel. Evol.* 50(1), article number 11.

Daigle, C. (2018). Parallels between postpartum disorders in humans and preweaning piglet mortality in sows. *Animals* 8(2), 22 (11 pp).

David, I., Canario, L., Combes, S. and Demars, J. (2019). Intergenerational transmission of characters through genetics, epigenetics, microbiota, and learning in livestock. *Front. Genet.* 10, 1058.

D'Eath, R. B., Roehe, R., Turner, S. P., Ison, S. H., Farish, M., Jack, M. C. and Lawrence, A. B. (2009). Genetics of animal temperament: aggressive behaviour at mixing is genetically associated with the response to handling in pigs. *Animal* 3(11), 1544–1554.

Dervishi, E., Zande, L., Reimert, I., Pramod, M. K., Knol, E. F. and Plastow, G. S. (2018). A pilot study to investigate the relationship between social behavior and serum metabolite concentration in finisher pigs. *J. Anim. Sci.* 96, Issue supplement 3. - ISSN 0021-8812 - p. 12 - 12.

Dervishi, E., van der Zande, L., Valente, S. T. da, Reimert, I., Pramod, M. K., Knol, E. F., Plastow, G. S. and Lopes, M. S. (2019). Combining metabolomics and genomics to elucidate physiological processes related to tail damage score in pigs. In: *Proceedings of the 37th International Society for Animal Genetics Conference (ISAG)*, pp. 147–147.

Desire, S., Turner, S. P., D'Eath, R. B., Doeschl-Wilson, A. B., Lewis, C. R. G. and Roehe, R. (2015). Genetic associations of short- and long-term aggressiveness identified by skin lesions with growth, feed efficiency, and carcass characteristics in growing pigs. *J. Anim. Sci.* 93(7), 3303–3312.

Desire, S., Turner, S. P., D'Eath, R. B., Doeschl-Wilson, A. B., Lewis, C. R. and Roehe, R. (2016). Prediction of reduction in aggressive behaviour of growing pigs using skin lesion traits as selection criteria. *Animal* 10(8), 1243–1253.

Ding, R., Quan, J., Yang, M., Wang, X., Zheng, E., Yang, H., Fu, D., Yang, Y., Yang, L., Li, Z., Liu, D., Cai, G., Wu, Z. and Yang, J. (2017). Genome-wide association analysis reveals genetic loci and candidate genes for feeding behavior and eating efficiency in Duroc boars. *PLoS ONE* 12(8), e0183244.

Duijvesteijn, N. (2014). Sociable Swine: prospects of indirect genetic effects for the improvement of productivity, welfare and quality. PhD thesis, Wageningen University, the Netherlands.

Gäde, S., Bennewitz, J., Kirchner, K., Looft, H., Knap, P. W., Thaller, G. and Kalm, E. (2008). Genetic parameters for maternal behaviour traits in sows. *Livest. Sci.* 114(1), 31–41.

Gilbert, C. L. (2001). Endocrine regulation of periparturient behaviour in pigs. *Reprod. Suppl.* 58, 263–266.

Gley, K., Murani, E., Trakooljul, N., Zebunke, M., Puppe, B., Wimmers, K. and Ponsuksili, S. (2019). Transcriptome profiles of hypothalamus and adrenal gland linked to haplotype related to coping behavior in pigs. *Sci. Rep.* 9(1), article number 13038.

Grandinson, K., Lund, M. S., Rydhmer, L. and Strandberg, E. (2002). Genetic parameters for the piglet mortality traits crushing, stillbirth and total mortality, and their relation to birth weight. *Acta Agric. Scand. A* 52(4), 167–173.

Guo, Y. M., Zhang, Z. Y., Ma, J. W., Ai, H. S., Ren, J. and Huang, L. S. (2015). A genomewide association study of feed efficiency and feeding behaviors at two fattening stages in a White Duroc × Erhualian F population. *J. Anim. Sci.* 93(4), 1481–1489.

Hellbrügge, B., Tölle, K.-H., Henze, C., Presuhn, U., and Krieter, J. (2007). A note on genetic parameters for human-approach tests on gilts. In: *Genetic Aspects of Piglet Losses and the Maternal Behaviour of Sows. Schriftenreihe Des Instituts Fur Tierzücht und Tierhaltung Der Christian-Albrechts-Universität zu Kiel*, Heft, 158.

Hellbrügge, B., Tölle, K. H., Bennewitz, J., Henze, C., Presuhn, U. and Krieter, J. (2008). Genetic aspects regarding piglet losses and the maternal behaviour of sows, Part 2. Genetic relationship between maternal behaviour in sows and piglet mortality. *Animal* 2(9), 1281–1288.

Hong, J. K., Kim, Y. M., Cho, K. H., Park, J. C. and Lee, D. H. (2017). Reproductive performance of sows selected for divergent social genetic effects for growth. *Anim. Reprod.* 14(Sup. 1), 1292–1297.

Hong, J. K., Kim, K. H.,Song, N. R., Choi, T., Hwang, H. S., Lee, J. K., Eom, T. K. and Rhim, S. J. (2018a). Social behavior and group growth of finishing pigs with divergent social breeding values. *Rev. Colom Cienc. Pecua* 31(4), 267–275.

Hong, J. K., Jeong, Y. D., Cho, E. S., Choi, T. J., Kim, Y. M., Cho, K. H., Lee, J. B., Lim, H. T. and Lee, D. H. (2018b). A genome-wide association study of social genetic effects in Landrace pigs. *Asian-Australas. J. Anim. Sci.* 31(6), 784–790.

Hong, J. K., Kim, Y. S., Cho, K. H., Lee, D. H., Min, Y. J.and Cho, E. S. (2019). Application of single-step genomic evaluation using social genetic effect model for growth in pig. *Asian-Australas. J. Anim. Sci.* 32(12), 1836–1843.

Jarvis, S., Reed, B. T., Lawrence, A. B., Calvert, S. K. and Stevenson, J. (2004). Peri-natal environmental effects on maternal behaviour, pituitary and adrenal activation, and the progress of parturition in the primiparous sow. *Anim. Welf.* 13(2), 171–181.

Jonas, E. and Rydhmer, L. (2018). Effect of candidate genes for maternal ability on piglet survival and growth. *Livest. Sci.* 207, 83–90.

Kavlak, A. T. and Uimari, P. (2019). Estimation of heritability of feeding behaviour traits and their correlation with production traits in Finnish Yorkshire pigs. *J. Anim. Breed. Genet.* 136(6), 484–494.

Kim, Y. M., Cho, E. S., Cho, K. H., Sa, S. J., Jeong, Y. D., Woo, J. S., Lee, I. J. and Hong, J. K. (2016). Social genetic effects on days to 90 kg in Duroc and Yorkshire pigs. Korean. *J. Agr. Sci.* 43(4), 595–602.

König von Borstel, U., Tönepöhl, B., Appel, A. K., Voß, B., Brandt, H., Naderi, S. and Gauly, M. (2018). Suitability of traits related to aggression and handleability for integration into pig breeding programmes: genetic parameters and comparison between Gaussian and binary trait specifications. *PLoS ONE* 13(12), 17 pp.

Koolhaas, J. M. and Boer, S. F. (2008). Coping styles and aggression: a biobehavioral approach. In: Vingerhoets, A. J. J. M., Nyklíček, I. and Denollet, J. (Eds), *Emotion Regulation: Conceptual and Clinical Issues*. Springer, Boston, US, pp. 12–26.

Lan, J., Lei, M. G., Zhang, Y. B., Wang, J. H., Feng, X. T., Xu, D. Q., Gui, J. F. and Xiong, Y. Z. (2009). Characterization of the porcine differentially expressed PDK4 gene and association with meat quality. *Mol. Biol. Rep.* 36(7), 2003–2010.

Løvendahl, P., Damgaard, L. H., Nielsen, B. L., Thodberg, K., Su, G. and Rydhmer, L. (2005). Aggressive behaviour of sows at mixing and maternal behaviour are heritable and genetically correlated traits. *Livest. Prod. Sci.* 93(1), 73–85.

Meunier-Salaün, M. C., Guérin, C., Billon, Y., Sellier, P., Noblet, J. and Gilbert, H. (2014). Divergent selection for residual feed intake in group-housed growing pigs: characteristics of physical and behavioural activity according to line and sex. *Animal* 8(11), 1898–1906.

Muir, W. M. (2005). Incorporation of competitive effects in forest tree or animal breeding programs. *Genetics* 170(3), 1247–1259.

Nielsen, H. M., Ask, B. and Madsen, P. (2018). Social genetic effects for growth in pigs differ between boars and gilts. *Genet. Sel. Evol.* 50(1), 4.

Ocepek, M., Rosvold, E. M., Andersen-Ranberg, I. and Andersen, I. L. (2017). Can we improve maternal care in sows? Maternal behavioral traits important for piglet survival in loose-housed sow herds. *J. Anim. Sci.* 95(11), 4708–4717.

Ocepek, M. and Andersen, I. L. (2018). Sow communication with piglets while being active is a good predictor of maternal skills, piglet survival and litter quality in three different breeds of domestic pigs (Sus scrofa domesticus). *PLoS ONE* 13(11), e0206128.

Peden, R. S. E., Turner, S. P., Boyle, L. A. and Camerlink, I. (2018). The translation of animal welfare research into practice: the case of mixing aggression between pigs. *Appl. Anim. Beh. Sci.* 204, 1–9.

Ponsuksili, S., Zebunke, M., Murani, E., Trakooljul, N., Krieter, J., Puppe, B., Schwerin, M. and Wimmers, K. (2015). Integrated Genome-wide association and hypothalamus eQTL studies indicate a link between the circadian rhythm-related gene PER1 and coping behavioral. *Sci. Rep.* 5, 16264.

Rauw, W. M., Johnson, A. K., Gomez-Raya, L. and Dekkers, J. C. M. (2017). A hypothesis and review of the relationship between selection for improved production efficiency, coping behavior, and domestication. *Front. Genet.* 8, 134.

Reimert, I., Rodenburg, T. B., Ursinus, W. W., Duijvesteijn, N., Camerlink, I., Kemp, B. and Bolhuis, J. E. (2013). Backtest and novelty behavior of female and castrated male piglets, with diverging social breeding values for growth. *J. Anim. Sci.* 91(10), 4589–4597.

Reimert, I. (2014). (Em)pathetic pigs?: the impact of social interactions on welfare, health and productivity. PhD thesis, Wageningen University, the Netherlands.

Reimert, I., Rodenburg, T. B., Ursinus, W. W., Kemp, B. and Bolhuis, J. E. (2014a). Responses to novel situations of female and castrated male pigs with divergent social breeding values and different backtest classifications in barren and straw-enriched housing. *Appl. Anim. Behav. Sci.* 151, 24–35.

Reimert, I., Rodenburg, T. B., Ursinus, W. W., Kemp, B. and Bolhuis, J. E. (2014b). Selection based on indirect genetic effects for growth, environmental enrichment and coping style affect the immune status of pigs. *PLoS ONE* 9(9), e108700.

Reyer, H., Shirali, M., Ponsuksili, S., Murani, E., Varley, P. F., Jensen, J. and Wimmers, K. (2017). Exploring the genetics of feed efficiency and feeding behaviour traits in a pig line highly selected for performance characteristics. *Mol. Genet. Genomics* 292(5), 1001–1011.

Rodenburg, T. B., Bijma, P., Ellen, E. D., Bergsma, R., de Vries, S., Bolhuis, J. E., Kemp, B. and van Arendonk, J. A. M. (2010). Breeding amiable animals? Improving farm animal welfare by including social effects in breeding programmes. *Anim. Welf.* 19(S), 77–82.

Rodenburg, T. B., Van Der Zande, L., De Haas, E. N., Košťál, L., Pichová, K., Piette, D., Tetens, J., Visser, B., DeKlerk, B., Van Der Sluis, M., Bennewitz, J., Siegford, J., Norton, T., Guzhva, O. and Ellen, E. E. (2019). Reduce damaging behaviour in laying hens and pigs by developing sensor technologies to inform breeding programs. In: Newberry, R. C. and Braastad, B. O. (Eds), *Animal Lives Worth Living Proceedings 53rd Congress of the International Society for Applied Ethology*, Norway.

Rohrer, G. A., Brown-Brandl, T., Rempel, L. A., Schneider, J. F. and Holl, J. (2013). Genetic analysis of behavior traits in swine production. *Livest. Sci.* 157(1), 28–37.

Rydhmer, L. and Jonas, E. (2016). Oxytocin candidate genes and maternal behaviour of sows. *Proceedings of the 67th Meeting of the European Federation of Animal Science, UK.* Session 60, p. 529.

Sheffler, K., Stamer, E., Traulsen, I. and Krieter, J. (2016). Estimation of genetic parameters for agonistic behaviour of pigs at different ages. *J. Agric. Sci.* 154, 732–741.

Sherwin, E., Bordenstein, S. R., Quinn, J. L., Dinan, T. G. and Cryan, J. F. (2019). Microbiota and the social brain. *Science* 366(6465).

Stratz, P., Just, A., Faber, H. and Bennewitz, J. (2016). Genetic analyses of mothering ability in sows using field-recorded observations. *Livest. Sci.* 191, 1–5.

Stukenborg, A., Traulsen, I., Stamer, E., Puppe, B., Presuhn, U. and Krieter, J. (2012). Heritabilities of agonistic behavioural traits in pigs and their relationships within and between different age groups. *Livest. Sci.* 149(1-2), 25-32.

Terenina, E., Bazovkina, D., Rousseau, S., Salin, F., D'Eath, R., Turner, S., Kulikov, A. and Mormède, P. (2012). Association entre polymorphisms de genes candidats et comportements agressifs chez le porc. *J. Rech. Porcine Fr.* 44, 45-46.

Turner, S. P., Roehe, R., D'Eath, R. B., Ison, S. H., Farish, M., Jack, M. C., Lundeheim, N., Rydhmer, L. and Lawrence, A. B. (2009). Genetic validation of postmixing skin injuries in pigs as an indicator of aggressiveness and the relationship with injuries under more stable social conditions. *J. Anim. Sci.* 87(10), 3076-3082.

Ursinus, W. (2014). A tale too long for a tail too short? Identification of characteristics in pigs related to tail biting and other oral manipulations directed at conspecifics. PhD thesis, Wageningen University, the Netherlands.

Ursinus, W. W., Van Reenen, C. G., Kemp, B. and Bolhuis, J. E. (2014). Tail biting behaviour and tail damage in pigs and the relationship with general behaviour: predicting the inevitable? *Appl. Anim. Beh. Sci.* 156, 22-36.

Van Eenennaam, A. L. (2017). Genetic modification of food animals. *Curr. Opin. Biotechnol.* 44, 27-34.

Vangen, O., Holm, B., Valros, A., Lund, M. S. and Rydhmer, L. (2005). Genetic variation in sows' maternal behaviour, recorded under field conditions. *Livest. Prod. Sci.* 93(1), 63-71.

Wallenbeck, A. and Keeling, L. J. (2013). Using data from electronic feeders on visit frequency and feed consumption to indicate tail biting outbreaks in commercial pig production. *J. Anim. Sci.* 91(6), 2879-2884.

Wilson, K., Zanella, R., Ventura, C., Johansen, H. L., Framstad, T., Janczak, A., Zanella, A. J. and Neibergs, H. L. (2012). Identification of chromosomal locations associated with tail biting and being a victim of tail-biting behaviour in the domestic pig (Sus scrofa domesticus). *J. Appl. Genet.* 53(4), 449-456.

Wu, P., Wang, K., Yang, Q., Zhou, J., Chen, D., Liu, Y.,Ma, J., Tang, Q., Jin, L., Xiao, W., Lou, P., Jiang, A., Jiang, Y., Zhu, L., Li, M., Li, X. and Tang, G. (2019). Whole-genome re-sequencing association study for direct genetic effects and social genetic effects of six growth traits in Large White pigs. *Scientific Reports* 9, article number 9667.

Wurtz, K. E., Siegford, J. M., Bates, R. O., Ernst, C. W. and Steibel, J. P. (2017). Estimation of genetic parameters for lesion scores and growth traits in group-housed pigs. *J. Anim. Sci.* 95(10), 4310-4317.

Zebunke, M., Nurnberg, G., Melzer, N. and Puppe, B. (2017). The backtest in pigs revisited–inter-situational behaviour and animal classification. *Appl. Anim. Beh. Sci.* 194, 7-13.

Chapter 2

Developmental influences on pig behaviour

Giuliana Miguel-Pacheco and Yolande M. Seddon, University of Saskatchewan, Canada

1 Introduction
2 Prenatal environmental effects on development of pig behaviour
3 Environmental effects during the early postnatal period on the development of behaviour
4 Conclusion
5 Future trends in research
6 Where to look for further information
7 References

1 Introduction

Intensive swine production has developed systems for rearing pigs with great production efficiency that can be adopted on a global scale. However, from building design to husbandry, many of the approaches give little consideration as to how a system may influence the behaviour of the pigs and the subsequent implications for their welfare. The development of behaviour arises from the interaction of intrinsic and extrinsic processes and the experiences of the individual within the environment. Whether it be stimulating or stressful, the factors experienced by the individual interact to modulate the emotional, autonomic and endocrine responses. These contribute to the development of individual differences, and influence emotionality, health and reproductive success, with implications for the ethics and efficiency of modern swine production. In the field of animal science, it has long been known that early-life influences shape the development of behaviour in diverse ways. Only in recent years has it been discovered exactly how far back this early life experience can impact the development of not only the individual but also the subsequent generations of offspring.

Covering the pre-conception, prenatal and early-postnatal period, this chapter brings together the current understanding of how factors occurring

http://dx.doi.org/10.19103/AS.2020.0081.02

within each of these periods contribute to the development of pig behaviour. The pre-conception period covers from gamete maturation to the point of conception or fertilization (zygote formation). The prenatal period covers the entirety of gestation, from conception to birth. The postnatal period covers the period from birth to the first four weeks of the piglets' life, which for modern swine production, covers the pre-weaning period, with some discussion of where effects may continue beyond this period included.

With global questions surrounding the rearing of farmed pigs, the sustainability of practices and the push for the diversification of rearing systems to improve animal welfare and meet environmental goals, this area presents an important and exciting research field that opens up a new era of discovery to understand how the management of farmed pigs influences ontogeny. With an increasing number of sows being reared in group-gestation systems and methods to enhance current housing and rearing systems in progress, the time is now to understand how these factors influence the development of offspring behaviour and the implications for pig welfare and productivity.

The aim of this chapter is to give the reader a comprehensive understanding of the current knowledge in this area, the implications for pig welfare and production and the important research questions that must be answered to progress this field.

2 Prenatal environmental effects on development of pig behaviour

Pigs and sows are exposed to external stressors throughout their lifetime; the way animals adapt and respond will depend on the type of stressor, the length of the event and how well-adapted the individual is to cope (Martínez-Miró et al., 2016). Research in humans (Mina and Reynolds, 2014) and various species (Rutherford et al., 2012; Sinclair et al., 2016), including pigs (Otten et al., 2015), has shown that the prenatal or in-utero period can modify the development of offspring physiology and behaviour, which potentially modifies their stress-coping mechanisms and, in turn, affecting the way they engage with their environment. Recent research has shown that the pre-conception environment could also alter foetal and offspring development (Velazquez et al., 2019; Boersma et al., 2014). The following sections explain the findings from research exploring how the stressors experienced by the parents prenatally, including both the pre-conception period and during gestation, influence the development of offspring behaviour, and the implications for pig welfare and production. Where research on pigs is not available, work from other species is explored.

2.1 Effects of the pre-conception environment on offspring behaviour

In modern swine production, it is common for breeding pigs to encounter various stressors during their lifetime. These can include early-life stressors such as an abrupt weaning, painful procedures, transportation and social mixing (Roldan-Santiago et al., 2013; Rioja-Lang et al., 2019; Telkänranta and Edwards, 2018) but also includes stressors during maturation before breeding and during gestation. These later-life stressors include close confinement housing (i.e. stalls), competition for resources (i.e. feeding or lying space), social stressors, feed availability, barren environments, heat stress and fear from rough handling and a poor human-animal relationship (European Food Safety Authority, 2007; Einarsson et al., 2008). The pre-conceptive stage could be defined as the period between the gamete maturation and fertilization (Velazquez et al., 2019), and has been identified as a sensitive period when external factors may alter parental physiology (e.g. oocyte characteristics) before conception, subsequently affecting offspring development. In human research, it has been proposed that stressors experienced in the weeks, months or even years before conception can have a trans-generational impact on offspring development (Pembrey et al., 2014). Due to a lack of research in non-human animals, it is more difficult to identify how far back, before conception, stressors may affect the subsequent offspring. In ruminants, environmental stressors experienced by the mother up to two months before conception influence offspring physiology and behaviour (Perry et al., 2019), whereas in rodents stressors experienced throughout the parental lifetime (from birth to conception) have resulted in changes to offspring behaviour (Cutuli et al., 2018; Rodgers et al., 2013; Chavatte-Palmer et al., 2016).

The changes in the offspring physiology and behaviour resulting from factors experienced in pre-conception environment are driven by epigenetic mechanisms at the maternal and paternal level and are linked to the type of stressor, duration and the period of parental development (e.g. adolescence) in which they occur (Evans et al., 2019; Hamada and Matthews, 2019; Cutuli et al., 2018). On the maternal side, nutritional factors during the pre-conception period can influence the oocyte quality and follicular fluid composition (Ashworth et al., 2009), while stress may alter oocyte DNA methylation of regulatory genes associated with hypothalamic pituitary adrenal (HPA) axis stress sensitivity (Chan et al., 2018). In the case of the paternal side, epigenetic changes to the sperm mechanism of non-coding RNAs are believed to be behind the epigenetic changes in their HPA stress responses seen in offspring (Chan et al., 2018; Evans et al., 2019; Rodgers et al., 2013). In both cases, definitive epigenetic paths are yet to be discovered.

Specifically, with respect to paternal effects, stress experienced during growth maturation (e.g. adolescence), or in the weeks before mating has been linked to increased basal corticosterone, anxiety and depressive-like behaviours (Dietz et al., 2011; Short et al., 2016), negative brain morphological alterations (Harker et al., 2015), a blunted corticosterone response during restraint stress (Rodgers et al., 2013) and reduced cognitive performance in rodent offspring (Yeshurun et al., 2017). It is important to mention that with regards to depressive-like behaviours interpretation of a forced swim test should be taken with caution. It has been proposed that a rodent giving up in the forced swim test may not indicate depression but may be a learnt response that the test will end upon cessation of swimming, and therefore could reflect a stress-coping strategy for survival (for more information, refer to Reardon, 2019; Molendijk and de Kloet, 2015; Commons et al., 2017). In the case of maternal effects, chronic (lasting several days) unpredictable (i.e. isolation, food and water deprivation) pre-conception stress in rodents resulted in offspring showing increased anxiety responses, depressive tendencies and impaired memory function (Li et al., 2010; Huang et al., 2010; Shachar-Dadon et al., 2009; Huang et al., 2013). In comparison, maternal nutritional stress applied before (up to 60 days before conception) and after conception (up to 30 days after insemination) increased anxiety behaviour in one-month-old lambs (returned faster to their mothers when separated), while reducing reactivity in older offspring (reduced cortisol response, locomotor behaviour and lack of escape attempts while in isolation between 4 and 18 months of age) (Kleemann et al., 2015; Donovan et al., 2013; Hernandez et al., 2010). In cattle, a low protein diet before conception (two months before insemination) increased feeding duration in newborn calves (Miguel-Pacheco et al., 2019). These findings in ruminants match those seen in rodents (e.g. Wright et al., 2011). On the other hand, research into positive factors has shown that enrichment might benefit offspring behavioural programming. Paternal or maternal environmental enrichment provision, for example, access to an exercise wheel or a novel environment, and social interactions throughout a rodent's lifetime or during the period before conception, have been found to improve offspring-complex motor abilities, cognitive performance and coping response to stressors, reducing anxiety while improving social behaviours (Cutuli et al., 2018).

There is a lack of research into how the pre-conception environment may alter offspring behaviour in pigs. However, pig nutritional studies looking at underfeeding or overfeeding before conception (days 14–21 of lactation) have observed effects on offspring intrauterine development (reduced embryonic and foetal growth rate, reduced weight in male and female foetuses) and survival (reduced number of piglets weaned per litter) (Ji et al., 2017). In gilts, provision of dietary fibre before mating (days 1–19 of the third post-pubertal oestrous cycle) improved oocyte quality and embryo development, with higher embryo survival resulting in more piglets born alive (Jarrett and Ashworth, 2018). These findings are similar to those observed in ruminants (Perry et al., 2019).

Additionally, epigenetic studies have shown reprogramming effects on muscle, uterine capacity and uterine environment, with one study observing carry-over effects that altered the fat metabolism pathway in following subsequent generations (up to F2 – the second generation) (Feeney et al., 2014).

Consequently, it is expected that stressors during the pre-conception period will affect pig behaviour similar to that observed in other species. Little consideration has been given towards understanding how the rearing and gestation environment could shape the behavioural development of the future breeding animals, and the trans-generational effects it could have on offspring. Research is required to evaluate exactly how both negative and positive factors experienced by designated breeding pigs within the pre-conception period may shape the development, characteristics and resilience of offspring, and the consequences for welfare and production performance. This information is important to support the development of optimal management systems for pigs; the effects must be understood for subsequent meat pigs but also the future generations of replacement breeding animals.

2.2 Effects of the post-conception environment on piglet behaviour

Contrary to the pre-conception environment, to date, various studies have been carried out to explore the different factors during the post-conception or prenatal period in pigs that modify offspring behaviour. During this period, the embryo and foetus are very susceptible to external factors that affect the health and welfare of the sow, whether they be negative (e.g. competition (Rutherford et al., 2012)), neutral (e.g. feed flavours (Figueroa et al., 2013)) or positive (e.g. gentle brushing (Tallet et al., 2016)). However, in typical intensive swine production environments, the number of negative factors likely outweighs the positive ones. Gilts and sows may encounter various stressors from the housing environment (e.g. space restriction, lack of enrichment), the social group (mixing, aggression, social hierarchy), nutrition (e.g. restricted feeding regime, competition to enter an electronic sow feeder) and management (e.g. fear due to rough handling, heat stress) (Bench et al., 2013a,b; Greenwood et al., 2014; D'Eath et al., 2018; Pedersen, 2018, Table 1). The combination of these factors alters offspring immune and neurological development with postnatal consequences (Weinstock, 2017; McGowan and Matthews, 2018; Monk et al., 2012). The changes or adaptions seen in the offspring due to the in-utero environment are 'predictive adaptive responses' that prepare the foetus for its life in the postnatal environment (Rutherford et al., 2012). However, when the postnatal environment does not match the factors experienced prenatally, the adaptions may cause health and welfare problems for the offspring (Rutherford et al., 2012; Sinclair et al., 2016). Maternal stress affects the foetal development

Table 1 Housing type, trimester and stressors that may be experienced by sows during gestation (Pedersen, 2018; Bench et al., 2013a,b)

Housing type	Gestation period (days)			Stressors
	1st trimester (<39)	2nd trimester (~39–76)	3rd trimester (>76)	
Breeding stall[a]	■			Restriction of movement and social contact Health concerns: reduced cardiovascular fitness and bone strength Heat stress Lack of bedding material and enrichment Flooring design – reducing lying and standing comfort Feeding restriction
Gestation stall[b]		■	■	Restriction of movement and social contact Health problems: reduced cardiovascular fitness and bone strength Heat stress Lack of bedding material and enrichment Flooring design – Reducing lying and standing comfort Feeding restriction
Group housing[c]		■	■	Social stress: aggression, mounting behaviour, dominance hierarchy Limited space allowance Dynamic group Feeding system Feeding restriction and competition Lack or insufficient enrichment Flooring design – reducing comfort during lying and movement. Health concerns: lameness, injury

[a]Up to 4 weeks after insemination in the European Union, except United Kingdom, Sweden and Norway where confinement in breeding stalls is prohibited.
[b]Banned in the European Union, but still largely used in various countries.
[c]Stressors may vary depending on the type of group housing: electronic sow feeder (EFS), free access feeding stalls, partly unprotected feeding.

through two possible channels: (1) an increase in maternal glucocorticoid concentrations that cross the placental barrier directly affecting and remodelling the foetal HPA axis and other tissues (e.g. gonads); or (2) altering the nutrient supply and placental function causing epigenetic changes (e.g. glucocorticoid receptor expression) (Fowden et al., 2006).

Of importance for pig development is the nature of the effect (e.g. social stress), the development stage at the time of the effect (e.g. pregnancy trimester) and the size of the effect (e.g. acute single event vs. chronic, intensive) (Sinclair et al., 2016; Rutherford et al., 2012; Prunier et al., 2019; Otten et al., 2015). From insemination to parturition, sows may go through different types of management depending on the accepted production systems in that region of the world; each production system variation will present different challenges or stressors to sows throughout the gestation, with these stressors varying in length (Table 1). For example, when sows are housed in a group gestation pen and sequentially fed with an electronic sow feeder (ESF), from 35 to 105 days of gestation, they may encounter repeated social stressors, in the form of subtle threats, bullying or physical aggression arising from competition to enter the ESF. This could be exacerbated if the number of sows on the ESF is increased to maximize capacity and throughput (Bench et al., 2013b). Another example is a dynamic gestation group, in which new groups of sows are added to an existing group of sows the group throughout gestation. This method of group gestation management can result in repeated aggression within the group, and increase risk of sow injury, and could create a chronic stressor for the sow throughout the gestation (Bench et al., 2013b). Sows that experience repeated social stress during mid-gestation (days 39–69) produce piglets with a more reactive behaviour phenotype (Jarvis et al., 2006; Ison et al., 2010), while social stress experienced during late gestation (days 77–105) affects the piglet's immune response, decreasing blood leukocytes and increasing blood lymphocyte proliferation, particularly in female offspring (Couret et al., 2009a,b).

In the case of the size of the effect, a study simulating chronic stress found that piglets born to sows that received weekly intravenous injections of ACTH between the 6th and 12th week of gestation while restrained by a nose snare performed less oral manipulation behaviours but had a greater plasma cortisol concentrations at mixing (60 days of age) than controls (no treatment or manipulation) which may indicate a piglet with a hyperactive HPA axis in response to stressful events (Haussmann et al., 2000). Though no apparent negative behavioural findings were observed in treatment piglets at mixing, biopsy punctures took longer to heal, indicating that the immune system of these animals was compromised by the maternal stress (Haussmann et al., 2000), which may be due to overproduction of cortisol. These potential contradictory findings may indicate that chronic stress could influence behaviour and physiological development differently: an animal that shows a lower behavioural reaction but a higher physiological reaction.

Offspring sex may also affect behaviour development and expression (Otten et al., 2015); however, behavioural studies looking at the differences between male and female piglets arising from prenatal stress are still limited. One study found that males seem to be more fearful during a novel object test (Kranendonk et al., 2007), but more studies are required in this area, with particular consideration to non-castrated males. To date, studies have been carried out on castrated males, in which the sex differences are difficult to interpret (Otten et al., 2015).

The following section will discuss how individual factors experienced by the sow during gestation can affect the offspring.

2.3 Prenatal housing conditions

2.3.1 Gestation stalls

Breeding or gestation stalls are still widely used around the world, with very few countries outside the EU with the legislation in place limiting or banning their use. Stall-housing presents specific stressors for sows, which includes the movement restriction and a lack of control over their environment (Baxter et al., 2017). Despite much debate on stalls versus group housing, research is limited to how stall-housing may affect offspring behaviour. When the behaviour of offspring from the stall and group-housed gilts has been compared, barrow piglets originating from stall-housed gilts appear to be more reactive, increasing vocalizations during an isolation test between 14 and 17 days of age, but being no different in other behaviours such as belly nosing (Sorrells et al., 2006). This suggests stall-housing may produce more emotionally reactive piglets. It is understood that stressors that increase maternal cortisol may produce an emotionally reactive piglet (Kranendonk et al., 2008; Otten et al., 2015). This is based on evidence from controlled studies in which piglets from sows treated with adrenocorticotropic hormone (ACTH) or hydrocorticosterone acetate (HCA) at different trimesters have been found to be more active and aggressive during social confrontation tests (Kranendonk et al., 2006), and more active during a novel environment situation (Otten et al., 2008). Repeated research has failed to find differences in cortisol response between stall and group-gestation housed sows (McGlone, 2013). However, this may indicate a failure of acute measures of cortisol to accurately identify chronic stress and may indicate that measuring offspring responses provides a sensitive measure to detect whether housing system stressors are influencing the sow. Further work is required in this area.

2.3.2 Group housing

Social stress is one of the predominant stressors experienced by sows in group gestation housing. Social mixing of sows to form gestation groups can occur as soon as gilts or sows are inseminated or 28 days after, upon confirmation of pregnancy. Sows are mixed in static or dynamic groups, and in the latter sows are exposed to the addition of new sows to the already formed group at weekly intervals or longer depending on the batch-farrowing intervals. This repeated introduction of new sows to the group has the potential to produce chronic social stress which could be exacerbated by other factors such as hierarchy or age (Bench et al., 2013b). The effects of the social mixing on offspring behaviour have been assessed, considering the different periods of gestation (Table 2). For pregnant sows, social mixing stress increases submissive behaviour, skin lesions and salivary cortisol, without evident effects on the sow's postnatal maternal behaviour (Jarvis et al., 2006). However, sows that have experienced social stress during gestation have been found to take longer to respond to the vocalizations of isolated piglets, suggesting a reduced maternal response (Ringgenberg et al., 2012). Additionally, the offspring from sows that have experienced social stress during gestation show reduced exploratory behaviour during lactation, increased anxiety-like behaviours after weaning, and after parturition sows originating from prenatally stressed mothers react negatively (e.g. increased risk of savaging) to their own piglets (Table 2; Jarvis et al., 2006).

Offspring behavioural changes are dependent on when the prenatal social mixing occurred – the sex of the offspring and the social rank of the sow (Kranendonk et al., 2008). The second trimester is the most sensitive period for foetal development, and it has been observed that the foetus releases cortisol in response to maternal ACTH, indicating that the foetal HPA axis function is established during this trimester (Otten et al., 2004). It is within this period that studies have found significant changes in offspring behaviour (Table 2). For example, piglets from mothers that had more skin lesions due to fighting at social mixing at mid-gestation (second trimester) spent less time sitting with fewer posture changes, and vocalized less at weaning and during the post-weaning period (days 28–59 of age), and were less aggressive after weaning (from day 31 onwards) (Table 2; Ison et al., 2010), while piglets from mothers that had more fighting lesions from social mixing at late gestation (third trimester) showed increased sitting behaviour and vocalizations on the day of weaning (Table 2; Ison et al., 2010). Offspring behavioural changes caused by late gestation mixing indicate a more stressful phenotype (e.g. increased sitting behaviour and vocalizations at weaning). However, Ison et al. (2010) interpret it that the lower behavioural reactivity of piglets from mid-gestation mixed sows could also indicate inhibition of behaviour, as has been observed in prenatally stressed rats (Harvey and Chevins, 1985). Further work is required to understand how previous sow experience and stage of foetal development may modulate behavioural changes in the offspring.

Lastly, the effect of social rank of the sow on offspring behaviour was studied by one paper which observed that piglets from low-social-rank sows that were group-housed during gestation were less reactive to novel objects,

Table 2 Studies on prenatal social mixing in female pigs and the effects on offspring behaviour

Reference	Gestation period (days)			Number of mixings per trimester	Duration of each mixing	Lactation	Social mixing on offspring behaviour (vs. control)
	1st trimester (<39)	2nd trimester (~39-76)	3rd trimester (>76)				
Jarvis et al. (2006)		39–45 and 59–65 days	77–83 and 97–103 days	One mix per week	Seven days		**Increased anxiety at and after weaning** Mid- and late-gestation mixing: ↓ long attacks, ↑ short retaliatory attacks and ↑ nosing of littermates. ↑ retreat frequency age when attacking. Mid-gestation mixing: ↑ short retaliatory attacks **Daughters with increased anxiety at their own parturition** Mid-gestation mixing: ↑ posture changes and frequency of substrate-directed behaviour during nesting ↑ frequency to bite and snap their piglets caused within 6 h post-parturition
Rutherford et al. (2009)		39–45 and 59–65 days		One mix per week	Seven days		**Increased pain response** ↑ pain score after tail docking (at 3 days of age) ↑ offspring pain score as maternal cortisol levels ↑ (1st day of each mixing week)
Ison et al. (2010)		39–45 and 59–65 days	77–83 and 97–103 days	One mix per week	Seven days		**Less reactive piglets at and after weaning from gilts with more lesions during mid-gestation mixing** ↓ activity and vocalizations ↑ aggression and ↓ activity from 31 days onwards **More reactive piglets at and after weaning from gilts with more lesions during late gestation mixing** ↑ activity and vocalizations

Sandercock et al. (2011)	39–45 and 59–65 days	One mix per week	Seven days	+ Piglets: Tail docking and sham (2–4-day-old)	**Increased pain response thresholds** ↑ noxious stimulation response, ↓ basal thresholds response and ↓ integrated responses to inflammatory challenges
Rutherford et al. (2014)	39–45 and 59–65 days	One mix per week	Seven days	[a]Farrowing crate or open pens	**Daughters with increased anxiety at their own parturition in both farrowing environments** ↑ ventral lying, restless behaviour, attention, and reaction to piglets when approaching her head **Increased piglet mortality when daughters housed in pen**
Ringgenberg et al. (2012)	39–45 and 59–65 days	One mix per week	Seven days	Enrichment or not at lactation (Cross-over design)	**Negative impact on sow maternal behaviour not counteracted by farrowing environment** ↑ latencies to respond vocally to isolated piglet vocalization Early lactation: ↑ lying time ventrally and ↓ time spent in the nesting straw area
Brajon et al. (2017)	39–45 and 59–65 days	One mix per week	Seven days	Enrichment or not at lactation (Cross-over design)	**Behaviour inhibition on day 14 and 27 of age (weaning at 21 days) not counteracted by farrowing environment** ↓ vocalizations frequency, ↑ lying and ↓ environmental exploration, play behaviour (whether it is locomotion play or fighting play) and mounting

[a]Lactation period of maternal stress offspring once they have themselves farrowed.

vocalizing less and staying closer to the novel object than piglets from high-ranking sows (Kranendonk et al., 2007). Offspring behavioural changes due to social mixing and social ranking of sows should be investigated further taking into consideration the type of feeding and grouping system, particularly as more producers consider moving gestating sows to group systems.

2.3.3 Environmental enrichment

A lack of enrichment in the gestation environment causes stress on sows, resulting in higher salivary cortisol levels during gestation and litters with higher pre-weaning mortality (Merlot et al., 2017, 2019). In group gestation systems, it is not only the availability of enrichment but also the quantity of enrichment and additionally the complexity of the pen environment that can be beneficial for the sows, reducing competition for resources allowing low-ranking sows to access the enrichment and pen resources (Bench et al., 2013b). Provision of appropriate enrichment is not only beneficial for sows but also has positive impacts on the behavioural development of their offspring. Piglets from sows provided with the daily straw substrate during the final trimester (from day 90) of gestation showed a reduction in aggressive and nosing behaviour towards pen mates, and female piglets were less fearful and more curious during a novel object test (Tatemoto et al., 2019a). This suggests a valuable role for enrichment to support the production of more sociable and resilient offspring, while also supporting the welfare of the sow throughout gestation through the provision of effective environmental enrichment.

2.4 Maternal nutrition

2.4.1 Feeding regime

Gestating sows are limit fed to prevent excessive weight gain during pregnancy. These restricted feeding practices are a source of stress as the sow is left with a feeling of hunger for many hours in the day. Restricted feeding practices may be more detrimental when combined with the effect of other environmental stressors; due to feeding restriction competitiveness increases in group sow housing, particularly in those with sequential or competitive feeding systems, which translate in higher aggression levels (Bench et al., 2013a). As a response to hunger and anxiety created by the restricted feeding practices, many sows display stereotypies such as vacuum or sham chewing (D'Eath et al., 2018). Studies investigating stereotypies in sows have observed a relationship between the rate and presence of this stereotypic behaviour in sows and piglet behaviour. In an open-field test, piglets from mothers performing a high rate of stereotypic behaviour walked more in the central and lateral quadrants and vocalized less in an open-field test (Tatemoto et al., 2020). When comparing piglets from sows that performed sham chewing (regardless of the rate) with control (non-sham chewing sows), piglets from sham chewing sows displayed more nosing in the first week of life, and during an open-field test piglets were more active and had a shorter latency to move than piglets from control sows (Tatemoto et al., 2019b). Also, sows that performed sham chewing had placentas containing greater amounts of cortisol than sows that did not perform sham chewing (Tatemoto et al., 2019b). Considering that performance of stereotypies may be a coping mechanism, the maternal stress may be reprogramming piglets to be able to cope better with an uncertain environment, therefore resulting in less fearful behaviours. Further investigation is needed to understand if this behavioural phenotype is of benefit for the welfare of the pig.

Selecting sows to produce a larger litter size has made intrauterine crowding a frequent occurrence, caused by the physiological pressure within the uterine space, and the increased blood supply due to a larger number of foetuses that the uterine physiology cannot easily support (Wang et al., 2017). This results in low-birthweight piglets and piglets with intrauterine growth retardation (IUGR) as they compete for nutrients in the womb. A larger surface area to volume ratio predisposes low birthweight piglets to hypothermia and hypoglycaemia, which reduces their ability to compete for a teat (Ward et al., 2020). Similarly, and more problematic, IUGR piglets are born not only weak and underweight but also physiologically immature (Ward et al., 2020; Wang et al., 2017). Studies show that undernutrition during the last trimester, particularly, a lower amount of specific nutrients (e.g. amino acids), can exacerbate the problem of intrauterine growth retardation, and benefits to piglet birthweight are observed when levels of these specific nutrients are improved in the maternal diet (Wang et al., 2017; Ward et al., 2020). Additionally, low-birthweight piglets (with or without IUGR) may also be more susceptible to stress, while piglets with IUGR may have impaired short-term memory (Schmitt et al., 2019). Interestingly, during the pre-weaning period offspring growth is modulated by gender; female piglets from restricted fed sows (50% of nutritional requirements) showed similar growth pattern as control piglets, while males from underfed sows were lighter than the control male offspring (Gonzalez-Bulnes et al., 2012). In other species undernutrition during gestation modifies appetite in the neonate through epigenetic changes at the hypothalamus level (Begum et al., 2012; Rocha et al., 2014); underweight offspring show an increased appetite (as observed through increased suckling bouts) mainly seen in female offspring (Stěhulová et al., 2013). The increase in appetite may modulate for increased aggression in piglets as they compete for access to the teat; however, a recent review on piglet birthweight and aggression failed to find this association (Prunier et al., 2019). This does not mean that piglets from undernourished mothers may not have an increased appetite; it is possible that they may develop strategies to reach a teat faster and feed without confrontation or fighting. Therefore, studies assessing the effect of low birthweight and IUGR on piglet behavioural development are required, in conjunction with research on the effect of sow undernutrition on offspring productivity and behavioural traits.

2.4.2 Diet formulation and supplementation

Gestating diets are formulated to provide sows with sufficient nutrients to maintain body condition while preventing excessive weight gain (Rosero et al., 2016). Additionally, formulations consider improvements to offspring survival, milk quality and quantity and the benefits for offspring long-term production traits (Tilton et al., 1999; Rosero et al., 2016; Peng et al., 2019). In other species, it has been observed that gestation diets poor in energy or protein modulate offspring behaviour (Rutherford et al., 2012; Sinclair et al., 2016). The only studies in pigs come from experimental medicine, where pigs have been used as animal models for human research (Table 3). Sows in these studies were fed a normal or experimental gestation diet high in fat and in sugar with 12% animal fat and 18.5% sucrose (HFHS). It has been identified that prenatal HFHS diets produce a less-aggressive piglet, but with a more anxious behavioural phenotype (Clouard et al., 2016a). Piglets showed fewer skin lesions and deep wounds due to aggression before weaning and continued to present a lower number of skin lesions in the middle part of the body up to three weeks after weaning. Additionally, piglets showed an increase in the duration of time walking and exploring feedstuffs and a tendency to increase exploration of their pen mates up to four weeks after weaning. Yet, these piglets tended to have decreased basal salivary cortisol levels at eight weeks of age which could be a blunted response to stress (Clouard et al., 2016a). However, HFHS diets can also positively affect the cognitive abilities of offspring. Piglets from HFHS mothers tested between 33 and 53 days of age using the holeboard test to evaluate spatial learning and reference memory consumed more rewards than control piglets but showed lower neurogenesis in brain areas associated with cognitive abilities and learning plasticity (Val-Laillet et al., 2017). When piglets were tested between 11 and 15 weeks of age (weaned at four weeks of age) using the same test, HFHS piglets had higher scores of working and reference memory (Clouard et al., 2016b). Long-term alterations have also been identified in the offspring of HFHS mothers; prenatal diets affect the cognitive abilities of offspring, with differences by sex, producing a more anxious behavioural phenotype with a lower glucose tolerance at 10 months of age (Gautier et al., 2018). Sex differences in the behavioural responses have been observed in the spatial holeboard discrimination task, with males visiting fewer bowls than females. When spatial cognition was tested without a food reward (alley maze test), HFHS piglets showed a longer test duration and increased immobility and exploration of the wall, which relates to anxiety or lower cognitive abilities.

Table 3 Studies exploring the effect of dietary supplementation, or flavour intervention on piglet behaviour: summarizing the stage of development (prenatal or postnatal), the duration of treatment application and the treatment

	Sow				Piglet		
Reference	1st trimester (<39 days)	2nd trimesters (~39–76 days)	3rd trimester (>76 days)	Lactation	Creep feed	Post-weaning	Diet supplementation/flavour intervention
Clouard et al. (2016a,b)			>59 days to end	Day 0–28		56 days (4 weeks)	HF (12%) HS (18.5%) diet vs. standard diet
Gautier et al. (2018); Val-Laillet et al. (2017)		28 to end		Day 0–28			HF (12%) HS (18.5%) diet vs. standard diet
Clouard et al. (2015)			61 days to end	Day 0–28			FO (DHA-rich) HOSF (high-oleic acid)
Gunnarsson et al. (2009)	All gestation period						LF (standard diet) HFS (6% fat, low PUFA) HFO (6%, high n-6 PUFA) HFL (6%, high n-3 PUFA)
Bernardino et al. (2016)	All gestation period						HFD (12.86% CF) vs. LFD (2.53% CF)
Blavi et al, (2016)			73 days to end	Day 28	>12 days old	For 35 days	Sows: FLA (Fluidarom) and control (no flavour) Piglets: FLA and lacto-vanilla
Figueroa et al. (2013)			Last 2 weeks				Anise flavour Milky-cheese flavour Control (no flavour)

(*Continued*)

Table 3 (*Continued*)

Reference	Sow				Piglet		Diet supplementation/flavour intervention
	1st trimester (<39 days)	2nd trimesters (~39–76 days)	3rd trimester (>76 days)	Lactation	Creep feed	Post-weaning	
Figueroa et al. (2019)			91 to 114		10–20 days old		Anise flavour Garlic flavour Control (no flavour)
Oostindjer et al. (2009)			98 to 108	Day 9–27			Anise flavour Control (soy oil)
Oostindjer et al. (2010a)			98 to 108	Day 6–24		28–42 days old	Anise flavour Control (soy oil)
Oostindjer et al. (2011a)			98 to 115	Day 2–25		25–39 days old	Anise flavour Control (soy oil)

HF = high fat, HS = high sugar; SD = standard diet; HE = high energy; DHA = docosahexaenoic acid; HFD = high fibre diet; LFD = low fibre diet; CF = crude fibre; vs. = versus; LF = low fat; HFS = increased saturated fat; HFO = high-fat oats (40%); HFL = high-fat oats and linseed oil (20%); PUFA = polyunsaturated fatty acids; FO = DHA-rich fish oil; HOSF = high-oleic acid sunflower oil.

Additionally, HFHS pigs showed altered brain glucose metabolism with the deactivation of brain areas related to reward sensitivity and cognitive control of eating behaviour, which could result in less-regulated appetite control and depression (Gautier et al., 2018). Behavioural changes observed in these studies are very likely to be linked with epigenetic modifications related to the maternal diet, as seen in other species. However, two things must be considered: (1) the milk of sows fed HFHS diets presented a higher lipid content, which may have supported better cognitive abilities at an early age, the effects of which were not seen when pigs were older (Leskanich and Noble, 1999); (2) tested diets had different crude fibre content (experimental diet 8% and standard diet 5.1%) which may have also influenced the results seen here.

Neonatal survival relies on piglets finding a teat as soon as possible following birth, to intake energy from colostrum, helping to support thermoregulation and to avoid being crushed by their mother. However, selection for larger litters has reduced the physiological maturity and birthweight of piglets, which affects their survival rate (Wang et al., 2017; Ward et al., 2020). Supplementation with long-chain polyunsaturated fatty acid (LC-PUFA) such as docosahexaenoic acid (DHA) can benefit the neurological development of piglets, contributing to their health and welfare (Lauritzen et al., 2016). Accumulation of LC-PUFAs in the foetal brain occurs predominantly in the last trimester, delivered through the placental blood supply. This accumulation continues through the neonatal period from lactation (Leskanich and Noble, 1999; Lauritzen et al., 2016). Piglets from sows supplemented with DHA given via salmon oil had higher survival rates, resulting from reduced incidences of deaths caused by sow crushing (Rooke et al., 2001). However, studies on how DHA influences piglet behaviour are inconclusive. One study evaluating the effect of sow diets containing high fat (HFS, increased saturated fat and low PUFA) or low-fat content (standard diet) and two types of PUFAs (HFO, increased fat high n-6 PUFA; or HFL, increased fat high n-3 PUFA) found that HFS in 2-day-old piglets remained close to their mother's faeces for a greater duration of time when tested in a T-maze and displayed shorter escape durations in comparison to piglets from mothers fed high-fat oats and linseed oil diets when tested in a modified backtest (Gunnarsson et al., 2009). These findings may indicate that low PUFA diets are conducive to a more fearful and reactive offspring phenotype. Comparisons of supplementing sow diets with either fish (FO) or sunflower oil (HSOF) from 61 days of gestation and throughout lactation identified that piglets from sows fed a FO diet displayed more social play behaviour and less manipulation of pen mates (Clouard et al., 2015). This latter study, however, failed to separate post-conception from postnatal effects. It is evident that DHA supplementation is beneficial to piglet survival potentially through modulation of their vigour and behaviour at birth; however, further research is needed to learn about the best period for supplementation and the type of fatty acid that would most greatly benefit the sows and piglets, while also considering the cost of dietary supplementation.

In some countries, dietary recommendations for gestating sows point towards the inclusion of fibre to bulk the content of the diet to modulate sow hunger (National Farm Animal Care Council (NFACC), 2014; European Union, 2008). There is evidence to suggest the provision of increased fibre in the gestation diets of sows beneficially influences piglet behaviour development. Piglets that fed a high-fibre diet (HFD; 35% soybean hulls) from mothers throughout the gestation had fewer skin lesions pre-weaning and at two days post-weaning following mixing with unfamiliar piglets (Bernardino et al., 2016). These findings could be related to sows experiencing a less stressful pregnancy, resulting in a less reactive piglet behaviour phenotype that is better at engaging with other piglets. It is clear that maternal diet supplementation holds great potential for supporting positive behavioural development in the piglet, and for certain diet supplementations, such as the addition of increased fibre content, there are also benefits for the sow. However, where the provision of fibre results in bulkier diets, it may also increase feeding time which may create challenges with increased competition between sows in group-housing systems with competitive feeding systems (Meunier-Salaün et al., 2001), and would reduce the number of sows that can be fed on one ESF feeder (Bench et al., 2013a). Further research is required in this area to separate these effects and to understand which is the best gestation period during which to supplement the maternal diet, what type of supplementation works best, such as type of fibre and fatty acid, what is the cost–benefit of the diet formulations and the effects for mother and offspring, and so recommendations can be made to producers.

2.4.3 Flavoured diets

Supplementation of prenatal diets with flavours provides an innovative way to reduce food neophobia, promote creep feed intake and encourage newly weaned piglets to increase feed intake at what is a very stressful transitional period in their lives (Weary et al., 2008). Studies in other species, including humans, have shown that exposing the foetuses to flavours during the last trimester, through supplementation of the maternal diet, increases the intake and preference of feed with these flavours in the postnatal environment (Simitzis et al., 2008; Bilkó et al., 1994; Bolhuis et al., 2009). However, studies in piglets have failed to observe both effects consistently.

Studies investigating the effect of prenatal flavoured diets on offspring food preferences have fed anise flavoured food or control diets to sows between 98 and 108 days of gestation. Immediately after birth piglets were cross-fostered and during lactation sows were fed either an anise flavoured or a control diet. Prenatally exposed piglets demonstrated recognition of the prenatal flavour through behavioural changes when they were re-exposed during lactation (Oostindjer et al., 2009). For example, piglets explored the anise-scented gate more and spend more time in the scented area of an arena during a gate preference test. In a novel environment test when the anise odour was present, piglets vocalized and tried to escape less frequently, while taking longer to initiate contact with another piglet (Oostindjer et al., 2009). At weaning these piglets were offered two diets (anise flavour and control), piglets showed an increase in food intake, but no clear preference for the control or anise-flavoured diet within the first three days after weaning (Oostindjer et al., 2010a). Interestingly, piglets explored the anise-flavoured diet for longer, indicating flavour recognition and showed less anxious profile as they had reduced oral manipulation and mounting of pen mates, with a longer time to initiate fights after weaning. In a similar study, sows were offered a flavoured or control gestation and lactation diet but without cross-fostering, and at and after weaning piglets were presented the maternal flavour in the food or in the air (Oostindjer et al., 2011a). Findings were similar to the previous studies, and flavour did not affect food intake or preference in any of the piglet groups (flavour or control) after weaning. However, flavour-exposed piglets had reduced latency to eat (without affecting feed intake or growth) and a lower salivary cortisol immediately after weaning. They also showed increased play behaviour and less aggressive behaviours in the first two weeks after weaning. The findings from these studies indicate a positive stress-modulation effect from the maternal flavour diet, but no lasting effect on feed preference in offspring.

Similarly, other studies using different flavours and experimental designs have not found positive results on food preference, although some still observed recognition of the maternal flavour. Figueroa et al. (2013) fed sows one of three diets in the last two weeks of gestation (Table 3). Piglets were tested for recognition of maternal flavour using strips impregnated with one of the two flavours used in maternal diets and the third strip with water in a triple-U-testing arena. Prenatally exposed piglets showed a preference for the strip infused with their mother's diet flavour, spending more time close to that strip and performing more oral-nasal contact with the strip. This preference was significantly consistent during lactation (14–21 days of age) with a reduction but still significant preference two days after weaning (26 days of age). To investigate feed preference, Figueroa et al. (2013) offered piglets creep feed with the same flavour as their mothers' prenatal diet and control creep feed during the last 6 days of lactation. Control piglets and prenatally exposed piglets preferred the control creep feed over the maternal flavour, which suggest a reduced neophobia in prenatally exposed piglets. In a follow-up study, prenatally exposed litters were either offered creep feed with their mother's prenatal diet flavour, an alien flavour or a control diet during lactation

(Figueroa et al., 2019). As in their previous study, the authors did not find an influence of prenatal flavour diet on feed intake, body weight or average daily gain. Only one study has found a preference for the flavoured feed on day two after weaning used a commercial flavour feed additive (Blavi et al., 2016). This product was added to sow diets from day 78 of gestation until day 28 of lactation. Piglets were offered creep feed similar to their mother's flavour diet from day 12 and flavoured weanling diet until day 35 after weaning. Average daily feed intake and average daily weight gain were influenced by the flavoured diet of the sow. In a second trial, only prenatally and postnatally exposed piglets that received a flavoured creep diet preferred the weanling flavour diet showing a high consumption when offered this and a control diet on day two after weaning (Blavi et al., 2016).

The increase in feed and preference for prenatal flavours is based on intrauterine sensory entrainment by the ingestion of the amniotic liquid and/or through the placental blood flow that crosses to the foetal nasal capillaries (Schaal et al., 2004; Figueroa et al., 2013; Blavi et al., 2016). This affinity for prenatal flavours is part of a trans-natal sensory continuity that would help the offspring for life after birth aiming to support recognition and selection of safe foods after weaning (Mellor, 2019). Some of the findings of these studies suggest a possible emotional association made by piglets with the flavours they were exposed to during gestation. Findings indicate that prenatally exposed piglets may modulate their behaviour based on the presence or absence of the aroma when tested and show a less aggressive or anxious phenotype particularly during lactation and immediately after weaning (Oostindjer et al., 2011a) Additionally, intrauterine aromas seem to reduce piglets' neophobia, evidenced by increased intake of unknown feed than the unexposed piglets (Figueroa et al., 2013). Feed intake seems to be promoted when flavour diet is provided for longer during the prenatal period (from 73 days) and throughout lactation and weaning to sows and piglets (Blavi et al., 2016). The prenatal exposure to flavours may not only elicit a possible preference for the maternal diet in other species but also in piglets; this preference may additionally be modulated with a psychological component through a desensitization or habituation process supported by the emotional association with the mother (Mellor, 2019). The lack of a strong and consistent preference for maternal flavour may be altered in pigs due to the various other emotional stressors that sows experience during gestation. Further research is needed to identify the gestation period when the foetus is more sensible to maternal flavours and the correct management for the use of flavour during and after the gestation period to encourage feed intake in weaned piglets. However, other lines of enquiry can also be assessed, such as the use of maternal flavours to provide a calmer post-weaning environment using aromas.

2.5 Heat stress

Prenatal heat stress (PHS) can be defined as foetal exposure to maternal hyperthermia, which occurs when the maternal body temperature increases above a set point (Hansen, 2009). The majority of global pork production takes place in countries with climates that are subject to reaching climatic, thermal extremes, resulting in gestating and lactating sows regularly experiencing heat stress at certain times of the year (Parois et al., 2018). During gestation, PHS alters foetal and placenta development, while also affecting nursing during lactation in all mammals, including pigs (Hansen, 2009; Johnson and Baumgard, 2019). Piglets born to PHS sows showed physiological changes such as elevated body temperature setpoints regardless of the ambient room temperature, altered production performance due to increase stress and inflammatory response during the nursery and growing stage, altering their production efficiency as feed intake and live weight are reduced (Johnson and Baumgard, 2019). When matured, boars from PHS sows had reduced sperm production and increased sperm abnormalities (Lugar et al., 2018), while gilts had reduced milk quality and lower number of piglets weaned (Lucy and Safranski, 2017). PHS impacts the neurodevelopmental process in the foetus, which is related to neurological and behavioural alterations in the offspring (Dowell et al., 2019). These changes are a potential response or re-programming associated with in-utero stress which alters the offspring HPA axis, and is not solely a reaction to the heat itself, as is seen in egg-laying species (Johnson and Baumgard, 2019).

In pigs, the only study looking at PHS effects on offspring behaviour exposed sows to heat stress (28°C night-time and 38°C daytime; 71.1 ± 0.2% relative humidity) between 30 and 60 days of gestation and assessed the behavioural reaction of piglets to transport immediately after weaning (~22 days). The study found that PHS piglets showed reduced cortisol and non-esterified fatty acid (NEFA) concentrations, increased drinking behaviour and a tendency for increased aggression in comparison to non-PHS piglets (Byrd et al., 2019). The unexpected reduction in blood cortisol concentration after weaning and transport could be related to a reduced stress reaction or on the contrary, adrenal fatigue, resulting from a chronic stress state due to the stress of multiple concurrent stressors, the transport, social mixing and weaning. Additionally, the reduced cortisol and NEFA concentrations may be the result of PHS in-utero programming; female heat-stress pigs showed reduced NEFA concentrations despite 40% feed intake reduction and even after an epinephrine challenge (Sanz Fernandez et al., 2015). Reduced NEFA concentrations in the PHS piglets translates into an animal with reduced ability to mobilize energy reserves at periods of reduced energy availability or stress, which has the potential to impact in pig performance. Increased drinking behaviour and aggression in PHS piglets upon arrival to the new barn can be associated to a reactive or, on the contrary, a fearful, behavioural phenotype responding to a frustrating environment after a feed-restriction period in a barren new environment, all these after a chain of stressful events.

The findings in pigs align with what has been observed in other mammalian species. For example, goat's kids from mothers exposed to heat stress from 12 days to 45 days of gestation showed reduced exploratory behaviour in the first month of age (Coloma-García et al., 2020), while dairy calves from cows exposed to heat stress at the end of the pregnancy showed differences in perception of thermal comfort, potentially caused by concomitant stress as calves reduced standing time even when having a higher body temperature after weaning (Laporta et al., 2017). Pregnant sows' exposure to heat stress is more likely in coming years due to global warming exacerbated by building characteristics that may not be prepared to mitigate the increased heat inside (Schauberger et al., 2019). Studies to develop methods to mitigate heat stress in sows have started showing promising results that benefit sow welfare and productivity (e.g. Parois et al., 2018). However, these studies should be complemented with observations on how these heat stress mitigation strategies affect or modify offspring physiological and behavioural profile.

2.6 Infectious diseases

Recent studies have identified that maternal immune activation (MIA) affects offspring neural development that relates to behaviour (Bergdolt and Dunaevsky, 2019; Nardone and Elliott, 2016). Similar to findings in rodents and non-human primates, Antonson et al. (2017) observed that piglets from sows infected with porcine reproductive and respiratory syndrome virus (PRRSv) on day 76 of gestation lacked pro-social behaviours at 28-day-old. In a three-chamber social behaviour testing paradigm, piglets from PRRSv-infected mothers did not show a preference for an empty compartment, the centre section or a compartment with a novel piglet, in contrast to the offspring of non-infected sows, which preferred to spend most of the time within compartments with a novel piglet (Antonson et al., 2017). Behavioural changes due to maternal infectious diseases may be caused by direct injury to specific sites of the brain during development or epigenetic alterations (Bergdolt and Dunaevsky, 2019). It is understood that cytokines can cross the placenta, potentially affecting brain development (Malkova et al., 2012; Smith et al., 2007). Antonson et al. (2018) observed a reduction in porcine foetal hippocampal neurons on near full-term foetuses and reduction of the major histocompatibility complex two (MHCII) proteins in microglia, which may explain the behavioural changes observed at 28 days of age. Though these anatomical changes were not observed in 28-day-old piglets (Antonson et al., 2017), it is still possible that other maternal pathways may have been altered (e.g. IL-6, IL-7 and astrocytes) contributing to behavioural changes (Gumusoglu et al., 2017; Nardone and Elliott, 2016). Studies in other species have shown alterations to microglia activity and behavioural changes of offspring in adulthood, and therefore early-life observations may not be enough to understand the full impact of maternal infection on offspring behaviour and long-term welfare (Antonson et al., 2017). Future studies should focus on longer-term observations of maternal infection effects on offspring behaviour to fully understand the impact on pig welfare and production.

2.7 Sow management

2.7.1 Handling

Depending on how much contact staff have with the sows in a swine operation, and the implementation of appropriate animal handling training, the human-animal relationship could be another source of stress for gestating sows. Animals that experience repeated negative interactions with humans may develop chronic stress, leading to a negative impact on their production parameters such as a reduced pregnancy rate and increased percentage of stillborn piglets (Tallet et al., 2018). Research into the effect of this type of prenatal stressor on piglets is limited; only a couple of studies have assessed the effect of rough handling. Piglets from mothers treated with rough handling (mild electric shock three times during herding) for 10 min during mid-gestation (42–77 days) did not show plasma cortisol concentration changes after weaning despite their mothers having higher plasma cortisol concentrations than control sows (Lay et al., 2008). While at two months of age, these piglets also showed less sickness behaviour when challenged with lipopolysaccharide but did not show behavioural changes (postures and aggression) during mixing at four months of age (Lay et al., 2011). The lack of physiological and behavioural changes found in these studies should not be taken that rough handling during gestation is not negative for offspring development, as it is very likely that the experimental treatments were not able to mimic real-life rough handling. Another way to interpret these findings is that piglets may have been programmed to better withstand stress or, on the contrary, that their HPA-axis response may have been blunted due to chronic stress started in-utero. Further studies are required to understand the full extent of rough handling on offspring behaviour in pigs.

2.7.2 Auditory cues

Studies in humans and other animal species reveal that auditory cues linked to emotional experiences may alter offspring behaviour (Lickliter, 2011). In pigs, Tallet et al. (2016) tested the theory that piglets could relate prenatal auditory cues to positive or negative sow experiences. Sows in the last month of gestation were positively (soft brush) and negatively (electric prod) handled for 10 min, once per week, while hearing a human voice or not. During the handling procedures, sows reacted to the negative experience with distress calls and behavioural changes suggestive of stress (e.g. withdrawal, aggressiveness), regardless of whether there was a human voice played in the background or not. However, treated sows seem to pay more attention to the voice cues, when compared to control sows that were never treated but heard the human voices. Tallet et al. (2016) expected that based on their in-utero experience, piglets would react negatively to the human voices associated with negative handling. Interestingly, piglets generalized the human voice (negative or positive, or neutral (unknown voice)) and reacted more calmly when isolated, with fewer distress calls than controls when the human voice was played. Even more, they were able to differentiate the voice with the negative valence, producing more distress calls when hearing the negative voice, in comparison to positive voice. In humans, evidence from studies in pregnant women and newborns suggests that foetuses learn sounds from their prenatal environment (Pino, 2016), and this learning can also be related to emotional voice tones (Mastropieri and Turkewitz, 1999). In those species that are mature at birth, the in-utero acoustic environment has an impact on the auditory development, particularly sounds that have different acoustic characteristics to the mother 'voice', eliciting reaction of cerebral areas linked to auditory recognition, increasing heart rate and musculoskeletal movements (Mellor, 2019). These foetus's reaction and posterior behavioural modification are based on what is known as trans-natal sensory continuity, where various foetal sensory receptors activate under a stimulation modifying neuronal pathways associated with the auditory development, in this case, preparing the future newborn to its life after birth (Mellor, 2019). It is already known that rough human handling increases fear and stress in sows with negative consequences for reproduction and piglet survival (Tallet et al., 2018), than the growing piglet environment with additional stressful cues (e.g. auditory) may exacerbate their reaction to other stressful factors. It is important for production systems to encourage positive cues during sow and pig handling that involve not only gentle physical management but also reduction in auditory cues (e.g. yelling). Further research is needed in this area to understand the extent of the impact of auditory cues on piglet development and production.

3 Environmental effects during the early postnatal period on the development of behaviour

3.1 Introduction

The mammalian brain continues to develop postnatally, with the organization of neural structures arising from an interaction between genetics and early-life experiences (Tierney & Nelson, 2009). In swine, the period of rapid brain development occurs within the first 12 weeks of life, with the maximum increase in brain volume occurring within the first four weeks following birth (Conrad et al., 2012). The large increase in total brain volume in the postnatal period suggests this is a critical period where disruption of developmental processes by environmental inputs can have long-lasting or permanent effects on brain structure and function (Conrad et al., 2012). Events that occur during a period of rapid brain growth shape the development of neural circuits, mediating the expression of behavioural, emotional, autonomic and endocrine responses to stress, and shaping how an individual responds to subsequent life events, contributing to individual differences in emotionality and stress responsiveness (Rinaman et al., 2011). Depending on the nature of the experience, this can lead to a positive or negative fitness outcome for the individual (Plotsky and Meaney, 1993).

The behaviour of the domestic pig mimics that of its ancestor, the wild boar (Stolba and Wood-Gush, 1989). The brain of the pig is also largely the same as that of the wild boar (Schmidt, 2015). The pig evolved to develop in a highly complex rearing environment, filled with sensory stimuli, with close and complex social bonds. For the development of healthy and normal processes in the brain, the domestic pig requires inputs similar to that which would be encountered in the evolutionary environment (Telkänranta and Edwards, 2018). In contrast, the early-life stimuli afforded to pigs in the typical intensive swine production environment is far removed. Rather than being complex, a global standard farrowing environment for intensively farmed pigs is a full or part-slatted pen, bedding-free, with the sow restrained in a farrowing crate for the duration of lactation. This creates a barren and static environment, with limited space and sensory input, restricted maternal contact with early separation from the sow, the occurrence of intermittent stressful events from human handling and painful procedures, abrupt early weaning at three or four weeks of age and mixing with unfamiliar pigs in a limited space, and depending on the production system, transportation. This implies that the early-life features of the modern swine production environment have major implications for pig behavioural development that will impact both the immediate and longer-term animal welfare and productivity.

The following sections shall review the knowledge on how the stimuli experienced by pigs in the early life modern indoor, intensive swine production environment influences the behavioural development of the pig, and what is known about how manipulations to the environment, or including management interventions influence behavioural development of the pig, and how this may relate to the long-term welfare outcomes for the pig.

3.2 Complexity of the pre-weaning environment

Due to concern over meeting behavioural needs and supporting the expression of innate behaviours, how the barren environment influences the pig, and how to overcome it and provide outlets for the expression of species-specific behaviour has received a considerable amount of research focus. This work has involved comparing pigs raised in highly contrasting environments, such as piglets born into outdoor environments to those in the barren farrowing pen with crated sows, and methods to increase the environmental complexity of indoor farrowing pens. The body of work can also be divided between that which has explored more biologically relevant substrate such as peat, straw, mushroom compost, to that using objects that have no biological relevance, and may or may not contain the properties known to be attractive to pigs (Van De Weerd et al., 2003).

Born into free-ranging, natural conditions piglets start to display exploratory behaviours (rooting, chewing, sniffing) from the first day of life, and direct their investigatory behaviours only to the environment, and not to other animals (Petersen, 1994). For indoor-raised piglets, a typical farrowing pen with a crated sow is a barren environment, devoid of complexity, that presents limitations on what the piglets can manipulate and explore. The space in the pre-weaning environment is also limited, with the typical area available to piglets being in the range from 4.3 m^2 to 4.8 m^2 for indoor farrowing pens containing crated sows (Guy et al., 2012; Baxter et al., 2012). Consequently, piglets direct their investigatory behaviours towards litter-mates, the sow, and pen fittings, and this is well-documented in literature from observations of piglets raised in barren intensive production environments in comparison to those born into outdoor farrowing systems (Cox and Cooper, 2001; Hötzel et al., 2004).

Raising pigs in enriched environments from birth to slaughter can produce pigs with quite divergent characteristics to their barren-raised counter parts. Pigs raised indoors from birth to slaughter in enriched pens (straw and peat provision, and increased space) produce pigs that are more active, show lower levels of aggression, are less fearful of novelty and produce better carcass characteristics than barren-raised pigs (Beattie et al., 2000a,b). Enriched raised pigs have been found to learn tasks quicker (Sneddon et al., 2000), although not consistently found (De Jong et al., 2000). However, the barren-raised pigs have shown impaired long-term memory to a maze test (De Jong et al., 2000), the evidence suggests that rearing in an enriched environment supports cognitive development. Taken together, there is evidence that the complexity of the rearing environment alters the cognition, reaction to novelty, social behaviour and also the stress response of pigs, which is likely responsible for the meat quality differences found by Beattie et al. (2000b).

In contrast, our understanding of the contribution of the early rearing environment to the development of the pig, and subsequent welfare and productivity outcomes is still being unravelled. A large number of the studies that have investigated effects of the pre-weaning environment have studied environments that contain more biologically relevant elements, for example, peat, loose contact with the sow and early socialization. A consistent finding is that a more complex pre-weaning environment supports the transition to weaning, reducing stress and improving productivity and health (Kutzer et al., 2009; Middelkoop et al., 2019; Luo et al., 2020; Ko et al., 2020). It is clear from a large array of studies that within the immediate environment, enrichment of the pre-weaning environment with biologically relevant rooting materials (e.g. peat) has strong effects on pig behaviour (e.g. Kutzer et al., 2009; Luo et al., 2020). Supporting more appropriate/normal foraging behaviour patterns is desirable, because redirected oral exploratory behaviour can decrease growth rate in the receiving pigs (Camerlink et al., 2012), and is considered a contributor to the development of tail biting (D'Eath et al., 2014), with increased tail biting occurring in pens of pigs that show greater levels of pig and pen-directed oral manipulation (Ursinus et al., 2014). Efforts to support the development of biting and foraging behaviours directed away from pen mates may help to reduce the risk for tail biting occurring in the first place. Far less is known about how the provision of non-biologically relevant enrichment, such as the provision of objects in the pre-weaning period, influences pig behaviour development; fewer studies have explored this, the majority providing object or point-source enrichments (those limited in size and contained within a specific area) to older pigs from the nursery stage onwards (Horback et al., 2016; Holling et al., 2017; e.g. see de Oliveira et al., 2020).

The type of enrichment in the pre-weaning environment is also important. Not all enrichment carries the same value for pigs (Van De Weerd et al., 2003), and, partly in consequence, not all enrichment can therefore elicit positive/ behaviour changing effects. Some enrichment may actually increase frustration (van de Weerd and Day, 2009), and there is evidence to suggest biologically relevant enrichment may elicit a greater magnitude of effects. Provision of straw (Oostindjer et al., 2011b) pre-weaning (Fig. 1) resulted in increases in creep intake post-weaning, but not the provision of chewable rope and paper pre-weaning (Telkänranta et al., 2014). Provision of object enrichment pre-weaning increased the expression of play, but only litters with rootable substrate showed a lower stress response to weaning (Yang et al., 2018). Considering the reluctance for fully slatted systems to add manipulable substrates such as straw, further work is warranted to understand what, and the extent of, benefits that object enrichment given in the pre-weaning period can provide for piglet development. Only through comparing these effects to those arising from the provision of more biologically relevant materials can the relative contribution of these environmental factors towards pig development be understood. This is important for evidence-based requirements for pig housing that can support appropriate swine development and promote improved welfare.

Play behaviour is an early-life behaviour that first appears in the days following birth, peaking between two and six weeks of age (Newberry et al.,

Figure 1 Provision of chopped straw to piglets in the farrowing environment: provision on a mat enables the provision in a slatted system and is suitable during the first week of life (a), and provision in a box helps to contain the straw better as piglets grow (b), photos provided by D. Pollock/Y. Seddon, University of Saskatchewan.

1988). Its expression is believed to largely be suggestive of good welfare, as the behaviour is typically expressed only in the absence of sickness, hunger, and so on and has important developmental effects including somatic control, development of social skills and preparing for the unexpected (Špinka et al., 2001; Held and Špinka, 2011) (Špinka et al., 2001). The complexity of the rearing environment greatly influences the expression of play. Provision of more space and straw stimulates pre-weaning play behaviour, and although aggression is observed as higher than barren reared pigs at weaning (Chaloupková et al., 2007), less chronic aggression has been found in piglets reared in enhanced pre-weaning environments (PigSafe pen, more space and straw substrate) (Martin et al., 2015), which is believed to be related to a greater expression of play behaviour in which socio-cognitive skills can develop. This may suggest a positive effect of the environment or simply that the space and facilities are available to perform the behaviour. The total amount of space available will influence locomotor play (Blackshaw et al., 1997). Locomotor play is strongly related to social play, and therefore in pens with limited space, the expression of these play behaviours may be suppressed. This was found by Yang et al. (2018), in which the provision of objects in a farrowing crate has been found to increase object play, but not locomotor or social play, a result that was believed to be related to the restrictive space. Higher welfare-farrowing pens provide an environment in which the expression of play can be increased in part because of the increase in space, increasing up to 9 m^2 for some loose lactation pen designs (Baxter et al., 2012; Guy et al., 2012). For the standard farrowing crate there are implications for how this is stunting the development of the pig through the restriction of play. How play relates to the development of social behaviour is discussed in Section 3.2.

However, whether the effects of the farrowing environment complexity can permanently shape behavioural development is harder to determine because the immediate environment has such a profound effect on the behaviour of the pig. For example, in the pre-weaning environment, piglets reared with the provision of peat display more foraging and play behaviour than barren-reared piglets but do not differ from barren-raised pigs in behaviour when moved for rearing in a barren nursery (Vanheukelom et al., 2011). When observed in the nursery phase, piglets reared with the provision of peat in both the pre-weaning and nursery phase, or provided with peat in the nursery phase only, also do not differ in behaviours, with both showing equal and greater levels of positive behaviours (play, eating) and fewer negative/stress-related behaviours (biting, fighting and sitting) than pigs that are raised in only a barren environment or received peat in the pre-weaning environment only (Vanheukelom et al., 2011). It is a consistent finding that moving piglets from an enriched pre-weaning environment to a barren post-weaning environment results in behavioural changes suggestive of poorer welfare and vice versa (Munsterhjelm et al., 2009; Oostindjer et al., 2011b). However, piglets that move from a barren pre-weaning environment to an enriched post-weaning environment have been observed to exhibit more pronounced exploration and play, described by Luo et al. (2020) as if observed to be going through a phase of catching up. Comparing the interaction between sow lactation housing (loose or crated) and enrichment (enriched vs. barren) in the pre- and post-weaning period, Oostindjer et al. (2010b) observed behavioural changes suggesting negative effects from switching environments to be greatest when piglets born into an enriched environment and a crated sow were weaned into a barren environment, suggesting that the sow housing pre-weaning has an effect on how the piglets value their post-weaning environment (Oostindjer et al., 2010b).

However, there is also evidence that there is a clear effect of the provision of enrichment materials in the pre-weaning environment and shaping of the behavioural development. Telkänranta et al. (2014) identified that providing enrichment in the form of sisal rope and paper from birth reduced oral manipulation of pen mates by pigs, and the severity of tail biting observed post-weaning, compared to the provision of these enrichment materials from the nursery stage.

Provision of enrichment in the pre-weaning environment influences the development of the circadian rhythm of cortisol, and this may have implications for behavioural changes. Pigs reared from birth to slaughter in enriched (straw-bedded) environments show a pronounced and established circadian rhythm of cortisol by 22 weeks of age, whereas, for barren pigs, this is blunted (De Jong et al., 2000). However, the work of Munsterhjelm et al. (2010) suggests providing the environmental complexity is increased during the early life period, the development of rhythm is achieved. Munsterhjelm et al. (2010) found the provision of enrichment (thin layer of wood shavings and chopped straw) from birth to nine weeks of age increased the odds of a cortisol rhythm developing, compared to barren housing, and a greater percentage of pigs had a pronounced circadian rhythm by 21 weeks of age when enrichment was provided from birth to four weeks of age. The significance of a blunted rhythm for pigs is still not uncovered, but relationships between pens where the focal pig had blunted rhythms showing increased tail-biting lesions in the fattening stage suggest it could relate to chronic stress (Munsterhjelm et al., 2010).

Similarly, the concentration of brain-derived neurotrophic factor (BDNF) was higher in piglets provided with a foraging block either pre-weaning or pre- and post-weaning, compared to barren-raised pigs or pigs provided with the enrichment only post-weaning (Rault et al., 2018). Due to its role in hypocampal neurogenesis, Rault et al. (2018) proposed that measures of BDNF may help to reflect the functional benefits of environmental enrichment, and can reflect how the enrichment supports neural development. Further understanding on how different enrichment types influence BDNF concentrations could be used as a marker of the effectiveness of enrichment, and to understand how the enrichment has influenced pig development and measures of resilience remains to be discovered. Establishing understanding of the physiological changes that occur can help increase knowledge as to how the environment is influencing the physiology, which may also present itself in behavioural changes.

3.3 The influence of the early social environment

Pigs are a highly social species and form complex social bonds (Focardi et al., 2015), the closest of which are between litter-mates and their mother (Newberry and Wood-Gush, 1986; Petersen et al., 1989). For the vast majority of piglets born into commercial production environments, the early social environment involves the litter-mates and the mother, whose ability to interact is somewhat restricted by the farrowing crate. Piglets will remain in their litters until weaning at around 21–28 days of age, at which point they are then mixed with unfamiliar piglets from a variety of different litters to form nursery pen groups. This mixing causes significant aggression, with increased stress and risk of injury (Ekkel et al., 1995; Puppe et al., 1997) and can have negative consequences for animal welfare and growth.

Communication from the sow to her piglets in the first few days of life, through vocalizations, grunts, sniffing and nudging supports the maternal-offspring bonding process, and attracts the piglets to stay close to the mother (Melišová et al., 2011). Sows that perform a greater amount of this communication when active in a loose lactation pen have been found to increase piglet survival (Melišová et al., 2011; Ocepek and Andersen, 2018), despite that the piglets respond to these calls by coming closer to the sow. While this suggests the role of the sow's maternal care and attention to her piglets, her communication may also be shaping the response of the piglets in ways that support a greater survival, and this has not been studied to date.

The social behaviour of pigs still mimics that of the wild boar (Stolba and Wood-Gush, 1989; Gabor et al., 1999). Having farrowed in an isolated nest away from the sounder (stable group of breeding females), the sow would lead her piglets out of the nest at around 10 days of age to rejoin the larger family group. At this point, piglets from other litters would be brought together to form one family group. Studied under free-ranging natural conditions, at this joining of litters, piglets will form social bonds with non-litter-mates peacefully, with no aggression (Petersen et al., 1989). This period of social integration takes place between two- and seven-weeks post-partum (Petersen et al., 1989), suggesting this is a sensitive time for learning of social behaviours to take place. That pigs can form social bonds peacefully has significant implications for the management of swine in production systems. There is considerable evidence that taking advantage of this socialization window and providing piglets with the opportunity to interact with non-litter-mates confer lasting benefits for the social development of pigs.

Piglets co-mingled in the pre-weaning period and then re-mixed with unfamiliar piglets at weaning start to fight quickly but resolve conflict and established a social hierarchy sooner (D'Eath, 2005), resulting in overall lower level of aggression (Weary et al., 1999) and fewer injuries (Salazar et al., 2018). When tested in matched pair tests at eight weeks of age, contests between pigs socialized pre-weaning result in a quicker resolution, with less cost to the animals (lower injury), with increased ritualized display behaviour indicative of mutual contest assessment (Camerlink et al., 2019). It is clear that the early-life socialization supports the development of social skills with the pigs having increased opportunities to understand the consequences of conflict and exercise assessment of the cost of getting involved in the contest, which requires cognitively advanced assessment. The behaviour changes resulting from socializing piglets pre-weaning are present six weeks after socializing (Camerlink et al., 2018, 2019); however, the influence of pre-weaning socialization on longer-term social behaviours of pigs has not yet received much research focus. As a first study to document the effects of pre-weaning socialization on pig social behaviour at a late stage of life, Ko et al. (2020)

observed that piglets reared under commercial conditions with co-mingling and objects within the farrowing environment continue to show altered social behaviours at pre-slaughter mixing. Skin lesions were suggestive of control animals (standard rearing conditions) engaging in more reciprocal fighting and early-life enriched pigs (socialisation through co-mingling and object provision) receiving more unilateral bullying (Ko et al., 2020). The aforementioned work could not fully separate the effects of treatment as control, and enriched pigs were mixed together pre-slaughter for practicality, and therefore further work with greater control over treatments and more detailed observations would be beneficial. There is value to investigate this, considering the implications for understanding the role of early-life management to support the development of breeding gilts for group gestation housing.

Play behaviour assists with the development of social skills (Špinka et al., 2001), particularly aggressive contest escalation and resolution behaviour, with divergent gender effects that may help to reduce escalated aggression (Weller et al., 2019). Relationships exist between early play fighting frequency and the development of different social assessment strategies in the pigs, and therefore there is potential that early-life play behaviour may be predictive of adult social behaviour (Weller et al., 2020). Therefore, supporting the full expression of play behaviour during the pre-weaning environment may be one method to support the development of social skills in pigs that has long-lasting implications. This might require a larger farrowing environment because locomotor and social play is increased in larger farrowing environments compared to the standard farrowing pen containing a crate, as is discussed by Yang et al. (2018). However, for piglets to be able to experience interactions with non-litter-mates during early life will further complement the social skills acquired through play as individuals can gain a superior understanding of their own resource holding potential (the ability to win a fight) (Weller et al., 2019).

Cross-fostering of piglets between litters is a practice performed within the first few days post-partum, in which piglets are taken from a sow, and placed within the litter of another. This practice is done to reduce competition within litters, for example, by removal of excess piglets, so there are an equal number of piglets to functional teats. It is also sometimes done to equalize litter size according to birthweight (Heim et al., 2012). Teat order and material recognition are not fully established until several days post-birth (De Passillé et al., 1988), and cross-fostering performed before this time has fewer behavioural consequences (Robert and Martineau, 2001; Heim et al., 2012). However, when performed beyond this, teat order is disrupted and aggression ensues as the newly formed litter must re-establish a teat order (Robert and Martineau, 2001). Consequently, when fostering occurs later than two days post-partum, piglets from fostered litters (containing both residents and fostered on piglets) have been observed to fight less at weaning, resulting in fewer skin lesions (Giroux et al., 2000), suggesting that the action of cross-fostered litters encountering and fighting with unfamiliar piglets has an early-life socialization effect, influencing subsequent social behaviours. While this is beneficial, there is evidence to suggest that cross-fostering is an early-life practice that contributes to the development of non-aggressive tail biting in pigs. This is reviewed by Prunier et al. (2020), and more data is needed to draw solid conclusions. However, the level of competition in the pre-weaning environment appears to shape later aggressive behaviour. Teat order, which reflects the level of competition that piglets experience pre-weaning, relates to the level of aggression pigs perform post-weaning (Skok et al., 2014). This has implications for the development of aggressive traits in pigs from large litter sizes, in which competition at the udder may be greater and more sustained throughout lactation if the sow is expected to raise an increased number of piglets to maximize teat use.

Weaning drastically changes the social environment of pigs, with the removal of the sow and the introduction of new conspecifics, which occur with the additional concomitant stressors or moving to a new environment and being moved from a liquid to a solid diet. In global pig production practices, weaning typically occurs at 21 or 28 days, but in some parts of the world, weaning is even earlier, and closer to 18 days. Not only is this far removed from the gradual weaning that piglets will undergo if left with the sow in free-ranging conditions, but within standard practice it also takes place during the period of rapid postnatal brain development, and therefore concerns remain over whether the practice of early weaning can create lasting developmental changes in the pig. While early weaning influences the behaviour of the piglets in the immediate term, most commonly resulting in an increase in oral-nasal manipulation behaviours (belly-nosing, sucking, chewing, nibbling, tail biting) (Bøe, 1993; Gonyou et al., 1998; Worobec et al., 1999). The effects have been observed to be transient, disappearing in growth stages beyond the nursery (Hohenshell et al., 2000) and when raised in pens with substrate enrichment (Bøe, 1993). However, it has also been observed that compared to piglets weaned at 21 days of age, very early weaned piglets (12 days of age) continue to show increased oral-nasal manipulation of pen mates and objects within the pen in the grow-finish stage (Gonyou et al., 1998). There is no clear evidence that early weaning practices contribute to the development of tail biting behaviour as reviewed by Prunier et al. (2019), and despite weaning being a severe stressor during a period when the HPA axis is still developing, a reprogramming of the HPA axis resulting from weaning at different ages has not been observed (Jarvis et al., 2008).

Artificial rearing of piglets fed by milk cups, which is applied when there are more piglets than sow teats in a farrowing batch, rendering cross-fostering impossible, also produces an increase in redirected oral-nasal facial behaviours (Schmitt et al., 2019). With increased push to produce more piglets per sow, the practice of artificial rearing could remain for the rearing of proportions of piglets. Due to limited data, the full impact of artificial rearing to shape behavioural development is still largely unclear. However, considering this involves separation from the sow and requires piglets to feed on a bowl cup rather than a nipple, there is a good potential that long-term behavioural changes could be observed, and may contribute to the development of tail biting behaviour (see review by Prunier et al. (2019)).

The sow also has a role in supporting the vertical transfer of information on appropriate foodstuffs. This latter topic is covered in Section 3.3, in which the development of ingestive behaviours is discussed.

3.4 Supporting development of appetitive and ingestive behaviours

Under natural conditions, weaning, resulting in the piglet transitioning from milk to solid food, is a gradual process. Piglets begin to explore their environment, rooting, biting and chewing from the first day of life, and sampling and ingestion of solid food has been observed to develop in the third week of life (Petersen, 1994), with the weaning transition completing between 8 and 22 weeks of age (Newberry and Wood-Gush, 1985; Jensen and Stangel, 1992; Petersen, 1994). Within the intensive production environment, the weaning process occurs much earlier, typically at 21 or 28 days of age, and is usually an abrupt switch from a liquid to solid diet. The opportunities for learning about solid foodstuffs and developing ingestive behaviours are typically limited by the management style. Creep feed is proposed for stimulating intake and supporting the transition onto solid food. However, problems exist and persist with stimulating the piglets to eat the creep; creep consumption is highly variable across litters; not all piglets within a litter will consume the creep, and those that do consume limited amounts, reducing the cost-effectiveness of providing creep. Consequently, creep provision is not a standard practice on all farms.

Supporting a gradual transition onto solid food is important for the maintenance of gut integrity. A drop in food intake in the period immediately following weaning plays a major role in altering gut integrity, increasing the risk of gastrointestinal disorders post-weaning (Hamps et al., 1994, cited in Montagne et al., 2007; Spreeuwenberg et al., 2001), reduced growth performance and increased need for supportive antibiotics. Therefore methods to stimulate the intake of feed pre-weaning, and help to smooth the transition to weaning are valuable. It is believed that early exploratory behaviour of the pigs is one means by which pigs gain information about their environment (Wood-Gush and Vestergaard, 1993), and therefore stimulating exploratory behaviour may support learning of novel foodstuffs in piglets. Stimulating exploratory behaviour around the creep has been investigated through the use of commercially available

(Kuller et al., 2010), creation of a more diverse 'play feeder' (Middelkoop et al., 2019), which involved simple modifications to a traditional round creep feeder design; rope, PVC spiral tubing and strips of canvas cloth was added to the inside of the feeder, to support increased exploration and foraging around the creep feed. In the work of Kuller et al. (2010), pigs raised with the play feeder had a greater disappearance of creep feed pre-weaning but did not have an increased weight gain, with no post-weaning measures collected. In the work of Middelkoop et al. (2019) the play feeder did not influence creep intake pre-weaning, but did attract more piglets to explore the creep feeder pre-weaning, resulting in increased ingestive behaviours, a 15% greater feed intake and growth between days 0 and 15 post-weaning, and a 5% higher body weight at day 15 post-weaning. Furthermore, the provision of a play feeder resulted in fewer ear and tail lesions and improved faecal consistency in the post-weaning period (Middelkoop et al., 2019). These profound benefits suggest that the play feeder may ease the weaning transition overall. In part, this may be because of the benefits that manipulable objects provided in the pre-weaning can offer. It may be that enrichment that promotes exploratory behaviour pre-weaning may support increased exploration in the post-weaning environment, which could result in increased ingestion of feed. However, it is a common finding that manipulable objects that provided pre-weaning have little to no effect on growth performance post-weaning (Telkänranta et al., 2014; Yang et al., 2018), or if they do, it is related to supporting the transition to weaning through reducing the stress or weaning, through mechanisms such as better preparing the piglets to encounter novelty and reducing the incidence of enteric disease, rather than increasing feed intake (Vanheukelom et al., 2011). Therefore the findings of Middelkoop et al. (2019) are interesting in that they appear to suggest that providing enrichment objects around the feeder could support piglets to make a beneficial association between the performance of object exploration/play and feed. Making such an association may also be related to why providing creep feed in pellets of a larger diameter results in increased feed intake and body weight gain post-weaning (van den Brand et al., 2014); it provides a greater opportunity for manipulation of the pellets and for making the association between oral manipulation and solid feed.

3.4.1 Interaction between the sow and piglets

Within the intensive production environment, the restriction of the sow to a farrowing crate throughout lactation drastically reduces the interaction between sow and piglets and results in the sow and piglets feeding in separate areas of the farrowing pen and on a different feed. But piglets take cues from the sow to learn about novel food, and observation, participation and stimulus enhancement appear to be important for the piglet to learn about safe food and develop ingestive behaviours (Oostindjer et al., 2011c). Piglets show a preference for the feed the sow consumes (Oostindjer et al., 2011c), and when provided with an option to feed with the sow or simply observe the sow eating, food neophobia is reduced, and intake increased (Oostindjer et al., 2011c,d). While piglets show a preference to feed in the same location as the sow, if forced to feed separately, piglets choose to eat in a separate location, where the food provided is the same as that eaten by the sow (Oostindjer et al., 2011c). Piglets housed with loose sows during lactation have been found to exhibit more chewing behaviour pre-weaning, supporting the development of muscles for mastication (Herring, 1985), and more food exploration post-weaning (Oostindjer et al., 2010b). This is powerful information that can be combined in designs for the farrowing environment, to support better piglet behavioural development that will support the adaptation of the piglet to weaning. The design of the family feeder as described by Oostindjer et al. (2014), in which the sow and piglets can both eat, illustrates the application of this concept in practice. It has been found to increase pre-weaning growth rate by 6% at no extra effort to the producer (Oostindjer et al., 2014). Oostindjer et al. (2014) provide a thorough review of how factors in the pre-weaning environment influence the development of eating behaviours in the pig; it is recommended reading for those wishing to learn more.

3.5 Human interaction

Early-life human contact may influence subsequent responses to humans. It provides a window of opportunity in which to shape how the pig will respond to subsequent encounters to humans, and this could be used to put in place management strategies that could result in lasting benefits for the animal and to improve the human-animal relationship.

Handling of piglets early in life (increased exposure to human presence, stroking and petting of pigs that approach a human) has been found to have lasting effects on the behavioural response to humans, reducing the fear of humans as measured by increased approaches to a stationary human up to 18–24 weeks of age (Hemsworth et al., 1986; Hemsworth and Barnett, 1992). If handling is applied later in life (from 11 to 22 weeks of age), the differences in positive- and negative-handled gilts that were evident at 25 weeks were absent when retested at 40 weeks (Hemsworth et al., 1981), suggesting the effect to reduce the fear response may only be short term, and that for a longer effect, a sensitive period may exist. The work of Weaver et al. (2000) further supports that a sensitive period exists. Daily handling of neonatal boars during the first 14 days of life (litter-mates caught and placed in a straw-lined box for 10 min) produced long-lasting changes to the behaviour and HPA axis of boars, detectable at seven months of age (Weaver et al., 2000). There is great value in identifying a sensitive period; it would enable recommendations for targeted management to be given within a specific window of time to modulate the pigs' response to humans and to beneficially shape the behaviour. To date, this has not been well researched.

In applying early-life handling, the nature of human interaction is also important. It is well-documented that positive human-pig interactions (i.e. gentle strokes and pats) reduce the fear of humans, whereas negative interactions (i.e. slapping and electric shocks) increase fear and avoidance of humans (Hemsworth et al., 1981). The beneficial effect of this early handling may be a result of positive conditioning (associating the human handling to a positive or negative stimulus), and additionally habituation/familiarization of the piglets to human handling, resulting in them being less fearful if the interaction is passive (Muns et al., 2015). Considering this, extremely brief, positive human contact given on the first day of life (caressing piglets and human talking during six suckling bouts on the first day of life) can immediately alter the behavioural response of piglets to subsequent stressful events, lowering the duration of escape behaviour during tail docking at 2 days of age (Muns et al., 2015). Over a period of 5 days, repeated interactions with weaned piglets of positive (gentle stroking and food rewards), neutral (passive presence) or negative (rough handling without physical stress) influenced subsequent behavioural responses to humans 5 weeks later; positive treatments reduced fear responses, passive supported habituation to human presence and negative interactions induced fear responses (Brajon et al., 2015).

How frequent or severe the nature of the early human handling has to be to produce long-lasting changes in behavioural reactivity and HPA axis functioning is not fully known. Initial work has identified that piglets tail-docked within the first 3 days of age are slower to make contact with a human when tested at 14 days of age, compared to undocked piglets (Tallet et al., 2019). Castration within the first week of life has also been found to influence the behavioural response of pigs to humans when measured at 11 weeks of age, with castrated pigs spending less time within close proximity to an unfamiliar human than entire males (Tallet et al., 2013). These works suggest that the routine painful husbandry procedures shape the human-animal relationship and may result in pigs generalizing negative experiences with humans early in life to subsequent interactions (Tallet et al., 2013). It is of value to understand how the administration of the range of routine painful husbandry procedures early in life (teeth clipping, castration, tail docking, iron injection and vaccination), influence the HPA-axis functioning and subsequent behavioural response of pigs to humans. It is important to consider whether the beneficial effects of positive early handling may be overridden by subsequent negative handling experiences, such as the aforementioned practices (Hemsworth and Barnett, 1992). This is not known but should be investigated. Otherwise, efforts to rebuild a more positive association for pigs to humans may be necessary to support ongoing benefits. The challenge is that the correct form and balance of handling must be employed, or else it can result in pigs that are so accustomed to human presence that they become harder to move/handle (Day et al., 2002).

The effects of human handling as tactile stimulation early in life appears to offer benefits beyond simply reducing fear responses to humans. Pigs that received positive early human tactile contact show a reduction in fear when tested in a novel environment (de Oliveira et al., 2015), and increased locomotor skills and social interactions (Zupan et al., 2016). When pigs were stroked by humans it also increased their salivary oxytocin concentrations (Lürzel et al., 2020). Together this suggests that early-life tactile stimulation from human contact can improve the response of pigs to a new situation and influence motor, cognitive, socio-emotional development, and influencing hormonal concentrations associated with social recognition and positive welfare states (Rault et al., 2017). This suggests the potential for a far greater role from early-life human tactile stimulation to benefit the development and welfare of the pig.

4 Conclusions

In recent years, swine production systems have gone through various management changes to provide for sow welfare. Depending on the production system, growing pigs and sows experience various management practices that expose them to stressors of different intensity and length. The body of work collated in the present chapter reveals that how the stressors the sow experiences over her life, and how she responds, could be modulated or programmed even before conception. The changes seen in the piglets' physiological and behavioural responses are likely to be based on epigenetic factors resulting from the stressors experienced by the parents during the prenatal environment and the pre-conception period.

Some of the studies presented here show that the prenatal conditions can produce (1) a better suited behavioural phenotype, and (2) that positive factors can improve sow and piglet welfare, and this should be researched further. However, it is important to consider how the postnatal life can alter further, for better or worse behavioural phenotype of piglets. Similarly, to understand how far following birth the beneficial effects that develop prenatally last, when considering their interaction with the postnatal environment. Research seems to show that piglet behaviour modified by maternal stress may be buffered by the postnatal environment. However, the long-term consequences of these behavioural changes for offspring adaption, and the consequences for health, welfare and productivity remain to be determined.

Early-life stimuli experienced by piglets in the postnatal environment elicit immediate behavioural changes, while also showing evidence of permanently altering the physiological development and conferring long-lasting changes. Simple changes that increase the complexity of the environment provide opportunities for increased and improved social contacts between the sow and piglets, increase the range of social partners and provide positive human interaction; delivered separately or together, support increased expression of innate behaviours, a reduction in damaging behaviours, improved social skills, cognitive development and modulation of stress reactivity, and demonstrates that piglets require key environmental inputs early in life to support a normal, and healthy behavioural development. There is substantial evidence that the most pronounced and longer-lasting benefits that appear to extend into the life of the growing pig arise when stimuli with biological relevance are provided. Combining this knowledge into modern pork production herein lies significant opportunities to increase the behavioural development, stress resilience, welfare and production qualities of pigs.

The sow is an important figure in the piglet's life and can support piglet survival and learning about appropriate feedstuffs, and with the majority of sows rearing piglets in farrowing crates, the role of the sow in supporting the development of the piglet is underutilized in modern production facilities. The research to date suggests that simple modifications to the environment can support the transfer of information from sow to piglets, and the pairing of behaviours with stimuli can bring about beneficial development changes, helping the piglet to associate one with another.

Considering the research presented herein, it highlights the importance of understanding exactly how modern production practices affect the animals before new production systems and management methods are adopted on a global scale and is especially important considering that the evolving pig production climate requires production with less antibiotic use and more resilient pigs.

5 Future trends in research

Current knowledge of pre-conception and prenatal environment effects is still limited. Areas of research interest could be the effect of single or combined stressors before and after conception, the effect of sex on the behaviour phenotype and the use of non-castrated pigs to understand gender effects. This latter is of importance as some countries have started to ban piglet castration, and it is applicable for the development of breeding boars. Long-term studies are needed to understand how these behavioural phenotypes cope or adapt in their postnatal environment and how management practices different from those experienced by the parents may benefit or negatively affect offspring welfare and productivity. Understanding how prenatal stressors influence piglet resilience traits based on behavioural adaptations should include individual traits influenced by genetics and personality.

In the postnatal period, future research should focus on identifying sensitive periods, the physiological changes that drive the behavioural effects that are reported, understanding the interplay between pre- and post-weaning environments, and the duration of effects to influence pig behaviour, physiology and the subsequent consequences for welfare and productivity. The majority of studies reviewing the early life period only apply treatments up to weaning, but given that rapid development of the pig brain continues until 12 weeks of age, more studies that are continuing to investigate how the behavioural development is enhanced and reinforced throughout this period would be beneficial. However, it is also important that future work identifies if there is an opposing negative effect of repeated stressful events that occur in this early life period.

There is a growing need for scientific evidence in the determinants of pig behaviour as this will facilitate evidence-based recommendations to farmers to improve their production systems while improving the welfare and productivity of pigs. The need for this research is even greater, now more than ever considering the increasing numbers of sows moving to group-housing systems, in which depending on the system and management, they can be exposed to an increased range of stressors throughout gestation. Further research is required to confirm that positive behaviour modifications gained in any of the production stages are long-term or negatively affected by different management practices later in life.

6 Where to look for further information

The European Reference Centre for Animal Welfare, Pigs provides factsheets summarizing the latest information on various pig welfare topics. Some of this material makes reference to provisions in the early life environment of importance to pig welfare.

https://www.eurcaw.eu/en/eurcaw-pigs/output/scientific-output.htm.

Below are some scientific reviews that can provide the reader with a basic understanding of the physiology behind pig behavioural development and the changes caused by prenatal and postnatal factors:

- Mellor, D. J. (2019). Preparing for life after birth: Introducing the concepts of intrauterine and extrauterine sensory entrainment in ammalian young. *Animals: An Open Access Journal from MDPI* 9. https://doi.org/10.3390/ani9100826.
- Mina Theresia, H. and Reynolds, R. M. (2014). Mechanisms linking in utero stress to altered offspring behaviour. In: Pariante Carmine, M. and Lapiz-Bluhm, M. D. (Eds), *Behavioral Neurobiology of Stress-Related Disorders*.

Springer Berlin Heidelberg, Berlin, Heidelberg, pp. 93–122. https://doi.org/10.1007/7854_2014_291.

- Oostindjer, M., Kemp, B., van den Brand, H. and Bolhuis, J.E. (2014). Facilitating 'learning from mom how to eat like a pig' to improve welfare of piglets around weaning. *Applied Animal Behaviour Science* 160: 19–30. https://doi.org/https://doi.org/10.1016/j.applanim.2014.09.006.
- Otten, W., Kanitz, E. and Tuchscherer, M. (2015). The impact of pre-natal stress on offspring development in pigs. *Journal of Agricultural Science* 153: 907–919. https://doi.org/10.1017/S0021859614001361.
- Prunier, A., Averos, X., Dimitrov, I., Edwards, S. A., Hillmann, E., Holinger, M., Ilieski, V., Leming, R., Tallet, C., Turner, S. P., Zupan, M. and Camerlink, I. (2020). Review: Early life predisposing factors for biting in pigs. *Animal* 14: 570–587. https://doi.org/10.1017/S1751731119001940.
- Špinka, M. (2017). *Advances in Pig Welfare*, Woodhead Publishing Series in Food Science, Technology and Nutrition. Elsevier Science. https://doi.org/10.1016/c2015-0-05682-9.

7 References

Antonson, A. M., Balakrishnan, B., Radlowski, E. C., Petr, G. and Johnson, R. W. (2018). Altered hippocampal gene expression and morphology in fetal piglets following maternal respiratory viral infection. *Developmental Neuroscience* 40(2): 104–119. doi:10.1159/000486850.

Antonson, A. M., Radlowski, E. C., Lawson, M. A., Rytych, J. L. and Johnson, R. W. (2017). Maternal viral infection during pregnancy elicits anti-social behavior in neonatal piglet offspring independent of postnatal microglial cell activation. *Brain, Behavior, and Immunity* 59: 300–312. doi:10.1016/j.bbi.2016.09.019.

Ashworth, C. J., Toma, L. M. and Hunter, M. G. (2009). Nutritional effects on oocyte and embryo development in mammals: implications for reproductive efficiency and environmental sustainability. *Philosophical Transactions of the Royal Society of London. Series B, Biological Sciences* 364(1534): 3351–3361. doi:10.1098/rstb.2009.0184.

Baxter, E. M., Andersen, I. L. and Edwards, S. A. (2017). Sow welfare in the farrowing crate and alternatives. In: Špinka, M. B. T.-A. (Ed.) *Advances in Pig Welfare*. Woodhead Publishing, Cambridge. pp. 27–71. doi:10.1016/B978-0-08-101012-9.00002-2.

Baxter, E. M., Lawrence, A. B. and Edwards, S. A. (2012). Alternative farrowing accommodation: welfare and economic aspects of existing farrowing and lactation systems for pigs. *Animal* 6(1): 96–117. doi:10.1017/S1751731111001224.

Beattie, V. E., O'Connell, N. E., Kilpatrick, D. J. and Moss, B. W. (2000a). Influence of environmental enrichment on welfare-related behavioural and physiological parameters in growing pigs. *Animal Science* 70(3): 443–450. doi:10.1017/S1357729800051791.

Beattie, V. E., O'Connell, N. E. and Moss, B. W. (2000b). Influence of environmental enrichment on the behaviour, performance and meat quality of domestic pigs. *Livestock Production Science* 65(1-2): 71–79. doi:10.1016/S0301-6226(99)00179-7.

Begum, G., Stevens, A., Smith, E. B., Connor, K., Challis, J. R., Bloomfield, F. and White, A. (2012). Epigenetic changes in fetal hypothalamic energy regulating pathways are associated with maternal undernutrition and twinning. *FASEB Journal* 26(4): 1694-1703. doi:10.1096/fj.11-198762.

Bench, C. J., Rioja-Lang, F. C., Hayne, S. M. and Gonyou, H. W. (2013a). Group gestation housing with individual feeding-I: how feeding regime, resource allocation, and genetic factors affect sow welfare. *Livestock Science* 152(2-3): 208-217. doi:10.1016/j.livsci.2012.12.021.

Bench, C. J., Rioja-Lang, F. C., Hayne, S. M. and Gonyou, H. W. (2013b). Group gestation sow housing with individual feeding-II: how space allowance, group size and composition, and flooring affect sow welfare. *Livestock Science* 152(2-3): 218-227. doi:10.1016/j.livsci.2012.12.020.

Bergdolt, L. and Dunaevsky, A. (2019). Brain changes in a maternal immune activation model of neurodevelopmental brain disorders. *Progress in Neurobiology* 175: 1-19. doi:10.1016/j.pneurobio.2018.12.002.

Bernardino, T., Tatemoto, P., Morrone, B., Mazza Rodrigues, P. H. and Zanella, A. J. (2016). Piglets born from sows fed high fibre diets during pregnancy are less aggressive prior to weaning. *PLoS ONE* 11(12): e0167363. doi:10.1371/journal.pone.0167363.

Bilkó, A., Altbäcker, V. and Hudson, R. (1994). Transmission of food preference in the rabbit: the means of information transfer. *Physiology and Behavior* 56(5): 907-912. doi:10.1016/0031-9384(94)90322-0.

Blackshaw, J. K., Swain, A. J., Blackshaw, A. W., Thomas, F. J. M. and Gillies, K. J. (1997). The development of playful behaviour in piglets from birth to weaning in three farrowing environments. *Applied Animal Behaviour Science* 55(1-2): 37-49. doi:10.1016/S0168-1591(97)00034-8.

Blavi, L., Solà-Oriol, D., Mallo, J. J. and Pérez, J. F. (2016). Anethol, cinnamaldehyde, and eugenol inclusion in feed affects postweaning performance and feeding behavior of piglets1. *Journal of Animal Science* 94(12): 5262-5271. doi:10.2527/jas.2016-0760.

Bøe, K. (1993). The effect of age at weaning and post-weaning environment on the behaviour of pigs. *Acta Agriculturae Scandinavica, Section A - Animal Science* 43(3): 173-180. doi:10.1080/09064709309410162.

Boersma, G. J., Bale, T. L., Casanello, P., Lara, H. E., Lucion, A. B., Suchecki, D. and Tamashiro, K. L. (2014). Long-term impact of early life events on physiology and behaviour. *Journal of Neuroendocrinology* 26(9): 587-602. doi:10.1111/jne.12153.

Bolhuis, J. E., Oostindjer, M., Van Den Brand, H., Gerrits, W. J. J. and Kemp, B. (2009). Voluntary feed intake in piglets: potential impact of early experience with flavours derived from the maternal diet. *Voluntary Feed Intake in Pigs*: 37-60. doi:10.3921/978-90-8686-689-2.

Brajon, S., Laforest, J. P., Bergeron, R., Tallet, C., Hötzel, M. and Devillers, N. (2015). Persistency of the piglet's reactivity to the handler following a previous positive or negative experience. *Applied Animal Behaviour Science* 162: 9-19. doi:10.1016/j.applanim.2014.11.009.

Brajon, S., Ringgenberg, N., Torrey, S., Bergeron, R. and Devillers, N. (2017). Impact of prenatal stress and environmental enrichment prior to weaning on activity and social behaviour of piglets (*Sus scrofa*). *Applied Animal Behaviour Science* 197: 15-23. doi:10.1016/j.applanim.2017.09.005.

Byrd, C. J., Anderson, N. C., Lugar, D. W., Safranski, T. J., Lucy, M. C. and Johnson, J. S. (2019). Evaluating the effects of in utero heat stress on piglet physiology and behavior following weaning and transport. *Animals: An Open Access Journal from MDPI* 9(4): 191. doi:10.3390/ani9040191.

Camerlink, I., Bijma, P., Kemp, B. and Bolhuis, J. E. (2012). Relationship between growth rate and oral manipulation, social nosing, and aggression in finishing pigs. *Applied Animal Behaviour Science* 142(1-2): 11-17. doi:10.1016/j.applanim.2012.09.004.

Camerlink, I., Farish, M., D'eath, R. B., Arnott, G. and Turner, S. P. (2018). Long term benefits on social behaviour after early life socialization of piglets. *Animals: An Open Access Journal from MDPI* 8(11): 192. doi:10.3390/ani8110192.

Camerlink, I., Turner, S. P., Farish, M. and Arnott, G. (2019). Advantages of social skills for contest resolution. *Royal Society Open Science* 6(5): 181456. doi:10.1098/rsos.181456.

Chaloupková, H., Illmann, G., Bartoš, L. and Špinka, M. (2007). The effect of pre-weaning housing on the play and agonistic behaviour of domestic pigs. *Applied Animal Behaviour Science* 103(1-2): 25-34. doi:10.1016/j.applanim.2006.04.020.

Chan, J. C., Nugent, B. M. and Bale, T. L. (2018). Parental advisory: maternal and paternal stress can impact offspring neurodevelopment. *Biological Psychiatry* 83(10): 886-894. doi:10.1016/j.biopsych.2017.10.005.

Chavatte-Palmer, P., Tarrade, A., Kiefer, H., Duranthon, V. and Jammes, H. (2016). Breeding animals for quality products: not only genetics. *Reproduction, Fertility, and Development* 28(1-2): 94-111. doi:10.1071/RD15353.

Clouard, C., Gerrits, W. J. J., Kemp, B., Val-Laillet, D. and Bolhuis, J. E. (2016a). Perinatal exposure to a diet high in saturated fat, refined sugar and cholesterol affects behaviour, growth, and feed intake in weaned piglets. *PLoS ONE* 11(5): e0154698. doi:10.1371/journal.pone.0154698.

Clouard, C., Kemp, B., Val-Laillet, D., Gerrits, W. J., Bartels, A. C. and Bolhuis, J. E. (2016b). Prenatal, but not early postnatal, exposure to a Western diet improves spatial memory of pigs later in life and is paired with changes in maternal prepartum blood lipid levels. *FASEB Journal* 30(7): 2466-2475. doi:10.1096/fj.201500208R.

Clouard, C., Souza, A. S., Gerrits, W. J. J., Hovenier, R., Lammers, A. and Bolhuis, J. E. (2015). Maternal fish oil supplementation affects the social behavior, brain fatty acid profile, and sickness response of piglets. *Journal of Nutrition* 145(9): 2176-2184. doi:10.3945/jn.115.214650.

Coloma-García, W., Mehaba, N., Llonch, P., Caja, G., Such, X. and Salama, A. A. K. (2020). Prenatal heat stress effects on gestation and postnatal behavior in kid goats. *PLoS ONE* 15(2): e0220221. doi:10.1371/journal.pone.0220221.

Commons, K. G., Cholanians, A. B., Babb, J. A. and Ehlinger, D. G. (2017). The rodent forced swim test measures stress-coping strategy, not depression-like behavior. *ACS Chemical Neuroscience* 8(5): 955-960. doi:10.1021/acschemneuro.7b00042.

Conrad, M. S., Dilger, R. N. and Johnson, R. W. (2012). Brain growth of the domestic pig (Sus scrofa) from 2 to 24 weeks of age: a longitudinal MRI study. *Developmental Neuroscience* 34(4): 291-298. doi:10.1159/000339311.

Couret, D., Jamin, A., Kuntz-Simon, G., Prunier, A. and Merlot, E. (2009a). Maternal stress during late gestation has moderate but long-lasting effects on the immune system of the piglets. *Veterinary Immunology and Immunopathology* 131(1-2): 17-24. doi:10.1016/j.vetimm.2009.03.003.

Couret, D., Prunier, A., Mounier, A. M., Thomas, F., Oswald, I. P. and Merlot, E. (2009b). Comparative effects of a prenatal stress occurring during early or late gestation on pig immune response. *Physiology and Behavior* 98(4): 498–504. doi:10.1016/j.physbeh.2009.08.003.

Cox, L. N. and Cooper, J. J. (2001). Observations on the pre- and post-weaning behaviour of piglets reared in commercial indoor and outdoor environments. *Animal Science* 72(1): 75–86. doi:10.1017/S1357729800055570.

Cutuli, D., Berretta, E., Laricchiuta, D., Caporali, P., Gelfo, F. and Petrosini, L. (2018). Pre-reproductive parental enriching experiences influence progeny's developmental trajectories. *Frontiers in Behavioral Neuroscience* 12: 254. doi:10.3389/fnbeh.2018.00254.

D'Eath, R. B. (2005). Socialising piglets before weaning improves social hierarchy formation when pigs are mixed post-weaning. *Applied Animal Behaviour Science* 93(3–4): 199–211. doi:10.1016/j.applanim.2004.11.019.

D'Eath, R. B., Arnott, G., Turner, S. P., Jensen, T., Lahrmann, H. P., Busch, M. E., Niemi, J. K., Lawrence, A. B. and Sandøe, P. (2014). Injurious tail biting in pigs: how can it be controlled in existing systems without tail docking? *Animal* 8(9): 1479–1497. doi:10.1017/S1751731114001359.

D'Eath, R. B., Jarvis, S., Baxter, E. M. and Houdijk, J. (2018). Mitigating hunger in pregnant sows. In: Špinka, M. B. T.-A. (Ed.) *Advances in Pig Welfare*. Woodhead Publishing, Cambridge. pp. 199–234. doi:10.1016/B978-0-08-101012-9.00007-1.

De Jong, I. C., Prelle, I. T., Van De Burgwal, J. A., Lambooij, E., Korte, S. M., Blokhuis, H. J. and Koolhaas, J. M. (2000). Effects of environmental enrichment on behavioral responses to novelty, learning, and memory, and the circadian rhythm in cortisol in growing pigs. *Physiology and Behavior* 68(4): 571–578. doi:10.1016/S0031-9384(99)00212-7.

de Oliveira, D., Paranhos da Costa, M. J. R., Zupan, M., Rehn, T. and Keeling, L. J. (2015). Early human handling in non-weaned piglets: effects on behaviour and body weight. *Applied Animal Behaviour Science* 164: 56–63. doi:10.1016/j.applanim.2015.01.002.

de Oliveira, G. F., Marcon, A. V., Martins, R. A., Crone, C., Garcia, R. G., Paz, I. C. L. A., Nieto, V. M. O. S., Odakura, A. M., Braz, J. M. and Caldara, F. R. (2020). Environmental enrichment strategies for nursery piglets and efficacy for maintaining interest. *Animal Production Science*. doi:10.1071/AN19351.

De Passillé, A. M. B., Rushen, J. and Hartsock, T. G. (1988). Ontogeny of teat fidelity in pigs and its relation to competition at suckling. *Canadian Journal of Animal Science* 68(2): 325–338. doi:10.4141/cjas88-037.

Dietz, D. M., Laplant, Q., Watts, E. L., Hodes, G. E., Russo, S. J., Feng, J., Oosting, R. S., Vialou, V. and Nestler, E. J. (2011). Paternal transmission of stress-induced pathologies. *Biological Psychiatry* 70(5): 408–414. doi:10.1016/j.biopsych.2011.05.005.

Day, J. E. L., Spoolder, H. A. M., Burfoot, A., Chamberlain, H. L. and Edwards, S. A. (2002). The separate and interactive effects of handling and environmental enrichment on the behaviour and welfare of growing pigs. *Applied Animal Behaviour Science* 75(3): 177–192. doi:10.1016/S0168-1591(01)00199-X.

Donovan, E. L., Hernandez, C. E., Matthews, L. R., Oliver, M. H., Jaquiery, A. L., Bloomfield, F. H. and Harding, J. E. (2013). Periconceptional undernutrition in sheep leads to decreased locomotor activity in a natural environment. *Journal of Developmental Origins of Health and Disease* 4(4): 296–299. doi:10.1017/S2040174413000214.

Dowell, J., Elser, B. A., Schroeder, R. E. and Stevens, H. E. (2019). Cellular stress mechanisms of prenatal maternal stress: heat shock factors and oxidative stress. *Neuroscience Letters* 709: 134368. doi:10.1016/j.neulet.2019.134368.

Einarsson, S., Brandt, Y., Lundeheim, N. and Madej, A. (2008). Stress and its influence on reproduction in pigs: a review. *Acta Veterinaria Scandinavica* 50(1): 48. doi:10.1186/1751-0147-50-48.

Ekkel, E. D., van Doorn, C. E., Hessing, M. J. and Tielen, M. J. (1995). The Specific-Stress-Free housing system has positive effects on productivity, health, and welfare of pigs. *Journal of Animal Science* 73(6): 1544–1551. doi:10.2527/1995.7361544x.

European Food Safety Authority (2007). Animal health and welfare aspects of different housing and husbandry systems for adult breeding boars, pregnant, farrowing sows and unweaned piglets - Scientific Opinion of the Panel on Animal Health and Welfare. *EFSA Journal* 5(10): 572. doi:10.2903/j.efsa.2007.572.

European Union (2008). Council Directive 2008/120/EC of 18 December 2008 laying down minimum standards for the protection of pigs. *Official Journal of the European Union* 316: 36–38.

Evans, J. P., Wilson, A. J., Pilastro, A. and Garcia-Gonzalez, F. (2019). Ejaculate-mediated paternal effects: evidence, mechanisms and evolutionary implications. *Reproduction* 157(4): R109–R126. doi:10.1530/REP-18-0524.

Feeney, A., Nilsson, E. and Skinner, M. K. (2014). Epigenetics and transgenerational inheritance in domesticated farm animals. *Journal of Animal Science and Biotechnology* 5(1): 48. doi:10.1186/2049-1891-5-48.

Figueroa, J., Marchant, I., Morales, P. and Salazar, L. C. (2019). Do prenatally-conditioned flavor preferences affect consumption of creep feed by piglets? *Animals: An Open Access Journal from MDPI* 9(11). doi:10.3390/ani9110944.

Figueroa, J., Solà-Oriol, D., Vinokurovas, L., Manteca, X. and Pérez, J. F. (2013). Prenatal flavour exposure through maternal diets influences flavour preference in piglets before and after weaning. *Animal Feed Science and Technology* 183(3–4): 160–167. doi:10.1016/j.anifeedsci.2013.04.023.

Focardi, S., Morimando, F., Capriotti, S., Ahmed, A. and Genov, P. (2015). Cooperation improves the access of wild boars (Sus scrofa) to food sources. *Behavioural Processes* 121: 80–86. doi:10.1016/j.beproc.2015.10.019.

Fowden, A. L., Giussani, D. A. and Forhead, A. J. (2006). Intrauterine programming of physiological systems: causes and consequences. *Physiology* 21(1): 29–37. doi:10.1152/physiol.00050.2005.

Gabor, T. M., Hellgren, E. C., Bussche, R. A. and Silvy, N. J. (1999). Demography, sociospatial behaviour and genetics of feral pigs (Sus scrofa) in a semi-arid environment. *Journal of Zoology* 247(3): 311–322. doi:10.1111/j.1469-7998.1999.tb00994.x.

Gautier, Y., Luneau, I., Coquery, N., Meurice, P., Malbert, C. H., Guerin, S., Kemp, B., Bolhuis, J. E., Clouard, C., Le Huërou-Luron, I., Blat, S. and Val-Laillet, D. (2018). Maternal Western diet during gestation and lactation modifies adult offspring's cognitive and hedonic brain processes, behavior, and metabolism in Yucatan minipigs. *FASEB Journal* 32(12): 6478–6494. doi:10.1096/fj.201701541.

Giroux, S., Robert, S. and Martineau, G.-P. (2000). The effects of cross-fostering on growth rate and post-weaning behavior of segregated early-weaned piglets. *Canadian Journal of Animal Science* 80(4): 533–538. doi:10.4141/A99-122.

Gonyou, H. W., Beltranena, E., Whittington, D. L. and Patience, J. F. (1998). The behaviour of pigs weaned at 12 and 21 days of age from weaning to market. *Canadian Journal of Animal Science* 78(4): 517–523.

Gonzalez-Bulnes, A., Ovilo, C., Lopez-Bote, C. J., Astiz, S., Ayuso, M., Perez-Solana, M. L., Sanchez-Sanchez, R. and Torres-Rovira, L. (2012). Gender-specific early postnatal catch-up growth after intrauterine growth retardation by food restriction in swine with obesity/leptin resistance. *Reproduction* 144(2): 269–278. doi:10.1530/REP-12-0105.

Greenwood, E. C., Plush, K. J., van Wettere, W. H. E. J. and Hughes, P. E. (2014). Hierarchy formation in newly mixed, group housed sows and management strategies aimed at reducing its impact. *Applied Animal Behaviour Science* 160(1): 1–11. doi:10.1016/j.applanim.2014.09.011.

Gumusoglu, S. B., Fine, R. S., Murray, S. J., Bittle, J. L. and Stevens, H. E. (2017). The role of IL-6 in neurodevelopment after prenatal stress. *Brain, Behavior, and Immunity* 65: 274–283. doi:10.1016/j.bbi.2017.05.015.

Gunnarsson, S., Pickova, J., Högberg, A., Neil, M., Wichman, A., Wigren, I., Uvnäs-Moberg, K. and Rydhmer, L. (2009). Influence of sow dietary fatty acid composition on the behaviour of the piglets. *Livestock Science* 123(2–3): 306–313. doi:10.1016/j.livsci.2008.12.002.

Guy, J. H., Cain, P. J., Seddon, Y. M., Baxter, E. and Edwards, S. (2012). Economic evaluation of high welfare indoor farrowing systems for pigs. *Animal Welfare* 21(Suppl. 1): 19–24. doi:10.7120/096272812X13345905673520.

Hamada, H. and Matthews, S. G. (2019). Prenatal programming of stress responsiveness and behaviours: progress and perspectives. *Journal of Neuroendocrinology* 31(3): e12674. doi:10.1111/jne.12674.

Hansen, P. J. (2009). Effects of heat stress on mammalian reproduction. *Philosophical Transactions of the Royal Society of London. Series B, Biological Sciences* 364(1534): 3341–3350. doi:10.1098/rstb.2009.0131.

Harker, A., Raza, S., Williamson, K., Kolb, B. and Gibb, R. (2015). Preconception paternal stress in rats alters dendritic morphology and connectivity in the brain of developing male and female offspring. *Neuroscience* 303: 200–210. doi:10.1016/j.neuroscience.2015.06.058.

Harvey, P. W. and Chevins, P. F. D. (1985). Crowding pregnant mice affects attack and threat behavior of male offspring. *Hormones and Behavior* 19(1): 86–97. doi:10.1016/0018-506X(85)90009-1.

Haussmann, M. F., Carroll, J. A., Weesner, G. D., Daniels, M. J., Matteri, R. L. and Lay, D. C. (2000). Administration of ACTH to restrained, pregnant sows alters their pigs' hypothalamic-pituitary-adrenal (HPA) axis. *Journal of Animal Science* 78(9): 2399–2411. doi:10.2527/2000.7892399x.

Heim, G., Mellagi, A. P. G., Bierhals, T., de Souza, L. P., de Fries, H. C. C., Piuco, P., Seidel, E., Bernardi, M. L., Wentz, I. and Bortolozzo, F. P. (2012). Effects of cross-fostering within 24h after birth on pre-weaning behaviour, growth performance and survival rate of biological and adopted piglets. *Livestock Science* 150(1–3): 121–127. doi:10.1016/j.livsci.2012.08.011.

Held, S. D. E. and Špinka, M. (2011). Animal play and animal welfare. *Animal Behaviour* 81(5): 891–899.

Hemsworth, P. H. and Barnett, J. L. (1992). The effects of early contact with humans on the subsequent level of fear of humans in pigs. *Applied Animal Behaviour Science* 35(1): 83–90. doi:10.1016/0168-1591(92)90018-7.

Hemsworth, P. H., Barnett, J. L. and Hansen, C. (1981). The influence of handling by humans on the behavior, growth, and corticosteroids in the juvenile female pig. *Hormones and Behavior* 15(4): 396–403. doi:10.1016/0018-506X(81)90004-0.

Hemsworth, P. H., Barnett, J. L., Hansen, C. and Gonyou, H. W. (1986). The influence of early contact with humans on subsequent behavioural response of pigs to humans. *Applied Animal Behaviour Science* 15(1): 55–63. doi:10.1016/0168-1591(86)90022-5.

Hernandez, C. E., Matthews, L. R., Oliver, M. H., Bloomfield, F. H. and Harding, J. E. (2010). Effects of sex, litter size and periconceptional ewe nutrition on offspring behavioural and physiological response to isolation. *Physiology and Behavior* 101(5): 588–594. doi:10.1016/j.physbeh.2010.08.020.

Herring, S. W. (1985). The ontogeny of mammalian mastication. *American Zoologist* 25(2): 339–350.

Hohenshell, L. M., Cunnick, J. E., Ford, S. P., Kattesh, H. G., Zimmerman, D. R., Wilson, M. E., Matteri, R. L., Carroll, J. A. and Lay, D. C. (2000). Few differences found between early- and late-weaned pigs raised in the same environment. *Journal of Animal Science* 78(1): 38–49. doi:10.2527/2000.78138x.

Holling, C., grosse Beilage, E., Vidondo, B. and Nathues, C. (2017). Provision of straw by a foraging tower – effect on tail biting in weaners and fattening pigs. *Porcine Health Management* 3(1): 4. doi:10.1186/s40813-017-0052-7.

Horback, K. M., Pierdon, M. K. and Parsons, T. D. (2016). Behavioral preference for different enrichment objects in a commercial sow herd. *Applied Animal Behaviour Science* 184: 7–15. doi:10.1016/j.applanim.2016.09.002.

Hötzel, M. J., Pinheiro Machado F°, L. C., Wolf, F. M. and Costa, O. A. D. (2004). Behaviour of sows and piglets reared in intensive outdoor or indoor systems. *Applied Animal Behaviour Science* 86(1–2): 27–39. doi:10.1016/j.applanim.2003.11.014.

Huang, Y., Chen, S., Xu, H., Yu, X., Lai, H., Ho, G., Huang, Q. and Shi, X. (2013). Pre-gestational stress alters stress-response of pubertal offspring rat in sexually dimorphic and hemispherically asymmetric manner. *BMC Neuroscience* 14: 67. doi:10.1186/1471-2202-14-67.

Huang, Y., Shi, X., Xu, H., Yang, H., Chen, T., Chen, S. and Chen, X. (2010). Chronic unpredictable stress before pregnancy reduce the expression of brain-derived neurotrophic factor and N-methyl-D-aspartate receptor in hippocampus of offspring rats associated with impairment of memory. *Neurochemical Research* 35(7): 1038–1049. doi:10.1007/s11064-010-0152-0.

Ison, S. H., D'Eath, R. B., Robson, S. K., Baxter, E. M., Ormandy, E., Douglas, A. J., Russell, J. A., Lawrence, A. B. and Jarvis, S. (2010). "Subordination style" in pigs? The response of pregnant sows to mixing stress affects their offspring's behaviour and stress reactivity. *Applied Animal Behaviour Science* 124(1–2): 16–27. doi:10.1016/j.applanim.2010.02.001.

Jarrett, S. and Ashworth, C. J. (2018). The role of dietary fibre in pig production, with a particular emphasis on reproduction. *Journal of Animal Science and Biotechnology* 9(1): 59. doi:10.1186/s40104-018-0270-0.

Jarvis, S., Moinard, C., Robson, S. K., Baxter, E., Ormandy, E., Douglas, A. J., Seckl, J. R., Russell, J. A. and Lawrence, A. B. (2006). Programming the offspring of the pig by prenatal social stress: neuroendocrine activity and behaviour. *Hormones and Behavior* 49(1): 68–80. doi:10.1016/j.yhbeh.2005.05.004.

Jarvis, S., Moinard, C., Robson, S. K., Sumner, B. E. H., Douglas, A. J., Seckl, J. R., Russell, J. A. and Lawrence, A. B. (2008). Effects of weaning age on the behavioural and neuroendocrine development of piglets. *Applied Animal Behaviour Science* 110(1–2): 166–181. doi:10.1016/j.applanim.2007.03.018.

Jensen, P. and Stangel, G. (1992). Behaviour of piglets during weaning in a seminatural enclosure. *Applied Animal Behaviour Science* 33(2–3): 227–238. doi:10.1016/S0168-1591(05)80010-3.

Ji, Y., Wu, Z., Dai, Z., Wang, X., Li, J., Wang, B. and Wu, G. (2017). Fetal and neonatal programming of postnatal growth and feed efficiency in swine. *Journal of Animal Science and Biotechnology* 8(1): 42. doi:10.1186/s40104-017-0173-5.

Johnson, J. S. and Baumgard, L. H. (2019). PHYSIOLOGY SYMPOSIUM: postnatal consequences of in utero heat stress in pigs. *Journal of Animal Science* 97(2): 962–971. doi:10.1093/jas/sky472.

Kleemann, D. O., Kelly, J. M., Rudiger, S. R., McMillen, I. C., Morrison, J. L., Zhang, S., MacLaughlin, S. M., Smith, D. H., Grimson, R. J., Jaensch, K. S., Brien, F. D., Plush, K. J., Hiendleder, S. and Walker, S. K. (2015). Effect of periconceptional nutrition on the growth, behaviour and survival of the neonatal lamb. *Animal Reproduction Science* 160: 12–22. doi:10.1016/j.anireprosci.2015.06.017.

Ko, H.-L., Chong, Q., Escribano, D., Camerlink, I., Manteca, X. and Llonch, P. (2020). Pre-weaning socialization and environmental enrichment affect life-long response to regrouping in commercially-reared pigs. *Applied Animal Behaviour Science* 229: 105044.

Kranendonk, G., Hopster, H., Fillerup, M., Ekkel, E. D., Mulder, E. J. and Taverne, M. A. (2006). Cortisol administration to pregnant sows affects novelty-induced locomotion, aggressive behaviour, and blunts gender differences in their offspring. *Hormones and Behavior* 49(5): 663–672. doi:10.1016/j.yhbeh.2005.12.008.

Kranendonk, G., Van Der Mheen, H., Fillerup, M. and Hopster, H. (2007). Social rank of pregnant sows affects their body weight gain and behavior and performance of the offspring. *Journal of Animal Science* 85(2): 420–429. doi:10.2527/jas.2006-074.

Kranendonk, G., Mulder, E. J. H., Parvizi, N. and Taverne, M. A. (2008). Prenatal stress in pigs: experimental approaches and field observations. *Experimental and Clinical Endocrinology and Diabetes* 116(7): 413–422. doi:10.1055/s-2008-1065335.

Kuller, W. I., Tobias, T. J. and Van Nes, A. (2010). Creep feed intake in unweaned piglets is increased by exploration stimulating feeder. *Livestock Science* 129(1–3): 228–231.

Kutzer, T., Bünger, B., Kjaer, J. B. and Schrader, L. (2009). Effects of early contact between non-littermate piglets and of the complexity of farrowing conditions on social behaviour and weight gain. *Applied Animal Behaviour Science* 121(1): 16–24. doi:10.1016/j.applanim.2009.08.004.

Laporta, J., Fabris, T. F., Skibiel, A. L., Powell, J. L., Hayen, M. J., Horvath, K., Miller-Cushon, E. K. and Dahl, G. E. (2017). In utero exposure to heat stress during late gestation has prolonged effects on the activity patterns and growth of dairy calves. *Journal of Dairy Science* 100(4): 2976–2984. doi:10.3168/jds.2016-11993.

Lauritzen, L., Brambilla, P., Mazzocchi, A., Harsløf, L. B., Ciappolino, V. and Agostoni, C. (2016). DHA effects in brain development and function. *Nutrients* 8(1). doi:10.3390/nu8010006.

Lay, D. C., Kattesh, H. G., Cunnick, J. E., Daniels, M. J., McMunn, K. A., Toscano, M. J. and Roberts, M. P. (2008). Prenatal stress effects on pig development and response to weaning. *Journal of Animal Science* 86(6): 1316–1324. doi:10.2527/jas.2007-0812.

Lay, D. C., Kattesh, H. G., Cunnick, J. E., Daniels, M. J., Kranendonk, G., McMunn, K. A., Toscano, M. J. and Roberts, M. P. (2011). Effect of prenatal stress on subsequent response to mixing stress and a lipopolysaccharide challenge in pigs. *Journal of Animal Science* 89(6): 1787–1794. doi:10.2527/jas.2010-3612.

Leskanich, C. O. and Noble, R. C. (1999). The comparative roles of polyunsaturated fatty acids in pig neonatal development. *British Journal of Nutrition* 81(2): 87–106. doi:10.1017/s0007114599000215.

Li, H., Zhang, L., Fang, Z., Lin, L., Wu, C. and Huang, Q. (2010). Behavioral and neurobiological studies on the male progeny of maternal rats exposed to chronic unpredictable stress before pregnancy. *Neuroscience Letters* 469(2): 278–282. doi:10.1016/j.neulet.2009.12.017.

Lickliter, R. (2011). The integrated development of sensory organization. *Clinics in Perinatology* 38(4): 591–603. doi:10.1016/j.clp.2011.08.007.

Lucy, M. C. and Safranski, T. J. (2017). Heat stress in pregnant sows: thermal responses and subsequent performance of sows and their offspring. *Molecular Reproduction and Development* 84(9): 946–956. doi:10.1002/mrd.22844.

Lugar, D. W., Proctor, J. A., Safranski, T. J., Lucy, M. C. and Stewart, K. R. (2018). In utero heat stress causes reduced testicular area at puberty, reduced total sperm production, and increased sperm abnormalities in boars. *Animal Reproduction Science* 192: 126–135. doi:10.1016/j.anireprosci.2018.02.022.

Luo, L., Reimert, I., Middelkoop, A., Bas, K. and Elizabeth, B. J. (2020). Effects of early and current environmental enrichment on behavior and growth in pigs. *Frontiers in Veterinary Science* 7: 268.

Lürzel, S., Bückendorf, L., Waiblinger, S. and Rault, J. L. (2020). Salivary oxytocin in pigs, cattle, and goats during positive human-animal interactions. *Psychoneuroendocrinology* 115: 104636. doi:10.1016/j.psyneuen.2020.104636.

Malkova, N. V., Yu, C. Z., Hsiao, E. Y., Moore, M. J. and Patterson, P. H. (2012). Maternal immune activation yields offspring displaying mouse versions of the three core symptoms of autism. *Brain, Behavior, and Immunity* 26(4): 607–616. doi:10.1016/j.bbi.2012.01.011.

Martin, J. E., Ison, S. H. and Baxter, E. M. (2015). The influence of neonatal environment on piglet play behaviour and post-weaning social and cognitive development. *Applied Animal Behaviour Science* 163: 69–79. doi:10.1016/j.applanim.2014.11.022.

Martínez-Miró, S., Tecles, F., Ramón, M., Escribano, D., Hernández, F., Madrid, J., Orengo, J., Martínez-Subiela, S., Manteca, X. and Cerón, J. J. (2016). Causes, consequences and biomarkers of stress in swine: an update. *BMC Veterinary Research* 12(1): 171. doi:10.1186/s12917-016-0791-8.

Mastropieri, D. and Turkewitz, G. (1999). Prenatal experience and neonatal responsiveness to vocal expressions of emotion. *Developmental Psychobiology* 35(3): 204–214. doi:10.1002/(SICI)1098-2302(199911)35:3<204::AID-DEV5>3.0.CO;2-V.

McGlone, J. J. (2013). REVIEW: updated scientific evidence on the welfare of gestating sows kept in different housing systems. *The Professional Animal Scientist* 29(3): 189–198. doi:10.15232/S1080-7446(15)30224-2.

McGowan, P. O. and Matthews, S. G. (2018). Prenatal stress, glucocorticoids, and developmental programming of the stress response. *Endocrinology* 159(1): 69–82. doi:10.1210/en.2017-00896.

Melišová, M., Illmann, G., Andersen, I. L., Vasdal, G. and Haman, J. (2011). Can sow pre-lying communication or good piglet condition prevent piglets from getting crushed? *Applied Animal Behaviour Science* 134(3–4): 121–129. doi:10.1016/j.applanim.2011.06.015.

Mellor, D. J. (2019). Preparing for life after birth: introducing the concepts of intrauterine and extrauterine sensory entrainment in mammalian young. *Animals: An Open Access Journal from MDPI* 9(10). doi:10.3390/ani9100826.

Merlot, E., Calvar, C. and Prunier, A. (2017). Influence of the housing environment during sow gestation on maternal health, and offspring immunity and survival. *Animal Production Science* 57(8): 1751–1758. doi:10.1071/AN15480.

Merlot, E., Pastorelli, H., Prunier, A., Père, M. C., Louveau, I., Lefaucheur, L., Perruchot, M. H., Meunier-Salaün, M. C., Gardan-Salmon, D., Gondret, F. and Quesnel, H. (2019). Sow environment during gestation: part I. Influence on maternal physiology and lacteal secretions in relation with neonatal survival. *Animal* 13(7): 1432–1439. doi:10.1017/S1751731118002987.

Meunier-Salaün, M. C., Edwards, S. A. and Robert, S. (2001). Effect of dietary fibre on the behaviour and health of the restricted fed sow. *Animal Feed Science and Technology* 90(1–2): 53–69. doi:10.1016/S0377-8401(01)00196-1.

Middelkoop, A., Costermans, N., Kemp, B. and Bolhuis, J. E. (2019). Feed intake of the sow and playful creep feeding of piglets influence piglet behaviour and performance before and after weaning. *Scientific Reports* 9(1): 16140. doi:10.1038/s41598-019-52530-w.

Miguel-Pacheco, G. G., Perry, V. E. A., Hernandez-Medrano, J. H., Wapenaar, W., Keisler, D. H. and Voigt, J. P. (2019). Low protein intake during the preconception period in beef heifers affects offspring and maternal behaviour. *Applied Animal Behaviour Science* 215: 1–6. doi:10.1016/j.applanim.2019.04.003.

Mina, T. H. and Reynolds, R. M. (2014). Mechanisms linking in utero stress to altered offspring behaviour. In: Pariante Carmine, M. and Lapiz-Bluhm, M. D. (Eds) *Current Topics in Behavioral Neurosciences*. Berlin, Heidelberg: Springer. pp. 93–122. doi:10.1007/7854_2014_291.

Molendijk, M. L. and de Kloet, E. R. (2015). Immobility in the forced swim test is adaptive and does not reflect depression. *Psychoneuroendocrinology* 62: 389–391. doi:10.1016/j.psyneuen.2015.08.028.

Monk, C., Spicer, J. and Champagne, F. A. (2012). Linking prenatal maternal adversity to developmental outcomes in infants: the role of epigenetic pathways. *Development and Psychopathology* 24(4): 1361–1376. doi:10.1017/S0954579412000764.

Montagne, L., Boudry, G., Favier, C., Le Huërou-Luron, I., Lallès, J. P. and Sève, B. (2007). Main intestinal markers associated with the changes in gut architecture and function in piglets after weaning. *British Journal of Nutrition* 97(1): 45–57.

Muns, R., Rault, J. L. and Hemsworth, P. (2015). Positive human contact on the first day of life alters the piglet's behavioural response to humans and husbandry practices. *Physiology and Behavior* 151: 162–167. doi:10.1016/j.physbeh.2015.06.030.

Munsterhjelm, C., Peltoniemi, O. A. T., Heinonen, M., Hälli, O., Karhapää, M. and Valros, A. (2009). Experience of moderate bedding affects behaviour of growing pigs. *Applied Animal Behaviour Science* 118(1–2): 42–53. doi:10.1016/j.applanim.2009.01.007.

Munsterhjelm, C., Valros, A., Heinonen, M., Hälli, O., Siljander-Rasi, H. and Peltoniemi, O. A. (2010). Environmental enrichment in early life affects cortisol patterns in growing pigs. *Animal* 4(2): 242–249.

Nardone, S. and Elliott, E. (2016). The interaction between the immune system and epigenetics in the etiology of autism spectrum disorders. *Frontiers in Neuroscience* 10: 329. doi:10.3389/fnins.2016.00329.

National Farm Animal Care Council (NFACC) (2014). *Code of Practice for the Care and Handling of Pigs*, Canadian Pork Council, Ottawa. ISBN 978-0-9936189-3-2.

Newberry, R. C. and Wood-Gush, D. G. M. (1985). The suckling behaviour of domestic pigs in a semi-natural environment. *Behaviour* 95(1-2): 11-25. doi:10.1163/156853985X00028.

Newberry, R. C. and Wood-Gush, D. G. M. (1986). Social relationships of piglets in a semi-natural environment. *Animal Behaviour* 34(5): 1311-1318. doi:10.1016/S0003-3472(86)80202-0.

Newberry, R. C., Wood-Gush, D. G. M. and Hall, J. W. (1988). Playful behaviour of piglets. *Behavioural Processes* 17(3): 205-216. doi:10.1016/0376-6357(88)90004-6.

Ocepek, M. and Andersen, I. L. (2018). Sow communication with piglets while being active is a good predictor of maternal skills, piglet survival and litter quality in three different breeds of domestic pigs (Sus scrofa domesticus). *PLoS ONE* 13(11): e0206128. doi:10.1371/journal.pone.0206128.

Oostindjer, M., Bolhuis, J. E., van den Brand, H. and Kemp, B. (2009). Prenatal flavor exposure affects flavor recognition and stress-related behavior of piglets. *Chemical Senses* 34(9): 775-787. doi:10.1093/chemse/bjp063.

Oostindjer, M., Bolhuis, J. E., van den Brand, H., Roura, E. and Kemp, B. (2010a). Prenatal flavor exposure affects growth, health and behavior of newly weaned piglets. *Physiology and Behavior* 99(5): 579-586. doi:10.1016/j.physbeh.2010.01.031.

Oostindjer, M., Bolhuis, J. E., Mendl, M., Held, S., Gerrits, W., van den Brand, H. and Kemp, B. (2010b). Effects of environmental enrichment and loose housing of lactating sows on piglet performance before and after weaning. *Journal of Animal Science* 88(11): 3554-3562. doi:10.2527/jas.2010-2940.

Oostindjer, M., Bolhuis, J. E., Mendl, M., Held, S., van den Brand, H. and Kemp, B. (2011a). Learning how to eat like a pig: effectiveness of mechanisms for vertical social learning in piglets. *Animal Behaviour* 82(3): 503-511. doi:10.1016/j.anbehav.2011.05.031.

Oostindjer, M., Bolhuis, J. E., Simon, K., van den Brand, H. and Kemp, B. (2011b). Perinatal flavour learning and adaptation to being weaned: all the pig needs is smell. *PLoS ONE* 6(10): e25318. doi:10.1371/journal.pone.0025318.

Oostindjer, M., van Den Brand, H., Kemp, B. and Bolhuis, J. E. (2011c). Effects of environmental enrichment and loose housing of lactating sows on piglet behaviour before and after weaning. *Applied Animal Behaviour Science* 134(1-2): 31-41. doi:10.1016/j.applanim.2011.06.011.

Oostindjer, M., Kemp, B., van den Brand, H. and Bolhuis, J. E. (2014). Facilitating "learning from mom how to eat like a pig" to improve welfare of piglets around weaning. *Applied Animal Behaviour Science* 160(1): 19-30. doi:10.1016/j.applanim.2014.09.006.

Oostindjer, M., Munoz, J. M., Van Den Brand, H., Kemp, B. and Bolhuis, J. E. (2011d). Maternal presence and environmental enrichment affect food neophobia of piglets. *Biology Letters* 7(1): 19-22. doi:10.1098/rsbl.2010.0430.

Otten, W., Kanitz, E., Tuchscherer, M., Schneider, F. and Brüssow, K. P. (2004). Effects of adrenocorticotropin stimulation on cortisol dynamics of pregnant gilts and their fetuses: implications for prenatal stress studies. *Theriogenology* 61(9): 1649-1659. doi:10.1016/j.theriogenology.2003.09.009.

Otten, W., Kanitz, E., Tuchscherer, M., Brüssow, K. P. and Nürnberg, G. (2008). Repeated administrations of adrenocorticotropic hormone during late gestation in pigs: maternal cortisol response and effects on fetal HPA axis and brain neurotransmitter systems. *Theriogenology* 69(3): 312–322. doi:10.1016/j.theriogenology.2007.09.025.

Otten, W., Kanitz, E. and Tuchscherer, M. (2015). The impact of pre-natal stress on offspring development in pigs. *Journal of Agricultural Science* 153(5): 907–919. doi:10.1017/S0021859614001361.

Parois, S. P., Cabezón, F. A., Schinckel, A. P., Johnson, J. S., Stwalley, R. M. and Marchant-Forde, J. N. (2018). Effect of floor cooling on behavior and heart rate of late lactation sows under acute heat stress. *Frontiers in Veterinary Science* 5: 223. doi:10.3389/fvets.2018.00223.

Pedersen, L. J. (2018). Overview of commercial pig production systems and their main welfare challenges. In: Špinka, M. B. T.-A. (Ed.) *Advances in Pig Welfare*. Woodhead Publishing, Cambridge. pp. 3–25. doi:10.1016/B978-0-08-101012-9.00001-0.

Pembrey, M., Saffery, R., Bygren, L. O., Network in Epigenetic Epidemiology and Network in Epigenetic Epidemiology (2014). Human transgenerational responses to early-life experience: potential impact on development, health and biomedical research. *Journal of Medical Genetics* 51(9): 563–572. doi:10.1136/jmedgenet-2014-102577.

Peng, X., Yan, C., Hu, L., Liu, Y., Xu, Q., Wang, R., Qin, L., Wu, C., Fang, Z., Lin, Y., Xu, S., Feng, B., Zhuo, Y., Li, J., Wu, D and Che, L. (2019). Effects of fat supplementation during gestation on reproductive performance, milk composition of sows and intestinal development of their offspring. *Animals: An Open Access Journal from MDPI* 9(4): 125. doi:10.3390/ani9040125.

Perry, V. E. A., Copping, K. J., Miguel-Pacheco, G. and Hernandez-Medrano, J. (2019). The effects of developmental programming upon neonatal mortality. *Veterinary Clinics of North America. Food Animal Practice* 35(2): 289–302. doi:10.1016/j.cvfa.2019.02.002.

Petersen, H. V., Vestergaard, K. and Jensen, P. (1989). Integration of piglets into social groups of free-ranging domestic pigs. *Applied Animal Behaviour Science* 23(3): 223–236. doi:10.1016/0168-1591(89)90113-5.

Petersen, V. (1994). The development of feeding and investigatory behaviour in free-ranging domestic pigs during their first 18 weeks of life. *Applied Animal Behaviour Science* 42(2): 87–98. doi:10.1016/0168-1591(94)90149-X.

Pino, O. (2016). Fetal memory: the effects of prenatal auditory experience on human development. *BAOJ Medical and Nursing* 2(2). doi:10.24947/baojmn/2/2/120.

Plotsky, P. M. and Meaney, M. J. (1993). Early, postnatal experience alters hypothalamic corticotropin-releasing factor (CRF) mRNA, median eminence CRF content and stress-induced release in adult rats. *Brain Research. Molecular Brain Research* 18(3): 195–200. doi:10.1016/0169-328X(93)90189-V.

Prunier, A., Averos, X., Dimitrov, I., Edwards, S. A., Hillmann, E., Holinger, M., Ilieski, V., Leming, R., Tallet, C., Turner, S. P., Zupan, M. and Camerlink, I. (2019). Review: Early life predisposing factors for biting in pigs. *Animal* 14(3): 570–587. doi:10.1017/S1751731119001940.

Puppe, B., Tuchscherer, M. and Tuchscherer, A. (1997). The effect of housing conditions and social environment immediately after weaning on the agonistic behaviour, neutrophil/lymphocyte ratio, and plasma glucose level in pigs. *Livestock Production Science* 48(2): 157–164. doi:10.1016/S0301-6226(97)00006-7.

Rault, J. L., Lawrence, A. J. and Ralph, C. R. (2018). Brain-derived neurotrophic factor in serum as an animal welfare indicator of environmental enrichment in pigs. *Domestic Animal Endocrinology* 65: 67–70. doi:10.1016/j.domaniend.2018.05.007.

Rault, J. L., van den Munkhof, M. and Buisman-Pijlman, F. T. A. (2017). Oxytocin as an indicator of psychological and social well-being in domesticated animals: a critical review. *Frontiers in Psychology* 8: 1521. doi:10.3389/fpsyg.2017.01521.

Reardon, S. (2019). Depression researchers rethink popular mouse swim tests. *Nature* 571(7766): 456–457. doi:10.1038/d41586-019-02133-2.

Rinaman, L., Banihashemi, L. and Koehnle, T. J. (2011). Early life experience shapes the functional organization of stress-responsive visceral circuits. *Physiology and Behavior* 104(4): 632–640. doi:10.1016/j.physbeh.2011.04.008.

Ringgenberg, N., Bergeron, R., Meunier-Salaün, M. C. and Devillers, N. (2012). Impact of social stress during gestation and environmental enrichment during lactation on the maternal behavior of sows. *Applied Animal Behaviour Science* 136(2–4): 126–135. doi:10.1016/j.applanim.2011.12.012.

Rioja-Lang, F. C., Brown, J. A., Brockhoff, E. J. and Faucitano, L. (2019). A review of swine transportation research on priority welfare issues: a Canadian perspective. *Frontiers in Veterinary Science* 6: 36. doi:10.3389/fvets.2019.00036.

Robert, S. and Martineau, G. P. (2001). Effects of repeated cross-fosterings on preweaning behavior and growth performance of piglets and on maternal behavior of sows. *Journal of Animal Science* 79(1): 88–93. doi:10.2527/2001.79188x.

Rocha, M. L. M., Fernandes, P. P., Lotufo, B. M., Manhães, A. C., Barradas, P. C. and Tenorio, F. (2014). Undernutrition during early life alters neuropeptide Y distribution along the arcuate/paraventricular pathway. *Neuroscience* 256: 379–391. doi:10.1016/j.neuroscience.2013.10.040.

Rodgers, A. B., Morgan, C. P., Bronson, S. L., Revello, S. and Bale, T. L. (2013). Paternal stress exposure alters sperm microRNA content and reprograms offspring HPA stress axis regulation. *Journal of Neuroscience* 33(21): 9003-9012. doi:10.1523/JNEUROSCI.0914-13.2013.

Roldan-Santiago, P., Martinez-Rodriguez, R., Yanez-Pizana, A. and Trujillo-Ortega, M. (2013). Stressor factors in the transport of weaned piglets: a review. *Veterinarni Medicina* 58(5): 241–251. doi:10.17221/6805-VETMED.

Rooke, J. A., Sinclair, A. G., Edwards, S. A., Cordoba, R., Pkiyach, S., Penny, P. C., Penny, P., Finch, A. M. and Horgan, G. W. (2001). The effect of feeding salmon oil to sows throughout pregnancy on pre-weaning mortality of piglets. *Animal Science* 73(3): 489–500. doi:10.1017/S135772980005846X.

Rosero, D. S., Boyd, R. D., Odle, J. and van Heugten, E. (2016). Optimizing dietary lipid use to improve essential fatty acid status and reproductive performance of the modern lactating sow: a review. *Journal of Animal Science and Biotechnology* 7(1): 34. doi:10.1186/s40104-016-0092-x.

Rutherford, K. M. D., Donald, R. D., Arnott, G., Rooke, J., Dixon, L., Mehers, J., Turnbull, J. and Lawrence, A. (2012). Farm animal welfare: assessing risks attributable to the prenatal environment. *Animal Welfare* 21(3): 419–429. doi:10.7120/09627286.21.3.419.

Rutherford, K. M. D., Piastowska-Ciesielska, A., Donald, R. D., Robson, S. K., Ison, S. H., Jarvis, S., Brunton, P. J., Russell, J. A. and Lawrence, A. B. (2014). Prenatal stress produces anxiety prone female offspring and impaired maternal behaviour in the domestic pig. *Physiology and Behavior* 129: 255–264. doi:10.1016/j.physbeh.2014.02.052.

Rutherford, K. M. D., Robson, S. K., Donald, R. D., Jarvis, S., Sandercock, D. A., Scott, E. M., Nolan, A. M. and Lawrence, A. B. (2009). Pre-natal stress amplifies the immediate behavioural responses to acute pain in piglets. *Biology Letters* 5(4): 452–454. doi:10.1098/rsbl.2009.0175.

Salazar, L. C., Ko, H. L., Yang, C. H., Llonch, L., Manteca, X., Camerlink, I. and Llonch, P. (2018). Early socialisation as a strategy to increase piglets' social skills in intensive farming conditions. *Applied Animal Behaviour Science* 206: 25–31. doi:10.1016/j.applanim.2018.05.033.

Sandercock, D. A., Gibson, I. F., Rutherford, K. M. D., Donald, R. D., Lawrence, A. B., Brash, H. M., Scott, E. M. and Nolan, A. M. (2011). The impact of prenatal stress on basal nociception and evoked responses to tail-docking and inflammatory challenge in juvenile pigs. *Physiology and Behavior* 104(5): 728–737. doi:10.1016/j.physbeh.2011.07.018.

Sanz Fernandez, M. V., Johnson, J. S., Abuajamieh, M., Stoakes, S. K., Seibert, J. T., Cox, L., Kahl, S., Elsasser, T. H., Ross, J. W., Isom, S. C., Rhoads, R. P. and Baumgard, L. H. (2015). Effects of heat stress on carbohydrate and lipid metabolism in growing pigs. *Physiological Reports* 3(2): e12315. doi:10.14814/phy2.12315.

Schaal, B., Hummel, T. and Soussignan, R. (2004). Olfaction in the fetal and premature infant: functional status and clinical implications. *Clinics in Perinatology* 31(2): 261–285. doi:10.1016/j.clp.2004.04.003.

Schauberger, G., Mikovits, C., Zollitsch, W., Hörtenhuber, S. J., Baumgartner, J., Niebuhr, K., Piringer, M., Knauder, W., Anders, I., Andre, K., Hennig-Pauka, I. and Schönhart, M. (2019). Global warming impact on confined livestock in buildings: efficacy of adaptation measures to reduce heat stress for growing-fattening pigs. *Climatic Change* 156(4): 567-587. doi:10.1007/s10584-019-02525-3.

Schmidt, V. (2015). *Comparative Anatomy of the Pig Brain - An Integrative Magnetic Resonance Imaging (MRI) Study of the Porcine Brain with Special Emphasis on the External Morphology of the Cerebral Cortex*. VVB Laufersweiler Verlag, Giessen.

Schmitt, O., O'Driscoll, K. and Baxter, E. M. (2019). Exploratory study of the effects of intra-uterine growth retardation and neonatal energy supplementation of low birth-weight piglets on their post-weaning cognitive abilities. *Animal Cognition* 22(3): 373–385. doi:10.1007/s10071-019-01251-8.

Shachar-Dadon, A., Schulkin, J. and Leshem, M. (2009). Adversity before conception will affect adult progeny in rats. *Developmental Psychology* 45(1): 9–16. doi:10.1037/a0014030.

Short, A. K., Fennell, K. A., Perreau, V. M., Fox, A., O'Bryan, M. K., Kim, J. H., Bredy, T. W., Pang, T. Y. and Hannan, A. J. (2016). Elevated paternal glucocorticoid exposure alters the small noncoding RNA profile in sperm and modifies anxiety and depressive phenotypes in the offspring. *Translational Psychiatry* 6(6): e837. doi:10.1038/tp.2016.109.

Simitzis, P. E., Deligeorgis, S. G., Bizelis, J. A. and Fegeros, K. (2008). Feeding preferences in lambs influenced by prenatal flavour exposure. *Physiology and Behavior* 93(3): 529–536. doi:10.1016/j.physbeh.2007.10.013.

Sinclair, K. D., Rutherford, K. M. D., Wallace, J. M., Brameld, J. M., Stöger, R., Alberio, R., Sweetman, D., Gardner, D. S., Perry, V. E., Adam, C. L., Ashworth, C. J., Robinson, J. E. and Dwyer, C. M. (2016). Epigenetics and developmental programming of welfare and production traits in farm animals. *Reproduction, Fertility, and Development* 28(10): 1443–1478. doi:10.1071/RD16102.

Skok, J., Prevolnik, M., Urek, T., Mesarec, N. and Škorjanc, D. (2014). Behavioural patterns established during suckling reappear when piglets are forced to form a new dominance hierarchy. *Applied Animal Behaviour Science* 161(1): 42–50. doi:10.1016/j.applanim.2014.09.005.

Smith, S. E. P., Li, J., Garbett, K., Mirnics, K. and Patterson, P. H. (2007). Maternal immune activation alters fetal brain development through interleukin-6. *Journal of Neuroscience* 27(40): 10695–10702. doi:10.1523/JNEUROSCI.2178-07.2007.

Sneddon, I. A., Beattie, V. E., Dunne, L. and Neil, W. (2000). The effect of environmental enrichment on learning in pigs. *Animal Welfare* 9(4): 373–383. Available at: https://www.ingentaconnect.com/content/ufaw/aw/2000/00000009/00000004/art00003.

Sorrells, A. D., Eicher, S. D., Scott, K. A., Harris, M. J., Pajor, E. A., Lay, D. C. and Richert, B. T. (2006). Postnatal behavioral and physiological responses of piglets from gilts housed individually or in groups during gestation. *Journal of Animal Science* 84(3): 757–766. doi:10.2527/2006.843757x.

Špinka, M., Newberry, R. C. and Bekoff, M. (2001). Mammalian play: training for the unexpected. *Quarterly Review of Biology* 76(2): 141–168. doi:10.1086/393866.

Spreeuwenberg, M. A. M., Verdonk, J. M., Gaskins, H. R. and Verstegen, M. W. (2001). Small intestine epithelial barrier function is compromised in pigs with low feed intake at weaning. *The Journal of Nutrition* 131(5): 1520–1527.

Stěhulová, I., Špinka, M., Šárová, R., Máchová, L., Kněz, R. and Firla, P. (2013). Maternal behaviour in beef cows is individually consistent and sensitive to cow body condition, calf sex and weight. *Applied Animal Behaviour Science* 144(3–4): 89–97. doi:10.1016/j.applanim.2013.01.003.

Stolba, A. and Wood-Gush, D. G. M. (1989). The behaviour of pigs in a semi-natural environment. *Animal Science* 48(2): 419–425. doi:10.1017/S0003356100040411.

Tallet, C., Brajon, S., Devillers, N. and Lensink, J. (2018). *Pig-Human Interactions: Creating a Positive Perception of Humans to Ensure Pig Welfare*. doi:10.1016/B978-0-08-101012-9.00008-3.

Tallet, C., Brilloüet, A., Meunier-Salaün, M.-C., Paulmier, V., Guérin, C. and Prunier, A. (2013). Effects of neonatal castration on social behaviour, human–animal relationship and feeding activity in finishing pigs reared in a conventional or an enriched housing. *Applied Animal Behaviour Science* 145(3–4): 70–83.

Tallet, C., Rakotomahandry, M., Guérin, C., Lemasson, A. and Hausberger, M. (2016). Postnatal auditory preferences in piglets differ according to maternal emotional experience with the same sounds during gestation. *Scientific Reports* 6: 37238. doi:10.1038/srep37238.

Tallet, C., Rakotomahandry, M., Herlemont, S. and Prunier, A. (2019). Evidence of pain, stress, and fear of humans during tail docking and the next four weeks in piglets (*Sus scrofa domesticus*). *Frontiers in Veterinary Science* 6: 462.

Tatemoto, P., Bernardino, T., Alves, L., Cristina de Oliveira Souza, A., Palme, R. and José Zanella, A. (2019a). Environmental enrichment for pregnant sows modulates HPA-axis and behavior in the offspring. *Applied Animal Behaviour Science* 220. doi:10.1016/j.applanim.2019.104854.

Tatemoto, P., Bernardino, T., Alves, L. and Zanella, A. J. (2019b). Sham-chewing in sows is associated with decreased fear responses in their offspring. *Frontiers in Veterinary Science* 6: 390. doi:10.3389/fvets.2019.00390.

Tatemoto, P., Bernardino, T., Morrone, B., Queiroz, M. R. and Zanella, A. J. (2020). Stereotypic behavior in sows is related to emotionality changes in the offspring. *Frontiers in Veterinary Science* 7: 79. doi:10.3389/fvets.2020.00079.

Telkänranta, H. and Edwards, S. A. (2018). Lifetime consequences of the early physical and social environment of piglets. In: Špinka, M. B. T.-A. (Ed.) *Advances in Pig Welfare*. Woodhead Publishing, Cambridge. pp. 101–136. doi:10.1016/B978-0-08-101012-9.00013-7.

Telkänranta, H., Swan, K., Hirvonen, H. and Valros, A. (2014). Chewable materials before weaning reduce tail biting in growing pigs. *Applied Animal Behaviour Science* 157: 14–22. doi:10.1016/j.applanim.2014.01.004.

Tierney, A. L. and Nelson, C. A. (2009). Brain development and the role of experience in the early years. *Zero to Three* 30(2): 9–13. Available at: http://www.ncbi.nlm.nih.gov/pubmed/23894221%0A, http://www.pubmedcentral.nih.gov/articlerender.fcgi?artid=PMC3722610.

Tilton, S. L., Miller, P. S., Lewis, A. J., Reese, D. E. and Ermer, P. M. (1999). Addition of fat to the diets of lactating sows: I. Effects on milk production and composition and carcass composition of the litter at weaning. *Journal of Animal Science* 77(9): 2491–2500. doi:10.2527/1999.7792491x.

Ursinus, W. W., Van Reenen, C. G., Kemp, B. and Bolhuis, J. E. (2014). Tail biting behaviour and tail damage in pigs and the relationship with general behaviour: predicting the inevitable? *Applied Animal Behaviour Science* 156: 22–36. doi:10.1016/j.applanim.2014.04.001.

van de Weerd, H. A. and Day, J. E. L. (2009). A review of environmental enrichment for pigs housed in intensive housing systems. *Applied Animal Behaviour Science* 116(1): 1–20. doi:10.1016/j.applanim.2008.08.001.

Van De Weerd, H. A., Docking, C. M., Day, J. E. L., Avery, P. J. and Edwards, S. A. (2003). A systematic approach towards developing environmental enrichment for pigs. *Applied Animal Behaviour Science* 84(2): 101–118. doi:10.1016/S0168-1591(03)00150-3.

van den Brand, H., Wamsteeker, D., Oostindjer, M., van Enckevort, L. C., van der Poel, A. F., Kemp, B. and Bolhuis, J. E. (2014). Effects of pellet diameter during and after lactation on feed intake of piglets pre- and postweaning. *Journal of Animal Science* 92(9): 4145–4153. doi:10.2527/jas.2014-7408.

Val-Laillet, D., Besson, M., Guérin, S., Coquery, N., Randuineau, G., Kanzari, A., Quesnel, H., Bonhomme, N., Bolhuis, J. E., Kemp, B., Blat, S., Le Huërou-Luron, I. and Clouard, C. (2017). A maternal Western diet during gestation and lactation modifies offspring's microbiota activity, blood lipid levels, cognitive responses, and hippocampal neurogenesis in Yucatan pigs. *FASEB Journal* 31(5): 2037–2049. doi:10.1096/fj.201601015R.

Vanheukelom, V., Driessen, B., Maenhout, D. and Geers, R. (2011). Peat as environmental enrichment for piglets: the effect on behaviour, skin lesions and production results. *Applied Animal Behaviour Science* 134(1–2): 42–47. doi:10.1016/j.applanim.2011.06.010.

Velazquez, M. A., Fleming, T. P. and Watkins, A. J. (2019). Periconceptional environment and the developmental origins of disease. *Journal of Endocrinology* 242(1): T33–T49. doi:10.1530/JOE-18-0676.

Wang, J., Feng, C., Liu, T., Shi, M., Wu, G. and Bazer, F. W. (2017). Physiological alterations associated with intrauterine growth restriction in fetal pigs: causes and insights for nutritional optimization. *Molecular Reproduction and Development* 84(9): 897–904. doi:10.1002/mrd.22842.

Ward, S. A., Kirkwood, R. N. and Plush, K. J. (2020). Are larger litters a concern for piglet survival or an effectively manageable trait? *Animals: An Open Access Journal from MDPI* 10(2): 309. doi:10.3390/ani10020309.

Weary, D. M., Jasper, J. and Hötzel, M. J. (2008). Understanding weaning distress. *Applied Animal Behaviour Science* 110(1-2): 24-41. doi:10.1016/j.applanim.2007.03.025.

Weary, D. M., Pajor, E. A., Bonenfant, M., Ross, S. K., Fraser, D. and Kramer, D. L. (1999). Alternative housing for sows and litters: 2. Effects of a communal piglet area on pre- and post-weaning behaviour and performance. *Applied Animal Behaviour Science* 65(2): 123-135. doi:10.1016/S0168-1591(99)00053-2.

Weaver, S. A., Aherne, F. X., Meaney, M. J., Schaefer, A. L. and Dixon, W. T. (2000). Neonatal handling permanently alters hypothalamic-pituitary-adrenal axis function, behaviour, and body weight in boars. *Journal of Endocrinology* 164(3): 349-359. doi:10.1677/joe.0.1640349.

Weinstock, M. (2017). Prenatal stressors in rodents: effects on behavior. *Neurobiology of Stress* 6: 3-13. doi:10.1016/j.ynstr.2016.08.004.

Weller, J. E., Camerlink, I., Turner, S. P., Farish, M. and Arnott, G. (2019). Playful pigs: early life play-fighting experience influences later life contest dynamics. *Animal Behaviour* 158: 269-279. doi:10.1016/j.anbehav.2019.09.019.

Weller, J. E., Turner, S. P., Farish, M., Camerlink, I. and Arnott, G. (2020). The association between play fighting and information gathering during subsequent contests. *Scientific Reports* 10(1): 1133. doi:10.1038/s41598-020-58063-x.

Wood-Gush, D. G. M. and Vestergaard, K. (1993). Inquisitive exploration in pigs. *Animal Behaviour* 45(1): 185-187.

Worobec, E. K., Duncan, I. J. H. and Widowski, T. M. (1999). The effects of weaning at 7, 14 and 28 days on piglet behaviour. *Applied Animal Behaviour Science* 62(2-3): 173-182. doi:10.1016/S0168-1591(98)00225-1.

Wright, T., Langley-Evans, S. C. and Voigt, J. P. (2011). The impact of maternal cafeteria diet on anxiety-related behaviour and exploration in the offspring. *Physiology and Behavior* 103(2): 164-172. doi:10.1016/j.physbeh.2011.01.008.

Yang, C. H., Ko, H. L., Salazar, L. C., Llonch, L., Manteca, X., Camerlink, I. and Llonch, P. (2018). Pre-weaning environmental enrichment increases piglets' object play behaviour on a large scale commercial pig farm. *Applied Animal Behaviour Science* 202: 7-12. doi:10.1016/j.applanim.2018.02.004.

Yeshurun, S., Rogers, J., Short, A. K., Renoir, T., Pang, T. Y. and Hannan, A. J. (2017). Elevated paternal glucocorticoid exposure modifies memory retention in female offspring. *Psychoneuroendocrinology* 83: 9-18. doi:10.1016/j.psyneuen.2017.05.014.

Zupan, M., Rehn, T., De Oliveira, D. and Keeling, L. J. (2016). Promoting positive states: the effect of early human handling on play and exploratory behaviour in pigs. *Animal* 10(1): 135-141. doi:10.1017/S1751731115001743.

Part 2

Management of behaviour in different production stages

Chapter 3

Optimising pig welfare in breeding and gestation

Paul H. Hemsworth, University of Melbourne, Australia

1 Introduction
2 Breeding and gestation of gilts and sows
3 Animal management
4 Conclusion and future trends
5 Where to look for further information
6 References

1 Introduction

The basic social group of wild boars is typically organised around a nucleus of closely related females and their most recent litters (Dardallion, 1988). These matriarchal groups consist of related subadult (younger than one year) and adult (older than one year) females and their piglets, and groups of up to 30–40 individuals have been reported in non-hunted areas (Leaper et al., 1999; Kramer-Schadt et al., 2007). The groups are spatially restricted to home ranges of about 5–20 km^2 (Leaper et al., 1999) and although the respective home ranges may overlap, these matriarchal groups do not appear to associate with other families, avoiding contact through a strategy of mutual avoidance (Dardallion, 1988; Gabor et al., 1999; Kramer-Schadt et al., 2007).

The dynamics of the matriarchal group include the isolation of pre-parturient females in spring, with their re-entry to the group with their new born, the entry of nulliparous females and the arrival of adult males during the rutting season with simultaneous departure of subadult males (Leaper et al., 1999). There is little or no aggression with the introduction of piglets to the matriarchal group (Jensen, 1986; Mauget, 1981). Subadult females split up and disperse with other subadult females to constitute a new social group, although sometimes they may return to their matriarchal group in autumn, together with their newborn piglets (Dardallion, 1988). However, it is suggested that only about 20% of yearling females may leave the matriarchal group and disperse (Kaminski et al., 2005). Adult males are frequently solitary, but subadult males

http://dx.doi.org/10.19103/AS.2020.0081.03

often roam together in small groups (Fernandez-Llario et al., 1996). Males fight for access to adult (older than one year) females and, having achieved matings, they move to another group.

Aggression is qualitatively similar in domestic pigs and their wild ancestors (Stolba, 1988). However, in contrast to domestic pigs in modern pig production systems, aggression between individuals of the wild boar in the wild and between individuals of the domestic pig returned to semi-natural settings outside of the breeding season is infrequent and rarely injurious (Stolba, 1988; Marchant-Forde and Marchant-Forde, 2005; Turner et al., 2006). Aggression is partially genetically determined and there is evidence that breeding programmes for improved growth may have resulted in increased aggression in the modern domestic pig. For example, pigs with a high breeding value for growth show more aggression when grouped with unfamiliar pigs (Rodenburg et al., 2010; Canario et al., 2012). There is also evidence that, genetically, young pigs that receive many skin injuries show improved growth performance compared to those with few injuries (Desire et al., 2015).

Social interactions are also obviously affected by the environment. Matriarchal groups of wild boar in the wild and groups of domestic pigs in semi-natural settings have a stable, linear hierarchy, based on sex and, within each sex group, the closely related factors of weight, size and strength (Mauget, 1981; Stolba and Wood-Gush, 1989). Aggression is infrequent and is generally associated with key resources such as food (Mauget, 1981; Jensen and Wood-Gush, 1984; Stolba and Wood-Gush, 1989). The inability of domestic pigs to form matriarchal groupings in modern pig production with its housing systems and management practices, such as early weaning, frequent mixing of unfamiliar pigs and penning that limits avoidance responses by pigs and dispersal and the display of appropriate submissive behaviour, obviously contributes to increased aggression. With the wide disparity between the 'natural' and 'farmed' situations, it is therefore not surprising that problems arise with aggression when domestic pigs are mixed in modern production systems (Marchant-Forde and Marchant-Forde, 2005; Marchant-Forde, 2009a).

2 Breeding and gestation of gilts and sows

2.1 Boar exposure, puberty stimulation and oestrus detection

Selected breeding gilts are generally exposed to boars for puberty stimulation at 150–180 days of age (Kraeling and Webel, 2015). Effective puberty stimulation has been recently described by several authors (Kraeling and Webel, 2015; Patterson and Foxcroft, 2019). A combination of tactile, visual, auditory and olfactory boar stimuli and daily, direct exposure to a rotation of mature boars (older than 10 months) for a minimum of 10–15 min maximizes the pubertal response in gilts. Both the stimulation of puberty and detection of puberty are facilitated by providing fence-line and direct contact (15 min daily) with multiple mature boars (Siswadi and Hughes, 1995; Patterson et al., 2016).

Direct contact with mature boars during gilt puberty stimulation and oestrus detection of both gilts and sows should be supervised to avoid stress as well as injury to gilts and sows. Based on fleeing and prolonged high-pitched vocalizations, subordinate sows in pair housing showed more fear in response to boar stimulation even when they were in oestrus, thus possibly impairing both oestrus detection and mating (Pedersen et al., 2003). Gilts and sows are also at risk of fractures or spinal cord injuries being mounted excessively by boars or other females in the pen (Levis, 2011). Rapid growth and increased body weight before mating and excessive body condition predispose gilts and young sows to lameness, and natural mating may further increase the risk of foot and leg injury (Gill, 2007). While there is no evidence, oestrus detection using full boar contact may also increase the risk of foot and leg injury in gilts and sows. Furthermore, the physical environment, particularly floor type, quality and maintenance, will affect the risk of foot and leg injury during oestrus detection and natural mating and poor natural mating conditions such as slippery floors, small floor area and obstructions such as drinkers may depress the sexual receptivity of gilts (Hemsworth et al., 1989a; Jongman et al., 1996).

2.2 Housing gilts and sows around mating and during gestation

Weaned and pregnant gilts and sows are housed under varying conditions indoors and outdoors (Marchant-Forde, 2009a; Kemp and Soede, 2012). They can be housed from weaning in gestation stalls full-time or until post-insemination or pregnancy confirmation when then housed in groups, or housed in groups from weaning. Group size can vary from small (4–5) to very large (up to 400), and the feeding systems can vary from feeding on the floor or in troughs, feeding stalls (free-access partial or free-access full-body length stalls), trickle feeders or electronic sow feeders (ESF), and thus pigs may feed simultaneously and sequentially. Space allowance can vary from 1.4 m^2/pig to higher space allowances, particularly in outdoors systems. Other features in group-housing systems may also differ such as flooring, the provision and type of bedding, temperature and feeding level, although gilts and sows are mostly restrictively fed during pregnancy. These design features of the housing system can affect the welfare of gilts and sows.

Aggression is inevitable when unfamiliar pigs are mixed. Agonistic behaviour and particularly aggression may lead to injuries such as claw and limb lesions, fractures, or spinal cord damage associated with incidents such as slips, falls, knocks, bites, head slashes and toes getting stuck in the gaps of concrete slats (Anil et al., 2005). Under commercial conditions, pigs exhibit dyadic dominance relationships but mixing unfamiliar pigs of similar weight and age elicits intense aggression, particularly during the first day and is likely to be exacerbated by a close similarity in dominance ability (see review by Turner and Edwards, 2004). Pigs in small groups of 10–30 exhibit dyadic dominance relationships that show stability in a range of commercial contexts, including changes to the group composition (Turner and Edwards, 2004; Verdon and Rault, 2018). Once established, social relationships are largely maintained by low-intensity aggression, threats and avoidance behaviour (Beilharz and Cox, 1967; Gonyou, 2001). Thus, once these dyadic dominance relationships are established, there is a reduced need to deliver aggression, resulting in an overall reduction in aggression in the group (Gonyou, 2001). In larger groups, there appears to be less reliance on aggression during the period immediately post-mixing than would be expected and possible changes in social behaviour in large group sizes that may be responsible are discussed in the section on group size. Furthermore, as reviewed by Verdon and Rault (2018), once dominance relationships are established, the characteristics of the aggression changes to shorter durations but more frequent aggression and generally occurs over access for, or defence of, a resource such as food or lying space. When gilts or sows have to compete for food, aggression can remain high for at least 3–9 days after mixing (Barnett, 1997; Arey, 1999; Hemsworth et al., 2013; Verdon et al., 2016). Interestingly while frequency of aggression in gilts and sows was observed to remain high for at least 9 days after mixing, Verdon et al. (2016) found that fresh injuries were lower at 2 days than at 9 days after mixing, perhaps as a consequence of the change in the characteristics of aggression from less-frequent but longer durations to more frequent but shorter durations as the dominance relationships are established. Thus pigs in groups are exposed to a range of incidents that may cause injury such as slips, falls, knocks, bites, head slashes, collisions with other pigs and pen features and toes getting stuck in the gaps of concrete slats.

In addition to agonistic behaviour associated with the establishment of dominance relationships, agonistic behaviour can be affected by a range of housing and management practices as well as pig characteristics such as experience and genetics. As a consequence of legislative, consumer and retailer pressure on the housing of gilts and sows in gestation stalls (Matthews and Hemsworth, 2012), many countries have moved to, or are moving to, group-housing systems for gestating sows. The following section on housing around mating and gestation will therefore focus on gilts and sows housed in groups.

2.2.1 Pen design

Floor space. Floor space is a key determinant of aggression, injuries and stress when mixing and housing unfamiliar pigs. In addition to spatial requirements for physical size and basic movement, pigs need to access key resources such as feed, water and lying space. They are also motivated to interact with other pigs and to explore particularly if hungry. Therefore, they need floor space not only to access the resources but also, if necessary, to distance themselves from others, including when accessing these resources (Verdon et al., 2015a).

Floor space is particularly relevant early after mixing since insufficient space, in terms of both quantity (amount) and quality (configuration, including physical and visual barriers), can prolong aggression by affecting a pig's ability to avoid or escape others and hence delay the formation of a stable hierarchy (Lindberg, 2001). However, defining floor space allowance in terms of minimising aggression can never be definitive because, for example, the animal's response is curvilinear and the level of aggression will depend on factors such as the quality of space, access to feed and parity variation in the group.

There is evidence of increased aggression and stress in both the short and long terms if space is insufficient for gilts. Reducing floor space for gilts within the range of 1.0–3.0 m^2/animal increases aggression and stress. Floor space of 1.0 m^2/gilt elevated plasma cortisol concentrations on weeks 2 and 10 after mixing in adult gilts compared to 2.0 or 3.0 m^2/gilt (Hemsworth et al., 1986) and increased aggression around feeding on days 2–30 after mixing elevated plasma cortisol concentration and increased the cortisol response to ACTH on weeks 5 and 7 after mixing pregnant gilts compared to 1.4 and 2.0 m^2/gilt (Barnett et al., 1992; Barnett, 1997). Aggression during the first 90 min after mixing pregnant gilts was similar in floor spaces of 1 and 2 m^2/gilt (Barnett et al., 1992). However, this is in contrast to a subsequent experiment which showed reduced aggression during the first 90 min of mixing unfamiliar adult gilts when floor space was reduced from 3.4 to 1.4 m^2/gilt (Barnett et al., 1993a). The feeding system in the experiment by Hemsworth and colleagues was floor feeding and the experiments by Barnett and colleagues used a factorial design to examine both space and provision of feeding stalls.

Reducing floor space has also been shown to increase sow aggression and stress. Aggression on days 1 and 2 after mixing unfamiliar sows was higher at 1.2 than at 2.0 m^2/sow in a floor-feeding system (Taylor et al., 1997). In an experiment involving established groups of pregnant sows, each studied in each of four floor space treatments for 7-day periods and confined in individual feeding stalls around feeding, aggression on days 6 and 7 of treatment was higher at 2.0 than at 2.4, 3.6 or 4.8 m^2/sow (Weng et al., 1998). While salivary cortisol concentrations at 2 and 26 h after mixing were not affected, reducing floor space from 3.0 to 2.25 m^2/sow increased aggression on days 1, 8, 15 and 22 after mixing in a dynamic group with an electronic sow-feeding system (Remience et al., 2008). Aggression on the first two days after mixing was higher at 4.1 than at 9.3 m^2/sow in elongated rectangular and circular pens, but not in rectangular or square pens (Docking et al., 2001). Furthermore, within the range of 1.4–3.0 m^2/sow, significant negative relationships were found between space allowance and both aggression at feeding and stress based on plasma cortisol concentrations at day 2 after mixing unfamiliar sows, but not at days 8–9 and 51 after mixing in a floor-feeding system (Hemsworth et al., 2013, 2016). Reducing floor space from 6.0 or 4.0 to 2.0 m^2/sow did not affect aggression on days 1–4 after mixing in a floor-feeding system, but salivary cortisol concentrations were surprisingly lower at 2.0 m^2/sow (Greenwood et al., 2016).

This effect of space on aggression and stress in sows found on day 2 after mixing, but not later in pregnancy (Hemsworth et al., 2013, 2016), is in contrast to that of gilts in which reducing floor space from 2.0 to 1.4 m^2/gilt increased aggression on 2–30 days after mixing and elevated plasma cortisol concentration on 5 and 7 weeks after mixing (Barnett et al., 1992; Barnett, 1997). While there is no obvious explanation for these conflicting effects, it has been suggested that sows that are more experienced with group housing may adapt more quickly to spatial restriction in groups than gilts (Hemsworth et al., 2013, 2016).

Reducing floor space has been shown to increase the prevalence of skin injuries in some studies within the range of 2.0–4.0 m^2/sow (Weng et al., 1998), 4.1-9.3 m^2/sow (Docking et al., 2001), 1.4–3.3 m^2/sow (Salak-Johnson et al., 2007) and 2.25–3.0 m^2/sow (Remience et al., 2008), but not in others within the range of 1.4–3.4 m^2/gilt (Barnett et al., 1993a), 1.4–3.0 m^2/sow (Hemsworth et al., 2013) and 2.0–6.0 m^2/sow (Greenwood et al., 2016). Differences between studies in feeding system, group size, parity and physical features in the pen may be responsible for these conflicting results on skin injuries. In an experiment examining the effects of floor space allowance (2.3, 2.8 and 3.2 m^2/sow) by varying group size from 11 to 31 sows, Séguin et al. (2006a) found that skin injuries at 24 h after mixing were not affected by floor space or group size. While there was no effect of floor space on lameness during the first 3 weeks after mixing, Salak-Johnson et al. (2007) found that a floor space of 3.3 m^2/sow was associated with a greater risk of lameness later in gestation than 1.4 or 2.3 m^2/sow and Maes et al. (2016) suggested that sows with increased floor space may have a greater opportunity for activity which may increase the risk of injuries leading to lameness.

Floor space of sows mixed early after insemination may affect reproductive performance since stress can impair reproduction (Turner et al., 2005; Spoolder et al., 2009). Within the range of 1.4–3.0 m^2/sow, a significant relationship has been shown between space allowance and farrowing rate in one experiment (Hemsworth et al., 2013), but not in another (Hemsworth et al., 2016). Although there was no time of year effect on aggression or stress in either experiment, the fertility of sows in summer in the earlier experiment was more susceptible to reduced space than in other replicates (Hemsworth et al., 2013). Unlike the earlier experiment, sows in the 2016 experiment, once confirmed pregnant at about day 27 after insemination, were housed in their treatment groups with a minimum of 1.8 m^2/sow. The authors of this 2016 experiment proposed that providing more space after confirmed pregnancy for sows initially housed at 1.45 and 1.61 m^2/sow in the experiment may have ameliorated spatial restriction effects on reproduction, particularly in summer. Furthermore, they suggested that a proportion of female pigs are expected to be resistant to the effects of prolonged stress or sustained increased cortisol (Turner et al., 2005).

These observations on the effects of space, particularly those on aggression and stress, indicate that a space allowance for gilts and sows of 1.4 m^2/animal is likely to be too small from an animal welfare perspective and that significant improvements in gilt and sow welfare are likely to be achieved with space allowances for gilts and sows in the range of 2.0–2.4 m^2/animal. Furthermore, the effects of space on aggression and stress are most pronounced soon after mixing (Hemsworth et al., 2013, 2016), highlighting the importance of floor space at mixing. With its animal welfare and economic implications, a strategy of staged-gestation penning, with more space immediately after mixing and less space later in gestation (Hemsworth et al., 2016), requires investigation. The implications of staged-gestation penning in which increased space is provided immediately after mixing and space is subsequently decreased once dominance relationships are established are considered in more detail in the section on specific-mixing pen.

Group size. There is a considerable body of research in young pigs indicating that aggression is not affected by group size ranging from 6 to 80 pigs (Turner et al., 1999, 2000; Samarakone and Gonyou, 2009). It is often assumed that housing pigs in large groups increases aggression, presumably on the basis that aggression is expected to increase in large groups because there are more dominance relationships to establish (Arey and Edwards, 1998) and that individual recognition may be difficult in large groups (Estevez et al., 2007). As reviewed by Turner and Edwards (2004), evidence suggests that in large groups there is less reliance on aggression during the immediate post-mixing phase than would be expected if the same proportion of dominance relationships is established using aggression in both large and small groups. The authors propose that this regulation of aggression in young pigs may simply be explained by the more effective avoidance of aggressive pigs that is possible in large pens with greater dimensions. However, Turner and Edwards (2004) propose that pigs may fight after mixing until a ceiling of fatigue or injury is reached or that they may become more selective over which group members they chose to fight, perhaps utilising status signals such as size.

Studies with sows indicate that while group size may not affect aggression or stress, injuries may be higher in large groups. Some limited research by Taylor et al. (1997) examining the effects of varying group sizes of 5, 10, 20 and 40 sows with a space allowance of 2.0 m^2/sow has shown that the prevalence of skin injuries measured at days 5 and 53 after mixing was higher in groups of 10 than 5, although neither of these two groups sizes differed from 20 and 40. Taylor and colleagues also found a tendency for aggression to be higher in large groups at days 1 and 2 after mixing. In a large experiment, housing sows in groups of 10, 30 and 80 did not affect aggression around feeding or plasma cortisol concentrations measured at days 2, 9 and 51 after mixing (Hemsworth et al., 2013). However, group size was related to skin injuries, with groups of 10 consistently having the lowest number of total skin injuries over this period. While aggression is likely to lead to skin injuries, the authors concluded that contact with pen features associated with avoidance of other sows, particularly with fast movement in large pens, may also increase the incidence of skin injuries. Olsson et al. (1994) reported increased injuries as group size increases, but group size was confounded by space, feeding system and the presence of bedding.

Thus, other factors such as floor space and quality and competition for feed or feeding space may have a greater impact on sow aggression and stress than group size. As discussed earlier, it has been proposed that in large groups, less aggression than expected may occur because of more total space available to effectively avoid aggressive pigs in large pens with greater dimensions, pigs may fight after mixing until a ceiling of fatigue and injury is reached, and/or they become more selective over which group members they chose to fight, utilizing status signals such as size (Turner and Edwards, 2004). However, the limited number of studies on adult gilts and sows indicates that smaller groups of pigs may have less skin injuries.

Feeding system. Providing escape areas allows subordinates to avoid dominant gilts and sows (Olsson and Samuelsson, 1993) and providing feeding stalls allows gilts and sows protection from group-mates at feeding. Aggression was reduced in pens with partial stalls (0.74 m depth) during 15–90 min after mixing, but not during the first 15 min (Barnett et al., 1992), however, in a second similar study, partial stalls had no effect on aggression during the first 90 min after mixing (Barnett et al., 1993a). Differences between the two studies in floor area behind the stalls may have been responsible for this inconsistency in the results.

In comparison to floor feeding, providing group-housed pregnant gilts with feeding stalls, particularly full body-length stalls, reduces aggression, injuries (in some studies) and plasma cortisol concentrations in the long term. Feeding stalls without rear gates reduced aggression during and at 40 min after feeding on days 2–15 after mixing and reduced baseline plasma cortisol concentration and the cortisol response to ACTH on weeks 4 and 7–8 after mixing pregnant gilts but did not affect prevalence of skin injuries compared to floor feeding (Barnett et al., 1992). In a similar experiment, feeding stalls reduced aggression measured for 40 min, 10 min after feeding commenced on days 2–30 after mixing and reduced baseline plasma cortisol concentration and the cortisol response to ACTH on weeks 5 and 8 after mixing pregnant gilts compared to floor feeding (Barnett, 1997). It is of interest that a floor space of 1.0 m^2/gilt, either total space or the space outside the feeding stalls, and independent of the feeding system, increases stress compared to 1.4 and 2.0 m^2/gilt (Barnett et al., 1992; Barnett, 1997), highlighting the importance of floor space. Body-length stalls reduced sow aggression in established groups compared to shoulder stalls or no stalls at the feeding trough, however, this effect only occurred with dry feed and not with wet feed (Andersen et al., 1999).

Furthermore, feeding stalls, particularly full body-length stalls without rear gates, improved cell-mediated immunity in pregnant gilts in the long term (Barnett et al., 1992; Barnett, 1997). However, increased vulva biting occurred in pens with full body-length feeding stalls without rear gates (Andersen et al., 1999), indicating that stall design may influence the nature as well as the amount of aggression. As recognised by several authors, although floor feeding is competitive, gaining access to feeding stalls can also lead to competition and aggression between group-housed sows (see review by Bench et al., 2013).

Allowing group-housed-gestating sows' continuous access to feeding stalls reduces aggression and skin injuries, but not salivary cortisol in comparison to allowing access only around feeding time (Wang and Li, 2016). Thus, providing stalls allows gilts and sows protection from group-mates particularly at feeding and facilitates both access to important resources, such as feed, and avoidance responses from aggressive pigs.

While sequential feeding systems, such as an electronic sow feeding system, allow control over individual gilts and sow intake, animals have to queue to access the stalls. Although feeding orders can develop (see Spoolder et al., 2009), this can result in aggression and the displacement of subordinates at the entrance of the feeder (Bench et al., 2013). The design of the pen and the placement of the feeding stalls may affect aggression since less sows per electronic sow feeding system, a long route from exit to entry of the electronic sow feeding system and positioning the electronic sow feeding system away from busy areas or other resources (e.g. drinker, hay rack), may assist in improving accessibility to the electronic sow feeding system and reduce aggression (see review by Verdon et al., 2015a).

Flooring including bedding. Musculoskeletal problems in commercial pigs are common, occurring in all stages of commercial production and these problems can involve all components of the musculoskeletal system such as the claw, foot, lower leg, thigh, shoulder and pelvis (see reviews by Nalon et al., 2013; Pluym et al., 2013). The general clinical sign of musculoskeletal problems is lameness, an impaired movement or deviation from normal gait (Cockram and Hughes, 2011) and the main causes of lameness in pigs range from inflammation and pain (Cockram and Hughes, 2011) and infected skin and claw lesions (Velarde, 2007) to broken bones (Marchant-Forde, 2009a). Among these, it has been proposed that osteochondrosis is a major factor contributing to lameness in pigs (see review by Supakorn et al., 2018). It has been estimated that the total annual culling and death rate from 2011 to 2016 was 8.3% to 13.0% for gilts and 45.7% to 47.4% for sows, and the reported culling rate due to locomotor problems of 15.2% in the United States, 16.9% in England, 9.7% in Belgium, 15.5% in Southern Mexico, 9–15% in Sweden, Finland and Denmark, 22.5% in Southern China and 37.4% in Thailand (Supakorn et al., 2018). While claw lesions are common in sows with a prevalence ranging between 59% and 99% of sows having one or more claw lesions, not all of these sows display lameness (Heinonen et al., 2013; Pluym et al., 2011, 2013).

Lameness is a multifactorial problem affected by genetics, housing and management practices such as herd health and nutrition (Nalon et al., 2013; Pluym et al., 2013). Most lameness cases during group housing develop shortly after the sows are introduced to groups (Supakorn et al., 2018). Genetic background is implicated. Disparity in claw size, which is associated with higher culling risk, is hereditable although there is a wide range of heritability values, depending on breed and statistical method (Pluym et al., 2013). Nutrition is an important predisposing factor of sow lameness, but information on the optimum feeding strategy and nutrient requirements for modern sows in contemporary housing are limited (Pluym et al., 2013; van Riet et al., 2013). Supakorn et al. (2018) have emphasised the need to fully understand the interaction between genetics and nutrition, relative to lameness and osteochondrosis.

Quality of flooring is essential to pig welfare as it is likely to have a direct effect on foot health and the culling rate from lameness (Barnett et al., 2001; Borell et al., 1997). The primary cause of lameness is pain (Cockram and Hughes, 2011) and consequently lameness is likely to be associated with some level of suffering in the animal (Dawkins, 1998). The review by Pluym et al. (2011) of research over the last 30 years on flooring and claw health in sows highlights the impact of floor quality and hygiene. For example, some of the main contributors to the development of claw lesions include group housing on fully or partly slatted floors, without bedding; wet and soiled dunging area; and old concrete flooring with rough, crumbled slat edges and enlarged gaps. Since slip-resistance, abrasiveness, hardness, surface profile and void ratio are the main floor characteristics contributing to claw health (Pluym et al., 2011, 2013; Supakorn et al., 2018), the authors recommended research on contemporary housing and flooring systems for sows to define optimum values for these floor characteristics. Foraging substrates can reduce injury caused by slipping (Heinonen et al., 2013), some abrasion is necessary for foot health (Webb and Nilsson, 1983) and the lack of natural wearing of the horn on solid floors with deep bedding may increase the risk of toe erosions, especially when straw is wet and soiled (Kilbride et al., 2009). The quality of flooring interacts with the thermal requirements and pen design on lameness in sow. For outdoor sows, site and soil type are significant factors affecting lameness (Thornton, 1990).

2.2.2 Specific mixing pen

The use of mixing pens when mixing gilts and sows has been proposed by many authors (e.g. Edwards et al., 1993; Arey and Edwards, 1998; Barnett et al., 2001; Verdon et al., 2015a; Verdon and Rault, 2018) on the basis that more floor space may provide the opportunity for avoidance by less aggressive pigs, while enabling the dominance relationships to form with less aggression, and subsequently less injuries and stress, at mixing.

The attributes of mixing pens that have been proposed to reduce aggression at mixing include: straw or rice hulls and feed, which may provide a distraction (straw or rice hulls also offers a foothold during aggression or fleeing); easy access to feed; absence of tightly confined areas in which a pig could be cornered and unable to escape from an aggressive pig; adequate space for pigs to turn around and for two pigs to easily pass side by side in all places; modifying pen size and shape since pigs require a minimum space in which they fight; the use of masking odours since anosmic pigs show reduced aggression; and sedation using pharmacological agents and grouping after dark to utilise this normal inactive time (Arey and Edwards, 1998; Barnett et al., 2001; Verdon et al., 2015a; Verdon and Rault, 2018). Although Barnett et al. (2001) concluded that all or some of these above suggestions may only be effective in postponing aggression, rather than reducing it, the aim when mixing gilts or sows should be to introduce them in a setting in which individuals can avoid aggressive ones when required with minimum risk to injury and stress while also allowing the dominance relationships to quickly form. Bedding such as straw while not likely to reduce aggression (see review by Verdon et al., 2015a) will reduce the risks of leg injury associated with aggression (see review by Spoolder et al., 2009).

As discussed in the section on floor space, one of the most obvious features of a mixing pen that will reduce aggression and stress is increased floor space (Barnett et al., 1992, 1993a). Furthermore, the sow's requirement for space may be less once the group is established (Hemsworth et al., 2013, 2016). A recent study in which the effects of housing sows at 4.0 or 6.0 m^2/sow for the first 4 days after mixing in comparison to 2.0 m^2/sow and then housing all sows at 2.0 m^2 per sow did not affect aggression, plasma cortisol concentrations or skin injuries in either period (Greenwood et al., 2016). Nevertheless, further research is required on this topic since Hemsworth et al. (2013, 2016) have shown that sows in static groups may adapt over time to a reduced space of

1.4 m^2/sow and thus a strategy of staged-gestation penning, with more space immediately after insemination and less space later in gestation once the dominance relationships have been established, may address both animal welfare and economic considerations. Therefore, further research is required, particularly on the long-term effects on the pigs when subsequently placed in their gestation group systems.

A visual barrier in the pen may reduce aggression after mixing by allowing retreating pigs to visually isolate themselves from aggressive ones (Marchant-Forde and Marchant-Forde, 2005). The provision of a partial barrier centrally placed along the length of the pen reduced sow aggression in the 12-h period after mixing but not skin injuries (Edwards et al., 1993). In contrast Luescher et al. (1990) found that the provision of barriers placed perpendicular to the back of the pen and head dividers in feeding troughs did not affect the time that previously unfamiliar gilts spent fighting during 3 days after mixing. Surprisingly gilts in pens with barriers had more skin injuries on front and cheeks at day 4 after mixing, possibly arising from contact with the partitions associated with avoidance of aggressive gilts.

As discussed in the section on feeding system, aggression was reduced in pens with partial stalls early after mixing. Furthermore, full-body length feeding stalls reduced aggression around feeding on days 2–30 after mixing (Barnett et al., 1992; Barnett, 1997).

The shape of the group pen may influence aggression at mixing. Aggression in growing pigs was found to be higher in circular pens and lower in square pens, with rectangular and triangular pens exhibiting intermediate levels (Wiegand et al., 1994), leading the authors to conclude that corners may act as hide areas, which have been shown to decrease aggression in newly grouped pigs (McGlone and Curtis, 1985). There is limited evidence that aggression in adult gilts at mixing is reduced in rectangular pens compared to square pens with a floor area of 1.4 m^2/gilt (Barnett et al., 1993a). However, the benefits were lost in larger pens providing 3.4 m^2/gilt. In a study of pen shape (rectangular, square, elongated rectangular and circular pens) and floor space (4.1 and 9.3 m^2/sow) in the first 28 hours after mixing, Docking et al. (2001) found the least aggression in the circular pen with 9.3 m^2/sow.

Another management technique that has been examined is boar presence. Boar presence in the pen at mixing reduced aggression during both the feeding and non-feeding periods, and skin injuries on the first day of mixing (Docking et al., 2001). Similarly, the presence of a boar in the pen reduced the duration of fighting on days 1 and 2 after mixing during both the feeding and non-feeding periods, but not the frequency of aggression or fights (Borberg and Hoy, 2009). However, there was no effect of boar presence on skin injuries at the day after mixing. In contrast, Séguin et al. (2006b) found that the frequency of aggressive interactions and duration of fighting during the floor feeding and non-feeding periods in the first 2 days after mixing were unaffected by the presence of a boar either in the pen or with fenceline contact, although fenceline contact increased aggression during the feeding period in the 25–48 h period after mixing. Sows mixed with a boar had fewer shoulder scratches than sows without boar contact. There were also no treatment effects on agonistic behaviour for two days after boars were removed 48 h after mixing or saliva cortisol concentrations on days 1, 2 and 6 after mixing (Séguin et al., 2006b). While two of these three studies indicate that mixing with a boar provides some welfare benefits in terms of reducing sow agonistic behaviour and skin injuries, these studies are confounded by group size and floor space (Marchant-Forde and Marchant-Forde, 2005).

Some chemical interventions have been shown to reduce aggression at mixing. 'Anti-aggressive' drugs such as amperozide (Barnett et al., 1993a,b, 1996) and azaperone (Luescher et al., 1990: Csermely and Wood-Gush, 1990) appear to reduce aggression in mixed gilts in the short term for as long as the drug has efficacy, but once the effects of the drug have worn off, aggression returns to levels seen in untreated gilts.

A synthetic formulation of a maternal pheromone, which closely matches the composition of skin secretions isolated from sows, has been shown to reduce aggression and injuries in pigs mixed post-weaning (McGlone and Anderson, 2002; Guy et al., 2009). While bouts of fighting decreased in the presence of a synthetic olfactory agonist over 8 days after mixing, the overall time the sows spent involved in aggressive behaviour remained unaffected, indicating that the treated sows displayed fewer, but longer aggressive bouts. Furthermore, there were no treatment effects on injuries, salivary cortisol concentrations or conception rate. As indicated by the authors, it is not possible to determine whether the observed effects on sow behaviour were due to the 'pheromone-like' properties of the synthetic olfactory agonist, or masking by the chemical odour that was emitted by the product since anosmic pigs show reduced aggression (Luescher et al., 1990).

2.2.3 Time of mixing

Gilts and sows can be mixed at weaning, after breeding, or after they are confirmed pregnant. Higher levels of aggression and higher cortisol concentrations on the day of mixing have been observed in sows mixed in the week after insemination (early gestation) than in 5–6 weeks after insemination (Stevens et al., 2015). These effects of timing of mixing were not seen at day 7 after mixing or later, day 91 of gestation. As with cortisol and aggression, sows mixed early in gestation had more skin injuries at day 7 after mixing than sows mixed later in gestation (Stevens et al., 2015), possibly because of declining aggression. Similarly, more skin injuries, more vulva injuries and a greater incidence of lameness were observed in sows mixed early after insemination than in those mixed later in gestation (Knox et al., 2014).

In contrast to the effects on aggression and stress found by Stevens et al. (2015), Strawford et al. (2008) and Knox et al. (2014) found that aggression early after mixing was similar for sows mixed at 2-9 days after insemination and those mixed 35-46 days after insemination and Knox et al. (2014) found that sows mixed early after insemination had a smaller increase in serum cortisol from a baseline measure than did sows mixed later in gestation. Stevens et al. (2015) found no treatment effects on skin injuries at 91 days after insemination, but Li and Gonyou (2013) found that sows mixed in either static or dynamic groups early in gestation had more skin injuries before farrowing than did those mixed later in gestation. These contradictory results may be due to differences between studies in sows, such as genetics, size and experience, design features, such as feeding system, pen design and group size, and timing of measurements relative to insemination. Interestingly conception rates were lower for sows mixed early in gestation than for those mixed later in gestation or those housed in stalls for the entire gestation (Knox et al., 2014), and farrowing rates were lower for sows mixed early in gestation than for those mixed later in gestation (Li and Gonyou, 2013; Knox et al., 2014). Increased stress when mixed early after insemination may be responsible for these reproductive effects.

With the move away from stall housing of sows, there is interest in the use of grouping sows at weaning. While group housing may facilitate sexual behaviour, there is evidence that grouping may stimulate the sexual behaviour of dominant sows but suppress that of subordinate sows (Pedersen et al., 1993, 2003). Furthermore, there is evidence that oestrus may be delayed and variation in the onset of oestrus may be increased with grouping of sows at weaning (Langendijk et al., 2000; Rault et al., 2014), although with high levels of boar stimulation, neither the detection nor duration of oestrus differed between weaned sows in stalls and those in groups (Langendijk et al., 2000). Cortisol concentrations on the day after weaning were higher for sows mixed at weaning than for those housed in stalls after weaning, but cortisol concentrations were similar after insemination when the two treatment of sows were regrouped or grouped for the first time, respectively (Rault et al., 2014). The higher stress in sows mixed at weaning could be due to either the stage of reproduction or the accumulation of various stressors when weaning sows into groups (Hemsworth, 2018).

While the literature is inconsistent on the effects of the timing of mixing post-insemination, possibly because of differences between studies in intrinsic sow factors, the design of the group systems and timing of measurements, the aim should be to minimise aggression and thus injuries and stress to safeguard both sow welfare and reproduction.

2.2.4 Static and dynamic groups

Static groups are mostly formed after insemination for the entire gestation, while sows previously unfamiliar in dynamic systems are frequently introduced to the group so that they experience between 3 and 12 mixings per gestation (Marchant-Forde, 2009a). It has been suggested that aggression may be greater in dynamic groups of sows because of the frequent introduction of unfamiliar sows to the group (Arey and Edwards, 1998; Barnett et al., 2001). The few studies that have been conducted have indicated that while no effects were found on aggression, cortisol concentrations, reproductive performance and longevity, skin injuries (cuts, swellings and wounds) and lameness were greater in dynamic groups (Anil et al., 2006; Strawford et al., 2008; Li and Gonyou, 2013).

2.2.5 Hunger in gestating sows

Sows are commonly restrictively fed during gestation for productivity and lameness reasons (Meunier-Salaün et al., 2001). It has been estimated that these levels are about 60–70% of ad libitum intake (D'Eath et al., 2018) and operant conditioning studies indicate that sows fed these restricted levels can be hungry for a considerable period of the day (Hutson, 1991; Lawrence and Terlouw, 1993).

Because conventional low-fibre, high-concentrate diets require little food-searching (appetitive) behaviour and consummatory behaviours like chewing, restrictive feeding result in unfulfilled motivations to perform these natural foraging activities, leading to increased oral stereotypies (such as oral stereotypic licking, bar-biting and sham-chewing or vacuum-chewing) (Bergeron et al., 2006) (Fig. 1).

Food restriction increases stereotypies in both group- and stall-housed sows (Terlouw et al., 1991; Terlouw and Lawrence, 1993; Bergeron et al., 2000). In addition to increasing feeding levels, increasing digestible energy in these diets can also reduce stereotypies (Bergeron and Gonyou, 1997). Feeding high-fibre diets that meet the sow's nutrient requirements increases feeding time and reduces stereotypies (e.g. Robert et al., 1993; Brouns et al., 1994; Ramonet et al., 1999), but the increased feeding time appears to account for much of this reduction in stereotypy level (Robert et al., 1997).

A recent comprehensive review on mitigating hunger in pregnant sows by D'Eath et al. (2018) also concluded that providing fibre in gestating sow diets, while still restricting energy intake, can reduce redirected oral behaviours. However, fibre does not completely satisfy the motivation of sows to access food in operant tasks. Furthermore, natural foraging behaviours such as rooting and chewing suitable substrates are increased as a result of restrictive feeding, but there are foraging opportunities with the use of straw bedding or outdoor

Figure 1 A sow displaying the oral stereotypy sham-chewing.

pastures (D'Eath et al., 2018). Straw that elicits foraging behaviour has been shown to reduce oral stereotypies in gestating sows (Spoolder et al., 1995; Bergeron et al., 2006).

2.2.6 Fully or partially slatted, non-bedded and non-enriched environments for breeding pigs

Indoor production systems are considered by some to provide barren environments for animals (Barnett et al., 2001) and some authors have raised concerns about the extensive use of fully or partially slatted, non-bedded and non-enriched environments for young and breeding pigs (e.g. Marchant-Forde, 2009b). The European Union Commission Directive 2001/93/EC (European Commission Directive, 2001, p. 37) states that 'pigs must have permanent access to a sufficient quantity of material to enable proper investigation and manipulation activities, such as straw, hay, wood, sawdust, mushroom compost or a mixture of such that does not compromise the health of animals'. However, the extent to which this provides effective enrichment has been questioned (Marchant-Forde, 2009b; Van de Weerd and Day, 2009). Although the effect of space is confounded by flooring, Merlot et al. (2017, 2019) found that gestating sows housed on slatted floors with 2.4 m^2/sow of floor space had higher cortisol concentrations, granulocyte counts and immunoglobulin G than gestating sows housed on deep straw bedding with 3.5 m^2/sow. Interestingly in both experiments, sows housed on deep straw bedding with more floor space had lower preweaning mortality, implicating prenatal stress effects on the developing offspring.

The topic of enrichment and pig welfare is reviewed elsewhere in this book.

2.2.7 Individual sow characteristics

Research on sows at mixing has shown that the heritability for severe aggressive behaviours is intermediate (h^2=0.24; Løvendahl et al., 2005). The feasibility and desirability of selection for reduced aggression in pigs is considered elsewhere in this book.

Individual sow aggressive behaviour within groups of unfamiliar animals is more predictable within a gestation than between gestations (Verdon et al., 2015b; Horback and Parsons, 2016). Verdon and colleagues concluded that while genetics is likely to contribute to individual sow aggression within groups of unfamiliar animals, the reduced strength of between-gestation relationships suggests that social experience and group composition may also influence the aggressive phenotype. In fact, when housed in groups, sows may show flexibility in their use of aggression depending on the physical characteristics or behaviour of their pen mates (see reviews by Turner and Edwards, 2004; Verdon et al., 2015a). Furthermore, while variation in age exists in groups of wild boar in the wild and groups of domestic pigs in semi-natural settings and this diversity may be integral to social learning and stabilisation of the social group, little is known of the importance of social experience particularly early in life on aggression in sows (see reviews by Verdon et al., 2015a; Telkanranta and Edwards, 2017). The importance of experience is demonstrated by the findings in several species that repeated encounter and operant performance studies that intraspecific aggression can be positively reinforcing for successful aggressors, whereas the punishing effects of defeat are likely to result in avoidance (see review by Potegal, 1979).

2.3 Boar

There has been very limited research conducted on the effects of housing and management on the welfare of breeding boars. In many countries housing boars in individual stalls where they cannot turn around is now either not recommended under domestic codes of practice for pig welfare or prohibited under the regulated standards of pig welfare. For example, the European Commission Directive (2008, pg. 11) and the New Zealand Code of Welfare (Anon, 2010, pg. 21) and Canadian Code of Practice (Anon, 2014, pg. 13) require for all holdings that boar pens must provide sufficient space to allow the boar to turn around. The Australian Model Code of Practice requires that boars accommodated individually in stalls must be able to stand, get up and lie down without being obstructed by the bars and fittings of the stall, lie with limbs extended to stretch and to be able to freely undertake such movements

(Anon, 2010, pg. 8). Furthermore, this code requires that if boars are housed continuously in stalls, they must be released for mating or exercise at least twice per week. Recommendations for floor space for boars in individual pens varies from 5.6 m^2 for partially or fully slatted floors (Anon, 2014, pg. 51), and 6.0 m^2 (European Commission Directive, 2008, pg. 11; Anon, 2010, pg. 8) and 7.4 m^2 for solid bedded floors (Anon, 2014, pg. 51). The New Zealand Code of Welfare (Anon, 2010) provides no minimum floor space, but requires that boars must be provided with sufficient space so that they can stand up, turn around and lie comfortably in a natural position.

Artificial insemination (AI) is widely used in modern commercial pig production and boars are mainly housed in stalls in AI centres, but pens are also used (Singleton, 2001; Knox, 2016). Surprisingly, some codes of practice or standards such as those in Australia (Anon, 2010) and the EU (European Commission Directive, 2008) do not specifically refer to housing of AI boars, and others such as in Canada (Anon, 2014) have a disclaimer that the code does not apply to associated industries such as AI boar stud farms.

Adult males of the wild ancestors of the domestic pig are frequently solitary in the wild, but during the breeding season fight to mate adult females in matriarchal groups and then move to another group (Fernandez-Llario et al., 1996). Thus, individual housing per se with sexual activity associated with regular contact with females when used for oestrus detection and mating in commercial production units or regular semen collection in AI centres, is unlikely to be stressful, although the effects of using boars solely for oestrus detection without mating on their sexual frustration and stress have not been examined.

The post-pubertal social environment of the boar appears to affect the sexual behaviour of boars. Isolation of mature boars from female pigs depresses their sexual behaviour (see review by Hemsworth and Tilbrook, 2007); however, this effect is not permanent since rehousing isolated boars near females restores their sexual behaviour within 4 weeks. Interestingly, the oestrous status of the females did not influence the effectiveness of females in maintaining the sexual behaviour of mature boars. The effect on sexual behaviour of using boars solely for oestrus detection in commercial production units or semen collection in AI centres is unknown, although low sexual motivation of AI boars may be less critical because of the moderate collection frequency generally required (Singleton, 2001; Knox, 2016). However, data from AI centres indicate that 15% of boars in three 250-head boar studs in the United States could not be trained to collection on a dummy (Flowers, 2008) and apart from boars with either poor semen or reduced for their semen, 21% of boars at a Polish AI centre from 1998 to 2013 were culled for poor sexual motivation.

While group housing of breeding boars is common in outdoor systems, boars are generally housed individually in indoor production systems. Boars housed in groups are reported to have better physical conditions, presumably because of improved space for exercise (Cordoba-Dominguez et al., 1991). In a study examining the effects of housing mature boars individually either in stalls (0.8 × 2.9 m) housed adjacent to a mating pen compared or in pens (2.6 × 3m) away from the mating pen, Levis et al. (1995) found no treatment effects on daytime baseline concentrations of plasma cortisol or the cortisol response to administration of ACTH. These physiological measures of stress provide no evidence that housing boars individually in stalls adjacent to a mating pen has any adverse effects on their welfare relative to housing boars individually in pens. However, in light of the effects of housing gilts and sows in stalls during gestation, it would be prudent to examine the welfare implications of stall housing of boars used for natural matings, boars solely used for puberty stimulation and oestrus detection and boars housed in AI centres. There is evidence that housing gestating gilts and sows in stalls depresses their welfare. Pregnant gilts in stalls experienced a chronic stress response, based on a sustained elevation of basal cortisol concentrations, similar to that seen in tethered gilts, and active avoidance of neighbours (see review by Barnett et al., 2001) as well as a higher incidence of lameness and an increased neutrophil to lymphocyte ratios, perhaps as a consequence of increased stress (Karlen et al., 2007).

Lameness and penile injuries may not physically allow the achievement of copulation or may inhibit copulation because of pain (Christensen, 1953), while injury sustained during copulation may produce a psychological effect for some time after physical recovery has occurred, again inhibiting copulation. Prevention of lameness and penile injuries should include attention to the design and maintenance of the accommodation and mating or semen collection areas, appropriate supervision and assistance at mating or semen collection, and selection for heritable conformational traits of the feet and legs (Hemsworth and Tilbrook, 2007). Obesity and skeletal defects can also affect the sexual behaviour of boars, often by affecting mobility and thus mating competency. Lameness or locomotor problems, and death and euthanasia account for 8–12% and 7–8%, respectively, of removals of breeding boars in commercial production units (D'Allaire and Leman, 1990; Koketsu and Sasaki, 2009) and lameness is one of the main reasons for culling boars from AI centres (Knecht et al., 2017). Lameness or locomotor injuries and death in AI and commercial production units are likely to be painful and thus a welfare concern (Whay et al., 2003).

While it does not appear to be widely viewed as a high priority based on stall housing of boars in some countries, research is required on the welfare implications of stall housing of boars used for natural matings, boars solely used for puberty stimulation and oestrus detection and boars housed in AI centres.

3 Animal management

3.1 Animal abuse and cruelty

It is indisputable that there have been cases of animal abuse and cruelty in livestock production, just as there have been cases involving companion animals, and these are likely to continue. Animal welfare is a highly emotive subject and the news media frequently report animal welfare abuses, often as a result of covert recording or filming by animal rights organisations. These stories generate considerable public interest and the message conveyed is that the livestock industry in question engages in poor animal welfare practices and that some aspect of it should be substantially changed or even banned.

Anecdotal experience in extensively studying human-animal interactions in the livestock industries is that the prevalence of deliberate mistreatment of animals appears to be low (Hemsworth and Coleman, 2011). That is not to say that there are occasions when an individual might mistreat an animal because of a particular set of circumstances, such as in response to the animal's behaviour or because the person is upset for another reason or is unwell (Coleman and Hemsworth, 2014). There also will be cases where a person is so insensitive to the state of the animal that their behaviour is persistently inappropriate. However, most of the instances of poor behaviour appear to arise because the management of animals has largely been learned 'on the job' and people may be unaware of the effect of their behaviour on the welfare of the animals under their care.

3.2 Stockmanship

Management affects animal welfare in several ways (Hemsworth and Coleman, 2009, 2011). At the level of farm management, human resource management practices, including employee selection and training, and animal management practices, such as best practice in housing and husbandry, and implementation of welfare protocols and audits, all impact on farm animal welfare. At the stockperson level, together with the opportunity to perform their tasks well, stockpeople require a range of well-developed husbandry skills and knowledge to effectively care for and manage farm animals.

Stockpeople clearly require a basic knowledge of both the requirements and behaviour of farm animals, and also must possess a range of well-developed husbandry and management skills to care for and manage their animals effectively. Other important job-related characteristics include job satisfaction, work motivation and motivation to learn new skills and knowledge about the animal. Since the 1980s there has been an ever-increasing body of evidence on the profound effects of stockperson interactions on farm animal fear and in turn, animal stress, welfare and productivity. Research in the livestock industries, including the pig industry and supported by handling studies in laboratories, demonstrates sequential relationships between stockperson attitudes to interacting and working with their animals, stockperson interactions and animal fear, stress and productivity (see Coleman, 2004; Hemsworth and Coleman, 2011). The specific attitudes identified in the field studies were stockperson attitudes towards their animals and working with them, their beliefs about other people's expectations of them, and their beliefs about the extent to which they have control over how they interact with and manage their animals. These, in turn, determine the nature and extent of stockperson interactions with their animals and, in situations in which stockperson behaviour is poor, animal fear and stress are elevated and thus both animal productivity and welfare are at risk. For example, it has been shown that positive attitudes among stockpeople to the use of patting and talking and to the avoidance of intense verbal and physical effort to handle gilts and sows are negatively correlated with the use of negative tactile interactions such as slaps, pushes and hits in handling pigs (Hemsworth et al., 1989b; Coleman et al., 1998). In turn, the frequent use of negative tactile interactions was positively associated with increased fear of humans. Furthermore, attitudes to handling pigs are also correlated with aspects of the job such as job satisfaction, work motivation and motivation to learn new skills and knowledge about the pig (Coleman et al., 1998). Therefore, attitudes influence not only the manner in which stockpeople handle pigs, but also their motivation to care for pigs.

3.3 Minimising handling stress

This research has shown that the way in which stockpeople routinely handle their animals is largely learned and that a major factor affecting stockperson behaviour is stockperson attitudes. These attitudes are based on the beliefs that stockpeople hold about the animals and are shaped by both ad hoc advice from co-workers and by direct and indirect experiences with animals. They are therefore learnt, and consequently training of stockpeople targeting the key attitudes and related behaviour that affect animal fear and stress should be effective in reducing animal fear and stress and consequently minimising risks to animal welfare through poor handling. In fact, when training programmes that are based on an understanding of both the beliefs that underpin behaviour and the appropriate ways of handling animals are used, there is a dramatic and sustained improvement in stockperson behaviour with a concomitant improvement in animal welfare as well as animal productivity (Hemsworth and Coleman, 2011; Coleman and Hemsworth, 2014). Indeed, studies in the pig industry (Hemsworth et al., 1994a; Coleman et al., 2000) have shown that cognitive behavioural training, in which these key attitudes and behaviour of stockpeople are targeted, can improve the attitudes and behaviour of stockpeople towards their pigs successfully, with consequent beneficial effects on gilt and sow fear and reproduction (Fig. 2).

Thus, training programmes that target the attitude and behaviour of stockpeople offer considerable opportunity to improve pig welfare. Improved human-animal interactions may also enhance job-related characteristics, such as job satisfaction, motivation and commitment, thereby potentially improving the stockperson's job performance and career prospects (Coleman and Hemsworth, 2014). Thus, including training targeting the attitudes and behaviours of stockpeople towards pigs in conjunction with the technical skills, and knowledge of stockpeople is likely to not only reduce the stress associated with handling and husbandry procedures involving humans, but also improve the motivation in stockpeople to learn new technical skills and knowledge and to apply these competencies in the management of the animals under their care.

Most industries have animal welfare policies in place, but policies alone do not change the behaviour of individual stockpeople unless there are major sanctions that are enforced but this requires constant surveillance which is rarely acceptable in the workplace. Direct interventions, in the form of targeted behaviour change training programmes, as research has shown, provide a proactive means to retain a skilled workforce that acts in the best interests of the animals under their care (Coleman and Hemsworth, 2019).

4 Conclusion and future trends

Aggression and consequently stress are inevitable when unfamiliar pigs are mixed. Agonistic behaviour and particularly aggression may also lead to injuries

Figure 2 Positive attitudes to handling pigs are associated with positive handling of pigs.

such as claw and limb lesions, fractures or spinal cord damage associated with incidents such as slips, falls, knocks, bites, head slashes and toes getting stuck in the gaps of concrete slats. Pen design is an important determinant of the welfare of group-housed gilts and sows. Floor space is particularly relevant early after mixing since insufficient space, in terms of both quantity (amount) and quality (configuration, including physical and visual barriers), can prolong aggression by affecting a pig's ability to avoid or escape other pigs and hence delay the formation of a stable hierarchy. Providing group-housed pregnant pigs with feeding stalls, particularly full body-length stalls, reduces the long-term aggression, plasma cortisol concentrations and often injuries. Quality of flooring in terms of bedding, dry dunging area, and well-maintained and designed slatted and solid floors is essential to pig welfare as it is likely to have a direct effect on foot health and the culling rate from lameness. While aggression, cortisol concentrations, reproductive performance and longevity are not affected, skin injuries and lameness are greater in dynamic than static groups.

While not studied extensively, particularly in sows, the use of mixing pens in which more space, free-access, full-body length stalls, visual barriers and bedding are provided is likely to provide the opportunity for avoidance by less-aggressive gilts and sows, enabling the dominance relationships to form with less aggression, and subsequently less injuries and stress. Further research however is required, particularly on the long-term effects on the pigs when subsequently placed in gestation group systems.

Although there is evidence in other species that early experience may affect social skills later in life, there are few studies on the effects of early social experience on aggressive behaviour of sows. Genetic selection also has the potential to reduce aggression, and therefore, continued research on the opportunity to genetically select against aggressiveness and its broader implications is required.

While it does not appear to be widely viewed as a high priority, research is required on the welfare implications of stall housing of boars used for natural matings, boars solely used for puberty stimulation and oestrus detection and boars housed in AI centres.

Lameness is a multifactorial problem affected by genetics, housing and management practices such as herd health and nutrition, and prevention of lameness from developing in group-housed sows is a major challenge since most lameness during group housing developing shortly after the sows are introduced to groups. The quality of flooring is essential to pig welfare as it is likely to have a direct effect on foot health and the culling rate from lameness. Slip-resistance, abrasiveness, hardness, surface profile and void ratio are the main floor characteristics contributing to lameness and research on contemporary housing and flooring systems is required to define optimum values for these floor characteristics. The interaction between genetics and nutrition, relative to lameness and osteochondrosis needs to be fully understood.

Direct contact with mature boars during gilt puberty stimulation and oestrus detection of both gilts and sows should be supervised to avoid stress as well as injury to gilts and sows. The physical environment, particularly floor type, quality and maintenance, will contribute to the risk of foot and leg injury during puberty stimulation, oestrus detection and natural mating.

Most industries have animal welfare policies in place, but policies alone do not change the behaviour of individual stockpeople. Direct interventions, in the form of targeted behaviour change training programmes, as research has shown, provide a proactive means to retain a skilled workforce that acts in the best interests of the animals under their care. It is likely that general community will place an increasing emphasis on ensuring the competency of stockpeople to manage the welfare of livestock.

In conclusion, animal welfare problems for pigs during breeding and gestation are generally less a function of the type of housing system than of how well it operates. The focus on different housing systems ignores many of the important factors that can affect animal welfare such as the design of the system per se and the quality of management, particularly stockmanship. As persuasively stated by Fraser (2005, pp. 8C, 9C):

> some of the most important determinants of animal welfare are not specific to any one type of housing and production system. Whether dairy cows have more health problems in tie-stalls or on pasture is a matter of debate, but there is no disagreement that their welfare is improved by staff who are skilled at detecting and treating disease. Whether sows are better off in stalls or group-housed systems is sometimes disputed, but there is no disagreement that good maintenance and functioning of equipment is important to their welfare. In fact, if we think of animal welfare as being influenced by such key factors such as staff time and skill, substrate, temperature, feed quality and disease prevention measures, then animal welfare problems may be less a function of the type of rearing system - confinement, semi-confinement or extensive - than of how well it operated.

5 Where to look for further information

The welfare of breeding pigs has been reviewed by Barnett et al. (2001), Marchant-Forde (2009a) and Verdon et al. (2015a). The topic of human-animal interactions in the livestock industries has been reviewed by Hemsworth and Coleman (2011) and Coleman and Hemsworth (2014), and details of the commercialized training programme targeting attitudinal and behavioural change in stockpeople ('ProHand') are available at: https://www.animalwelfare-science.net/uploads/1/2/3/2/123202832/prohand_background_jul19.pdf.

6 References

Andersen, I. L., Bøe, K. E. and Kristiansen, A. L. (1999). The influence of different feeding arrangements and food type on competition at feeding in pregnant sows. *Appl. Anim. Behav. Sci.* 65(2): 91–104.

Anil, L., Anil, S. S., Deen, J., Baidoo, S. K. and Walker, R. D. (2006). Effect of group size and structure on the welfare and performance of pregnant sows in pens with electronic sow feeders. *Can. J. Vet. Res.* 70(2): 128–136.

Anil, S. S., Anil, L., Deen, J., Baidoo, S. K. and Wheaton, J. E. (2005). Characterization of claw lesions in gestating sows. In: 2005 Allen D. Leman Swine Conference, UDA, Minneapolis, MN, pp. 193–199.

Anon. (2010). *Animal Welfare (Pigs), Code of Welfare: A Code of Welfare Issued under the Animal Welfare Act 1999*. National Animal Welfare Advisory Committee, New Zealand.

Anon. (2014). *Code of Practice for the Care and Handling of Pigs, Canada*. National Farm Animal Care Council (NFACC), Canada.

Arey, D. S. (1999). Time course for the formation and the disruption of social organisation in group-housed sows. *Appl. Anim. Behav. Sci.* 62(2–3): 199–207.

Arey, D. S. and Edwards, S. A. (1998). Factors influencing aggression between sows after mixing and the consequences for welfare and production. *Livest. Prod. Sci.* 56(1): 61–70.

Barnett, J. L. (1997). Modifying the design of group pens with individual feeding places affects the welfare of pigs. In: Bottcher, R. W. and Hoff, S. J. (Eds). Proceedings of the 5th International Symposium American Society of Agricultural Engineers, St. Joseph, MI, pp. 613–618.

Barnett, J. L., Hemsworth, P. H., Cronin, G. M., Jongman, E. C. and Hutson, G. D. (2001). A review of the welfare issues for sows and piglets in relation to housing. *Aust. J. Agric. Res.* 52(1): 1–28.

Barnett, J. L., Hemsworth, P. H., Cronin, G. M., Newman, E. A., McCallum, T. H. and Chilton, D. (1992). Effects of pen size, partial stalls and method of feeding on welfare-related behavioural and physiological responses of group-housed pigs. *Appl. Anim. Behav. Sci.* 34(3): 207–220.

Barnett, J. L., Cronin, G. M., McCallum, T. H. and Newman, E. A. (1993a). Effects of pen size/shape and design on aggression when grouping unfamiliar adult pigs. *Appl. Anim. Behav. Sci.* 36(2–3): 111–122.

Barnett, J. L., Cronin, G. M., McCallum, T. H. and Newman, E. A. (1993b). Effects of 'chemical intervention' techniques on aggression and injuries when grouping unfamiliar adult pigs. *Appl. Anim. Behav. Sci.* 36(2–3): 135–148.

Barnett, J. L., Cronin, G. M., McCallum, T. H., Newman, E. A. and Hennessy, D. P. (1996). Effects of grouping unfamiliar adult pigs after dark, after treatment with amperozide and by using pens with stalls, on aggression, skin lesions and plasma cortisol concentrations. *Appl. Anim. Behav. Sci.* 50(2): 121–133.

Beilharz, R. G. and Cox, D. F. (1967). Social dominance in swine. *Anim. Behav.* 15(1): 117–122.

Bench, C. J., Rioja-Lang, F. C., Hayne, S. M. and Gonyou, H. W. (2013). Group gestation sow housing with individual feeding–II: how space allowance, group size and composition, and flooring affect sow welfare. *Livest. Sci.* 152(2–3): 218–227.

Bergeron, R., Badnell-Waters, A. J., Lambton, S. and Mason, G. (2006). Stereotypic oral behaviour in captive ungulates: foraging, diet and gastrointestinal function. In:

Mason, G. and Rushen, J. (Eds). *Stereotypic Animal Behaviour: Fundamentals and Applications to Welfare* (2nd edn.). CABI Publishing, Wallingford, UK, pp. 19-27.

Bergeron, R., Bolduc, J., Ramonet, Y., Meunier-Salaün, M. C. and Robert, S. (2000). Feeding motivation and stereotypies in pregnant sows fed increasing levels of fibre and/or food. *Appl. Anim. Behav. Sci.* 70(1): 27-40.

Bergeron, R. and Gonyou, H. W. (1997). Effects of increasing energy intake and foraging behaviours on the development of stereotypies in pregnant sows. *Appl. Anim. Behav. Sci.* 53(4): 259-270.

Borberg, C. and Hoy, S. (2009). Mixing of sows with or without the presence of a boar. *Livest. Sci.* 125(2-3): 314-317.

Borell, E., von, B. D. M., Csermely, D., Dijkhuizen, A. A., Edwards, S. A., Jensen, P., Madec, F. and Stamataris, C. (1997). The welfare of intensively kept pigs. In: Report of the Scientific Veterinary Committee, European Union, Brussels, Belgium.

Brouns, F., Edwards, S. A. and English, P. R. (1994). Effect of dietary fibre and feeding system on activity and oral behaviour of group housed gilts. *Appl. Anim. Behav. Sci.* 39(3-4): 215-223.

Canario, L., Turner, S. P., Roehe, R., Lundeheim, N., D'Eath, R. B., Lawrence, A. B., Knol, E., Bergsma, R. and Rydhmer, L. (2012). Genetic associations between behavioural traits and direct-social effects of growth rate in pigs. *J. Anim. Sci.* 90(13): 4706-4715.

Christensen, N. O. (1953). Impotentia coeundi in boars due to arthrosis deformans. In: 15th International Veterinary Congress, Part I, Stockholm, Sweden, Vol. 2, pp. 742-745; Part II, pp. 332-333.

Cockram, M. S. and Hughes, B. (2011). Health and disease. In: Appleby, M. C., Hughes, B. and Mench, J. (Eds). *Animal Welfare*. CABI Publishing, Wallingford, UK, pp. 120-137.

Coleman, G. J. (2004). Personnel management in livestock industries. In: Benson, G. J. and Rollin, B. E. (Eds). *The Well-Being of Farm Animals: Challenges and Solutions*. Blackwell Publishing, Ames, IA, pp.167-181.

Coleman, G. J. and Hemsworth, P. H. (2014). Training to improve stockperson beliefs and behaviour towards livestock enhances welfare and productivity. *Scientific and Technical Review of the Office International des Epizooties* 33: 131-137.

Coleman, G. J. and Hemsworth, P. H. (2019). Scientific commentary: animal abuse and cruelty in livestock production. Animal Welfare Science Centre. Available at: https://www.animalwelfare-science.net/news/scientific-commentary-animal-abuse-and-cruelty-in-livestock-production; https://www.animalwelfare-science.net/ (Accessed 20th September 2019).

Coleman, G. J., Hemsworth, P. H., Hay, M. and Cox, M. (1998). Predicting stockperson behaviour towards pigs from attitudinal and job-related variables and empathy. *Appl. Anim. Behav. Sci.* 58(1-2): 63-75.

Coleman, G. J., Hemsworth, P. H., Hay, M. and Cox, M. (2000). Modifying stockperson attitudes and behaviour towards pigs at a large commercial farm. *Appl. Anim. Behav. Sci.* 66(1-2): 11-20.

Cordoba-Dominguez, J., Dunne, J., Cliff, A., MacPherson, O., Menaya, C., Vidal, J. and English, P. (1991). Evaluation of the behaviour of mature boars housed in groups of four. In: Proceedings of the Winter Meeting British Society of Animal Production, Scarborough, UK, pp. 65.

Csermely, D. and Wood-Gush, D. G. M. (1990). Agonistic behaviour in grouped sows. II. How social rank affects feeding and drinking behaviour. *Ital. J. Zool.* 57(1): 55-58.

D'Allaire, S. and Leman, A. D. (1990). Boar culling in swine breeding herds in Minnesota. *Can. Vet. J.* 31(8): 581–583.

Dardallion, M. (1988). Wild boar social groupings and their seasonal changes in the Camargue, southern France. *Z. Säugetierkd.* 53: 22–30.

Dawkins, M. S. (1998). Evolution and animal welfare. *Q. Rev. Biol.* 73(3): 305–328.

D'Eath, R. B., Jarvis, S., Baxter, E. M. and Houdijk, J. (2018). Mitigating hunger in pregnant sows. In: Spinka, M. (Ed.). *Advances in Pig Welfare, Woodhead Publishing Series in Food Science, Technology and Nutrition*. Woodhead Publishing, Duxford, UK.

Desire, S., Turner, S. P., D'Eath, R. B., Doeschl-Wilson, A. B., Lewis, C. R. G. and Roehe, R. (2015). Genetic associations of short- and long-term aggressiveness identified by skin lesions with growth, feed efficiency and carcass characteristics in growing pigs. *J. Anim. Sci.* 93(7): 3303–3312.

Docking, C. M., Kay, R. M., Day, J. E. L. and Chamberlain, H. L. (2001). The effect of stocking density, group size and boar presence on the behaviour, aggression and skin damage of sows mixed in a specialized mixing pen at weaning. In: Proceedings of the British Society of Animal Science, Scarborough, UK, p. 46 (Abst).

Edwards, S. A., Mauchline, S. and Stewart, A. H. (1993). Designing pens to minimise aggression when sows are mixed. *Farm Build. Prog.* 113: 20–23.

Estevez, I., Andersen, I. and Naevdal, E. (2007). Group size, density and social dynamics in farm animals. *Appl. Anim. Behav. Sci.* 103(3–4): 185–204.

European Commission Directive (2008). Council Directive 2008/120/EC of 18 December 2008 laying down minimum standards for the protection of pigs. *Official Journal* 18.2.2009 L 47/5-13.

Fernandez-Llario, P., Carranza, J. and Hidalgo de Trucios, S. J. (1996). Social organization of the wild boar (*Sus scrofa*) in Donana National Park. *Misc. Zool.* 19: 9–18.

Fraser, D. (2005). Animal welfare and the intensification of animal production. An alternative interpretation. *FAO Readings in Ethics* 2. Editorial Production and Design Group, Publishing Management Service, FAO, Rome. ISBN 92-5-105386-3.

Gabor, T. M., Hellgren, E. C., van den Bussche, R. A. and Silvy, N. J. (1999). Demography, sociospatial behaviour and genetics of feral pigs (*Sus scrofa*) in a semi-arid environment. *J. Zool.* 247(3): 311–322.

Gill, P. (2007). Nutritional management of the gilt for lifetime productivity – feeding for fitness or fatness? In: *Today's Challenges. Tomorrow's Opportunities*, London Swine Conference Proceedings, London, Ontario, Canada, pp. 83–99.

Gonyou, H. W. (2001). The social behaviour of pigs. In: Keeling, L. J. and Gonyou, H. W. (Eds). *Social Behaviour in Farm Animals*. CAB International, Wallingford, UK, pp. 147–168.

Greenwood, E. C., Plush, K. J., van Wettere, W. H. E. J. and Hughes, P. E. (2016). Group and individual sow behaviour is altered in early gesation by space allowance in the days immediately following grouping. *J. Anim. Sci.* 94(1): 385–393.

Guy, J. H., Burns, S. E., Barker, J. M. and Edwards, S. A. (2009). Reducing post-mixing aggression and skin lesions in weaned pigs by application of a synthetic maternal pheromone. *Anim. Welf.* 18: 249–255.

Heinonen, M., Peltoniemi, O. and Valros, A. (2013). Impact of lameness and claw lesions in sows on welfare, health and production. *Livest. Sci.* 156(1–3): 2–9.

Hemsworth, P. H. (2018). Key determinants of pig welfare: implications of animal management and housing design on livestock welfare. *Anim. Prod. Sci.* 58(8): 1375–1386.

Hemsworth, P. H., Barnett, J. L., Coleman, G. J. and Hansen, C. (1989b). A study of the relationships between the attitudinal and behavioural profiles of stockpersons and the level of fear of humans and reproductive performance of commercial pigs. *Appl. Anim. Behav. Sci.* 23(4): 301–314.

Hemsworth, P. H., Barnett, J. L., Hansen, C. and Winfield, C. G. (1986). Effects of social environment on welfare status and sexual behaviour of female pigs. II. Effects of space allowance. *Appl. Anim. Behav. Sci.* 16(3): 259–267.

Hemsworth, P. H. and Coleman, G. J. (2009). Animal welfare and management. In: Smulders, F. J. M. and Algers, B. (Eds). *Food Safety Assurance and Veterinary Public Health Volume 5. Welfare Productions Animals: Assessment and Management Risks*. Wageningen Academic Publishers, the Netherlands, pp. 133–147.

Hemsworth, P. H. and Coleman, G. J. (2011). *Human-Livestock Interactions: The Stockperson and the Productivity and Welfare of Farmed Animals* (2nd edn.). CAB International, Oxon, UK.

Hemsworth, P. H., Coleman, G. J. and Barnett, J. L. (1994). Improving the attitude and behaviour of stockpersons towards pigs and the consequences on the behaviour and reproductive performance of commercial pigs. *Appl. Anim. Behav. Sci.* 39(3–4): 349–362.

Hemsworth, P. H., Hansen, C. and Winfield, C. G. (1989a). The influence of mating conditions on the sexual behaviour of male and female pigs. *Appl. Anim. Behav. Sci.* 23(3): 207–214.

Hemsworth, P. H., Morrison, R. S., Tilbrook, A. J., Butler, K. L., Rice, M. and Moeller, S. J. (2016). Effects of varying floor space on aggressive and cortisol concentrations in group housed sows. *J. Anim. Sci.* 94(11): 4809–4818.

Hemsworth, P. H., Rice, M., Nash, J., Giri, K., Butler, K. L., Tilbrook, A. J. and Morrison, R. S. (2013). Effects of group size and floor space allowance on grouped sows: aggression, stress, skin injuries and reproductive performance. *J. Anim. Sci.* 91(10): 4953–4964.

Hemsworth, P. H. and Tilbrook, A. J. (2007). Sexual behavior of male pigs. *Horm. Behav.* 52(1): 39–44.

Horback, K. M. and Parsons, T. D. (2016). Temporal stability of personality traits in group-housed gestating sows. *Animal* 10(8): 1351–1359.

Hutson, G. D. (1991). A note on hunger in the pig: sows on restricted rations will sustain an energy deficit to gain additional food. *Anim. Sci.* 52(1): 233–235.

Jensen, P. (1986). Observations on the maternal behaviour of free-ranging domestic pigs. *Appl. Anim. Behav. Sci.* 16(2): 131–142.

Jensen, P. and Wood-Gush, D. G. M. (1984). Social interactions in a group of free-ranging sows. *Appl. Anim. Behav. Sci.* 12(4): 327–337.

Jongman, E. C., Hemsworth, P. H. and Galloway, D. B. (1996). The influence of conditions at the time of mating on the sexual behaviour of male and female pigs. *Appl. Anim. Behav. Sci.* 48(3–4): 143–150.

Kaminski, G., Brandt, S., Baubet, E. and Baudoin, C. (2005). Life-history patterns in female wild boars (Sus scrofa): mother-daughter postweaning associations. *Can. J. Zool.* 83(3): 474–480.

Karlen, G. A. M., Hemsworth, P. H., Gonyou, H. W., Fabrega, E., Strom, A. D. and Smits, R. J. (2007). The welfare of gestating sows in conventional stalls and large groups on deep litter. *Applied Animal Behaviour Science* 105(1–3): 87–101.

Kemp, B. and Soede, N. M. (2012). Reproductive issues in welfare-friendly housing systems in pig husbandry: a review. *Reprod. Domest. Anim.* 47(Suppl. 5): 51–57.

Kilbride, A., Gillman, C., Ossent, P. and Green, L. (2009). Impact of flooring on the health and welfare of pigs. *In Practice* 31(8): 390–395.

Knecht, D., Jankowska-Mąkosa, A. and Duziński, K. (2017). Analysis of the lifetime and culling reasons for AI boars. *J. Anim. Sci. Biotechno.* 8: 49.

Knox, R., Salak-Johnson, J., Hopgood, M., Greiner, L. and Connor, J. (2014). Effect of day of mixing gestating sows on measures of reproductive performance and animal welfare. *J. Anim. Sci.* 92(4): 1698–1707.

Knox, R. V. (2016). Artificial insemination in pigs today. *Theriogenology* 85(1): 83–93.

Koketsu, Y. and Sasaki, Y. (2009). Boar culling and mortality in commercial swine breeding herds. *Theriogenology* 71(7): 1186–1191.

Kraeling, R. R. and Webel, S. K. (2015). Current strategies for reproductive management of gilts and sows in North America. *J. Anim. Sci. Biotechnol.* 6(1): 3.

Kramer-Schadt, S., Fernández, N. and Thulke, H.-H. (2007). Potential ecological and epidemiological factors affecting the persistence of classical swine fever in wild boar *Sus scrofa* populations. *Mamm. Rev.* 37(1): 1–20.

Langendijk, P., Soede, N. M. and Kemp, B. (2000). Effects of boar contact and housing conditions on estrus expression in weaned sows. *J. Anim. Sci.* 78(4): 871–878.

Lawrence, A. B. and Terlouw, E. M. C. (1993). A review of behavioral factors involved in the development and continued performance of stereotypic behaviors in pigs. *J. Anim. Sci.* 71(10): 2815–2825.

Leaper, R., Massei, G., Gorman, M. L. and Aspinall, R. (1999). The feasibility of reintroducing wild boar (Sus scrofa) to Scotland. *Mamm. Rev.* 29(4): 239–258.

Levis, D. G., Barnett, J. L., Hemsworth, P. H. and Jongman, E. C. (1995). The effect of breeding facility and sexual stimulation on plasma cortisol in boars. *J. Anim. Sci.* 73(12): 3705–3711.

Li, Y. Z. and Gonyou, H. W. (2013). Comparison of management options for sows kept in pens with electronic feeding stations. *Can. J. Anim. Sci.* 93(4): 445–452.

Lindberg, A. C. (2001). Group life. In: Keeling, L. J. and Gonyou, H. W. (Eds). *Social Behaviour in Farm Animals*. CABI Publishing, Oxfordshire, UK, pp. 37–58.

Løvendahl, P., Damgaard, L. H., Nielsen, B. L., Thodberg, K., Su, G. and Rydhmer, L. (2005). Aggressive behaviour of sows at mixing and maternal behaviour are heritable and genetically correlated traits. *Livest. Prod. Sci.* 93(1): 73–85.

Luescher, U. A., Friendship, R. M. and McKeown, D. B. (1990). Evaluation of methods to reduce fighting among regrouped gilts. *Can. J. Anim. Sci.* 70(2): 363–370.

Maes, D., Pluym, L. and Peltoniemi, O. (2016). Impact of group housing of pregnant sows on health. *Porcine Health Manag.* 2: 17.

Marchant-Forde, J. N. (2009a). Welfare of dry sows. In: Marchant-Forde, J. N. (Ed). *The Welfare of Pigs*. Springer Science and Business Media, New York City, pp. 95–139.

Marchant-Forde, J. N. (2009b). Future perspectives of the welfare of pigs. In: Marchant-Forde, J. N. (Ed). *The Welfare of Pigs*. Springer Science and Business Media, New York City, pp. 331–342.

Marchant-Forde, J. N. and Marchant-Forde, R. M. (2005). Minimizing inter-pig aggression during mixing. *Pig News Inf.* 26: 63N–71N.

Matthews, L. R. and Hemsworth, P. H. (2012). Drivers of change: law, international markets, and policy. *Anim. Front.* 2(3): 40–45.

Mauget, R. (1981). Behavioural and reproductive strategies in wild forms of *Sus scrofa* (European wild boar and feral pigs). In: Sybesma, W. (Ed). *The Welfare of Pigs* (vol. 11). Martinus Nijhoff, The Hague, pp. 3–13.

McGlone, J. J. and Anderson, D. L. (2002). Synthetic maternal pheromone stimulates feeding behavior and weight gain in weaned pigs. *J. Anim. Sci.* 80(12): 3179–3183.

McGlone, J. J. and Curtis, S. E. (1985). Behavior and performance of weanling pigs in pens equipped with hide areas. *J. Anim. Sci.* 60(1): 20–24.

Meunier-Salaün, M. C., Edwards, S. A. and Robert, S. (2001). Effect of dietary fibre on the behaviour and health of the restricted sow. *Anim. Feed Sci. Technol.* 90(1–2): 53–69.

Nalon, E., Conte, S., Maes, D., Tuyttens, F. A. M. and Devillers, N. (2013). Assessment of lameness and claw lesions in sows. *Livest. Sci.* 156(1–3): 10–23.

Olsson, A. C. and Samuelsson, O. V. (1993). Grouping studies of lactating and newly weaned sows. In: Collins, E. and Boom, C. (Eds). Proceedings of the 4th International Symposium American Society of Agricultural Engineers, Coventry, UK, pp. 475–482.

Olsson, A. C., Svendsen, J. and Reese, D. (1994). Housing of gestating sows in long narrow pens with liquid feeding: function studies and grouping routines in five sow pools. *Swed. J. Agric. Res.* 24: 131–141.

Patterson, J. and Foxcroft, G. (2019). Gilt management for fertility and longevity. *Animals* 9(7): 434.

Patterson, J., Triemert, E., Gustafson, B., Werner, T., Holden, N., Pinilla, J. C. and Foxcroft, G. (2016). Validation of the use of exogenous gonadotropins (PG600) to increase the efficiency of gilt development programs without affecting lifetime productivity in the breeding herd. *J. Anim. Sci.* 94(2): 805–815.

Pedersen, L. J., Heiskanen, T. and Damm, B. I. (2003). Sexual motivation in relation to social rank in pair-housed sows. *Anim. Reprod. Sci.* 75(1–2): 39–53.

Pedersen, L. J., Rojkittikhun, T., Einarsson, S. and Edqvist, L. E. (1993). Postweaning grouped sows: effects of aggression on hormonal patterns and estrus behaviour. *Appl. Anim. Behav. Sci.* 38(1): 25–39.

Pluym, L., Van Nuffel, A., Dewulf, J., Cools, A., Vangroenweghe, F., Van Hoorebeke, S. and Maes, D. (2011). Prevalence and risk factors of claw lesions and lameness in pregnant sows in two types of group housing. *Veterinarni Medicina* 56: 101–109.

Pluym, L., Van Nuffel, A. and Maes, D. (2013). Treatment and prevention of lameness with special emphasis on claw disorders in group-housed sows. *Livest. Sci.* 156(1–3): 36–43.

Potegal, M. (1979). The reinforcing value of several types of aggressive behavior: a review. *Aggr. Behav.* 5(4): 353–373.

Ramonet, Y., Meunier-Salaun, M. C. and Dourmad, J. Y. (1999). High-fiber diets in pregnant sows: digestive utilization and effects on the behavior of the animals. *J. Anim. Sci.* 77(3): 591–599.

Rault, J. L., Morrison, R. S., Hansen, C. F., Hansen, L. U. and Hemsworth, P. H. (2014). Effects of group housing after weaning on sow welfare and sexual behavior. *J. Anim. Sci.* 92(12): 5683–5692.

Remience, V., Wavreille, J., Canart, B. V., Meunier-Salaün, M. C., Prunier, A., Bartiaux-Thill, N., Nicks, B. and Vandenheede, M. (2008). Effects of space allowance on the welfare of dry sows kept in dynamic groups and fed with electronic sow feeder. *Appl. Anim. Behav. Sci.* 112(3–4): 284–296.

Robert, S., Matte, J. J., Farmer, C., Girard, C. L. and Martineau, G. P. (1993). High-fibre diets for sows: effects on stereotypies and adjunctive drinking. *Appl. Anim. Behav. Sci.* 37(4): 297–309.

Robert, S., Rushen, J. and Farmer, C. (1997). Both energy content and bulk of feed affect stereotypic behaviour, heart rate and feeding motivation of female pigs. *Appl. Anim. Behav. Sci.* 54(2–3): 161–171.

Rodenburg, T. B., Bijma, P., Ellen, E. D., Bergsma, R., de Vries, S., Bolhuis, J. E., Kemp, B. and Van Arendonk, J. A. M. (2010). Breeding amiable animals? Improving farm animal welfare by including social effects in breeding programmes. *Anim. Welf.* 19(S): 77–82.

Salak-Johnson, J. L., Niekamp, S. R., Rodriguez-Zas, S. L., Ellis, M. and Curtis, S. E. (2007). Space allowance for dry, pregnant sows in pens: body condition, skin lesions and performance. *J. Anim. Sci.* 85(7): 1758–1769.

Samarakone, T. S. and Gonyou, H. W. (2009). Domestic pigs alter their social strategy in response to social group size. *Appl. Anim. Behav. Sci.* 121(1): 8–15.

Séguin, M. J., Barney, D. and Widowski, T. M. (2006a). Assessment of a group-housing system for gestating sows: effects of space allowance and pen size on the incidence of superficial skin lesions, changes in body condition, and farrowing performance. *J. Swine Health Prod.* 14: 89-96.

Séguin, M. J., Friendship, R. M., Kirkwood, R. N., Zanella, A. J. and Widowski, T. M. (2006b). Effects of boar presence on agonistic behaviour, shoulder scratches, and stress response of bred sows at mixing. *J. Anim. Sci.* 84(5): 1227–1237.

Singleton, W. L. (2001). State of the art in artificial insemination of pigs in the United States. *Theriogenology* 56(8): 1305–1310.

Siswadi, R. and Hughes, P. E. (1995). The efficacy of the boar effect when conducted in a modified detection-mating area (DMA). *Aust. J. Agric. Res.* 46(8): 1517–1523.

Spoolder, H. A. M., Burbidge, J. A., Edwards, S. A., Simmins, P. H. and Lawrence, A. B. (1995). Provision of straw as a foraging substrate reduces the development of excessive chain and bar manipulation in food restricted sows. *Appl. Anim. Behav. Sci.* 43(4): 249–262.

Spoolder, H. A. M., Geudeke, M. J., Van der Peet-Schwering, C. M. C. and Soede, N. M. (2009). Group housing of sows in early pregnancy: a review of success and risk factors. *Livest Sci.* 125: 1–14.

Stevens, B. G., Karlen, G. M., Morrison, R., Gonyou, H. W., Butler, K. L., Kerswell, K. J. and Hemsworth, P. H. (2015). Effects of stage of gestation at mixing on aggression, injuries and stress in sows. *Appl. Anim. Behav. Sci.* 165: 40–46.

Stolba, A. (1988). Ethograms of the domestic pig and European wild boar. *The Libr. Assoc. Study Anim. Behav.*: 287–298.

Stolba, A. and Wood-Gush, D. G. M. (1989). The behaviour of pigs in a semi-natural environment. *Anim. Sci.* 48(2): 419–425.

Strawford, M. L., Li, Y. Z. and Gonyou, H. W. (2008). The effect of management strategies and parity on the behaviour and physiology of gestating sows housed in an electronic sow feeding system. *Can. J. Anim. Sci.* 88(4): 559–567.

Supakorn, C., Stock, J. D., Garay, E., Johnson, A. K. and Stalder, K. J. (2018). Lameness: a principle problem to sow longevity in breeding herds. *CAB Rev.* 13(23): 1–14.

Taylor, I. A., Barnett, J. L. and Cronin, G. M. (1997). Optimum group size for pigs. In: Bottcher, R. W. and Hoff, S. J. (Eds). Proceedings of the 5th International Symposium American Society of Agricultural Engineers, St. Joseph, MI, pp. 965-971.

Telkanranta, H. and Edwards, S. A. (2017). Lifetime consequences of the early physical and social environment of piglets. In: Spinka, M. (Ed.). *Advances in Pig Welfare*. Woodhead Publishing Series in Food Science, Technology and Nutrition, Duxford, United Kingdom, pp. 101–136.

Terlouw, E. M. C. and Lawrence, A. B. (1993). Long-term effects of food allowance and housing on development of stereotypies in pigs. *Appl. Anim. Behav. Sci.* 38(2): 103–126.

Terlouw, E. M. C., Lawrence, A. B. and Illius, A. W. (1991). Influences of feeding level and physical restriction on development of stereotypies in sows. *Anim. Behav.* 42(6): 981–991.

Thornton, K. (1990). *Outdoor Pig Production*. Farming Press Ltd., Ipswich, UK.

Turner, S. P. and Edwards, S. A. (2004). Housing immature domestic pigs in large social groups: implications for social organisation in a hierarchical society. *Appl. Anim. Behav. Sci.* 87(3–4): 239–253.

Turner, S. P., Edwards, S. A. and Bland, V. C. (1999). The influence of drinker allocation and group size on the drinking behaviour, welfare and production of growing pigs. *Anim. Sci.* 68(4): 617–624.

Turner, S. P., Ewen, M., Rooke, J. A. and Edwards, S. A. (2000). The effect of space allowance on performance, aggression and immune competence of growing pigs housed on straw deep-litter at different group sizes. *Livest. Prod. Sci.* 66(1): 47–55.

Turner, S. P., Farnworth, M. J., White, I. M. S., Brotherstone, S., Mendl, M., Knap, P., Penny, P. and Lawrence, A. B. (2006). The accumulation of skin lesions and their use as a predictor of individual aggressiveness in pigs. *Appl. Anim. Behav. Sci.* 96(3–4): 245–259.

Turner, A. I., Hemsworth, P. H. and Tilbrook, A. J. (2005). Susceptibility of reproduction in female pigs to impairment by stress or elevation of cortisol. *Domest. Anim. Endocrinol.* 29(2): 398–410.

Van de Weerd, H. A. and Day, J. E. L. (2009). A review of environmental enrichment for pigs housed in intensive housing systems. *Appl. Anim. Behav. Sci.* 116(1): 1–20.

van Riet, M. M. J., Millet, S., Aluwé, M. and Janssens, G. P. J. (2013). Impact of nutrition on lameness and claw health in sows. *Livest. Sci.* 156(1–3): 24–35.

Velarde, A. (2007). Agonistic behaviour. In: Verlarde, A. and Geers, R. (Eds). *On Farm Monitoring of Pig Welfare*. Wageningen Academic Press, Wageningen, the Netherlands, pp. 53–56.

Verdon, M., Hansen, C. F., Rault, J. L., Jongman, E., Hansen, L. U., Plush, K. and Hemsworth, P. H. (2015a). Effects of group-housing on sow welfare: a review. *J. Anim. Sci.* 93(5): 1999–2017.

Verdon, M., Morrison, R. S. and Hemsworth, P. H. (2015b). Short and long-term repeatability of individual sow aggressiveness. *Anim. Prod. Sci.* 55(12): 1512 (abstract).

Verdon, M., Morrison, R. S., Rice, M. and Hemsworth, P. H. (2016). Individual variation in sow aggressive behaviour and its relationship with sow welfare. *J. Anim. Sci.* 94(3): 1203–1214.

Verdon, M. and Rault, J. -L. (2018). *Aggression in Group Housed Sows and Fattening Pigs*. In: Spinka, M. (Eds.). *Advances in Pig Welfare*. Woodhead Publishing Series in Food Science, Technology and Nutrition, Duxford, United Kingdom, pp. 235–251.

Wang, L. H. and Li, Y. Z. (2016). Effect of continuous access to feeding stalls during mixing on behavior, welfare, and performance of group-housed gestating sows in different social ranks. *Can. J. Anim. Sci.* 96(3): 386–396.

Webb, N. G. and Nilsson, C. (1983). Flooring and injury–an overview. In: Baxter, S. H., Baxter, M. R. and MacCormack, J. A. D. (Eds). *Farm Animal Housing and Welfare*. Martinus Nijhoff, The Hague, The Netherlands, pp. 226–259.

Weng, R. C., Edwards, S. A. and English, P. R. (1998). Behaviour, social interactions and lesion score of group-housed sows in relation to floor space allowance. *Appl. Anim. Behav. Sci.* 59(4): 307–316.

Whay, H. R., Main, D. C. J., Green, L. E. and Webster, A. J. F. (2003). Animal-based measures for the assessment of welfare state of dairy cattle, pigs and laying hens: consensus of expert opinion. *Anim. Welf.* 12: 205–217.

Wiegand, R. M., Gonyou, H. W. and Curtis, S. E. (1994). Pen shape and size: effects on pig behaviour and performance. *Appl. Anim. Behav. Sci.* 39(1): 49–61.

Chapter 4

Optimising sow and piglet welfare during farrowing and lactation

Emma M. Baxter, Animal Behaviour and Welfare Team, Animal and Veterinary Sciences Research Group, SRUC, UK; and Sandra Edwards, Newcastle University, UK

1 Introduction
2 Welfare challenges during farrowing and lactation
3 Mitigating welfare challenges
4 Conclusions
5 Future trends in research
6 Where to look for further information
7 References

1 Introduction

One of the most contentious and persistent welfare issues in livestock production is the use of housing systems involving confinement. In the pig industry, these include gestation stalls and farrowing crates that together house sows during much of their reproductive life. The welfare detriments to the sow include physical and behavioural restriction leading to physiological and psychological stress. There is growing pressure, expressed through consumer demand, societal opinion and government legislation, to abolish confinement systems such as the farrowing crate (Eurobarometer, 2016). However, farmers have valid concerns regarding such change; they need to achieve good animal performance (i.e. high piglet survival), in systems with acceptable capital, running and labour costs, and which facilitate efficient labour routines and safeguard the operator. Originally, farrowing crates were introduced to achieve these aims, especially to reduce piglet mortality attributed to overlaying by the sow. While reductions in this type of mortality were achieved, piglets also die from other causes (Edwards, 2002) and their mortality is a persistent welfare and economic problem (Baxter and Edwards, 2018). Narrow production-focussed breeding goals exacerbating many of the pre-disposing risk factors for piglet mortality have hindered improvements in survival and further challenge both piglet and sow welfare during farrowing and lactation.

http://dx.doi.org/10.19103/AS.2020.0081.04

There are also welfare concerns for piglets relating to painful husbandry procedures such as tooth resection, tail docking and castration. Furthermore, certain management strategies to rear surplus piglets from large litters pose new risks for piglet welfare. This chapter will highlight some of these welfare challenges, before describing various mitigation strategies to optimise welfare of both sows and piglets in the farrowing environment. Various examples of best practice will be detailed, with a case study describing the development and successful implementation of a high-welfare alternative farrowing and lactation system ('PigSAFE') that attempts to reconcile the 'triangle of needs' relating to the farmer, the sow and her litter.

2 Welfare challenges during farrowing and lactation

The most prominent welfare concerns during farrowing and lactation focus on systems imposing both physical and behavioural restriction on the sow. This is an area of continued scientific and public debate. 'Naturalness' and 'freedom to express natural behaviour' are facets of animal welfare that citizens value highly (Verbeke, 2009; Sørensen and Fraser, 2010; Boogaard et al., 2011) and surveys repeatedly emphasise that sow confinement is in direct opposition to these values (Boogaard et al., 2011; Grunert et al., 2018), with respondents wanting 'spacious farrowing pens with unfixed sows' (Boogaard et al., 2011). Despite cultural differences in attitudes towards animals (D'Silva and Turner, 2012), dissatisfaction with restrictive systems appears to be globally shared. European concerns are well documented (Eurobarometer, 2016) and have resulted in total or partial prohibition of some confinement systems such as sow gestation stalls. Studies in the USA and Canada (Ryan et al., 2015) have shown that North American consumers also oppose the use of stalls and therefore it is likely that they would raise similar objections to 'fixation' of the sow at any period in her life. Brazilian citizens preferred cage-free or free-range systems based on naturalness, animals' freedom to move, and ethics (Yunes et al., 2017). Although animal welfare is a relatively new concept in China, there are indications that the consumers here are also becoming increasingly interested in improving the rearing conditions of farm livestock (You et al., 2014; van de Weerd and Ison, 2019), especially if this impacts upon food safety. Younger consumers, in particular, value environmentally friendly and organic products (de Barcellos et al., 2013; Thøgersen and Zhou, 2012; Chen and Lobo, 2012), suggesting increasing pressure on confinement systems in the future.

While progress has been made worldwide in restricting the use of stalls in gestation (Buller et al., 2018), the predominant maternity system used globally continues to be the farrowing crate. It persists because its principal reason for development persists - high levels of pre-weaning piglet mortality, along with the perception that the majority of these deaths are attributable to overlying by the sow (Edwards, 2002). The farrowing crate reduces crushing mortality by significantly limiting the sow's freedom of movement and provides a safe and cost-effective working environment for farm staff. However, piglets can still die from other causes and the farrowing crate may not be the only solution. Average piglet pre-weaning mortality levels in UK commercial outdoor systems, where the sow experiences zero-confinement throughout farrowing and lactation, are on a par with those in indoor crated systems (12.20% vs. 12.19% respectively), with stillborn mortality actually lower in outdoor units (3.3% vs. 5.8%, AHDB Pork[1]). Furthermore, there is a growing body of evidence demonstrating that the reduction in sow welfare associated with confinement in a farrowing crate actually results in impaired welfare outcomes for her piglets. Confined sows show increased piglet-directed aggression (Jarvis et al., 2006), and greater restlessness during farrowing which increases crushing risk and reduces safe access to the udder (Ocepek and Andersen, 2017). Conversely, in loose-housing environments where sow welfare is improved, piglet advantages include increased maternal carefulness (Ocepek and Andersen, 2017), improved suckling success, as measured by increased IgG levels (Yun et al., 2014), and increased weaning weight (Pedersen et al., 2011; Nowland et al., 2019).

2.1 Piglet welfare challenges

The main welfare challenges for piglets are the pre-disposing factors for mortality. Piglet mortality during farrowing and lactation averages 16-20% per litter (Baxter and Edwards, 2018). Their most vulnerable period is during the first 72h of life, when they can suffer asphyxia, hypothermia, crushing and starvation. These early causes of live-born mortality are often interlinked, and many piglets will become chilled, fail to compete at the udder for vital colostrum and lack energy to move away from the sow when she changes posture (Edwards, 2002). Piglets can also suffer from disease pre-weaning, especially if they fail to suckle colostrum from which they gain passive immunity. Thus, many piglets that die pre-weaning are likely to be exposed to some degree of either pain, hunger and/or fear and stress which could be either acute or chronic. Piglets that are born dead are unlikely to have reached a conscious state and therefore stillbirth is less of a welfare issue for the piglet itself (for a more detailed discussion see Baxter and Edwards, 2018). However, with piglet mortality of approximately 7% attributed to this condition (annual benchmarking figures from commercial herds in USA, UK and DK for 2018, sourced from PigCHAMP, AHDB Pork, SEGES), it is certainly an economic and ethical concern. Many of

1 https://porktools.ahdb.org.uk/prices-stats/costings-herd-performance/ data acquired January 2020.

the pre-disposing risk factors for mortality have been exacerbated by narrow breeding goals focussing on production traits such as hyperprolificacy and lean tissue growth rate, which reduce piglet maturity at birth and increase within-litter competition both pre- and post-natally. A particular concern for piglet welfare is the increased number of piglets born with intrauterine growth retardation (IUGR) (Edwards et al., 2019a). This not only increases neonatal mortality risk, but also has detrimental long-term consequences for physiology and behaviour.

If piglets do survive early life, they will commonly undergo a number of painful husbandry procedures including tooth resection, tail docking and (for males) castration. All of these procedures are performed within the first 7 days of life, usually within 24h post-partum. Piglets may also receive vaccinations and injections of iron, vitamins and antibiotics. While each of these procedures is carried out with some longer-term welfare justification, all cause acute pain and can give rise to additional medium- and long-term welfare detriments.

Piglets can experience further behavioural detriments as a result of barren housing environments, with no access to environmental enrichment or structurally complex surroundings that are known to have positive effects on social and cognitive development (De Jonge et al., 1996; Martin et al., 2015), adaptation to weaning (Oostindjer et al., 2011a), growth rate (Brown et al., 2015; Lawrence et al., 2018), immune responses (van Dixhoorn et al., 2016) and stress regulation mechanisms (Fox et al., 2006). Piglets reared in barren environments often develop poor social skills, display abnormal behaviours both pre- and post-weaning, and lack behavioural flexibility to cope with challenges later in life (for a review see Telkänranta and Edwards, 2018).

2.2 Sow welfare challenges

In conventional pig production, sows are moved to farrowing crates approximately five days before they are due to give birth. A typical farrowing crate has tubular metal bars running horizontally along its length, with additional bars positioned above the sow to prevent escape by jumping or climbing. It measures approximately 2.00m in length and is between 0.45-0.65m wide (Table 4). Flooring is typically fully or partially slatted (plastic or metal) to allow easy removal of waste into slurry pits below. The farrowing crate restricts movements, allowing the sow only enough space to stand up and lie down but not to turn around. The flooring and manure management system generally prohibit the provision of substrate required to help fulfil behavioural needs, such as the performance of highly motivated nest-building behaviour (Wischner et al., 2009). Nest-building is a behavioural pattern typically initiated by sows from 16-24h before they give birth. Its performance reflects a strong evolutionary survival value since, in the wild, these nests protect the young from predators and inclement weather. Nest-building is therefore intrinsically motivated; triggered by endogenous hormones (Algers and Uvnas-Moberg, 2007) and stimulated by extrinsic environmental factors in a sequence including finding a nest-site, rooting the ground, finding, carrying and arranging suitable substrate (Jensen, 1986, 1993). The time sequence in which nest-building behaviour ceases and farrowing starts is influenced by the performance of these activities and the effects that they have on hormone levels that prepare the sow for farrowing. A satisfactory phase of active nest-building occurring 3-7h pre-farrowing coincides with increases in oxytocin level. The sow then becomes less active, lies down and goes into a 'quiet phase' before farrowing begins.

A sow will attempt to perform nest-building behaviour no matter what her environment. Sows kept in farrowing crates or pens with no nest-building material redirect these behaviours to the pen equipment and perform bar-biting, manipulate the drinker and root and paw at the floor (Lawrence et al., 1994). A common misconception is that modern domestic breeds of sow do not nest-build as they have no need, experiencing no risk from predators or inclement weather when housed indoors, and because the behaviour has been bred out of them. This is incorrect and nest-building remains functionally important. Despite years of domestication and selective breeding, and despite protective, warm environments provided to piglets, the modern sow continues to be highly motivated to perform the behavioural patterns of nest-building (Jensen, 2002; Yun and Valros, 2015). When sows are unable to fully perform nest-building due to a lack of space and/or substrate, they show increased stress (Lawrence et al., 1994, 1997; Jarvis et al., 1997, 2001). This adversely affects various aspects of maternal behaviour (for a review see Yun and Valros, 2015) including the progress of farrowing, which has been shown in some (Oliviero et al., 2010), but not all (Jarvis et al., 2001; Hales et al., 2015), studies to increase stillborn mortality. Stress can activate brain opioid pathways, resulting in an inhibition of oxytocin release which is essential to the farrowing process (Lawrence et al., 1992). Opioids play an important role in preventing pain during farrowing (Jarvis et al., 1998; Ison et al., 2018) and may also promote sow passivity towards piglets during the early post-natal period (Jarvis et al., 1999). Disruption of the hormonal control of this passive response may trigger piglet-directed aggression (Ahlström et al., 2002), which is more prevalent when sows are housed in farrowing crates (Jarvis et al., 2004; Ison et al., 2015). Furthermore, reductions in positive maternal behaviours including sustained lateral lying, carefulness when changing posture, including pre-lying sow-piglet communication and responsiveness to piglets are also reported in restrictive systems (Jarvis et al., 1999; Andersen et al., 2014; Yun and Valros, 2015).

Other physiological and physical impacts of confinement on the sow include a reduced ability to thermoregulate (Quiniou and Noblet, 1999; Phillips et al., 2000; Muns et al., 2016), increased risk of hoof, leg and shoulder lesions (Boyle et al., 2002; Leeb et al., 2001) and reduced muscle mass due to prolonged reduction in movement (Barnett et al., 2001). Restrictive housing and the impairment of thermoregulation leads to reduction in feed intake (Pajor, 1998; Muns et al., 2016; Black et al., 1993), which contributes to the development of shoulder lesions/ulcers (Bonde, 2008) and adversely affects milk production and piglet growth rate (Black et al., 1993; Renaudeau and Noblet, 2001). The modern hyperprolific sow is at greater risk of developing injuries when housed in farrowing crates due to her significant increase in size; she is now substantially heavier and longer than her equivalent of 40 years ago (Moustsen et al., 2011). If used as a nurse sow to rear surplus piglets or piglets weaned at an older age, an extended lactation length also results in increased incidence of leg bursae and udder lesions (Ladewig et al., 1984; Sørensen et al., 2016). Research shows that prolonged confinement in a farrowing crate leads to chronic stress in sows (Cronin et al., 1991; Jarvis et al., 2006). This appears to arise from enforced subjection to increasingly persistent piglet attention (Weary et al., 2002) and a lack of ability to express the exploratory and foraging behaviours seen under more extensive conditions as the sow increasingly spends more time away from the nest (Hötzel et al., 2004). Piglets housed in commercial farrowing crates spend more time interacting with their dam, including suckling and udder massage, than piglets reared outdoors (Hötzel et al., 2004). Therefore, it is likely that a lack of exploratory opportunities within conventional farrowing crates also contributes to welfare problems associated with this system.

3 Mitigating welfare challenges

Genetic selection programmes targeting piglet survival traits, including increased robustness and resilience of the neonate, as well as selection for good maternal traits (e.g. udder quality - Balzani et al., 2016; maternal behaviour - Ocepek and Andersen, 2017) would help to tackle some of the welfare detriments described previously. These strategies are discussed elsewhere (e.g. Turner et al., 2018). This chapter will concentrate on optimising nutritional inputs to help combat piglet mortality, as well as describing how changes to the farrowing and lactation environment can improve welfare for both the sow and piglets, while being mindful of practical considerations for the farmers.

3.1 Optimising nutritional inputs

3.1.1 Sow nutrition

There are basic nutritional requirements for sows and piglets that should be met to ensure, at least, provision of the nutrients required for maintenance and production outputs. However, research suggests that the nutritional needs of hyperprolific sows are not always met. Breeding for very large litter sizes imposes more demands on sows, and a need to update nutrient requirements (Strathe et al., 2017). For example, Danish hyperprolific sows have required increases in levels of lysine (by 28%) and crude protein (by 14%) in their rations (Tybirk et al., 2012; Tybirk et al., 2015 - cited in Pedersen et al., 2019). There is also evidence that modern hyperprolific sows have a compromised energy status at the time of farrowing (Feyera et al., 2018), in part due to a prolonged farrowing duration. Farrowing is now reported to last, on average, 7h (Hales et al., 2015), representing an increase of 150 minutes compared to their 'standard' counterparts producing 'normal' sized litters (Oliviero et al., 2010). Protracted farrowing can lead to dystocia, increased stillbirths and piglets born with hypoxia (Björkman et al., 2017; Oliviero et al., 2010), pain for the sow (Mainau et al., 2012; Ison et al., 2016), as well as reduced colostrum yield (Hasan et al., 2019). Given this extended farrowing period, it is likely that sows are suffering from maternal fatigue. Feyera et al. (2018) demonstrated that the time delay between a sow's last meal and the onset of farrowing affected farrowing duration and the number of stillborn piglets. If the time between the two events was below 3h there was no effect on farrowing duration or stillborn mortality, but if the delay was above 6h both were significantly affected. The risk of this occurring is high, given that sows prefer to farrow undisturbed overnight when staff are absent. Feyera et al. (2018) went on to show that the energy status of the sow was depleted as farrowing progressed and recommended that sows were fed three meals a day to reduce the inter-meal interval. They also demonstrated benefits from high-fibre diets, fed from two weeks prior to farrowing and over the transition period, which provided a more sustained energy release after feeding. Feeding high-fibre diets also provides sows with greater gut fill and feelings of satiety (D'Eath et al., 2009), and also relieves constipation (Oliviero et al., 2010), and has benefits for reproductive efficiency when fed post-weaning (for a review, see Jarrett and Ashworth, 2018). Providing sows with pain relief could also mitigate some of the welfare detriments associated with problematic farrowings. Mainau et al. (2012) demonstrated that sows treated with a non-steroidal anti-inflammatory drug (NSAID - meloxicam) spent less time lying inactive on the third day following parturition. They also demonstrated benefits for piglets (increased IgG concentration in serum and enhanced pre-weaning growth) when multiparous sows were given oral meloxicam at the beginning of farrowing (Mainau et al., 2016). Similarly, Homedes et al. (2014) saw a reduction in piglet mortality when sows were given the NSAID ketoprofen.

Other nutritional interventions at the sow level can indirectly influence piglet welfare. These include late gestational feeding of supplements to reduce the impacts of low birth weight on piglet vitality (e.g. palm oil distallate - Amdi et al., 2013a; fatty acids - Rooke et al., 2001; Bontempo and Jiang, 2015;

L-glutamine - Wu et al., 2011). The evidence for improvements in colostrum yield from dietary supplements is conflicting (Theil et al., 2014), but there appears to be greater success in altering colostrum composition via sow nutrition to provide more fat and energy to the piglet (e.g. supplementary leucine metabolite - Nissen et al., 1994; increased dietary fibre - Loisel et al., 2013). Introducing fats and oils into late gestation and lactation diets as a high-energy supplement has been shown to increase sow milk yield, improve neonatal growth and development, and alter sow metabolism (Bontempo and Jiang, 2015). Peng et al. (2019) showed that sows fed a gestation diet supplemented with fat (2% soybean oil) had greater plasma prolactin concentrations, which was thought to be the mechanism benefiting their lactation capabilities. The same study demonstrated increased colostral protein concentration, with others also showing soy oil supplementation in sow lactation diets increases protein concentration in milk (Jones et al., 2002).

3.1.2 Piglets: ensuring optimum colostrum intake

Piglets are completely reliant on their mother for the provision of nutrition in the first few days of life. In order to survive they must quickly get to the udder, acquire a functional teat and suckle colostrum. This not only aids thermoregulation and the acquisition of nutrients and immunoglobulins, but also initiates gut closure, which then reduces the risk of pathogens entering the piglet's systemic circulation (Gaskin and Kelley, 1995). With large litter size, the risk of failing to acquire enough colostrum for survival, growth and development is high (Quesnel et al., 2012; Decaluwé et al., 2014) due to competition at the udder and/or compromised birth weight status (Declerck et al., 2015; Hasan et al., 2019; Amdi et al., 2013b). Piglets need to ingest a minimum of 200g colostrum for survival, but at least 250g is required for normal growth and development (Devillers et al., 2011; Hasan et al., 2019). There is also evidence that, for successful transfer of lymphocytes (B and T cells), cytokines, nucleotides, various growth factors (Bandrick et al., 2011) and other important biochemical signalling factors (Power and Schulkin, 2013), piglets need to suckle colostrum from their own mother. This has implications for optimal farrowing management, and limits the success of early nutritional interventions providing colostrum supplements (e.g. Muns et al., 2014). It is likely that the energy piglets acquire from any supplements has more effect on piglet survival than the provision of immunoglobulins (Thorup et al., 2015; Muns et al., 2015, 2017). This is because the newborn piglet's body reserves do not cover its high-energy needs (e.g. to ensure thermoregulation, locate the udder, compete for a teat and escape potentially dangerous movements of the sow). Providing piglets with extra energy shortly after birth to boost activity could be a means to promote suckling (i.e. the acquisition of sufficient amounts of colostrum) and, consequently, neonatal survival and growth. However, it is difficult to suggest one protocol describing how best to deliver supplements, of which type and to whom, as studies vary in all these factors (e.g. Muns et al., 2014, 2015, 2017; Schmitt et al., 2019a; Englesmann et al., 2019). It is likely that, to be effective, energy supplements should be administered very early in the piglet's life and in not too great a volume in any one meal. This has implications for farrowing management and how best to design a farrowing environment that allows such targeted interventions while still safeguarding sow welfare. Certainly, maximising sow lactational output should be a priority and changes to the farrowing environment to influence colostrum quality (Yun et al., 2014) and provide easy access to the udder (Pedersen et al., 2011) have demonstrated positive effects on suckling success and piglet outcomes.

It has long been established that optimum birth weight is key to piglet survival (Tuchscherer et al., 2000; Edwards, 2002; Baxter et al., 2008) and the long-term prospects for piglets born under that optimum weight are poor because of their impaired suckling success. Roehe (1999) demonstrated that 1.6 kg was optimum, with piglets below this weight suffering a three-fold increase in mortality risk. Twenty years later, selection for hyperprolificacy has increased litter size and reduced average piglet birth weight, and it appears that there has been a shift in this optimum. A multi-study project, measuring birth weight of over 4000 piglets from 394 litters at four different commercial farms, concluded that piglets born below 1.11 kg suffered a six-fold increase in mortality risk (Feldpausch et al., 2019). The shift in average birth weight accompanying hyperprolificacy has not resulted in small but robust piglets more capable of survival. It is more likely that increased husbandry interventions for large litters have influenced the survival prospects of these smaller piglets. This is supported by recent evidence suggesting that, for piglets born under 1.8 kg in weight, it is their weaning weight that can best predict their life-time growth rate (Douglas et al., 2013), indicating that such piglets can display catch-up growth if provided with the correct resources. Therefore, ensuring high milk intake during the pre-weaning period for smaller piglets may help them to reach their longer-term growth potential. Supplementary milk provision in the farrowing pen, in addition to a stable suckling environment, may aid this endeavour. However, it should be noted that these interventions concentrate on growth and survival. Long-term health and welfare implications for piglets born into large litters and with low birth weight are yet to be fully understood (Edwards et al., 2019a).

3.1.3 Large litter management and ensuring optimum milk intake

If piglets are born into a very large litter and there are more piglets than functional teats, spilt suckling can be performed to ensure all piglets get as much colostrum as possible from their own mother (Baxter et al., 2013). This involves splitting the litter into two groups, usually based on their weight and/or vitality (Kyriazakis and Edwards, 1986; Donovan and Dritz, 2000). The lightest/weakest piglets are allowed access to the udder first, while the heaviest/strongest ones are enclosed in a heated creep area or a designated box. The piglet groups are then alternated after a few successful sucklings and this continues until fostering opportunities present themselves.

Cross-fostering can ensure a more even number and weight distribution of piglets on the udder to reduce competition and, if performed correctly in the first 48h of life (after a recommended 6h minimum with their own mother), it can be successful in promoting greater survival rates and growth of low birthweight piglets (Alexopoulos et al., 2018; Douglas et al., 2014). However, if performed too early (before 6h post-farrowing), too late (after 48h post-farrowing - Price et al., 1994; Straw et al., 1998; Horrell and Bennett, 1981) or over-performed it can be disruptive, stressful and counterproductive (Robert and Martineau, 2001; Straw et al., 1998). Piglets develop a teat order within the first few days of life, where they become faithful to an individual teat and will vigorously defend it (Puppe and Tuchscherer, 1999). Teat order stability comes from cohesive group suckling behaviour (Skok and Škorjanc, 2014) which promotes growth and development of piglets. Cross-fostering can disrupt this stability, and the introduction of new piglets into a litter that has already established its teat order can lead to fights causing facial injuries to piglets, sow udder injuries and disrupted milk ejections, which in turn can impact growth and development of both fostered and resident piglets (for a review, see Baxter et al., 2013).

Simple cross-fostering cannot solve the problem of supernumerary piglets in a whole farrowing batch, which is now commonplace in hyperprolific herds. More complex strategies need to be adopted to deal with these extra piglets throughout lactation. These include nurse sow strategies (i.e. use of sows that have had their own piglets early weaned in order to free up spare teats to rear surplus piglets) or artificial rearing of surplus piglets. Both practices are reviewed in detail in Baxter et al. (2013, 2018) and have implications for both piglet and sow welfare (Sørensen et al., 2016; Schmitt et al., 2019b). For nurse sow strategies, some studies (Kobek-Kjeldager et al., 2020) but not all (Schmitt et al., 2019c) show detrimental impacts on piglet weaning weight. Schmitt et al. (2019c) did find significant weight reduction in the first week after fostering when comparing piglets moved to nurse sows with piglets left with their mother, but by weaning the differences were no longer significant. They selected the heaviest and most robust piglets for fostering and chose nurse sows with good temperament and udder quality, which are likely to be factors in how successful these strategies can be. It should be noted that one major threat to piglet health is introduced by use of nurse sows; that is, the compromise of biosecurity by breaking batch integrity in the farrowing system (Calderón Díaz et al., 2017).

Minimising exposure of suckling piglets to pathogens is an integral part of controlling pre-weaning mortality and 'all-in-all-out' (AIAO) management of farrowing batches is the key to this. As nurse sow strategies involve some form of early weaning (at least relative to EU regulations), there are also implications for piglet welfare because of the known stressors involved in early separation from the dam (Drake et al., 2008).

It can be argued that artificial rearing is worse for piglet welfare than nurse sow strategies, since it typically involves very early separation from the mother and removal to a more barren environment. Here piglets are placed in specialised enclosures, usually located in a separate room or sitting above the farrowing crate, where they will be fed milk replacer until weaning age (usually at 28 days) (Schmitt et al., 2019d). The enclosures also contain a heat lamp to ensure thermal comfort of the piglets, milk and water cups that can be activated by nudging with their snout, and solid 'creep' food. The fact that piglets are fed ad libitum in a controlled environment, where the risk of crushing is removed, is quite attractive to farmers who may not be able to implement nurse sow strategies. Additionally, some studies have shown artificially reared piglets to have higher weaning weights (van Beirendonck et al., 2015; Cabrera et al., 2010). However, others have observed short-term (De Vos et al., 2014) or sustained (Schmitt et al., 2019d) impairments of growth in artificially-reared piglets as well as behavioural abnormality.

3.1.4 Pre-weaning solid feed intake

Piglets kept under natural or semi-natural conditions begin to forage for solid feed from about two weeks of age and gradually increase their intake as the sow's milk production declines. Introducing piglets to solid feed pre-weaning can reduce weaning stress, which can otherwise contribute to compromised gut and immune function, reduced feed-intake, growth rate, health and welfare (Weary et al., 2008). While early introduction of a starter diet is typically practiced on commercial farms from the second week of life, further optimisation could consider the palatability and presentation of the feed provided. In a series of experiments by Oostindjer et al. (reviewed in Oostindjer et al., 2014) it was demonstrated that post-weaning feed intake was improved when piglets could eat together with their mothers pre-weaning, and that flavour learning was also effective in promoting growth and reducing weaning stress. In these studies, the authors designed a shared feeding station allowing piglets to eat with their mother (Oostindjer et al., 2011b) and gave sows flavoured gestational feed before presenting piglets with the same flavour in both the feed and the air post-weaning (Oostindjer et al., 2011c). Middelkoop et al. (2018) also studied the impact of altering food presentation and flavour on piglet feed intake pre- and post-weaning. As early as two days post-farrowing, one group of piglets was presented with a choice of two diets differing in production method, size, flavour, ingredient composition and nutrient profile, smell, texture and colour (i.e. dietary diversity – DD). The other group had feed that was changed in flavour every six days (i.e. flavour novelty – FN). The DD piglets were more interested in feed and had higher intake than the FN group and it was thought that the diversity stimulated feed exploration and therefore intake. The same authors also demonstrated that the type of creep feeder influenced feed intake (Middelkoop et al., 2019). A round conventional feeder was less attractive to the piglets than a play-feeder, whereby the same round design was adorned with enrichment materials to stimulate rooting and exploratory behaviour.

Providing an enriched neonatal environment can also be beneficial for pre-weaning growth rate. Oostindjer et al. (2010) showed that piglets developed feeding habits better in a loose-housed farrowing and lactation environment compared to a conventional crate system. This is supported by evidence that an enriched neonatal environment that stimulates play behaviour can improve growth of litters (Brown et al., 2015). A combination of nutritional and environmental enrichment may be particularly valuable in mitigating the effects of being born with low birth weight. Understanding the importance of environmental enrichment is one aspect of a larger discussion about providing for the biological needs of the animals to optimise their welfare and enhance performance.

3.2 Biological specifications for optimal housing design

Although the definition of animal welfare has historically separated the animal's physical health and well-being from its psychological health (i.e. the 'biological functioning' vs. the 'feelings' schools of thought – Duncan, 2005), providing for both is imperative for good animal welfare. If a 'need' is denied, this results in a negative welfare state (Jensen and Toates, 1993). Where a need is life-sustaining (e.g. food and water), prolonged failure to make provision for it would be fatal. Therefore, hierarchies of need have been proposed (Duncan, 2005) with these basic life-sustaining needs understandably prioritised. Within livestock production, behavioural needs are often demoted or ignored within the hierarchy, perhaps because of a failure to recognise (or for animal welfare scientists to communicate) their importance for biological functioning. This may actually be counter-productive for animal productivity, as it is well established that performance of species-typical behaviour contributes to an animal's biological fitness (Hamilton, 1964a,b). Behaviours (e.g. nest-building) that have persisted despite centuries of domestication and selection pressure for production traits are those that remain biologically significant.

Thus, biological needs should include physiological, physical and behavioural necessities, and providing for these should promote welfare, health and performance. Meeting biological needs in the farrowing and lactation environment requires the identification of what is most important for both piglets and sows, and then determining the best way to make provision. When designing a new housing system, this means understanding that the animals themselves are active participants in the process and not just passive receivers of the husbandry systems they are put into, as typically practiced since the industrialisation of farming (Bos and Koerkamp, 2009). Many of the health, welfare and environmental problems facing livestock sectors today arise from the dominant post-war policies of modernisation, which stem, understandably, from the focus on ensuring domestic food security at an affordable price. Re-designing the environment with the occupants in mind should redress the balance.

In the case of farrowing and lactation, Baxter et al. (2011a) attempted this approach by first describing each physiological and behavioural need, as indicated by sound biological evidence, and then determining which of these needs are sensitive to the physical environment. They subsequently translated this information into design criteria for a maternity unit built to the biological specifications of the pigs (as summarised in Table 1). This initial task formed the basis for re-designing the farrowing and lactation environment to maximise welfare (The PigSAFE project – see Section 3.4).

3.2.1 Space

The sow needs space for finding and creating a nest, giving birth (parturition), suckling and interacting with her piglets, feeding, drinking, defaecation as well as space for re-integration with her social group. The piglet's spatial needs include space in which to be born, search for the udder, find a teat and suckle. It needs space to rest and keep warm, interact with its siblings and mother, as well as to avoid her during posture changes, and to grow and develop. With current genetics, this means providing space for at least 14 piglets (or the likely maximum number of piglets a sow could support) until weaning at 28 days post-partum.

Quantifying the spatial needs of the animals involves measuring the size of the modern-day domestic sow and her piglets (at weaning) based on the 95th percentile (i.e. largest) individuals. Then, determining both the static and dynamic space requirements for different postures and posture changes, turning around, piglet gathering, separation of functional areas, etc. (Baxter and Schwaller, 1983; Petherick, 1983, Gonyou et al., 2006). Allometry, based on the equation $A = k*W^{0.667}$ (where A=area, k= a constant and W= live-weight), can be used to estimate the space an animal occupies as a consequence of its mass (Baxter and Schwaller, 1983). This equation is very general and assumes all animals are the same shape and that this is consistent over time. In general,

Table 1 Summary of sows' and piglets' biological specifications for a farrowing and lactation system and estimated 'values' required to meet their needs. More detailed descriptions are given in Baxter et al. (2011a, 2018) and any new information published since these manuscripts has been incorporated. Where no new information has materialised and values for components cannot be given, 'further research' is noted

Component of system	Sows	Value of required specification at minimum level to meet needs	Piglets	Value of required specification at minimum to ideal level
Space	Increased activity for nest-site seeking	$4.9m^2$	Parturition	$2.79m^2$
	Hygiene – separate dunging space from feeding and lying area	Separate dunging area from nest and feed sites.	Udder access for suckling throughout lactation	$2.79m^2$
	Feeding and foraging	Separate feeding area from nest and dung sites	Protection, safe lying area for parturition and nest-occupation	Separate space inaccessible to the sow e.g. $0.8m^2$ per 10–12 neonates
	Turn-around nest space for piglet inspection and gathering behaviour	Floor space = $2.44m^2$, planar space = $3.17m^2$. Further research needed		
	Lateral lying and parturition	$2.79m^2$	Protected lying area during lactation	$0.96m^2$ for 14, four-week-old piglets
	Thermal comfort via posture changes	$2.44m^2$	Area for feed trough to introduce starter diet and area for supplying supplementary nutrition/energy (separate from the sow)	Provide in the creep, interacts with above
	Nest-departure	Separate area from nest site		
	Gradual separation from piglets and sow-controlled nursing	Separate space inaccessible to piglets	Hygiene	Separate area for dunging, interacts with flooring
	Social contact with other sows	Separate space inaccessible to piglets to allow contact between sows, but if sows fully reintegrated before weaning larger space to allow body language assessment and/or fighting to establish dominance – further research needed to determine minimum space per sow.		

Substrate	Nest-building – carrying and manipulating	2 kg long-stemmed straw	Foraging, nutritional development	Earth-like materials (e.g. peat)
	Complete nest	2 kg long-stemmed straw and branches	Enrichment, social and cognitive development	Further research needed on quantity. Novelty requires fresh input daily. Complex materials (e.g. branches) preferred.
	Udder comfort	Further research needed, interaction with floor properties	Thermal comfort during parturition	2.5cm of straw, interacts with flooring
	Thermal comfort during nest-building and parturition	2 kg long-stemmed straw in nest site – interaction with flooring and room temperature.	Physical comfort	Further research needed, interacts with thermal comfort and flooring properties
	Foraging material	Further research needed	Protection	Deep bedding – 10–12cm, interacts with flooring
Walls	Enclosure/Isolation of nest	3 solid-sided walls (cul-de-sac)	Protection from sow posture changes	Sloped wall or protection bars
	Visual and physical contact with non-litter pigs	Vertical barred area with void wide enough to allow at least nasal contact between pigs	Social contact (visual and physical)	Vertical barred area
		Solid at base with separation between pens	Hygiene	Solid walls (at least at bottom of penning) separating other litters
	Supported posture changes	Solid sloped or vertical walls	Thermal comfort	Solid walls with thermal resistance properties to limit heat loss via radiation – interacts with substrate and flooring
	Lack of disturbance	Further research needed		

(*Continued*)

Table 1 (*Continued*)

Component of system	Sows	Value of required specification at minimum level to meet needs	Piglets	Value of required specification at minimum to ideal level
Flooring	Nest-building - digging, rooting and hollowing	Malleable (e.g. earthen) or solid to accommodate deep substrate	Thermal comfort during parturition and first 24h of life	High thermal resistance – e.g. rubber matting or deep substrate or under-floor/localised heating
	Nest-building and parturition	Solid to accommodate substrate	Thermal comfort during lactation	High thermal resistance – e.g. rubber matting or deep substrate (see above) or under-floor/ localised heating (see general)
	Thermal comfort during nest-building, parturition and lactation	Temperature differentials in separate areas allowing choice. High thermal resistance e.g. rubber matting or deep substrate. Low thermal resistance e.g. metal.	Physical comfort – avoiding injury, promoting suckling behaviour Protection from fatal crushing by the sow	Solid flooring with minimal abrasiveness and well-maintained (e.g. rubber matting or specialised screed with non-slip properties), interacts with substrate
	Physical comfort – avoiding injury, promoting suckling behaviour	Non-slip surface e.g. rubber matting or plastic-coated metal Minimal abrasiveness (interacts with substrate). Solid to avoid teat injuries	Hygiene	Slatted flooring with void width no more than 10mm and rounded edges. Interacts with temperature (see general)
	Hygiene	Slatted area Gradation of floor with slope away from lying area. E.g. 10% slope for drainage		

General	Thermal comfort	Ambient temperature 12–22°C, interactions with substrate and flooring	Health treatment for injuries, vaccines, etc.	Safe area for handling required, interacts with space
	High feed intake		Promote weaning, reduce nutritional stress and encourage increased feed and water intake	Suitable solid food, inaccessible to the sow – interacts with space and substrate. Provide feed tray and sufficient space to allow social facilitation
	Low light in nest-site	See space and thermal comfort	Thermal comfort	Localised heat source set at thermo-neutral temperature (e.g. 34°C at birth) – interacts with substrate
			Hygiene	Temperature differentials to encourage dunging outside of nest site – interacts with flooring

because these equations err on the side of caution, they are still relevant today for calculating static space requirements (Gonyou et al., 2006). However, many systems are so fixed in their pen dimensions that they have provided no flexibility for coping with changes in the animal's size and productivity over time, or changes to husbandry practices that impact on space. The farrowing crate is such a system. Because the modern-day sow is substantially larger than her counterparts of the 1980s (Moustsen et al., 2011) and she rears a larger litter for a longer lactation length, many farrowing pen designs do not allow sufficient space for hyperprolific litters (Pedersen et al., 2013). Piglets can neither rest together in a thermally comfortable area nor have unobstructed suckling. EU Council Directive 2008/120/EC states, 'When a farrowing crate is used, the piglets must have sufficient space to be able to suckle without difficulties'. The length of a four-week-old piglet (approximately 0.5 m - Moustsen and Poulsen, 2004), therefore needs to be provided at each side of the crate to avoid fighting caused by blocking of piglets' access to preferred teats. Suckling success is promoted when such constraints are not in place, with greater milk let-down reported when sows show better udder access in loose housing (Pedersen et al., 2011). The detailed body measurements provided by Moustsen and colleagues allow calculations for the necessary creep dimensions, suckling space and space for fixtures and fittings within a pen to prevent injury (e.g. creep bar widths). These data can then be applied to determine the detailed design dimensions for any new farrowing and lactation system (Table 1).

3.2.2 Substrate

Although there is strong evidence that space is more important than substrate for allowing the behavioural expression of nest-building (Jarvis et al., 2002; Cronin et al., 1994; Hartsock and Barczewski, 1997), substrate is still very important for sows and piglets. Given that the majority of commercial sows are housed in conventional crated environments (>90% in selected European countries, >85% in Australia and New Zealand - Baxter and Edwards, 2016[2]), it is important to investigate more fully which substrates are of relevance to the sow before and after farrowing in order to improve her welfare, maternal behaviour and the welfare of her piglets. In the periparturient period, the sow needs to collect and arrange suitable substrate for nest-building, and to achieve thermal and physical comfort for herself and her piglets. Substrate also facilitates foraging and nutritional development, providing enrichment which is important for the social and cognitive development of piglets.

2 Proceedings of Free Farrowing Workshop 2016, Belfast NI (eds, Baxter, E. M. and Edwards, S. A.). Proceedings available at https://www.freefarrowing.org/info/2/research/45/free_farrowing_workshops.

The type of substrate provided should have properties which satisfy the biological needs. For nest-building, the material needs to be suitable to provide feedback to the sow signalling that the nest is complete. Without this, some sows may continue to be motivated to nest-build even during farrowing (Thodberg et al., 1999; Damm et al., 2000), which constitutes a risk for the newborn piglets. Substrate manipulation to allow arrangement, as well as a sense of enclosure, appears to give a sense of completion and, in preference tests, sows chose pre-formed nest-sites that also offered a sufficient quantity of straw to satisfy nest-building behaviour (Arey, 1992). 'Sufficient quantity' is a rather open description but it is one found in the European legislation outlining provision of enrichment material at all stages of production. Council Directive 2008/120/EC states that 'pigs must have permanent access to a sufficient quantity of material to enable proper investigation and manipulation activities'. For nest-building, studies have suggested as little as 2 kg of long-stemmed straw to be sufficient (on solid floors – L. J. Pedersen *personal communication*), whereas voluntary use of as much as 255 kg of mixed substrate has been reported (outdoor semi-natural systems - Zanella and Zanella, 1993). Practicality is important and interactions with the flooring and waste management system are factors requiring consideration. EU legislation says that all sows must have material for nest-building but provide possibilities for derogation, stating that 'unless the slurry system makes provision unfeasible'. Straw is widely considered as the gold-standard material for its manipulable as well as thermal properties (Rosvold et al., 2018; Mount, 1967; Wathes and Whittemore, 2006). Westin et al. (2014) demonstrated that, if used strategically, a slurry system could cope with large quantities of straw at the time of nest-building. They provided 15–20 kg of long-stemmed straw when sows moved in to the farrowing accommodation and did not subsequently remove (unless soiled) or add to this, allowing the straw to gradually fall through the slatted part of the pen and be removed with the excreta. Alternative nesting materials have also been studied (e.g. cloth tassels - Widowski and Curtis, 1990; burlap sacks – Bolhuis et al., 2018; Plush et al., 2019; lucerne hay - Edwards et al., 2019b; peat - Rosvold and Andersen, 2019; newspaper, sawdust, shredded paper - Swan et al., 2018). Something as simple as a cloth tassel tied to pen fittings that animals can pull, tear and manipulate may provide welfare benefits, even though the material cannot result in the building of a suitable nest. This is particularly relevant for animals kept in crates, as most of the tassel remains attached to the front of the crate where sows can continue to access it. In contrast, substrate on the floor often gets pushed to the rear of the crate or out of reach during nest-building, with potentially frustrating consequences. Swan et al. (2018) specifically looked at nest-building in crates and compared six materials, either attached to the farrowing crate or in amounts of 1–2 L placed in front of the sow. These included point-source objects offered on the ground or to the side, wood shavings, straw, shredded paper and whole newspaper. Sows showed nest-building behaviour with all materials, but a functionality assessment saw straw, wood shavings and newspaper tested further. There were different benefits for the selected options: the newspaper group performed more nest-building and fewer bar-biting activities, while piglet mortality during the entire lactation period was lower in the straw group than the other groups. This study demonstrates benefits to the piglets of maternal nest-building and is supported by studies, both in crates and loose-housing, that show providing for nest-building can improve suckling success and growth rate in piglets (e.g. crates - Edwards et al., 2019b; loose housing - Yun et al., 2014; Pedersen et al., 2011; Plush et al., 2019) and can reduce stillbirths (e.g. Rosvold and Andersen, 2019; Edwards et al., 2019b). Edwards et al. (2019b) also showed that sows in crates continued to interact with provided substrate (in the form of lucerne hay) throughout lactation, suggesting they still found the enrichment rewarding after the nest-building phase.

Enriching the environment throughout lactation will also benefit the piglets. Studies have shown that providing substrates, point-source objects, opportunities for play and socialisation improves cognitive and social skills (Martin et al., 2015), growth rate (Brown et al., 2015), gut health (Oostindjer et al., 2010) and immunity (van Dixhoorn et al., 2016), and can reduce weaning stress. Which material and how much to provide to best achieve these benefits is still difficult to quantify but, as a general rule of thumb, it appears that diversity of the environment will accrue the most benefits to different aspects of the piglets' welfare. It is fast becoming recognised that an optimised gut microbiota can enhance disease resistance (Patil et al., 2020) and, as efforts are being made to reduce antibiotic use, finding alternative methods to achieve such outcomes are being sought. Increasing diet complexity, by introducing foraging opportunities to piglets pre-weaning, can increase microbial diversity and the presence of beneficial microbes, reducing the risk of post-weaning diarrhoea. Reduction of stress is also important for healthy gut development, with early-life stressful experiences demonstrated, in rodents, to lead to dysfunction of the intestinal barrier (O'Mahony et al., 2009). This is relevant for piglets, which may experience a number of acutely stressful husbandry procedures within the first few days of life.

3.2.3 Walls

The benefits of suitable walls in a farrowing environment include imparting a sense of enclosure for nest-building, providing a solid surface to support sow posture changes and thereby providing protection for piglets and also providing opportunities for both seclusion from conspecifics and interactions with them. When sows choose nest-sites they prefer a 'cul-de-sac' arrangement, providing protection on three sides, and a nest-entrance with a view, presumably for vigilance against potential threats. The degree of enclosure afforded by the nest-site will influence disturbance, and disturbance is known to adversely affect oxytocin levels (Lawrence et al., 1992). Providing sows with a supportive surface to lean against during their descent when lying reduces crushing (Baxter, 1991, Marchant et al., 2001; Damm et al., 2006) and sows prefer sloped or vertical walls to lie down against rather than farrowing rails (Damm et al., 2006). Manufacturers of outdoor farrowing huts or arks recognise that having angled walls is protective for piglets, as they prevent the sow from fully contacting the lower walls when changing posture and therefore trapping piglets. Recreating these protective elements in indoor environments requires quantification of the best dimensions/angles for the sloped walls to provide piglet escape zones, while being an attractive prospect to encourage supported lying behaviour by sows. Moustsen (2006 – cited in Pedersen et al., 2013) provided these details and alternative systems have successfully incorporated such features (Baxter et al., 2015).

While enclosure is important in the periparturient period, as lactation progresses nest-departure and re-integration of the sow and litter into the social group occur in natural conditions when the piglets are about two weeks old. Sows leave the nest to increase their foraging territory and increase feed intake during the metabolically demanding lactation phase. Piglets also benefit from this foraging activity. At the time of reintegration, prior familiarity with other group members minimises aggressive interactions that might impair fitness. Under natural or semi-natural conditions, pigs maintain established groups where aggression is regulated via an 'avoidance order', with specific behavioural patterns reducing risk of attacks by dominant individuals (Jensen, 2002). When a sow leaves the group to farrow, the longer she remains isolated, the more challenging it is for her and her litter to re-integrate. Early socialisation with other litters during lactation benefits piglets by reducing the impact of weaning (Pajor et al., 1999; Hessel et al., 2006), particularly the effects of mixing aggression, and improving piglet social skills post-weaning (Morgan et al., 2014). Alternative farrowing systems which include group housing or multisuckling allow this more natural reintegration; this was popular in the early 1980s, especially in Sweden. They are a cheaper alternative to individual housing throughout lactation and, if the benefits outlined above can be realised, they seem like a sensible choice. However, in general, group housing systems have returned poor production figures for piglet mortality (see Baxter et al., 2012 for a review). More recent research in Germany (Bohnenkamp et al., 2013), the Netherlands (van Nieuwamerongen et al., 2015) and Australia (Verdon et al., 2019; Greenwood et al., 2019) has revisited these systems, partly as a way to increase lactation length and reduce weaning stress, and thus reduce the post-weaning reliance on antibiotic interventions to maintain piglet health. However very careful management is required to ensure success of such systems (van Nieuwamerongen et al., 2014; Thomsson et al., 2016). Facilitating some form of social contact between different sows and litters in individual housing systems involves designing partitions with areas to allow fence-line contact between neighbours. Although this does not allow for full reintegration before weaning, it has been shown to benefit post-weaning outcomes such as reducing aggression upon mixing (Martin et al., 2015). Whether walls are solid or barred will also impact on dunging behaviour (Moustsen and Jensen, 2008) and the microclimate within the pen. Providing cooler (better ventilated) areas will reduce the risk of heat stress, particularly for sows during lactation (Muns et al., 2016), although thermal comfort also depends on substrate provision and floor type.

3.2.4 Flooring

Suitable flooring is one of the topics that requires significant attention in pig production systems, especially when it comes to providing for biological needs. Flooring qualities for locomotion and lying behaviour are based on many different aspects: friction, abrasiveness, hardness, surface profile and thermal properties (Lensink et al., 2013). For all livestock, floors should provide physical and thermal comfort when lying, should not lead to injury or slipping when standing and walking, but should also not be too abrasive. To promote good hygiene, they should facilitate easy cleaning, which will reduce transmission of infectious diseases (Lensink et al., 2013). There are consequently many trade-offs to consider; deciding on the optimal flooring in the farrowing and lactation environment is particularly complicated because of the diverse needs of the sows and piglets at different stages (i.e. nest-building, parturition, lactation) and because flooring interacts with many other design specifications when trying to meet those needs. For nest-building, a sow would ideally want a malleable floor to dig a hollow by rooting the ground. While this is achievable with soil under natural conditions, a proxy indoors involves provision of plentiful substrate, which means that the flooring needs to be solid to retain this substrate. This is also the case for substrate provided as enrichment material and to improve comfort by reducing the risk of development of shoulder ulcers in sows during prolonged lying bouts (Rolandsdotter et al., 2009; Rioja-Lang et al., 2018) and leg injuries in piglets during suckling (Zoric et al., 2009; Lewis et al., 2005; Mouttotou and Green, 1999). However, excreta build-up on solid floors can increase the risk of disease, and self-cleaning slatted systems tend to promote good pen hygiene (as well as reducing labour) (Rantzer and Svendsen, 2001). Good hygiene can be further promoted by clear differentiation between lying areas and areas designed for defaecation. Providing part-slatted flooring in pens helps to designate these areas, and also to provide temperature differentials as slatted floors have different thermal properties to solid floors. If given the correct cues, pigs will actively separate their dunging, resting and feeding sites (Randall et al., 1983). They will also seek corners in which to defecate (Wiegand et al., 1994), which has implications for pen design.

If plentiful substrate is not provided, soft surfaces, such as rubber matting, may allow some deformation of the surface. This reduces contact pressure and mechanical stress on the body, and can aid healing if sows have shoulder sores (Zurbrigg, 2006). Rubber mats can also reduce carpal lesions of the piglets (Courboulay et al., 2000). However, the quality and abrasiveness of the mat are important to prevent excessive wear and tear, as well as to prevent slipping. Some studies suggest rubber flooring is less slippery (Boylet et al., 2000), but others report a 'film of slurry' when mats were used in farrowing houses (Calderón Díaz et al., 2013) which will increase the slip risk for sows and impair hygiene. Flooring is a very difficult system component to perfect and, when designing a new indoor system, providing different floor types in different functional areas within the same pen is likely to be the best way to meet all biological needs.

3.2.5 Thermal comfort

The thermo-neutral zones of sows and newborn piglets are markedly different. A sow's evaporative critical temperature is the upper limit of the thermo-neutral zone and represents the temperature at which evaporative heat loss begins to increase and heat stress develops. This depends on a variety of factors associated with the sow's ability to thermoregulate and is heavily influenced by components of the housing system and ambient temperature. It has been postulated that peri-parturient sows will actively choose a nest-site based on the most thermally resistant flooring (Pedersen et al., 2006). In a study by Hunt and Petchey (1987), sows given the choice of either a concrete floor or a rubber floor with varying amounts of straw chose the mat with the greatest amount of straw. However this choice might have been dictated by comfort or substrate availability, as subsequent studies investigating nest-site choice have shown no difference between heated or room-temperature flooring with the same properties and substrate availability (Pedersen et al., 2006; Baxter et al., 2015). It is generally thought that peri-parturient sows can tolerate slightly higher temperatures than their typical thermal comfort zone during lactation of 10–20°C (Black et al., 1993) and it could be postulated that this tolerance enhances biological fitness as piglets are extremely cold sensitive. This tolerance may be limited, however, as Muns et al. (2016) found that sows in farrowing crates experiencing 25°C around parturition altered their postural behaviour relative to those kept at 20°C. They reacted to the thermal challenge with higher respiration rate, but both their rectal and udder temperatures were elevated, indicating that they were not able to compensate for the higher ambient temperature. When sows are loose-housed they have more possibility to regulate their temperature by altering their postures and choosing different areas in their environment. A study investigating three room temperatures (15°C, 20°C and 25°C) for lactating sows kept loose in pens with a partly slatted concrete floor showed that sows used the cooler slatted floor for behavioural thermoregulation by resting in this zone in between daily activity bouts (Malmkvist et al., 2012). During lactation, sows are more at risk from heat stress due to the higher metabolic activity associated with elevated feed intake and milk synthesis (Williams et al., 2013). Sows unable to thermoregulate will experience hyperthermia/heat stress, which can impact milk production and therefore piglet growth rate and survival. Thus, loose housing with different thermal zones may increase sow thermal comfort and positively affect piglet growth (e.g. Pedersen et al., 2011; Oostindjer et al., 2010).

The thermal challenge for piglets is avoiding hypothermia. At the time of birth they experience a sudden drop in ambient temperature (of approximately 15–20°C – Herpin et al., 2002). Consequently, the lower limit of the thermoneutral zone (the lower critical temperature – approximately 34°C Mount, 1968), is exceeded, resulting in chilling. It is impossible to keep the piglets in their thermoneutral zone by raising room temperature, as heat stress for the sow could be fatal. Thus, providing a suitable microclimate in a designated piglet zone is a necessity, and substrate, flooring and the degree of enclosure in the nest-site all influence the effectiveness of this. Deep bedding slows heat loss, having a thermal resistance 11 times greater than that of concrete slats and 22 times greater than solid, wet concrete flooring (Wathes and Whittemore, 2006). Mount (1967) demonstrated that piglets in contact with a concrete floor lost 40% more heat than those in contact with 2.5cm of straw. This is, in part, why straw management is such an important factor in outdoor pig production. In the absence of deep bedding, an artificially heated creep area of suitable size (Table 1) can provide an effective microclimate, provided that piglets can be attracted to use this area at an early age.

3.2.6 General system components

Although there are some advances to be made in applying nutritional strategies (see Section 3.1), it is assumed that the fundamental physiological requirements for food and water are generally well catered for. However, when sows are loose-housed, placement of feeders and drinkers can influence the spatial organisation of behaviour, especially excretory behaviour (Moustsen and Jensen, 2008; Andersen and Pedersen, 2011; Ocepek et al., 2018), and could influence farrowing location (Baxter et al., 2015). Sows will typically eat and then turn away from their feeding area to defecate, often orientating themselves so that their head faces away from their feed site (Moustsen et al., 2007; Moustsen and Jensen, 2008; Andersen and Pedersen, 2011). Incorporating barred partitions between pens in the designated dunging area, and a solid dividing wall between designated dunging and nesting sites, can improve pen hygiene (Moustsen and Jensen, 2008). However, there is large individual variation in dunging patterns in loose-housed sows (Bøe et al., 2016) and maintaining hygiene continues to be a subject needing research activity (Hansen, 2018). This is not just because good pen hygiene reduces the risk of disease transmission, but it would also encourage greater acceptance of loose-housing systems. Cleaning out farrowing pens and providing new bedding was estimated to require 33% of total daily work time for Swedish farmers on 35 commercial pig farms (Mattsson et al., 2004). The stockperson is a key contributor to animal welfare and should be considered as an important participant in system design. Stockperson intervention to maximise piglet survival has increased in the wake of superprolific breeding programs, where the number of piglets born regularly surpasses the sow's ability to rear them. Large litter size management involves handling of piglets and sows, cross-fostering, split-suckling, establishment of nurse sows, and provision of supplementary nutrition (Baxter et al., 2013, 2018). This represents a challenge in any system but could be regarded as more challenging in loose-housing conditions (Rosvold et al., 2017). Facilitating stockperson interventions while safeguarding the welfare of the sow is therefore an important element of system design. Several Danish studies have evaluated loose-housing systems and pinpoint large litter size as the main risk factor in allowing sows to be loose all the time (i.e. zero-confinement) (e.g. Hales et al., 2015). As such, there has been sustained interest in developing 'alternatives' that retain some form of restraint of the sow (i.e. temporary crating).

The increased research and development activity in alternative systems is dominated by temporary crating designs (for reviews see Baxter et al., 2018; Glencorse et al., 2019; www.freefarrowing.org). Temporary crating systems (e.g. Table 4) facilitate loose-lactation but cannot support all the biological needs of the sow, especially her functionally important nest-building behaviour (Hansen et al., 2017). These systems are therefore a compromise; they are generally the most economically attractive choice out of all the alternatives (bar outdoor production – Guy et al., 2012). They are often built on the same spatial footprint as a conventional crate system, with fully or partly slatted flooring, and maintain most of the stockperson advantages that conventional systems impart (i.e. easy and safe intervention, easy dung removal). Where pledges to phase out farrowing crates have been made (e.g. Denmark and Austria - Hansen, 2018; Heidinger et al., 2018), there has been great investment in evaluating these compromise systems, including identifying when sows should be let out of their crates post-partum. The 'critical time window' for piglet survival has been suggested as four days post-partum (Heidinger et al., 2018). It can, of course, be argued that any reduction of restraint is an improvement in animal welfare, but temporary crating systems are far from optimal for sows or piglets. Many of the benefits of allowing for more satisfaction of nest-building motivation are not realised for the sow or her piglets, and ensuring the sow is released after a short period post-partum will be based on the stockperson's discretion. There can also be spikes in piglet mortality at this time if opening protocols are sub-optimal (King et al., 2019), or if pens are operated as zero-confinement when they have been designed for temporary crating. This is repeatedly seen in studies where the same design of temporary confinement system is compared when kept in an open position all of the time versus varying lengths of confinement (e.g. Hales et al., 2014; Condous et al., 2016; Lohmeier et al., 2020). As these systems rarely have all the design features that promote good maternal behaviour (as described above), it is almost inevitable that they will fail when the sow is completely free. It is understandable that temporary crating options may be perceived by farmers as a good introduction to loose-housing, with the crate offering an 'insurance' policy to return to conventional methods if there is no uptake either by staff or the supply chain. However, even temporary crating systems represent an investment - the most basic ones (i.e. those occupying the same footprint as conventional systems) cost ~1.6 times more than farrowing crates (Guy et al., 2012) and such investment might be costly if not future-proofed for consumer demands. This was evident in the poultry industry, where the introduction of the 'enriched (or furnished) cage' to replace battery cages did not fully appreciate public perception that 'a cage is still a cage' and therefore a push for alternatives which resulted in a surge in the free range market and development of barn egg and aviary alternatives.

Thus, it is still important to determine whether optimal welfare can be attained in a farrowing and lactation system where the needs of all stakeholders are taken into account (i.e. sow, piglet, stockperson and consumer).

3.4 Case study: PigSAFE – a zero confinement farrowing and lactation system

A project commissioned by the UK government aimed to develop a suitable alternative to the farrowing crate that provides for the maximal sow and piglet welfare that can be achieved under commercial conditions. This required the re-designing of a system that could reconcile a 'triangle of needs' relating to the farmer, the sow and her litter. However, as the needs of the litter and the farmer are often closely aligned, both prioritising improved piglet survival and growth, the main conflict to be resolved lies between the sow and the farmer. This requires methods to provide the appropriate level of environmental enrichment to meet the biological 'needs' of the farrowing sow that are also consistent with good piglet survival, and other management and business constraints. A literature review was undertaken to establish/confirm the biological principles (including ethological understanding) underlying sow and piglet needs in the perinatal period. From this, a list of biological specifications (or design criteria) for zero-confinement farrowing systems was derived (as discussed in summary above and in Table 1, and in detail in Baxter et al., 2011a). A second critical review was undertaken to gather information on all known alternative farrowing and lactation systems (including outdoor systems), to identify the extent to which each system addressed biological specifications and its practical success (Baxter et al., 2012; updated in Baxter et al., 2018). From this platform, and in collaboration with representatives from industry, non-governmental organisations, biologists, economists and engineers, a prototype system was developed - PigSAFE (Piglet and Sow Alternative Farrowing Environment). This was tested on two research farms to 'model' potential implementation scenarios (i.e. conversion of existing farrowing crate accommodation (Farm A) and a new-build (Farm B)) and to rapidly de-bug obvious problems.

3.4.1 The PigSAFE design

PigSAFE is a 'designed pen'. This terminology is used for a pen that has different areas to fulfil different functions (Baxter et al., 2012, 2018) and specific pen features designed to stimulate good animal behaviour (i.e. good maternal behaviours: correct farrowing location, careful lying behaviours, appropriate interactions; good hygiene: use of provided dunging sites). The PigSAFE designs on Farm A and B are shown in Figs 1a and 2a, respectively. They have similar features, but Farm A has a narrower dunging passage and is built above a slurry pit with plastic slotted and slatted flooring. Both pens have a nest area, with solid flooring to allow provision of nesting material, and sloping walls to control the stand-to-lie posture changes of the sow and lower the risk of piglets being trapped and killed. A heated creep area has easy access from the nest for the piglets and from the passageway for stockperson inspection. A separate slatted dunging area is bounded by walls with barred panels to adjacent pens, to discourage farrowing outside the nest and allow sow-to-sow visual and oral-nasal contact. A sow feeding crate (solid-sided) is included at one side of the pen, where the sow can be locked in to allow safe inspection or treatment of the piglets, but is not wide enough to allow the sow to be locked

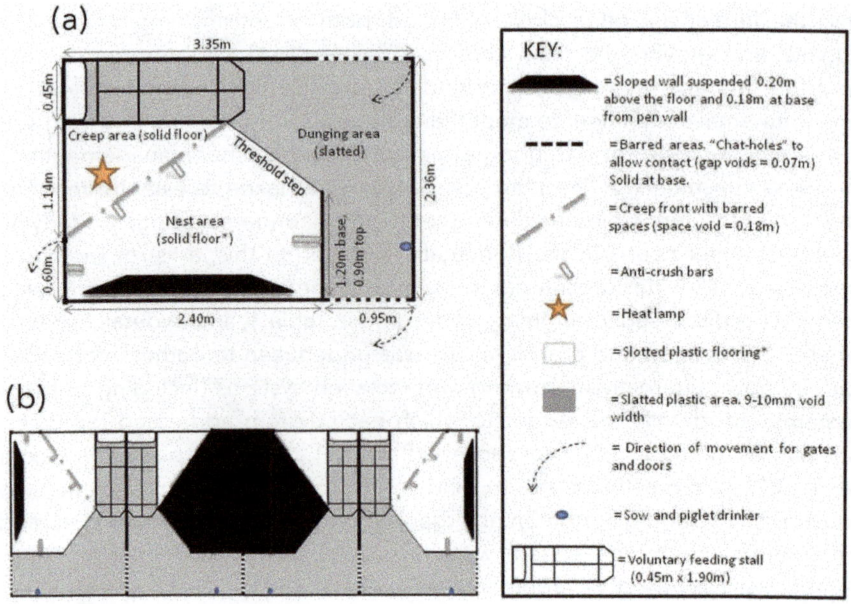

Figure 1 (a) Prototype PigSAFE pen (not for scale) and (b) Four pens were constructed in one room: in each room, two pens were provided with sound proofing material and a temporary roof fitted (height 1.2m) for the QUIET treatment.

in for farrowing. Farm B's new-build design incorporates additional dynamic features such as the ability to open up the nest area after a designated time (~7 days post-farrowing) to improve hygiene and simulate nest departure. The pen is partially slatted, with a scraper system under the metal slats in the dunging area.

The first iterations of both designs tested some gaps in the knowledge regarding the effects of a 'quiet' nest-site and quantification of substrate needed in the nest-site (Farm A - Edwards et al., 2012), as well as the effects of under-floor heating in the nest-site and quantification of nest size (Farm B - Baxter et al., 2015).

3.4.2 Initial trials

A concern when sows farrow loose is that they will not farrow in the desired location. In the PigSAFE design, that location was the nest-site where substrate was present and sloped walls provided protection and get-away sites for piglets, as well as a large heated creep. Initial trials were focussed on determining whether we could provide the correct stimuli to achieve the desired farrowing location and acceptable piglet survival (defined as comparable to farrowing crates and national herd averages).

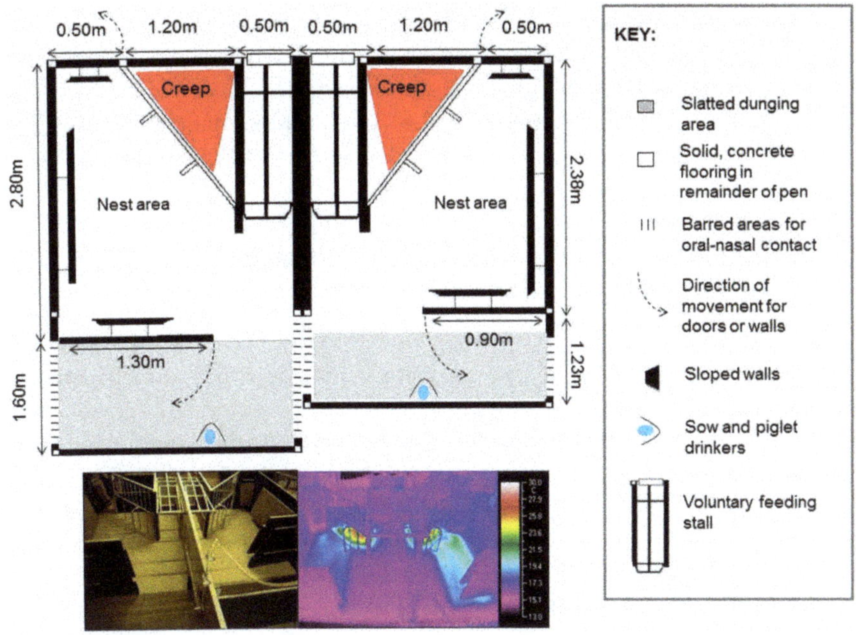

Figure 2 PigSAFE design showing LARGE (left) and SMALL (right) pen designs and thermal image showing dispersion of heat in the nest-site when under-floor heating was applied for treatments.

3.4.2.1 The influence of sound-proofing and substrate on farrowing location and performance

For the trials on Farm A, it was hypothesised that 1) 'ad libitum' substrate provided in the nest area would encourage nest-building and farrowing in the correct location, and 2) a quiet nest-site would encourage farrowing in the correct location and benefit piglet survival. The treatments were applied within the PigSAFE pen design (Fig. 1a) (overall pen size = 2.36m x 3.35m = 7.9m^2). Sound-suppression material and a roof over the nest-site were used to create a 'quiet' nest (Fig. 1b), and long-stemmed straw was provided at two levels (MAX - 4 kg and MIN - 2 kg) before farrowing until two days post-farrowing. A 2 × 2 factorial design saw 99 Large White × Landrace sows randomly assigned to quiet (i.e. quiet or control) and substrate (MAX or MIN) treatments. Linear mixed models analysed data. All the sows chose to start farrowing in the nest and 99% of all piglets were farrowed in the nest. While there were numerical trends in the expected direction, neither the degree of nest enclosure/soundproofing nor the quantity of nesting substrate supplied had a statistically significant influence on piglet survival (Live-born mortality, Quiet = 14% vs. Control = 17%, sem = 1.52, P = 0.16; Min = 17% vs. Max = 14%, sem 1.52, P = 0.11).

The overall design successfully promoted farrowing in the desired location, irrespective of substrate amount or sound-suppression. This allowed the more cost-effective and practical pen design (i.e. no covered nest and sound proofing) and substrate amount (i.e. 2 kg long-stemmed straw provided in nest-site on entry and maintained at that level until farrowing) to be taken forward for the next iterations and 'commercial' trials.

3.4.2.2 The influence of temperature and nest size on farrowing location and performance

At Farm B, two hypotheses to optimise farrowing location and improve piglet survival were tested: 1) a heated nest-site would be more attractive to the farrowing sow, and 2) greater space would improve maternal behaviour and therefore benefit piglet performance. PigSAFE (Fig. 2) was adapted: the LARGE treatment measured 9.7m^2 in total, with a nest area of 4.0m^2, and the SMALL treatment (same design but smaller) measured 7.9m^2 in total with a nest area of 3.3m^2. The nest floor was heated to either 30°C (T30) or 20°C (T20) from 48h before, until 24h after farrowing. Room temperature was kept at 18°C.

A 2 × 2 factorial design saw 88 Large White × Landrace sows randomly assigned to space and temperature treatments. Generalised linear mixed models analysed data. The overall design successfully promoted farrowing in the desired location, irrespective of nest size and floor temperature (97% of all piglets born were born in the nest). There was no significant effect of floor temperature on performance. However, space influenced mortality, with significantly greater live-born mortality when sows were afforded a larger farrowing space (LARGE=18.1% vs. SMALL=10.9% P=0.028). Behavioural analysis suggested that the greater mortality was a result of a larger nest area in which the sow could lie down unsupported (full results of this study can be found in Baxter et al., 2015). These results suggest the larger nest size was less protective for the piglets and thus a smaller nest would be recommended. The pen features with nesting substrate provided enough stimuli, regardless of floor temperature, to attract sows into the nest.

Conclusion from the initial trials: the results indicated that the PigSAFE pen provided the correct stimuli (enclosed nest-site, substrate) to attract the sow to farrow in the desired location and that piglet survival levels were acceptable at both sites within the constraints of the experiment. The final prototypes would adopt the best designs at both sites (which were also the most cost-effective) to run 'commercially' using farm rather than research staff.

3.4.3 PigSAFE 'commercial' trials and performance

For this phase, farm staff at each site were responsible for operating both PigSAFE and farrowing crates in parallel, given minimal experimental constraints.

The optimised designs were used at each site and only basic instructions were given, including the best practice for substrate provision based on initial trials. Data presented are on pig performance.

3.4.3.1 Piglet survival

Combined data from the two sites (304 sows, n=164 in Crates n=140 in PigSAFE, approximately 10% gilts) are shown in Table 2 against the national average and top third data at the time of the trial.

When the data at each farm were looked at separately, there were interesting trends suggesting farm differences. At Farm A, the PigSAFE system performed very well, with no significant difference compared to crates (Fig. 3) and both systems performing better than the industry top third. At Farm B, although there were no significant differences in pig performance, mortality in the PigSAFE pen was closer to the industry indoor average.

At Farm B, pig performance in PigSAFE pens showed an improvement with every batch farrowed (Fig. 4), highlighting that stock-person training is an important aspect when adopting a new system. In contrast with Farm A, where the manager had previous experience with loose-farrowing outdoor sows, the manager at Farm B had previously worked with sows in crates for over 25 years and thus was unfamiliar with loose farrowing and lactating sows. The learning curve is indicative of potential scenarios when implementing a new system on a commercial farm where staff have previously had little or no experience of animals farrowing and lactating while loose.

Table 2 Performance of PigSAFE pens and farrowing crates for the period up to August 2011, with national average and top third farrowing crate data for comparison (AHDB Pork, 2010[1])

	Total born (BA+BD)	Total weaned	%Total mortality	%Live-born mortality	%Stillborn mortality*
PigSAFE	13.0	10.7[a]	13.3[b]	9.6[c]	4.1[b]
Farrowing crates	12.8	10.8[a]	12.0[b]	9.0[c]	3.4[b]
National indoor average	12.3	10.1	17.6	12.3	6.1
National indoor top third	13.4	11.2	16.5	10.3	5.6

[a]Adjusted for net fostering; [b]This figure includes both stillborn piglets and those dying before weaning and is adjusted for litter size; [c]Adjusted for litter size post fostering. BA = Born alive, BD = Born dead/Stillborn. *Stillborn% = percentage of total litter size that were born dead.

[1] https://porktools.ahdb.org.uk/prices-stats/costings-herd-performance/indoor-breeding-herd/.

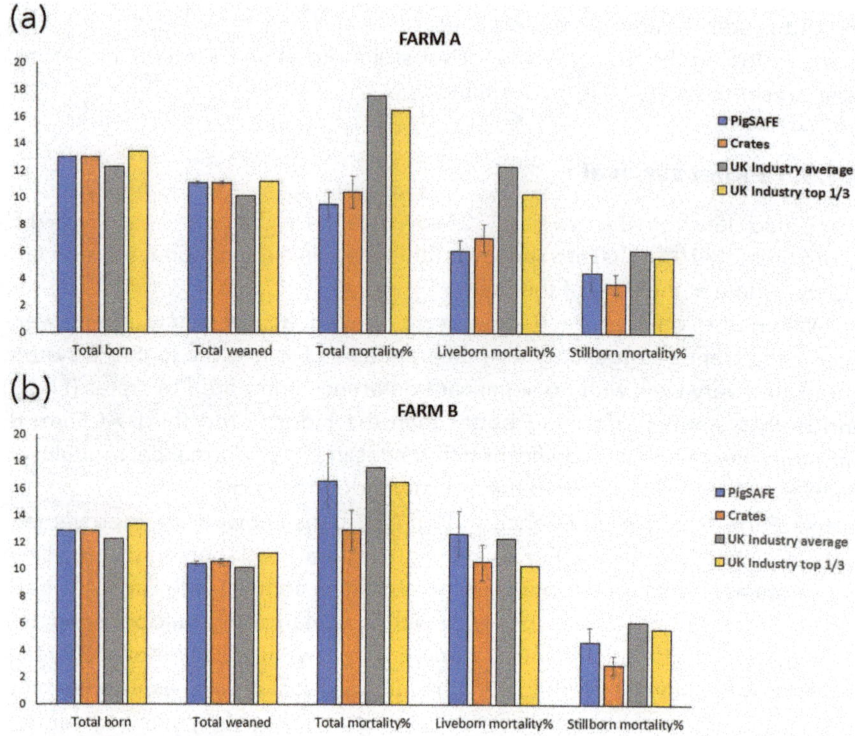

Figure 3 Performance at Farm A (a) and Farm B (b). Descriptive comparison against industry average and top third at the time (AHDB Pork, 2010, https://porktools.ahdb.org.uk/prices-stats/costings-herd-performance/indoor-breeding-herd/). Mortality data are adjusted for net fostering (similar in both systems), parity and litter size.

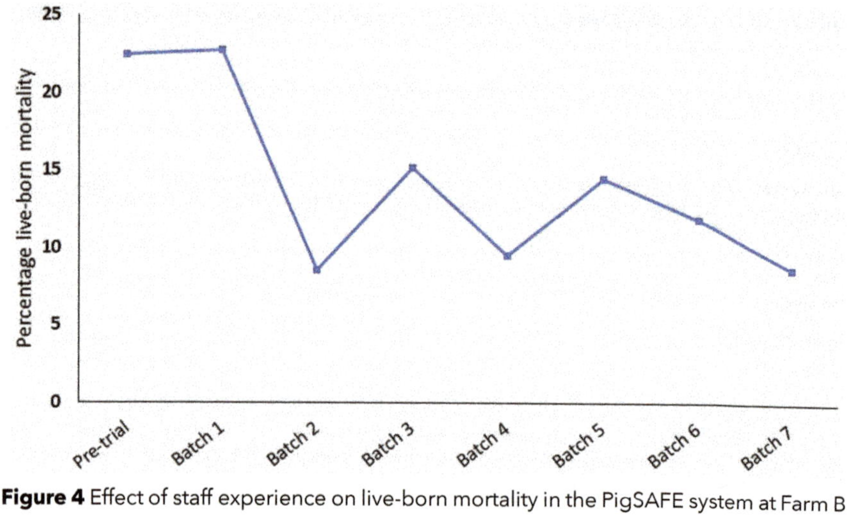

Figure 4 Effect of staff experience on live-born mortality in the PigSAFE system at Farm B.

3.4.3.2 Sow and piglet weight and condition

Body weight (kg) and back-fat depth were measured on sows pre-farrowing and post-weaning, and piglets were weighed at weaning. Table 3 shows that although there were no significant differences between housing systems at Farm A, at Farm B piglets weaned from PigSAFE pens were significantly heavier than those from crates (average individual weight 8.8 kg vs. 8.5 kg).

At Farm B there was a tendency for sows housed in PigSAFE pens to eat more than those housed in crates (7.27 kg vs. 6.44 kg per sow per day $F_{1,6}=3.45$ $P=0.088$). At Farm A there were no significant differences in feed intake between farrowing systems ($P=0.621$). It is possible that the difference in feed intake is only seen at Farm B as a result of potentially beneficial effects of the building in which the PigSAFE pens were built. The 'new build' scenario implemented at Farm B involved a large building shell with high roof space, and thus a more airy environment for the sows. It is possible that the higher piglet weaning weights seen in PigSAFE at Farm B relate to this sow feed intake, but it could also be that the more open pens (at 7 days post-farrowing the dividing wall opened up) improved udder access and milk let-down. This has been seen in other work demonstrating the benefits of space on suckling success as a result of easier udder access (Pedersen et al., 2011). These performance advantages are important, as the capital investment required to install a PigSAFE pen has been shown to be more than a conventional crate (Guy et al., 2012).

Conclusions from commercial-type trials: Results from the commercial scale-up phase showed no significant differences in performance between PigSAFE pens and conventional crates within farms. The most promising results from Farm A showed that the PigSAFE system can perform better than the industry top third producers. A key aspect of any new system is the ability and willingness of stockpersons to adapt to a new system and this effect was clearly

Table 3 Influence of farrowing accommodation on sow body condition and piglet weaning weight (adjusted for litter size, weaned litter size and weaning age). Sow weight loss includes the birth of the litter. For statistical purposes, systems within sites were compared.

	Sow condition		Litter condition
	Weight loss (kg)	Back-fat loss (mm)	Average litter weaned weight (kg)
Farm A PigSAFE	38.81	4.18	78.05
Farm A Farrowing crates	38.54	4.28	76.14
Farm B PigSAFE	27.68	4.45	91.33[a]
Farm B Farrowing crates	29.71	4.18	87.77[b]

Subscripts with different letters indicate figures are significantly different at $P < 0.05$.

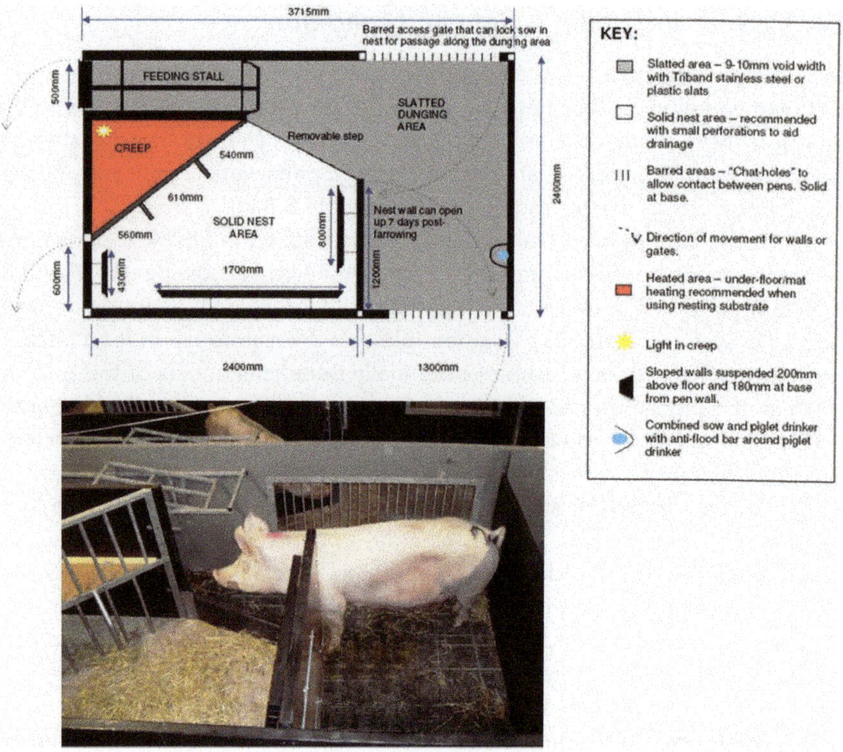

Figure 5 Recommended design and dimensions for building a PigSAFE pen.

seen at Farm B. There is also some indication of benefits in weaning weights and sow feed intake when using PigSAFE pens compared with conventional crates. The recommended best design as a result of the research is shown in Fig. 5, and full details of all components can be found at 'www.freefarrowing.org'.

Throughout the project it was emphasised that the detail of the design is very important in its performance and therefore any deviation from the recommended details is done at the farmer's own risk.

3.5 Commercial developments in alternative farrowing and lactation systems

Since the publication of the initial review manuscripts (Baxter et al., 2011a, 2012), there has been a great deal of research and industry activity in the development of alternative systems (Table 4). These have included large national studies in Austria (Pro-SAU) and Denmark (SEGES showroom) and plans for reform to phase out farrowing crates in these countries. Thus Baxter

Table 4 Examples of different farrowing and lactation systems (Specific information on alternative systems can be found at www.freefarrowing.org)

System			Descriptions	
Crates	Conventional	Crate	Tubular metal bars run horizontally along the length of the crate, with additional bars positioned above the sow to prevent escape via climbing or jumping. Crate prevents turning around. Crates are fixed width or can be adjusted at the back for larger sows. Space available to sow: 0.90–1.23 m^2. Total space available to piglets: ~3.5–4.0 m^2. A water and food trough is situated at the front with piglet water drinker within pen. Flooring: fully or part-slatted with pens usually having a solid floor or mat for the piglet resting area, which includes a heat source.	Fixed-width farrowing crate with front covered and heated creep, solid floors with cast-iron slatted area at rear for waste removal. Photo courtesy of E. M. Baxter.  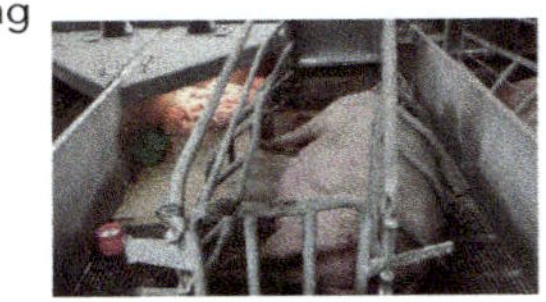 Adjustable farrowing crate with capability to widen but prevent turning. Corner, heated creep, partially slatted floors (concrete/cast-iron). Photo courtesy of V. A. Moustsen.

(*Continued*)

Table 4 (*Continued*)

System			Descriptions	
	Temporary crating	Hinged/ swing-side for example, 360° MPP, Combi-flex, Combi-flex hexel, Winged	Farrowing crate structure within pen as in conventional systems. Spatial footprint and flooring same as conventional system. Sow is crated on entry to farrowing house until approximately 5–7 days post-farrowing. Crate can be opened up or swung-open for lactation allowing the sow to turn around. An additional heated and solid area (typically a heat-mat) inaccessible to the sow is provided for the piglets.	 360° MPP temporary crate in 'open' (left) and 'closed' (right) position. Crate typically opened after 4–7 days. Heat mat for piglets on fully slatted plastic floor. Photo courtesy of E. M. Baxter. 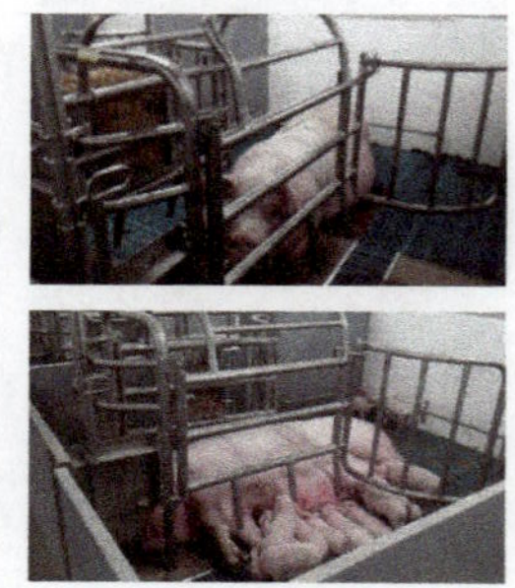 'Winged' temporary crating option. Crate opens up at the back to allow the sow to reverse out and turn in the dunging area. Photo courtesy of V. A. Moustsen.

		SWAP	The SWAP system (Sow Welfare And Piglet Protection) is designed to allow nest-building and then confinement post-farrowing until approximately four days post-farrowing (Moustsen et al., 2013). Sow should be able to turn around for nesting and lactation but crated for farrowing and immediately post-partum. Larger footprint than the crate – 6 m^2 built on part-slatted flooring. Separate large corner heated and covered area for piglets.	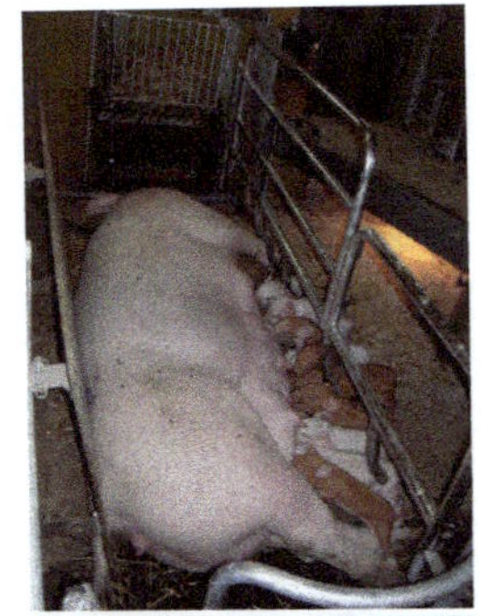  SWAP in closed and open positions. Photo courtesy of V. A. Moustsen.
Pens	Simple	For example, simple pen, mushroom pen, sloped floor	Same spatial footprint as conventional crate. No crate. Sow able to turn around at all times. Fully or part-slatted flooring. Possibility of modified floor design. Separate heated area for piglets.	 Simple pen on the same spatial footprint as a conventional farrowing crate. Fully slatted flooring. Photo courtesy of V. A. Moustsen.

(*Continued*)

Table 4 Examples of different farrowing and lactation systems (Specific information on alternative systems can be found at www.freefarrowing.org) (*Continued*)

System		Descriptions	
Designed pen	For example, Comfort Sow, Danish FF, FATs, PigSAFE, WelCon, Werribee	Zero-confinement. Sow can turn around. Defined areas provided in the pen for feeding, dunging and lying/nesting. Size varies (5–8.5 m^2). Additional pen 'furniture' such as rails or sloped walls to assist sow posture changes and protect piglets. Solid or part-slatted floor with generous lying area to provide substrate for nest-building. A separate, heated creep area for the piglets, inaccessible to the sow is provided. Some systems have the potential for separating the sow from piglets to allow farmers to perform husbandry procedures safely.	 Danish Free Farrower Zero confinement pen with part-slatted flooring and a corner creep. Photo courtesy of V. A. Moustsen. 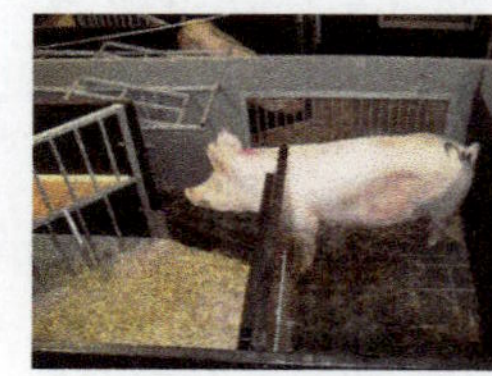 PigSAFE zero confinement pen with part slatted flooring, corner creep and separate feeding, nesting and dunging areas. Photo courtesy of E. M. Baxter.

Group	Zero-confinement group	Separate or free-access nests then group	Systems allow sows and litters to mix before weaning, typically 10–21 days into lactation. Based on multi-suckling accommodation (e.g. Swedish systems – Ljungström, Thorstensson). Variable but large amounts of space for both sows and piglets. For farrowing, sows are initially individually housed in single pens, but are integrated with their litter into groups in larger multi-suckling pens between 10–21 days post-farrowing Alternatively, sows are already grouped prior to farrowing and have free access to individual nest boxes for farrowing.	 Typical Swedish multi-suckling system with deep-straw bedding. Photo courtesy of J. H. Pedersen.  Group housing system for lactating sows at Swine Innovation Center Sterksel in the Netherlands. Photo courtesy of C. van der Peet-Schwering, Wageningen University and Research Centre.

(Continued)

Table 4 Examples of different farrowing and lactation systems (Specific information on alternative systems can be found at www.freefarrowing.org) (*Continued*)

System			Descriptions	
	Temporary crating then group	Crated then grouped	Sows are conventionally crated and then grouped 10–14 days post-partum to allow litter integration.	
Kennels	Kennels	Kennel with run, for example, 'Solari'	An outside space is intended for dunging and feeding by the individually housed sow, with an indoor kennel for farrowing. Floors are solid to facilitate provision of deep substrate. A heated, creep space may be provided within the kennel for the piglets.	 Outdoor kennel and run. Photo courtesy of E. M. Baxter.
Outdoor	Outdoor	Outdoor	These are systems with low capital investment and running costs, where sows and their piglets are housed individually, outdoors in farrowing arks or huts, with access to individual or group paddocks. There are different ark and hut designs available and described in detail elsewhere (e.g. Honeyman et al., 1998; Baxter et al., 2009).	 Outdoor farrowing huts and paddocks. Photos courtesy of S.-L. A. Schild (top) and E. M. Baxter (bottom).

et al. (2018) updated their earlier scientific review to include new information in the state-of-the-art. In addition, Baxter and Edwards developed a website to provide a more farmer-friendly interface for those interested in building alternative systems (www.freefarrowing.org).

Many of the alternative systems available commercially are temporary crating systems, as farmers seek to retain greater control over the sow's movements, particularly around the first few days post-partum when interventions to promote piglet survival and piglet husbandry procedures are performed. Farmers have also said that they want an 'insurance policy' of being able to resort to the conventional crated system if they do not succeed in adopting free farrowing or loose lactation, either because of stockperson reluctance, performance problems or a failure in the market to recognise the additional financial requirements of higher welfare systems. Although there has been a lot of work on pen design, less attention has been given to optimising management and genetic selection of the best animals for loose indoor systems. It is likely that all three 'Ps' (Pens-People-Pigs) will need to be optimised to achieve consistent performance and reduce the barriers to uptake.

4 Conclusions

Determining the basic needs of all actors (animals, farmers) that are involved in a system is a key starting point for designing higher welfare, sustainable alternatives to conventional farrowing crate systems that are known to impose welfare challenges for both sows and piglets.

Allowing the animals to be more active in the control of their environment is a positive aspect of animal welfare that appears to be an important part of their behavioural needs. When given the chance to be an active participant in her own environment, the sow will fulfil her own needs but also contribute to the goals of others. Maximising piglet survival increases her biological fitness and also satisfies the main needs of the piglets and farmer. The current challenge is what happens when the biological limits of the sow have been exceeded and she can no longer fully control the outcomes of all her offspring without the contribution of another actor. Designing systems that can allow stockperson interventions without imposing any further constraints on sow welfare is an important goal for research and development.

5 Future trends in research

This chapter has concentrated on the managerial and environmental interventions required to optimise animal welfare around farrowing. Research on the development of alternative farrowing and lactation systems is reasonably mature; however, greater innovation is required to optimise system components. For example, designing suitable flooring and waste management systems that allow the provision of environmental enrichment without compromising hygiene is still a challenge.

Optimising the biological components of farrowing and lactation systems will involve breeding for improved maternal behaviour (Grandison, 2005; Gäde et al., 2008; Baxter et al., 2011b) and investigating strategies for breeding a more robust piglet. Both have potential to improve piglet survival and therefore the potential to promote greater confidence with loose housing. However, breeding strategies seeking short-term gains in production, such as selecting for increased numbers of piglets born beyond the biological limits of the sow to rear them, threaten welfare. They also challenge system design to incorporate features that help cater for the increasingly specialised needs of piglets born into large litters. Finally, a social science element meriting further research is understanding the barriers to uptake of alternative systems, including stockperson behaviour and willingness to change. Greater attention to all these aspects should improve sow and piglet welfare while addressing farmer concerns.

6 Where to look for further information

In addition to the peer-reviewed literature provided in the reference section, web-based information can be found detailing practical ways to optimise the farrowing and lactation environment.

- www.freefarrowing.org.
- http://practicalpig.ahdb.org.uk/.

7 References

Ahlström, S., Jarvis, S. and Lawrence, A. B. 2002. Savaging gilts are more restless and more responsive to piglets during the expulsive phase of parturition. *Applied Animal Behaviour Science* 76(1), 83–91.

Alexopoulos, J. G., Lines, D. S., Hallett, S. and Plush, K. J. 2018. A review of success factors for piglet fostering in lactation. *Animals* 8(3), 38.

Algers, B. and Uvnäs-Moberg, K. 2007. Maternal behavior in pigs. *Hormones and Behavior* 52(1), 78–85.

Amdi, C., Hansen, C. F., Krogh, U., Oksbjerg, N. and Theil, P. K. 2013a. Less brain sparing occurs in severe intrauterine growth-restricted piglets born to sows fed palm fatty acid distillate. In: *Manipulating Pig Production XIV*. APSA Biennial Conference, Melbourne, Victoria, Australia, 24th to 27th November, p. 125.

Amdi, C., Krogh, U., Flummer, C., Oksbjerg, N., Hansen, C. F. and Theil, P. K. 2013b. Intrauterine growth restricted piglets defined by their head shape ingest insufficient amounts of colostrum. *Journal of Animal Science* 91(12), 5605–5613.

Andersen, H. M. L. and Pedersen, L. J. 2011. The effect of feed trough position on choice of defecation area in farrowing pens by loose sows. *Applied Animal Behaviour Science* 131(1–2), 48–52.

Andersen, I. L., Vasdal, G. and Pedersen, L. J. 2014. Nest building and posture changes and activity budget of gilts housed in pens and crates. *Applied Animal Behaviour Science* 159, 29–33.

Arey, D. S. 1992. Straw and food as reinforcers for prepartal sows. *Applied Animal Behaviour Science* 33(2–3), 217–226.

Balzani, A., Cordell, H. J., Sutcliffe, E. and Edwards, S. A. 2016. Heritability of udder morphology and colostrum quality traits in swine. *Journal of Animal Science* 94(9), 3636–3644.

Bandrick, M., Pieters, M., Pijoan, C., Baidoo, S. K. and Molitor, T. W. 2011. Effect of cross-fostering on transfer of maternal immunity to Mycoplasma hyopneumoniae to piglets. *Veterinary Record* 168(4), 100.

Barnett, J. L., Hemsworth, P. H., Cronin, G. M., Jongman, E. C. and Hutson, G. D. 2001. A review of the welfare issues for sows and piglets in relation to housing. *Australian Journal of Agricultural Research* 52(1), 1–28.

Baxter, E. M. and Edwards, S. A. 2018. Piglet mortality and morbidity: inevitable or unacceptable? In: Špinka, M. (Ed.), *Advances in Pig Welfare*, pp. 73–100. Woodhead Publishing, Cambridge, UK.

Baxter, E. M., Jarvis, S., D'eath, R. B., Ross, D. W., Robson, S. K., Farish, M., Nevison, I. M., Lawrence, A. B. and Edwards, S. A. 2008. Investigating the behavioural and physiological indicators of neonatal survival in pigs. *Theriogenology* 69(6), 773–783.

Baxter, E. M., Jarvis, S., Sherwood, L., Robson, S. K., Ormandy, E., Farish, M., Smurthwaite, K. M., Roehe, R., Lawrence, A. B. and Edwards, S. A. 2009. Indicators of piglet survival in an outdoor farrowing system. *Livestock Science* 124(1-3), 266–276.

Baxter, E. M., Lawrence, A. B. and Edwards, S. A. 2011a. Alternative farrowing systems: design criteria for farrowing systems based on the biological needs of sows and piglets. *Animal: An International Journal of Animal Bioscience* 5(4), 580–600.

Baxter, E. M., Jarvis, S., Sherwood, L., Farish, M., Roehe, R., Lawrence, A. B. and Edwards, S. A. 2011b. Genetic and environmental effects on piglet survival and maternal behaviour of the farrowing sow. *Applied Animal Behaviour Science* 130(1-2), 28–41.

Baxter, E. M., Lawrence, A. B. and Edwards, S. A. 2012. Alternative farrowing accommodation: welfare and economic aspects of existing farrowing and lactation systems for pigs. *Animal* 6(1), 96–117.

Baxter, E. M., Rutherford, K. M. D., d'Eath, R. B., Arnott, G., Turner, S. P., Sandøe, P., Moustsen, V. A., Thorup, F., Edwards, S. A. and Lawrence, A. B. 2013. The welfare implications of large litter size in the domestic pig II: management factors. *Animal Welfare* 22(2), 219–238.

Baxter, E. M., Adeleye, O. O., Jack, M. C., Farish, M., Ison, S. H. and Edwards, S. A. 2015. Achieving optimum performance in a loose-housed farrowing system for sows: the effects of space and temperature. *Applied Animal Behaviour Science* 169, 9–16.

Baxter, E. M., Andersen, I. L. and Edwards, S. A. 2018. Sow welfare in the farrowing crate and alternatives. In: Špinka, Marek (Ed.), *Advances in Pig Welfare*, pp. 27–72. Woodhead Publishing, Cambridge, UK.

Baxter, M. R. and Schwaller, C. E. 1983. Space requirements for sows in confinement. In: Baxter, S. H., Baxter, M. R. and MacCormack, J. (Eds), *Farm Animal Housing and Welfare*, pp. 181–195. Martinus Nijhoff Publishers, The Hague, The Netherlands.

Baxter, M. R. 1991. The Freedom farrowing system. *Farm Building Progress* 104, 9–15.

Björkman, S., Oliviero, C., Rajala-Schultz, P. J., Soede, N. M. and Peltoniemi, O. A. T. 2017. The effect of litter size, parity and farrowing duration on placenta expulsion and retention in sows. *Theriogenology* 92, 36–44.

Black, J. L., Mullan, B. P., Lorschy, M. L. and Giles, L. R. 1993. Lactation in the sow during heat stress. *Livestock Production Science* 35(1–2), 153–170.

Bøe, K. E., Kvaal, I., Hall, E. J. S. and Cronin, G. M. 2016. Individual differences in dunging patterns in loose-housed lactating sows. *Acta Agriculturae Scandinavica, Section A–Animal Science* 66(4), 221–230.

Bohnenkamp, A. L., Traulsen, I., Meyer, C., Müller, K. and Krieter, J. 2013. Comparison of growth performance and agonistic interaction in weaned piglets of different weight classes from farrowing systems with group or single housing. *Animal: An International Journal of Animal Bioscience* 7(2), 309–315.

Bolhuis, J. E., Raats-van den Boogaard, A. M. E., Hoofs, A. I. J. and Soede, N. M. 2018. Effects of loose housing and the provision of alternative nesting material on peri-partum sow behaviour and piglet survival. *Applied Animal Behaviour Science* 202, 28–33.

Bonde, M. 2008. Prevalence of decubital shoulder lesions in Danish sow herds. Internal Report 12, University of Aarhus, Faculty of Agricultural Sciences, p. 8.

Bontempo, V. and Jiang, X. R. 2015. Feeding various fat levels in sows: effects on immune status and performance of sows and piglets. In: Farmer, C. (Ed.), *The Gestating and Lactating Sow*, pp. 357–375. Wageningen Academic Publishers, Netherlands.

Boogaard, B. K., Boekhorst, L. J. S., Oosting, S. J. and Sørensen, J. T. 2011. Socio-cultural sustainability of pig production: citizen perceptions in the Netherlands and Denmark. *Livestock Science* 140(1–3), 189–200.

Bos, B. and Koerkamp, P. G. 2009. Synthesising needs in system innovation through structured dresign: a methodical outline of the role of needs in reflexive interactive design (RIO). Sustainable agriculture and food chains in peri-urban areas. In: Poppe, K. J., Termeer, K. and Slingerland, M. (Eds), *Transitions Towards Sustainable Agriculture and Food Chains in Peri-Urban Areas*, Chapter 12p. 219–237. Wageningen Academic Publishers.

Boyle, L. A., Leonard, F. C., Lynch, P. B. and Brophy, P. 2002. Effect of gestation housing on behaviour and skin lesions of sows in farrowing crates. *Applied Animal Behaviour Science* 76(2), 119–134.

Boyle, L. A., Regan, D., Leonard, F. C., Lynch, P. B. and Brophy, P. 2000. The effect of mats on the welfare of sows and piglets in the farrowing house. *Animal Welfare* 9(1), 39–48.

Brown, S. M., Klaffenböck, M., Nevison, I. M. and Lawrence, A. B. 2015. Evidence for litter differences in play behaviour in pre-weaned pigs. *Applied Animal Behaviour Science* 172, 17–25.

Buller, H., Blokhuis, H., Jensen, P. and Keeling, L. 2018. Towards farm animal welfare and sustainability. *Animals* 8(6), 81.

Cabrera, R. A., Boyd, R. D., Jungst, S. B., Wilson, E. R., Johnston, M. E., Vignes, J. L. and Odle, J. 2010. Impact of lactation length and piglet weaning weight on long-term growth and viability of progeny. *Journal of Animal Science* 88(7), 2265–2276.

Calderón Díaz, J. A., Fahey, A. G., KilBride, A. L., Green, L. E. and Boyle, L. A. 2013. Longitudinal study of the effect of rubber slat mats on locomotory ability, body, limb and claw lesions, and dirtiness of group housed sows. *Journal of Animal Science* 91(8), 3940–3954.

Calderón Díaz, J. A., Diana, A., Boyle, L. A., Leonard, F. C., McElroy, M., McGettrick, S., Moriarty, J. and García Manzanilla, E. 2017. Delaying pigs from the normal production flow is associated with health problems and poorer performance. *Porcine Health Management* 3(1), 13.

Chen, J. and Lobo, A. 2012. Organic food products in China: determinants of consumers' purchase intentions. *The International Review of Retail, Distribution and Consumer Research* 22(3), 293–314.

Condous, P. C., Plush, K. J., Tilbrook, A. J. and Van Wettere, W. H. E. J. 2016. Reducing sow confinement during farrowing and in early lactation increases piglet mortality. *Journal of Animal Science* 94(7), 3022–3029.

Council Directive 2008/120/EC of 18 December 2008 laying down minimum standards for the protection of pigs.

Courboulay, V., le Roux, A., Collin, F., Dutertre, C. and Rousseau, P. 2000. Incidence du type de sol en maternite sue le confort de la truie et des porcelets. *Journees Recherche Porcine en France* 32, 115–122.

Cronin, G. M., Barnett, J. L., Hodge, F. M., Smith, J. A. and McCallum, T. H. 1991. The welfare of pigs in two farrowing/lactation environments: cortisol responses of sows. *Applied Animal Behaviour Science* 32(2–3), 117–127.

Cronin, G. M., Smith, J. A., Hodge, F. M. and Hemsworth, P. H. 1994. The behaviour of primiparous sows around farrowing in response to restraint and straw bedding. *Applied Animal Behaviour Science* 39(3–4), 269–280.

Damm, B. I., Vestergaard, K. S., Schroder-Petersen, D. L. and Ladewig, J. 2000. The effect of branches on prepartum nest building in gilts with access to straw. *Applied Animal Behaviour Science* 69(2), 113–124.

Damm, B. I., Moustsen, V., Jørgensen, E., Pedersen, L. J., Heiskanen, T. and Forkman, B. 2006. Sow preferences for walls to lean against when lying down. *Applied Animal Behaviour Science* 99(1–2), 53–63.

D'Eath, R. B., Tolkamp, B. J., Kyriazakis, I. and Lawrence, A. B. 2009. 'Freedom from hunger'and preventing obesity: the animal welfare implications of reducing food quantity or quality. *Animal Behaviour* 77(2), 275–288.

de Barcellos, M. D., Grunert, K. G., Zhou, Y., Verbeke, W., Perez-Cueto, F. J. A. and Krystallis, A. 2013. Consumer attitudes to different pig production systems: a study from mainland China. *Agriculture and Human Values* 30(3), 443–455.

Decaluwé, R., Maes, D., Wuyts, B., Cools, A., Piepers, S. and Janssens, G. P. J. 2014. Piglets' Colostrum intake associates with daily weight gain and survival until weaning. *Livestock Science* 162, 185–192.

Declerck, I., Dewulf, J., Piepers, S., Decaluwé, R. and Maes, D. 2015. Sow and litter factors influencing colostrum yield and nutritional composition. *Journal of Animal Science* 93(3), 1309–1317.

De Jonge, F. H., Bokkers, E. A. M., Schouten, W. G. P. and Helmond, F. A. 1996. Rearing piglets in a poor environment: developmental aspects of social stress in pigs. *Physiology and Behavior* 60(2), 389–396.

Devillers, N., Le Dividich, J. and Prunier, A. 2011. Influence of colostrum intake on piglet survival and immunity. *Animal: An International Journal of Animal Bioscience* 5(10), 1605–1612.

De Vos, M., Huygelen, V., Willemen, S., Fransen, E., Casteleyn, C., Van Cruchten, S., Michiels, J. and Van Ginneken, C. 2014. Artificial rearing of piglets: effects on small intestinal morphology and digestion capacity. *Livestock Science* 159, 165–173.

Donovan, T. S. and Dritz, S. S. 2000. Effect of split nursing on variation in pig growth from birth to weaning. *Journal of the American Veterinary Medical Association* 217(1), 79-81.

Douglas, S. L., Edwards, S. A., Sutcliffe, E., Knap, P. W. and Kyriazakis, I. 2013. Identification of risk factors associated with poor lifetime growth performance in pigs. *Journal of Animal Science* 91(9), 4123-4132.

Douglas, S. L., Edwards, S. A. and Kyriazakis, I. 2014. Management strategies to improve the performance of low birth weight pigs to weaning and their long term consequences. *Journal of Animal Science* 92(5), 2280-2288.

Drake, A., Fraser, D. and Weary, D. M. 2008. Parent-offspring resource allocation in domestic pigs. *Behavioral Ecology and Sociobiology* 62(3), 309-319.

D'Silva, J. and Turner, J. (Eds) 2012. *Animals, Ethics and Trade: The Challenge of Animal Sentience*. Earthscan, London, UK.

Duncan, I. J. 2005. Science-based assessment of animal welfare: farm animals. *Revue Scientifique et Technique* 24(2), 483-492.

Edwards, S. A. 2002. Perinatal mortality in the pig: environmental or physiological solutions? *Livestock Production Science* 78(1), 3-12.

Edwards, S. A., Brett, M., Seddon, Y. M., Ross, D. and Baxter, E. M. 2012. Evaluation of nest design and nesting substrate options for the PigSAFE free farrowing pen. *Proceedings of the British Society of Animal Science*, 7.

Edwards, S. A., Matheson, S. M. and Baxter, E. M. 2019a. Genetic influences on intra-uterine growth retardation of piglet and management interventions for low birth weight piglets. In: González, E. R. (Ed.), *Nutrition of Hyperprolific Sows*, pp. 141-166. Novus International, Inc., Spain.

Edwards, L. E., Plush, K. J., Ralph, C. R., Morrison, R. S., Acharya, R. Y. and Doyle, R. E. 2019b. Enrichment with Lucerne hay improves sow maternal behaviour and improves piglet survival. *Animals* 9(8), 558.

Engelsmann, M. N., Hansen, C. F., Nielsen, M. N., Kristensen, A. R. and Amdi, C. 2019. Glucose injections at birth, warmth and placing at a nurse sow improve the growth of IUGR piglets. *Animals* 9(8), 519.

Eurobarometer 2016. *Special Eurobarometer* 442./Wave EB84.4 - TNS Opinion & Social. Attitudes of Europeans towards Animal Welfare, Available at: http://ec.europa.eu/COMMFrontOffice/PublicOpinion/index.cfm/ResultDoc/download/Document.

Feldpausch, J. A., Jourquin, J., Bergstrom, J. R., Bargen, J. L., Bokenkroger, C. D., Davis, D. L., Gonzalez, J. M., Nelssen, J. L., Puls, C. L., Trout, W. E. and Ritter, M. J. 2019. Birth weight threshold for identifying piglets at risk for preweaning mortality. *Translational Animal Science* 3(2), 633-640.

Feyera, T., Pedersen, T. F., Krogh, U., Foldager, L. and Theil, P. K. 2018. Impact of sow energy status during farrowing on farrowing kinetics, frequency of stillborn piglets, and farrowing assistance. *Journal of Animal Science* 96(6), 2320-2331.

Fox, C., Merali, Z. and Harrison, C. 2006. Therapeutic and protective effect of environmental enrichment against psychogenic and neurogenic stress. *Behavioural Brain Research* 175(1), 1-8.

Gäde, S., Bennewitz, J., Kirchner, K., Looft, H., Knap, P. W., Thaller, G. and Kalm, E. 2008. Genetic parameters for maternal behaviour traits in sows. *Livestock Science* 114(1), 31-41.

Gaskin, H. R. and Kelly, K. W. 1995. Immunology and neonatal Mortality. In: Varley, M. A. (Ed.), *The Neonatal Pig: Development and Survival*, pp. 39-56. CAB International, UK.

Glencorse, D., Plush, K., Hazel, S., D'Souza, D. and Hebart, M. 2019. Impact of non-confinement accommodation on farrowing performance: A systematic review and meta-analysis of farrowing crates Versus pens. *Animals* 9(11), 957.

Gonyou, H. W., Brumm, M. C., Bush, E., Deen, J., Edwards, S. A., Fangman, T., McGlone, J. J., Meunier-Salaun, M., Morrison, R. B., Spoolder, H., Sundberg, P. L. and Johnson, A. K. 2006. Application of broken-line analysis to assess floor space requirements of nursery and grower-finisher pigs expressed on an allometric basis. *Journal of Animal Science* 84(1), 229–235.

Grandinson, K. 2005. Genetic background of maternal behaviour and its relation to offspring survival. *Livestock Production Science* 93(1), 43–50.

Greenwood, E. C., van Dissel, J., Rayner, J., Hughes, P. E. and van Wettere, W. H. E. J. 2019. Mixing sows into alternative lactation housing affects sow aggression at mixing, future reproduction and piglet injury, with marked differences between multisuckle and sow separation systems. *Animals* 9(9), 658.

Grunert, K. G., Sonntag, W. I., Glanz-Chanos, V. and Forum, S. 2018. Consumer interest in environmental impact, safety, health and animal welfare aspects of modern pig production: results of a cross-national choice experiment. *Meat Science* 137, 123–129.

Guy, J. H., Cain, P. J., Seddon, Y. M., Baxter, E. M. and Edwards, S. A. 2012. Economic evaluation of high welfare indoor farrowing systems for pigs. *Animal Welfare* 21(1) (Suppl.1), 19–24.

Hales, J., Moustsen, V. A., Nielsen, M. B. F. and Hansen, C. F. 2014. Higher preweaning mortality in free farrowing pens compared with farrowing crates in three commercial pig farms. *Animal* 8(1), 113–120.

Hales, J., Moustsen, V. A., Devreese, A. M., Nielsen, M. B. F. and Hansen, C. F. 2015. Comparable farrowing progress in confined and loose housed hyper-prolific sows. *Livestock Science* 171, 64–72.

Hamilton, W. D. 1964a. The genetical evolution of social behaviour. I. *Journal of Theoretical Biology* 7(1), 1–16.

Hamilton, W. D. 1964b. The genetical evolution of social behaviour. II. *Journal of Theoretical Biology* 7(1), 17–52.

Hansen, C. F., Hales, J., Weber, P. M., Edwards, S. A. and Moustsen, V. A. 2017. Confinement of sows 24 h before expected farrowing affects the performance of nest building behaviours but not progress of parturition. *Applied Animal Behaviour Science* 188, 1–8.

Hansen, L. U. 2018. Test of 10 different farrowing pens for loose-housed sows. Report no. 1803. SEGEs, Denmark.

Hartsock, T. G. and Barczewski, R. A. 1997. Prepartum behaviour in swine: effects of pen size. *Journal of Animal Science* 75(11), 2899–2904.

Hasan, S., Orro, T., Valros, A., Junnikkala, S., Peltoniemi, O. and Oliviero, C. 2019. Factors affecting sow colostrum yield and composition, and their impact on piglet growth and health. *Livestock Science* 227, 60–67.

Heidinger, B., Stinglmayr, J., Maschat, K., Oberer, M., Kuchling, S. and Baumgartner, J. 2018. Summary of the Austrian project "pro-SAU": evaluation of novel farrowing systems with possibility for the sow to move 2410:Eval. AFB Supplement zum Abschlussbericht des Projekts Pro-SAU.

Herpin, P., Damon, M. and Le Dividich, J. 2002. Development of thermoregulation and neonatal survival in pigs. *Livestock Production Science* 78(1), 25–45.

Hessel, E. F., Reiners, K. and Van den Weghe, H. F. A. 2006. Socializing piglets before weaning: effects on behavior of lactating sows, pre-and postweaning behavior, and performance of piglets. *Journal of Animal Science* 84(10), 2847-2855.

Homedes, J., Salichs, M., Sabaté, D., Sust, M. and Fabre, R. 2014. Effect of ketoprofen on pre-weaning piglet mortality on commercial farms. *The Veterinary Journal* 201(3), 435-437.

Honeyman, M. S., Roush, W. B. and Penner, A. D. 1998. Pig crushing mortality by hut type in outdoor farrowing. Annual Progress Report, Iowa State University, Ames, pp. 16-17.

Horrell, I. and Bennett, J. 1981. Disruption of teat preferences and retardation of growth following cross-fostering of 1-week-old pigs. *Animal Science* 33(1), 99-106.

Hötzel, M. J., Pinheiro Machado F°, L. C., Wolf, F. M. and Costa, O. A. D. 2004. Behaviour of sows and piglets reared in intensive outdoor or indoor systems. *Applied Animal Behaviour Science* 86(1-2), 27-39.

Hunt, K. and Petchey, A. M. 1987. A study of the environmental preferences of sows around farrowing. *Farm Building Progress* 89, 11-14.

Ison, S. H., Wood, C. M. and Baxter, E. M. 2015. Behaviour of pre-pubertal gilts and its relationship to farrowing behaviour in conventional farrowing crates and loose-housed pens. *Applied Animal Behaviour Science* 170, 26-33.

Ison, S. H., Clutton, R. E., Di Giminiani, P. and Rutherford, K. M. 2016. A review of pain assessment in pigs. *Frontiers in Veterinary Science* 3, 108.

Ison, S. H., Jarvis, S., Hall, S. A., Ashworth, C. J. and Rutherford, K. M. D. 2018. Periparturient behavior and physiology: further insight into the farrowing process for primiparous and multiparous sows. *Frontiers in Veterinary Science* 5, 122.

Jarrett, S. and Ashworth, C. J. 2018. The role of dietary fibre in pig production, with a particular emphasis on reproduction. *Journal of Animal Science and Biotechnology* 9(1), 1-11.

Jarvis, S., Lawrence, A. B., McLean, K. A., Deans, L. A., Chirnside, J. and Calvert, S. K. 1997. The effect of environment on behavioural activity, ACTH, b-endorphin and cortisol in pre-parturient gilts. *Animal Science* 65(3), 465-472.

Jarvis, S., Lawrence, A. B., McLean, K. A., Chirnside, J., Deans, L. A. and Calvert, S. K. 1998. The effect of environment on plasma cortisol and b-endorphin in the parturient pig and the involvement of endogenous opioids. *Animal Reproduction Science* 52(2), 139-151.

Jarvis, S., McLean, K. A., Calvert, S. K., Deans, L. A., Chirnside, J. and Lawrence, A. B. 1999. The responsiveness of sows to their piglets in relation to the length of parturition and the involvement of endogenous opioids. *Applied Animal Behaviour Science* 63(3), 195-207.

Jarvis, S., van der Vegt, B. J., Lawrence, A. B., McLean, K. A., Deans, L. A., Chirniside, J. and Calvert, S. K. 2001. The effect of parity and environmental restriction on behavioural and physiological responses of pre-parturient pigs. *Applied Animal Behaviour Science* 71(3), 203-216.

Jarvis, S., Calvert, S. K., Stevenson, J., van Leeuwen, N. and Lawrence, A. B. 2002. Pituitary-adrenal activation in pre-parturient pigs (Sus scrofa) is associated with behavioural restriction due to lack of space rather than nesting substrate. *Animal Welfare* 11, 371-384.

Jarvis, S., Reed, B. T., Lawrence, A. B., Calvert, S. K. and Stevenson, J. 2004. Peri-natal environmental effects on maternal behaviour, pituitary and adrenal activation, and the progress of parturition in the primiparous sow. *Animal Welfare* 13(2), 171-181.

Jarvis, S., D'Eath, R. B., Robson, S. K. and Lawrence, A. B. 2006. The effect of confinement during lactation on the hypothalamic-pituitary-adrenal axis and behaviour of primiparous sows. *Physiology and Behavior* 87(2), 345-352.

Jensen, P. 1986. Observations on the maternal behaviour of free-ranging domestic pigs. *Applied Animal Behaviour Science* 16(2), 131-142.

Jensen, P. 1993. Nest building in domestic sows: the role of external stimuli. *Animal Behaviour* 45(2), 351-358.

Jensen, P. 2002. Behaviour of pigs. In: P Jensen (Eds), *The Ethology of Domestic Animals*, pp. 159-172. CABI Publishing, Wallingford, UK.

Jensen, P. and Toates, F. M. 1993. Who needs 'behavioural needs'? Motivational aspects of the needs of animals. *Applied Animal Behaviour Science* 37(2), 161-181.

Jones, G. M., Edwards, S. A., Sinclair, A. G., Gebbie, F. E., Rooke, J. A., Jagger, S. and Hoste, S. 2002. The effect of maize starch or soya-bean oil as energy sources in lactation on sow and piglet performance in association with sow metabolic state around peak lactation. *Animal Science* 75(1), 57-66.

King, R. L., Baxter, E. M., Matheson, S. M. and Edwards, S. A. 2019. Temporary crate opening procedure affects immediate post-opening piglet mortality and sow behaviour. *Animal: An International Journal of Animal Bioscience* 13(1), 189-197.

Kobek-Kjeldager, C., Moustsen, V. A., Theil, P. K. and Pedersen, L. J. 2020. Effect of litter size, milk replacer and housing on production results of hyper-prolific sows. *Animal* 14(4), 824-833.

Kyriazakis, I. and Edwards, S. A. 1986. The effects of split suckling on behaviour and performance of piglets. *Applied Animal Behaviour Science* 16(1), 92.

Ladewig, J., Kloeppel, P. and Kallweit, E. 1984. A case of "reversed cannibalism" the piglets damaging the sow [vulvar lesions]. *Annales de Recherches Veterinaires (France)*. Annals of veterinary research, 15(2), pp.275-277.

Lawrence, A. B., McLean, K. A., Jarvis, S., Gilbert, C. L. and Petherick, J. C. 1997. Stress and parturition in the pig. *Reproduction in Domestic Animals* 32(5), 231-236.

Lawrence, A. B., Petherick, J. C., McLean, K. A., Deans, L. A., Chirnside, J., Gaughan, A., Clutton, E. and Terlouw, E. M. C. 1994. The effect of environment on behaviour, plasma cortisol and prolactin in parturient sows. *Applied Animal Behaviour Science* 39(3-4), 313-330.

Lawrence, A. B., Petherick, J. C., Mclean, K. and Gilbert, C. 1992. Mediation of stress-induced inhibition of oxytocin in farrowing sows by endogenous opioids. *Proceedings of the British Society of Animal Production* 1992, 56.

Leeb, B., Leeb, C., Troxler, J. and Schuh, M. 2001. Skin lesions and callosities in group-housed pregnant sows: animal-related welfare indicators. *Acta Agriculturae Scandinavica, Section A – Animal Science* 51(S30), 82-87.

Lensink, B. J., Ofner-Schröck, E., Ventorp, M., Zappavigna, P., Flaba, J., Georg, H. and Bizeray-Filoche, D. 2013. In Andres Aland, Thomas Banhazi (Eds) Lying and walking surfaces for cattle, pigs and poultry and their impact on health, behaviour and performance. In: *Livestock Housing: Modern Management to Ensure Optimal Health and Welfare of Farm Animals*, pp. 2509-2514. Wageningen Academic Publishers, Wageningen, The Netherlands.

Lewis, E., Boyle, L. A., Brophy, P., O'Doherty, J. V. and Lynch, P. B. 2005. The effect of two piglet teeth resection procedures on the welfare of sows in farrowing crates. Part 2. *Applied Animal Behaviour Science* 90(3-4), 251-264.

Lohmeier, R. Y., Grimberg-Henrici, C. G. E., Büttner, K., Burfeind, O. and Krieter, J. 2020. Farrowing pens used with and without short-term fixation impact on reproductive traits of sows. *Livestock Science* 231, 103889.

Loisel, F., Farmer, C., Ramaekers, P. and Quesnel, H. 2013. Effects of high fiber intake during late pregnancy on sow physiology, colostrum production, and piglet performance. *Journal of Animal Science* 91(11), 5269–5279.

Mainau, E., Ruiz-de-la-Torre, J. L., Dalmau, A., Salleras, J. M. and Manteca, X. 2012. Effects of meloxicam (Metacam®) on post-farrowing sow behaviour and piglet performance. *Animal: An International Journal of Animal Bioscience* 6(3), 494–501.

Mainau, E., Temple, D. and Manteca, X. 2016. Experimental study on the effect of oral meloxicam administration in sows on pre-weaning mortality and growth and immunoglobulin G transfer to piglets. *Preventive Veterinary Medicine* 126, 48–53.

Malmkvist, J., Pedersen, L. J., Kammersgaard, T. S. and Jørgensen, E. 2012. Influence of thermal environment on sows around farrowing and during the lactation period. *Journal of Animal Science* 90(9), 3186–3199.

Marchant, J. N., Broom, D. M. and Corning, S. 2001. The influence of sow behaviour on piglet mortality due to crushing in an open farrowing system. *Animal Science* 72(1), 19–28.

Martin, J. E., Ison, S. H. and Baxter, E. M. 2015. The influence of neonatal environment on piglet play behaviour and post-weaning social and cognitive development. *Applied Animal Behaviour Science* 163, 69–79.

Mattsson, B., Susic, Z. and Lundeheim, N. 2004. Time for care and comfort. In Proceedings of 18th IPVS Congress, Hamburg, Germany, p. 800.

Middelkoop, A., Choudhury, R., Gerrits, W. J. J., Kemp, B., Kleerebezem, M. and Bolhuis, J. E. 2018. Dietary diversity affects feeding behaviour of suckling piglets. *Applied Animal Behaviour Science* 205, 151–158.

Middelkoop, A., Costermans, N., Kemp, B. and Bolhuis, J. E. 2019. Feed intake of the sow and playful creep feeding of piglets influence piglet behaviour and performance before and after weaning. *Scientific Reports* 9(1), 16140.

Morgan, T., Pluske, J., Miller, D., Collins, T., Barnes, A. L., Wemelsfelder, F. and Fleming, P. A. 2014. Socialising piglets in lactation positively affects their post-weaning behaviour. *Applied Animal Behaviour Science* 158, 23–33.

Mount, L. E. 1967. The heat loss from new born pigs to the floor. *Research in Veterinary Science* 8(2), 175–186.

Mount, L. E. 1968. *The Climatic Physiology of the Pig*. Edward Arnold, London, UK.

Moustsen, V. A. and Poulsen, H. L. 2004. Anbefalinger vedr. dimensioner på fareboks og kassesti. *Landsudvalget for Svin, Danske Slagterier, Notat* nr. 414.

Moustsen, V. and Jensen, T. 2008. *Inventar til forbedring af hygiejne i stier til løsgående farende og diegievende søer*. Notat, Nr. 809, Dansk Svineproduktion.

Moustsen, V. A., Pedersen, L. J. and Jensen, T. 2007. Afprøvning af stikoncepter til løsgående farende og diegivende søer. Meddelelse (805): 1–25. Dansk Svineproduktion.

Moustsen, V. A., Lahrmann, H. P. and D'Eath, R. B. 2011. Relationship between size and age of modern hyper-prolific crossbred sows. *Livestock Science* 141(2–3), 272–275.

Moustsen, V. A., Hales, J., Lahrmann, H. P., Weber, P. M. and Hansen, C. F. 2013. Confinement of lactating sows in crates for 4 days after farrowing reduces piglet mortality. *Animal* 7(4), 648–654.

Mouttotou, N. and Green, L. E. 1999. Incidence of foot and skin lesions in nursing piglets and their association with behavioural activities. *Veterinary Record* 145(6), 160–165.

Muns, R., Silva, C., Manteca, X. and Gasa, J. 2014. Effect of cross-fostering and oral supplementation with colostrums on performance of newborn piglets. *Journal of Animal Science* 92(3), 1193–1199.

Muns, R., Manteca, X. and Gasa, J. 2015. Effect of different management techniques to enhance colostrum intake on piglets' growth and mortality. *Animal Welfare* 24(2), 185–192.

Muns, R., Malmkvist, J., Larsen, M. L. V., Sørensen, D. and Pedersen, L. J. 2016. High environmental temperature around farrowing induced heat stress in crated sows. *Journal of Animal Science* 94(1), 377–384.

Muns, R., Nuntapaitoon, M. and Tummaruk, P. 2017. Effect of oral supplementation with different energy boosters in newborn piglets on pre-weaning mortality, growth and serological levels of IGF-I and IgG. *Journal of Animal Science* 95(1), 353–360.

Nissen, S., Faidley, T. D., Zimmerman, D. R., Izard, R. and Fisher, C. T. 1994. Colostral milk fat percentage and pig performance are enhanced by feeding the leucine metabolite β-hydroxy-β-methyl butyrate to sows. *Journal of Animal Science* 72(9), 2331-2337.

Nowland, T. L., van Wettere, W. H. E. J. and Plush, K. J. 2019. Allowing sows to farrow unconfined has positive implications for sow and piglet welfare. *Applied Animal Behaviour Science* 221, 104872.

Ocepek, M. and Andersen, I. L. 2017. What makes a good mother? Maternal behavioural traits important for piglet survival. *Applied Animal Behaviour Science* 193, 29–36.

Ocepek, M., Goold, C. M., Busančić, M. and Aarnink, A. J. A. 2018. Drinker position influences the cleanness of the lying area of pigs in a welfare-friendly housing facility. *Applied Animal Behaviour Science* 198, 44–51.

Oliviero, C., Heinonen, M., Valros, A. and Peltoniemi, O. 2010. Environmental and sow-related factors affecting the duration of farrowing. *Animal Reproduction Science* 119(1–2), 85–91.

O'Mahony, S. M., Marchesi, J. R., Scully, P., Codling, C., Ceolho, A. M., Quigley, E. M., Cryan, J. F. and Dinan, T. G. 2009. Early life stress alters behavior, immunity, and microbiota in rats: implications for irritable bowel syndrome and psychiatric illnesses. *Biological Psychiatry* 65(3), 263–267.

Oostindjer, M., Bolhuis, J. E., Mendl, M., Held, S., Gerrits, W., Van den Brand, H. and Kemp, B. 2010. Effects of environmental enrichment and loose housing of lactating sows on piglet performance before and after weaning. *Journal of Animal Science* 88(11), 3554–3562.

Oostindjer, M., van den Brand, H., Kemp, B. and Bolhuis, J. E. 2011a. Effects of environmental enrichment and loose housing of lactating sows on piglet behaviour before and after weaning. *Applied Animal Behaviour Science* 134(1–2), 31–41.

Oostindjer, M., Bolhuis, J. E., Mendl, M., Held, S., van den Brand, H. and Kemp, B. 2011b. Learning how to eat like a pig: effectiveness of mechanisms for vertical social learning in piglets. *Animal Behaviour* 82(3), 503–511.

Oostindjer, M., Bolhuis, J. E., Simon, K., van den Brand, H. and Kemp, B. 2011c. Perinatal flavour learning and adaptation to being weaned: all the pig needs is smell. *PLoS ONE* 6(10), e25318.

Oostindjer, M., Kemp, B., van den Brand, H. and Bolhuis, J. E. 2014. Facilitating 'learning from mom how to eat like a pig' to improve welfare of piglets around weaning. *Applied Animal Behaviour Science* 160, 19–30.

Pajor, E. A. 1998. Parent-offspring conflict and its implications for maternal housing systems in domestic pigs. Doctoral dissertation, McGill University Libraries.

Pajor, E. A., Weary, D. M., Fraser, D. and Kramer, D. L. 1999. Alternative housing for sows and litters: 1. Effects of sow-controlled housing on responses to weaning. *Applied Animal Behaviour Science* 65(2), 105–121.

Patil, Y., Gooneratne, R. and Ju, X. H. 2020. Interactions between host and gut microbiota in domestic pigs: a review. *Gut Microbes*, 11(3), 310–334.

Pedersen, L. J., Jørgensen, E., Heiskanen, T. and Damm, B. I. 2006. Early piglet mortality in loose-housed sows related to sow and piglet behaviour and to the progress of parturition. *Applied Animal Behaviour Science* 96(3–4), 215–232.

Pedersen, L. J., Malmkvist, J. and Andersen, H. M. L. 2013. Housing of sows during farrowing: a review on pen design, welfare and productivity. In: Andres Aland and Thomas Banhazi (Eds), *Livestock Housing: Modern Management to Ensure Optimal Health and Welfare of Farm Animals*, pp. 285–297. Wageningen Academic Publishers. Wageningen, The Netherlands.

Pedersen, M. L., Moustsen, V. A., Nielsen, M. B. F. and Kristensen, A. R. 2011. Improved udder access prolongs duration of milk letdown and increases piglet weight gain. *Livestock Science* 140(1–3), 253–261.

Pedersen, T. F., Van Vliet, S., Bruun, T. S. and Theil, P. K. 2019. Feeding sows during the transition period–is a gestation diet, a simple transition diet, or a lactation diet the best choice? *Translational Animal Science* 4(1), 34–48.

Peng, X., Yan, C., Hu, L., Liu, Y., Xu, Q., Wang, R., Qin, L., Wu, C., Fang, Z., Lin, Y., Xu, S., Feng, B., Zhuo, Y., Li, J., Wu and Che, L. 2019. Effects of fat supplementation during gestation on reproductive performance, milk composition of sows and intestinal development of their offspring. *Animals* 9(4), 125.

Petherick, J. C. 1983. A biological basis for the design of space in livestock housing. In: Baxter, S. H., Baxter, M. R. and MacCormack, J. A. S. C. (Eds), *Farm Animal Housing and Welfare*, pp. 103–120. Martinus Nijoff Publisher, Boston, MA.

Phillips, P. A., Fraser, D. and Pawluczuk, B. 2000. Floor temperature preference of sows at farrowing. *Applied Animal Behaviour Science* 67(1–2), 59–65.

Plush, K., McKenny, L., Nowland, T., van Wettere, W. and Terry, R. 2019. *Reducing Sow Stress around Farrowing*. Report prepared for the Co-operative Research Centre for High Integrity Australian Pork. *(Pork CRC) project # 1C-114)*.

Power, M. L. and Schulkin, J. 2013. Maternal regulation of offspring development in mammals is an ancient adaptation tied to lactation. *Applied and Translational Genomics* 2, 55–63.

Price, E. O., Hutson, G. D., Price, M. I. and Borgwardt, R. 1994. Fostering in swine as affected by age of offspring. *Journal of Animal Science* 72(7), 1697–1701.

Puppe, B. and Tuchscherer, A. 1999. Developmental and territorial aspects of suckling behaviour in the domestic pig (Sus scrofa f. domestica). *Journal of Zoology* 249(3), 307–313.

Quesnel, H., Farmer, C. and Devillers, N. 2012. Colostrum intake: influence on piglet performance and factors of variation. *Livestock Science* 146(2–3), 105–114.

Quiniou, N. and Noblet, J. 1999. Influence of high ambient temperatures on performance of multiparous lactating sows. *Journal of Animal Science* 77(8), 2124–2134.

Randall, J. M., Armsby, A. W. and Sharp, J. R. 1983. Cooling gradients across pens in a finishing piggery: II. *Journal of Agricultural Engineering Research* 28(3), 247–259.

Rantzer, D. and Svendsen, J. 2001. Slatted versus solid floors in the dung area of farrowing pens: effects on hygiene and pig performance, birth to weaning. *Acta Agriculturae Scandinavica, Section A - Animal Science* 51(3), 167-174.

Renaudeau, D. and Noblet, J. 2001. Effects of exposure to high ambient temperature and dietary protein level on sow milk production and performance of piglets. *Journal of Animal Science* 79(6), 1540-1548.

Rioja-Lang, F. C., Seddon, Y. M. and Brown, J. A. 2018. Shoulder lesions in sows: a review of their causes, prevention, and treatment. *Journal of Swine Health and Production* 26(2), 101-107.

Robert, S. and Martineau, G. P. 2001. Effects of repeated crossfosterings on preweaning behavior and growth performance of piglets and on maternal behavior of sows. *Journal of Animal Science* 79(1), 88-93.

Roehe, R. 1999. Genetic determination of individual birth weight and its association with sow productivity traits using Bayesian analyses. *Journal of Animal Science* 77(2), 330-343.

Rolandsdotter, E., Westin, R. and Algers, B. 2009. Maximum lying bout duration affects the occurrence of shoulder lesions in sows. *Acta Veterinaria Scandinavica* 51(1), 44.

Rooke, J. A., Sinclair, A. G., Edwards, S. A., Cordoba, R., Pkiyach, S., Penny, P. C., Penny, P., Finch, A. M. and Horgan, G. W. 2001. The effect of feeding salmon oil to sows throughout pregnancy on pre-weaning mortality of piglets. *Animal Science* 73(3), 489-500.

Rosvold, E. M. and Andersen, I. L. 2019. Straw vs. peat as nest-building material-the impact on farrowing duration and piglet mortality in loose-housed sows. *Livestock Science* 229, 203-209.

Rosvold, E. M., Kielland, C., Ocepek, M., Framstad, T., Fredriksen, B., Andersen-Ranberg, I., Næss, G. and Andersen, I. L. 2017. Management routines influencing piglet survival in loose-housed sow herds. *Livestock Science* 196, 1-6.

Rosvold, E. M., Newberry, R. C., Framstad, T. and Andersen, I. L. 2018. Nest-building behaviour and activity budgets of sows provided with different materials. *Applied Animal Behaviour Science* 200, 36-44.

Ryan, E. B., Fraser, D. and Weary, D. M. 2015. Public attitudes to housing systems for pregnant pigs. *PLoS ONE* 10(11), e0141878.

Schmitt, O., Baxter, E. M., Lawlor, P. G., Boyle, L. A. and O'Driscoll, K. 2019a. A single dose of fat-based energy supplement to light birth weight pigs shortly after birth does not increase their survival and growth. *Animals* 9(5), 227.

Schmitt, O., Baxter, E. M., Boyle, L. A. and O'Driscoll, K. 2019b. Nurse sow strategies in the domestic pig: I. Consequences for selected measures of sow welfare. *Animal* 13(3), 580-589.

Schmitt, O., Baxter, E. M., Boyle, L. A. and O'Driscoll, K. 2019c. Nurse sow strategies in the domestic pig: II. Consequences for piglet growth, suckling behaviour and sow nursing behaviour. *Animal* 13(3), 590-599.

Schmitt, O., O'Driscoll, K., Boyle, L. A. and Baxter, E. M. 2019d. Artificial rearing affects piglets pre-weaning behaviour, welfare and growth performance. *Applied Animal Behaviour Science* 210, 16-25.

Skok, J. and Škorjanc, D. 2014. Group suckling cohesion as a prelude to the formation of teat order in piglets. *Applied Animal Behaviour Science* 154, 15-21.

Sørensen, J. T. and Fraser, D. 2010. On-farm welfare assessment for regulatory purposes: issues and possible solutions. *Livestock Science* 131(1), 1-7.

Sørensen, J. T., Rousing, T., Kudahl, A. B., Hansted, H. J. and Pedersen, L. J. 2016. Do nurse sows and foster litters have impaired animal welfare? Results from a cross-sectional study in sow herds. *Animal* 10(4), 681–686.

Strathe, A. V., Bruun, T. S. and Hansen, C. F. 2017. Sows with high milk production had both a high feed intake and high body mobilization. *Animal* 11(11), 1913–1921.

Straw, B. E., Dewey, C. E. and Bürgi, E. J. 1998. Patterns of crossfostering and piglet mortality on commercial US and Canadian swine farms. *Preventive Veterinary Medicine* 33(1–4), 83–89.

Swan, K. M., Peltoniemi, O. A. T., Munsterhjelm, C. and Valros, A. 2018. Comparison of nest-building materials in farrowing crates. *Applied Animal Behaviour Science* 203, 1–10.

Telkänranta, H. and Edwards, S. A. 2018. Lifetime consequences of the early physical and social environment of piglets. In: Špinka, M. (Ed.), *Advances in Pig Welfare* pp. 101–136. Woodhead Publishing, Cambridge, UK.

Theil, P. K., Lauridsen, C. and Quesnel, H. 2014. Neonatal piglet survival: impact of sow nutrition around parturition on fetal glycogen deposition and production and composition of colostrum and transient milk. *Animal: An International Journal of Animal Bioscience* 8(7), 1021–1030.

Thodberg, K., Jensen, K. H., Herskin, M. S. and Jorgensen, E. 1999. Influence of environmental stimuli on nest building and farrowing behaviour in domestic sows. *Applied Animal Behaviour Science* 63(2), 131–144.

Thøgersen, J. and Zhou, Y. 2012. Chinese consumers' adoption of a "green" innovation– the case of organic food. *Journal of Marketing Management* 28(3–4), 313–333.

Thomsson, O., Sjunnesson, Y., Magnusson, U., Eliasson-Selling, L., Wallenbeck, A. and Bergqvist, A. S. 2016. Consequences for piglet performance of group housing lactating sows at one, two, or three weeks post-farrowing. *PLoS ONE* 11(6), e0156581.

Thorup, F., Wedel-Müller, R. L., Hansen, C. F., Kanitz, E. and Tuchscherer, M. 2015. Neonatal mortality in piglets is more due to lack of energy than lack of immunoglobulins. In: *International Conference on Pig Welfare: Improving Pig Welfare-What Are the Ways Forward?*, p. 84. Wageningen Academic Publishers; Copenhagen, Denmark

Tuchscherer, M., Puppe, B., Tuchscherer, A. and Tiemann, U. 2000. Early identification of neonates at risk: traits of newborn piglets with respect to survival. *Theriogenology* 54(3), 371–388.

Turner, S. P., Camerlink, I., Baxter, E. M., D'Eath, R. B., Desire, S. and Roehe, R. 2018. Breeding for pig welfare: opportunities and challenges. In: Špinka, M. (Ed.), *Advances in Pig Welfare*, pp. 399–414. Woodhead Publishing, Cambridge, UK.

Tybirk, P., Sloth, N. M. and Jorgensen, L. 2012. *Danish Nutrient Requirement Standards (in Danish: Normer for Naringsstoffer)* (17th rev. edn.). SEGES Pig Research Centre, Axelborg, Denmark.

Tybirk, P., Sloth, N. M., Sonderby, T. B.. and Kjeldsen, N. 2015. *Danish Nutrient Requirement Standards (in Danish Normer for Naringsstoffer)* (22th rev. edn.). SEGES Pig Research Centre, Axelborg, Denmark.

van Beirendonck, S., Schroijen, B., Bulens, A., Van Thielen, J. and Driessen, B. 2015. A solution for high production numbers in farrowing units? In: *International Conference on Pig Welfare: Improving Pig Welfare-What Are the Ways Forward,* pp. 85–85. Wageningen Academic Publishers, Copenhagen, Denmark.

van de Weerd, H. and Ison, S. 2019. Providing effective environmental enrichment to pigs: how far have we come? *Animals* 9(5), 254.

van Dixhoorn, I. D., Reimert, I., Middelkoop, J., Bolhuis, J. E., Wisselink, H. J., Koerkamp, P. W. G., Kemp, B. and Stockhofe-Zurwieden, N. 2016. Enriched housing reduces disease susceptibility to co-infection with porcine reproductive and respiratory virus (PRRSV) and Actinobacillus pleuropneumoniae (A. pleuropneumoniae) in young pigs. *PLoS ONE* 11(9), e0161832.

van Nieuwamerongen, S. E., Bolhuis, J. E., Van der Peet-Schwering, C. M. C. and Soede, N. M. 2014. A review of sow and piglet behaviour and performance in group housing systems for lactating sows. *Animal* 8(3), 448–460.

van Nieuwamerongen, S. E., Soede, N. M., van der Peet-Schwering, C. M. C., Kemp, B. and Bolhuis, J. E. 2015. Development of piglets raised in a new multi-litter housing system vs. conventional single-litter housing until 9 weeks of age. *Journal of Animal Science* 93(11), 5442–5454.

Verbeke, W. 2009. Stakeholder, citizen and consumer interests in farm animal welfare. *Animal Welfare* 18(4), 325–333.

Verdon, M., Morrison, R. S. and Rault, J. L. 2019. Sow and piglet behaviour in group lactation housing from 7 or 14 days post-partum. *Applied Animal Behaviour Science* 214, 25–33.

Wathes, C. and Whittemore, C. T. 2006. Environmental management of pigs. In: Kyriazakis, I. and Whittemore, C. T. (Eds), *Whittemore's Science and Practice of Pig Production*, pp. 533–592. Blackwell Publishing, Oxford, UK.

Weary, D. M., Pajor, E. A., Bonenfant, M., Fraser, D. and Kramer, D. L. 2002. Alternative housing for sows and litters. Part 4. *Applied Animal Behaviour Science* 76(4), 279–290.

Weary, D. M., Jasper, J. and Hötzel, M. J. 2008. Understanding weaning distress. *Applied Animal Behaviour Science* 110(1–2), 24–41.

Westin, R., Holmgren, N., Hultgren, J. and Algers, B. 2014. Large quantities of straw at farrowing prevents bruising and increases weight gain in piglets. *Preventive Veterinary Medicine* 115(3–4), 181–190.

Widowski, T. M. and Curtis, S. E. 1990. The influence of straw, cloth tassel, or both on the prepartum behaviour of sows. *Applied Animal Behaviour Science* 27(1–2), 53–71.

Wiegand, R. M., Gonyou, H. W. and Curtis, S. E. 1994. Pen shape and size: effects on pig behavior and performance. *Applied Animal Behaviour Science* 39(1), 49–61.

Williams, A. M., Safranski, T. J., Spiers, D. E., Eichen, P. A., Coate, E. A. and Lucy, M. C. 2013. Effects of a controlled heat stress during late gestation, lactation, and after weaning on thermoregulation, metabolism, and reproduction of primiparous sows. *Journal of Animal Science* 91(6), 2700–2714.

Wischner, D., Kemper, N. and Krieter, J. 2009. Nest-building behaviour in sows and consequences for pig husbandry. *Livestock Science* 124(1–3), 1–8.

Wu, G., Bazer, F. W., Johnson, G. A., Knabe, D. A., Burghardt, R. C., Spencer, T. E., Li, X. L. and Wang, J. J. 2011. Triennial Growth Symposium: important roles for L-glutamine in swine nutrition and production *Journal of Animal Science* 89(7), 2017–2030.

You, X., Li, Y., Zhang, M., Yan, H. and Zhao, R. 2014. A survey of Chinese citizens' perceptions on farm animal welfare. *PLoS ONE* 9(10), e109177.

Yun, J., Swan, K. M., Vienola, K., Kim, Y. Y., Oliviero, C., Peltoniemi, O. A. T. and Valros, A. 2014. Farrowing environment has an impact on sow metabolic status and piglet colostrum intake in early lactation. *Livestock Science* 163, 120–125.

Yun, J. and Valros, A. 2015. Benefits of prepartum nest-building behaviour on parturition and lactation in sows–a review. *Asian-Australasian Journal of Animal Sciences* 28(11), 1519–1524.

Yunes, M. C., von Keyserlingk, M. A. G. and Hötzel, M. J. 2017. Brazilian citizens' opinions and attitudes about farm animal production systems. *Animals* 7(10), 75.

Zanella, A. J. and Zanella, E. L. 1993. Nesting material used by free-range sows in Brazil. In: Nichelmann, M., Wierenga, H. K. and Braun, S. (Eds), *Proceedings of the 3rd Joint Meeting of the International Congress on Applied Ethology*, p. 411. Humboldt-Universitaet, Berlin.

Zoric, M., Nilsson, E., Lundeheim, N. and Wallgren, P. 2009. Incidence of lameness and abrasions in piglets in identical farrowing pens with four different types of floor. *Acta Veterinaria Scandinavica* 51(1), 23.

Zurbrigg, K. 2006. Sow shoulder lesions: risk factors and treatment effects on an Ontario farm. *Journal of Animal Science* 84(9), 2509–2514.

Chapter 5

Optimising pig welfare at the weaning and nursery stage

Nicole Kemper, University of Veterinary Medicine Hannover, Germany

1 Introduction
2 Minimising the impact of weaning
3 Optimised management: the human key factor
4 Conclusion
5 Future trends in research
6 Where to look for further information
7 References

1 Introduction

1.1 Background

The most frequently mentioned key production parameters in pig farming are litter size and daily weight gain in the fattening period. However, between farrowing and fattening, the rearing of the weaned piglets in the nursery unit represents a very important and delicate period in the life of a pig with a number of risks for their welfare. Also at this bridging stage, optimal care and profound knowledge about pigs are essential for enabling both pig health and welfare in the young animals and laying the foundation for successful fattening. Since weaning itself probably represents the most stressful event in the life of a pig, maintaining their welfare is of enormous importance in this period to prevent stress and serious drawbacks, such as weight check or diarrhoea. Weaning leads to a stress response with increased levels of corticotrophin-releasing factor (CRF) and cortisol after activation of the hypothalamic pituitary adrenal axis (Moeser et al., 2017). Together with activation of the immune system and compromised digestive and absorptive capacities of the gut due to physiological changes in structure and function, a reduction in feed efficiency and lean tissue deposition can be observed (Campbell et al., 2013; Spurlock, 1997). Weight losses of approximately 100–250 g on the first day post-weaning are caused by this stress, and undernutrition can be induced by the change from milk to solid feed (Hennig-Pauka et al., 2019; Le Dividich and Sève, 2000).

http://dx.doi.org/10.19103/AS.2020.0081.05

This loss is usually recovered by about four days post-weaning if no further adverse effects occur (Le Dividich and Sève, 2000).

The following statements refer mainly to commercial production systems with separate farrowing housings, after which weaned piglets are transported to another barn or compartment, the nursery unit. The particular challenge in the nursery stage is the provision of a good start to the weaners from the beginning. In contrast to weaning in wild boars or pigs in a (semi-) natural environment, which happens progressively until the 9th and 16th weeks of life (Jensen, 1986; Newberry and Wood-Gush, 1985), weaning in pig production is an abrupt procedure, usually occurring at an age around four weeks of life or less with several extreme changes. Over the last several decades, the weaning age has decreased significantly, as described for French piglets from about eight weeks in the 1950s to about 25 days today (Prunier et al., 2010). In the European Union, legal regulations demand a weaning age of at least 28 days of age with an exception for piglets that are moved into specialised compartments (emptied, thoroughly cleaned and disinfected, and separated from sow housings), which can be weaned seven days earlier to ease all-in-all-out batch management (Council Directive 2008/120/EC, 2008). In organic systems, piglets are allowed to suckle longer until weaning starts between six and eight weeks of age. The actual weaning age in commercial production is a compromise between producing a high number of piglets per sow and per year via short intervals between successive farrowing on one hand, versus having piglets robust enough to cope with the weaning stress on the other. While it has been shown that three to four weeks are the optimal weaning time to maximise sow performance (Levis, 1997), the weaned piglets are still immature at this age and very sensitive to weaning stress. Weaning stress is caused by abrupt changes in the piglets' life, including, among others, separation from the mother sow, dietary changes, exposure to a new environment and mixing with unacquainted pen mates. It can impair piglet welfare seriously if no appropriate counter or mitigation measures are applied. Weaned piglets in commercial conditions face concurrent stressors, especially directly after weaning. Each single stressor that can be minimised is helpful in reducing negative impacts on piglet welfare. The most obvious effect of decreased welfare in weaners, especially in barren environments, is the occurrence of abnormal behaviours; with no or only few physical- and mental-stimulating activities and too little space allowance, stereotypes or injurious behaviours such as tail- or ear-biting can occur (Beattie et al., 1995; Stafford, 2010). The etiology of these behaviours is often multifactorial; thus the provision of an adequate environment and the reduction of possible stressors are even more important.

The process of weaning and the following period demand high adaptive capabilities to help the piglets cope with stress. Even if the first critical days in the nursery stage are overcome, the stockperson has to have a careful eye on the growing pigs to ensure they become healthy fatteners without any declines in health and welfare. In general, several factors are attributed to pig welfare and have to be addressed carefully in the nursery unit: (1) space allowance; (2) structuring the housing environment in functional areas; (3) availability of food, water and rooting material; (4) social contacts; (5) cleanliness; (6) efficient regulation of the barn climate; and (7) management in order to enable a stable social hierarchy (Bracke et al., 2002a,b; Scipioni et al., 2009). The more elements from the farrowing unit can be offered in a similar way in the nursery, the smoother the abrupt post-weaning changes will be resolved.

1.2 The importance of enabling optimised welfare in nursery pigs and the key challenges

Welfare is a basic precondition for healthy pigs showing a high performance, most apparent in good growth and feed efficiency, at the nursery stage. Even more important, the opportunity to manage pig welfare is required by law in many countries and is an indicator for professional pig farming following the best practices. In the nursery, the foundation is laid for successful fattening, and, under commercial aspects, the aims are to maximise growth performance and reduce morbidity and mortality. From a welfare point of view, the weaning process with all of the accompanying changes in the lives of piglets combined with their very young age represents a potential tremendous strain. Welfare impairments can be classified as environmental, nutritional and social stress (Lallès et al., 2004), and subsequent health problems (Fig. 1). Keeping these negative impacts as minimal as possible is urgently needed because otherwise the piglets are more prone to diseases, grow less and are difficult to manage, especially when lagging behind the rest of the group. These shortcomings in the nursery stage usually cannot be made up later on in the piglets' life; bodyweight at weaning, and weight gain directly post-weaning can affect growth until the end of the fattening period (Pajor et al., 1991).

A major challenge is the sensitivity of the young piglets at weaning and in the first weeks after that. Weaners are vulnerable and immature piglets, and require far more care than finishing pigs. These piglets are fragile and inexperienced. Before weaning, the sow's milk is the single nutrient and energy source for the suckling piglets (Bøe, 1991). The separation from their mothers leaves them without a milk supply, warmth and motherly care, including touching, comforting and pheromones (Niekamp et al., 2007). Because of this separation and transport to an unknown environment, they become mentally and physically stressed, even before arriving at the nursery barn. The nursery barn represents a new physical and microbial environment. Until then, the farrowing pen was the only environment they knew. The new surroundings, together with the changes in barn climate and microclimate, represent

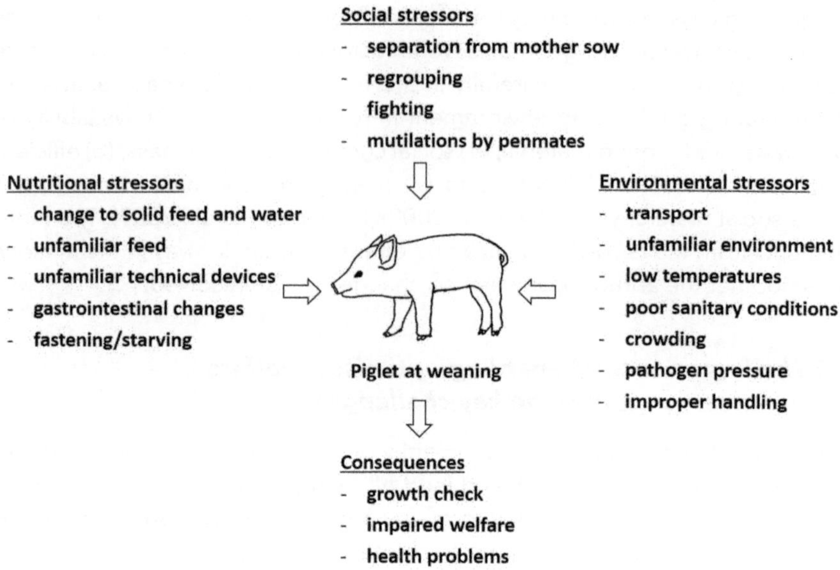

Figure 1 Potential stressors of different origins and possible consequences for piglets at weaning.

environmental stress. Nutritional stress results from the abrupt change from milk to solid feed, which goes along with a decreased nutrient intake and as a consequence, lower body heat. In turn, this decrease exacerbates environmental stress, especially in a very cold environment. Social stress is caused both by the separation from their mother and mixing with unknown piglets from other litters that can result in hierarchical fights. Moreover, weaning and movement to the nursery unit also represent a challenge to piglet health. Because health is one prerequisite to good welfare, it is of the utmost importance to prevent infections. Infections such as diarrhoea or respiratory diseases at weaning age are caused by pathogens present in the environment after insufficient cleaning and disinfection. These two factors are facilitated by not yet been fully developed immune response in the young animals and their lack of effective antibody formation. The immune system of weaners has not matured at this point, and even worse, the usual period of weaning falls within a critical lack of protective immunoglobulins in weaner blood. Their immature immune system is not yet capable of producing specific immunoglobulins (antibodies), and the immunoglobulins acquired via colostrum from their mother (maternal immunity) are already declining (Moeser et al., 2017). This lack of immunity appears between the second and fourth week of life, which is the exact time of weaning in commercial pig production systems. Negative effects, such as infectious diseases, especially post-weaning diarrhoea and even deaths, have been well described (Lallès et al., 2004; Weary et al., 2008). These negative effects can be aggravated when piglets are weaned with low weaning weights as could happen in large, uneven litters with a certain percentage of low-weighing and sometimes underdeveloped piglets (Hales et al., 2013; Koketsu et al., 2017). Only a very low percentage of piglets weighing less than 1 kg at birth were still alive at weaning in a study by Quiniou et al. (2002). In accordance, in another study, 86% of piglets weighing less than 0.80 kg at birth did not survive to weaning (Gondret et al., 2005). Birth weight and weaning weight correlate with one another, and as stated by Smith et al. (2007): 'Pigs that are heavier at birth are worth more money because they are likely to be heavier at the end of the nursery phase.' Therefore, special attention should be given to low-weight piglets from litters of highly prolific sows, not only in the farrowing unit (as reviewed by De Vos et al., 2014), but also in the subsequent weaning steps and while in the nursery stage.

2 Minimising the impact of weaning

2.1 The weaning process: preparation of procedures and facilities

Piglets are usually weaned at a certain age in order to empty the farrowing compartment and make room for the next farrowing batch. However, weaning is not only based on pure planning data, but also on the actual appearance of the piglet and its weight. Low birthweight piglets are defined in two ways; first as newborns with a birthweight below the 10th percentile of the mean birthweight of the litter, and second, with birthweight less than the mean birthweight minus two times the standard deviation (Cooper, 1975). Similarly, low-weaning weight piglets can be defined with the mean weaning weight of the batch as reference. Those piglets are at special risk for post-weaning welfare and health impairments.

The weaned piglets should be uniform and at a comparable weight level. All weaning preparations for sows and to where to move the single sow (slaughter or service) and the weaned piglets have to be completed in due time. The impact of weaning can be reduced by a careful planning of all processes and a thorough preparation of the facilities. The preparation of the nursery facilities in such a way that the piglets' environmental needs are adequately reflected is essential for giving them a good start in life. Good preparation includes calculations of the space and respective number of pens needed for the whole batch of weaned piglets. An adequate space allowance is important for weaner pigs, taking into account their expected weights when leaving the nursery unit for the calculation of stocking density. Legal requirements have to be met even though it has been shown that they often fail to provide a high amount of free space for different species-specific behaviours (Fels et al., 2019).

At the time of weaning in most production systems, the piglets are moved from the farrowing barns into the nursery unit. Only on some farms, the sows are removed from the farrowing pen, and the piglets stay there until they are brought to the fattening units. The following details refer mainly to the first situation. In general, experienced and trained farm personnel form an indispensable prerequisite for a smooth weaning process by keeping the negative impacts on the piglets as low as possible. When chasing piglets, a driving board should be used and all movements should be performed in calm manner.

Another point to consider in organising the weaning process is animal transportation. In the United States, Canada and other countries, transportation of weaned piglets to separate production facilities is a common practice. Transportation duration should be as short as possible, however, for the United States, duration up to 24 h in trailers without supplemental heat, feed and water are described (Lewis and Berry, 2006; Sutherland et al., 2009). Each transport affects the welfare of the newly weaned piglets as the animals experience stress, as reviewed comprehensively by Sutherland et al. (2014). The extent of stress and impaired welfare, especially by the concurrent weaning and transport, is influenced by season, respectively temperature, space allowance and duration of the transport and further aspects such as mixing with unfamiliar conspecifics. In addition, handling at loading, vibrations, noise, acceleration, water and feed withdrawal contribute to the extent of stress (Sutherland et al., 2014). In Northern America, segregated early weaned piglets are typically 14–21 days of age (Lewis and Berry, 2006), while in European countries and other regions of the world, they are usually older when weaned and transported, and therefore more robust than at younger age. Concerning mortality in weaners at transport, only limited data is available. Averós et al. (2010) stated that in 13.8% of 109 European transports, including 58 682 piglets, deaths occurred, and 0.07% of all transported piglets were found dead on arrival. However, these data did not refer to freshly weaned piglets, but to animals with 85–100 days of age, which were weaned between 21 and 28 days of age (Averós et al., 2010). Nevertheless, the identified risk factors, duration of journey and mean outside temperature with significant effects on mortality are most probably also true for freshly weaned piglets, as also shown in transport simulations by Berry and Lewis (2001). An overview of the special situation in transport of early weaned piglets is given by Lewis (2008), stressing the importance of ambient temperature and duration as well. Regarding ambient temperature, both summer (Lewis and Berry, 2006) and winter (Lewis and Berry, 2006; Wamnes et al., 2006) were described to negatively affect weaners' growth check and recovery time. Besides ambient conditions, the genetic background of the piglets might also influence the actual effect of transport on welfare, as stated by Averós et al. (2009). Sutherland et al. (2014) recommended to provide 0.06

or 0.07 m^2/pig at short (< 1 h) transportation trips in summer, and concluded that transports of up to 6 h might not cause more stress than weaning alone, but further studies on the short- and long-term effects of transport on weaners' welfare are needed to develop detailed recommendations. Each transport has an effect on piglet welfare, and therefore, should be kept as short as possible. Transportation vehicles, whether motorised or simple trolleys, should be equipped with adequate bedding to avoid physical damages to or cooling of the piglets.

In order to minimise the negative impact of weaning on pig welfare from environmental changes, the nursery barn must be prepared carefully before the freshly weaned piglets arrive. These preparations include measures to ensure both proper health and welfare by minimising infection risks and providing an adequate environment by meeting housing requirements of young piglets. Housing requirements of weaners are summarized also by Madec et al. (2003).

Independent of the housing system and the exact design, such as flooring or bedding, the facilities must be thoroughly cleaned and disinfected before the piglets arrive. The cleaning is crucial to remove all possible pathogenic microorganisms originating from the preceding batch of nursery piglets. Proper hygiene management based on all-in/all-out management and cleaning and disinfection of all housing facilities has been shown to be of the utmost importance for the pigs' health (Madec et al., 1998). As shown by Le Floch et al. (2006), piglets weaned with 28 days of life and housed in a housing facility of high sanitary status showed significantly higher final body weight 45 days after weaning compared to piglets kept in poor sanitary conditions; however, they also received antibiotic supplementation via standard diet and were kept in single cages. Sanitation measures are often unpopular tasks, but the importance of their careful performance cannot be stressed enough. There are many possible pitfalls in cleaning and disinfection, but the most important points to consider in the nursery are listed:

- thoroughly remove faecal soiling from the previous batch of piglets by soaking and cleaning as soon as the previous group of pigs is moved to the fattening unit;
- dust should be removed, including from equipment that is not easy to reach, such as fans, air ducts, and feed and water pipes above head-height, because pathogens attach to dust and can survive and stay infectious over extremely long periods of time (Schulz et al., 2016);
- use effective disinfectants with a broad spectrum, and, if needed, also specifically against known, identified pathogens;
- avoid gaps in effectiveness of disinfection caused by protein failures, soap failures or low temperature failures; and

- also clean and disinfect technical equipment, such as trolleys on which the piglets are held or moved.

These cleaning and disinfection procedures should be incorporated into a general farm biosecurity concept, including, among others, proper batch management with all-in/all-out, boot hygiene, shower facilities at the entrance, and rodent control. One important point to consider after cleaning and disinfection is the drying of the barn. Drying is essential for a proper disinfection. After cleaning and before application of the disinfectant drying avoids dilution effects of the disinfectant by puddles, and after disinfection drying prevents the animals from drinking and ingesting disinfectant residues. No wet spots should be present anymore when the piglets arrive.

Even with the most careful sanitising, exposures of the pigs to environmental pathogens cannot be avoided. To a certain degree, this exposure is even desirable, because it can lead to a resilient immunity when the immune system of the pigs is fully developed. However, due to the lack of immune competence in weaners at the age of two–four weeks, this option is not applicable in the early nursery stage, and instead, everything has to be done to ensure the piglets are strong enough to cope with all stressors that occur immediately post-weaning. Besides the reduction in pathogen concentrations and possible transmission routes in the weaners' environment, the immediate surroundings should be prepared and equipped in such a way that help the pigs to adjust and, at a minimum, does not represent a source of further health and welfare risks.

First, the temperature of the pens is the key; piglets at weaning experience cold stress because they are removed from the farrowing environment providing a heated creep area for them as well as from their mothers, which also provided warmth and other nurturing functions. Piglets lack thermoregulation capacities, meaning that they cannot regulate their body temperature sufficiently because they do not possess the necessary insulating fat reserves at this point and additionally, cannot consume satisfactory amounts of feed as an energy source in order to maintain their body temperatures high enough, especially after the sudden change from milk to solid feed. The temperature in the pen at the animals' body height in the first week after weaning should be between 28°C and 32°C for piglets weighing about 5 kg; especially in the lying area this temperature is required (Mavromichalis, 2001; Muirhead et al., 2013). For weaners, the lower critical temperature ranges from 26°C to 28°C during the first week and from 23°C to 24°C during the second week after weaning (Noblet et al., 2001). The nursery unit must be heated up and prewarmed to these temperatures before the weaners arrive. Keeping this temperature at the animals' body height depends significantly on the design and types of equipment in the pen. In straw-based systems, a sufficient amount of bedding material should be provided to keep the piglets warm even without an additional heat source.

Floor design is of special importance; in slatted floor systems, drafts can develop easily, and depending on the floor type, ammonia emission rates can increase (Aarnink et al., 1997). Lying mats (made of rubber or plywood), pen divisions, and shelters or covers can help to establish segments with an optimal microclimate. In this way, structuring the pen into separate dunging and lying areas is supported. Regarding flooring, a detailed overview was provided by the EFSA Panel on Animal Health and Welfare (2005). The floor has to be designed in such a way that the risk of claw and leg injuries due to inappropriate slat and gap widths and rough materials is minimized. In a British study, KilBride et al. (2009) compared the impact of different floor types on the occurrence of foot and limb lesions, and showed that the presence of bursitis, capped hock and calluses was lowest in rearing pigs between 6 and 22 weeks of age housed on soil (outdoor rearing), followed by solid concrete with a mixture of deep and sparse bedding, and highest in pigs housed on fully slatted floors. The positive effect of bedding was confirmed by Munsterhjelm et al. (2009).

Concerning ventilation for freshly weaned pigs, an air renewal rate of not more than 0.1 m^2/h/kg of body weight is recommended (Mavromichalis, 2001). As appropriate for pigs at all ages, relative air humidity should range between 60% and 80%. In a two-year-experiment, Done et al. (2005) found no significant effects of ammonia levels of up to 37 ppm and dust concentrations of up to 9.9 mg/m^3 (inhalable fraction) on general pathology, respiratory tract pathology and microbiological findings in lungs, tracheae and nasal cavities in weaners kept under these conditions for 37 days. The authors conclude that such high concentrations are unlikely to be associated with respiratory diseases in weaners under the conditions of a good all-in-all-out management with thorough cleaning and disinfection, a high standard of nutrition and husbandry and a special attention to pigs' welfare (Done et al., 2005). However, as weaned piglets show an aversion against ammonia at levels of and above 10 ppm (Jones et al., 1996), ammonia concentrations in the nursery should be kept below this concentration to promote welfare.

Before taking the weaners to the new facility, the equipment, especially feeders and drinking nipples or other sources of water must be checked and be in a proper, functioning condition. The piglets must acclimate to the water source, especially when a different type of water cup or nipple was provided in the farrowing unit. Sufficient amounts of fresh water, without any deficiencies in quality should be accessible ad libitum, with recommended water flow rates at nipple drinkers of 500 mL/min (Scipioni et al., 2009). Besides the drinkers' efficiency, the number and the position of any watering unit are highly relevant. Before the piglets arrive, fresh, clean water should be poured into cups or should trickle from the nipples to help them to become familiar with these devices. All other equipment, such as manipulable materials, should be clean at arrival. Bedding or enrichment material should be dry and located in the assigned areas and at room temperature. In addition, the hospital pens should also be prepared before arrival.

2.2 Environmental stress: strategies to reduce possible welfare impacts

For the provision of a suitable environment, not only the technical set-up has to be animal-friendly, but also the climate is of key importance. As already mentioned, focus should be laid on the optimal barn temperature for the respective age of the piglets. In the first two weeks post-weaning, piglets need, in addition to the above-mentioned temperatures at animals' body height, a room temperature of about 25–30°C in order to maintain thermal neutrality, depending on their weight and the floor type (Muirhead et al., 2013). Even though it was shown in experiments that piglets weaned at three weeks of age can adapt to long-term cold exposures (12°C for three weeks), the physiological consequences as changes in glucose turnover, thyroid hormones and catecholamines have significant negative impacts especially on weight gain, but possibly also on health (Herpin et al., 1987). Le Dividich and Herpin (1994) summarized the effects of climatic conditions on performance, metabolism and health of weaned pigs in a review. With increasing weight and in bedded systems, temperatures can be lower. In preference experiments, weaners aged between four and eight weeks preferred straw-bedded areas for resting at ambient temperatures of 18–21°C, while at 25–27°C, concrete floors were selected (Fraser, 1985). This indicates that the weaners' preference for type of flooring depends strongly on temperature. Instead of only checking and comparing the detailed measured values of ambient climate via different climate data loggers, the animals and their behaviour should be observed at each tour through the compartments. Their behaviour is the ultimate indicator of climatic comfort and welfare in general. Healthy pigs in their climatic comfort zone do not huddle together or spread out as much as possible. They do not pant. It has to be stressed that pigs in conventional nurseries without any outdoor access are kept in a manmade environment for 24 h a day. Their possibilities to choose or avoid unfavourable environmental conditions are very limited. For instance, if the underfloor heating is too hot, piglets will avoid it as far as possible, but with that avoidance, available space for free movements is limited (Fig. 2). The responsibility for the provision of an adequate animal-friendly environment lies with the farmer and the stockmen on the farm.

Pen cleanliness has to be checked and controlled daily, and bedding and enrichment material has to be renewed if necessary. From a technical point of view, all devices, such as fans, heaters, lights, feeders and drinkers have to be

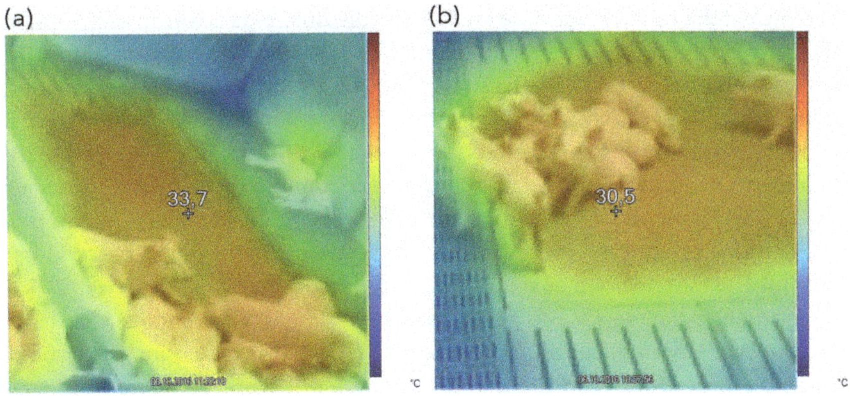

Figure 2 Piglets avoid the over-heated heating panel (a) and use the correctly heated one (b) (Photo with thermal imaging camera: ITTN, TiHo).

checked daily for proper function. Even with careful preparation of the facilities, movement into a new, unknown environment is a stressful situation for the pigs. In order to avoid health and welfare impairments, not only a good preparation of the pure infrastructure is necessary, but the pens should also be large enough and equipped with enrichment to smooth negative impacts caused by other stressors, especially social stress. If the space is limited, high animal density and crowding impacts animal welfare and animal health, and with those negative effects, also influences performance. Several studies showed the negative effect of decreased space allowance on feed intake and weight gain in weaners (Brumm et al., 2001; Stojanac et al., 2014). Besides significantly lower daily weight gains and higher feed conversion rates, Stojanac et al. (2014) detected a significantly increased occurrence of *Salmonella* species, higher mortalities and higher culling rates in weaners kept at the highest stocking rate of 0.16 m^2 available space per pig, compared to 0.25 m^2 and 0.32 m^2 per pig. These findings emphasise the importance of adequate space allowance not only for production parameters, but also for the safety of pig meat for human nutrition. In a South Korean study, reduced space allowance affected not only daily weight gain negatively, but resulted in a linear increase in plasma-cortisol and Interleukin-1ß in weaners (Oh et al., 2010). From their results, the authors suggest a space allocation between 0.30 m^2 and 0.43 m^2 per animal for weaners until 15 kg of bodyweight to maximize growth performance and immune response (Oh et al., 2010). In the study by Smith et al. (2004), increasing space allowance from 0.23 m^2 per pig to 0.35 m^2 was associated with increased body weight at approximately ten weeks of age. Regarding covered floor spaces by freshly weaned piglets, an individual piglet of 7 kg bodyweight covered 793 cm^2 in standing position, 761 cm^2 in sitting posture and 838 cm^2 in a recumbent position (Fels et al., 2018a). Covered space increased linearly with weight, and at the end of the rearing period with 25 kg, individual pigs covered 1.704 cm^2 in a standing position, 1.687 cm^2 sitting and 1.798 cm^2 in a recumbent posture, on average (Fels et al., 2018a). These static space data can be useful in determining the amount of free space, in relation to the offered pen size, which remains per pig after deducting the space needed by the pig's body. In more detail, the influence of space allowance is discussed by Brumm and Gonyou (2001) and by the EFSA Panel on Animal Health and Welfare (2005). Especially, when pigs are crowded, the gut microflora can be altered and pathogenic bacteria can proliferate, with diarrhoea as a consequence (Jayaraman and Nyachoti, 2017). Management practices for promoting a good gut health in weaners have been reviewed by Jayaraman and Nyachoti (2017) in detail, emphasising the need to avoid crowding stress in addition to feeding strategies and sanitation.

Environmental enrichment after weaning has been shown to improve feed efficiency and reduce the incidence of diarrhoea (Oostindjer et al., 2010). Compared with a barren environment, moderate bedding has positive impacts not only on welfare and in the reduction of post-weaning diarrhoea, but also in increasing growth of the weaners (Munsterhjelm et al., 2009). Increased space allowance alone does not result in decreased aggression or other harmful behaviour, but combined with enrichment, positive effects were shown (Beattie et al., 1996). Martin et al. (2015) examined the influence of an enriched environment in the farrowing unit on play behaviour and on post-weaning social and cognitive development in weaners, and concluded that a constant enrichment is needed to regularly stimulate play behaviour in order to facilitate animal welfare. As it is highly beneficial to provide structural and nutritional elements both in the farrowing and in the rearing unit, the provision of such enrichment materials should be started at a very young age. Such continuous enrichment can yield positive effects on post-weaning performance (Vanheukelom et al., 2011), but it should be noted that a pre-weaning provision has to be followed by a post-weaning provision, because any later loss of enrichment is assumed to lead to compromised welfare in the weaned piglets that were used to prior enrichment (Oostindjer et al., 2014). A decrease in mutual oral manipulations and skin lesions was detected in suckling piglets when enrichment with straw towers, paper dispensers, chew-bones and ropes was offered starting in the farrowing unit, in which the piglets began to explore the material starting on their second day of life and continuing (Fels et al., 2016). However, manipulable materials at this early stage do not seem to reduce aggressive interactions after weaning and when mixing piglets from different litters, but positive long-term effects can be assumed (Rauterberg et al., 2018). Interestingly, the use of enrichment material at this early age seems also to be linked to feed intake, because the interaction with substrates increases the flexibility towards new objects, including food, and allows the piglets to develop a broader explorative and feeding-related behaviour (Oostindjer et al., 2014). Moreover, it was shown that enrichment increases the behavioural flexibility of pigs in general and decreases fear of novel objects (Tönepöhl et al., 2012). After weaning, enrichment material can distract the piglets from performing stress-related abnormal behaviour (Fraser et al., 1991). Therefore, enrichment materials are also one set of the main preventive and emergency tools in cases of tail biting. Summarising the facts about enrichment, even the provision of small quantities of material was shown to produce positive effects (Fig. 3). Enrichment is an effective tool for increasing piglet welfare, but the substrates have to be in a proper hygienic quality because their potential to carry and transfer pathogens and toxins has already been shown; peat and material originating from maize can especially present a hygienic risk because they possibly contain mycobacteria and mycotoxins (Wagner et al., 2018).

2.3 Case study: implementation of elevated platforms in the nursery unit

One practical and inexpensive method for increasing space allowance in existing nursery pens is implementation of elevated platforms. In this way, split-level housing systems are established, which not only offer more space, but also more structure in the piglets' environment. This idea was first mentioned in the scientific literature about thirty years ago (Fraser et al., 1986; Phillips and Fraser, 1987). However, it has never been practiced on a larger scale, but has recently

Figure 3 Provision of linen straw as enrichment material in a trough, one possibility to provide substrates in nurseries with fully slatted floors (Photo: ITTN, TiHo).

drawn attention because of the ongoing debate concerning animal welfare in some European countries and the potential benefit that it possibly imparts to the piglets. Therefore, we examined the use of elevated platforms for weaners on farm and evaluated possible advantages and limitations in more detail (Fels et al., 2018b,c). In the experimental nursery unit, a platform of 5 m^2 was not only installed to provide more space per animal, but also to offer enrichment by nine different manipulable materials and toys installed on the upper level of the platform aimed to create an activity area (Fig. 4). Moreover, it was hypothesised that the additional structural elements provided by the elevated platform lead to the establishment of different functional areas, that is, for resting, activity, feeding and dunging. Forty piglets were weaned and grouped together in the two-level-pen for 33 days.

Scientific analyses included behavioural observations via video-recordings using the scan-sampling method to determine the location of the piglets within the two-level pen and their body postures (standing/sitting and lying) (Fels et al., 2018c). Moreover, ten focal animals were chosen per pen and followed continuously during the housing period to assess if they actually use the platform, and to reveal individual differences in space utilisation (Fels et al., 2018b). In order to test possible impacts on barn climate and the general hygiene, floor soiling was documented, and ammonia and air velocity were measured, too (Fels et al., 2018c). Our results showed that the piglets used the offered space to establish an activity area on the platform, a lying area under the platform, and dunging and feeding areas. This structuring led to the desired dung concentration in the dunging area; however, the animals in this study also defecated in all other areas, but to a lower extent, however, no effects of the elevated platform on the barn climate were detected (Fels

Figure 4 Enriched two-level pen in the nursery (Photo: ITTN, TiHo).

et al., 2018c). The elevated platform was used continuously during the housing period (Fig. 5), and especially in the piglets' active hours in the afternoon (Fels et al., 2018c). Each of the focal animals used the platform during the housing period; however, individual differences became apparent (Fels et al., 2018b).

The studies clearly showed that a two-level pen was a good option to increase the available space, as well as availability of enrichment and structure in existing nurseries and with those additions, piglet welfare. In our studies, the platform was offered as additional space and was not included in stocking density calculations (m^2 per animal). The legally required space was covered already by the space at the base level, and further studies might evaluate whether some recognition of the platform to fulfil (legal) spatial requirements can be accepted based on animal welfare aspects to increase the farmers' willingness to equip their nurseries with two-level pens. Full recognition of the additional space should not be recommended based on the present studies, because behavioural observations revealed that the piglets rested at night mainly under the platform and therefore, the additional space was not used in periods of inactivity (Fels et al., 2018b).

2.4 Nutritional stress: optimising feed and water provision

Sufficient water and nutrient intake is essential for welfare; otherwise, the piglet becomes impaired by thirst, hunger, and cold. At the usual age of

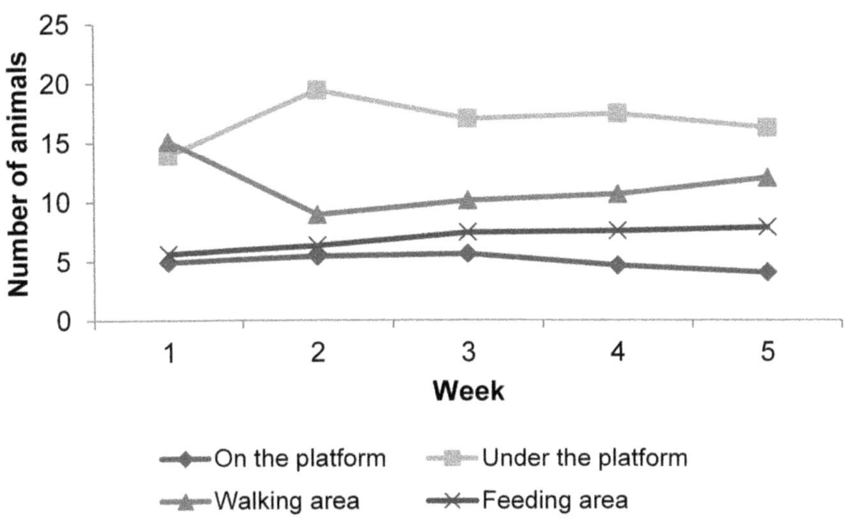

Figure 5 Mean number of animals per five minutes in the different pen areas during week 1-5, total number of 40 animals in the two-level pen (reprinted from Fels et al., Journal of Applied Animal Welfare Science 21: 267-282, 2018; with the permission of Taylor and Francis Group, www.tandfonline.com).

weaning at three to four weeks of age, sow's milk represents the main source of nutrient intake. In the farrowing unit, the sow nurses her whole litter at frequent intervals, providing them with both liquid and solid nutrients. After weaning, the piglets have to face the facts that there is no sow and no milk anymore, and they have to satisfy hunger and thirst via different, hitherto mostly unknown, resources. It takes some time to get the piglets accustomed to these new supplies, and voluntary nutrient intake decreases during that time. Bruininx et al. (2001) showed in an experiment without creep feed provision during the suckling period, that after weaning, single piglets fasted up to 50 h before they start eating, while 50% of the piglets started eating within 4 h after weaning. In another experiment, piglets without access to creep feed took longer to start feeding, up to 76 h after weaning, than piglets that got creep feed during lactation with 20 h longest before initial feed intake after weaning (Bruininx et al., 2002). Decreasing the number of animals per feeder hole from 7.5 freshly weaned piglets per feeder hole to 3.75 resulted in a decrease of average time to the onset of feed intake from 39.16 h to 31.87 h and increased feed consumption (Laskoski et al., 2019). During the critical period after weaning, in addition to changes in the gastrointestinal tract, such as crypt hyperplasia and villous atrophy, a weight loss can be observed in nearly each pig (Le Dividich and Sève, 2000; Pluske et al., 1997). Usually after two to three days longest, piglets should start feeding, but even then, they are still not gaining weight. This is caused by the fact that besides the adaptation to the technical facilities, another drawback is the slow habituation of the porcine digestive tract to the new feed composition that contains ingredients, which have to be digested by enzymes (Pluske et al., 1997). Before weaning, piglets were used to sow's milk being rich in animal protein and liquids, and their enzyme expression has to be increased in order to digest the new diet. To facilitate this process, creep feed should already be provided during the suckling period. At the time of weaning, the digestive system of the piglets is still immature, characterised by the mentioned reduced activity of enzymes, changes in intestinal morphology and reduced digestion in the small intestine (Jayaraman and Nyachoti, 2017). The metabolised energy intake (ME) in the first week post-weaning is estimated to be around 60–70% of the pre-weaning milk intake, increasing to 100% only after approximately 2–3 weeks post-weaning (Campbell et al., 2013; Pluske et al., 1997). In these first weeks in the nursery unit, the piglets' digestive system is not only limited in absorptive and digestive capacities, but also in its intestinal barrier function (Moeser and Blikslager, 2007). Therefore, cellular permeability is increased, and pathogens and toxins can cross the epithelium and cause further problems such as infections, inflammation and growth reduction (Campbell et al., 2013; Moeser and Blikslager, 2007; Oostindjer et al., 2014). This process is exacerbated by immune system activation at weaning, for instance, an upregulation of pro-inflammatory cytokines during weaning (Campbell et al., 2013). Thus, piglets are very susceptible to diarrhoea during this period, and it is no coincidence that diarrhoea is often observed during the second and third weeks post-weaning (Madec et al., 1998). Diarrhoea is not only a health, but also a welfare problem, and should therefore be prevented as much as possible.

One of the most important prevention measures is to smoothen the abrupt nutritional changes after weaning. Piglets often show food neophobia, increased by the general stressful situation at weaning (Oostindjer et al., 2014). In order to familiarise the piglets to solid feed and to trigger enzyme expression, creep feed should be offered already in the farrowing pen, as already mentioned. Piglets under commercial conditions eat small amounts of creep feed from an age of 12 days onwards (Pajor et al., 1991). At this young age, the provision of creep feed in the farrowing unit does not necessarily lead to a high intake of solid feed by the young piglets, varying individually between <1% and 20% of the daily metabolic intake (Pluske et al., 1995). However, provision of creep feed helps the piglets become accustomed to solid feed, and a positive correlation between pre- and post-weaning feed intake was shown under commercial conditions (Bruininx et al., 2002; Sulabo et al., 2010). A promising way to increase feed intake in suckling piglets, and, subsequently, in weaners is to allow the piglets to interact more with the sow and to learn to eat solid food by olfactory, visual and auditory cues learned from her (Oostindjer et al., 2014). This social learning is best when the sow and her piglets eat the same, or similar in taste feed from the same location (Oostindjer et al., 2014). Piglets interacting more with the sow in loose-housing systems pre-weaning showed more playing behaviour and less damaging behaviour in the first two weeks after weaning (Oostindjer et al., 2011)

After weaning, the same creep feed should be given in the nursery unit in order to offer familiar food to the piglets and to avoid abrupt changes. If possible, synchronal feeding should be enabled, where all piglets can access feed at the same time. Besides the provision via the feeders, small amounts of pellets should be offered on floor mats or on solid floor on the first day after weaning (Mavromichalis, 2001). In order to offer the feed via different options, mats and troughs, for instance, help to increase feed intake, because the pigs' curiosity is addressed, and they can discover different possibilities. Small quantities of pellets should be given up to three times per day to stimulate feed intake. Another approach for increasing feed intake is to offer food out of specially designed feeders that aim at the explorative behaviour in the piglets. In this way, it was shown that by offering creep feed via a 'playfeeder', which stimulated exploration, feed intake increased (Kuller et al., 2010). Placing mats in proximity to the feeder helps the piglets to get accustomed to the feeder and consume more feed in this way (Mavromichalis, 2001). These mats should be equipped with a rim to avoid rolling, pushing and wasting of the pellets. After two to three days post-weaning, the piglets should become accustomed to the feeders. It has to be kept in mind that heavy piglets at weaning are often slower in adapting to dry diets than their lighter litter mates, because they were used to drink large quantities of milk instead of eating creep feed offered in the farrowing pen. Same attention with regard to diet has to be given to all piglets. Another point to consider is that provision of a high amount of milk is important for a high feed intake capacity even later in life; therefore, supplementing the sow's diet during late gestation and lactation helps to improve piglets' performance and provides them a good start in life (Kirkden et al., 2013). Other approaches for reducing nutritional stress by the use of special feed ingredients such as prebiotics and probiotics, n-3 long chain polyunsaturated fatty acids, n-6 linoleic acid or glutamine still need to be investigated (Campbell et al., 2013). Moreover, the correlation between feed intake and other important environmental factors is of importance. Feed intake and ambient temperature are negatively correlated; however, with the weaners' higher requirements for temperature, a negative impact on feed intake, which is directly related to high ambient temperature, is rarely observed, except in very hot summer months (Dong and Pluske, 2007). Other than temperature, relative humidity does not seem to influence the performance of weaners (Noblet et al., 2001).

Concerning water provision, the piglets have to learn that after weaning water is provided from a source other than the solid feed, unlike the situation at the sow, and moreover, the way in which to use drinkers. Sufficient intake of clean, fresh and safe water is of utmost importance, affecting feed intake and therewith the growth rate. Eating and drinking times in newly weaned pigs were shown to be positively associated (Dybkjaer et al., 2006). However, it can take more than one week after weaning to reach the same intake level of fluids in the pigs as observed before weaning (Brooks and Tsourgiannis, 2003). Nipples, especially, can cause problems because their mode of use is not in line with the natural drinking position of pigs, which is done with their heads in a low position. Moreover, the delivery rate from nipple drinkers is often too low. Barber et al. (1989) described best effects on both feed intake and growth rate with water delivery rates of 0.700 L/min. Bowls and open water sources with a large reservoir capacity offer a more convenient and obvious way to drink and meet the needs of the weaners (Barber et al., 1989). As for feed, several options of water sources help the piglet to increase their intake of liquids. Concerning the numbers of drinkers per weaner, Dong and Pluske (2007) recommended a minimum of one drinker per 8–10 pigs. Drinkers for piglets should also be offered in the farrowing pen to promote water intake after weaning.

2.5 Social stress: approaches to lower welfare impacts

Social stress in weaned piglets is caused by separation from their mother and by regrouping with unfamiliar pen mates. Directly after separation, the stress caused by the separation from the mother and litter mates is mainly expressed vocally, which is in most cases only temporary. Longer-lasting is the occurrence of agonistic interactions in the new social group; directly after weaning, piglets fight to establish a new social hierarchy. Aggressive fighting is one of the most important stressors at weaning (Kutzer et al., 2009), resulting in social and physiological stress which potentially influences growth rate, causes injuries and even leads to death (Arey and Edwards, 1998). Mixing piglets, especially in combination with other adverse factors, such as high stocking density, causes stress, and if this social stress becomes chronic, it not only results in decreased growth, but also in immunosuppression and impairment of the epithelial barrier as shown in 10-week-old grower piglets (Li et al., 2017). Usually, the new rank order in a mixed group of piglets is established within 48–72 h (Fels et al., 2014a). The main reason for this increase in restriction of aggressive behaviours can be explained by the avoidance strategies of the subdominant animals (Jensen, 1982). The dominant animals are the ones that engage in more frequent fighting and more often initiate fights than the aggressors (Fels and Kemper, 2017; Rhim et al., 2015). Moreover, they often show higher weight gains by having easier access to resources (Schoenfelder, 2005). In this way, the rank order shows a direct effect on animal health and welfare. Once a social hierarchy is formed, it usually remains stable if no further animals join the group and if conditions are favourable. However, aggressive behaviours, such as dominating and threatening behaviour, can be observed even with an established rank order, when feed is restricted, animal-feeder ratio is limited, stocking density is adverse, or retreat possibilities are lacking (Fels et al., 2014b). With regard to stocking density, most fighting occurs in intermediate stocking densities, when animals cannot avoid one another, and the dominant ones can benefit from gaining access to resources by expelling other animals (Weary et al., 2008).

In order to reduce the impact of aggressive behaviours caused by post-weaning mixing and the encroaching welfare impairments, several strategies were tested, mostly without any success. These strategies only delayed the occurrence of agonistic behaviours but did not lead to suppression (Fels et al., 2014a). Distractive feeding and straw provision did not cause a steady reduction in aggression in pigs (Arey and Edwards, 1998; Arey and Franklin, 1995). Other approaches, such as the application of sedatives or deodorant substances or the reduction of light, only led to a delay in occurrence of aggression (Fels et al., 2014a). Approaches that are more promising aim at shifting mixing to an earlier stage. Under semi-natural conditions, piglets leave the nest and mix with piglets from other litters at an age of ten days (Jensen and Redbo, 1987) with only low levels of aggression. Subsequently, social relations between the piglets are established by playful, aggressive interactions, resulting in positive social integration (Petersen et al., 1989). Under farm conditions, it was shown that aggression decreases when piglets from different litters are allowed to co-mingle before weaning because the social hierarchy is formed more easily with less fighting and consequently, with fewer injuries (Morgan et al., 2014; Schrey et al., 2019; and as reviewed by Weary et al., 2008). At the age between five and twelve days, the establishment of the social hierarchy was found to be 'especially good', as reported by D'Eath (2005). These observations led to several technical approaches to enable piglets to mix with non-litter mates in the suckling period before weaning, all showing positive effects resulting in decreased aggressive behaviours and injuries, not only directly after weaning, but also longer-lasting over the fattening period (Olsson et al., 1999; Parratt et al., 2006). Such early socialised piglets seem to be better prepared for social and non-social challenges at weaning (Hillmann et al., 2003), which emphasises the importance of the early life experience for long-lasting animal welfare. Technically, the easiest and cheapest way to enact socialisation is a piglet door between neighbouring pens, thus allowing the piglets to mix and move around in a larger area than just the direct farrowing pen. Kutzer et al. (2009) showed a reduction of stress after weaning and of skin lesions due to agonistic interactions as well as a better performance in piglets with early social contact starting 10 days post-partum, via piglet doors between two adjacent pens. Another possibility is to offer a communal piglet area in which piglets of different litters can mix. Finally, group housing of lactating sows in multi-suckling systems also allows contact between the litters. Several studies have shown reduced aggression at weaning in the piglets born and raised in these systems compared to piglets from conventional single-housing pens (Bohnenkamp et al., 2013; van Nieuwamerongen et al., 2014; Schrey et al., 2019).

If a pre-weaning socialisation cannot be realised, at least mixing piglets from different litters should be avoided as far as possible. With larger group sizes in the nursery unit, this mixing becomes especially more difficult. It was shown that after weaning and mixing, aggressive behaviour is mainly observed between non-litter mates and increases with higher numbers of litters included in the group (Hwang et al., 2016; Fels and Kemper, 2017). Mixing should be minimised to decrease agonistic behaviours, but also to reduce the risk of pathogen transmission as mixing has been described as very significant risk factor for *Streptococcus suis* (Dee et al., 1993) and *Actinobacillus pleuropneumoniae* infections (Tobias et al., 2014). This point also has to be taken into consideration when pre-weaning mingling of piglets is performed.

Other behaviours directed to the pen mates include massaging, belly-nosing, mounting, or tail-biting and indicate a high level of stress (Dudink et al.,

2006). With these behaviours, compromises in welfare follow not only for the animal performing the behaviour because of high stress levels, but also for the target animal (Oostindjer et al., 2014). Concerning the time points and period of occurrence, these can vary depending on the weaning age. The younger the piglets are at weaning, the more marked the behavioural disturbances are (Prunier et al., 2010). For instance, in Germany, in which piglets are usually weaned at three or four weeks of age, peaks of tail-biting behaviour are usually seen in the second and third week after weaning (Veit et al., 2017), whereas in Sweden, tail-biting is assessed later and at lower incidences mainly in the fattening unit (Wallgren et al., 2016). In this context, the technology of segregated early weaning has to be mentioned, which is common in large pig farm enterprises with multi-site production with all-in all-out pig flow across North America. It is characterised by weaning piglets between 7 and 21 days of age (mostly between days 12–16), which are transported to isolated housings in nurseries units. Segregated early weaning aims at breaking the infection chain by utilising acquired maternal (passive) immunity from the dam before piglets develop their own active immunity in response to pathogens (von Borell, 2000). However, it was shown that increasing weaning age from 12 to 21.5 days in a multi-site swine production system significantly improved wean-to-finish growth, with weight at weaning having more influence on the post-weaning growth than changing the feeding regime (Main et al., 2004). Worobec et al. (1999) investigated the effects of very early weaning at 7 days on piglets' behaviour and performance, and found behaviour problems such as nosing and chewing pen-mates related to welfare, and lower weights at 6 weeks of age compared to piglets weaned at 14 or 28 days. They concluded that piglets do not adapt well to very early weaning at 7 days, with slightly better adaption to weaning at 14 days and significantly better adaption to weaning at 28 days (Worobec et al., 1999). Another point to consider is that weaning very early in lactation is also more difficult for the sow (Robert et al., 1999). Due to the high impact of segregated early weaning on the welfare of piglets, also including management of low-weight piglets and transportation, weaning piglets closer to 21 days of age was recommended to find both production and welfare advantages (Robert et al., 1999).

It was shown that piglets from loose-housed sows expressed less belly nosing and other manipulative post-weaning behaviours (Oostindjer et al., 2014). Of course, in this context, the environment plays a key role; piglet-directed behaviours, such as mutual oral manipulation, occur especially in barren environments and are very low in piglets raised and weaned outdoors or in other complex environments (Weary et al., 2008). Therefore, an effective way to react on or to prevent behavioural disturbances in the nursery is the provision of adequate enrichment or bedding material. Bedding, especially straw, has been shown to be a recreational stimulus with both representing a stimulus for rooting and a source of thermal and physical comfort (Fraser et al., 1991).

3 Optimised management: the human key factor

Management during the nursery stage includes batch management strategies, nutrition practices, maintenance of hygiene and biosecurity standards, and disease prevention protocols in addition to animal welfare considerations. Even if the latter are often not explicitly mentioned, careful optimisation of the other factors form a solid foundation for achieving high welfare standards if a few further above-mentioned aspects are included. However, even if all these aspects are considered, the success of a nursery, which enables animal welfare, depends on one key factor: the stockperson. The stockman is of utmost importance, because without an excellent staff, it is impossible to realise piglet welfare at the weaning and nursery stage even if all other factors are at a good level. The first weeks in the nursery unit are one of the most critical periods in a pig's life and demand most attention by the farm staff because most of the welfare problems in the nursery arise during that period. Because the sensitive piglets experience stress from different factors, as explained above, any other stressful procedures, for instance, vaccinations, should be avoided in the period around weaning to minimise additional stressors with negative impacts on health and welfare. Besides the preparations and implementations of measures previously mentioned, the stockman has to handle the piglets, to keep a careful eye on them, and decide whether and when actions have to be taken to maintain pig welfare. During and after weaning, piglets undergo handling procedures that they have previously not experienced (Orihuela et al., 2018). Thus, concerning handling, especially in weaners, all animals should be treated calmly, carefully and gently. Any kind of lesion has to be avoided by picking them up only when needed using correct picking form, which is done by the back legs and if only caught by one leg, support is needed under the chest. Moreover, they should be let down also very carefully. Having a careful eye on the piglets and their facilities is essential; feed and water intake should be checked daily in the nursery unit in addition to symptoms of diseases, such as diarrhoea and any signs of irregularities. When taking care of the piglets, attention must be given to the details, which presents a challenging and demanding task, especially for beginners. Therefore, appropriate staff training on a regular basis and an exchange of experiences are essential. Even though new trends, especially in the field of digitalisation, offer innovative ways to monitor the animals, including camera systems that detect behavioural patterns, such as increased movements, at this delicate age in the nursery, human care cannot be replaced.

4 Conclusion

Optimising piglet welfare at the nursery stage is one of the most difficult tasks in modern pig farming. In the existing systems and management structures, the young age of the animals at weaning in most countries represents a big challenge, and the interaction of different approaches has to work smoothly to fulfil the most urgent needs of the piglets. In order to implement weaner welfare, the farmers have to be aware that every step towards best welfare, including health, is always also one step towards good performance. Any drawbacks, but also any benefits, are reflected in the balance sheet. Weaners are future capital and have to be treated in a respectful manner by doing everything possible to avoid or at least reduce environmental, nutritional and social stress. In the myriads of scientific studies, manifold aspects of improving weaners' performance and welfare in existing systems have been elucidated. The challenge lies more in the actual combination and transfer of knowledge into practice, including scientific support to analyse the outcome. This can only be realised by on-farm research projects, a highly motivated and skilled farm staff, and incentives for those farms that successfully improve welfare. Whether changes in the whole pig farming framework towards more animal friendly systems, for instance, with longer lactation periods and higher age at weaning, as already realised in some countries, will happen on a broader base in the near future is doubtful. It is therefore all the more important to make every effort within the realms of possibility and considering all the important points mentioned in this chapter to enable good pig welfare during the nursery stage.

5 Future trends in research

Even though a large number of studies have focussed on weaners' welfare and possible ways to improve it in commercial pig production over the last decades, there is still room for improvements. Future trends in research might depend on the demands and settings in the respective countries. In Northern and Middle European countries, where animal welfare is a highly relevant topic, and changes in the husbandry systems can be expected in future, open research questions are, for instance, the evaluation of 'enriched' nurseries with more structures, bedded floors and outdoor access, and the possible consequences on welfare and production parameters. In other countries, where the existing systems are not expected to change, other topics such as the optimised transportation of freshly, maybe early, weaned piglets may have priority. As a general topic, the use of advanced computer-based, digitalized technologies is most probably a huge future research trend in the nursery, as well. Automated techniques, for example, for detecting aggressive episodes in weaned pigs (e.g. Chen et al., 2020), may be beneficial for increasing weaners' welfare. Concluding, research on optimising welfare in nurseries has to consider many factors from different areas, including behaviour studies, process management and animal hygiene, and therefore, is a major challenge.

6 Where to look for further information

- Homepage of the EU-Project FareWellDock (tail docking, tail biting, health and enrichment in pigs): https://farewelldock.eu/info/.
- Homepage of the EU-Project GroupHouseNet (synergy for preventing damaging behaviour in group-housed pigs and chickens): https://www.grouphousenet.eu/.
- Homepage of the EU PiG Innovation group: https://www.eupig.eu/.
- Homepage of the European Reference Centre for Animal Welfare Pigs (EURCAW-Pigs): https://www.eurcaw.eu/en/eurcaw-pigs.htm.

7 References

Aarnink, A. J. A., Swierstra, D., van den Berg, A. J. and Speelman, L. (1997). Effect of type of slatted floor and degree of fouling of solid floor on ammonia emission rates from fattening piggeries. *Journal of Agricultural Engineering Research* 66(2): 93–102.

Arey, D. S. and Edwards, S. A. (1998). Factors influencing aggression between sows after mixing and the consequences for welfare and production. *Livestock Production Science* 56(1): 61–70.

Arey, D. S. and Franklin, M. F. (1995). Effects of straw and unfamiliarity on fighting between newly mixed growing pigs. *Applied Animal Behaviour Science* 45(1–2): 23–30.

Averós, X., Herranz, A., Sánchez, R. and Gosálvez, L. F. (2009). Effect of the duration of commercial journeys between rearing farms and growing-finishing farms on the physiological stress response of weaned piglets. *Livestock Science* 122(2–3): 339–344.

Averós, X., Knowles, T. G., Brown, S. N., Warriss, P. D. and Gosálvez, L. F. (2010). Factors affecting the mortality of weaned piglets during commercial transport between farms. *Veterinary Record* 167(21): 815–819.

Barber, J., Brooks, P. H. and Carpenter, J. L. (1989). The effects of water delivery rate on the voluntary food intake, water use and performance of early-weaned pigs from 3 to 6 weeks of age. *BSAP Occasional Publication* 13: 103–104.

Beattie, V., Walker, N. and Sneddon, I. (1995). Effects of environmental enrichment on behaviour and productivity of growing pigs. *Animal Welfare* 4: 207–220.

Beattie, V. E., Walker, N. and Sneddon, I. A. (1996). An investigation of the effect of environmental enrichment and space allowance on the behaviour and production of growing pigs. *Journal of Applied Animal Behaviour Science* 48(3–4): 151–158.

Berry, R. J. and Lewis, N. J. (2001). The effect of duration and temperature of simulated transport on the performance of early-weaned piglets. *Canadian Journal of Animal Science* 81(2): 199–204.

Bøe, K. (1991). The process of weaning in pigs: when the sow decides. *Applied Animal Behaviour Science* 30(1–2): 47–59.

Bohnenkamp, A. L., Traulsen, I., Meyer, C., Müller, K. and Krieter, J. (2013). Comparison of growth performance and agonistic interaction in weaned piglets of different weight classes from farrowing systems with group or single housing. *Animal* 7(2): 309–315.

Bracke, M. B. M., Metz, J. H. M., Spruijt, B. M. and Schouten, W. G. P. (2002a). Decision support system for overall welfare assessment in pregnant sows B: validation by expert opinion. *Journal of Animal Science* 80(7): 1835–1845.

Bracke, M. B. M., Spruijt, B. M., Metz, J. H. M. and Schouten, W. G. P. (2002b). Decision support system for overall welfare assessment in pregnant sows A: model structure and weighting procedure1. *Journal of Animal Science* 80(7): 1819–1834.

Brooks, P. and Tsourgiannis, C. A. (2003). Factors affecting the voluntary feed intake of the weaned pig. In: Pluske, J., Le Dividich, J. and Verstegen, M. W. A. (Eds), *Weaning the Pig: Concepts and Consequences*. Wageningen Academic Publishers, Wageningen, The Netherlands, 81–116.

Bruininx, E. M. A. M., Binnendijk, G. P., van der Peet-Schwering, C. M. C, Schrama, J. W., den Hartog, L. A., Everts, H. and Beynen, A. C. (2002). Effect of creep feed consumption on individual feed intake characteristics and performance of group-housed weanling pigs. *Journal of Animal Science* 80: 1413–1418.

Bruininx, E. M. A. M., van der Peet-Schwering, C. M. C., Schrama, J. W., Vereijken, P. F. G., Vesseur, P. C., Everts, H., den Hartog, L. A. and Beynen, A. C. (2001). Individually measured feed intake characteristics and growth performance of group-housed weanling pigs: effects of sex, initial body weight, and body weight distribution within groups. *Journal of Animal Science* 79: 301-308.

Brumm, M. C., Ellis, M., Johnston, L. J., Rozeboom, D. W., Zimmerman, D. R. and NCR-89 Committee on Swine Management (2001). Interaction of swine nursery and grow-finish space allocations on performance. *Journal of Animal Science* 79: 1967–1972.

Brumm, M. C. and Gonyou, H. W. (2001). Effects of facility design on behavior and feed and water intake. In: Lewis, A. J. and Southern, L. L. (Eds), *Swine Nutrition* (2nd edn). CRC Press LLC Press, Boca Raton, FL, 499–517.

Campbell, J. M., Crenshaw, J. D. and Polo, J. (2013). The biological stress of early weaned piglets. *Journal of Animal Science and Biotechnology* 4(1): 19.

Chen, C., Zhu, W., Steibel, J., Siegford, J., Wurtz, K., Han, J. and Norton, T. (2020). Recognition of aggressive episodes of pigs based on convolutional neural network and long short-term memory. *Computers and Electronics in Agriculture* 169: 105166.

Cooper, J. E. (1975). The use of the pig as an animal model to study problems associated with low birthweight. *Laboratory Animals* 9(4): 329–336.

Council Directive 2008/120/EC (2008). *Of 18 December 2008 Laying down Minimum Standards for the Protection of Pigs*. https://eur-lex.europa.eu/legal-content/EN/ALL/?uri=CELEX%3A32008L0120. Accessed November 14, 2019.

D'Eath, R. B. (2005). Socialising piglets before weaning improves social hierarchy formation when pigs are mixed post-weaning. *Applied Animal Behaviour Science* 93(3–4): 199–211.

Dee, S. A., Carlson, A. R., Winkelman, N. L. and Corey, M. M. (1993). Effect of management practices on the Streptococcus suis carrier rate in nursery swine. *Journal of the American Veterinary Medical Association* 203(2): 295–299.

De Vos, M., Che, L., Huygelen, V., Willemen, S., Michiels, J., Van Cruchten, S. and Van Ginneken, C. (2014). Nutritional interventions to prevent and rear low birth weight piglets. *Journal of Animal Physiology and Animal Nutrition* 98(4): 609–619.

Done, S. H., Chennells, D. J., Gresham, A. C. J., Williamson, S., Hunt, B., Taylor, L. L., Bland, V., Jones, P., Armstrong, D., White, R. P., Demmers, T. G. M., Teer, N. and Wathes, C. M. (2005). Clinical and pathological responses of weaned pigs to atmospheric ammonia and dust. *Veterinary Record* 157(3): 71-80.

Dong, G. Z. and Pluske, J. R. (2007). The low feed intake in newly-weaned pigs: problems and possible solutions. *Asian-Australasian Journal of Animal Sciences* 20(3): 440-452.

Dudink, S., Simonse, H., Marks, I., de Jonge, F. H. and Spruijt, B. M. (2006). Announcing the arrival of enrichment increases play behaviour and reduces weaning-stress-induced behaviours of piglets directly after weaning. *Applied Animal Behaviour Science* 101(1-2): 86-101.

Dybkjaer, L., Jacobsen, A. P., Togersen, F. A. and Poulsen, H. D. (2006). Eating and drinking activity of newly weaned piglets: effects of individual characteristics, social mixing, and addition of extra zinc to the feed. *Journal of Animal Science* 84(3): 702-711.

EFSA Panel on Animal Health and Welfare (2005). *EFSA Journal* 3(10): 268, doi:10.2903/j.efsa.2005.268.

Fels, M., Giersberg, M., Bill, J., Gillandt, K. and Kemper, N. (2016). Effects of a uniform environmental enrichment during lactation and after weaning on the behaviour and skin lesion score of piglets. *Züchtungskunde* 88: 241-253.

Fels, M., Hartung, J. and Hoy, S. (2014a). Social hierarchy formation in piglets mixed in different group compositions after weaning. *Applied Animal Behaviour Science* 152: 17-22.

Fels, M., Thays Sonoda, L., Rauterberg, S., Hartung, J. and Kemper, N. (2014b). Aggressive behaviour of piglets after mixing - causes, factors and control measures. *Der Praktische Tierarzt* 95: 952-960.

Fels, M. and Kemper, N. (2017). Fighting activity and establishment of a litter-associated dominance order in weaned piglets after mixing. *Berliner und Münchener Tierärztliche Wochenschrift* 130: 306-313.

Fels, M., Konen, K., Hessel, E. and Kemper, N. (2018a). Determination of static space occupied by individual weaner and growing pigs using an image-based monitoring system. *The Journal of Agricultural Science* 156(2): 282-290.

Fels, M., Lüthje, F., Bill, J., Aleali, K. and Kemper, N. (2018b). Elevated platforms for weaners: do pigs use the extra space? *Veterinary Record* 183(7): 222-222.

Fels, M., Lüthje, F., Faux-Nightingale, A. and Kemper, N. (2018c). Use of space and behavior of weaned piglets kept in enriched two-level housing system. *Journal of Applied Animal Welfare Science* 21(3): 267-282.

Fels, M., Konen, K., Hessel, E. and Kemper, N. (2019). Biometric measurement of static space required by weaned piglets kept in groups of eight during 6 weeks. *Animal Production Science* 59(7): 1327-1335.

Fraser, D. (1985). Selection of bedded and unbedded areas by pigs in relation to environmental temperature and behaviour. *Applied Animal Behaviour Science* 14(2): 117-126.

Fraser, D., Phillips, P. A. and Thompson, B. K. (1986). A test of a free-access two-level pen for fattening pigs. *Journal of Animal Science* 42(2): 269-274.

Fraser, D., Phillips, P. A., Thompson, B. K. and Tennessen, T. (1991). Effect of straw on the behaviour of growing pigs. *Applied Animal Behaviour Science* 30(3-4): 307-318.

Gondret, F., Lefaucheur, L., Louveau, I., Lebret, B., Pichodo, X. and Le Cozler, Y. (2005). Influence of piglet birth weight on postnatal growth performance, tissue lipogenic capacity and muscle histological traits at market weight. *Livestock Production Science* 93(2): 137–146.

Hales, J., Moustsen, V. A., Nielsen, M. B. F. and Hansen, C. F. (2013). Individual physical characteristics of neonatal piglets affect preweaning survival of piglets born in a noncrated system. *Journal of Animal Science* 91(10): 4991–5003.

Hennig-Pauka, I., Menzel, A., Boehme, T. R., Schierbaum, H., Ganter, M. and Schulz, J. (2019). Haptoglobin and C-reactive protein–non-specific markers for nursery conditions in swine. *Frontiers in Veterinary Science* 6: 92.

Herpin, P., Bertin, R., Le Dividich, J. and Portet, R. (1987). Some regulatory aspects of thermogenesis in cold-exposed piglets. *Comparative Biochemistry and Physiology. A, Comparative Physiology* 87(4): 1073–1081.

Hillmann, E., von Hollen, F., Bünger, B., Todt, D. and Schrader, L. (2003). Farrowing conditions affect the reactions of piglets towards novel environment and social confrontation at weaning. *Applied Animal Behaviour Science* 81(2): 99–109.

Hwang, H. S., Lee, J. K., Eom, T. K., Son, S. H., Hong, J. K., Kim, K. H. and Rhim, S. J. (2016). Behavioral characteristics of weaned piglets mixed in different groups. *Asian-Australasian Journal of Animal Sciences* 29(7): 1060–1064.

Jayaraman, B. and Nyachoti, C. M. (2017). Husbandry practices and gut health outcomes in weaned piglets: a review. *Animal Nutrition* 3(3): 205–211.

Jensen, P. (1982). An analysis of agonistic interaction patterns in group-housed dry sows–aggression regulation through an "avoidance order". *Journal of Applied Animal Ethology* 9(1): 47–61.

Jensen, P. (1986). Observations on the maternal behaviour of free-ranging domestic pigs. *Applied Animal Behaviour Science* 16(2): 131–142.

Jensen, P. and Redbo, I. (1987). Behavior during nest-leaving in free-ranging domestic pigs. *Applied Animal Behaviour Science* 110: 355–362.

Jones, J. B., Burgess, L. R., Webster, A. J. F. and Wathes, C. M. (1996). Behavioural responses of pigs to atmospheric ammonia in a chronic choice test. *Animal Science* 63(3): 437–445.

KilBride, A., Gillman, C., Ossent, P. and Green, L. (2009). Impact of flooring on the health and welfare of pigs. *In Practice* 31(8): 390–395.

Kirkden, R. D., Broom, D. M. and Andersen, I. L. (2013). Invited review: piglet mortality: management solutions. *Journal of Animal Science* 91(7): 3361–3389.

Koketsu, Y., Tani, S. and Iida, R. (2017). Factors for improving reproductive performance of sows and herd productivity in commercial breeding herds. *Porcine Health Management* 3: 1.

Kuller, W. I., Tobias, T. J. and Van Nes, A. (2010). Creep feed intake in unweaned piglets is increased by exploration stimulating feeder. *Journal of Livestock Science* 129(1–3): 228–231.

Kutzer, T., Bünger, B., Kjaer, J. B. and Schrader, L. (2009). Effects of early contact between non-littermate piglets and of the complexity of farrowing conditions on social behavior and weight gain. *Applied Animal Behaviour Science* 121(1): 16–24.

Lallès, J. P., Boudry, G., Favier, C., Le Floch, N., Luron, I., Montagne, L., Oswald, I. P., Pié, S., Piel, C. and Sève, B. (2004). Gut function and dysfunction in young pigs: physiology. *Animal Research* 53(4): 301–316.

Laskoski, F., Faccin, J. E. G., Vier, C. M., Gonçalves, M. A. D., Orlando, U. A. D., Kummer, R., Mellagi, A. P. G., Bernardi, M. L., Wentz, I. and Bortolozzo, F. P. (2019). Effects of pigs per feeder hole and group size on feed intake onset, growth performance, and ear and tail lesions in nursery pigs with consistent space allowance. *Journal of Swine Health and Production* 27(1): 12–18.

Le Dividich, J. and Herpin, P. (1994). Effects of climatic conditions on the performance, metabolism and health status of weaned piglets: a review. *Livestock Production Science* 38(2): 79–90.

Le Dividich, J. and Sève, B. (2000). Effects of underfeeding during the weaning period on growth, metabolism, and hormonal adjustments in the piglet. *Domestic Animal Endocrinology* 19(2): 63–74.

Le Floch, N., Jondreville, C., Matte, J. J. and Seve, B. (2006) Importance of sanitary environment for growth performance and plasma nutrient homeostasis during the post-weaning period in piglets. *Archives of Animal Nutrition* 60(1): 23–34, doi:10.1080/17450390500467810.

Levis, D. G. (1997). Effect of early weaning on sow reproductive performance - a review. Nebraska Swine Reports 204. http://digitalcommons.unl.edu/coopext_swine/204.

Lewis, N. J. (2008). Transport of early weaned piglets. *Applied Animal Behaviour Science* 110(1–2): 128–135.

Lewis, N. J. and Berry, R. J. (2006). Effects of season on the behaviour of early-weaned piglets during and immediately following transport. *Applied Animal Behaviour Science* 100(3–4): 182–192.

Li, Y., Song, Z., Kerr, K. A. and Moeser, A. J. (2017). Chronic social stress in pigs impairs intestinal barrier and nutrient transporter function, and alters neuro-immune mediator and receptor expression. *PLoS ONE* 12(2): e0171617.

Madec, F., Bridoux, N., Bounaix, S. and Jestin, A. (1998). Measurement of digestive disorders in the piglet at weaning and related risk factors. *Preventive Veterinary Medicine* 35(1): 53–72.

Madec, F., Le Dividich, J., Pluske, J. R. and Verstegen, M. W. A. (2003). Environmental requirements and housing of the weaned pig. In: Pluske, J., Le Dividich, J. and Verstegen, M. W. A. (Eds), *Weaning the Pig: Concepts and Consequences*. Wageningen Academic Publishers, Wageningen, The Netherlands, 337–360.

Main, R. G., Dritz, S. S., Tokach, M. D., Goodbandt, R. D. and Nelssen, J. L. (2004). Increasing weaning age improves pig performance in a multisite production system. *Journal of Animal Science* 82(5): 1499–1507.

Martin, J. E., Ison, S. H. and Baxter, E. M. (2015). The influence of neonatal environment on piglet play behaviour and post-weaning social and cognitive development. *Applied Animal Behaviour Science* 163: 69–79.

Mavromichalis, I. (2001). Management of the nursery pig. *Illinois Porknet Papers*. http://livestocktrail.illinois.edu/uploads/porknet/papers/ManagementOfTheNurseryPig.PDF. Accessed October 12, 2019.

Moeser, A. J. and Blikslager, A. T. (2007). Mechanisms of porcine diarrheal disease. *Journal of the American Veterinary Medical Association* 231(1): 56–67.

Moeser, A. J., Pohl, C. S. and Rajput, M. (2017). Weaning stress and gastrointestinal barrier development: implications for lifelong gut health in pigs. *Animal Nutrition* 3(4): 313–321.

Morgan, T., Pluske, J., Miller, D., Collins, T., Barnes, A. L., Wemelsfelder, F. and Fleming, P. A. (2014). Socialising piglets in lactation positively affects their post-weaning behaviour. *Journal of Applied Animal Behaviour Science* 158: 23–33.

Muirhead, M. R., Alexander, T. J. L. and Carr, J. (2013). *Managing Pig Health: A Reference for the Farm* (2nd edn). 5M PUB, Sheffield, England.

Munsterhjelm, C., Peltoniemi, O. A. T., Heinonen, M., Hälli, O., Karhapää, M. and Valros, A. (2009). Experience of moderate bedding affects behaviour of growing pigs. *Applied Animal Behaviour Science* 118(1-2): 42-53.

Newberry, R. C. and Wood-Gush, D. G. M. (1985). The suckling behaviour of domestic pigs in a semi-natural environment. *Behaviour* 95(1-2): 11-25.

Niekamp, S. R., Sutherland, M. A., Dahl, G. E. and Salak-Johnson, J. L. (2007). Immune response of piglets to weaning stress: impacts of photoperiod. *Journal of Animal Science* 85(1): 93-100.

Noblet, J., Le Dividich, J. and Van Milgen, J. (2001). Thermal environment and swine nutrition. In: Lewis, A. J. and Southern, L. L. (Eds), *Swine Nutrition* (2nd edn). CRC Press LLC Press, Boca Raton, FL, 519-544.

Oh, H. K., Choi, H. B., Ju, W. S., Chung, C. S. and Kim, Y. Y. (2010). Effects of space allocation on growth performance and immune system in weaning pigs. *Livestock Science* 132(1-3): 113-118.

Olsson, I. A. S., de Jonge, F. H., Schuurman, T. and Helmond, F. A. (1999). Poor rearing conditions and social stress in pigs: repeated social challenge and the effect on behavioural and physiological responses to stressors. *Behavioural Processes* 46(3): 201-215.

Oostindjer, M., Bolhuis, J. E., Mendl, M., Held, S., Gerrits, W., Van den Brand, H. and Kemp, B. (2010). Effects of environmental enrichment and loose housing of lactating sows on piglet performance before and after weaning. *Journal of Animal Science* 88(11): 3554-3562.

Oostindjer, M., Kemp, B., van den Brand, H. and Bolhuis, J. E. (2014). Facilitating 'learning from mom how to eat like a pig' to improve welfare of piglets around weaning. *Applied Animal Behaviour Science* 160: 19-30.

Oostindjer, M., van den Brand, H., Kemp, B. and Bolhuis, J. E. (2011). Effects of environmental enrichment and loose housing of lactating sows on piglet behaviour before and after weaning. *Journal of Applied Animal Behaviour Science* 134(1-2): 31-41.

Orihuela, A., Mota-Rojas, D., Velarde, A., Strappini, A., de la Vega, L., Borderas, F. and Alonso-Spilsbury, M. (2018). Environmental enrichment to improve behaviour in farm animals. CABI. England. *CAB Reviews Perspectives in Agriculture Veterinary Science Nutrition and Natural Resources* 13: 1-25.

Pajor, E. A., Fraser, D. and Kramer, D. L. (1991). Consumption of solid food by suckling pigs: individual variation and relation to weight gain. *Applied Animal Behaviour Science* 32(2-3): 139-155.

Parratt, C. A., Chapman, K. J., Turner, C., Jones, P. H., Mendl, M. T. and Miller, B. G. (2006). The fighting behaviour of piglets mixed before and after weaning in the presence or absence of a sow. *Journal or Applied Animal Behavioural Science* 101(1-2): 54-67.

Petersen, H. V., Vestergaard, K. and Jensen, P. (1989). Integration of piglets into social groups of free-ranging domestic pigs. *Applied Animal Behaviour Science* 23(3): 223-236.

Phillips, P. and Fraser, D. (1987). Design, cost and performance of a free-access, two-level pen for growing-finishing pigs. *Canadian Agricultural Engineering* 29.

Pluske, J. R., Hampson, D. J. and Williams, I. H. (1997). Factors influencing the structure and function of the small intestine in the weaned pig: a review. *Livestock Production Science* 51(1-3): 215-236.

Pluske, J. R., Williams, I. H. and Aherne, F. X. (1995). Nutrition of the neonatal pig. In: Varlex, M. A. (Ed), *The Neonatal Pig: Development and Survival*. CAB Int., Wallingford (UK), 187-235.

Prunier, A., Heinonen, M. and Quesnel, H. (2010). High physiological demands in intensively raised pigs: impact on health and welfare. *Animal* 4(6): 886-898.

Quiniou, N., Dagorn, J. and Gaudré, D. (2002). Variation of piglets' birth weight and consequences on subsequent performance. *Livestock Production Science* 78(1): 63-70.

Rauterberg, S., Kemper, N. and Fels, M. (2018). Influence of environmental enrichment during lactation and after weaning on aggressive behaviour and skin lesion score of piglets. *Berliner und Münchener Tierärztliche Wochenschrift* 131(1-2): 12-19.

Rhim, S. J., Son, S. H., Hwang, H. S., Lee, J. K. and Hong, J. K. (2015). Effects of mixing on the aggressive behavior of commercially housed pigs. *Asian-Australasian Journal of Animal Sciences* 28(7): 1038-1043.

Robert, S., Weary, D. M. and Gonyou, H. (1999). Segregated early weaning and welfare of piglets. *Journal of Applied Animal Welfare Science* 2(1): 31-40.

Schoenfelder, A. (2005). The effect of rank order on feed intake and growth of fattening pigs. *DTW. Deutsche Tierarztliche Wochenschrift* 112(6): 215-218.

Schrey, L., Kemper, N. and Fels, M. (2019). Behaviour and skin injuries of piglets originating from a novel group farrowing system before and after weaning. *Agriculture*, MDPI 9: 14.

Schulz, J., Ruddat, I., Hartung, J., Hamscher, G., Kemper, N. and Ewers, C. (2016). Antimicrobial-resistant Escherichia coli survived in dust samples for more than 20 years. *Frontiers in Microbiology* 7: 866.

Scipioni, R., Martelli, G. and Antonella Volpelli, L. (2009). Assessment of welfare in pigs. *Italian Journal of Animal Science* 8(sup1): 117-137.

Smith, A. L., Stalder, K. J., Serenius, T. V., Baas, T. J. and Mabry, J. W. (2007). Effect of piglet birth weight on weights at weaning and 42 days post weaning. *Journal of Swine Health and Production* 15(4): 213-218.

Smith, L. F., Beaulieu, A. D., Patience, J. F., Gonyou, H. W. and Boyd, R. D. (2004). The impact of feeder adjustment and group size-floor space allowance on the performance of nursery pigs. *Journal of Swine Health and Production* 12(3): 111-118.

Spurlock, M. E. (1997). Regulation of metabolism and growth during immune challenge: an overview of cytokine function. *Journal of Animal Science* 75(7): 1773-1783.

Stafford, K. J. (2010). Tail biting: an important and undesirable behaviour of growing pigs. *The Veterinary Journal* 186(2): 131-132.

Stojanac, N., Stevančević, O., Potkonjak, A., Savić, B., Stančić, I. and Vračar, V. (2014). The impact of space allowance on productivity performance and Salmonella spp. shedding in nursery pigs. *Livestock Science* 164: 149-153.

Sulabo, R. C., Tokach, M. D., Dritz, S. S., Goodband, R. D., DeRouchey, J. M. and Nelssen, J. L. (2010). Effects of varying creep feeding duration on the proportion of pigs consuming creep feed and neonatal pig performance. *Journal of Animal Science* 88(9): 3154-3162.

Sutherland, M. A., Backus, B. L. and McGlone, J. J. (2014). Effects of transport at weaning on the behavior, physiology and performance of pigs. *Animals* 4(4): 657-669; doi:10.3390/ani4040657.

Sutherland, M. A., Bryer, P. J., Davis, B. L. and McGlone, J. J. (2009). Space requirements of weaned pigs during a 60 minute transport in summer. *Journal of Animal Science* 87(1): 363–370.

Tobias, T. J., Bouma, A., van den Broek, J., van Nes, A., Daemen, A. J. J. M., Wagenaar, J. A., Stegeman, J. A. and Klinkenberg, D. (2014). Transmission of Actinobacillus pleuropneumoniae among weaned piglets on endemically infected farms. *Preventive Veterinary Medicine* 117(1): 207–214.

Tönepöhl, B., Appel, A. K., Welp, S., Voss, B., König von Borstel, U. and Gauly, M. (2012). Effect of marginal environmental and social enrichment during rearing on pigs' reactions to novelty, conspecifics and handling. *Applied Animal Behaviour Science* 140(3–4): 137–145.

van Nieuwamerongen, S. E., Bolhuis, J. E., van der Peet-Schwering, C. M. and Soede, N. M. (2014). A review of sow and piglet behaviour and performance in group housing systems for lactating sows. *Animal* 8(3): 448–460.

Vanheukelom, V., Driessen, B., Maenhout, D. and Geers, R. (2011). Peat as environmental enrichment for piglets: the effect on behaviour, skin lesions and production results. *Applied Animal Behaviour Science* 134(1–2): 42–47.

Veit, C., Büttner, K., Traulsen, I., Gertz, M., Hasler, M., Burfeind, O., große Beilage, E. and Krieter, J. (2017). The effect of mixing piglets after weaning on the occurrence of tail-biting during rearing. *Livestock Science* 201: 70–73.

von Borell, E. (2000). Welfare assessment of segregated early weaning (SEW) - a review. *Archives of Animal Breeding* 43: 337–345.

Wagner, K. M., Schulz, J. and Kemper, N. (2018). Examination of the hygienic status of selected organic enrichment materials used in pig farming with special emphasis on pathogenic bacteria. *Porcine Health Management* 4: 24–24.

Wallgren, T., Westin, R. and Gunnarsson, S. (2016). A survey of straw use and tail biting in Swedish pig farms rearing undocked pigs. *Acta Veterinaria Scandinavica* 58(1): 84.

Wamnes, S., Lewis, N. J. and Berry, R. J. (2006). The performance of early-weaned piglets following transport: effect of season and weaning weight. *Canadian Journal of Animal Science* 86(3): 337–343.

Weary, D. M., Jasper, J. and Hötzel, M. J. (2008). Understanding weaning stress. *Applied Animal Behaviour Science* 110(1–2): 24–41.

Worobec, E. K., Duncan, I. J. H. and Widowski, T. M. (1999). The effects of weaning at 7, 14 and 28 days on piglet behaviour. *Applied Animal Behaviour Science* 62(2–3): 173–182.

Chapter 6

Optimizing pig welfare in the growing and finishing stage

Arlene Garcia and John J. McGlone, Texas Tech University, USA

1 Introduction
2 Pig behavioral issues that impact pig welfare
3 Production systems
4 Group size
5 Living conditions in indoor systems
6 Animal health and animal care
7 Conclusion and future trends in research
8 Where to look for further information
9 References

1 Introduction

The US pig inventory has increased by 15% in the past 3 years, with a pig population greater than 71 million (NASS, 2017). Global pork production has increased fourfold over the past 50 years, driven by population growth and dietary transition toward more animal protein per capita (Lassaletta et al., 2014, 2019; Bai et al., 2018a). The US markets over 100 million pigs per year (about 10% of the world production).

In 2018, China had about half of the world's numbers of pigs. In that year, China and other Asian countries broke with African Swine Fever (ASF). Prior to this outbreak, pork was the meat consumed in the highest amount in the world, mainly due to a high level of consumption in Asia. Estimates are that at least half of the pigs in China have been culled due to ASF. This means that the world's pig population decreased in a short period by over 25%. ASF has also broken out in Southeast Asia, Russia, Poland and Belgium. It is thought to be carried among farms by feral or wild pigs in Europe. If this is so, ASF may spread to other European countries. When ASF breaks in a country, that country is not allowed to export pig meat. This changes the economics of the infected country. The concern is that when pig prices are low, or when large numbers of pigs must be euthanized, that animal welfare may suffer. Currently,

http://dx.doi.org/10.19103/AS.2020.0081.06

in 2020 a pandemic that began in China (in December 2019) has spread around the world, and many countries and regions have 'stay-at-home' orders. How this pandemic will impact the world's swine business is yet to be determined. If farm and slaughter plant workers do not work, then the pork supply (and all food) will decrease. Additionally, the inability of employees to go to work negatively impacts the welfare of pigs on farms. The extent of a pandemic on pig welfare will most likely be studied later.

ASF virus (ASFV) has a complex transmission cycle involving African wild suid species, soft ticks, and domestic pigs (Sánchez-Vizcaíno et al., 2019). ASFV is transmitted, most often directly via contact between sick and healthy animals, including domestic pigs and wild boars (Sánchez-Vizcaíno et al., 2019). As the disease progresses, animals may stop eating and become increasingly moribund, presenting bloody diarrhea, vomiting and possibly abortion (Dixon et al., 2019). Pathological changes associated with vasculitis include skin erythema, pulmonary edema, hyperemic splenomegaly, hemorrhagic lymphadenitis and petechial hemorrhages in the kidneys, lungs and urinary bladder (Dixon et al., 2019). Dixon et al. (2019) reported that infection is associated with very high levels of virus in the blood (up to 109 TCID50/ml) and tissues, particularly the spleen and lymph nodes. ASF is a hemorrhagic disease that causes illness quickly with a very high mortality rate (Dixon et al., 2019). The usual practice is to euthanize all the pigs on the infected farms and on all farms near the infected farm. ASF causes two main welfare problems. First, infected animals suffer a horrible death. Second, euthanizing so many pigs is a challenge, even in countries with sophisticated animal welfare rules.

As a result of the Asian ASF outbreak, pig meat is being exported from many exporting countries to China. About 25% of the pork in the USA is exported. In 2019, the USA and China were in a trade war. This results in little US pork going to China. This lowers the pig price in the USA since there is an oversupply without the Chinese market. The rest of the world has high pig prices due to the increased demand and because China must import more pork than in the past. The ASF outbreak will increase the world's pork prices until the disease is eradicated, yet there is no vaccine for ASF. To meet this demand, new farms will be built using current intensive technologies. Smaller farms were hit harder with ASF than modern-style, intensive farms. The regrowth in the Chinese pig industry, which is under way now, will be primarily large, intensive pig units. ASF accelerated the move away from small, low-input farms to larger, intensive pig units. Small, backyard farms and large intensive farms have very different welfare concerns.

The majority of pigs in pork-exporting countries are raised in intensive production systems. The trend is for the world industry to continue to restructure with more intensive farms and fewer small farms. ASF accelerates this trend greatly in infected and pork-exporting countries (USA, Brazil, Germany, Spain, Denmark, The Netherlands and other EU countries).

In intensive farms in the USA, pigs are weaned from the sow around 3 weeks of age (while European directives do not allow weaning until 28 days of age) and moved to a nursery (or a wean-to-finish barn, bypassing the nursery) where they will stay for 6–8 weeks, and then to a grow-finishing unit where they will spend 16–17 weeks until they reach market weight (National Pork Board, NPB, 2019). When pigs are moved to the finishing barn, they are termed grow-finishing pigs and are fed there until they reach market weight (at about 280 lbs/127 kg in the USA); then they are termed market pigs.

Intensive production systems are where animals are kept and raised within fewer production units, with a large number of animals within these units (Fraser, 2005). Such intensification can have negative effects if not managed properly, including increased animal aggression, competition over resources, disease, injuries and death, and over-all poor animal welfare. However, intensive production systems can be well controlled (Clark et al., 2019) and have fewer animal welfare issues if they are properly managed.

Societal concerns about the well-being of livestock animals are a driving factor for the swine industry to adopt practices and systems that are pig welfare friendly. As the growing demand for animal protein grows, and hence a shortage of labor, there is more need for automated systems, but caretakers play a vital role in ensuring these systems work properly. Caretakers are responsible for the day-to-day care of pigs and identifying problems within the barn. Currently, we ultimately rely on human observation for the welfare of pigs.

Real-time monitoring of pigs in large-scale settings has become crucial (Nasirahmadi et al., 2017), as the direct observation of every animal is intermittent and impractical (Millman, 2007; Weary et al., 2009). New smart barns may allow the automation of animal welfare checks, among other benefits.

The main objectives of this chapter are to define animal welfare and to identify where intensive production systems may put finishing pigs at risk of poor animal welfare.

The first challenge in assessing animal welfare is to define the term. Its definition varies, which adds to the complexity of this issue. The OIE (Terrestrial Code) animal welfare code defines animal welfare as 'the physical and mental state of an animal in relation to the conditions in which it lives and dies' (OIE, 2010). The OIE further guides its definition of animal welfare through the 'guiding principles', known as the 'Five Freedoms'. The Five Freedoms were proposed in 1965 and describe society's expectations for the conditions animals should experience under human control (OIE, 2010). The Five Freedoms encompass the following:

- Freedom from hunger, malnutrition and thirst;
- Freedom from fear and distress;

- Freedom from heat stress and physical discomfort;
- Freedom from pain, injury and disease; and
- Freedom to express normal patterns of behavior.

However, animal welfare can be defined in a number of ways, but, whatever the definition is, some authors believe it must include three elements (Velarde et al., 2015): the biological function of the animal (or ability to cope with its environment), its emotional state, and its ability to show normal patterns of behavior (Manteca et al., 2009).

Animal welfare is perceived as a 'public good' (Miele and Evans, 2010) and is a necessary element of sustainable animal production (McGlone, 2001; Broom, 2010). Sustainable meat production is 'economically sound, economically viable, socially just, and humane' (Appleby, 2004). Further, animal welfare is a matter of ethics and an essential tool to gain and maintain markets, and husbandry that benefits sustainability and maximizes the welfare of animals, but avoids its potential impairment (Velarde et al., 2015).

2 Pig behavioral issues that impact pig welfare

Welfare issues may arise in pig production when there is a mismatch between a pig's instincts or behavior and its environment (Kittawornrat and Zimmerman, 2011). Behavioral impulses may be expressed inappropriately when instinctual behavior is thwarted (Kittawornrat and Zimmerman, 2011). In the wild, pigs are active at night and spend the majority of their awake time in foraging-related activities, including rooting, grazing and exploring with their snout (Caley, 1997), while outdoor domestic pigs are active during the day (Stolba and Wood-Gush, 1984). Pigs in commercial settings are active in the day and continue to express exploratory behaviors (Kittawornrat and Zimmerman, 2011).

2.1 Oral-nasal-facial/oral behaviors

Oral-nasal-facial (ONF) behaviors are behaviors associated with the snout, nose and face of the pig. Pigs in both indoor and outdoor systems express ONF behaviors (Dailey and McGlone, 1997). Some people call them oral-nasal behaviors, but the detailed work by Dailey and McGlone (1997) showed that the face is also used to interact with substrates. Pigs kept outdoors have a much higher rate of ONF behaviors than pigs kept indoors. ONF behaviors include chew/bite grass, chew/bite fence/bars, chew rocks/soil and rooting the ground or trough. Indoor pigs have less material to manipulate than pigs on a pasture or wooded area. This lack of material to manipulate may be frustrating to indoor pigs, which can lead to what we call abnormal behaviors, which may not be abnormal if these are appropriate behaviors for the environment. Certainly, pork producers would like to minimize damaging ONF behaviors. Providing rooting materials may reduce unwanted damage toward animals or penning and direct these behaviors to rootable substrates. Dailey and McGlone (1997) concluded that ONF-stereotyped behaviors may be natural pre- and post-feeding appetitive and consummatory chewing. Rooting activities in sequences form by the available substrates.

2.2 Tail biting

Tail biting is often an indication that welfare is low in the pen as a whole (Fritschen and Hogg, 1996). There are a variety of management practices that can reduce tail biting outbreaks, such as the provision of enrichment (Beattie et al., 2001), removal of the tail biter or removal of the victim (Zonderland et al., 2008), but some farms at certain times cannot control tail biting. Further research in this area is needed to find one or more solutions.

2.3 Belly nosing

Belly nosing is the act of rhythmic up-and-down movement of one piglet rubbing the belly of another with its snout (Fraser, 1978). This behavior can lead to skin lesions and ulcerations on the belly and flank of the victim (Straw and Bartlett, 2001).

When piglets are raised with the sow, piglets orient toward the udder and massage the sow's underline to induce milk letdown as an example of ONF behaviors expressed within a few minutes of birth (Frei et al., 2018). Piglets raised without a sow develop belly nosing behaviors (Weary et al., 1999; Rzezniczek et al., 2015), and the prevalence increases with decreasing weaning age (Metz and Gonyou, 1990; Worobec et al., 1999). Studies suggest that belly nosing is not an indicator of stress (Gardner et al., 2001a), not influenced by diet or diet quality or the presence of milk in the diet (Gardner et al., 2001b), but is more closely related to social interactions than with eating or drinking (Li and Gonyou, 2002); thus it is a type of redirected ONF sucking behavior (Fraser, 1978). Further, piglets reared with the sow do not show belly nosing behaviors (Rzezniczek et al., 2015), which may indicate the behavioral needs of piglets for a specific kind of ONF behavior are not met in artificial rearing systems (Frei et al., 2018).

Commercial environments are often barren and prevent pigs from displaying natural behaviors, such as rooting; thus an environment with poor stimuli can contribute to the development of what are called abnormal behaviors (Cooper et al., 2001; Bolhuis et al., 2005), but may be perfectly normal given the pig's development and inappropriate environment. The inability to express species-specific behavior in foraging in a barren environment can increase aggression by enhancing competition, leading to more negative agonistic behavior (Docking et al., 2008), as well as to frustration, a stress factor itself (Casal-Plana et al., 2017).

2.4 Aggression

During a pig's life cycle, there may be several aggressive interactions with other pigs, which can result in injuries and stress (Hemsworth et al., 2015) and detrimentally affect productivity (McGlone et al., 1980; Verdon and Rault, 2018), at least temporarily. Aggression can be observed during the mixing of pigs, as a social hierarchy is established within the group (McGlone, 1986) and typically declines after the hierarchy is established (Langbein and Puppe, 2004).

Mixing of unfamiliar pigs leads to aggression and can negatively impact physiology and welfare (Arey and Edwards, 1998; Einarsson et al., 1996; Knox et al., 2014; Rault et al., 2014; Verdon et al., 2015) due to fighting when pigs form social hierarchies and during feeding (Arey and Edwards, 1998; Li et al., 2012). These aggressive encounters can result in injury, lameness and increased cortisol concentrations (Anil et al., 2006; Arey and Edwards, 1998; Li et al., 2012) and can affect weight gain (Mormède et al., 2007).

When unfamiliar pigs are mixed in intensive units, they fight. At a minimum, intensive pigs are mixed at weaning and when they are transported to slaughter. Some pigs are also mixed between growing and finishing phases and in late finishing when regrouping smaller finishing pigs, as heavier pigs in the barn are marketed.

When outdoor pigs are weaned and mixed, they do not fight, while indoor pigs do fight when weaned and mixed (Sarignac et al., 1997). Outdoor pigs mingle among different litters during lactation. Social behaviors are practiced, and this behavioral practice is thought to reduce fighting later because piglets learn appropriate social skills. Pigs kept with a single sow in a farrowing crate may lack the social skills needed to prevent fighting. In this thinking, the indoor system causes pig aggression due to a lack of social skills in developing piglets, which is directly caused by the intensive production system.

A commonly overlooked factor that leads to aggression in the herd is the increase in pig weights over time. Increased pig weight and weight gain can lead to increased motivation to eat and chew and can cause stocking density issues toward the end of the finishing phase, leading to competition over resources. Market weight has linearly increased by 5.8 kg every 10 years for the past 4 decades, driven by the dilution of fixed cost over more weight per pig, improved genetics and nutrition resulting in more efficient pigs (Brumm, 2012). Higher pig weights are essentially driven by packer demands for carcasses that can provide more kg of lean meat to improve packing plant throughput. Producers consistently respond to market demands, and one of the market demands has been for greater market weights that produce carcasses with greater kg of lean meat per carcass.

Although pig market weights vary depending on regions and cultural backgrounds, as of the beginning of 2017 in the USA, the average market weight was reported to be 129 kg, but varied between 124 kg and 130 kg in 2016 (NASS, 2017). The design and management of commercial group housing are complex and with wide disparity (Grandin and Whiting, 2018). Proper facility management can decrease pig aggression and lead to a positive impact on pig welfare (Grandin and Whiting, 2018). Such management strategies should focus on controlling aggression and protecting vulnerable pigs from aggression. This may include minimizing changes in the social structure (i.e. mixing), modifying pen design to allow escape from aggressors (McGlone and Curtis, 1985), incorporating distractions at regrouping, reducing stocking density and facilitating access to resources such as feed and water (Marchant-Forde and Marchant-Forde, 2005).

3 Production systems

Pigs are typically weaned at 21 days and moved to a nursery. Confinement systems for pigs are generally confronted by skepticism and concern from the public (Sørensen and Schrader, 2019) because of the negative impact they can have on animal welfare. However, any poorly managed facility can negatively affect animal welfare.

Most confined systems follow the all-in/all-out (AIAO) management system. The AIAO strategy has many advantages for pig production, if it is adhered to, which many times does not happen. The benefits of AIAO include improved biosecurity/minimized disease transmission, health, and growth performance (Scheidt et al., 1995; Owsley et al., 2013). Pigs in the AIAO system are matched by age and move forward through the production stages in the same groups without re-mixing pigs of different ages (Owsley et al., 2013). The rooms these pigs occupied are, therefore, left empty and are cleaned and disinfected for the next group of pigs (Diana et al., 2019). Adhering to an AIAO management strategy is reported to have major constraints. These constrains include the lack of facilities to house slow-growing pigs/sick pigs (Diana et al., 2019), leading to the practice of sorting pigs by body weight into the next production stage (Calderón Díaz et al., 2017) to achieve uniformity in slaughter weights (Brumm et al., 2004). Slow-growing/sick pigs disrupt the normal flow by being delayed for several weeks from moving to the next stage and are then re-mixed with pigs of their own size, but not age (Diana et al., 2017). This practice is reported to increase disease transmission between different age groups, can have an adverse effect on pig performance, can cause pigs to be 10 kg lighter at slaughter and have increased risk of diseases such as pleurisy and pericarditis compared to the pigs in the 'normal' flow (Calderón Díaz et al., 2017). The stress associated with re-mixing these slow-growing/sick pigs leads to the development of damaging behaviors such as fighting wounds and ear and tail biting (Schrøder-Petersen and Simonsen, 2001; Martínez-Miró et al., 2016), reflecting poor welfare (Schrøder-Petersen and Simonsen, 2001).

3.1 Grow-finish pigs in alternative housing

Indoor, mechanically ventilated systems are generally chosen over other systems such as hoop and pasture systems due to their lifetime use of 15–20 years or longer (Kliebenstein et al., 2003), the ability to have environmental control (ventilation, misters, heating system), increased biosecurity and other factors that can be controlled. Under year-round pork production, confinement systems generally have superior pig performance and lower variable costs (such as feed costs) compared to hoop or pasture systems (Larson et al., 2002, 2003).

Due to the increased concern about animal welfare, there is an interest in alternative production systems, such as straw-based systems, straw yards and outdoor production systems (Guy et al., 2002). In a study by Guy et al. (2002), genotype and finishing systems were compared to identify an interaction. The welfare of progeny from outdoor/semiintensive-type sows and indoor sows was evaluated, in addition to pig welfare in straw yards (considered a less-intensive system), outdoor paddocks and conventional, fully slatted pens. They found that the interaction between the housing system and genotype for health parameters was not significant. However, Guy et al. (2002) reported that outdoor paddocks and straw yards enhanced pig health and welfare compared to conventional, fully slatted pens, shown by the levels of stomach ulcers, lung damage, bursitis and body damage.

Most pigs in developed countries are finished indoors in pens with concrete slats and mechanical ventilation. The intensity of indoor housing has been increasing in recent years, with the introduction of what is called filter barns (Reicks, 2009). In these barns, the incoming air is filtered to prevent microbes from entering the building. This means the barns must be sealed to any outside air, and mechanical ventilation is the only way air is moved in and out. We have not seen studies that attempt to compare the welfare of pigs in a filter barn with a conventional barn. However, today the non-filtered grow-finish barn with concrete slats and mechanical ventilation is the industry standard. With the advent of highly infectious diseases like ASF or porcine reproductive and respiratory syndrome (PRRS), filter barns are likely to be more common in the future.

An alternative system must be different from the standard indoor building we have today. While people suggest that alternative systems are more humane, the main reasons to use an alternative barn are often the lower cost to build or use the structure. Three common alternative systems (from most to least intensive) are bedded poultry barns, hoop structure and free range with a shelter. While there are many others, these three have at least some published data.

Twenty years ago, Jeff Hill (2000) reviewed deep-bedded finishing barns and hoops. He identified that there should be considerations of animal performance, environmental issues, marketing options, initial facility investments, and animal welfare issues. This section will touch on these issues but focus on animal welfare and on more recent literature. Most lay people observing a well-managed bedded or outdoor system would prefer that system over a more intensive system. However, alternate systems sometimes have serious welfare issues if they are not managed properly. We must also consider other society issues when a system is considered (McGlone, 2001). At least we need to consider animal welfare, food safety and environmental impacts of any outdoor or indoor-bedded facility.

One society issue has trumped all others in the recent disfavor of alternative systems: food safety. In the past (15–20 years ago), we could keep *Salmonella* out of indoor pigs, but we could not keep it out of the outdoor system, especially the wallows (Callaway et al., 2005). More recently, porcine epidemic diarrhea virus (PEDv) and ASF have moved among outdoor herds freely and unpredictably by birds, wild pigs and rodents. These infectious diseases (and probably more) are deal-killers for outdoor systems in the USA and parts of Europe and Asia. Still, outdoor and bedded facilities are favored at first glance by consumers.

3.2 Flooring

Slatted concrete floors are the most commonly used type of floor in intensive units due to their effective drainage of manure (Devillers et al., 2019). However, slatted floors can have negative effects (such as lameness), but limited studies have been conducted to identify the effect of slatted concrete flooring design on pig performance and welfare (Devillers et al., 2019).

Slat design has been studied for manure drainage and ammonia emissions, with slat width based on the efficiency of manure drainage (Aarnink et al., 1996, 1997; Vermeij et al., 2009; Ye et al., 2009). In North America, slats usually measure 5 inches and gaps measure 1 inch in width, with a permeability of 17% for a fully slatted floor (Devillers et al., 2019).

Rahse and Hoy (2007) found a tendency for more severe claw lesions in fattening pigs kept on slatted floors with a gap wider than 20 mm (0.78 inches), whereas Falke et al. (2018) found no effect on the slat width on leg and foot lesions between 153 mm (6.02 inches) and 190 mm (7.5 inches) slats, both with gaps of 15 mm (0.59 inches). The relatively high prevalence of toe lesions may be linked to pigs being predominately housed on concrete floors (Cameron,

2012; Penny et al., 1965; Schulenburg et al., 1986). When pigs are housed on slatted floors, an increased risk of sole and heel erosions has been reported and these lesions' prevalence is higher compared to pigs housed on solid floors with straw bedding (Jørgensen, 2003). Gentry et al. (2002a) examined feet and lungs (and meat quality) of indoor pigs raised on concrete slats with pigs raised on bedding. Pigs on bedding had fewer overall foot lesions, but the percentage of severe foot lesions was three times higher on bedding than for pigs on concrete slats. When pigs on bedding get a foot lesion, the lesion is more likely to develop into a severe foot lesion.

Webb (1984) reported that pigs walk mainly on their outer digits and also use their inner toe to some extent; however, the compressive strength of the hoof wall does not increase in heavier pigs. The peak pressure on the foot is almost independent of weight, and the total contact area varies almost linearly, which gives weight peak hoof pressures and hoof compressive strengths a safety factor for the foot (Webb, 1984). Gap recommendations on slat and gap width vary based on the size of the animal, mainly due to the size of the foot in proportion to the body weight and the void percentage of the slatted floor as a function of live weight (Jensen et al., 1997; Broom et al., 2005). Matching the pig foot size with the slats is important so that the manure flows, and the pigs do not get their feet stuck in the slats. Anatomical measurement of claw widths in sows is 25–33 mm (0.98–1.30 inches), and toes measure 45–51 mm (1.77–2 inches), on average (Broom et al., 2005; Sasaki et al., 2015). A study by Tubbs (1988) gave a recommendation of 13–16 mm (0.51–0.63 inches) rather than 19–25 mm (0.75–0.98 inches) to minimize injuries to toes and dew claws, which would then be more consistent with having slat width measure at least the sole length or twice the claw width (an average of more than 60 mm (2.40 inches); Devillers et al., 2019).

3.3 Bedded indoor and outdoor systems

A bedded indoor system is a way for pork producers to have pigs on bedding with a relatively low-cost building. Usually, these buildings are a re-purposed building that is converted for finishing pigs. These buildings became known when, in Missouri and surrounding states, the US packers needed more pigs and the finishing spaces were not enough in numbers. Figure 1 shows an example of a converted Turkey building in a region that has a large amount of fescue hull bedding from farms that produce fescue for seed. In the northern states, they use corn stalks, wheat, or barley straw for bedding. Because shipping bedding is expensive, producers use bedding that is available locally.

An important topic to consider with bedded facilities is the management of large amounts of wet or soiled bedding. This chapter does not review this issue, but it must be considered in practice.

Figure 1 Wide view (a) and close view (b) of finishing pigs near the end of the grow-finish cycle. Pigs are on fescue hull bedding, which is common in Missouri. The barn started with 2000 pigs, bedded, with natural ventilation (curtains go up and down). This barn was used to grow turkeys and re-purposed into a 'natural' finishing facility with bedding rather than concrete slats.

Two features of most bedded indoor facilities are the flooring (concrete slats or bedding) and group size (usually more than 500 pigs per barn, often 1000–2000 pigs).

The performance, health and welfare of pigs in bedded buildings versus conventional indoor slatted-floor systems have previously been studied. Gentry and others (Gentry et al., 2002a) examined pig growth, pork composition, and pork quality for pigs finished indoors on slats, and indoors on deep bedding (Fig. 2), or outdoor housing (Fig. 3). Pigs born and raised outdoors in the summer had higher weight gain and heavier carcasses and redder meat color than pigs in a more conventional indoor slatted barn. However, loins in a simulated retail display from the outdoor system in the summer had more discoloration and browning 4 days after display compared with indoor pigs. In another study, during the Texas winter, no differences were found in pig performance or in most meat quality traits. In the next comparison, cohorts of pigs were finished in either a conventional indoor barn or a deep-bedded barn with 1500 pigs in a single pen. Pigs on bedding grew faster and had

Figure 2 A row of hoop structures for growing-finishing pigs. Each barn held about 500 pigs. A tractor with a feed wagon was used to re-fill feeders.

Figure 3 An outdoor pig nursery (a) in Texas. Pigs will need to be moved when they get bigger to maintain their health and welfare. A Brazilian low-cost, very extensive system (b). As the pigs grow they will need to be moved to a larger space with more ground cover to improve their health and maintain the land. An outdoor finishing pen (c) in Texas. These pigs have a concrete slat section under the waterer to prevent pigs from making a wallow (there is a wallow in those pens away from the waterer). And the grass is green and growing in this field due to the low pig density.

heavier carcasses than indoor-reared pigs. More directly related to welfare, the authors measured foot and lung lesions from pigs on bedding and slats. The conclusion about the foot lesions is that, overall, bedding reduces foot lesions; however, if any pig begins to develop foot lesions on bedding, it will be much worse than if the pigs were on slats. For lung lesions, pigs on bedding had more overall lung lesions but fewer severe lung lesions than pigs on slatted floors.

Gentry et al. (2002a) further explored the mechanisms behind the improvements observed for pigs in the outdoor system. In their study, pigs born to contemporary sows which farrowed either indoors or outdoors were moved either indoors or outdoors. Overall, the outdoor-born pigs had higher growth and heavier carcasses than indoor-born pigs, regardless of being finished indoors or outdoors.

The same group examined muscle fiber types for indoor and outdoor-born pigs (Gentry et al., 2004). Loin chops from outdoor-born pigs had more type I and type IIA muscle fibers and a lower number of type IIB/X fibers than indoor pigs. Thus, the outdoor farrowing system changes muscle development compared to indoor-reared pigs in Texas. A similar study in Switzerland by Bee et al. (2004) performed in the winter did not find the same results as Gentry et al. (2004), but they measured different traits. Meat from outdoor-reared pigs had more drip loss and had fewer fast glycolytic fibers and different fatty acid composition than indoor pigs. These two investigators' findings did not actually conflict because the positive meat quality results by Gentry were found mostly in the summer.

To consider if the results above between housing systems were due to exercise among pigs in large pens or outdoors, Gentry et al. (2002b) used very narrow and long pens with the feeder on one end and the water on the other end. These pigs spent more time walking than pigs in a conventional finishing pen. Pigs forced to exercise did not have many differences in pig performance or meat quality. These results (Gentry et al., 2002b) cannot be explained simply by pigs in alternative systems getting more exercise. Other, yet-to-be-determined, factors are responsible for the differences in pig growth meat quality.

Clearly, more is to be learned about housing system effects on muscle fat/lean composition, metabolism and muscle fiber types. The effects of rearing systems on muscle fibers and collagen were reviewed 15 years ago by Gondret et al. (2005). Additionally, we need to focus on pig factors and less on growth and meat quality. Food safety, pig health issues and environmental issues should be considered and can be challenging. For pig factors related to animal welfare, the bedded facilities have a positive view from casual observers and there is some evidence that pigs are more active and more social in larger groups with bedding or outdoors.

3.4 Hoop buildings

A hoop building is a special type of bedded facility (Fig. 3). The structure can be made of metal or different kinds of heavy plastic. The inside is bedded, although we have seen an inside with slats. In this section, we refer to bedded hoop buildings. Some research was done on this housing system 15–25 years ago. Little current research can be found in the literature.

In an attempt to directly assess pig welfare in hoops and slatted facilities, Lay et al. (2000) reported that compared to pigs on slats, pigs in hoops had lower cortisol, fewer aberrant behaviors, more play and fewer injuries. We speculate that the hoop system was well managed. If the bedding is wet, especially in the winter, we may find different effects. A report from Portugal by Araújo et al. (2016) found no meaningful differences in pig welfare for a special Portuguese breed in hoop buildings, but the study had an unclear control treatment group.

Honeyman (2005) summarized his work in a 2005 review. This author has conducted the most extensive work on hoop structures in the USA. Honeyman (2005) concluded that pork quality was only slightly different for pigs in hoops compared to conventional indoor, slatted-floor facilities. He reported that pigs in hoops tended to be fatter (as we found with pigs in bedded indoor facilities). This may indicate that pigs in bedded facilities need to be fed differently. Also, this finding may indicate that the bedding creates a warmer microenvironment, and with a similar feed intake, they would be fatter. It is interesting that Honeyman reports smaller carcass weight in hoop buildings, while in our bedded poultry facility, the carcass weights were heavier than in indoor slatted facilities. Part of this may be explained by differential effects depending on the season.

A common difference for pigs in hoops is that they are often sold in a 'natural' pork scheme of various types. As such, they are often not fed antibiotics. The comparison of indoor slated facilities with hoop or bedded or outdoor facilities needs to have similar antibiotic applications and a common cohort of pigs (same genetics and previous experience). For the most part, some data show similar performance and health issues in a variety of systems if the study has controlled for known factors (temperature, genetics, feed, health, time, etc.). The main issues in the USA are the significant food safety and animal health challenges with alternative systems – these were enough to limit the use of alternative systems except in small, specialized niche markets.

4 Group size

Marginal profit increases with the size of pig operations (Martin and Kruja, 2000); the swine industry is shifting toward larger production units. Thus, housing large numbers of similarly aged animals and housing pigs in large groups can reduce housing costs (Schmolke et al., 2003).

It has been suggested that as the group size increases, variation in performance also increases (English, 1988). Housing large groups of pigs has long been reported to have a negative impact that may lead to poor performance, increased variation in body weight (BW) and a higher incidence of behavioral vices (English, 1988). Group size can affect the performance of pigs, although welfare is not necessarily compromised when increasing the number of individuals within a group for a given space allowance per pig (Turner et al., 2003; Turner and Edwards, 2004). Group size and space allowance per pig are reported to often be confounded in many studies, making the interpretation of these two factors difficult (Averós et al., 2010).

Schmolke et al. (2003) found that average daily gain (ADG) in groups of 40 pigs was 12% lower than pigs in groups of 10 during the initial 2 weeks of their study. Spoolder et al. (1999) reported a 5% reduction in the growth rate in groups of 40–80 pigs up to 65 kg but not during the finishing period, compared with pigs in groups of 20. Wolter et al. (2001) conducted a wean-to-finish study where he found that growth rates were poorer in groups of 50 and 100 than in groups of 25 between the weights of 6 kg and 35 kg. However, Wolter et al. (2001) did not find differences in the final BW and overall growth rate among the group sizes. Schmolke et al. (2003) did not observe an increase in variation in performance as the group size increased. Petherick et al. (1989) reported that pigs housed in groups of 36 had greater initial variation in BW than groups of 6 or 18 pigs; however, the variation in BW was similar among group sizes when the study concluded.

It has been suggested that one of the main sources of stress in large groups is the complexity of the social hierarchy (Stricklin and Mench, 1987; Moore et al., 1996). If social stress is significantly greater in pens of larger group sizes, one could expect a breakdown in social stability and hence an increase in the variability of animals (Stricklin and Mench, 1987).

One strategy that pigs could use to reduce the complexity would be to form social subgroups that avoid interacting with other subgroups (Stricklin and Mench, 1987; Gonyou and Keeling, 2001). The result of such a strategy would be that the pigs remain in a limited area of the pen and avoid other sections (Schmolke et al., 2003). Additionally, the probability of forming linear hierarchies is relatively low, even for moderate group sizes of seven to eight individuals (Mesterton-Gibbons and Dugatkin, 1995).

The increase in pig weights can lead to increased stocking densities toward the end of the finishing phase, increasing feeder and water competition, causing aggression among pen mates that can lead to an increase in injuries, lameness and other problems.

Resources such as food, water and attractive lying places are often limited in space, creating a competitive environment, which leads to aggression and social stress, even when the amount of resources is sufficient (McGlone, 1985; Mendl et al., 1992). Due to the differences in competitive ability, some individuals will monopolize resources, while others will refrain from competing and use the resources when the other group members are resting (Andersen et al., 2004). Resource monopolization is reported to be reduced in larger groups since a high rate of intrusions reduces the effectiveness of aggression in controlling a resource (Davies and Houston, 1981) and increases the time spent, energy expended and injury (Andersen et al., 2004). As group size increases, more individuals will benefit from not engaging in fights and will use alternative non-aggressive strategies to acquire resources (Andersen et al., 2004). Pagel and Dawkins (1997) suggested that the likeliness of encountering the same individual will decline with increasing group size. However, larger groups imply a more complex social environment, and usually only those individuals with high resource-holding potential are involved in serious fights (Andersen et al., 2004). Thus, it is assumed that there are large individual differences in aggression within groups (Mendl et al., 1992; Hessing et al., 1993; Erhard et al., 1997; Andersen et al., 2000).

5 Living conditions in indoor systems

5.1 Temperature

Confined livestock buildings are usually equipped with mechanical ventilation systems that will (1) provide sufficient air quality during the winter season, in combination with air temperatures that are close to the thermoneutral zone of the animals, despite restricted ventilation rates and (2) removal of sensible heat from animals by high ventilation rates to avoid high house temperatures during the summer (Vitt et al., 2017). The ventilation system must remove water and noxious gases in the barn, but limitations result from economic constraints due to high investment, energy and maintenance costs (Vitt et al., 2017). Newer filtered barns are being developed for increased biosecurity. These barns consume more energy and, if not managed right, can have poor air quality.

Pigs growing near their thermoneutral zone have better feed efficiency and improved welfare (Jackson et al., 2018; Cecchin et al., 2019) compared to when pigs are cold. Pigs in temperatures outside their thermoneutral zone may experience behavioral and physiological changes, consequentially affecting performance (Kiefer et al., 2009). Compared to other livestock species reared for meat production, pigs have a relatively low level of insulation in their coat hair, making them more susceptible to temperatures outside their thermoneutral zone within their housing environment (Jackson et al., 2018). Feral pigs can change their microenvironment by manipulating the material to provide insulation while indoor pigs cannot.

In cold conditions, pigs' thermoregulatory heat production increases, and energy retention from consumed nutrients is depressed, causing worse feed efficiency (Lopez et al., 1991). In warm weather, pigs will begin to reduce feed intake in order to decrease their metabolic heat production (Jackson et al., 2018), leading to a significant impact on pig productivity (Renaudeau et al., 2011).

Pigs have underdeveloped sweat glands, and, thus, it is more likely that pigs will experience heat stress in their housing environment than other farm animals. Pig housing designs should include water misting (Santonja et al., 2017), as this type of cooling system can aid in heat transfer through evaporation and convection (Jackson et al., 2018). Evaporation will transform water droplets into vapor, increasing humidity, reducing the surrounding air temperature, while convection created by an adequate velocity of air passing over the pigs will remove heat and vapor before the pigs get wet (Jackson et al., 2018).

In a study by Vitt et al. (2017), earth heat exchangers (those that use the earth for heat storage) showed the best performance compared to direct evaporative cooling by cooling pads and indirect evaporative cooling by cooling pads in combination with a heat exchanger by totally avoiding heat stress. Earth inlet heat exchangers were effective at damping short-term temperature fluctuations and heating the inlet air temperature during the winter, whereas direct evaporative cooling pads reduced temperature depending on the relative humidity of the outdoor temperature (but moistened bedding materials), and indirect evaporative cooling by cool pads in combination with a heat exchanger reduced the inlet air temperature by evaporation without humidification (Vitt et al., 2017).

5.2 Air quality

Gentry et al. (2002a) studied the lungs of pigs kept indoors on either slats or indoors on bedding. The lungs of pigs kept indoors on bedding, and slats were similar on average. However, pigs in the indoor slatted system had two times the severe lung lesions than pigs on bedding indoors. Air quality is a challenge in both systems; in indoor slatted systems it is ammonia and other gases, while, on bedding, it is dust. Apparently poor gas air quality is worse for pig lungs than the dust from the bedding.

Pig production is one of the major contributors to ammonia production (Groot Koerkamp et al., 1998; Aneja et al., 2000), methane and nitrous oxide (Blanes-Vidal et al., 2008). These gases are produced by direct emission from the digestive system of animals or from the decomposition of animal waste (Blanes-Vidal et al., 2008). Among the different swine buildings, ammonia concentrations are highest in finishing buildings (Groot Koerkamp et al., 1998). Ammonia concentration typically ranges from 0 ppm to 40 ppm (Heber et al., 2005), and the concentration is usually higher at night and in winter when building ventilations are low due to lower outside temperatures (Ni et al., 2018).

Ammonia is toxic and irritates the respiratory tract at concentrations exceeding 15 ppm (Banhazi et al., 2008). High ammonia concentrations can have negative effects (Fig. 4) on human health (Urbain et al., 1994) and the health of pigs (Gerber et al., 1991).

In a study by Jones et al. (1996), pigs were given access to compartments with different ammonia concentrations (0, 10, 20 and 40 ppm). Pigs spent a greater amount of their time in the compartments with lower ammonia concentration than the compartments with higher concentration, 0 ppm (54.3%), 10 ppm (26.9%), 20 ppm (7·1%) and 40 ppm (5.1%), respectively. Pigs visited areas with higher concentrations of ammonia less often and for shorter periods of time. However, this group reported that the aversion to areas with higher

Ammonia concentration (ppm)	Symptoms
10	Some negative effects at long term exposure
15	Smell threshold for human beings
20	Eye irritation for broilers
20–40	Increase of respiratory diseases
25–35	Stockmen feel uncomfortable
50	Disturbance of productive capacity; Water flows from the eyes
50–150	Decrease of young pig growth by 12– 29%
70	Reduced daily gain and poor feed conversion
100–200	Irritation and anorexia
5000	Deadly within a few minutes

Figure 4 Ammonia concentration and the effects/symptoms on humans and animals. Adapted from Ni et al. (2018).

concentrations of ammonia was not immediate and movement to other areas with lower ammonia concentrations may have been due to a sense of malaise that was possibly developed in a polluted atmosphere. The pigs also chose to rest, feed and forage in areas where there was lower ammonia concentration.

5.3 Space allowance

Space allowance is of interest to animal producers, policy makers and animal welfare groups and plays an important role in public concerns associated with intensive farming (Estevez, 2007; SVC, 1996, 1997). Space allowance can affect the welfare and productivity of pigs (Weng et al., 1998; Spoolder et al., 1999). New pig diseases that are endemic in many countries (Chae, 2004; Baekbo et al., 2012) may be exacerbated by social stress and crowding (Sutherland et al., 2007; Alarcon et al., 2011). Space allowance has been extensively studied throughout the years, with a negative effect of high stocking densities being found on different behavioral and physiological indicators associated with impaired welfare (Meunier-Salaun et al., 1987, 2007). Thus, recommendations to producers include reduced stocking density, that is, providing more space (Hassing and Bækbo, 2004).

The formula that relates body weight to surface areas is known as the k-value (NFACC, 2014). When multiplied by a pig's body weight (kg), k-value gives the floor surface area in m^2: $A = k \times BW^{0.667}$, where A = floor surface area in m^2; k-value = floor space allowance coefficient; BW = pig body weight in kg (NFACC, 2014). Allometric space allowance provides a basis for the determination of the spatial needs for pigs (BW; Petherick, 1983; Baxter, 1984) and the homogenization of all the available information according to BW (Gonyou et al., 2006).

In a study by Thomas et al. (2017), different space allowances (0.84, 0.74 or 0.65 m^2 per pig) were tested using finishing pigs. Thomas et al. (2017) found a reduction in ADFI as space allocation decreased, starting at a mean BW of 80.3 kg and that ADFI and ADG decreased linearly, starting at a mean BW of 74 kg, as space allocation decreased. This group concluded that decreasing space allocation resulted in poorer ADG driven by a reduction in ADFI. In another study by Carpenter et al. (2018), over a 71-day period, provided a space allowance of 0.91, or 0.63 m^2 per pig, with either a gate adjustment (on days 28, 45 and 62) or removal of the heaviest pig from the pen (on days 28 and 45) to provide more space and keep pigs in accordance with their predicted minimum space requirement ($(m^2) = 0.0336 \times (BW, kg)^{0.667}$). The results of their study were that pigs provided 0.91 m^2 had increased ADG compared with those allowed 0.63 m^2 with pigs provided space adjustments. Furthermore, pigs with 0.91 m^2 of space grew faster and consumed more feed than pigs restricted in space. As pigs grew to the predicted space requirement and were provided with more space, performance for pigs with adjustments was greater than those provided just 0.63 m^2 but less than those allowed 0.91 m^2. Carpenter et al. (2018) concluded that the industry-accepted critical k-value (of 0.0336) may not be adequate for optimal pig performance across multiple BW ranges. Flohr et al. (2018) used data from 30 publications on the effects of floor space allowance on the growth of finishing pigs using alternative prediction equations for ADG, average daily feed intake (ADFI) and gain:feed ratio (G:F). The optimum equations to predict ADG, ADFI and G:F were ADG, g = 337.57 + (16 468 × k) − (237 350 × k2) − (3.1209 × initial BW (kg)) + (2.569 × final BW (kg)) + (71.6918 × k × initial BW (kg)); ADFI, g = 833.41 + (24 785 × k) − (388 998 × k2) − (3.0027 × initial BW (kg)) + (11.246 × final BW (kg)) + (187.61 × k× initial BW (kg)); G:F = predicted ADG/predicted ADFI. The results of the meta-analysis indicated that BW was a predictor of ADG and ADFI even after computing the constant coefficient k, which used the final BW in its calculation. Flohr et al. (2018) suggested that including initial and final BW improves the prediction over using k as a predictor alone and that G:F of finishing pigs is influenced by floor space allowance.

Small space allowances can adversely affect productivity but also pen hygiene (Jensen et al., 2012). Pigs naturally avoid defecating in their lying areas, and generally hygiene will deteriorate as pigs get larger (Hacker et al., 1994; Rossi et al., 2008), as they will not be able to separate their dunning and lying areas with reduced space (Jensen et al., 2012). According to the EFSA the equation to calculate space allowance should be $A = 0.036 \times BW^{0.67}$ for pigs up to 110 kg, but pigs above 110 kg should have a k = 0.047; this equation is based solely on lying behavior and does not incorporate space for fundamental behaviors (such as feeding, drinking, excretion and exploration; Vermeer et al., 2014). A group of international scientists reviewed the scientific literature to determine the k-value for finishing pigs based on the actual published weight gain and other objective measures. Gonyou et al. (2006) concluded the k-value should be between 0.0317 and 0.0348. This range is based on science, whereas the EFSA value is based on the judgment of its authors.

5.4 Lying comfort

Growing-finishing pigs spend a large part of their time budget lying down (Ruckebusch, 1972), and, therefore, an adequate lying comfort seems important for their welfare (Tuyttens, 2005). Data relative to the allometric space required to satisfy pigs' different lying postures already exists (Petherick, 1983; Ekkel et al., 2003), but the quantitative relationships between housing factors and the percentage of time spent lying remain largely unknown (Averós et al., 2010). Averós et al. (2010) conducted a meta-analysis of 22 studies, in which they found a significant interaction between the k-value and the floor type, showing that the relationship between space allowance and lying behavior is affected by the presence or absence of slats. Averós et al. (2010) also suggested that the ability to rest as space availability decreases may be compromised before a reduced performance becomes apparent.

The concept of used and free spaces was first published by McGlone and Newby (1994). First, pigs utilize space differently depending on the time of day. Floor space occupied by standing and lying pigs is least in the middle of the night and most during active periods when more pigs are standing. The authors developed regression equations for free floor space over a day and found the point in which free space was the least. This could be the space requirement – the time when most pigs are lying down. Imagine a pen of pigs sleeping or lying down – the floor space not occupied is the free space. The authors provided experimental evidence for two conclusions. First, considering the USA required space allowance per pig, when 50% of the free space is removed by reducing pen size, then there is no performance set back as there is in crowded pigs. Second, as the group size increased; that is, assume you give heavy finishing pigs 1 m^2/pig. With 10 pigs, there is a certain, small amount of free space. But if you give 40 pigs 1 m^2/pig, there is a greater amount of free space. With very large group sizes (e.g. 1000 pigs in a pen), the total space can be reduced because there is so much unused, free space. Other authors have not been able to demonstrate this concept using other methods. However, the second point is a simple case of geometry. The first conclusion may work only with very large group sizes (100 or more pigs per pen).

Nannoni et al. (2019) reported that increasing the space allowance to 1.3 m^2/head for Italian heavy pigs had positive effects on animal welfare, as it increased the possibility of rest and improved productive parameters and reduced pen floor exploration. When there is insufficient space, group members time-share space; thus the space is dependent not only on space allowance per individual but also on group size (Petherick, 2007). Allowing all the pigs to lay down all at the same time may prevent sleep disruption by other pen mates and allow longer and synchronized sleeping bouts (Nannoni et al., 2019). Synchronized lying behavior has been reported as an indicator of improved welfare (Vermeer et al., 2014) in contrast to Gonyou et al. (2006) who based their conclusion on the point in lowering space allowances in which animal performance was negatively impacted. One cannot know the consequences of changes in lying behavior, but pig feed intake and growth are parameters that are very sensitive to space, especially too little space.

5.5 Lighting

Inappropriate lighting has a negative effect on the welfare of animals, and lighting influences an animal's ocular, physical and neural development and behavior (Taylor et al., 2006). Taylor et al. (2006) placed pigs in four chambers with different illuminance (2.4, 4, 40 and 400 lux) and found that pigs preferred the dimmest lighting and spent the least time in the brightest lighting. Pigs rested and slept in the dimmest lighting and defecated in the brightest lighting. Taylor et al. (2006) suggested that pigs should be provided at least 6 h of dim light (at 2.4 lux). Literature is limited to determine lighting schedules for pigs. Most of the literature reports lighting recommendations for low-stress handling (Grandin, 2003, 2014).

5.6 Stockmanship

The attitude of stockmen/caretakers toward animals is directly associated with their behavior during handling, and rough practices can negatively affect animal welfare (Hemsworth and Coleman, 2010). When animals experience positive interactions with humans, they become less fearful, which, in turn, facilitates handling (Schmied et al., 2010; Probst et al., 2012).

In modern husbandry production, human–animal relationships are an important component of farm animal welfare (Hosey and Melfi, 2014). Poor stockmanship can increase animal anxiety, leading to difficulty in handling and management (Wang et al., 2020). Pigs with a high degree of fear of humans have restricted growth and production performance due to a chronic stress responses (Hemsworth et al., 1981, 1986, 1987; Gonyou et al., 1986). A low-stress response of pigs toward people, and increased willingness to approach and make contact with people is an indicator of good human–animal relationship (Waiblinger et al., 2006). Gentle handling has been reported to improve the daily feed intake (Day et al., 2002), feed conversion rate (Hemsworth and Barnett, 1991) and growth efficiency (Hemsworth et al., 1987) of weaned piglets. Additionally, gentle handling can affect pigs' stress before slaughtering and can lead to improved meat quality (D'Souza et al., 1998; Hemsworth et al., 2002).

Handling practices are usually learned from working 'on the job', and it is common for stockmen to believe that their behaviors are harmless to animals (Hemsworth, 2007). Caretakers' behavior during livestock handling can be improved by carefully selecting personnel and developing training programs designed to reduce animal stress during handling (Boivin et al., 2007). Pig studies have shown that cognitive/behavioral interventions can improve workers' attitudes and consequently their behavior (Coleman et al., 2000; Hemsworth et al., 2002). However, training efficacy depends on the interactions between all trained and non-trained individuals on the same farm and on whether or not a common culture toward animal handling can be developed (Coleman and Hemsworth, 2014).

Coleman et al. (2000) found that training improved stockpeople's attitudes toward animals. In pig production, daily handling often occurs immediately following training sessions, and if done regularly after training, stockpeople can experience the benefits of training by observing the consequences of their behaviors on animals' reactions (Ceballos et al., 2018). Stockpeople's attitude toward the animals could be improved by a feedback loop since the nature and frequency of those behaviors determine the quality of human-animal relationships (Ceballos et al., 2018).

Ceballos et al. (2018) suggests, based on their study, that better managing style within the farm with regular trainings can favor the development of positive stockpeople attitudes toward animals or can reinforce the beneficial effects of training. Additionally, Ceballos et al. (2018) suggested that stockpeople who were not trained but worked on the same team or in direct contact with trained people performed more positive behaviors, which proposes possible social facilitation, imitation or transfer of these types of behaviors. Ceballos et al. (2018) stated that the presence of someone who has been trained has the potential to change the normal context of animal handling on the farm and affects the behavior of the non-trained people, possibly influencing them to behave more properly.

6 Animal health and animal care

Animal health management involves the farm veterinarian, the client or those working for the client/farm caretakers. Early detection of health problems and compromised welfare can increase treatment success, contain problems and enhance pig welfare and sustainability (Matthews et al., 2016). However, early detection of problems is challenging as human observation can be subjective and subclinical disease shows no signs (Matthews et al., 2016).

Inflammatory responses evoke changes in behavior, including increased thermoregulatory activities and sleep, reduced social exploration and appetite and altered food preferences (Millman, 2007). The term 'malaise' is commonly used to describe sickness feelings such as lethargy, depression and pain (Millman, 2007). Expression of sickness behavior would include huddling, shivering, dehydration, fever, anorexia, increased thirst, sleepiness, reduced grooming and exploration, uncoordinated body movements and increased pain and sensitivity (Millman, 2007). Other clinical signs used to identify sick pigs include diarrhea, vomiting, isolating themselves from the social group, seeking warmer environments and increasing lying behavior to minimize heat loss (Millman, 2007). Sickness behavior in pigs may be difficult to identify by the caretaker but also hard for pigs to express, especially toward the end of the finishing phase, where space allowance is reduced due to their group housing environments (group housing, commonly used in commercial settings).

Respiratory disease is also a common problem in late-stage finishing pigs. With increasing body weights, sanitation programs must be taken into consideration as pigs over 110 kg have increased risks of outbreaks of ileitis and mycoplasma pneumonia (Kim et al., 2005). *Mycoplasma hyopneumoniae* is the primary etiologic agent of enzootic pneumonia (a multifactorial disease) of pigs and is a key pathogen in the porcine respiratory disease complex (PRDC) (Thacker and Minion, 2012). *Actinobacillus pleuropneumonia* (APP) is one of the main pathogens associated with pleuritis in pigs (Merialdi et al., 2012). Both of these pathogens are associated with increased body weights, seen toward the end of the finishing phase, and are primarily seen in the USA in larger finishing pigs (Kim et al., 2005). Pigs in the finishing stage are usually affected by these pathogens, which can cause chronic coughing, growth retardation, low mortality but high morbidity (Maes et al., 1996).

6.1 Hospital/sick pens

Properly operated hospital pens (those that offer extra warmth, good footing, easy access to feed and water, have established protocols for treatment, culling and euthanasia) provide an economic return, improve worker morale and improve the welfare of compromised pigs (Blackwell and Wellington Place, 2005). Hospital pens are designed to house compromised pigs where they are supposed to be doctored and nursed back to health, but currently most hospital pens are actually sick pens that are more like holding areas where animals in reality are not commonly treated (Blackwell and Wellington Place, 2005). When a pig dies in a hospital pen, it should always be a surprise. If workers are saying to themselves that they are happy a certain pig finally died, euthanasia decision rules are not adequate and caretakers are neglecting part of their responsibilities (Blackwell and Wellington Place, 2005). As a rule, animals should be culled when they are no longer profitable and euthanized when it is inhumane to let the animal live, but deciding when it is uneconomic and whether to treat or euthanize is difficult for farm managers (Morrow et al., 2006).

The Pork Quality Assurance Plus Program notes that a pig in a hospital pen that does not show signs of improvement after 2 days of intensive care, is severely injured or non-ambulatory with the inability to recover, or any pig that is non-ambulatory with a body condition score (BCS) of 1 should be euthanized.

6.2 Culling

Culling is defined as early pig removal during production (Stein et al., 1990). Culling is important to prevent additional economic losses. The loss of investment at the late stage of production affects the producer's profitability (Jensen et al., 2012). Pigs are typically identified by the caretaker to be sold or are isolated or placed in a separate pen (commonly with other pigs that are compromised known as a sick or hospital pen). Compromised pig segregation allows for their condition to improve by providing them with a nurturing environment that is less competitive, by treating them, or euthanized if the animal does not improve.

6.3 Euthanasia

Subjective guidelines in companion animals look at the ability of the animal to enjoy food, breathe freely and without difficulty, eat and drink without pain and respond to their owner and family (Morrow et al., 2006), whereas objective guidelines evaluate weight loss, weakness, infection, organ failure and injuries (Duncan, 1988). In farm animals, euthanasia considers welfare aspects and economics, but few comprehensive guidelines have been created for swine producers (Morrow et al., 2006). The NPB gives practical recommendations on swine euthanasia, but the On Farm Euthanasia of Swine: Recommendations for the Producer document only describes approved euthanasia methods and not recommendations on etiologies that would require pigs to be euthanized.

Not only is timely euthanasia a matter of good animal welfare, but it is also required by packers and customers. The Common Swine Industry Audit (CSIA) was developed by producers, veterinarians, animal scientists, packers, processors and retail and food-service representatives to establish criteria that any on-farm swine audit must include to be comprehensive and credible (CSIA, 2019). The CSIA includes 27 key aspects of swine care and pre-harvest pork safety through all phases of production. The CSIA tool has five critical questions that, if failed, will cause the farm to fail the entire audit. Question 1 pertains to willful acts of abuse or neglect, and questions 2–5 involve humane euthanasia. Humane euthanasia requires timely euthanasia for animals that have been treated for 2 consecutive days and fail to show signs of improvement. Additional critical questions address the humane handling of animals to be euthanized and the confirmation of insensibility and death. As mentioned earlier, not following these stringent requirements will result in the automatic failure of the audit, leading to a corrective action, which may include being dropped by the packer if repetitive violations occur or the farm fails to meet the required scores. However, despite the criticality of these issues in the industry, neither the CSIA nor Pork Quality Assurance Plus Program recommendations on timely euthanasia have been verified with scientific studies producing empirical evidence.

7 Conclusion and future trends in research

Pig inventory is much greater for pigs from weaning to market than for breeding stock. Therefore, opportunities to improve animal welfare in wean-to-finish pigs are significant if one wants to help many animals. This chapter summarizes areas of animal welfare concern on commercial farms and offers solutions if they are available. The future will bring more automated technologies to improve the health and welfare of pigs.

On many farrow-to-finish farms where 25 pigs per sow per year are typically produced, a 1000-sow farm would have more than 25 000 finishing pigs. Welfare problems in this group impact more individuals than any item in the breeding herd.

In the authors' view, the largest problem in the swine industry is the absence of skilled labor. Many larger farms operate with less than full staff. Having an inadequate number of people directly relates to reduced pig welfare. The solution to this problem is to automate this (and other) stage of production. On many farms, they assign one person to look at eight finishing barns of 1200 pigs each or 9600 pigs in total. This means the worker spends 1 h, with 1200 pigs in a day. That equates to 20 pigs per minute or 3 s per pig, assuming all the time is spent in pig observation - which it is not - workers must check feeders and waters and deal with sick or dead pigs. All of this can be automated using smart barn technologies. Cameras and sensors can monitor temperature, humidity, air flow, air quality, coughs, pig health and pig weights. Algorithms can determine pigs that are out of compliance and alter caretakers to a specific animal to attend to.

One of the most common criticisms of activists is when they find injured or ill pigs. Unattended 'compromised' pigs are common at a low percentage on all farms. Finding these animals quickly and attending to them or conducting timely euthanasia is critical, but when farms are short of labor, it is difficult to get the tasks accomplished. Smart barns may eventually solve these problems.

All the other issues in pig production are less of a problem, but they still warrant attention. These include painful practices, early detection and treatment of pig diseases, space, transport and enrichment. Funding bodies can direct resources toward the most important problems first. One caution is that we should not try to find solutions to problems that are caused by our production system (e.g. tail biting, aggression and others). We should instead change the production system. The finishing system has not changed much in the past 50 years. Current stressors, including concrete slatted floors, no use of bedding, aggression, heat stress, and others merit solutions to prevent the stressor rather than looking for a way to reduce the problem. Prevention is far more effective than treatment (all of which do not work well).

Two other take-home messages are (1) any change in animal welfare for the better must not include losing ground on other society issues (McGlone, 2001) and (2) from an economic point of view, any change to improve welfare cannot increase production costs unless the market pays for the change. Finally, when one country or region implements what they believe as positive welfare changes, it often results in a negative economic outcome. Forcing people to change practices to accommodate views held by other cultures may not be the solution to improving animal welfare.

8 Where to look for further information

The science of pig welfare is a relatively new discipline. While farm animal practices on the farm and at processing plants have been a concern at least since 1906 when Upton Sinclair wrote The Jungle, actual scientific research about animal welfare really started only in the 1960s in Europe and shortly thereafter in North America. Today, swine scientists study pig welfare in every major country that produces and/or consumes pork. The scientific literature about pig (and other species) welfare has grown and continues to grow at a rapid rate. Scientific literature search engines will identify studies published over the last few decades. However, interested readers should also examine publications over the past 200 years or more for information on pig care and husbandry. Even in 1883, Joseph Harris highlighted the importance of good pig care in the second paragraph of his excellent book on the pig (Harris on the Pig, 1883, reprinted 1999 by Lyons Press, NY, NY). The historical literature often provides insights into pig care, especially when production technology is changing.

Government bodies, especially in Europe, have seemingly ever-evolving animal welfare rules. The EU has EU-wide rules about pig welfare, but each member country can have more strict rules about pig care. In the USA and Canada, there are few animal welfare laws, but the industry has published guidelines. Many food companies that purchase pork have animal welfare rules that their suppliers must follow. Guidelines and company rules are used as a basis for auditing pig welfare on farms and at processing plants. While some countries may have government inspectors, this is not the case in the USA. However, detailed on-farm audits are likely to be effective or more effective than government inspectors. In the EU, some countries adhere to EU and country laws via on-farm inspections, while other EU countries do not perform on-farm inspections. Because on-farm audits are required to market pork, when this system is in place, every farm must be audited. Market access requires compliance without any government oversight. Pig trade associations in the USA produce and refine guidelines over time. One can look at the websites of the National Pork Board (www.pork.org) and the National Pork Producers Council (www.nppc.org) for animal welfare information including audit instruments. In addition, the Professional Animal Auditor Association (PAACO; www.animalauditor.org) certifies animal welfare auditors and audit instruments. PAACO provides open access to audit tools.

9 References

Aarnink, A. J. A., Van Den Berg, A. J., Keen, A., Hoeksma, P. and Verstegen, M. W. A. (1996). Effect of slatted floor area on ammonia emission and on the excretory and lying behaviour of growing pigs. *Journal of Agricultural Engineering Research* 64(4), 299–310.

Aarnink, A. J. A., Swierstra, D., Van den Berg, A. J. and Speelman, L. (1997). Effect of type of slatted floor and degree of fouling of solid floor on ammonia emission rates from fattening piggeries. *Journal of Agricultural Engineering Research* 66(2), 93–102.

Alarcon, P., Velasova, M., Mastin, A., Nevel, A., Stärk, K. D. and Wieland, B. (2011). Farm level risk factors associated with severity of post-weaning multi-systemic wasting syndrome. *Preventive Veterinary Medicine* 101(3-4), 182–191.

Andersen, I. L. and Bøe, K. E. (1999). Straw bedding or concrete floor for loose-housed pregnant sows: consequences for aggression, production and physical health. *Acta Agriculturae Scandinavica, Section A - Animal Science* 49(3), 190–195.

Andersen, I. L., Andenæs, H., Bøe, K. E., Jensen, P. and Bakken, M. (2000). The effects of weight asymmetry and resource distribution on aggression in groups of unacquainted pigs. *Applied Animal Behaviour Science* 68(2), 107–120.

Andersen, I. L., Nævdal, E., Bakken, M. and Bøe, K. E. (2004). Aggression and group size in domesticated pigs, Sus scrofa: 'when the winner takes it all and the loser is standing small'. *Animal Behaviour* 68(4), 965–975.

Aneja, V. P., Chauhan, J. P. and Walker, J. T. (2000). Characterization of atmospheric ammonia emissions from swine waste storage and treatment lagoons. *Journal of Geophysical Research: Atmospheres* 105(D9), 11535–11545.

Anil, L., Anil, S. S., Deen, J., Baidoo, S. K. and Walker, R. D. (2006). Effect of group size and structure on the welfare and performance of pregnant sows in pens with electronic sow feeders. *Canadian Journal of Veterinary Research = Revue Canadienne de Recherche Veterinaire* 70(2), 128–136.

Appleby, M. C. (2004). Alternatives to conventional livestock production methods. In: Benson, G. J. and Rollin, B. E. (Eds) *The Well-Being of Farm Animals*, pp. 339-350. John Wiley & Sons.

Araújo, J. P., Amorim, I., Silva, J. S., Pires, P. and Cerqueira, J. (2016). Outdoor housing systems for Bísaro pig breed with a hoop barn: some effects on welfare. In: *Food Futures: Ethics, Science and Culture*, pp. 185–242. Wageningen Academic Publishers, Wageningen, The Netherlands.

Arey, D. S. and Edwards, S. A. (1998). Factors influencing aggression between sows after mixing and the consequences for welfare and production. *Livestock Production Science* 56(1), 61-70.

Averós, X., Brossard, L., Dourmad, J. Y., De Greef, K. H., Edge, H. L., Edwards, S. A. and Meunier-Salaün, M. C. (2010). Quantitative assessment of the effects of space allowance, group size and floor characteristics on the lying behaviour of growing-finishing pigs. *Animal: an International Journal of Animal Bioscience* 4(5), 777–783.

Baekbo, P., Kristensen, C. S. and Larsen, L. E. (2012). Porcine circovirus diseases: a review of PMWS. *Transboundary and Emerging Diseases* 59(Suppl. 1), 60–67.

Bai, Z., Ma, W., Ma, L., Velthof, G. L., Wei, Z., Havlík, P., Oenema, O., Lee, M. R. F. and Zhang, F. (2018). China's livestock transition: driving forces, impacts, and consequences. *Science Advances* 4(7), eaar8534.

Banhazi, T. M., Seedorf, J., Rutley, D. L. and Pitchford, W. S. (2008). Identification of risk factors for sub-optimal housing conditions in Australian piggeries: Part 2. Airborne pollutants. *Journal of Agricultural Safety and Health* 14(1), 21–39.

Baxter, S. (1984). *Intensive Pig Production: Environmental Management and Design*. Granada Technical Books, London.

Beattie, V. E., Sneddon, I. A., Walker, N. and Weatherup, R. N. (2001). Environmental enrichment of intensive pig housing using spent mushroom compost. *Animal Science* 72(1), 35–42.

Bee, G., Guex, G. and Herzog, W. (2004). Free-range rearing of pigs during the winter: adaptations in muscle fiber characteristics and effects on adipose tissue composition and meat quality traits. *Journal of Animal Science* 82(4), 1206–1218.

Blackwell, T. and Wellington Place, R. R. (2005). Effective treatment and handling of poor doing pigs in the finishing barn. In: Conference Proceedings, London Swine Conference, London, Ontario, p. 167.

Blanes-Vidal, V., Hansen, M. N., Pedersen, S. and Rom, H. B. (2008). Emissions of ammonia, methane and nitrous oxide from pig houses and slurry: effects of rooting material, animal activity and ventilation flow. *Agriculture, Ecosystems and Environment* 124(3–4), 237–244.

Bolhuis, J. E., Schouten, W. G. P., Schrama, J. W. and Wiegant, V. M. (2005). Behavioural development of pigs with different coping characteristics in barren and substrate-enriched housing conditions. *Applied Animal Behaviour Science* 93(3–4), 213–228.

Boivin, X., Marcantognini, L., Boulesteix, P., Godet, J., Brulé, A. and Veissier, I. (2007). Attitudes of farmers towards Limousin cattle and their handling. *Animal Welfare Potters Bar Then Whethampstead* 16(2), 147.

Broom, D. M. (2010). Animal welfare: an aspect of care, sustainability, and food quality required by the public. *Journal of Veterinary Medical Education* 37(1), 83–88.

Broom, D. M., Gunn, M., Edwards, S., Wechsler, B., Algers, B., Spoolder, H., Madec, F., Von Borell, E. and Olsson, O. (2005). The Welfare of Weaners and Rearing Pigs: Effects of Different Space Allowances and Floor Types. EFSA-Q-2004-077, p. 129.

Brumm, M. (2012). Impact of heavy market weights on facility and equipment needs. Proceedings of the Allen D. Leman Swine Conference. St Paul, Minnesota, pp. 165–168.

Brumm, M. C., Harmon, J. D., Honeyman, M. S., Kliebenstein, J. B., Lonergan, S. M., Morrison, R. and Richard, T. (2004). *Hoop Barns for Grow-Finish Swine*. MidWest Plan Service, Ames, IA.

Calderón Díaz, J. A., Diana, A., Boyle, L. A., Leonard, F. C., McElroy, M., McGettrick, S., Moriarty, J. and García Manzanilla, E. (2017). Delaying pigs from the normal production flow is associated with health problems and poorer performance. *Porcine Health Management* 3(1), 13.

Caley, P. (1997). Movements, activity patterns and habitat use of feral pigs (Sus scrofa) in a tropical habitat. *Wildlife Research* 24(1), 77–87.

Callaway, T. R., Morrow, J. L., Johnson, A. K., Dailey, J. W., Wallace, F. M., Wagstrom, E. A., McGlone, J. J., Lewis, A. R., Dowd, S. E., Poole, T. L., Edrington, T. S., Anderson, R. C., Genovese, K. J., Byrd, J. A., Harvey, R. B. and Nisbet, D. J. (2005). Environmental prevalence and persistence of Salmonella spp. in outdoor swine wallows. *Foodborne Pathogens and Disease* 2(3), 263–273.

Cameron, R. (2012). Integumentary system, skin, hoof and claw. In Zimmerman, J. J. (Ed.), *Diseases of Swine*, pp. 251–263. Wiley-Blackwell, West Sussex, UK.

Carpenter, C. B., Holder, C. J., Wu, F., Woodworth, J. C., DeRouchey, J. M., Tokach, M. D., Goodband, R. D. and Dritz, S. S. (2018). Effects of increasing space allowance by removing a pig or gate adjustment on finishing pig growth performance. *Journal of Animal Science* 96(7), 2659–2664.

Casal-Plana, N., Manteca, X., Dalmau, A. and Fàbrega, E. (2017). Influence of enrichment material and herbal compounds in the behaviour and performance of growing pigs. *Applied Animal Behaviour Science* 195, 38–43.

Ceballos, M. C., Sant'Anna, A. C., Boivin, X., Costa, FdO., Carvalhal, M. V. and Paranhos da Costa, M. J. R. (2018). Impact of good practices of handling training on beef cattle welfare and stockpeople attitudes and behaviors. *Livestock Science* 216, 24–31.

Cecchin, D., Ferraz, P. F. P., Campos, A. T., Sousa, F. A., Amaral, P. I. S., Castro, J. O., Conti, L. and da Cruz, V. M. F. (2019). Thermal comfort of pigs housed in different installations. *Agronomy Research* 17(2), 378–384.

Chae, C. (2004). Postweaning multisystemic wasting syndrome: a review of aetiology, diagnosis and pathology. *The Veterinary Journal* 168(1), 41–49.

Clark, B., Panzone, L. A., Stewart, G. B., Kyriazakis, I., Niemi, J. K., Latvala, T., Tranter, R., Jones, P. and Frewer, L. J. (2019). Consumer attitudes towards production diseases in intensive production systems. *PLoS ONE* 14(1), e0210432.

Coleman, G. J., Hemsworth, P. H., Hay, M. and Cox, M. (2000). Modifying stockperson attitudes and behaviour towards pigs at a large commercial farm. *Applied Animal Behaviour Science* 66(1–2), 11–20.

Coleman, G. J. and Hemsworth, P. H. (2014). Training to improve stockperson beliefs and behaviour towards livestock enhances welfare and productivity. *Revue Scientifique et Technique* 33(1), 131–137.

Common Swine Industry Audit (2019). Available at: https://www.pork.org/production/tools/common-swine-industry-audit/common-swine-industry-audit-materials/.

Cooper, J. J., Cox, L. N. and Whitworth, C. (2001). Early environmental experience and transferable skills in the weaned piglet. *Animal Welfare-Potters Bar* 10, S238–S238.

Dailey, J. W. and McGlone, J. J. (1997). Oral/nasal/facial and other behaviors of sows kept individually outdoors on pasture, soil or indoors in gestation crates. *Applied Animal Behaviour Science* 52(1–2), 25–43.

Davies, N. B. and Houston, A. I. (1981). Owners & satellites: the economics of territory defence in the pied wagtail, Motacilla alba. *Journal of Animal Ecology* 50(1), 157–180.

Day, J. E. L., Spoolder, H. A. M., Burfoot, A., Chamberlain, H. L. and Edwards, S. A. (2002). The separate and interactive effects of handling and environmental enrichment on the behaviour and welfare of growing pigs. *Applied Animal Behaviour Science* 75(3), 177–192.

Devillers, N., Janvier, E., Delijani, F., Méthot, S., Dick, K. J., Zhang, Q. and Connor, L. (2019). Effect of slat and gap width of slatted concrete flooring on sow gait using kinematics analysis. *Animals: an Open Access Journal from MDPI* 9(5), 206.

Diana, A., Boyle, L. A., García Manzanilla, E., Leonard, F. C. and Calderón Díaz, J. A. (2019). Ear, tail and skin lesions vary according to different production flows in a farrow-to-finish pig farm. *Porcine Health Management* 5(1), 19.

Dixon, L. K., Sun, H. and Roberts, H. (2019). African swine fever. *Antiviral Research* 165, 34–41.

Docking, C. M., Van de Weerd, H. A., Day, J. E. L. and Edwards, S. A. (2008). The influence of age on the use of potential enrichment objects and synchronisation of behaviour of pigs. *Applied Animal Behaviour Science* 110(3–4), 244–257.

D'Souza, D. N., Leury, B. J., Dunshea, F. R. and Warner, R. D. (1998). Effect of on-farm and pre-slaughter handling of pigs on meat quality. *Australian Journal of Agricultural Research* 49(6), 1021-1025.

Duncan, J. C. (1988). Careers in veterinary medicine. New York, New York: Rosen Publishing Group.

EFSA Panel on Animal Health and Welfare (2007). Scientific opinion of the Panel on Animal Health and Welfare on a request from Commission on the Risks Associated with Tail Biting in Pigs and Possible Means. *The EFSA Journal* 611, 1-13.

EFSA Panel on Animal Health and Welfare (AHAW) (2012). Statement on the use of animal-based measures to assess the welfare of animals. *EFSA Journal* 10(6), 2767.

EFSA Panel on Animal Health and Welfare (AHAW) (2014). Scientific opinion concerning a multifactorial approach on the use of animal and non-animal-based measures to assess the welfare of pigs. *EFSA Journal* 12(5), 3702.

Einarsson, S., Madej, A. and Tsuma, V. (1996). The influence of stress on early pregnancy in the pig. *Animal Reproduction Science* 42(1-4), 165-172.

Ekkel, E. D., Spoolder, H. A. M., Hulsegge, I. and Hopster, H. (2003). Lying characteristics as determinants for space requirements in pigs. *Applied Animal Behaviour Science* 80(1), 19-30.

English, P. R. (1988). *The Growing and Finishing Pig: Improving Efficiency (No. SF 396.9. G76).*

Erhard, H. W., Mendl, M. and Ashley, D. D. (1997). Individual aggressiveness of pigs can be measured and used to reduce aggression after mixing. *Applied Animal Behaviour Science* 54(2-3), 137-151.

Estevez, I. (2007). Density allowances for broilers: where to set the limits? *Poultry Science* 86(6), 1265-1272.

Falke, A., Friedli, K., Gygax, L., Wechsler, B., Sidler, X. and Weber, R. (2018). Effect of rubber mats and perforation in the lying area on claw and limb lesions of fattening pigs. *Animal: an International Journal of Animal Bioscience* 12(10), 2130-2137.

Flohr, J. R., Dritz, S. S., Tokach, M. D., Woodworth, J. C., DeRouchey, J. M. and Goodband, R. D. (2018). Development of equations to predict the influence of floor space on average daily gain, average daily feed intake and gain: feed ratio of finishing pigs. *Animal: an International Journal of Animal Bioscience* 12(5), 1022-1029.

Fraser, D. (1978). Observations on the behavioural development of suckling and early-weaned piglets during the first six weeks after birth. *Animal Behaviour* 26, 22-30.

Fraser, D. G. (2005). *Animal Welfare and the Intensification of Animal Production: An Alternative Interpretation*, Vol. 2. Food & Agriculture Org.

Frei, D., Würbel, H., Wechsler, B., Gygax, L., Burla, J. B. and Weber, R. (2018). Can body nosing in artificially reared piglets be reduced by sucking and massaging dummies? *Applied Animal Behaviour Science* 202, 20-27.

Fritschen, R. D. and Hogg, A. (1996). *Preventing Tail Biting in Swine (Anti-Comfort Syndrome)*. Cooperative Extension, Institute of Agriculture and Natural Resources, University of Nebraska--Lincoln.

Gardner, J. M., De Lange, C. F. M. and Widowski, T. M. (2001a). Belly-nosing in early-weaned piglets is not influenced by diet quality or the presence of milk in the diet. *Journal of Animal Science* 79(1), 73-80.

Gardner, J. M., Duncan, I. J. H. and Widowski, T. M. (2001b). Effects of social "stressors" on belly-nosing behaviour in early-weaned piglets: is belly-nosing an indicator of stress? *Applied Animal Behaviour Science* 74(2), 135-152.

Gentry, J. G., McGlone, J. J., Blanton, J. R. and Miller, M. F. (2002a). Alternative housing systems for pigs: influences on growth, composition and pork quality. *Journal of Animal Science* 80(7), 1781–1790.

Gentry, J. G., McGlone, J. J., Blanton, J. R., Jr. and Miller, M. F. (2002b). Impact of spontaneous exercise on performance, meat quality, and muscle fiber characteristics of growing/finishing pigs. *Journal of Animal Science* 80(11), 2833–2839.

Gentry, J. G., McGlone, J. J., Miller, M. F. and Blanton, J. R., Jr. (2004). Environmental effects on pig performance, meat quality, and muscle characteristics. *Journal of Animal Science* 82(1), 209–217.

Gerber, D. B., Veenhuizen, M. A. and Shurson, G. C. (1991). Ammonia, carbon monoxide, carbon dioxide, hydroden sulfide, and methane in swine confinement facilities. *The Compendium on Continuing Education for the Practicing Veterinarian (USA)* 13, 1483–1488.

Gondret, F., Combes, S., Lefaucher, L. and Lebret, B. (2005). Effects of exercise during growth and alternative rearing systems on muscle fibers and collagen properties. *Reproduction, Nutrition, Development* 45(1), 69–86.

Gonyou, H. W., Hemsworth, P. H. and Barnett, J. L. (1986). Effects of frequent interactions with humans on growing pigs. *Applied Animal Behaviour Science* 16(3), 269–278.

Gonyou, H. W. and Keeling, L. J. (Eds) (2001). *Social Behavior in Farm Animals*. CABI Publishing, Wallingford, UK.

Gonyou, H. W., Brumm, M. C., Bush, E., Davies, P., Deen, J., Edwards, S. A., Fangman, T., McGlone, J. J., Meunier-Salaun, M., Morrison, R. B., Spoolder, H., Sundberg, P. L. and Johnson, A. K. (2006). Application of broken line analysis to assess floor space requirements of nursery and grow/finish pigs expressed on a allometric basis. *Journal of Animal Science* 84, 229–235.

Grandin, T. (2003). Transferring results of behavioral research to industry to improve animal welfare on the farm, ranch and the slaughter plant. *Applied Animal Behaviour Science* 81(3), 215–228.

Grandin, T. (2014). Animal welfare and society concerns finding the missing link. *Meat Science* 98(3), 461–469.

Grandin, T. and Whiting, M. (Eds) (2018). *Are We Pushing Animals to Their Biological Limits?: Welfare and Ethical Implications*. CABI Publishing, Wallingford, UK.

Groot Koerkamp, P. W. G., Metz, J. H. M., Uenk, G. H., Phillips, V. R., Holden, M. R., Sneath, R. W., Short, J. L., White, R. P. P., Hartung, J., Seedorf, J., Schröder, M., Linkert, K. H., Pedersen, S., Takai, H., Johnsen, J. O. and Wathes, C. M. (1998). Concentrations and emissions of ammonia in livestock buildings in northern Europe. *Journal of Agricultural Engineering Research* 70(1), 79–95.

Guy, J. H., Rowlinson, P., Chadwick, J. P. and Ellis, M. (2002). Health conditions of two genotypes of growing-finishing pig in three different housing systems: implications for welfare. *Livestock Production Science* 75(3), 233–243.

Hacker, R. R., Ogilvie, J. R., Morrison, W. D. and Kains, F. (1994). Factors affecting excretory behavior of pigs. *Journal of Animal Science* 72(6), 1455–1460.

Hassing, A.-G. and Bækbo, P. (2004). PMWS manual. Videncenter for Svineproduction. Centre for Pig Production, Copenhagen, Denmark. Available at: http://vsp.lf.dk/Publikationer/Manualer/PMWS%20manual.aspx?450 full=1.

Heber, A. J., Tao, P. C., Ni, J. Q., Lim, T. T. and Schmidt, A. M. (2005). Air emissions from two swine finishing building with flushing: ammonia characteristics. In: *Livestock*

Environment VII, 18–20 May 2005, Beijing, China, p. 436. American Society of Agricultural and Biological Engineers.

Hemsworth, P. H. (2007). Ethical stockmanship. *Australian Veterinary Journal* 85(5), 194–200.

Hemsworth, P. H. and Barnett, J. L. (1991). The effects of aversively handling pigs, either individually or in groups, on their behaviour, growth and corticosteroids. *Applied Animal Behaviour Science* 30(1–2), 61–72.

Hemsworth, P. H. and Coleman, G. J. (2010). *Human-Livestock Interactions: The Stockperson and the Productivity of Intensively Farmed Animals*. CABI Publishing, Wallingford, UK.

Hemsworth, P. H., Barnett, J. L. and Hansen, C. (1981). The influence of handling by humans on the behavior, growth, and corticosteroids in the juvenile female pig. *Hormones and Behavior* 15(4), 396–403.

Hemsworth, P. H., Barnett, J. L. and Hansen, C. (1986). The influence of handling by humans on the behaviour, reproduction and corticosteroids of male and female pigs. *Applied Animal Behaviour Science* 15(4), 303–314.

Hemsworth, P. H., Barnett, J. L. and Hansen, C. (1987). The influence of inconsistent handling by humans on the behaviour, growth and corticosteroids of young pigs. *Applied Animal Behaviour Science* 17(3–4), 245–252.

Hemsworth, P. H., Barnett, J. L., Hofmeyr, C., Coleman, G. J., Dowling, S. and Boyce, J. (2002). The effects of fear of humans and pre-slaughter handling on the meat quality of pigs. *Australian Journal of Agricultural Research* 53(4), 493–501.

Hemsworth, P. H., Mellor, D. J., Cronin, G. M. and Tilbrook, A. J. (2015). Scientific assessment of animal welfare. *New Zealand Veterinary Journal* 63(1), 24–30.

Hessing, M. J. C., Hagelsø, A. M., van Beek, J. A. M., Wiepkema, R. P., Schouten, W. G. P. and Krukow, R. (1993). Individual behavioural characteristics in pigs. *Applied Animal Behaviour Science* 37(4), 285–295.

Hill, J. (2000). Deep bed swine finishing. *5 o Seminário Internacional de Suinocultura 27 e 28 de setembro de 2000*. Expo Center Norte, SP. Available at: http://www.cnpsa.embrapa.br/sgc/sgc_publicacoes/anais0009_hill.pdf.

Honeyman, M. S. (2005). Extensive bedded indoor and outdoor pig production systems in USA: current trends and effects on animal care and product quality. *Livestock Production Science* 94(1–2), 15–24.

Hosey, G. and Melfi, V. (2014). Human-animal interactions, relationships and bonds: a review and analysis of the literature. *International Journal of Comparative Psychology* 27(1), 117–142.

Jackson, P., Guy, J. H., Sturm, B., Bull, S. and Edwards, S. A. (2018). An innovative concept building design incorporating passive technology to improve resource efficiency and welfare of finishing pigs. *Biosystems Engineering* 174, 190–203.

Jensen, P., Von Borell, E.,Broom, D. M., Csermely, D., Dijkhuizen, A. A., Hylkema, S., Edwards, S. A., Madec, F. and Stamataris, C. (1997). The Welfare of Intensively Kept Pigs. Doc XXIV/B3/ScVC/0005/1997, p. 190. Commission of the European Communities; Scientific Veterinary Committee, Brussels, Belgium.

Jensen, T., Nielsen, C. K., Vinther, J. and D'Eath, R. B. (2012). The effect of space allowance for finishing pigs on productivity and pen hygiene. *Livestock Science* 149(1–2), 33–40.

Jones, J. B., Burgess, L. R., Webster, A. J. F. and Wathes, C. M. (1996). Behavioural responses of pigs to atmospheric ammonia in a chronic choice test. *Animal Science* 63(3), 437–445.

Jørgensen, B. (2003). Influence of floor type and stocking density on leg weakness, osteochondrosis and claw disorders in slaughter pigs. *Animal Science* 77(3), 439–449.

Kiefer, C., Meignen, B. C. G., Sanches, J. F. and Carrijo, E. A. S. (2009). Response of growing swine maintained in different thermal environments. *Archivos de Zootecnia* 58(221), 55–64.

Kim, Y. S., Kim, S. W., Weaver, M. A. and Lee, C. Y. (2005). Increasing the pig market weight: world trends, expected consequences and practical considerations. *Asian-Australasian Journal of Animal Sciences* 18(4), 590–600.

Kittawornrat, A. and Zimmerman, J. J. (2011). Toward a better understanding of pig behavior and pig welfare. *Animal Health Research Reviews* 12(1), 25–32.

Kliebenstein, J., Larson, B., Honeyman, M. and Penner, A. (2003). *A Comparison of Production Costs, Returns and Profitability of Swine Finishing Systems*. Iowa State University, Ames, IA.

Knox, R., Salak-Johnson, J., Hopgood, M., Greiner, L. and Connor, J. (2014). Effect of day of mixing gestating sows on measures of reproductive performance and animal welfare. *Journal of Animal Science* 92(4), 1698–1707.

Langbein, J. and Puppe, B. (2004). Analysing dominance relationships by sociometric methods—a plea for a more standardised and precise approach in farm animals. *Applied Animal Behaviour Science* 87(3–4), 293–315.

Larson, B., Kliebenstein, J. B., Honeyman, M. S. and Penner, A. D. (2002). An economic analysis of pork production in hoop and confinement facilities: a winter comparison. *Animal Industry Report* 1(1).

Larson, B., Kliebenstein, J. B., Honeyman, M. S. and Penner, A. (2003). Economics of finishing pigs in hoop structures and confinement: a summer group under different space restrictions. *Animal Industry Report* 1(1).

Lassaletta, L., Billen, G., Romero, E., Garnier, J. and Aguilera, E. (2014). How changes in diet and trade patterns have shaped the N cycle at the national scale: Spain (1961-2009). *Regional Environmental Change* 14(2), 785–797.

Lassaletta, L., Estellés, F., Beusen, A. H. W., Bouwman, L., Calvet, S., Van Grinsven, H. J. M., Doelman, J. C., Stehfest, E., Uwizeye, A. and Westhoek, H. (2019). Future global pig production systems according to the Shared Socioeconomic Pathways. *Science of the Total Environment* 665, 739–751.

Lay Jr., D. C., Haussmann, M. F., Daniels, M. J., Harmon, J. D. and Richard, T. L. (2000). Swine housing impacts on environment and behavior: a comparison between hoop structures and total environmental control. In: *Swine Housing, Proc. First Int. Conf.* (October 9–11, 2000, Des Moines, Iowa), pp. 49–55. ASAE. Pub., St. Joseph, MI.

Li, Y. and Gonyou, H. W. (2002). Analysis of belly nosing and associated behaviour among pigs weaned at 12–14 days of age. *Applied Animal Behaviour Science* 77(4), 285–294.

Li, Y. Z., Wang, L. H. and Johnston, L. J. (2012). Sorting by parity to reduce aggression toward first-parity sows in group-gestation housing systems. *Journal of Animal Science* 90(12), 4514–4522.

Lopez, J., Jesse, G. W., Becker, B. A. and Ellersieck, M. R. (1991). Effects of temperature on the performance of finishing swine: I. Effects of a hot, diurnal temperature on average daily gain, feed intake, and feed efficiency. *Journal of Animal Science* 69(5), 1843–1849.

Maes, D., Verdonck, M., Deluyker, H. and de Kruif, A. (1996). Enzootic pneumonia in pigs. *Veterinary Quarterly* 18(3), 104–109.

Manteca, X., Velarde, A. and Jones, B. (2009). Animal welfare components. In: Smulders, F. and Algers, B. (Eds) *Welfare of Production Animals: Assessment and Management of Risks*, pp. 61-77. Wageningen Academic Publishers, The Netherlands.

Marchant-Forde, J. N. and Marchant-Forde, R. M. (2005). Minimizing inter-pig aggression during mixing. *Pig News and Information* 26(3), 63N.

Martin, L. and Kruja, Z. (2000). The western Canada advantage. *Advances in Pork Production* 11, 17-36.

Martínez-Miró, S., Tecles, F., Ramón, M., Escribano, D., Hernández, F., Madrid, J., Orengo, J., Martínez-Subiela, S., Manteca, X. and Cerón, J. J. (2016). Causes, consequences and biomarkers of stress in swine: an update. *BMC Veterinary Research* 12(1), 171.

Matthews, S. G., Miller, A. L., Clapp, J., Plötz, T. and Kyriazakis, I. (2016). Early detection of health and welfare compromises through automated detection of behavioural changes in pigs. *The Veterinary Journal* 217, 43-51.

McGlone, J. J. (1985). A quantitative ethogram of aggressive and submissive behaviors in recently regrouped pigs. *Journal of Animal Science* 61(3), 559-565.

McGlone, J. J. (1986). Influence of resources on pig aggression and dominance. *Behavioural Processes* 12(2), 135-144.

McGlone, J. J. (2001). Farm animal welfare in the context of other society issues: toward sustainable systems. *Livestock Production Science* 72(1-2), 75-81.

McGlone, J. J. and Curtis, S. E. (1985). Behavior and performance of weanling pigs in pens equipped with hide areas. *Journal of Animal Science* 60(1), 20-24.

McGlone, J. J. and Newby, B. E. (1994). Space requirements for finishing pigs in confinement: behavior and performance while group size and space vary. *Applied Animal Behaviour Science* 39(3-4), 331-338.

McGlone, J. J., Kelley, K. W. and Gaskins, C. T. (1980). Lithium and porcine aggression. *Journal of Animal Science* 51(2), 447-455.

Mendl, M., Zanella, A. J. and Broom, D. M. (1992). Physiological and reproductive correlates of behavioural strategies in female domestic pigs. *Animal Behaviour* 44(6), 1107-1121.

Merialdi, G., Dottori, M., Bonilauri, P., Luppi, A., Gozio, S., Pozzi, P., Spaggiari, B. and Martelli, P. (2012). Survey of pleuritis and pulmonary lesions in pigs at abattoir with a focus on the extent of the condition and herd risk factors. *The Veterinary Journal* 193(1), 234-239.

Mesterton-Gibbons, M. and Dugatkin, L. A. (1995). Toward a theory of dominance hierarchies: effects of assessment, group size, and variation in fighting ability. *Behavioral Ecology* 6(4), 416-423.

Metz, J. H. M. and Gonyou, H. W. (1990). Effect of age and housing conditions on the behavioural and haemolytic reaction of piglets to weaning. *Applied Animal Behaviour Science* 27(4), 299-309.

Meunier-Salaun, M. C., Vantrimponte, M. N., Raab, A. and Dantzer, R. (1987). Effect of floor area restriction upon performance, behavior and physiology of growing-finishing pigs. *Journal of Animal Science* 64(5), 1371-1377.

Meunier-Salaun, M. C., Bizeray, D., Colson, V., Courboulay, V., Lensink, J., Prunier, A., Remience, V. and Vandenheede, M. (2007). The welfare of farmed pigs. *Productions Animales* 20, 73-80.

Miele, M. and Evans, A. (2010). When foods become animals: ruminations on ethics and responsibility in care-full practices of consumption. *Ethics, Place and Environment* 13(2), 171-190.

Millman, S. T. (2007). Sickness behaviour and its relevance to animal welfare assessment at the group level. *Animal Welfare* 16, 123–125.

Moore, C. M., Zhou, J. Z., Stricklin, W. R. and Gonyou, H. W. (1996). Influence of group size and floor area space on social organization of growing-finishing pigs. In: Proceedings of the 30th International Congress of the International Society for Applied Ethology, 14–17 August, 1996, Guelph, Ontario, Canada. Colonel KL Campbell Centre for the Study of Animal Welfare, Guelph, Ontario.

Mormède, P., Andanson, S., Aupérin, B., Beerda, B., Guémené, D., Malmkvist, J., Manteca, X., Manteuffel, G., Prunet, P., van Reenen, C. G., Richard, S. and Veissier, I. (2007). Exploration of the hypothalamic-pituitary-adrenal function as a tool to evaluate animal welfare. *Physiology and Behavior* 92(3), 317–339.

Morrow, W. M., Meyer, R. E., Roberts, J. and Lascelles, D. (2006). Financial and welfare implications of immediately euthanizing compromised nursery pigs. *Journal of Swine Health and Production* 14(1), 25.

Nannoni, E., Martelli, G., Rubini, G. and Sardi, L. (2019). Effects of increased space allowance on animal welfare, meat and ham quality of heavy pigs slaughtered at 160Kg. *PLoS ONE* 14(2), e0212417.

Nannoni, E., Sardi, L., Vitali, M., Trevisi, E., Ferrari, A., Barone, F., Bacci, M. L., Barbieri, S. and Martelli, G. (2016). Effects of different enrichment devices on some welfare indicators of post-weaned undocked piglets. *Applied Animal Behaviour Science* 184, 25–34.

Nasirahmadi, A., Edwards, S. A. and Sturm, B. (2017). Implementation of machine vision for detecting behaviour of cattle and pigs. *Livestock Science* 202, 25–38.

NASS (2017). Hogs: commercial slaughter average liveweight by month and year, US. Available at: https://www.nass.usda.gov/Charts_and_Maps/Livestock_Slaughter/hglvwgx6.php.

National Farm Animal Care Council (NFACC) (2014). Available at: https://www.nfacc.ca/pdfs/codes/pig_code_of_practice.pdf.

National Pork Board (2019). Available at: https://www.pork.org/facts/pig-farming/life-cycle-of-a-market-pig/.

Ni, J. Q., Heber, A. J. and Lim, T. T. (2018). Ammonia and hydrogen sulfide in swine production. In: *Air Quality and Livestock Farming*, pp. 69–88. CRC Press, London, UK.

OIE (2010). Available at: https://www.oie.int/en/animal-welfare/animal-welfare-at-a-glance/.

Owsley, F., Rodning, S. and Floyd, J. (2013). *Scheduling All-In/All-Out Swine Production, vol ANR-0847. Alabama Cooperative Extension System.*

Pagel, M. and Dawkins, M. S. (1997). Peck orders and group size in laying hens:futures contracts' for non-aggression. *Behavioural Processes* 40(1), 13–25.

Penny, R. H., Osborne, A. D., Wright, A. I. and Stephens, T. K. (1965). Foot-rot in pigs, observations on the clinical disease. *The Veterinary Record* 77(38), 1101–1108.

Petherick, J. C. (1983). A biological basis for the design of space in livestock housing. *Farm Animal Housing and Welfare*, 103–120.

Petherick, J. C. (2007). Spatial requirements of animals: allometry and beyond. *Journal of Veterinary Behavior* 2(6), 197–204.

Petherick, J. C., Beattie, A. W. and Bodero, D. A. V. (1989). The effect of group size on the performance of growing pigs. *Animal Science* 49(3), 497–502.

Probst, J. K., Neff, A. S., Leiber, F., Kreuzer, M. and Hillmann, E. (2012). Gentle touching in early life reduces avoidance distance and slaughter stress in beef cattle. *Applied Animal Behaviour Science* 139(1-2), 42-49.

Rahse, E. and Hoy, S. (2007). Investigations on frequency and severity of different claw lesions in fattening pigs with regard to housing conditions. *Praktische Tierarzt* 88(1), 40-47.

Rault, J. L., Morrison, R. S., Hansen, C. F., Hansen, L. U. and Hemsworth, P. H. (2014). Effects of group housing after weaning on sow welfare and sexual behavior. *Journal of Animal Science* 92(12), 5683-5692.

Reicks, D. (2009). Application of air filtration systems in swine operations. *Advances in Pork Production* 20, 163-171.

Renaudeau, D., Gourdine, J. L. and St-Pierre, N. R. (2011). A meta-analysis of the effects of high ambient temperature on growth performance of growing-finishing pigs. *Journal of Animal Science* 89(7), 2220-2230.

Rossi, R., Costa, A., Guarino, M., Laicini, F., Pastorelli, G. and Corino, C. (2008). Effect of group size-floor space allowance and floor type on growth performance and carcass characteristics of heavy pigs. *Journal of Swine Health and Production* 16(6), 304-311.

Ruckebusch, Y. (1972). The relevance of drowsiness in the circadian cycle of farm animals. *Animal Behaviour* 20(4), 637-643.

Rzezniczek, M., Gygax, L., Wechsler, B. and Weber, R. (2015). Comparison of the behaviour of piglets raised in an artificial rearing system or reared by the sow. *Applied Animal Behaviour Science* 165, 57-65.

Sánchez-Vizcaíno, J. M., Laddomada, A. and Arias, M. L. (2019). African swine fever virus. In: Zimmerman, J. J., Karriker, L. A., Ramirez, A., Schwartz, K. J., Stevenson, G. W., Zhang, J. (Eds) *Diseases of Swine*, 11th edition, pp. 443-452. Wiley, New York.

Santonja, G. G., Georgitzikis, K., Scalet, B. M., Montobbio, P., Roudier, S. and Sancho, L. D. (2017). *Best available Techniques (BAT) Reference Document for the Intensive Rearing of Poultry or Pigs*. Publications Office of the European Union, Luxembourg.

Sarignac, C., Signoret, J. P. and McGlone, J. J. (1997). Relation mere-jeune, comportement et performances en function du Systeme de logement et de l'environnement social. Sow and piglet performance and behavior in either intensive outdoor or indoor units with litters managed as individuals or as small social groups. *Journees Rech. Porcine en France* 29, 123-128.

Sasaki, Y., Ushijima, R. and Sueyoshi, M. (2015). Field study of hind limb claw lesions and claw measures in sows. *Animal Science Journal* 86(3), 351-357.

Scheidt, A. B., Cline, T. R., Clark, L. K., Mayrose, V. B., Van Alstine, W. G., Diekman, M. A. and Singleton, W. L. (1995). The effect of all-in-all-out growing-finishing on the health of pigs. *Journal of Swine Health and Production* 3(5), 202-205.

Schmied, C., Boivin, X., Scala, S. and Waiblinger, S. (2010). Effect of previous stroking on reactions to a veterinary procedure: behaviour and heart rate of dairy cows. *Interaction Studies* 11(3), 467-481.

Schmolke, S. A., Li, Y. Z. and Gonyou, H. W. (2003). Effect of group size on performance of growing-finishing pigs. *Journal of Animal Science* 81(4), 874-878.

Schrøder-Petersen, D. L. and Simonsen, H. B. (2001). Tail biting in pigs. *The Veterinary Journal* 162(3), 196-210.

Schulenburg, A., von der Meyer, K. and Dierks-Meyer, B. (1986). Orthopedic and microstructural studies on claw horn of fattening pigs in various types of housing. *Veterinary Medicine* 125, 130–132.

Sørensen, J. T. and Schrader, L. (2019). Labelling as a tool for improving animal welfare–the pig case. *Agriculture* 9(6), 123.

Spoolder, H. A. M., Edwards, S. A. and Corning, S. (1999). Effects of group size and feeder space allowance on welfare in finishing pigs. *Animal Science* 69(3), 481–489.

Stolba, A. and Wood-Gush, D. G. M. (1984). The identification of behavioural key features and their incorporation into a housing design for pigs. *Annales de Recherches Veterinaires. Annals of Veterinary Research* 15(2), 287–299.

Straw, B. E. and Bartlett, P. (2001). Flank or belly nosing in weaned pigs. *Journal of Swine Health and Production* 9(1), 19–23.

Stricklin, W. R. and Mench, J. A. (1987). Social organization. Farm Animal Behavior. E. O. Price (Ed.). *Veterinary Clinics of North America. Food Animal Practice* 3(2), 307–322.

Sutherland, M. A., Niekamp, S. R., Johnson, R. W., Van Alstine, W. G. and Salak-Johnson, J. L. (2007). Heat and social rank impact behavior and physiology of PRRS-virus-infected pigs. *Physiology and Behavior* 90(1), 73–81.

SVC (1996). Report on the welfare of laying hens. European Commission, the Scientific Veterinary Committee Animal Welfare Section. Available at: http://ec.uropa.eu/food/fs/sc/oldcomm4/out33_en.pdf.

SVC (1997). The welfare of intensively kept pigs. Report of the Scientific Veterinary Committee of the EU. Available at: http://ec.europa.eu/food/animal/welfare/farm/out17_en.pdf.

Taylor, N., Prescott, N., Perry, G., Potter, M., Le Sueur, C. L. and Wathes, C. (2006). Preference of growing pigs for illuminance. *Applied Animal Behaviour Science* 96(1–2), 19–31.

Thacker, E. L. and Minion, F. C. (2012). Mycoplasmosis. In: Zimmermann, J. J., Karriker, L. A., Ramirez, A., Schwartz, K. J. and Stevenson, G. W. (Eds), *Diseases of Swine*, pp. 779–798. WileyBlackwell, Ames, IA.

Thomas, L. L., Goodband, R. D., Woodworth, J. C., Tokach, M. D., DeRouchey, J. M. and Dritz, S. S. (2017). Effects of space allocation on finishing pig growth performance and carcass characteristics. *Translational Animal Science* 1(3), 351–357.

Tubbs, R. C. (1988). Lameness in sows: solving a preventable problem. *Veterinary Medicine*.

Turner, S. P., Allcroft, D. J. and Edwards, S. A. (2003). Housing pigs in large social groups: a review of implications for performance and other economic traits. *Livestock Production Science* 82(1), 39–51.

Turner, S. P. and Edwards, S. A. (2004). Housing immature domestic pigs in large social groups: implications for social organisation in a hierarchical society. *Applied Animal Behaviour Science* 87(3–4), 239–253.

Tuyttens, F. A. M. (2005). The importance of straw for pig and cattle welfare: a review. *Applied Animal Behaviour Science* 92(3), 261–282.

Urbain, B., Gustin, P., Prouvost, J. F. and Ansay, M. (1994). Quantitative assessment of aerial ammonia toxicity to the nasal mucosa by use of the nasal lavage method in pigs. *American Journal of Veterinary Research* 55(9), 1335–1340.

Velarde, A., Fàbrega, E., Blanco-Penedo, I. and Dalmau, A. (2015). Animal welfare towards sustainability in pork meat production. *Meat Science* 109, 13–17.

Verdon, M. and Rault, J. L. (2018). Aggression in group housed sows and fattening pigs. In: Spinka, M. (Ed.) *Advances in Pig Welfare*, pp. 235-260. Woodhead Publishing, Sawston, UK.

Verdon, M., Hansen, C. F., Rault, J. L., Jongman, E., Hansen, L. U., Plush, K. and Hemsworth, P. H. (2015). Effects of group housing on sow welfare: a review. *Journal of Animal Science* 93(5), 1999-2017.

Vermeer, H. M., De Greef, K. H. and Houwers, H. W. J. (2014). Space allowance and pen size affect welfare indicators and performance of growing pigs under Comfort Class conditions. *Livestock Science* 159, 79-86.

Vermeij, I., Enting, J. and Spoolder, H. A. M. (2009). *Effect of Slatted and Solid Floors and Permeability of Floors in Pig Houses on Environment, Animal Welfare and Health and Food Safety: A Review of Literature (No. 186).* Animal Sciences Group.

Vitt, R., Weber, L., Zollitsch, W., Hörtenhuber, S. J., Baumgartner, J., Niebuhr, K., Piringer, M., Anders, I., Andre, K., Hennig-Pauka, I., Schönhart, M. and Schauberger, G. (2017). Modelled performance of energy saving air treatment devices to mitigate heat stress for confined livestock buildings in Central Europe. *Biosystems Engineering* 164, 85-97.

Waiblinger, S., Boivin, X., Pedersen, V., Tosi, M. V., Janczak, A. M., Visser, E. K. and Jones, R. B. (2006). Assessing the human-animal relationship in farmed species: a critical review. *Applied Animal Behaviour Science* 101(3-4), 185-242.

Wang, C., Chen, Y., Bi, Y., Zhao, P., Sun, H., Li, J., Liu, H., Zhang, R., Li, X. and Bao, J. (2020). Effects of long-term gentle handling on behavioral responses, production performance, and meat quality of pigs. *Animals: an Open Access Journal from MDPI* 10(2), 330.

Weary, D. M., Appleby, M. C. and Fraser, D. (1999). Responses of piglets to early separation from the sow. *Applied Animal Behaviour Science* 63(4), 289-300.

Weary, D. M., Huzzey, J. M. and Von Keyserlingk, M. A. G. (2009). Board-invited review: using behavior to predict and identify ill health in animals. *Journal of Animal Science* 87(2), 770-777.

Webb, N. G. (1984). Compressive stresses on, and the strength of the inner and outer digits of pigs' feet, and the implications for injury and floor design. *Journal of Agricultural Engineering Research* 30, 71-80.

Weng, R. C., Edwards, S. A. and English, P. R. (1998). Behaviour, social interactions and lesion scores of group-housed sows in relation to floor space allowance. *Applied Animal Behaviour Science* 59(4), 307-316.

Wolter, B. F., Ellis, M., Curtis, S. E., Augspurger, N. R., Hamilton, D. N., Parr, E. N. and Webel, D. M. (2001). Effect of group size on pig performance in a wean-to-finish production system. *Journal of Animal Science* 79(5), 1067-1073.

Worobec, E. K., Duncan, I. J. H. and Widowski, T. M. (1999). The effects of weaning at 7, 14 and 28 days on piglet behaviour. *Applied Animal Behaviour Science* 62(2-3), 173-182.

Ye, Z., Zhang, G., Seo, I.-H., Kai, P., Saha, C. K., Wang, C. and Li, B. (2009). Airflow characteristics at the surface of manure in a storage pit affected by ventilation rate, floor slat opening, and headspace height. *Biosystems Engineering* 104(1), 97-105.

Zonderland, J. J., Wolthuis-Fillerup, M., Van Reenen, C. G., Bracke, M. B., Kemp, B., Den Hartog, L. A. and Spoolder, H. A. (2008). Prevention and treatment of tail biting in weaned piglets. *Applied Animal Behaviour Science* 110(3-4), 269-281.

Chapter 7

Optimising pig welfare during transport, lairage and slaughter

Luigi Faucitano, Agriculture and Agri-Food Canada, Canada; and Antonio Velarde, Institute of Agrifood Research and Technology, Spain

1 Introduction
2 Welfare during transport
3 Welfare in lairage
4 Welfare during stunning and slaughter
5 Animal welfare audit protocols
6 Conclusion and future trends
7 Where to look for further information
8 References

1 Introduction

Stress imposed on pigs during transport, in lairage and at slaughter is both an animal welfare and a meat quality issue. Studies revealed that poor handling practices and transport conditions and wrong facility design at the slaughter plant both, individually and/or additively, can contribute to the loss of profits due to animal losses (dead-on-arrival or DOA and downers), reduction in carcass value due to weight losses and skin bruises and meat quality defects due to abnormal *post-mortem* muscle acidification (Schwartzkopf-Genswein et al., 2012; Faucitano, 2018; Rioja-Lang et al., 2019).

The responsibility for animal losses during transport may either be equally shared among the producer, who must guarantee proper preparation of animals (i.e. feed withdrawal) prior to transport and handling of pigs at loading, the trucker and the abattoir (in the case of integrated production systems) or be bore alone by the trucker, whereas the abattoir is responsible alone for the optimisation of lairage and slaughter conditions (layout, ambient control and handling systems) in order to let pigs recover from the stress of handling and transport and ensure optimal and uniform carcass and meat quality.

Training of handling crews and the application of rewards and fines have become of paramount importance to improve handling and reduce animal

http://dx.doi.org/10.19103/AS.2020.0081.07
Published by Burleigh Dodds Science Publishing Limited, 2021.

losses (Correa, 2011; Rocha et al., 2016; Dalla Costa et al., 2019). A Canadian integrated company reported a decrease from 0.3% to 0.01% in the incidence of fatigued pigs on arrival at the abattoir by applying a new programme of payment with incentives to handlers and truckers as a reward for their willingness to slow down the loading/unloading speed rate to 100 pigs/h (Correa, 2011). In contrast, fines up to $6000 under Canadian Food Inspection Agency regulations are applied to Canadian truckers for having –three to four dead pigs in the truck load (Faucitano, 2018). Grandin (2018) recently reported that the decision of an abattoir management in the US to apply a $25 handling fee for each pig arriving in a non-ambulatory condition resulted in a great reduction in the occurrence of downers.

This chapter will focus on the handling practices to be applied during the transportation of pigs (market weight and culled sows) to slaughter, as this phase is considered as the most stressful of the whole preslaughter period, and in lairage and at stunning/slaughter at the abattoir, aiming at limiting their impact on animal losses and pork quality.

2 Welfare during transport

The situation when a pig is in transit is considered a major stressor and may have deleterious effects on its health, its well-being, carcass yield and ultimately pork quality. Major sources of stress during transport are, among others, vehicle design, the time spent in the truck and the space allowed to lie down.

2.1 Vehicle design

Vehicle design features that may influence the welfare of pigs during transport include the loading/unloading system (ramps or hydraulic decks), the compartment location and the microclimate control (Faucitano and Goumon, 2018; Rioja-Lang et al., 2019).

The presence of fixed decks and ramps within the vehicle makes the procedures of loading and unloading more difficult, resulting in a higher risk of animal losses (Barton-Gade et al., 2007; Correa et al., 2013) and poor pork quality, that is, either pale, soft and exudative (PSE) or pale or dark, firm and dry (DFD) pork (Guàrdia et al., 2004; Correa et al., 2013, 2014). The North American pot-belly (PB) trailer is an example of the vehicle design featuring two to five, often steep (up to 32° slope), ramps (see Fig. 1) and 180° turns that pigs have to negotiate during the loading and unloading process. These trailer features have been associated with a lower ease of handling as showed by the greater use of electric prods and longer handling time (Ritter et al., 2008; Torrey et al., 2013a,b; Weschenfelder et al., 2013b), eventually resulting in a greater proportion of DOA and fatigued pigs on arrival at the abattoir compared with other trailer models featuring hydraulic decks (Ritter et al., 2008; Sutherland

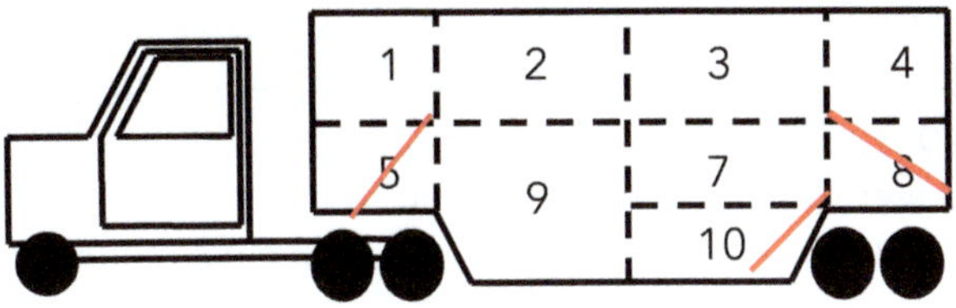

Figure 1 Position of compartments and internal ramps (solid red lines) inside a pot-belly trailer (modified from Correa et al., 2014). The represented model is a dual-purpose pot-belly trailer (for cattle and pig transport).

et al., 2009; Kephart et al., 2010; Weschenfelder et al., 2012; Correa et al., 2013).

The presence of ramps is one of the major contributors to the handling problems reported during loading and unloading in PB trailers (see Figs. 1 and 2; Faucitano and Goumon, 2018; Rioja-Lang et al., 2019). It is well known, in fact, that negotiating a ramp slope from 0° to 45° represents a significant physical and psychological challenging experience for pigs, as showed by the increased heart rate, frequency of turn backs and baulking, skin bruises and

Figure 2 Unloading from a pot-belly trailer through an internal ramp (L. Faucitano, AAFC, Canada).

handling time (Van Putten and Elshof, 1978; Warriss et al., 1991; Dalla Costa et al., 2007; Goumon et al., 2013b; Garcia and McGlone, 2015). Recently, Dalla Costa et al. (2019) compared ramp slopes of < 20° versus > 20° and reported a four-fold greater risk for DOA when steeper ramps were used. Furthermore, steep ramps (up to 26°) are also challenging for the handler, as showed by the increased heart rate and difficulty to handle (Goumon et al., 2013b). Current guidelines do not recommend to use ramps steeper than 20° for fixed ramps or than 25° for adjustable ramps (SCAHAW, 2011; TQA, 2016). However, as market pigs have become heavier (from 113 kg to 130 kg from 2000 to 2010 in Canada; Correa, 2011) and difficult to move (Bertol et al., 2011; Rocha et al., 2016), recent works recommend the maximum ramp slope to be reduced to 15° (Grandin, 2012; Goumon and Faucitano, 2017).

2.2 Animal position in the truck

The impact of the deck and/or compartment position on the vehicle pigs' welfare during transport and meat quality can be explained by the differences in the ease of handling (due to the presence of ramps), airflow rate, vibration rate and loading order (Faucitano and Goumon, 2018). European and Canadian studies showed that, when compared with the middle deck, pigs transported in the top and/or bottom decks presented higher body temperature, exsanguination blood cortisol levels, dehydration rates, carcass lesion score and incidence of PSE and exudative or pale/dark pork (Lambooij et al., 1985; Lambooij and Engel, 1991; Barton-Gade et al., 1996; Correa et al., 2013). The front and rear top and bottom rear compartments can be also critical, as showed by the increased core body temperatures after loading and during transport and higher heart rate after loading and longer unloading time (Goumon et al., 2013a; Torrey et al., 2013b; Conte et al., 2015). A greater risk for DFD pork production has been reported in pigs transported in the middle front compartment of a PB trailer (Correa et al., 2014), especially after long transportation (18 vs. 6 and 12 h) in winter (Scheeren et al., 2014).

2.3 Control of the microclimate within the transport vehicle

Generally, greater animal losses are reported during summer hauls (Vecerek et al., 2006; Werner et al., 2007; Haley et al., 2008; Correa et al., 2013; Vitali et al., 2014), with greater risk being recorded at ambient temperatures either between 17°C and 20°C (Warriss and Brown, 1994; Sutherland et al., 2009; Haley et al., 2008, 2010; Kephart et al., 2010) or between 29°C and 33°C (Peterson et al., 2017).

Within a vehicle the internal temperature can increase by almost 1°C for each 1°C increase in the environmental temperature (Dewey et al., 2009). The internal temperature increases more, up to 3–4°C rise in 5 min at a rate of approximately 1°C/min, in passively ventilated vehicles held in a stationary situation (Lambooij, 1988; Xiong et al., 2015). However, when passively ventilated trailers are compared, the environment is warmer in the PB trailer compared with the triple-deck flat-deck (FD) trailer while stationary or moving (Ritter et al., 2008; Weschenfelder et al., 2012; Faucitano and Goumon, 2018). The variation in the internal thermal conditions between the two trailer models has been explained by the difference in the side openings type (punch for the PB vs. slatted for the FD trailer) influencing the air flow inside the vehicle (Weschenfelder et al., 2012). More specifically, within the PB trailer the highest temperature peak (30.3°C) has been recorded in the bottom deck during transport (Brown et al., 2011), while the greater gradient (up to +11°C) between internal trailer and external ambient temperatures has been recorded in the middle and bottom front compartments during stops (Weschenfelder et al., 2012, 2013; Fox et al., 2014).

To ensure thermal comfort and reduce animal losses during vehicle stops, pigs should be cooled off by fan-assisted ventilation, water sprinkling/misting or the two systems combined. Research showed that when applied for 5 min after loading and before leaving the farm and at the end of the wait before unloading at the abattoir at ambient temperatures of 20°C and greater, water sprinkling reduced pigs' fatigue (as assessed by the blood lactate concentration) at slaughter and pork exudation, especially in pigs transported in the middle front and rear compartments (Nannoni et al., 2014). However, water sprinkling combined with insufficient ventilation can also result in an increased difference in humidity levels (up to +7.5%) between the trailer interior and the external environment (Fox et al., 2014), preventing efficient evaporative cooling in pigs. Pereira et al. (2018) showed that the use of fan-misting bank installed in the abattoir yard (see Fig. 3) for 30 min under warm ambient conditions (up to 28.1°C) was efficient to improve the quality of the internal environment of a PB trailer (lower temperature-humidity index or THI) and the pigs' thermal comfort (lower body heat loss) while waiting before unloading.

A greater proportion of DOAs and non-ambulatory pigs on arrival at the plant has been also reported in winter than in summer (Guàrdia et al., 1996; Sutherland et al., 2009), with a higher risk of death at 4–10°C than at 12–26°C (Peterson et al., 2017). Likely causes of the greater animal losses in winter compared with other seasons may be the more difficult animal handling through the icy internal ramps at loading and unloading (Torrey et al., 2013a,b) and insufficient bedding of the trailer floor, resulting in more pigs standing during transport to avoid the contact with the cold aluminium floor surface (Goumon et al., 2013a). The presence of slippery ramps can result in more slips and falls at loading and unloading (Torrey et al., 2013b), greater heart rates during transport and unloading (Goumon et al., 2013a) and increased blood creatine kinase (CK) and lactate concentrations at slaughter (Correa et al., 2014).

Figure 3 Operation of a fan-misting bank on pigs kept in a pot-belly trailer during their wait before unloading (L. Faucitano, AAFC, Canada).

In winter, the thermal comfort of pigs in the truck can be controlled by partially closing the ventilation openings (boarding) in order to reduce air flow and by adding 5-cm layer of Styrofoam to the vehicle top ceiling (Gonyou and Brown, 2012) and bedding on the truck floor (TQA, 2016). Within the air temperature range between 15°C and ≤ −12°C, trucks should be utilising from 25% to 95% boarding to prevent DOAs (McGlone et al., 2014a; TQA, 2016).

In winter, no more than six bales of bedding should be sufficient to avoid pigs to be in contact with the cold floor surface of the trailer during transport and reduce the rate of DOA or downers and the number of carcass lesions, including frostbites (Goumon et al., 2013a; McGlone et al., 2014b; Scheeren et al., 2014). The number of bales should be reduced to three in summer (McGlone et al., 2014b).

2.4 Journey time

Both long and short journeys can affect the welfare of slaughter pigs and mortality rate during transport (Rioja-Lang et al., 2019). Some studies reported an increased risk of animal and body weight losses (Warriss et al., 1990; Sutherland et al., 2009), fatigue and dehydration, as shown by the higher blood glucose, lactate and haematocrit levels at slaughter (Brown et al., 1999; Mota-Rojas et al., 2006; Becerril-Herrera et al., 2010; Sommavilla et al., 2017), and DFD pork production (Leheska et al., 2003; Mota-Rojas et al., 2006) with transport time longer than 4 h and up to 24 h. In contrast, in other studies shorter journeys (≤2 h) resulted in a lower ease of handling at the abattoir, greater concentrations of cortisol and lactate in exsanguination blood and a higher risk of PSE pork production (Faucitano and Lambooij, 2019). A possible reason for the detrimental effect of short journeys is that the pigs could not recover from the stress of handling experienced at loading upon arrival at the abattoir (Weschenfelder et al., 2013).

However, the effects of travel time on pig welfare can be additive to that of other concurrent factors, such as the time on fasting, season of the year, type of vehicle, loading density and pig genetic background (Rioja-Lang et al., 2019). Fasting pigs on farm for 12–18 h before loading reduces the risk of mortality during transport, regardless of the journey duration (up to 24 h), while in unfasted pigs this risk increases for journeys up to 8 h (Averós et al., 2008). In summer increased risk for DOA has been reported in hauls longer than 2 h (Vitali et al., 2014), while greater core body temperature, heart rate and incidence of DFD pork production have been recorded in pigs transported for 18 h in winter than in summer (Goumon et al., 2013a; Scheeren et al., 2014). The latter results may be caused by the longer exposure of pigs to cold stress. In transportation trials of different durations (45 min and 7 h), using both PB and FD trailers, Weschenfelder et al. (2013) reported an increased level of fatigue (based on exsanguination blood lactate concentration) at the time of slaughter in stress-susceptible pigs (Hal^{Nn}) hauled for a shorter time with the PB trailer compared with a FD trailer equipped with semi-hydraulic middle and top decks.

2.5 Space allowance

The recommended loading density for pigs during transport involves a trade-off between the economic pressure to increase the loading density in order to maximise profit from a single journey and the welfare of animals during transport (Rioja-Lang et al., 2019). However, particular attention must be paid to this transport variable considering its significant contribution to animal losses. In a US survey of more than 12 000 loads, Fitzgerald et al. (2009) reported that trailer density accounted for the largest portion of the variation in the total animal losses, that is, dead, fatigued and injured pigs, compared to other variables, such as the driver, handling crew, THI, wind speed, loading duration and wait time before unloading.

The EU legislation (Council Regulation (EC) No 1099/2009, 2009) is based on the evidence that when the loading density is higher than 235 kg/m^2 or lower than 0.425 m^2/100 kg not all pigs are able to lie down to rest and cannot rest as they are pushed to continually change their position (Lambooij et al., 1985; Lambooij, 2014). This uncomfortable situation has been associated with increased mortality rates and a higher number of non-ambulatory pigs on arrival at the plant (Ritter et al., 2006). Fitzgerald et al. (2009) reported that increasing the density from 212.4 kg/m^2 to 338.6 kg/m^2 corresponded to a 7.5-fold increase in animal losses on arrival at the slaughter plant. However, both low (0.50 m^2/100 kg) and high density (0.33 m^2/100 kg) during transport can cause physical stress, resulting in muscle fatigue and glycogen depletion, and a greater risk of DFD pork production (Lambooij et al., 1985; Guàrdia et al., 2005). In a low-density situation, physical stress is caused by the attempt of pigs to maintain their balance and cope with the unexpected and sudden movements of the truck or by fighting due to their greater freedom to move around in the compartment (Barton-Gade and Christensen, 1998; Guàrdia et al., 2005). In contrast, providing less space may fatigue pigs due to the frequent disturbance of lying animals by those seeking a place to rest and the difficulty of standing pigs to maintain their balance while the vehicle is moving (Lambooij and Engel, 1991).

The application of loading densities should be adjusted according to the body weight, ambient conditions and travel time. A recent Italian study reported a greater risk of DOAs when the EU-recommended density (235 kg/m^2) is applied for the transport of 160 kg pigs, which would correspond to a loading density of 0.7 pigs/m^2 (Nannoni et al., 2016). Because of their different physical needs and greater susceptibility to heat stress, due to the decreased heat dissipation rate (Renaudeau et al., 2011), the minimum recommended truck space required by heavier-market-weight pigs (from 114 kg to 182 kg) should be increased from 0.40 m^2/pig to 0.61 m^2/pig in winter and from 0.46 m^2/pig to 0.65 m^2/pig in summer (Grandin, 2017b). If the load size is simply determined by the number of animals that need to be shipped, then heavier pigs may be too packed on the truck (Zurbrigg et al., 2017). During hauls of 140-kg weight pigs, the reduction of the number of pigs from 3/m^2 to 2/m^2 has been associated with a decreased risk of animal losses from 1% to 0.2% (Fitzgerald et al., 2009).

During hot and humid ambient conditions, it is recommended to provide 15% to 25% more space (CARC, 2001; https://ec.europa.eu/food/animals/welfare/practice/transport_en). Indeed, transport losses are lower when more truck floor space (0.46 m^2/100 kg vs. 0.39 m^2/100 kg) is provided in summer (Ritter et al., 2012). However, greater space allowance (0.50 m^2/100 kg vs. 0.25 m^2/100 kg) may increase the risk for skin lesions (+28.2%) in this season (Guàrdia et al., 2009). In winter, the risk of skin lesions is higher (9.7%) at high loading density (0.25 m^2/100 kg) due to the effect of huddling to keep the body temperature and fighting/mounting to seek for space to lie down (Guàrdia et al., 2009).

Pilcher et al. (2011) showed that the increase of floor space (from 0.40–0.49 m^2/100 kg to 0.52 m^2/100 kg) helps reduce the incidence of fatigued pigs on arrival at the plant after short transport (<1 h) compared with longer journeys (3 h) as pigs can get adapted more to a long journey (550 km) as showed by the higher proportion of pigs lying on the truck floor and reduced body temperature and heart rate (Gerritzen et al., 2013). However, short journeys (1 h vs. 3 h) at higher loading densities (0.25 m^2/100 kg vs. 0.50 m^2/100 kg) have been also associated with a decreased risk of PSE pork production (Guàrdia et al., 2004). Based on this evidence, in order to prevent this meat quality defect,

Published by Burleigh Dodds Science Publishing Limited, 2021.

the EU-recommended space allowance of 0.425 m^2/100 kg should be only applied for journeys longer than 3 h. These results may be explained by the fact that at the departure from the farm pigs do not lie down immediately but still stand exploring and getting adapted to the novel environment. They remain so for the first few kilometres (0.5 h) to better cope with the fear and stress (as showed by the increased heart rate up to 220 beats/min) caused by the frequent truck accelerations/decelerations and vibrations generated by driving on the initial country unpaved roads (Chevillon, 2001). It can take up to 2 h of transport for the pigs to settle down in the truck (Lambooij et al., 1985; Barton-Gade and Christensen, 1998). In these conditions, providing pigs with less space would help them keep their balance by holding each other while the vehicle negotiates bends or poor road surfaces (Barton-Gade and Christensen, 1998).

2.6 Transport of culled sows

In pig production, sows are most likely culled due to poor body condition, lameness or failure to rebreed (Baloyh et al., 2015; Zhao et al., 2015; Grandin, 2016a). Such defects, along with greater mortality, are mostly reported at the largest farms (Koketsu, 2000).

Because of poor body and ambulatory conditions, cull sows may hardly walk and be loaded onto the truck, resulting in a greater risk of arriving fatigued, lame and even with worst body conditions, or dead, at the destination (i.e. buying station or slaughter plant) compared with market-weight pigs (Malena et al., 2007; McGee et al., 2016; Peterson et al., 2017). In a recent observational study on 47 loads of sows transported from the farm to slaughter, Thodberg et al. (2019) observed a deterioration of clinical signs, such as vulva and udder ulcers, skin lesions, wounds and torn hoofs, on arrival at the slaughter plant. These observations support the recommendation for producers to select fit sows and boars for transportation (Bench et al., 2008; Grandin et al., 2016a).

The effects of transportation on sow welfare can be exacerbated by the very long distance cull sows have to travel from the farm or assembly yard to the abattoir and ambient conditions. A recent study surveying the cull sow market network in the US reported that sows could be marketed between 24 and 120 h and travel straight to slaughter for an average of 1057 km prior to harvest (Blair and Lowe, 2019). Peterson et al. (2017) reported that for cull sows the risk to die during transport may be 1.93 and 0.81 times higher at outdoor ambient temperatures ranging from 29°C to 33°C and from 4°C to 10°C, respectively, compared to 12°C to 26°C.

Transport duration (4 h on average, ranging from <1 to 8 h), temperature in the truck (14°C on average, ranging from 3°C to 26°C) and duration of stops and wait before unloading at the slaughter plant (0.5 h on average for both, ranging from 0 h to 3 h and from 0 min to 75 min, respectively) have also been identified as the most important causing factors for the deterioration of clinical signs in cull sows on arrival at the abattoir (Thodberg et al., 2019).

3 Welfare in lairage

The purpose of lairage is to give stressed or fatigued animals an opportunity to recover from the stress of transport and previous handling in order to produce better meat quality (Faucitano, 2010, 2018; Gallo et al., 2016).

The recovery rate of pigs in lairage and the related economic losses due to poor carcass and meat quality depend on lairage time, the quality of the handling systems and ambient control (Faucitano, 2010; Gallo et al., 2016).

3.1 Lairage time

Two to three hours is usually the recommended rest time to allow the pigs to recover their physiological condition after transport and handling and to ensure the production of good-quality pork (Warriss, 2003). Unless required by the adversive ambient conditions, such as too high ambient temperature (>30°C; Fraqueza et al., 1998) or ammonia levels (>10 ppm; Weeks, 2008), short lairage intervals (<1 h) should not be applied as they may result in a higher muscle temperature and lactate level at slaughter, resulting in an increased incidence of PSE pork production (Fraqueza et al., 1998; Shen et al., 2006). However, while longer lairage time helps reduce the risk of PSE pork production (Guàrdia et al., 2005), extending lairage time from 3 h to overnight increases the risk of DFD pork production (Guàrdia et al., 2005) due to the depletion of the muscle glycogen content at slaughter caused by the combined effect of fasting and fighting within mixed groups of unfamiliar pigs (Nanni Costa et al., 2002; Guàrdia et al., 2009; Dalla Costa et al., 2016). Fighting-related skin lesion scores, in fact, increase with the lairage time (Warriss, 1996; Faucitano, 2001, 2010; Bottaccini et al., 2018), with almost a two-fold higher risk in pigs kept in lairage for 15 h compared to 3 h (Guàrdia et al., 2009).

3.2 Mixing

Under commercial conditions, mixing groups of unfamiliar pigs in lairage is a common practice because of the difference in size between the truck compartment and lairage pen, despite the risk of fighting, in terms of biting, pushing and head knocks (Rabaste et al., 2007) between pigs, and carcass value downgrading due to severe skin lesions (Faucitano, 2001). However, some strategies can be applied to limit fighting in lairage, such as keeping pigs in smaller groups (10 pigs/group vs. 30 pigs/group; Rabaste et al., 2007) and reducing the space allowance (0.85 m^2/pig to 0.42 m^2/pig; Moss, 1978; Geverink et al., 1996; Weeks, 2008). Anecdotally, some European plants managed to reduce fighting by enriching the pen with some corn kernels scattered on the floor that may distract pigs or sprinkling the back of pigs with vinegar that apparently mask the individual smell of each pig (Eyes on Animals, 2019).

The intensity and duration of fighting in lairage can be also aggravated by pig gender, with boars being more aggressive than gilts and barrows (Warriss and Brown, 1985) and immunocastrates fighting two-fold more than surgical castrates, especially when fed ractopamine during the last 28 days of the finishing period (Rocha et al., 2013). In a more recent study, Wesoly et al. (2015) reported an inverse relationship between testicular function and skin lesion score within mixed groups of boars, with high-ranking boars presenting lower skin lesions compared with low-ranked ones that are mounted more frequently (Rydhmer et al., 2006).

3.3 Ambient control

The control of temperatures between 15°C and 18°C and relative humidities (RH) between 59% and 65% is recommended to ensure the thermal comfort of pigs in lairage (Honkavaara, 1989). When these environmental conditions are not respected, pigs can either suffer from cold stress (shivering and huddling), which may result in DFD meat due to muscle energy depletion to maintain a constant body temperature (Knowles et al., 1998) or heat stress, as showed by increased panting, especially when they are kept at hot (>30°C) and humid (RH > 80%) conditions (Santos et al., 1997). According to Grandin (2012), death losses in lairage, also called 'dead-in-pen', almost doubled at temperatures above 32°C compared with 16°C.

Water spraying on pigs in the lairage pen helps reduce the heat-related respiration rate (Huynh et al., 2006), the number of dead-in-pen (Vitali et al., 2014) and the incidence of PSE pork production through a 2°C drop in the muscle mass temperature (Long and Tarrant, 1990). However, at temperatures below 5°C, showering is not recommended as it causes animal shivering and may lead to DFD pork due to muscle energy depletion to maintain a constant body temperature (Knowles et al., 1998). Ivn order to remove the excessive humidity produced by the application of sprinkling/misting systems as well as to control the concentration of noxious gases, for example, ammonia (Weeks, 2008), this practice must be combined with a proper ventilation (135 m^2 h^{-1}; Brent, 1986; Weeks, 2008). Vitali et al. (2014) reported a lower risk of death in lairages characterised, among others, by efficient ventilation through large open windows on the roof and side walls.

When compared to other livestock species (i.e. cattle and sheep), pig lairages are the loudest ones (Weeks et al., 2009), with the average noise level ranging from 76 dB to 108 dB and the highest peaks (120 dB) being recorded in the *peri-mortem* area (Talling et al., 1996; Rabaste et al., 2007). Excessive lairage noise produces a fear response in pigs, as showed by the number of pigs huddling in the pen looking for protection or escaping from the source of sound (Geverink et al., 1998), the increased heart rate and greater blood lactate, CK and cortisol levels at slaughter (Faucitano, 2010), all resulting in an increased production of PSE pork (Warriss et al., 1994; Van de Perre et al., 2010). Keeping the sound level lower than 85 dB in the *peri-mortem* area appears to reduce the risk for PSE meat (Vermeulen et al., 2015).

Lairage noise is mostly caused by gates clanging, operating machinery, echoes and pig vocalisation (Weeks, 2008; Weeks et al., 2009), although pigs appear to be more stressed by industrial sounds than the sounds of conspecifics (Geverink et al., 1998). To reduce the ambient sound level, some European plants replaced metal gates and fencing with plastic ones and modified the ceiling by decreasing the height and installing sound-absorbing materials (Eyes on Animals, 2019).

The movement of pigs in the alleys of the slaughter plant may be influenced by the lighting of these areas. Pigs are less reluctant to move from a dark area to a brightly lighted area (Van Putten and Elshof, 1978; Grandin, 1990; Tanida et al., 1996), making the use of moving tools less necessary. Grandin (2010) observed that the insufficient light (less than 160–215 lux) at the entrance of the stunning area increased the use of electric prods by 34%. The alleys must be consistently lighted to avoid shadows and contrasts in colour on the floor. It has been recently observed that the use of green lighting reduces shadows on the floor and improves the ease of handling through the alleys (Eyes on Animals, 2019; see Fig. 4).

3.4 Driving pigs to slaughter

The combination of a higher slaughter speed, poorly designed handling systems and the progressive passage from a free-moving group situation to a single line of aligned and restrained individuals during the short period of time between the exit from the lairage pen and stunning may result in a greater proportion of slips, jamming, backing-up and vocalisation (Warriss et al., 1994; Edwards et al., 2010, 2011; Van de Perre et al., 2010; Vermeulen et al., 2015; Rocha et al., 2016), and increased use of electric prods (Rocha et al., 2016). These behavioural responses have been associated with increased heart rate (up to 240 beats/min; Chevillon, 2001; Correa et al., 2010), blood lactate and CK levels at slaughter (Hambrecht et al., 2005; Edwards et al., 2010; Rocha et al., 2015), skin lesions scores (Rabaste et al., 2007) and exudative pork (Van der Wal et al., 1999; Hambrecht et al., 2005; Rabaste et al., 2007; Dokmanović et al., 2014; Rocha et al., 2016).

In the *peri-mortem* area critical factors are the entrance into the stun chute and the 'stop-start' forward movement of pigs towards the stunner,

Figure 4 Green lighting reduces the presence of shadows on the alley floor preventing stops and backups slowing down the flow of pigs towards slaughter (Photo courtesy of Eyes on Animals, The Netherlands).

which are both observed in raceways feeding electrical and CO_2 stunners and which result in the frequent use of electric prods to encourage the pig movement (Faucitano, 2010). The batch flow system or curved raceway has been observed to work well, easing the flow of pigs into the single/double electric stun raceway, as showed by the lower number of reluctant-to-move-forward pigs and the use of electric prods compared with the single file chute entrance (Edwards et al., 2010). However, this system must be well designed and managed, in that it must not have an abrupt entrance and should not be filled with pigs for more than 75% of its capacity (ideally 50%) to prevent jamming and overlaps in the crowd pen (Grandin, 2017b; see Fig. 5). The entrance of pigs into a CO_2 gas stunner has been significantly improved by the group-wise stunning system (Christensen and Barton-Gade, 1997), where pigs are moved forwards in groups of 15 pigs using pushing gates and are loaded in sub-groups of 5 into the cradle. This system proved to reduce the frequency of PSE (−3%) and blood-splashed pork (−5%) and bruised carcasses (−4%) due to reduced electric prodding and physical stress (Christensen and Barton-Gade, 1997).

4 Welfare during stunning and slaughter

The last stage of pig production is the slaughter for human consumption. The slaughter process includes the restraining of the animal, the stunning application and the exsanguination.

4.1 Restrainer types

The purpose of restraint is to facilitate the correct application of the stunning and bleeding. The restrainer type varies depending of the stunning method. In electrical stunning, restraint is needed to facilitate the correct placement of the electrodes between the eye and the ear and to facilitate the uninterrupted flow of current. However, incorrect restraint may lead to ineffective stunning or bleeding but can also cause pain and distress in its own right. The restrainer should match the size of the pigs. It should be narrow enough that the animals cannot move backwards and forwards or turn around. Insecure restraint may cause struggling or escape attempts (Grandin, 2016b). On the other hand, a too narrow device will provide excessive force to the animal and will be painful and may cause injuries. Prolonged restraint time may cause fear and exacerbate insecure or excessive restraint. Therefore, no pig should enter the restrainer until equipment and personnel are ready to slaughter that animal.

Several automatic electrical stunning methods are currently available. One device consists of a V-type restrainer where each pig makes contact with the electrodes and receives the stunning current (Fig. 6). Both sides must run at the

Figure 5 Filling the crowd pen with pigs for more than 75% of its capacity prevents easy handling into the stunning chute (L. Faucitano, AAFC, Canada).

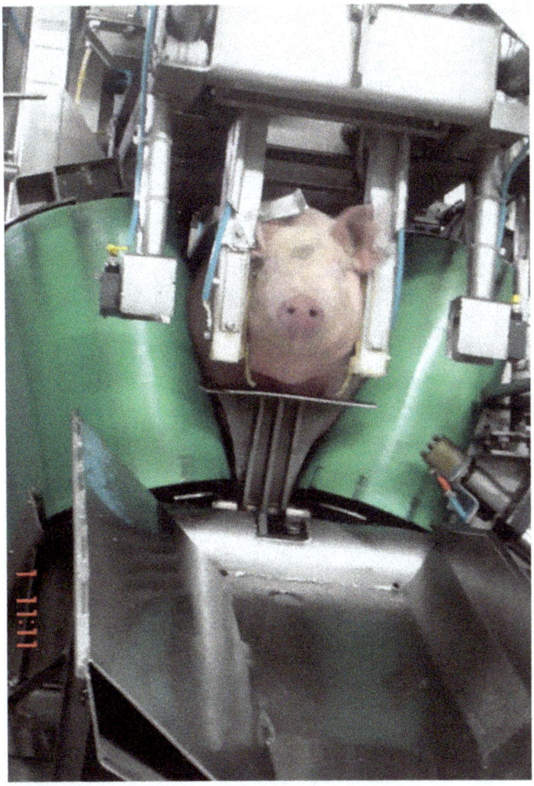

Figure 6 V-type restrainer (A. Velarde, IRTA, Spain).

same speed (Grandin, 2016b). When the two conveyors run at different speeds and their angle is too steep, the animals may struggle. A second method uses a conveyor belt system (Fig. 7). In this system, at the end of the restrainer, the nose of the pigs interrupts a beam of light, which activates the electrodes. In both systems, the animals are turned out and fall onto a table after the stunning.

Animal-based measures to assess pain, fear or distress of the pigs in the restrainer include slips or falls, struggling, escape attempts and vocalisation (Velarde and Dalmau, 2012; Grandin, 2016b).

Comparative studies showed that pigs restrained in the conveyor belt presented a lower heart rate (180 heartbeats/min vs. 220 heartbeats/min) and proportion of PSE pork production compared with those conveyed with the V-type restrainer (Griot et al., 2000).

On the other hand, the exposure to CO_2 at a high concentration does not require restraint as animals are stunned in groups (Velarde et al., 2000). In this case, gondolas for the gas stunning of pigs should not be overloaded, and animals should be able to stand without being on top of each other. Gondolas

Figure 7 Conveyor belt system (A. Velarde, IRTA, Spain).

overloading is a criterion in animal welfare auditing protocols at slaughter plants (Grandin, 2017b) based on the observation of 25–50 gondolas, depending on the plant size. Audit scores are given on a gondola basis and range from excellent (no gondolas are overloaded) to serious problem (the person moving the animals forces one or more pigs to jump on top of the other pigs in the gondolas with an electric prod or by hitting, shoving or kicking).

4.2 Stunning methods

To spare any avoidable pain, distress or suffering, stunning before bleeding is a common practice around the world. Stunning is defined as any intentionally induced process which causes the loss of consciousness and sensibility without pain, including any process resulting in instantaneous death (Regulation No. 1099/2009). During the loss of consciousness, the animal is unable to perceive external stimuli and control its voluntary mobility and, therefore, does not respond to normal stimuli, including pain (EFSA, 2004). In order to fulfil the humane slaughter requirements, the duration of unconsciousness must be longer than the sum of time that lapses between the end of stun and the time to the onset of death due to bleeding. Therefore, unconsciousness should be monitored at different stages of the process: (1) during shackling and hoisting, (2) before sticking and (3) during bleeding (EFSA, 2013). The EFSA AHAW panel has developed toolboxes of welfare indicators for developing monitoring procedures at slaughterhouses for pigs (EFSA, 2013). The most commonly used methods for stunning pigs at slaughter are electrical stunning and exposure to CO_2 at high concentrations.

Electrical stunning (Fig. 7) involves the application of an electric current of sufficient magnitude to the brain such that a generalised epileptiform activity is induced similar to that recorded in humans during grand mal epileptic seizures (Croft, 1952; Hoenderken, 1978). This seizure-like state, immediately followed by an exhausted state, is suggestive of an immediate loss of consciousness and appears to be associated with a lack of sensory awareness, which lasts a finite period of time (Anil, 1991). An effective stun is characterised by the presence of all of tonic–clonic seizures, the loss of posture, apnoea and the absence of corneal reflex (Velarde and Dalmau, 2012). The main hazards preventing effective stunning are incorrect electrode placement, poor contact, dirty or corroded electrode, too low voltage/current or high frequency (EFSA, 2004).

Exposure to CO_2 at high concentrations is also used for the stunning of pigs. In this case, the loss of consciousness is not immediate (Rodríguez et al., 2008). Since CO_2 does not induce immediate loss of consciousness, inhalation of concentrations greater than 30% of CO_2 by volume in the atmospheric air causes aversion, irritation of the mucous membranes (that can be painful) and respiratory distress during the induction phase (Velarde et al., 2007). During CO_2 exposure, pigs show signs of aversion such as retreat attempts, headshaking, sneezing, breathlessness, freezing, escape attempts, gasping (a very deep breath through a gasping-open mouth, indicative of breathlessness; Raj and Gregory, 1996) and vocalisations (Holst, 2001; Velarde et al., 2007).

The main hazards causing increased distress during induction to unconsciousness are irritant or aversive gas mixtures, low gas temperature or humidity.

The exposure to anoxic gases (argon or nitrogen) with less than 2% by volume of residual oxygen in air is non-aversive and does not cause sense of breathlessness before the loss of consciousness occurs. However, the time to induce unconsciousness when exposed to anoxia is longer than when exposed to hypercapnia (Raj et al., 1997) and might not be commercially feasible. Raj and Gregory (1996) reported that the addition of up to 30% CO_2 to an anoxic atmosphere reduces the time needed to induce unconsciousness with minimal aversion. Argon (Ar) has a low presence in the atmosphere (0.9% by volume), and its availability for commercial stunning practices might be limited. On the other hand, the relative density of nitrogen (N_2) is slightly lower than air concentrations and cannot be sustained within a pit at a higher concentration than 94% by volume (Dalmau et al., 2010). Nevertheless, this stability could be improved when nitrogen and CO_2 are combined (Dalmau et al., 2010).

The EU Regulation (EC 1099/2009) has approved the low atmospheric pressure stunning (LAPS) for broiler chickens. It consists of the exposure to gradual decompression with reduction in the available oxygen to less than 5% (Martin et al., 2016). Current research is carried out in pigs to assess the effectiveness in inducing unconsciousness without aversion.

An effective stun is characterised by the presence of the loss of posture, apnoea, absence of corneal reflex and absence of muscle tone (Velarde et al., 2007). The main hazards causing ineffective stunning are incorrect gas concentration or short gas exposure time (EFSA, 2004). In case of ineffective stunning or recovery, re-stun immediately using a backup system taking into account the causes of failure or recovery.

Comparative studies reported the positive effects of CO_2 stunning on the pork quality when compared with electrical stunning (Velarde et al., 2000, 2001). However, the effects of gas stunning on animal welfare and pork quality depend on the gas concentration and exposure time. Better results, in terms of the percentage of clinical reflexes and risk for PSE pork, were obtained after stunning with a higher CO_2 concentration (90% vs. 80%) and longer gas exposure time (100 vs. 70 sec; Nowak et al., 2007).

4.3 Exsanguination techniques

Exsanguination consists of the severance of the arteries supplying oxygenated blood to the brain. Pigs are bled by chest sticking, severing the brachiocephalic trunk and the major blood vessels which arise from the heart (Fig. 8). With an adequate incision, pigs lose between 40% and 60% of their total blood volume, and within 30 sec, the amount of blood lost is between 70% and 80% of the total amount of blood which will be lost (Warriss and Wilkins, 1987). The time to lose brain responsiveness (based on the reduction in visual evoked responses) ranges between 14 sec and 23 sec (mean 18 ± 63) and the development of an isoelectric electrocorticogram (ECoG) between 22 sec and 30 sec (Wotton and Gregory, 1986).

The main welfare concern at the time of bleeding following stunning is the recovery of consciousness due to prolonged stun-to-stick interval or due to incomplete severance of the main blood vessels.

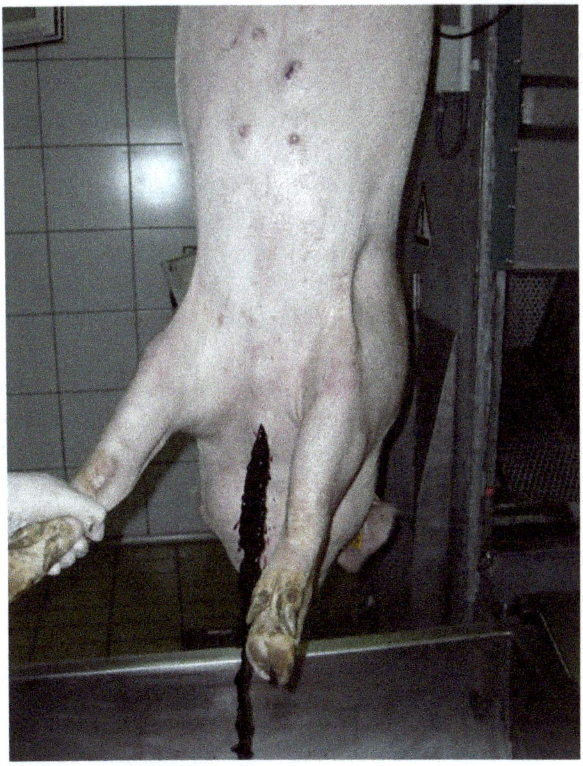

Figure 8 Chest sticking (A. Velarde, IRTA, Spain).

5 Animal welfare audit protocols

Nowadays, there is an increasing need for credible assessment systems to determine the welfare status of animals prior to slaughter. The application of animal welfare auditing protocols allows the evaluation of handling and slaughter practices, resulting in a significant improvement in handling practices, facilities design and quality of work, besides reassuring concerned consumers and increasing market opportunities due to the availability of certified products (Ballantyne, 2006; Grandin, 2017a).

Because of these benefits, animal welfare auditing has been increasingly implemented during transport and at slaughter since 1999 in the US, Australia, New Zealand and Europe, and in Canada since 2001 (Grandin and Smith, 2004).

Presently, the two main protocols used to assess animal welfare during transport and lairage and at slaughter are the audit protocols developed by Temple Grandin for the North American Meat Institute (NAMI; Grandin, 2017b) and by researchers of the European project Welfare Quality® (WQ®, 2009).

The NAMI audit protocol is based on the 'Recommended animal handling guidelines and audit guide' and has the objective to evaluate animal welfare during transport and at the slaughter plant through the assessment of seven core criteria based on the observation of animal behaviour (falls and vocalisation), assessment of facilities design, electric prod use, wilful acts of abuse, efficiency of stunning systems (electrical and CO_2 gas), access to water and insensibility on the bleed rail (Grandin, 2017b). The respect of the NAMI guidelines at the slaughter plant is a requirement to comply with the standards of the Humane Farm Animal Care and for the pork product to get the Certified Humane Raised and Handled® label (Humane Farm Animal Care, 2018).

The WQ® protocol (WQ®, 2009) has developed an integrated and standardised welfare assessment system based on 12 welfare criteria grouped into four main principles (good feeding, good housing, good health and appropriate behaviour) according to how they are experienced by animals (Dalmau et al., 2009). The innovation of the WQ® assessment system is its greater focus on animal-based measures (e.g. body condition, health, injuries and behaviour) rather than on resource and management-based measures (e.g. space allowance, number of drinkers and truck management). The WQ® welfare assessment starts in the unloading area, where general fear, thermoregulation behaviours, slipping and falling, sickness and dead animals are observed (Velarde and Dalmau, 2012). In lairage, five criteria are taken into consideration, namely the absence of thirst, hunger and disease; thermal comfort; and comfort while resting. The absence of thirst is calculated based on the number of drinking points in each pen and drinker functionality and cleanliness. The availability of feed for animals that have been held more than 12 h in the holding pen is the measure for the absence of hunger, while the absence of disease is measured by the presence of dead animals in the pen. Thermal comfort is evaluated by behavioural thermoregulation measures, such as huddling, shivering or panting scores. The last, but not least, criterion is comfort while resting in the pen, which is evaluated based on the space allowance. During movement from the lairage pen to the stunning area, good human–animal relationship is assessed by recording the incidence of high-pitched vocalisations. Afterwards, stunning effectiveness is assessed by observing the signs of the absence of pain, such as corneal and righting reflex, breathing and vocalisation, immediately after stunning and before sticking. After slaughter, skin lesion scoring provides valuable information regarding the management of animals on the farm of origin, during transport or in the lairage pen. Furthermore, the health status of the animals on the farm of origin is also assessed after slaughter by inspecting the internal organs for the presence of pleurisy and pneumonia in the lungs, pericarditis in the heart and white spots in the liver.

The efficiency of the WQ® programme as a tool for animal welfare monitoring has been scientifically validated at the abattoir (Dalmau et al., 2009). In 2014, a certification scheme of animal welfare based on the WQ® protocols has been developed by the Animal Welfare Subprogram of the IRTA (Monells, Spain) and the Spanish Association for Standardisation and Certification (AENOR). In 2017, already several pig abattoirs had been certified. The final objective of this programme is to include the certificate in the pork meat product label, named as 'Animal Welfare. AENOR Conform' (Dalmau, 2017). Recently, the IRTA has become scheme owner for the Animal Welfare Certification 'based on Welfare Quality®'.

However, neither the WQ® nor the NAMI assessment protocols cover the production process as a whole (from farm to slaughter). Rocha et al. (2016) developed a novel protocol for the animal welfare assessment of the whole pork chain by merging the WQ protocol with those of the Canadian Animal Care Assessment (CPC, 2011), which is a more resource- and management-based measurement of animal welfare at the farm level (including loading), and that of the American Meat Institute (AMI; Grandin, 2012) and validated it under Canadian commercial conditions. The objective of this study was to assess the relationship between the audit scores obtained at animal welfare-improved (AWI) and conventional farms using the WQ and CPC protocols and the variation in pig behaviours at loading, at unloading and in lairage as assessed by the WQ and AMI audit protocols. In this study, the frequency of slips used as an animal welfare audit criterion in the AMI protocol at the slaughter plant showed to be a good predictor of drip loss variation in the loin muscle, with muscle exudation being significantly related to slips during unloading ($r = 0.63$) and immediately before stunning ($r = 0.74$).

6 Conclusion and future trends

This chapter overviewed the effects of the stressors experienced by pigs during the transport, lairage and slaughter on animal losses; behavioural and physiological responses to stress; and carcass and meat quality.

Research provided the evidence that the use of truck models featuring hydraulic ramps or decks can help reduce the workload of handlers and improve the welfare of pigs during transport. However, more research on the truck design is needed with a study of the air flow patterns and vibration rate and insulation and cooling systems under different ambient conditions where temperature control becomes more critical and physiological heat maintenance and dissipation in pigs of different market weights become less effective. So far, most swine transportation research focused on slaughter weight pigs (up to 130 kg), but there is a considerable lack of information for the transportation of heavier pigs, including cull sows and boars. More specifically, information on their thermal and physical needs during transport and relative science-based guidelines for space allowance and travel time are needed.

Lairage and slaughter are extremely important for the pork chain economy as mistakes made at these points have irreversible effects on the welfare of pigs and carcass and meat quality and may offset all efforts made by the production sector to improve performance and animal welfare. Precautions must be taken to ensure proper handling and ambient control to keep the benefits of lairage as a resting area, allowing pigs to recover from the stress of transport and previous handling. The correct management and monitoring of critical areas in lairage is becoming paramount in the light of the increasing need for commercial abattoirs to obtain animal welfare audit approval and certification for their meat products.

7 Where to look for further information

7.1 Further reading

- Comprehensive and up-to-date reviews on swine transportation and animal welfare are the chapters authored by Faucitano and Goumon (2018) and Faucitano and Lambooij (2019) and published in the books *Advances in Pig Welfare* and *Livestock Transport and Handling*.
- An excellent source of science-based guidelines for the best handling practices at the abattoir (from reception to slaughter) in multiple species is the book *Animal Welfare at Slaughter: a Practitioner Guide* by Raj and Velarde (2016).
- Guidelines and recommendations on preslaughter handling and animal welfare can be found in the Temple Grandin web page: http://www.grandin.com/.
- A complete introduction to the relationship between animal welfare preslaughter and meat science is represented by the books *Animal Welfare and Meat Science* (1998) and *Animal Welfare and Meat Production* (2007) by Neville Gregory.
- A catalogue of training handbooks, guidelines and technical sheets in relation to transport and slaughter are available in the web page of the Humane Slaughter Association: https://www.hsa.org.uk/.

7.2 Key journals/conferences

- Special issues on pig transportation in the *Animals* journal (2014/2016): https://www.mdpi.com/journal/animals/special_issues/pig-trans-2016; https://www.mdpi.com/journal/animals/special_issues/pig-trans.
- Proceedings of special sessions on preslaughter handling and animal welfare included in the programme of the annual congresses of the

International Society of Applied Ethology (ISAE; https://www.applied-ethology.org/), the International Congress of Meat Science & Technology (ICoMST; http://www.icomst.helsinki.fi/index.htm) and American and Canadian Societies of Animal Science (ASAS-CSAS) Meeting & Trade Show; https://www.asas.org).

7.3 Major international research projects

- The Canadian Agricultural Partnership (http://www.agr.gc.ca/eng/about-us/key-departmental-initiatives/canadian-agricultural-partnership/?id=1461767369849) and Swine Innovation Porc (http://www.swineinnovationporc.ca/) are supporting projects on swine transportation.

8 References

Anil, M. H. (1991), 'Studies on the return of physical reflexes in pigs following electrical stunning', *Meat Sci.*, 30, 13–21.

Averós, X., Knowles, T. G., Brown, S. N., Warriss, P. D. and Gosálvez, L. F. (2008), 'Factors affecting the mortality of pigs being transported to slaughter', *Vet. Rec.*, 163, 386-90.

Ballantyne, W. (2006), 'A proactive approach to animal welfare', *Adv. Pork Prod.*, 17, 139–44.

Baloyh, P., Kapelankis, W., Jankowiak, H., Nagy, L., Kovacs, S., et al. (2015), 'The productive lifetime of sows on two farms and reasons for culling', *Ann. Anim. Sci.*, 15, 747–58.

Barton-Gade, P. and Christensen, L. (1998), 'Effect of different loading density during transport on welfare and meat quality in Danish slaughter pigs', *Meat Sci.*, 48, 237–47.

Barton-Gade, P., Christensen, L., Brown, S. N. and Warriss, P. D. (1996), 'Effect of tier and ventilation during transport on blood parameters and meat quality in slaughter pigs', *EU-Seminar: New Information on Welfare and Meat Quality of Pigs as Related to Handling, Transport and Lairage Conditions*, Landbauforschung Völkenrode, Kulmbach, Germany, 1996, vol. 166, 101–16.

Barton-Gade, P., Christensen, L., Baltzer, M. and Petersen, L. (2007), 'Causes of pre-slaughter mortality in Danish slaughter pigs with special emphasis on transport', *Anim. Welf.*, 16, 459–70.

Becerril-Herrera, M. M., Alonso-Spilsbury, M., Trujillo-Ortega, M. E., Guerrero-Legarreta, I., Ramirez-Necoechea, R., et al. (2010), 'Changes in blood constituents of swine transported for 8 or 16 h to an abattoir', *Meat Sci.*, 86, 945–8.

Bench, C., Schaefer, A. L. and Faucitano, L. (2008), 'The welfare of pigs during transport', *in* L. Faucitano and A. L. Schaefer (Eds), *Welfare of Pigs: From Birth to Slaughter*, Wageningen, The Netherlands: Wageningen Academic Publishers, pp. 161–95.

Bertol, T. M., Ellis, M., Ritter, M. J. and McKeith, F. K. (2011), 'Effect of feed withdrawal and handling intensity on longissimus muscle glycolytic potential and blood measurements in slaughter weight pigs', *J. Anim. Sci.*, 83, 1536–42.

Blair, B. and Lowe, J. (2019), 'Describing the cull sow market network in the US: a pilot project', *Prev. Vet. Med.*, 162, 107–9.

Bottaccini, M., Scollo, A., Edwards, S., Contiero, B., Veloci, M., et al. (2018), 'Skin lesion monitoring at slaughter on heavy pigs (170 kg): welfare indicators and ham defects', *PLoS ONE* (open access). doi: 10.1371/journal.pone.0207115.

Brent, G. (1986), *Housing the Pig*, Ipswich, UK: Farming Press Ltd.

Brown, S. N., Knowles, T. G., Edwards, J. E. and Warriss, P. D. (1999), 'Relationship between food deprivation before transport and aggression in pigs held in lairage before slaughter', *Vet. Rec.*, 145, 630–34.

Brown, J., Samarakone, T. S., Crowe, T., Bergeron, R., Widowski, T. M., et al. (2011), 'Temperature and humidity conditions in trucks transporting pigs in two seasons in eastern and western Canada', *Trans. ASABE*, 54, 1–8.

CARC. (2001), 'Recommended code of practice for the care and handling of farm animals - transportation', Canadian Agri-Food Research Council, Ottawa, Canada, 63p.

Chevillon, P. (2001), 'Pig welfare during pre-slaughter and stunning', *1st International Virtual Conference on Pork Quality*, Brazil, pp. 145–58.

Christensen, L. and Barton-Gade, P. (1997), 'New Danish developments in pig handling at abattoirs', *Fleischwirt.*, 77, 604–7.

Conte, S., Faucitano, L., Bergeron, R., Torrey, S., Gonyou, H. W., et al. (2015), 'Effects of season, truck type and location within truck on gastrointestinal tract temperature of market-weight pigs during transport', *J. Anim. Sci.*, 93, 5840–8.

Correa, J. A. (2011), 'Effect of farm handling and transport on physiological response, losses and meat quality of commercial pigs', *Adv. Pork Prod.*, 22, 249–56.

Correa, J. A., Torrey, S., Devillers, N., Laforest, J. P., Gonyou, H. W. and Faucitano, L. (2010), 'Effects of different moving devices at loading on stress response and meat quality in pigs', *J. Anim. Sci.*, 88, 4086–93.

Correa, J. A., Gonyou, H. W., Torrey, S., Widowski, T., Bergeron, R., et al. (2013), 'Welfare and carcass and meat quality of pigs being transported for 2 hours using two vehicle types during two seasons of the year', *Can. J. Anim. Sci.*, 93, 43–55

Correa, J. A., Gonyou, H. W., Torrey, S., Widowski, T., Bergeron, R., et al. (2014), 'Welfare of pigs being transported for long distance using a pot-belly trailer during winter and summer', *Animals*, 4, 200–13.

Council Regulation (EC) 1099/2009 (2009), 'Council Regulation No 1099/2009 on the protection of animals at the time of killing', *OJEU*, L303, 1–30.

CPC (2011), 'Animal care assessment', Canadian Pork Council, Ottawa, Canada. http://www.cqa-aqc.com/aca/documents/ACA-Animal-Care-Assessment.pdf.

Croft, P. G. (1952), 'Problem of electrical stunning', *Vet. Rec.*, 64, 255–8.

Dalla Costa, O. A., Faucitano, L., Coldebella, A., Ludke, J. V., Peloso, J. V., et al. (2007), 'Effects of the season of the year, truck type and location on truck on skin bruises and meat quality in pigs', *Livest. Sci.*, 107, 29-36.

Dalla Costa, F. A., Devillers, N., Paranhos da Costa, M. J. R. and Faucitano, L. (2016), 'Effects of applying preslaughter feed withdrawal at the abattoir on behaviour, blood parameters and meat quality in pigs', *Meat Sci.*, 119, 89-94.

Dalla Costa, O. A., Dalla Costa, F. A., Feddern, V., Dos Santos Lopes, L., et al. (2019), 'Risk factors associated with pig pre-slaughtering losses', *Meat Sci.*, 155, 61–8.

Dalmau, A. (2017), 'Development of a certification schema on animal welfare based on Welfare Quality protocols', *Newsl. Welf. Qual. Netw.*, 5, 2–3.

Dalmau, A., Temple, D., Rodríguez, P., Llonch, P. and Velarde, A. (2009), 'Application of the Welfare Quality® protocol at pig slaughterhouses', *Anim. Welf.*, 18, 497–505.

Dalmau, A., Llonch, P., Rodríguez, P., Ruíz-de-la-Torre, J. L., et al. (2010), 'Stunning pigs with different gas mixtures. Part 1: gas stability', *Anim. Welf.*, 19, 315–23.

Dewey, C., Haley, C., Widowski, T., Poljak, Z. and Friendship, R. (2009), 'Factors associated with in-transit losses of fattening pigs', *Anim. Welf.*, 18, 355–61.

Dokmanović, M., Velarde, A., Tomović,V., Glamočlija, N., Marković, R., et al. (2014), 'The effects of lairage time and handling procedure prior to slaughter on stress and meat quality parameters in pigs', *Meat Sci.*, 98, 220–6.

Edwards, L. N., Grandin, T. A., Engle, T. E., Porter, S. P., Ritter, M. J., et al. (2010), 'Use of exsanguination blood lactate to assess the quality of preslaughter pig handling', *Meat Sci.*, 86, 384–90.

Edwards, L. N, Engle, T. E., Grandin, T., Ritter, M. J., Sosnicki, A., et al. (2011), 'The effects of distance traveled during loading, lairage time prior to slaughter, and distance traveled to the stunning area on blood lactate concentration of pigs in a commercial packing plant', *Prof. Anim. Sci.*, 27, 485–91.

EFSA (2004), 'Welfare aspects of animal stunning and killing methods', Scientific Report of the Scientific Panel for Animal Health and Welfare on request from the Commission related to welfare aspects of animal stunning and killing methods (Question No. EFSA-Q-2003-093). European Food Safety Authority (AHAW 04-027).

EFSA (2013), 'Scientific Opinion on monitoring procedures at slaughterhouses for pigs', *EFSA J.*, 11(12), 3523.

Eyes on Animals (2019), 'Improving animal welfare in pig slaughterhouses'. http://www.eyesonanimals.com/wp-content/uploads/2016/06/Animal-welfare-in-pig-slaughterhouses-how-to-reduce-stress-suffering-and-ease-handling-aanp-1.pdf (accessed on 14 May 2019).

Faucitano, L. (2001), 'Causes of skin damage to pig carcasses', *Can. J. Anim. Sci.*, 81, 39–45.

Faucitano, L. (2010), 'Effects of lairage and slaughter conditions on animal welfare and pork quality', *Can. J. Anim. Sci.*, 90, 461–9.

Faucitano, L. (2018), 'Preslaughter handling practices and their effects on animal welfare and pork quality', *J. Anim. Sci.*, 96, 728–38.

Faucitano, L. and Goumon, S. (2018), 'Transport to slaughter and associated handling', *in* M. Špinka (Ed.), *Advances in Pig Welfare*, London, UK: Woodhead Publishing, pp. 261–93.

Faucitano, L. and Lambooij, E. (2019), 'Transport of pigs', *in* T. Grandin (Ed.), *Livestock Transport and Handling* (5th edn.), Wallingford, UK: CABI Publishing, pp. 302–27.

Fitzgerald, R. F., Stalder, K. J., Matthews, J. O., Schultz-Kaster, C. M. and Johnson, A. K. (2009), 'Factors associated with fatigued, injured, and dead pig frequency during transport and lairage at a commercial abattoir', *J. Anim. Sci.*, 87, 1156–66.

Fox, J., Widowski, T., Torrey, S., Nannoni, E., Bergeron, R., et al. (2014), 'Water sprinkling market pigs in a stationary trailer. 1. Effects on pig behaviour, gastrointestinal tract temperature and trailer micro-climate', *Livest. Sci.*, 160, 113–23.

Fraqueza, M. J., Roseiro, L. C., Almeida, J., Matias, E., Santos, C., et al. (1998), 'Effects of lairage temperature and holding time on pig behavior and on carcass and meat quality', *Appl. Anim. Behav. Sci.*, 60, 317–30.

Gallo, C., Faucitano, L. and Gerritzen, M. (2016), 'Effects of preslaughter handling on carcass and meat quality', in M. Raj and A. Velarde (Eds), *Animal Welfare at Slaughter: a Practitioner Guide*, Sheffield, UK: 5m Publishing, pp. 251–69.

Garcia, A. and McGlone, J. J. (2015), 'Loading and unloading finishing pigs: effects of bedding types, ramp angle, and bedding moisture', *Animals*, 5, 13–26.

Gerritzen, M. A., Hindle, V. A., Steinkamp, K. and Reimert, H. G. M. (2013), 'The effect of reduced loading density on pig welfare during long distance transport', *Anim.* 7, 1849–57.

Geverink, N. A., Engel, B., Lambooij, E. and Wiegant, V. M. (1996), 'Observations on behavior and skin damage of slaughter pigs and treatment during lairage', *Appl. Anim. Behav. Sci.*, 50, 1–13.

Geverink, N. A., Buhnemann, A., Van de Burgwal, J. A., Lambooij, E., Blokhuis, H. J., et al. (1998), 'Responses of slaughter pigs to transport and lairage sounds', *Physiol. Behav.*, 63, 667–73.

Gonyou, H. W. and Brown, J. (2012), 'Reducing stress and improving recovery from handling during loading and transport of market pigs', Final Report submitted to Alberta Livestock and Meat Agency, Edmonton, Canada, 40p.

Goumon, S. and Faucitano, L. (2017), 'Influence of loading handling and facilities on the subsequent response to pre-slaughter stress in pigs', *Livest. Sci.*, 200, 6–13.

Goumon, S., Brown, J. A., Faucitano, L., Bergeron, R., Widowski, T. M., et al. (2013a), 'Effects of transport duration on maintenance behavior, heart rate and gastrointestinal tract temperature of market-weight pigs in 2 seasons', *J. Anim. Sci.*, 91, 4925–35.

Goumon, S., Faucitano, L., Bergeron, R., Crowe, T., Connor, M. L., et al. (2013b), 'Effect of ramp configuration on easiness of handling, heart rate and behavior of near-market pigs at unloading', *J. Anim. Sci.*, 91, 3889–98.

Grandin, T. (1990), 'Design of loading facilities and holding pens', *Appl. Anim. Behav. Sci.*, 28, 187–201.

Grandin, T. (2010), 'The importance of measurement to improve the welfare of livestock, poultry and fish', in T. Grandin (Ed.), *Improving Animal Welfare: a Practical Approach*, Wallingford, UK: CABI Publishing, pp. 1–20.

Grandin, T. (2012), *Recommended Animal Handling Guidelines and Audit Guide: a Systematic Approach to Animal Welfare*, Washington, DC: American Meat Institute Foundation, 108p.

Grandin, T. (2016a), 'Transport fitness of cull sows and boars: a comparison of different guidelines of fitness for transport', *Animals*, 6, 77 (open access). doi: 10.3390/ani8070124.

Grandin, T. (2016b), 'Practical methods to improve animal handling and restraint', *in* A. Velarde and M. Raj (Eds), *Animal Welfare at Slaughter*, Sheffield, UK: 5M Publishing, pp. 71–90.

Grandin, T. (2017a), 'How to work with large meat buyers to improve animal welfare', *in* P. P. Purslow (Ed.), *New Aspects of Meat Quality: from Genes to Ethics*, London, UK: Woodhead Publishing, pp. 569–79.

Grandin, T. (2017b), *Recommended Animal Handling Guidelines and Audit Guide: Systematic Approach to Animal Welfare*, Washington DC: North American Meat Institute Foundation, 108p.

Grandin, T. (2018), 'Welfare problems in cattle, pigs, and sheep that persist even though scientific research clearly shows how to prevent them', *Animals*, 8, 124 (open access). doi: 10.3390/ani8070124.

Grandin, T. and Smith, G. C. (2004), 'Animal welfare and humane slaughter'. http://www.grandin.com/references/humane.slaughter.html (accessed on 23 September 2019).

Griot, B., Boulard, J., Chevillon, P. and Kerisit, R. (2000), 'Des restrainers à bande pour le bienêtre et la qualité de la viande', *Viandes et Produits Carnés*, 3, 91–7.

Guàrdia, M. D., Gispert, M. and Diestre, A. (1996), 'Mortality rates during transport and lairage in pigs for slaughter', *Meat Focus Inter.*, 10, 362–6.

Guàrdia, M. D., Estany, J., Balasch, S., Oliver, M. A., Gispert, M., et al. (2004), 'Risk assessment of PSE condition due to pre-slaughter conditions and RYR1 gene in pigs', *Meat Sci.*, 67, 471–8.

Guàrdia, M. D, Estany, J., Balasch, S., Oliver, M. A., Gispert, M., et al. (2005), 'Risk assessment of DFD meat due to pre-slaughter conditions in pigs', *Meat Sci.*, 70, 709–16.

Guàrdia, M. D., Estany, J., Balasch, S., Oliver, M. A., Gispert, M., et al. (2009), 'Risk assessment of skin damage due to pre-slaughter conditions and RYR1 gene in pigs', *Meat Sci.*, 81, 745–51.

Haley, C., Dewey, C. E., Widowski, T. and Friendship, R. (2008), 'Association between in-transit losses, internal trailer temperature, and distance travelled by Ontario market hogs', *Can. J. Vet. Res.*, 72, 385–9.

Haley, C., Dewey, C. E., Widowski, T. and Friendship, R. (2010), 'Relationship between estimated finishing-pig space allowance and in transit loss in a retrospective survey for 3 packing plant in Ontario in 2003', *Can. J. Vet. Res.*, 74, 178–84.

Hambrecht, E., Eissen, J. J., Newman, D. J., Smits, C. H., Verstegen, M. W., et al. (2005), 'Preslaughter handling effects on pork quality and glycolytic potential in two muscles differing in fiber type composition', *J. Anim. Sci.*, 83, 900–7.

Hoenderken, R. (1978), 'Electrical stunning of pigs for slaughter', Doctoral thesis, University of Utrecht, NL.

Holst, S. (2001), 'Carbon dioxide stunning of pigs for slaughter – practical guidelines for good animal welfare', *47th International Congress of Meat Science and Technology*, Krakow, Poland, pp. 48–54.

Honkavaara, M. (1989), 'Influence of lairage on blood composition of pig and on the development of PSE pork', *J. Agric. Sci. Finland*, 61, 433–40.

Humane Farm Animal Care (2018), 'Humane farm animal care standards', Humane Farm Animal Care, Middleburg, VA, 31p. www.certifiedhumane.org.

Huynh, T. T. T., Aarnink, A. J. A., Truong, C. T., Kemp, B. and Verstegen, M. W. A. (2006), 'Effects of tropical climate and water cooling on growing pigs' responses', *Livest. Sci.*, 104, 278–91.

Kephart, K. B., Harper, M. T. and Raines, C. R. (2010), 'Observations of market pigs following transport to a packing plant', *J. Anim. Sci.*, 88, 2199–203.

Knowles, T. G., Brown, S. N., Edwards, J. E. and Warriss, P. D. (1998), 'Ambient temperature below which pigs should not be continuously showered in lairage', *Vet. Rec.*, 143, 575–8.

Koketsu Y. (2000), 'Factors associated with increased sow mortality in North America', *Proceedings of the American Association of Swine Practitioners*, pp. 419–20.

Lambooij, E. (1988), 'Road transport of pigs over a long distance: some aspects of behaviour, temperature and humidity during transport and some effects of the last two factors', *Anim. Prod.*, 46, 257–63.

Lambooij, E. (2014), 'Transport of pigs', *in* T. Grandin (Ed.), *Livestock Handling and Transport*, Wallingford, UK: CABI Publishing, pp. 280–97.

Lambooij, E. and Engel, B. (1991), 'Transport of slaughter pigs by road over a long distance: some aspects of loading density and ventilation', *Livest. Prod. Sci.*, 28, 163–74.

Lambooij, E., Garssen, G. J., Walstra, P., Mateman, F. and Merkus, G. S. M. (1985), 'Transport of pigs by car for two days: some aspects of watering and loading density', *Livest. Prod. Sci.*, 13, 289–99.

Leheska, J. M., Wulf, D. M. and Maddock, R. J. (2003), 'Effects of fasting and transportation on pork quality development and extent of postmortem metabolism', *J. Anim. Sci.*, 80, 3194–202.

Long, V. P. and Tarrant, P. V. (1990), 'The effect of pre-slaughter showering and post-slaughter rapid chilling on meat quality in intact pork sides', *Meat Sci.*, 27, 181–95.

Malena, M., Voslárová, E., Kozák, A., Belobrádek, P., Bedánová, I., et al. (2007), 'Comparison of mortality rates in different categories of pigs and cattle during transport for slaughter', *Acta Vet. Brno*, 76, 109–16.

Martin, J. E., Christensen, K., Vizzier-Thaxton, Y. and McKeegan, D. E. F. (2016), 'Effects of analgesic intervention on behavioural responses to Low Atmospheric Pressure Stunning', *Appl. Anim. Behav. Sci.*, 180, 157–65.

McGee, M., Johnson, A. K., O'Connor, A. M., Tapper, K. R. and Millman, S. T. (2016), 'An assessment of swine marketed through buying stations and development of fitness for transport guidelines', *J. Anim. Sci.*, 94, 9.

McGlone, J. J., Johnson, A. K., Sapkota, A. and Kephart, R. K. (2014a), 'Establishing trailer ventilation (boarding) requirements for finishing pigs during transport', *Animals*, 4, 515–23.

McGlone, J. J., Johnson, A. K., Sapkota, A. and Kephart, R. K. (2014b), 'Transport of market pigs: improvements in welfare and economics', *in* T. Grandin (Ed.), *Livestock Handling and Transport*, Wallingford, UK: CABI Publishing, pp. 298–314.

Moss, B. W. (1978), 'Some observations on the activity and aggressive behavior of pigs when penned prior to slaughter', *Appl. Anim. Ethol.*, 4, 323–39.

Mota-Rojas, D., Becerril, M., Lemus, C., Sánchez, P., González, M., Olmos, S. A., et al. (2006), 'Effects of mid-summer transport duration on pre- and post-slaughter performance and pork quality in Mexico', *Meat Sci.*, 73, 404–12.

Nanni Costa, L., Lo Fiego, D. P., Dall'Olio, S., Davoli, R. and Russo, V. (2002), 'Combined effects of pre-slaughter treatments and lairage time on carcass and meat quality in pigs of different halothane genotype', *Meat Sci.*, 61, 41–7.

Nannoni, E., Widowski, T. M., Torrey, S., Fox, J., Rocha, L. M., et al. (2014), 'Water sprinkling market pigs in a stationary trailer. 2. Effects on selected exsanguination blood parameters and carcass and meat quality variation', *Livest. Sci.*, 160, 124–31.

Nannoni, E., Liuzzo, G., Serraino, A., Giacometti, F., Martelli, G., et al. (2016), 'Evaluation of pre-slaughter losses of Italian heavy pigs', *Anim. Prod. Sci.* (open-access). doi: 10.1071/AN15893.

Nowak, B., Mueffling, T. V. and Hartung, J. (2007), 'Effect of different carbon dioxide concentrations and exposure times in stunning of slaughter pigs: impact on animal welfare and meat quality', *Meat Sci.* 75, 290–8.

Pereira, T., Titto, E. A., Conte, S., Devillers, N., Sommavilla, R., et al. (2018), 'Use of fan-misters bank for cooling pigs kept in a stationary trailer before unloading: effects on trailer microclimate, and pig behavior and physiological response', *Livest. Sci.*, 216, 67–74.

Peterson, E., Remmenga, M., Hagerman, A. D. and Akkina, J. E. (2017), 'Use of temperature, humidity, and slaughter condemnation data to predict increases in transport losses in three classes of swine and resulting foregone revenue', *Front. Vet. Sci.*, 4, 67 (open access). doi: 10.3389/fvets.2017.00067.

Pilcher, C. M., Ellis, M., Rojo-Gomez, A., Curtis, S. E., Wolter, B. F., et al. (2011), 'Effects of floor space during transport and journey time on indicators of stress and transport losses in market weight pigs', *J. Anim. Sci.*, 89, 3809–18.

Rabaste, C., Faucitano, L., Saucier, L., Foury, D., Mormède, P., et al. (2007), 'The effects of handling and group size on welfare of pigs in lairage and its influence on stomach weight, carcass microbial contamination and meat quality variation', *Can. J. Anim. Sci.*, 87, 3-12.

Raj, A. B. M. and Gregory, N. G. (1996), 'Welfare implications of the gas stunning of pigs. Stress of induction of anaesthesia', *Anim. Welf.*, 5, 71-8.

Raj, A. B. M. and Velarde, A. (2016), *Animal Welfare at Slaughter: A Practitioner Guide*, Sheffield, UK: 5m Publishing.

Raj, A. B. M., Johnson, S. P., Wotton, S. B. and McInstry, J. L. (1997), 'Welfare implications of gas stunning of pigs 3. Time to loss of somatosensory evoked potentials and spontaneous electrocorticogram of pigs during exposure to gases', *Br. Vet. J.*, 153, 329-40.

Renaudeau, D., Gourdine, J. and St-Pierre, N. (2011), 'A meta-analysis of the effects of high ambient temperature on growth performance of growing-finishing pigs', *J. Anim. Sci.*, 89, 2220-30.

Rioja-Lang, F. C., Brown, J. A., Brockhoff, E. J. and Faucitano, L. (2019), 'A review of swine transportation research on priority welfare issues: a Canadian perspective', *Front. Vet. Sci.*, 6, 36 (open access). doi: 10.3389/fvets.2019.00036.

Ritter, M. J., Ellis, M., Brinkmann, J., DeDecker, J. M., Keffaber, K. K., et al. (2006), 'Effect of floor space during transport of market-weight pigs on the incidence of transport losses at the packing plant and the relationships between transport conditions and losses', *J. Anim. Sci.*, 84, 2856-64.

Ritter, M. J., Ellis, M., Bowman, R., Brinkmann, J., Curtis, S. E., et al. (2008), 'Effects of season and distance moved during loading on transport losses of market-weight pigs in two commercially available types of trailer', *J. Anim. Sci.*, 86, 3137-45.

Ritter, M., Rincker, P. and Carr, S. (2012), 'Pig handling and transportation strategies utilized under U.S. commercial conditions', London Swine Conference, London, Canada, pp. 109-20.

Rocha, L. M., Bridi, A. M., Foury, A., Mormède, P., Weschenfelder, A. V., et al. (2013), 'Effects of ractopamine administration and castration method on the response to pre-slaughter stress and carcass and meat quality in pigs of two Piétrain genotypes', *J. Anim. Sci.*, 91, 3965-77.

Rocha, L. M., Dionne, A., Saucier, L., Nannoni, E. and Faucitano, L. (2015), 'Hand-held lactate analyzer as a tool for the real-time measurement of physical fatigue before slaughter and pork quality prediction', *Anim.* 9, 707-14.

Rocha, L. M., Velarde, A., Dalmau, A., Saucier, L. and Faucitano, L. (2016), 'Can the monitoring of animal welfare parameters predict pork meat quality variation through the supply chain (from farm to slaughter)?', *J. Anim. Sci.*, 94, 359-76.

Rodríguez, P., Dalmau, A., Ruiz-de-la-Torre, J. L., Manteca, X., Jensen, E. W., et al. (2008), 'Assessment of unconsciousness during carbon dioxide stunning in pigs', *Anim. Welf.*, 17, 341-9.

Rydhmer, L., Zamaratskaia, G., Andersson, H. K., Algers, B., Guillemet, R., et al. (2006), 'Aggressive and sexual behavior of growing and finishing pigs reared in groups, without castration', *Acta Agric. Scand., Sect. A Anim. Sci.*, 56, 109-19.

Santos, C., Almeida, J. M., Matias, E. C., Fraqueza, M. J., et al. (1997), 'Influence of lairage environmental conditions and resting time on meat quality in pigs', *Meat Sci.*, 45, 253-62

SCAHAW (Scientific Committee on Animal Health and Animal Welfare) (2011), 'Scientific opinion concerning the welfare of animals during transport', *EFSA J.*, 9, 1966.

Scheeren, M. B., Gonyou, H. W., Brown, J., Weschenfelder, A. V. and Faucitano, L. (2014), 'Effects of transport time and location within truck on skin bruises and meat quality of market weight pigs in two seasons', *Can. J. Anim. Sci.*, 94, 71–8.

Schwartzkopf-Genswein, K. S., Faucitano, L., Dadgar, S., Shand, P., Gonzàlez, L. A., et al. (2012), 'Road transportation of cattle, swine and poultry in North America and its impact on animal welfare, carcass and meat quality: a review', *Meat Sci.*, 92, 227–43.

Shen, Q. W., Means, W. J., Thompson, S. A., Underwood, K. R., Zhu, M. J., et al. (2006), 'Pre-slaughter transport, AMP-activated protein kinase, glycolysis, and quality of pork loin', *Meat Sci.*, 74, 388–95.

Sommavilla, R., Faucitano, L., Gonyou, H. W., Seddon, Y., Bergeron, R., et al. (2017), 'Season, transport duration and trailer compartment effects on blood stress indicators in pigs: relationship to environmental, behavioural and other physiological factors, and pork quality traits', *Animals*, 7, 8 (open access). doi: 10.3390/ani7020008.

Sutherland, M. A., McDonald, A. and McGlone, J. J. (2009), 'Effects of variations in the environment, length of journey and type of trailer on the mortality and morbidity of pigs being transported to slaughter', *Vet. Rec.*, 165, 13–18.

Talling, J. C., Waran, N. K., Wathes, C. M. and Lines, J. A. (1996), 'Behavioural and physiological responses of pigs to sound', *Appl. Anim. Behav. Sci.*, 48, 187–202.

Tanida, H., Miura, A., Tanaka, T. and Yoshimoto, T. (1996), 'Behavioral responses of pig to darkness and shadows', *Appl. Anim. Behav. Sci.*, 49, 173–83.

Thodberg, K., Fogsgaard, K. K. and Herskin, M. S. (2019), 'Transportation of cull sows – deterioration of clinical condition from departure and until arrival at the slaughter plant', *Front. Vet. Sci.*, 6, 28 (open access). doi: 10.3389/fvets.2019.00028.

Torrey, S., Bergeron, R., Gonyou, H. W., Widowski, T. M., Lewis, N., et al. (2013a), 'Transportation of market-weight pigs: 1. Effect of season and truck type on behavior with a 2-hour transport', *J. Anim. Sci.*, 91, 2863–71.

Torrey, S., Bergeron, R., Faucitano, L., Widowski, T. M., Lewis, N., et al. (2013b), 'Transportation of market-weight pigs: 2. Effect of season and animal location in the truck on behavior with an 8-hour transport', *J. Anim. Sci.*, 91, 2872–8.

TQA (2016), '*National Pork Board Transport Quality Assurance Handbook*', Version 6. http://www.pork.org/tqa-certification/tqa-program-materials/ (accessed on 13 May 2019).

Van de Perre, V., Permentier, L., de Bie, S., Verbecke, G. and Geers, R. (2010), 'Effect of unloading, lairage, pig handling, stunning and season on pH of pork', *Meat Sci.*, 86, 931–7.

Van der Wal, P. G., Engel, B. and Reimert, H. G. M. (1999), 'The effect of stress, applied immediately before stunning, on pork quality', *Meat Sci.*, 53, 101–6.

Van Putten, G. and Elshof, W. J. (1978), 'Observations on the effect of transportation on the well-being and lean quality of slaughter pigs', *Anim. Reg. Stud.*, 1, 247–71.

Vecerek, V., Malena, M., Malena Jr., M., Voslarova, E. and Chloupek, P. (2006), 'The impact of the transport distance and season on losses of fattened pigs during transport to the slaughterhouse in the Czech Republic in the period from 1997 to 2004', *Vet. Med.*, 51, 21–8.

Velarde, A. and Dalmau, A. (2012), 'Animal welfare assessment at slaughter in Europe: moving from inputs to outputs', *Meat Sci.*, 92, 244–51.

Velarde, A., Gispert, M., Faucitano, L., Manteca, X. and Diestre, A. (2000), 'Survey of the effectiveness of stunning procedures used in Spanish pig abattoirs', *Vet. Rec.*, 146, 65–8.

Velarde, A., Gispert, M., Faucitano, L., Alonso, P., Manteca, X., et al. (2001), 'Effects of the stunning procedure and the halothane genotype on meat quality and incidence of haemorrhages in pigs', *Meat Sci.* 58, 313–19.

Velarde, A., Cruz, J., Gispert, M., Carrión, D., Ruiz-de-la-Torre, J. L., et al. (2007), 'Aversion to carbon dioxide stunning in pigs: effect of the carbon dioxide concentration and the halothane genotype', *Anim. Welf.*, 16, 513–22.

Vermeulen, L., Van de Perre, V., Permentier, L., De Bie, S., Verbeke, G., et al. (2015), 'Sound levels above 85 dB pre-slaughter influence pork quality', *Meat Sci.*, 100, 269–74.

Vitali, A., Lana, E., Amadori, M., Bernabucci, U., Nardone, A., et al. (2014), 'Analysis of factors associated with mortality of heavy slaughter pigs during transport and lairage', *J. Anim. Sci.*, 92, 5134–41.

Warriss, P. D. (1996), 'The consequences of fighting between mixed groups of unfamiliar pigs before slaughter', *Meat Focus Int.*, 5, 89–92.

Warriss, P. D. (2003), 'Optimal lairage times and conditions for slaughter pigs: a review', *Vet. Rec.*, 153, 170–6.

Warriss, P. D. and Brown, S. N. (1985), 'The physiological responses of fighting between in pigs and the consequences for meat quality', *J. Sci. Food Agric.*, 36, 87–92.

Warriss, P. D. and Brown, S. N. (1994), 'A survey of mortality in slaughter pigs during transport and lairage', *Vet. Rec.*, 134, 513–15.

Warriss, P. D. and Wilkins, L. J. (1987), 'Exsanguination in meat animals', *Seminar Preslaughter Stunning of Food Animals*, Brussels, Belgium, pp. 150–8.

Warriss, P. D., Brown, S. N., Bevis, E. A. and Kestin, S. C. (1990), 'The influence of pre-slaughter transport and lairage on meat quality in pigs of two genotypes', *Anim. Prod.*, 50, 165–72.

Warriss, P. D., Bevis, E. A., Edwards, J. E., Brown, S. N. and Knowles, T. G. (1991), 'Effect of the angle of slope on the ease with which pigs negotiate loading ramps', *Vet. Rec.*, 128, 419–21.

Warriss, P. D., Brown, S. N., Adams, S. J. M. and Corlett, I. K. (1994), 'Relationships between subjective and objective assessments of stress at slaughter and meat quality in pigs', *Meat Sci.*, 38, 329–40.

Weeks, C. A. (2008), 'A review of welfare in cattle, sheep, and pig lairages, with emphasis on stocking rates, ventilation and noise', *Anim. Welf.*, 17, 275–84.

Weeks, C. A., Brown, S. N., Warriss, P. D., Lane, S., Heasman, L., et al. (2009), 'Noise levels in lairages for cattle, sheep and pigs in abattoirs in England and Wales', *Vet. Rec.*, 165, 308–14.

Werner, C., Reiners, K. and Wicke, M. (2007), 'Short as well as long transport duration can affect the welfare of slaughter pigs', *Anim. Welf.*, 16, 385–9.

Weschenfelder, A. V., Torrey, S., Devillers, N., Crowe, T., Bassols, A., et al. (2012), 'Effects of trailer design on animal welfare parameters and carcass and meat quality of three Pietrain crosses being transported over a long distance', *J. Anim. Sci.*, 90, 3220–4676.

Weschenfelder, A. V., Torrey, S., Devillers, N., Crowe, T., Bassols, A., et al. (2013), 'Effects of trailer design on animal welfare parameters and carcass and meat quality of three Pietrain crosses being transported over a short distance', *Livest. Sci.*, 157, 234–44.

Wesoly, R., Jungbluth, I., Stefanski, V. and Weiler, U. (2015), 'Pre-slaughter conditions influence skatole and androstenone in adipose tissue of boars', *Meat Sci.*, 99, 6067.

WQ® (2009), 'Welfare Quality assessment protocol for pigs', Welfare Quality Consortium, Lelystad, the Netherlands, 122p.

Wotton, S. B. and Gregory, N. G. (1986), 'Pig slaughtering procedures: time to loss of brain responsiveness after exsanguination or cardiac arrest', *Res. Vet. Sci.*, 40, 148–51.

Xiong, Y, Green, A. and Gates, R. S. (2015), 'Characteristics of trailer thermal environment during commercial swine transport managed under U.S. industry guidelines', *Animals*, 5, 226–44.

Zhao, Y., Lu, X., Mo, D., Chen, Q. and Chen, Y. (2015), 'Analysis of reasons for sow culling and seasonal effects on reproductive disorders in southern China', *Anim. Rep. Sci.*, 159, 191–7.

Zurbrigg, K., van Dreumel, T., Rothschild, M. F., Alves, D., Friendship, R., et al. (2017), 'Pig-level risk factors for in-transit losses in swine: a review', *Can. J. Anim. Sci.*, 97, 339–46.

Published by Burleigh Dodds Science Publishing Limited, 2021.

Part 3

Current welfare issues

Chapter 8

Evidence of pain in piglets subjected to invasive management procedures

Armelle Prunier and Céline Tallet, PEGASE, INRAE, Institut Agro, France; and Dale Sandercock, Scotland's Rural College (SRUC), UK

1 Introduction
2 The why and how of invasive management procedures in piglets
3 Definition and mechanisms of pain
4 Neural evidence of pain due to invasive management procedures in piglets
5 Hormonal and metabolic evidence of pain due to invasive management procedures in piglets
6 Behavioural evidence of pain due to invasive management procedures in piglets
7 Conclusion and future trends
8 Summary
9 Where to look for further information
10 References

1 Introduction

Welfare of animals includes both physical and mental aspects and implies that animals should be free from pain, injury or disease (FAWC, 1992). Therefore, it is crucial for the welfare of pigs to identify all management techniques that are sources of pain and apply to them the '3S' approach accounting for 'Suppress, Substitute and Soothe' (Guatteo et al., 2012). First, a painful procedure that brings no obvious advantage to the animals or the producers or only small benefits exceeded by the negative effects should be suppressed. A good example is the stopping of tail docking in dairy cows (Guatteo et al., 2012). Second, when a painful procedure is unavoidable because strong negative effects are expected from its ending, a technique highly painful should be substituted by a less painful one. Third, a pharmacological treatment should be performed to soothe the pain derived from the unavailable less painful procedure. For example, it is

http://dx.doi.org/10.19103/AS.2020.0081.08

preferable to perform dehorning of dairy cows at a very young age by cauterising the horn bud and applying a pain relief treatment (Guatteo et al., 2012).

To identify painful procedures applied to pigs, it is necessary to use biological evidence based on scientific work. A prerequisite is the definition of what is pain. In human, it is defined as 'an unpleasant sensory and emotional experience associated with actual or potential tissue damage' (Williams and Craig, 2016). From this definition, we can deduce that pain warns the individual of a tissue damage and elicits physiological and behavioural reactions to stop, recover from or prevent the damage. This wide range of reactions can be used to assess the existence of pain derived from management procedures that are commonly used in pig farms. In the present chapter, we examine these reactions in pigs submitted to tissue-damaging procedures. They are grouped into three parts: (1) neural, (2) hormonal and metabolic and (3) behavioural consequences. We focus on surgical castration, tail docking and tooth resection, which are very common in the pig industry.

2 The why and how of invasive management procedures in piglets

2.1 Surgical castration

Most male piglets are still submitted to surgical castration across Europe (De Briyne et al., 2016). The main reasons are rearing less aggressive animals, having no risk of pregnancy if housed with females and, more importantly, avoiding the off-odours and off-flavours of entire male pig meat (boar taint) (EFSA, 2004). Surgical castration of male piglets is usually performed without any anaesthesia/analgesia during the first days or weeks of age. Directive 2001/93/EC stipulates: 'if castration is practised after the seventh day of life, it shall only be performed under anaesthetic and additional prolonged analgesia by a veterinarian'. Some pig producers carry out castration at birth or the day after, together with tail docking, iron injection and, in many cases, tooth resection. Surgical castration is a very rapid process that may take less than 30 s including the time for catching animals. Piglets are restrained during castration to minimise any movement, being held between the handler's legs with the head down, held on a flat bench, or restrained in a v-trough or in a commercial device. The scrotum is incised on one or two sides, depending on the producer, with a sharp scalpel. The incision(s) in the scrotum is approximately 2 cm in length, depending on the size of testes. Additional tissue separation is realised to free each testicle from the surrounding tissue, especially the gubernaculum. The testes are extracted and usually removed by cutting the cord (funiculus spermaticus) with a scalpel or an emasculator that clamps and crimps the cord for several seconds to limit bleeding. An antiseptic is often applied to the open wound. Piglets are rapidly returned to their pen.

From what is known in mature boars or other mammals, it can be assumed that the scrotum, testes and associated organs and structures concerned by surgical castration (skin, testes, epididymes, ductus deferens, fascial and muscular contributions from the abdominal wall and skin such as tunica and fascial sheaths, blood vessels, lymphatics) are highly innervated in piglets (Setchell et al., 1994). Sensory and motor innervations (sacral and lumbar nerves) are supplied to the skin of the scrotum and to the tissues that it contains. There are also sensory sympathetic nerves that can detect pain from the testes and associated structures, and that innervate the superficial muscle of the scrotum (tunica dartos) and the blood vessels. These innervations stem from both lumbar and sacral nerves and nerve plexi (nerve groupings as an identifiable structure). There are also sensory nerves to the testes that run within the cord. Therefore, all the tissues associated with castration are highly innervated and the tissue damage caused by surgical castration is likely to generate painful stimuli.

2.2 Tail docking

Even though it should not be performed on a routine basis (2001/93/EC, 2001), most piglets in commercial European farms are tail docked in order to reduce the risk for tail biting (EFSA, 2007). Indeed, it clearly reduces the risk of tail biting even if it is not fully effective to eliminate its occurrence and measures to improve the environment must be preferred (D'Eath et al., 2016; Lahrmann et al., 2017; Thodberg et al., 2018). Tail docking is carried out with scalpels, scissors/wire cutters or by cautery with a hot iron. As a general rule, no anaesthetic nor analgesic treatments are performed to reduce the pain. The proportion of the tail that is removed by docking is variable: from only the tip of the tail to up to three-fourth of the tail, or more. Docking itself is likely to be a source of pain since the tail is innervated already in neonatal pigs. Indeed, histological observations from Simonsen et al. (1991) have demonstrated the existence of peripheral nerves to the tip of tails in 1-day old piglets. In addition, long-term pain is suspected due the development of traumatic neuromas (random proliferation of regenerating axons and glial support cells) at the tip of the docked tail (Simonsen et al., 1991; Done et al., 2003; Herskin et al., 2015).

2.3 Tooth resection

Piglets are born with an incisor and a canine tooth present in each side of the upper and lower jaw. These eight teeth are very sharp and considered as 'weapons' that can easily injure sows or other piglets (Fraser and Thompson, 1991). They are milk teeth that are spontaneously shed between 2 and 4 months of age (Hay et al., 2004). A second milk incisor grows shortly after birth on each half jaw between 2 and 32 days of age with an average age of 9.5 days for the upper jaw and 5 days for the lower jaw (Tucker and Widowski, 2009). The third milk incisor begins to grow well after birth, being present at 35 days of age on the lower jaw in about 5% of piglets (Tucker and Widowski, 2009). Milk premolars also grow in most piglets after 35 days of age, although they can be observed from birth in a minority of piglets (Tucker and Widowski, 2009).

Even though it should not be performed on a routine basis (2001/93/EC, 2001), resection of teeth is practiced in piglets shortly after their birth in many European farms in order to limit lesions intact teeth might cause to other piglets or to sows' udders, to improve maternal behaviour and growth of piglets and to reduce piglet mortality. Reduction in the number and severity of skin lesions in piglets has been repeatedly demonstrated whereas the effect on sow teat lesions is inconsistent (Fu et al., 2019; Weary and Fraser, 1999; Holyoake et al., 2004; Gallois et al., 2005; Lewis et al., 2005a). Regarding maternal behaviour, no effect was demonstrated in the first two days following farrowing by Lewis et al. (2005a), Prunier et al. (2004) or overall lactation by Fu et al. (2019), but some benefit of tooth clipping was observed (Lewis et al., 2005a) on day 4 (less suckling bouts interrupted by the sow) and days 21 or 26 (less dog-sitting interpreted as avoidance of piglets at teats). The influence of tooth resection on piglet growth and mortality varies across studies with a reduction (growth: Weary and Fraser, 1999; Bataille et al., 2002; mortality: Holyoake et al., 2004), no significant effect (growth and mortality: Gallois et al., 2005; Lewis et al., 2005b; Fu et al., 2019) or an increase (growth: Holyoake et al., 2004) in piglets with resected teeth.

Tooth resection is performed in the days following birth, usually in conjunction with other procedures such as iron injection, tail cutting and sometimes castration in male piglets. It concerns the canines and incisors (= corners) of the two jaws that are present, that is, eight teeth in total. Teeth are typically resected with cutting pliers or an electric grinder (tooth abrasion with a stone). The proportion of the tooth removed varies between teeth (the longest teeth such as the upper incisors are proportionally more shortened) and the person carrying out the procedure. For example, in French commercial farms, Gallois et al. (2005) showed that depending on the tooth, an average of 10–40% of the tooth were removed leaving on average from 2.3 mm (inferior incisive) to 4 mm (superior incisive) of tooth above the gum line. In nearly all cases, tooth resection leads to the breaching of the tooth pulp chamber and exposure of the dental pulp (Hay et al., 2004; Hutter et al., 1994) that is highly innervated. Therefore, acute pain is expected. Long-term pain is also expected due to the development of inflammation and abscesses (Hay et al., 2004; Hutter et al., 1994).

3 Definition and mechanisms of pain

3.1 Definition of pain

As previously stated, pain is an unpleasant sensory and emotional experience associated with actual or potential tissue damage, or described in terms of such damage (Williams and Craig, 2016). Pain is always a subjective experience that is influenced to varying degrees by biological, psychological and social factors. It should be recognised that the inability to communicate verbally does not negate the possibility that an individual is experiencing pain. It is generally accepted that stimuli which cause pain are liable to damage tissues. Therefore, pain is most often reported in relation to experiences associated with actual or potential tissue damage. Pain is unequivocally a sensation in a part or parts of the body, but it is also always unpleasant and therefore also an emotional experience. In humans, pain is often reported in the absence of tissue damage or any likely pathophysiological cause, usually this happens for psychological reasons. It can be difficult to distinguish the experience of psychological pain from that attributable to tissue damage. Activity induced in the nociceptor and nociceptive pathways by a noxious stimulus is not pain, which is always a psychological state, even though it is acknowledged that pain most often has a proximate physical cause. More recently it has been proposed defining pain as an aversive sensory and emotional experience typically caused by, or resembling that caused by, actual or potential tissue injury (Williams and Craig, 2016).

3.2 Pain terminology

There are several broad categories of pain, linked to function, involvement with inflammation as well as anatomical location of the tissue damage (Cervero, 2012). Table 1 lists the main types of pain. These distinctions are recognised across animal species and are central to understanding the welfare consequence of pain in animals.

3.2.1 Nociceptive pain

Nociception (from Latin nocere 'to harm or hurt') is the neural process of detecting, transducing and encoding noxious stimuli by peripheral or central

Table 1 Characteristics of distinct types of pain based on duration and presence or absences of associated tissue inflammation (adapted from Cervero, 2012)

	Nociceptive pain	Inflammatory pain	Chronic pain
Description	Short-acting pain that arises from actual or threatened damage to tissue that is due to the activation of nociceptors	Pain attributable to inflammatory processes at the site of tissue damage that resolves once tissues have healed	A state of pain that persists long after normal healing and lacks the acute warning function of physiological nociception
Biological function	To protect the organism from injury	To protect the healing tissue	Non-adaptive or maladaptive

neurons of the somatosensory system (Sherrington, 1906). The transformation of a nociceptive stimulus into pain perception involves a highly specialised pathway with both structural and functional elements (Basbaum and Jessell, 2000). This system comprises nociceptors (first-order sensory neurons) in the periphery with specific molecular properties for differential coding of noxious modalities (e.g. thermal, mechanical and chemical), ascending and descending tracts that control the input into the dorsal horn of the spinal cord as well as supraspinal processing that regulates the integration of nociceptive information with other sensory modalities and autonomic function (Dubin and Patapoutain, 2010). The sensory pathway for the neural transmission from peripheral nociceptors is shown in Fig. 1. The first-order nociceptive neurons synapse with association neurons in the dorsal horn of the spinal cord. These association neurons synapse with second-order neurons that cross to the opposite side of the spinal cord where they ascend upward (as part of the spinothalamic tract) through the brain stem to the ventral posterolateral nucleus in the thalamus, where awareness of pain begins. From this region, signals are relayed by third-order neurons to the insula, anterior cingulate cortex (ACC) and primary (S1) and secondary (S2) somatosensory cortices.

3.2.2 Inflammatory pain

In contrast to acute nociceptive pain, inflammatory pain is long-lasting and arises from inflammatory processes at the site of tissue damage (Basbaum et al., 2009). The duration of inflammatory pain is usually linked to the time taken for wound healing. As such it has a protective function during tissue healing and its duration depends on the extent of the tissue damage and the conditions of healing, including the development of an infection. A key characteristic of inflammatory pain is that the somatosensory nervous system moves to a more excitable state, which continues to last for the duration of the healing process thereby leading to amplification of neural signalling from the inflamed tissues. Indeed, responses to noxious stimuli may be enhanced (hyperalgesia) or normal innocuous stimuli may produce pain (allodynia) (Kidd and Urban, 2001).

3.2.3 Chronic/persistent pain

Chronic pain is generally defined as pain that persists past normal healing and therefore lacks the acute warning function of physiological nociception and the protective function of inflammatory pain (Treede et al., 2015). Typically pain is regarded as chronic in humans when it lasts or recurs for more than 3–6 months. Persistent pain associated with injury or disease involves changes in the nervous system that are pathological. This pain is maladaptive offering no protective or survival advantage.

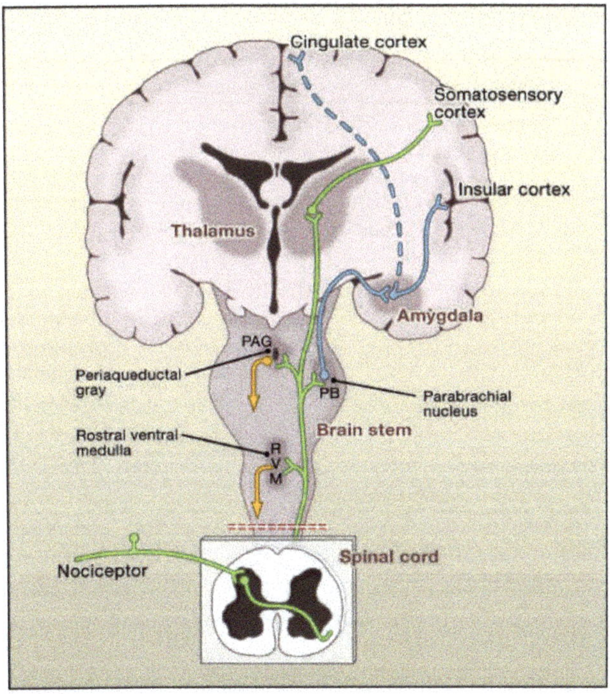

Figure 1 Anatomy of the pain pathway: primary afferent nociceptors convey noxious information to projection neurons within the dorsal horn of the spinal cord. A subset of these projection neurons transmits information to the somatosensory cortex via the thalamus, providing information about the location and intensity of the painful stimulus. Other projection neurons engage the cingulate and insular cortices via connections in the brainstem (parabrachial nucleus) and amygdala, contributing to the affective component of the pain experience. This ascending information also accesses neurons of the rostral ventral medulla and midbrain periaqueductal grey to engage descending feedback systems that regulate the output from the spinal cord. Figure reprinted from Cellular and Molecular Mechanisms of Pain (Basbaum et al., 2009) with permission from Elsevier.

3.3 Characterisation of pain based on anatomical location

Irrespective of its intensity, pain is often defined by its anatomical location which infers certain characteristics such as how pain is experienced and expressed. Typically anatomical locations are divided into the soma (e.g. skin, muscles, joints and bones), viscera (e.g. inner organs including reproductive organs) and the nervous system. The characteristics of these three types of pain are described in Table 2.

3.3.1 Somatic pain

The term 'somatic pain' is derived from the Greek *soma* meaning 'body'. Somatic pain is pain evoked by nociceptive information arising from any of the tissues

Table 2 Characteristics of distinct types of pain based on the anatomical location of tissue damage

	Somatic pain	Visceral pain	Neuropathic pain
Description	Pain arising from damage to skin, joints, muscles and bones. Dependent on activation of somatic afferent neurons	Pain form inner organs including reproductive organs. Dependent upon activation of visceral afferent neurons	Pain caused by a lesion or disease to the somatosensory nervous system
Characteristics	Well-described, easy to localise	Identification difficult to localise, diffuse. Pain felt in sites distant from source (referred pain)	Associated with sensory abnormalities (sensory deficits, dysesthesia). Exaggerated pain (hyperalgesia, allodynia), can be continual and/or episodic

that constitute the structure of the body (Murphy, 2007). This includes skin, joints, muscles, ligaments, tendons and bones, of the spine, trunk, and limbs and includes the skull, the meningeal membranes of the brain and spinal cord. More explicitly, somatic pain is used to distinguish pain that does not arise from the viscera (i.e. internal organs of the body). Somatic pain in humans is typically described as aching or stabbing and localised to the area of tissue injury or stimulation and may follow the distribution of a nerve root or peripheral nerve. A key feature of somatic pain is that it is caused by the stimulation of the nerve endings of the peripheral nerves that innervate the tissues that are the source of pain. In this way, this feature distinguishes somatic pain from neuropathic pain in which the source of pain is attributable to the affected nerves (Cervero and Laird, 1999).

3.3.2 Visceral pain

Visceral pain is, by definition, pain sensed as arising from the internal organs of the body (Cervero, 2009). It is often described as deep, dull, squeezing and sickening. Moreover, some organs are more sensitive to visceral pain than others. Diseases or disorders affecting solid organs such as the liver, lungs or kidneys are normally not associated with overt symptoms of pain per se but mainly with symptoms that are due to altered functioning of the organ itself (Giamberardino, 1999). Conversely, hollow organs such as the stomach, bladder and ureters are far more sensitive to damage and can elicit excruciating pain (Robinson and Gebhart, 2008). There can be multiple aetiologies for pain sensed in the internal organs linked to inflammation (acute and chronic), disruption of normal mechanical processes, neoplasms (benign or malignant), alterations in neurotransmission from the viscera and ischemia. Visceral pain is intriguing in that pain is commonly felt in sites distant from the location of the organ itself. This is known as referred pain and is a key feature of visceral pain. Visceral pain is often accompanied by motor and autonomic responses, such as vomiting and nausea (Cervero and Laird, 1999).

3.3.3 Neuropathic pain

The International Association for the Study of Pain (IASP) defines neuropathic pain as pain caused by a lesion or disease of the somatosensory system. It should be recognised that neuropathic pain is a clinical description (and not a diagnosis) which requires the presence of a lesion or disease that satisfies established neurological diagnostic criteria (Costigan et al., 2009). Neuropathic pain is typically subdivided into peripheral and central neuropathic pain. Peripheral neuropathic pain is attributable to lesions to the peripheral nervous system (PNS) caused by tissue trauma, diseases, chemical neurotoxicity, infection or tumours and involves multiple pathophysiological changes both within the PNS and in the CNS (Dworkin et al., 2003; Woolf and Mannion, 1999). Central neuropathic pain most commonly results from spinal cord injury, stroke or neuroimmune disease (Ducreux et al., 2006). In humans, the conventional approach in alleviating neuropathic pain has been to classify and treat it according to the underlying disease (Dworkin et al., 2003). However, this approach often fails to address a key feature of neuropathic pain, which is the manifestation of maladaptive plasticity in the nervous system. The primary disease and the neural damage it causes are only the initiators of a cascade of changes that lead to and sustain neuropathic pain (Costigan et al., 2009). Neuropathic pain is known to have severe and long-term consequences on the health-related quality of life in humans (Jensen et al., 2007). Neuropathic pain has been extensively investigated in a large number of clinical animal models of neuropathic pain (Jaggi et al., 2011). Neuropathic pain is also acknowledged as a major issue in the field of veterinary medicine and often represents an under-recognised and, therefore, untreated condition in veterinary patients (Moore, 2016).

3.4 Anatomy and physiology of the pig pain system

3.4.1 Somatosensory nervous system

The overall anatomy and physiology of the somatosensory system essential for the detection, perception and modulation of pain is common to most mammalian species, and is well described in pigs. The pig is a key model species in biomedical research as pigs have close anatomical, physiological and neurological similarities to humans (Rukwied et al., 2008; Castel et al.,

2016; Saalfrank et al., 2016; Rice et al., 2018). As previously discussed, primary nociceptive neurons convey information to projection neurons within the dorsal horn of the spinal cord that then transmit information to the somatosensory cortex via the thalamus, providing information about the location and intensity of the painful stimulus. Other projection neurons engage the cingulate and insular cortices via connections in the brainstem and amygdala, contributing to the affective component of the pain experience (Basbaum et al., 2009). In the first instance, the detection of actual or potential tissue damage is dependent on two specific types of primary sensory neurons, namely mylinated A-delta (Aδ) and unmyelinated C fibres (Julius and Basbaum, 2001). These fibres are found in various somatic tissues (see Section 3.3.1). The two fibre types are characterised by their conduction properties, which relate to axonal diameter and degree of myelination (Dubin and Patapoutain, 2010). Both properties affect nerve conduction velocities and pain quality and determine the perception of 'first' sharp (Aδ) and 'second' slow, dull (C-fibre) pain. Nociceptors are typically silent and transmit all-or-none action potentials only when sensitised (Gold and Gebhart, 2010).

An increasing number of studies have demonstrated that neural pain pathways in pigs are very similar to those in humans (Lynn et al., 1995, 1996; Rukwied et al., 2008, 2010; Obreja and Schmelz, 2013; Castel et al., 2014, 2016). In terms of their relative Aδ and C-fibre distribution and functional properties, there is strong homology between human and porcine nociceptive neurons (Obreja and Schmelz, 2010, 2013; Obreja et al., 2010; Kozłowska et al., 2017)

3.4.2 Autonomic nervous system (ANS)

The processing of pain in mammalian species is not exclusive to just the somatosensory system. The autonomic nervous system (ANS) has a key regulatory function in maintaining homeostasis by adapting and adjusting physiological processes following any disruption, such as a painful event. The ANS contributes to the modulation of pain pathways through excitatory sympathetic and inhitory parasympathetic mechanisms. Afferent nociceptive and efferent autonomic systems can interact in many ways (Schlereth and Birklein, 2008). Pain is regarded as a stressor, and as such activates the sympathetic nervous system; such activation can suppress pain via cortical pain control and descending inhibition of pain (Millan, 2002; Tracey and Mantyh, 2007), or in pathological states serve to exacerbate pain (Torebjork et al., 1995; Ali et al., 2000). Changes in heart rate and blood pressure reflecting ANS responses to painful events (e.g. surgical procedures) are routinely carried out in human patients as a way of monitoring and ensuring an appropriate level of anesthesia (Bantel and Trapp, 2011). Other indirect measurements of ANS responses to pain include changes in skin temperature, facial flushing, salivation, sweating, respiratory rate, piloerection, pupillary dilation and nausea, vomiting and bladder or bowel evacuation (Benarroch, 2001). Indirect assessment of sympathetic responses to painful management practices have been investigated in pigs and are discussed in more detail in Section 5.

3.5 Pain in neonates

Insensitivity to pain in very young animals compared to older animals has historically been used as an argument to justify carrying out painful procedures during this period of development. However, the weight of scientific evidence to the contrary makes this justification factually and ethically questionable even though variation with age in sensitivity to nociceptive stimuli may exist as suggested by a lack of electroencephalographic (EEG) response to an acute nociceptive stimulus in 1-day-old pigs whereas more typical responses to nociceptive stimuli were observed at 10 days of age and intermediate ones between 5 and 7 days of age (Kells et al., 2019 and section 4.2).

Processing of nociceptive stimuli requires functional peripheral sensory receptors, afferent and efferent sensory and motor pathways and subcortical and cortical neural integration of the related impulse traffic. Pigs as well as cattle, sheep, goats and horses are considered as precocial species, that is, their foetuses are neurologically matured at the time of birth even though brain continues to develop after birth (EFSA, 2017). Indeed, when pigs with normal birth weight were submitted to a battery of neurological tests the day after birth, they successfully completed (most) proprioception tests showing postural reactions to maintain their balance when challenged and nearly all of them displayed a withdrawal reflex after application of a noxious stimulus on the hind limb (Roelofs et al., 2019). It has been widely reported there are several placental-derived suppressors that act in utero to inhibit neural activity (awareness) in the foetus (Mellor et al., 2005). These sleep-inducing placental suppressors include adenosine, progesterone-neuroactive metabolites - allopregnanolone and pregnanolone, prostaglandin D2 and neuropeptide Y. The actions and the availability of these placental neuro-inhibitors to the newborn are clear before and during parturition with little or no carry over into post-partum period (Mellor et al., 2005).

4 Neural evidence of pain due to invasive management procedures in piglets

4.1 Castration

Electroencephalographic (EEG) responses to painful stimuli are similar in conscious and anesthetised humans and animals (Murrell et al., 2003). Electrical activity in the brain of piglets undergoing surgical castration under isoflurane anaesthesia is modified as shown by variations in the EEG signal (Haga et al., 2001; Haga and Ranheim, 2005). These variations, especially a decrease in the absolute theta power, are less marked in piglets receiving local anaesthesia with lidocaine before surgery.

The expression of the protein c-fos in neurons of the spinal cord, which are likely to transmit the nociceptive stimuli originating from the perineal region to the brain, has been studied in pigs submitted to surgical castration (Nyborg et al., 2000). It was shown that the number of activated neurons was three times lower in pigs treated with a local anaesthetic before castration than in pigs receiving an injection of saline.

Recently, neuroanatomical studies by Bengtsdotter et al. (2019) investigating the effects of castration in horses have demonstrated the presence of traumatic neuromas in the remnant testicular nerves at the site of castration. It is suggested that inguinal pain, unexplained hind limb lameness, back pain and behavioural issues in geldings may be attributable to painful traumatic neuromas that develop as a consequence of crushing and severing the testicular nerves during castration. Evidence of the presence of traumatic neuromas in the testicular nerves at the site of castration has yet to be confirmed in pigs.

4.2 Tail docking

Changes in EEG median frequency (F50) and 95% spectral edge frequency (F95), both indices of nociception, have been reported in lightly anaesthetised piglets undergoing tail docking by side cutting pliers or cautery iron at 2 or 20 days of age (Kells et al., 2017a,b). Following tail docking, F50 decreased briefly in 2-day-old piglets, but no differences in the reduction in the F50 in relation to the docking method were observed. Reduction in F50 was overall lower in 2-day-old docked piglets compared to those docked at 20 days of age. While reductions in F50 were observed in 2-day-old docked piglets, no significant changes in F95 or EEG total power (P_{TOT}) were observed. By contrast, pigs docked at 20 days of age exhibited sustained increases in P50 and decreases in P_{TOT}, consistent with expected response to nociceptive pain. Overall the data reported in this and in a more recent publication (Kells et al., 2019) imply that docking within the first days of birth may be less acutely painful than docking at a later age and that tail docking using clippers may be more painful than using a cautery iron. In addition there may be developmental qualitative differences in pain perception based on evidence of an increase in cortical responsiveness to tail docking injury with increasing age.

It has long been recognised that post-operative pain due to tissue injury, and the development and maintenance of on-going chronic pain (including neuropathic pain) is associated with marked changes in gene and associated peptide expression in affected neural and non-neural tissues (Hökfelt et al., 1994; Colburn et al., 1997). In a recent study by Sandercock et al. (2019), transcriptomic analysis of caudal dorsal root ganglia neurons from two groups of pigs docked at 3-days of age in line with commercial practice or at 63-days of age (a surgical model of tail biting in later life) revealed significant and sustained changes (4 months after tail amputation injury) in 185 inflammatory and neuropathic pain-associated genes. Gene ontology (GO) enrichment analysis of gene expression clusters identified gene family members of ions channels (e.g. voltage-gated potassium channels) and receptors (e.g. GABA receptors) that were significantly downregulated in both tail amputation groups. Neuronal membrane potential is largely determined by potassium channels and reducing or losing these channels can account for changes in neuronal excitability linked to ongoing pain (Ishikawa et al. 1999; Kim et al., 2002; Ocana 2004; Tsantoulas et al., 2012; Pollema-Mays et al., 2013). In addition, reductions in GABAergic inhibitory control can lead to a sustained increase in pain sensitivity (Fukuoka et al., 1988; Moore et al., 2002; Torsney and MacDermott, 2006). It has been proposed that the broader functional consequences of the down-regulation of these key pain neuromediators serve to increase and maintain peripheral neuronal excitability after tail amputation injury (Sandercock et al., 2019). This assertion is supported by histological evidence of traumatic neuroma development after tail amputation injury (Sandercock et al., 2016) and the presence of long-term mechanical sensitisation in tail stumps in tail-amputated pigs (Di Giminiani et al., 2017a).

4.3 Tooth resection

Neural evidence of the painful consequences of tooth resection in pigs is lacking in the scientific literature, although it is widely reported that traumatic dental injury in humans and other species (e.g. monkey, cat, dog) involving loss of pulp cavity integrity and dental pulp exposure (with or without nerve injury), activation of and infiltration by inflammatory cells and abscess formation are associated with severe dental pain (see review by Byers and Narhi, 1999). Similar features of traumatic tooth injury have been reported by Hay et al. (2004) in piglets subjected to tooth shortening by clipping and grinding, and it is therefore highly likely these procedures cause intense pain in piglets too. Recent preliminary studies by Sinclair et al. (2018) investigating gene expression in the dental pulp of needle teeth shortened by clipping and grinding have demonstrated sustained up-regulation (still evident 6 weeks after tooth injury) of the pro-inflammatory chemokine CXCL8 in exposed dental pulp by both methods, but found no difference between the method used. CXCL8 is involved in the NFκB signalling pathway, resulting in monocyte chemotaxis towards the site of injury (Enzerink et al., 2009). Its overexpression and its cognate receptor are critical in the development of neuro-inflammation and the establishment of on-going pain (White et al., 2005). Data from Sinclair et al. (2018) support evidence of a prolonged localised inflammatory state in the tooth pulp induced by both methods of tooth resection.

5 Hormonal and metabolic evidence of pain due to invasive management procedures in piglets

Pain is a powerful stressor that acts on the brain and stimulates the release of hormones from the hypothalamo-pituitary-adrenal (HPA) and sympathetic axes (Chapman et al., 2008), among them CRH (corticotropin-releasing hormone from the hypothalamus), ACTH (adrenocorticotropic hormone from the pituitary), cortisol (from the adrenal cortex) and catecholamines (from the adrenal medulla or neurons from the brain or autonomic nerves). These hormones/neuromediators stimulate nutrient release (e.g. glucose, lactate and free fatty acids), increase heart and respiratory rates, favour constriction of blood vessels in parts of the body or vasodilatation in others such as muscles, brain, lungs and heart. As a consequence, nutrient and oxygen supply to organs involved in fighting or fleeing behaviours is increased whereas it is decreased in other parts of the body. Another possible consequence is a decrease in peripheral skin temperature (Busnardo et al., 2010). Tissue damage activates the immune system and the release of numerous inflammatory mediators which, in turn, may activate the adrenal axis (Turnbull and Rivier, 1999; Chapman et al., 2008). For example, the pro-inflammatory cytokine interleukin-1, released by immune cells after a tissue damage, stimulates the release of ACTH and cortisol, and plays an active role in transmitting nociceptive stimuli to the brain (Turnbull and Rivier, 1999; Ren and Torres, 2009). Therefore, any sign of activation of the adrenal and sympathetic axes after application of a farm procedure can be used to suspect the existence of pain. However, to really prove it, it is necessary to check that the physiological response is higher after applying the true procedure than after a sham procedure which only includes the handling stress and that it is reduced by the use of an analgesic treatment (e.g. anaesthetic treatment and/or a non-steroidal anti-inflammatory drug = NSAID).

5.1 Castration

Compared to the pre-surgical period, heart rate increases in piglets castrated between 8 and 24 days of age but this increase is abolished by pre-treatment with the local anaesthetic lidocaine (White et al., 1995). However, castration and lidocaine effects are not seen in piglets castrated between 1 and 4 days of age probably because the heart rate is naturally very high in newborns and close to the maximum possible one. A very rapid and transient increase in plasma adrenaline followed by a longer-lasting increase in plasma noradrenaline is also present (Prunier et al., 2006, see Fig. 2). Adrenaline is probably of adrenal medullary origin and noradrenaline from peripheral sources. As a consequence of the catecholamine stimulation, muscle glycogen is mobilised, leading to a transient increase in plasma lactate (Prunier et al., 2005). In addition, skin temperature around the groin area is decreased immediately after castration but not 20 min later (Bonastre et al., 2016).

Shortly after surgical castration, measurement of hormones in plasma clearly indicates an activation of the adrenal axis (Carroll et al., 2006; Prunier et al., 2005; Prunier et al., 2006; Marchant-Forde et al., 2014; Sutherland et al., 2012). A 40-fold increase in plasma ACTH, peaking 5 min after surgery, is followed by a 3-fold increase in plasma cortisol, peaking from 15 to 30 min after surgery (Prunier et al., 2005, see Fig. 3). A major proportion of cortisol circulates in blood being bound to the cortico-binding globulin (CBG) that protects cortisol from being metabolised and avoids excessive action of cortisol on target tissues. By measuring cortisol and CBG, Carroll et al. (2006) calculated the free cortisol index (FCI) and observed that it is increased at 0.5 and 1.5 h after castration. Also, data from Carroll et al. (2006) indicate that plasma levels of cortisol and FCI are no longer elevated between 24 and 48 h after castration. If local anaesthesia with lidocaine is performed before surgical castration, the duration and amplitude of the post-surgical peaks of ACTH and cortisol are lower (see Fig. 3). However, when piglets are not catheterised and are sampled only once or twice in the first two hours after surgical castration, no reduction in the post-surgical peaks of ACTH or cortisol by local anaesthesia under lidocaine or procaine could be clearly demonstrated (Courboulay et al., 2010b; Kluivers-Poodt et al., 2012; Zols et al., 2006; Bonastre et al., 2016). When a NSAID such

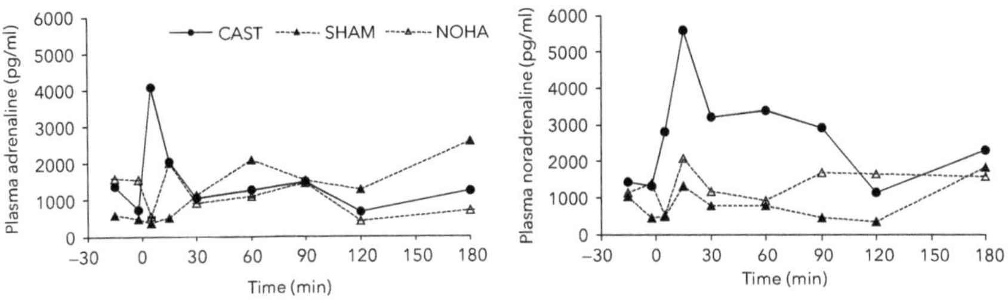

Figure 2 Comparison of neuroendocrine responses to surgical castration at time 0 without anaesthesia (CAST), sham-castration (SHAM) and no-handling (NOHA) in pigs castrated at 7–8 days of age (Prunier et al., 2006). Figure reproduced with permission from Universities Federation for Animal Welfare.

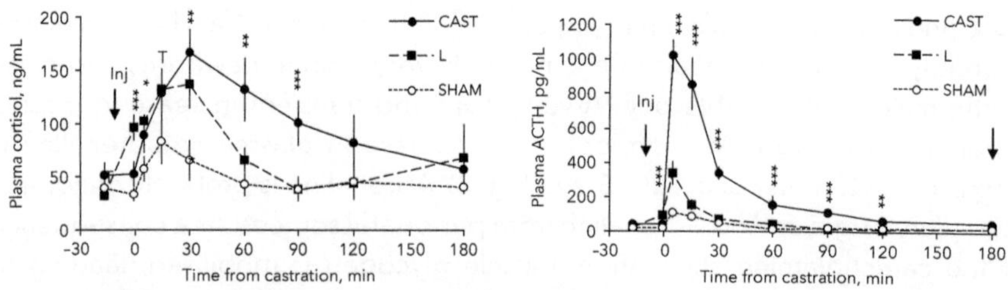

Figure 3 Plasma cortisol and ACTH in 7–8 day-old pigs around castration. They were submitted either to surgical castration (CAST, n = 9), castration + local anaesthesia (L, ~7.5 mg lidocaine/kg, 1/3 intratestis, 2/3 within the scrotal sac around the funiculus spermaticus injected 15 min before castration group L, n = 7), or sham-castration (SHAM, n = 7). Within the time of sampling, significant differences between treatments are indicated by stars (*** $P < 0.001$, ** $P < 0.01$, * $P < 0.05$, T $P < 0.07$) (Prunier, unpublished data).

as meloxicam, ketofen, or flunixin is injected intramuscularly before castration, the post-surgical increase in cortisol is dampened in numerous studies (e.g. in Courboulay et al., 2010a; Courboulay et al., 2010b; Langhoff et al., 2009; Sutherland et al., 2012; Bonastre et al., 2016), even if it is not always the case (Kluivers-Poodt et al., 2012; Bonastre et al., 2016). In piglets receiving meloxicam via the milk (sows were administered orally meloxicam), there was also no clear effect of meloxicam on cortisol measured 1 h after castration (Bates et al., 2014).

5.2 Tail docking

In 3-day-old piglets, an increase in heart rate was observed during docking by iron cautery but it was not accompanied by a decrease in peripheral skin temperature near the ear root (Fu et al., 2019). In 1-day-old piglet, no increase in plasma lactate was observed from 2 to 180 min after tail docking by iron cautery (Prunier et al., 2005). In the same study, no clear changes in plasma profiles of cortisol and ACTH were depicted. However, in a more recent experiment using 2- or 5-day-old piglets that were sampled once at 30 min after treatment, Courboulay et al. (2015) observed an increase in plasma cortisol in docked (hot iron) compared to sham-handled piglets and this increase was slightly mitigated by meloxicam injection before treatment. Using piglets docked at 6 days of age, Sutherland et al. (2008) also observed an increase in plasma cortisol at 60 min after docking with pliers but surprisingly not with hot iron cautery. In a more recent experiment, using piglets submitted to tail docking with pliers at about 3 days of age, the same authors observed higher plasma cortisol in docked than in sham-docked piglets at 30 min, but not at 60 min, after treatment.

From these results, it can be concluded that tail docking induces an activation of the HPA and sympathetic axes but of moderate amplitude, lower than after surgical castration. As a consequence, variations in physiological indicators are not always depicted after applying the practice.

5.3 Tooth resection

In 3-day-old piglets, an increase in heart rate and a decrease in peripheral skin temperature was observed during clipping teeth to the gum line with cutting pliers (Fu et al., 2019). In 1-day-old piglets, plasma concentrations of cortisol, ACTH and lactate were not affected by tooth resection with cutting pliers or an electric grinder from 2 min to 180 min after the procedure (Prunier et al., 2005). Similarly, Marchant-Forde et al. (2009) did not find a cortisol increase in 2-3-day-old piglets, sampled at 45 min, 4, 48 h, 1 and 2 weeks after tooth resection with both methods. Tooth resection was less severe in the studies from Prunier et al. (2005) and Marchant-Forde et al. (2009) leaving some millimetres of tooth above the gum line.

Overall, signs of activation of the HPA and sympathetic axes are scarce but the number of studies evaluating the physiological consequences of only tooth resection is very low.

6 Behavioural evidence of pain due to invasive management procedures in piglets

Numerous postural and behavioural indicators of pain have been described in mammals (Prunier et al., 2013; Mellor et al., 2000). They can be classified into five main categories. Three of them aim directly or indirectly to avoid or alleviate the painful stimulus: (1) avoidance and defensive behaviours in order to escape the nociceptive stimulus, (2) behaviours directed towards the painful areas with alleviating effects, (3) postures and behaviours aiming at reducing stimulation of the painful area. The fourth category is vocalisations that have a wide range of possible aiming to keep others away or, on the contrary, to attract them, in search of social support. The last category is related to general changes in activity, being motionless or agitated, feeding and drinking less, avoiding social contacts and reducing grooming behaviours. These changes are largely similar to those observed during sickness that were shown to highly depend on the stimulation of the immune system, especially on cytokine release (Dantzer et al., 2008). Based on apparent evolutionary conservation of facial expressions and on consistent association between pain and facial expression in human (Prkachin, 1992), it was demonstrated that facial expression can also be used to assess pain in rodents (e.g. in mice: Langford et al., 2010) and in farm animals (e.g. in horses: Dalla Costa et al., 2014). The methodology in pigs can be more (Di Giminiani et al., 2016) or less (Viscardi et al., 2017) complicated and consists in observing fine changes in facial muscle tensions around the nostrils, eyes, lips, ears and/or cheeks. The quality of the pictures used for the evaluation is crucial for a good assessment.

All these behavioural and postural changes can be used to suspect the existence of pain. As for physiological indicators, the level of evidence is highly strengthened when it is demonstrated that the behavioural consequences of the procedure are higher than after applying a sham-procedure and that they are reduced by the use of an analgesic treatment.

6.1 Castration

During castration, most piglets vocalise. High-frequency calls (> 1000 Hz) are attributable, at least in part, to the surgery of the animals because they are more frequent, of higher intensity and frequency in castrated than in sham-castrated pigs (Marx et al., 2003; Taylor and Weary, 2000; Weary et al., 1998; Hansson et al., 2011; Marchant-Forde et al., 2009). Marx et al. (2003) identified three types of vocalisations during castration: grunts, squeals and screams and showed that the rate of screams is highly increased (about 2–3 times) in surgical-castrated compared to sham-castrated piglets. Using an automatic system of detection of stress calls (STREMODO: Schon et al., 2001, 2004), numerous authors demonstrated that the occurrence of stress calls is higher in surgical-castrated than in sham-castrated piglets when the treatment is applied (Leidig et al., 2009; Sutherland et al., 2010, 2012, 2017; Backus and McGlone, 2018). Comparing the rate of high-frequency calls (> 1000 Hz), Taylor and Weary (2000) showed that it was highest during testis extraction and severing the spermatic cord and suggested that this part of the procedure is the most painful. Comparing the two methods of severing the cord, cutting with a sharp scalpel or by pulling/tearing, Taylor and Weary (2000) did not show any difference in the number of high-frequency calls. When a local anaesthesia with lidocaine was applied a couple of minutes before surgical castration, the rate of screams (Marx et al., 2003), of stressed calls (Leidig et al., 2009), the loudness of the vocalisations (Hansson et al., 2011; Kluivers-Poodt et al., 2012; Courboulay et al., 2010b) and their pureness (Kluivers-Poodt et al., 2012) were lower during the surgery.

In addition to vocalisations, surgical castration induced more defence movements (movements of legs often accompanied by body movements) and tail flicks (Leidig et al., 2009; Marchant-Forde et al., 2009; Sutherland et al., 2017; Courboulay et al., 2010b, 2010a) than sham-castration during treatment application. If local anaesthesia was applied by intra-testicular injection of lidocaine or procaine at least 3 min before castration, these movements were reduced (Horn et al., 1999; Leidig et al., 2009; Hansson et al., 2011; Courboulay et al., 2010b). As a consequence, surgical castration was easier to perform and took less time to be performed (Courboulay et al., 2010b).

During the first hours following castration, castrated pigs spent less time at the mammary glands, massaging and/or suckling (McGlone and Hellman, 1988; Hay et al., 2003; McGlone et al., 1993) (see Fig. 4). They were more inactive while awake (Hay et al., 2003), explored less (Courboulay et al., 2010b), showed more pain-related behaviours (prostration, stiffness, trembling, spasms, huddled-up and scratching the rump) (Kluivers-Poodt et al., 2013; Hay et al., 2003; Llamas Moya et al., 2008) and tail wagging (Hay et al., 2003; Viscardi and Turner, 2018; Wemelsfelder and van Putten, 1985). However, postures (ventral and lateral lying, sitting and standing) and location in the crate (at the sow's udder or sow's back, at heat lamp) were not altered (Hay et al., 2003; Llamas Moya et al., 2008). Castrated pigs were frequently isolated, being more frequently far from other littermates (Hay et al., 2003) or lying less frequently in contact with the sow and/or littermates (Llamas Moya et al., 2008; Sutherland et al., 2017, 2012). Behaviour seemed more often desynchronised than in their littermates (Hay et al., 2003). Using facial expression analysis to assess pain, Viscardi and Turner (2018) showed that castrated pigs had a higher grimace score than uncastrated ones but concluded that the method was less effective to detect pain in castrated pigs than the classical observation of pain-related behaviours (trembling, stiffness, spasms, tail wagging, rump scratching). When a local anaesthetic was applied, lying in isolation was decreased to the level of sham-castrated pigs (Sutherland et al., 2017) or close to it (Courboulay et al., 2010b). However, no clear effect of local anaesthesia was observed on pain-related behaviours (Sutherland et al., 2017; Kluivers-Poodt et al., 2013). When piglets were treated with NAIDs before surgery, isolation (Courboulay et al.,

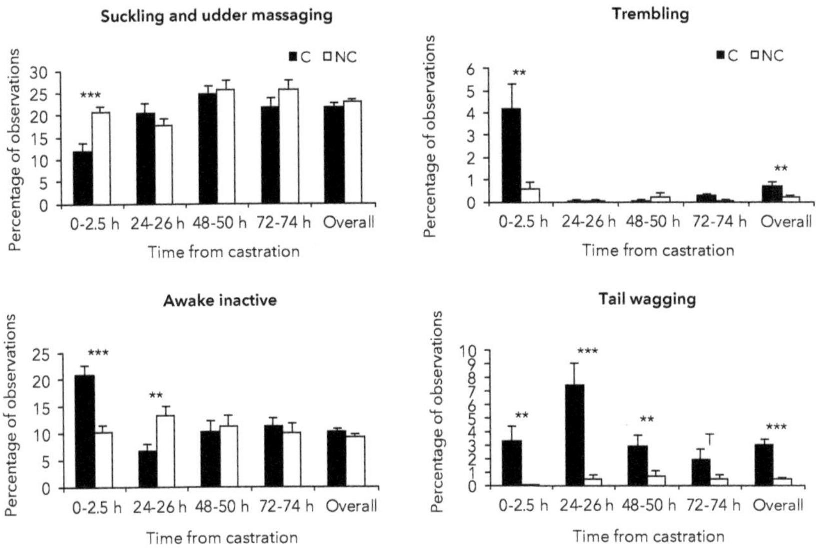

Figure 4 Comparison of behaviour in castrated and non-castrated piglets at different periods following castration (mean ± SEM; *** $P < 0.001$, ** $P < 0.01$, * **P** < 0.05; T P < 0.1) redrawn from Prunier et al. (2006) with permission from Universities Federation for Animal Welfare.

2010b) and some other pain-related behaviours (Kluivers-Poodt et al., 2013; Keita et al., 2010) were reduced but not all (Courboulay et al., 2010b; Kluivers-Poodt et al., 2013) and some authors could not demonstrate clear effect despite thorough evaluation (Viscardi and Turner, 2018; Courboulay et al., 2010a).

The day after surgery, walking was decreased (Hay et al., 2003) whereas scratching the rump and tail wagging (Hay et al., 2003; Courboulay et al., 2010b) were increased in castrated compared to uncastrated pigs. Some of these alterations in behaviour were observable until 3 days after castration: less walking, more scratching the rump and tail wagging (Wemelsfelder and van Putten, 1985; Hay et al., 2003). In addition, dog-sitting was decreased on the third day after castration (Llamas Moya et al., 2008). Increased tail wagging after surgical castration was attenuated by local anaesthesia with lidocaine but fully suppressed by ketofen treatment whereas increased rump scratching was influenced only by ketofen treatment with a full reversion (Courboulay et al., 2010b). The influence of surgical castration on combined pain-related behaviours was partly reversed by treatment with meloxicam but fully reversed when local anaesthesia was also applied (Hansson et al., 2011). However some authors failed to demonstrate any influence of surgical castration and pain treatment on behaviour the day after surgery despite thorough evaluation (Kluivers-Poodt et al., 2013).

In general, behavioural alterations after castration are of low or moderate amplitude but allow a reduction in the stimulation of the painful area by a direct effect (e.g. more huddling, less locomotion and dog-sitting) or by the avoidance of littermates (e.g. isolation and desynchronisation). They are more marked on the day of castration but some alterations are still visible 3 days later. The increase of scratching the rump seems paradoxical but this behaviour may inhibit the activation of nociceptive receptors through the simultaneous activation of mechanoreceptors as suggested by Hay et al. (2003).

6.2 Tail docking

During the intervention, the number and rate of high-pitched vocalisations (including screams and squeals) as well as their loudness were higher in pigs docked by cutting pliers or hot cautery than in sham-docked ones (Prunier et al., 2001; Backus and McGlone, 2018; Marchant-Forde et al., 2009; Tallet et al., 2019; Courboulay et al., 2015; Noonan et al., 1994; Herskin et al., 2016). On their side, Di Giminiani et al. (2017b) did not find a difference in the call rate, the maximum and the mean frequency of the calls, but demonstrated higher mean call energy and loudness in docked than in sham-docked pigs and a lower peak-to-peak pressure level in docked than in sham-docked pigs with a stronger effect in cold than in cautery-docked pigs. The percentage of piglets squealing during docking increased with the severity of docking and was decreased by local anaesthesia (subcutaneous injection of lidocaine 15 min before, see Fig. 5) but not by NSAID treatment (intramuscular injection of meloxicam 45 min before, see Fig. 5). More defence movements and attempts to escape were also observed in docked than in sham-docked pigs (Prunier et al., 2001; Marchant-Forde et al., 2009; Tallet et al., 2019; Courboulay et al., 2015; Herskin et al., 2016). These vigorous startling movements of body and one or more limbs were not influenced by the severity of docking (Herskin et al., 2016). They were reduced by local anaesthesia but not by NSAID (Herskin et al., 2016). Similarly, Courboulay et al. (2015) did not find a clear effect of NSAID treatment on piglet movements.

Just after treatment, while piglets were isolated in a cart, tail was seen more often immobile and ears more perpendicular to the head and moving in docked than in sham-docked pigs (Tallet et al., 2019). During the same period, grunts and high-pitched vocalisations (Noonan et al., 1994), rapid tail movements(= tail wagging) (Noonan et al., 1994; Prunier et al., 2001) and the tail pressed against the body (Prunier et al., 2001) were more frequent in docked than in sham-docked pigs. In the following minutes and hours, some behavioural differences were depicted: more tail trembling and less ample tail movements (Courboulay et al., 2015), more pain behaviours defined as scooting plus huddling and tail jammed

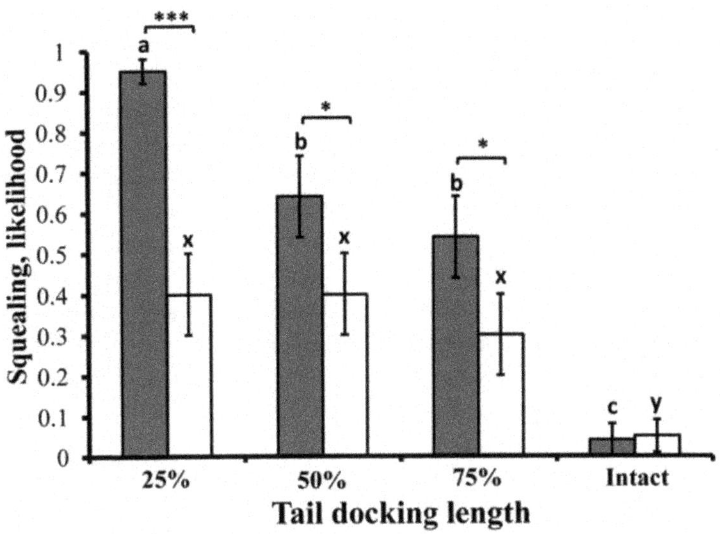

Figure 5 Influence of the severity of tail docking (25% vs. 50% vs. 75% of the tail were left) on the likelihood of piglet squealing during the procedure when either injected with lidocaine (open columns) or not injected (dark columns). Asterisks mark the significant effects of lidocaine injection within docking length treatment. Different letters a, b and c indicate significant differences ($P < 0.05$) for the non-injected pigs at the different docking lengths, whereas the letters x and y indicate the significant difference ($P < 0.05$) between docking lengths within the lidocaine-treated piglets (Herskin et al., 2016). Figure reproduced with permission from Elsevier.

between legs (Backus and McGlone, 2018), more lying (Tallet et al., 2019) and a preference for lying in the heated creep area (Herskin et al., 2016) were observed in docked than in sham-docked pigs. When pain alleviation was tested, no effect of local anaesthesia or NSAID was depicted (Herskin et al., 2016).

On the next day and the following ones, some differences were still depicted: more tail trembling and less ample tail movements (Courboulay et al., 2015) and a preference for lying in the heated creep area, more time lying alone and less time fighting/playing (Fu et al., 2019), the tail being more often immobile and in a horizontal position, and more behavioural reaction to tail touching (e.g. escape, vocalisation, aggression) (Tallet et al., 2019) were observed in docked than in sham-docked pigs. However, no effect of local anaesthesia of treatment with NSAID or both treatments combined could be detected (Courboulay et al., 2015; Herskin et al., 2016).

Tail sensitivity was explored by measuring the behavioural evoked response to mechanical and/or cold stimulation. In pigs docked with surgical cutters during the first week of age and tested close to the root of the tail at about 4–5 and 8 weeks of age, no effect of docking was depicted (Sandercock et al., 2011). Similar results were obtained with pigs docked by heat cautery and tested at the tip of the remaining tail at about 17 weeks of age, regardless the severity (one- or two-third of the tail removed) of docking (Di Giminiani et al., 2017b). In an experimental model of biting, tail amputation was performed at 9 or 17 weeks of age and animals showed signs of hypersensitivity at the tail tip at 1, 8 and 16 weeks after docking, regardless of the severity of docking (Di Giminiani et al., 2017a). Therefore, older animals are more prone to develop hypersensitivity when tail is amputated. However, the existence of hypersensitivity cannot be excluded in early docked animals (Di Giminiani et al., 2017b).

Overall, literature provides clear indications of procedural pain due to tail docking based on behaviour. Procedural pain seems more acute with cold cutting than with hot cautery and when the severity of docking is increased. After surgery, behavioural alterations are of low or moderate amplitude but seem to persist at least for 3 weeks after docking (e.g. tail posture and movement in Tallet et al., 2019). However, it should be mentioned that significant behavioural differences between docked and sham-docked pigs were not always depicted (e.g. in Sutherland et al., 2012) and when some behavioural differences were depicted, numerous other behavioural measures did not differ between groups (e.g. isolation and prostration in Courboulay et al., 2015).

6.3 Tooth resection

During treatment, squeal rate was lower in processed pigs than in sham ones when tooth resection was performed by grinding with a rotary grinder

(Marchant-Forde et al., 2009) whereas it was similar when tooth resection was performed with clippers (Marchant-Forde et al., 2009; Noonan et al., 1994). However, using vocalisations as a tool to evaluate pain during tooth resection is questionable as the process itself surely modifies the possibility to vocalise and vocal features (Marchant-Forde et al., 2009). More escape/defence movements per second were observed in processed pigs than in sham ones regardless of the method of tooth resection (Marchant-Forde et al., 2009).

Just after treatment, clipped piglets showed more teeth champing (movement of the jaws against each other, mouth empty) than sham processed ones (Bataille et al., 2002; Noonan et al., 1994; Lewis et al., 2005b; Sinclair et al., 2019), but the difference was less marked in ground piglets (Bataille et al., 2002) and the difference between ground and sham treatments was even not significant in two studies (Lewis et al., 2005b; Sinclair et al., 2019). However, it should be mentioned that in the two last studies the resection was more severe with clipping than with grinding (resection to the gum line vs. the tip of the teeth in Lewis et al. (2005b), about 1/3 versus 1/4 in Sinclair et al. (2019)). While isolated, clipped piglets showed less wood-shaving exploration than sham ones and ground piglets were intermediate (Sinclair et al., 2019).

During the first 10 min (Bataille et al., 2002) or 30 min (Lewis et al., 2005b) after tooth resection, the only significant difference that was depicted between processed and sham-processed pigs was less time active on a heatpad in processed pigs (Lewis et al., 2005b). Bataille et al. (2002) prolonged the observations for 12 h looking at the posture of the piglets, suckling and teat massaging behaviours but did not find any significant difference between treatments. More detailed observations by Lewis et al. (2005b) at 1, 4, 8, 14 and 21 days after the intervention showed some variations in the percentage of active piglets but these were minor and fluctuated from one stage to another. Contrarily, with a severe tooth resection to the gum line, Fu et al. (2019) showed that clipping increased the time that piglets spent by lying alone, and decreased the time spent by playing and fighting compared to sham procedure with observations at 1, 2, 3, 7, 14 and 21 days of age.

In total, the behavioural signs of pain after tooth resection are limited, but it should be remembered that it is based on only a few studies and that pigs are a 'prey' species that show little pain rendering its detection difficult.

7 Conclusion and future trends

Pigs suffer from acute pain during the process of castration, tail docking and tooth resection as well as shortly after. Pain lasts probably for several days, or even weeks after the practice is applied even if external signs are not always detected. Damage to peripheral nerves due to the application of these procedures can cause long-term pain lasting months after injury. This chronic pain has still been little studied in pigs and more data are needed for it to be more recognised and hence taken into account. Omics technics should be used to identify markers of chronic pain along the pathways of transmission of pain signals. In addition, consequences of chronic pain on learning, memory and other cognitive capacities and on the adaptation to rearing conditions should be investigated to eventually show that not only pain is not acceptable, but pain has probably co-lateral negative consequences on the physiological and behavioural development of the young. The existence of negative consequences of painful procedures on the quality of the human-animal relationship has already been pointed out but more results are needed to ensure the phenomenon.

Efforts should already be made to avoid these invasive management procedures. To achieve that, the environment of the pigs should be improved to better meet their behavioural needs. This is particularly crucial for avoiding tail docking. In addition, genetic lines should be modified either by choosing to rear existing lines that are less at risk for tail biting, boar taint and competition between piglets during lactation or by selecting on genetic goals including lower propensity for tail biting and boar taint and a better match between litter size at birth and sow maternal abilities. Besides working on solutions to avoid invasive management procedures in piglets, efforts should also be made to set up methods for easy identification of painful situations using non-invasive tools such as automatic scoring of facial expression or vocalisations. This will enable farmers to treat the pain more rapidly and efficiently.

8 Summary

Most piglets are still submitted to surgical castration, tail docking and tooth resection across Europe. Each of them is motivated by zootechnical reasons to prevent: boar taint (castration), tail biting (tail docking), lesions on the sow's teat and other piglets (tooth resection). However, they cause damage to living, highly innervated tissues and hence are potential sources of pain. The present review of the literature based on neural, neuroendocrine and behavioural evidence confirms the existence of pain during and well after applying each of the three procedures. Neuroendocrine and behavioural alterations are particularly evident during and/or in the first hours following surgical castration suggesting intense acute pain. They are less clear after tail docking and tooth resection, whereas neural evidence indicates long-term pain due to these practices. Therefore, all three practices should be avoided by applying preventive measures based on management, environment and genetics.

9 Where to look for further information

9.1 Further reading

- The suckling and weaned Piglet (2020), editor: Chantal Farmer.
- *Advances in Pig Welfare* (2017), editor: Marek Špinka.
- *The Welfare of Pigs* (2009), editor: Marchant-Forde, Jeremy N.

9.2 Key websites, projects and conferences

- ANIWHA Era-Net project *Ending tail docking and tail biting in the EU - Hazard characterization and exposure assessment of a major pig welfare Problem (FareWellDock)*, http://farewelldock.eu/about/.
- COST action *Synergy for preventing damaging behaviour in group housed pigs and chickens (GroupHouseNet)*, https://www.grouphousenet.eu/.
- European Food Safety Authority - Animal Health and Welfare, https://efsa.onlinelibrary.wiley.com/results/fdcaead5-2be1-4c72-bf46-18a428c3b60e/d98c76ac-8d4f-49aa-96b3-4eb760cb6b65.
- Farm Animal Welfare Education Centre, https://www.fawec.org/en/about-us.
- International Association for the Study of Pain (IASP), https://www.iasp-pain.org/.
- IASP World Congress of Pain, https://www.iaspworldcongress.org/.

10 References

2001/93/EC CD (2001). Minimum standards for the protection of pigs. In: E. Commission (Ed.), http://eur-lex.europa.eu/LexUriServ/LexUriServ.do?uri=CELEX:32001L0093:en:NOT, p. 3. Brussels, Belgium.

Ali, Z., Raja, S. N., Wesselmann, U., Fuchs, P. N., Meyer, R. A. and Campbell, J. N. (2000). Intradermal injection of norepinephrine evokes pain in patients with sympathetically maintained pain. *Pain* 88(2): 161–168.

Backus, B. L. and McGlone, J. J. (2018). Evaluating environmental enrichment as a method to alleviate pain after castration and tail docking in pigs. *Applied Animal Behaviour Science* 204: 37–42.

Bantel, C. and Trapp, S. (2011). The role of the autonomic system in acute surgical pain processing - what do we know? *Anaesthesia* 66(7): 541–544.

Basbaum, A. I. and Jessell, T. (2000). The perception of pain. In: Kendel, E. R., Schwartz, J. and Jessell, T. (Eds), *Principles of Neuroscience*. New York: Appleton and Lange.

Basbaum, A. I., Bautista, D. M., Scherrer, G. and Julius, D. (2009). Cellular and molecular mechanisms of pain. *Cell* 139(2): 267–284.

Bataille, G., Rugraff, Y., Meunier-Salaün, M. C., Bregeon, A. and Prunier, A. (2002). Conséquences comportementales, zootechniques et physiologiques de l'épointage des dents chez le porcelet. En Title: influence of tooth resection on behaviour, performance and physiology of piglets. *Journées de la Recherche Porcine* 34: 203–209.

Bates, J. L., Karriker, L. A., Stock, M. L., Pertzborn, K. M., Baldwin, L. G., Wulf, L. W., Lee, C. J., Wang, C. and Coetzee, J. F. (2014) Impact of transmammary-delivered meloxicam on biomarkers of pain and distress in piglets after castration and tail docking. *PLoS ONE* 9(12): e113678. doi: 10.1371/journal.pone.0113678.

Benarroch, E. E. (2001). Pain-autonomic interactions: a selective review. *Clinical Autonomic Research* 11(6): 343–349.

Bengtsdotter, E. A., Ekman, S. and Andersen, P. H. (2019). Neuromas at the castration site in geldings. *Acta Veterinaria Scandinavica* 61(1): 43. doi: 10.1186/s13028-019-0479-8.

Bonastre, C., Mitjana, O., Tejedor, M. T., Calavia, M., Yuste, A. G., Úbeda, J. L. and Falceto, M. V. (2016). Acute physiological responses to castration-related pain in piglets: the effect of two local anesthetics with or without meloxicam. *Animal* 10(9): 1474–1481.

Busnardo, C., Tavares, R. F., Resstel, L. B. M., Elias, L. L. K. and Correa, F. M. A. (2010). Paraventricular nucleus modulates autonomic and neuroendocrine responses to acute restraint stress in rats. *Autonomic Neuroscience: Basic and Clinical* 158(1–2): 51–57.

Byers, M. R. and Närhi, M. V. O. (1999). Dental injury models: experimental tools for understanding neuroinflammatory interactions and polymodal nociceptor functions. *Critical Reviews in Oral Biology and Medicine* 10(1): 4–39.

Carroll, J. A., Berg, E. L., Strauch, T. A., Roberts, M. P. and Kattesh, H. G. (2006). Hormonal profiles, behavioral responses, and short-term growth performance after castration of pigs at three, six, nine, or twelve days of age. *Journal of Animal Science* 84(5): 1271–1278.

Castel, D., Sabbag, I., Brenner, O. and Meilin, S. (2016). Peripheral neuritis trauma in pigs: a neuropathic pain model. *Journal of Pain* 17(1): 36–49.

Castel, D., Willentz, E., Doron, O., Brenner, O. and Meilin, S. (2014). Characterization of a porcine model for post-operative pain. *European Journal of Pain* 18: 495–505.

Cervero, F. (2009). Visceral versus somatic pain: similarities and differences. *Digestive Diseases* 27(Suppl. 1): 3–10.

Cervero, F. (2012). *Understanding Pain: Exploring the Perception of Pain*. Cambridge, MA: The MIT Press.

Cervero, F. and Laird, J. M. (1999). Visceral pain. *Lancet* 353(9170): 2145–2148.

Chapman, C. R., Tuckett, R. P. and Song, C. W. (2008). Pain and stress in a systems perspective: reciprocal neural, endocrine, and immune interactions. *Journal of Pain* 9(2): 122–145.

Colburn, R. W., DeLeo, J. A., Rickman, A. J., Yeager, M. P., Kwon, P. and Hickey, W. F. (1997). Dissociation of microglial activation and neuropathic pain behaviors following peripheral nerve injury in the rat. *Journal of Neuroimmunology* 79(2): 163–175.

Costigan, M., Scholz, J. and Woolf, C. J. (2009). Neuropathic pain: a maladaptive response of the nervous system to damage. *Annual Review of Neuroscience* 32: 1–32.

Courboulay, V., Hémonic, A., Gadonna, M., Meunier-Salaün, M. C. and Prunier, A. (2010a). Comparaison des effets d'une anesthésie locale (1ml de lidocaïne 1%) ou d'un traitement anti-inflammatoire sur la douleur due à la castration. Comparison between local anesthesia (1 ml lidocaïne 1% per piglet) and analgesia (ketoprofene) on pain associated with piglet castration. *Journées de la Recherche Porcine* 42: 35–36.

Courboulay, V., Hémonic, A., Gadonna, M. and Prunier, A. (2010b). Castration avec anesthésie locale ou traitement anti-inflammatoire: quel impact sur la douleur des porcelets et quelles conséquences sur le travail en élevage? Effect of local anesthesia or anti-inflammatory treatment on pain associated with piglet castration and on labour demand. *Journees de la Recherche Porcine en France* 42: 27–33.

Courboulay, V., Gillardeau, M., Meunier-Salaün, M. C. and Prunier, A. (2015). La Prise en charge de la douleur lors de la caudectomie et de la castration des porcelets. Pain relief at piglet tail docking and castration. *Journées Recherche Porcine* 47: 235–240.

Dalla Costa, E., Minero, M., Lebelt, D., Stucke, D., Canali, E. and Leach, M. C. (2014). Development of the Horse Grimace Scale (HGS) as a pain assessment tool in horses undergoing routine castration. *PLoS ONE* 9(3): e92281. doi: 10.1371/journal.pone.0092281.

Dantzer, R., O'Connor, J. C., Freund, G. G., Johnson, R. W. and Kelley, K. W. (2008). From inflammation to sickness and depression: when the immune system subjugates the brain. *Nature Reviews Neuroscience* 9(1): 46–56.

De Briyne, N., Berg, C., Blaha, T. and Temple, D. (2016). Pig castration: will the EU manage to ban pig castration by 2018? *Porcine Health Management* 2: 29.

D'Eath, R. B., Niemi, J. K., Ahmadi, B. V., Rutherford, K. M. D., Ison, S. H., Turner, S. P., Anker, H. T., Jensen, T., Busch, M. E., Jensen, K. K., Lawrence, A. B. and Sandøe, P. (2016). Why are most EU pigs tail docked? Economic and ethical analysis of four pig housing and management scenarios in the light of EU legislation and animal welfare outcomes. *Animal* 10(4): 687–699.

Di Giminiani, P., Brierley, V., Scollo, Gottardo, F., Malcolm, E. M., Edwards, S. A. and Leach, M. C. A. (2016). The assessment of facial expressions in piglets undergoing tail docking and castration: toward the development of the piglet grimace scale. *Frontiers in Veterinary Science* 3: 100. doi: 10.3389/fvets.2016.00100.

Di Giminiani, P., Edwards, S. A., Malcolm, E. M., Leach, M. C., Herskin, M. S. and Sandercock, D. A. (2017a). Characterization of short- and long-term mechanical sensitisation following surgical tail amputation in pigs. *Scientific Reports* 7(1): 4827. doi: 10.1038/s41598-017-05404-y.

Di Giminiani, P., Nasirahmadi, A., Malcolm, E. M., Leach, M. C. and Edwards, S. A. (2017b). Docking piglet tails: how much does it hurt and for how long? *Physiology and Behavior* 182: 69–76.

Done, S. H., Guise, J. and Chennells, D. (2003). Tail biting and tail docking in pigs. In: Proceedings of the Pig Veterinary Society Meeting, 14–15 November 2002, Burleigh Court, Loughborough, UK, 136–154.

Dubin, A. E. and Patapoutain, A. (2010). Nociceptors the sensors of the pain pathway. *Journal of Clinical Investigation* 120(11): 3760–3772.

Ducreux, D., Attal, N., Parker, F. and Bouhassira, D. (2006). Mechanisms of central neuropathic pain: a combined psychophysical and fMRI study in syringomyelia. *Brain* 129(4): 963–976.

Dworkin, R. H., Backonja, M., Rowbotham, M. C., Allen, R. R., Argoff, C. R., Bennett, G. J., Bushnell, M. C., Farrar, J. T., Galer, B. S., Haythornthwaite, J. A., Hewitt, D. J., Loeser, J. D., Max, M. B., Saltarelli, M., Schmader, K. E., Stein, C., Thompson, D., Turk, D. C., Wallace, M. S., Watkins, L. R. and Weinstein, S. M. (2003). Advances in neuropathic pain: diagnosis, mechanisms, and treatment recommendations. *Archives of Neurology* 60(11): 1524–1534.

EFSA (2004). Scientific report of the Scientific Panel for Animal Health and Welfare on a request from the Commission related to welfare aspects of the castration of Piglets. *EFSA Journal* 91: 1–118.

EFSA (2007). Scientific report on the risks associated with tail biting in pigs and possible means to reduce the need for tail docking considering the different housing and husbandry systems. *Annex to the EFSA Journal* 611: 1-98.

EFSA (2017). Animal welfare aspects in respect of the slaughter or killing of pregnant livestock animals (cattle, pigs, sheep, goats, horses). *EFSA Journal* 15(5): e04782.

Enzerink, A., Salmenperä, P., Kankuri, E. and Vaheri, A. (2009). Clustering of fibroblasts induces proinflammatory chemokine secretion promoting leukocyte migration. *Molecular Immunology* 46(8-9): 1787-1795.

FAWC (1992). Five freedoms. Available at: https://webarchive.nationalarchives.gov.uk/20121010012427/http://www.fawc.org.uk/freedoms.htm; Accessed 20 November, 2019.

Fraser, D. and Thompson, B. K. (1991). Armed sibling rivalry among suckling piglets. *Behavioral Ecology and Sociobiology* 29(1): 9-15.

Fu, L. L., Zhou, B., Li, H. Z., Liang, T. T., Chu, Q. P., Schinckel, A. P., Li, Y. and Xu, F. L. (2019). Effects of tail docking and/or teeth clipping on behavior, lesions, and physiological indicators of sows and their piglets. *Animal Science Journal* 90(9): 1320-1332.

Fukuoka, T., Tokunaga, A., Kondo, E., Miki, T., Tachbana, T. and Noguchi, K. (1988). Change in mRNAs for neuropeptide and the GABAA receptor in the dorsal root ganglion neurons in a rat experimental neuropathic pain model. *Pain* 78: 13-26.

Gallois, M., Cozler, Y. L. and Prunier, A. (2005). Influence of tooth resection in piglets on welfare and performance. *Preventive Veterinary Medicine* 69(1-2): 13-23.

Giamberardino, M. A. (1999). Recent and forgotten aspects of visceral pain. *European Journal of Pain* 3(2): 77-92.

Gold, M. S. and Gebhart, G. F. (2010). Nociceptor sensitization in pain pathogenesis. *Nature Medicine* 16(11): 1248-1257.

Guatteo, R., Levionnois, O., Fournier, D., Guémené, D., Latouche, K., Leterrier, C., Mormède, P., Prunier, A., Servière, J., Terlouw, C. and Le Neindre, P. (2012). Minimising pain in farm animals: the 3S approach - 'Suppress, Substitute, Soothe'. *Animal* 6(8): 1261-1274.

Haga, H. A. and Ranheim, B. (2005). Castration of piglets: the analgesic effects of intratesticular and intrafunicular lidocaine injection. *Veterinary Anaesthesia and Analgesia* 32(1): 1-9.

Haga, H. A., Tevik, A. and Mooerch, H. (2001). Electroencephalographic and cardiovascular indicators of nociception during isoflurane anaesthesia in pigs. *Veterinary Anaesthesia and Analgesia* 28(3): 126-131.

Hansson, M., Lundeheim, N., Nyman, G. and Johansson, G. (2011). Effect of local anaesthesia and/or analgesia on pain responses induced by piglet castration. *Acta Veterinaria Scandinavica* 53: 34.

Hay, M., Vulin, A., Génin, S., Sales, P. and Prunier, A. (2003). Assessment of pain induced by castration in piglets: behavioral and physiological responses over the subsequent 5 days. *Applied Animal Behaviour Science* 82(3): 201-218.

Hay, M., Rue, J., Sansac, C., Brunel, G. and Prunier, A. (2004). Long-term detrimental effects of tooth clipping or grinding in piglets: a histological approach. *Animal Welfare* 13: 23-32.

Herskin, M. S., Thodberg, K. and Jensen, H. E. (2015). Effects of tail docking and docking length on neuroanatomical changes in healed tail tips of pigs. *Animal* 9(4): 677-681.

Herskin, M. S., Di Giminiani, P. and Thodberg, K. (2016). Effects of administration of a local anaesthetic and/or an NSAID and of docking length on the behaviour of piglets during 5 h after tail docking. *Research in Veterinary Science* 108: 60-67.

Hökfelt, T., Zhang, X. and Wiesenfeld-Hallin, Z. (1994). Messenger plasticity in primary sensory neurons following axotomy and its functional implications. *Trends in Neurosciences* 17(1): 22–30.

Holyoake, P. K., Broek, D. J. and Callinan, A. P. L. (2004). The effects of reducing the length of canine teeth in sucking pigs by clipping or grinding. *Australian Veterinary Journal* 82(9): 574–576.

Horn, T., Marx, G. and von Borell, E. (1999). Behaviour of piglets during castration with and without a local anaesthesia. *Deutsche Tierarztliche Wochenschrift* 106(7): 271–274.

Hutter, S., Heinritzi, K., Reich, E. and Ehret, W. (1994). Efficacité de différentes méthodes de résection des dents chez le porcelet non sevré. The effect of different methods of tooth resection in piglets. *Revue de Médecine Vétérinaire* 145: 205–213.

Ishikawa, K., Tananka, M., Black, J. A. and Waxman, S. G. (1999). Changes in expression of voltage-gated potassium channels in dorsal root ganglion neurons following axotomy. *Muscle Nerve* 22: 502–507.

Jaggi, A. S., Jain, V. and Singh, N. (2011). Animal models of neuropathic pain. *Fundamental and Clinical Pharmacology* 25(1): 1–28.

Jensen, M. P., Chodroff, M. J. and Dworkin, R. H. (2007). The impact of neuropathic pain on health-related quality of life: review and implications. *Neurology* 68(15): 1178–1182.

Julius, D. and Basbaum, A. I. (2001). Molecular mechanisms of nociception. *Nature* 413(6852): 203–210.

Keita, A., Pagot, E., Prunier, A. and Guidarini, C. (2010). Pre-emptive meloxicam for postoperative analgesia in piglets undergoing surgical castration. *Veterinary Anaesthesia and Analgesia* 37(4): 367–374.

Kells, N. J., Beausoleil, N. J., Chambers, J. P., Sutherland, M. A., Morrison, R. S. and Johnson, C. B. (2017a). Electroencephalographic responses of anaesthetized pigs (Sus scrofa) to tail docking using clippers or cautery iron performed at 2 or 20 days of age. *Veterinary Anaesthesia and Analgesia* 44(5): 1156–1165.

Kells, N. J., Beausoleil, N. J., Johnson, C. B., Sutherland, M. A., Morrison, R. S. and Roe, W. (2017b). Comparison of neural histomorphology in tail tips from pigs docked using clippers or cautery iron. *Animal* 11(7): 1222–1227.

Kells, N. J., Beausoleil, N. J., Satherland, M. A. and Johnson, C. B. (2019). Post-natal development of EEG responses to noxious stimulation in pigs (Sus scrofa) aged 1–15 days. *Animal Welfare* 28(3): 317–329.

Kidd, B. L. and Urban, L. A. (2001). Mechanisms of inflammatory pain. *British Journal of Anaesthesia* 87(1): 3–11.

Kim, D. S., Choi, J. O., Rim, H. D. and Cho, H. J. (2002). Downregulation of voltage-gated potassium channel α gene expression in dorsal root ganglia following chronic constriction injury in the rat sciatic nerve. *Molecular Brain Research* 105(1–2): 146–152.

Kluivers-Poodt, M., Houx, B. B., Robben, S. R. M., Koop, G., Lambooij, E. and Hellebrekers, L. J. (2012). Effects of a local anaesthetic and NSAID in castration of piglets, on the acute pain responses, growth and mortality. *Animal* 6(9): 1469–1475.

Kluivers-Poodt, M., Zonderland, J. J., Verbraak, J., Lambooij, E. and Hellebrekers, L. J. (2013). Pain behaviour after castration of piglets; effect of pain relief with lidocaine and/or meloxicam. *Animal* 7(7): 1158–1162.

Kozłowska, A., Mikołajczyk, A., Adamiak, Z. and Majewski, M. (2017). Distribution and chemical coding of sensory neurons innervating the skin of the porcine hindlimb. *Neuropeptides* 61: 1–14.

Lahrmann, H. P., Busch, M. E., D'Eath, R. B., Forkman, B. and Hansen, C. F. (2017). More tail lesions among undocked than tail docked pigs in a conventional herd. *Animal* 11(10): 1825–1831.

Langford, D. J., Bailey, A. L., Chanda, M. L., Clarke, S. E., Drummond, T. E., Echols, S., Glick, S., Ingrao, J., Klassen-Ross, T., LaCroix-Fralish, M. L., Matsumiya, L., Sorge, R. E., Sotocinal, S. G., Tabaka, J. M., Wong, D., van den Maagdenberg, A. M., Ferrari, M. D., Craig, K. D. and Mogil, J. S. (2010). Coding of facial expressions of pain in the laboratory mouse. *Nature Methods* 7(6): 447-449.

Langhoff, R., Zols, S., Barz, A., Palzer, A., Ritzmann, M. and Heinritzi, K. (2009). Investigation about the use of analgesics for the reduction of castration-induced pain in suckling piglets. *Berliner Und Munchener Tierarztliche Wochenschrift* 122(9–10): 325–332.

Leidig, M. S., Hertrampf, B., Failing, K., Schumann, A. and Reiner, G. (2009). Pain and discomfort in male piglets during surgical castration with and without local anaesthesia as determined by vocalisation and defence behaviour. *Applied Animal Behaviour Science* 116(2–4): 174–178.

Lewis, E., Boyle, L. A., Brophy, P., O'Doherty, J. V. and Lynch, P. B. (2005a). The effect of two piglet teeth resection procedures on the welfare of sows in farrowing crates. Part 2. *Applied Animal Behaviour Science* 90(3–4): 251–264.

Lewis, E., Boyle, L. A., Lynch, P. B., Brophy, P. and O'Doherty, J. V. (2005b). The effect of two teeth resection procedures on the welfare of piglets in farrowing crates. Part 1. *Applied Animal Behaviour Science* 90(3–4): 233–249.

Llamas Moya, S. L., Boyle, L. A., Lynch, P. B., Arkins, S. (2008). Effect of surgical castration on the behavioural and acute phase responses of 5-day-old piglets. *Applied Animal Behaviour Science* 111(1–2): 133–145.

Lynn, B., Faulstroh, K. and Pierau, F. K. (1995). The classification and properties of nociceptive afferent units from the skin of the anaethetized pig. *European Journal of Neuroscience* 7(3): 431–437.

Lynn, B., Schütterle, S. and Pierau, F. K. (1996). The vasodilator component of neurogenic inflammation is caused by a special subclass of heat-sensitive nociceptors in the skin of the pig. *Journal of Physiology* 494(2): 587–593.

Marchant-Forde, J. N., Lay, D. C., Jr., McMunn, K. A., Cheng, H. W., Pajor, E. A. and Marchant-Forde, R. M. (2009). Postnatal piglet husbandry practices and well-being: the effects of alternative techniques delivered separately. *Journal of Animal Science* 87(4): 1479–1492.

Marchant-Forde, J. N., Lay, D. C., Jr., McMunn, K. A., Cheng, H. W., Pajor, E. A. and Marchant-Forde, R. M. (2014). Postnatal piglet husbandry practices and well-being: the effects of alternative techniques delivered in combination. *Journal of Animal Science* 92(3): 1150–1160.

Marx, G., Horn, T., Thielebein, J., Knubel, B. and von Borell, E. (2003). Analysis of pain-related vocalization in young pigs. *Journal of Sound and Vibration* 266(3): 687–698.

McGlone, J. J. and Hellman, J. M. (1988). Local and general anesthetic effects on behavior and performance of two- and seven-week-old castrated and uncastrated piglets. *Journal of Animal Science* 66(12): 3049–3058.

McGlone, J. J., Nicholson, R. I., Hellman, J. M. and Herzog, D. N. (1993). The development of pain in young pigs associated with castration and attempts to prevent castration-induced behavioral changes. *Journal of Animal Science* 71(6): 1441–1446.

Mellor, D. J., Cook, C. J. and Stafford, K. J. (2000). Quantifying some responses to pain as a stressor. In: Moberg, G. P. and Mench, J. A. (Eds) *The Biology of Animal Stress: Basic Principles and Implications for Welfare*. Wallingford: CAB International, 171–198.

Mellor, D. J., Diesch, T. J., Gunn, A. J. and Bennet, L. (2005). The importance of 'awareness' for understanding fetal pain. *Brain Research. Brain Research Reviews* 49(3): 455–471.

Millan, M. J. (2002). Descending control of pain. *Progress in Neurobiology* 66(6): 355–474.

Moore, K. A., Kohno, T., Karchewski, L. A., Scholz, J., Baba, H. and Woolf, C. J. (2002). Partial peripheral nerve injury promotes a selective loss of GABAergic inhibition in the superficial dorsal horn of the spinal cord. *Journal of Neuroscience* 22(15): 6724–6731.

Moore, S. A. (2016). Managing neuropathic pain in dogs. *Frontiers in Veterinary Science* 3: 12. doi: 10.3389/fvets.2016.00012.

Murphy, P. (2007). Somatic pain. In: Schmidt, R. and Willis, W. (Eds), *Encyclopedia of Pain*. Berlin, Heidelberg, Germany: Springer.

Murrell, J. C., Johnson, C. B., White, K. L., Taylor, P. M., Haberham, Z. L. and Waterman-Pearson, A. E. (2003). Changes in the EEG during castration in horses and ponies anaesthetised with halothane. *Veterinary Anaesthesia and Analgesia* 30(3): 138–146.

Noonan, G. J., Rand, J. S., Priest, J., Ainscow, J. and Blackshaw, J. K. (1994). Behavioural observations of piglets undergoing tail docking, teeth clipping and ear notching. *Applied Animal Behaviour Science* 39(3–4): 203–213.

Nyborg, P. Y., Srig, A., Lykkegaard, K. and Svendsen, O. (2000). Nociception after castration of juvenile pigs determined by quantitative estimation of c-Fos expressing neurons in the spinal cord dorsal horn. *Dansk Veterinartidsskrift* 83: 16–17.

Obreja, O., Ringkamp, M., Namer, B., Forsch, E., Klusch, A., Rukied, R., Petersen, M. and Schmelz, M. (2010). Patterns of activity-dependent conduction velocity changes differentiate classes of unmyelinated mechano-insensitive afferents including cold nociceptors, in pigs and in human. *Pain* 148(1): 59–69.

Obreja, O. and Schmelz, M. (2010). Single-fibre recordings of unmyelinated afferents in pig. *Neuroscience Letters* 470(3): 175–179.

Obreja, O. and Schmelz, M. (2013). The porcine nerve as a model of human pathology. In: Handwerker, H. O. and Arendt-Nielsen, L. (Eds), *Pain Models - Translational Relevance and Applications*. Washington, DC: IASP Press, 227–235.

Ocana, M. (2004). Potassium channels and pain: present realities and future opportunities. *European Journal of Pharmacology* 500(1–3): 203–219.

Pollema-Mays, S. L., Centeno, M. V., Ashford, C. J., Apkrian, A. V. and Martina, M. (2013). Expression of background potassium channels in rat DRG is cell-specific and down-regulated in a neuropathic pain model. *Molecular and Cellular Neurosciences* 57: 1–9.

Prkachin, K. M. (1992). The consistency of facial expressions of pain: a comparison across modalities. *Pain* 51(3): 297–306.

Prunier, A., Bataille, G., Meunier-Salaun, M. C., Bregeon, A. and Rugraff, Y. (2001). Consequences comportementales, zootechniques et physiologiques de la caudectomie realisee avec ou sans "insensibilisation" locale chez le porcelet. Influence of tail docking, with or without a cold analgesic spray, on the behaviour, performance and physiology of piglets. *Journees de la Recherche Porcine en France* 33: 313–318.

Prunier, A., Gallois, M., Klouytten, A. and Le Cozler, Y. (2004). Effets de l'épointage des dents sur les performances, les lésions cutanées et le comportement des truies et des porcelets. Influence of tooth resection on performance, skin lesions and behaviour of sows and piglets. *Journées de la Recherche Porcine* 36: 379-388.

Prunier, A., Mounier, A. M. and Hay, M. (2005). Effects of castration, tooth resection, or tail docking on plasma metabolites and stress hormones in young pigs. *Journal of Animal Science* 83(1): 216-222.

Prunier, A., Bonneau, M., Von Borell, E. H., Cinotti, S., Gunn, M., Frediksen, B., Giersing, M., Morton, D. B., Tuyttens, F. A. M. and Velarde, A. (2006). A review of the welfare consequences of surgical castration in piglets and the evaluation of non-surgical methods. *Animal Welfare* 15: 277-289.

Prunier, A., Mounier, L., Le Neindre, P., Leterrier, C., Mormède, P., Paulmier, V., Prunet, P., Terlouw, C. and Guatteo, R. (2013). Identifying and monitoring pain in farm animals: a review. *Animal* 7(6): 998-1010.

Ren, K. and Torres, R. (2009). Role of interleukin-1 beta during pain and inflammation. *Brain Research Reviews* 60(1): 57-64.

Rice, F. L., Castel, D., Ruggiero, E., Dockum, M., Houk, G., Sabbag, I., Albrecht, P. J. and Meilin, S. (2018). Human-like cutaneous neuropathologies associated with a porcine model of peripheral neuritis: a translational platform for neuropathic pain. *Neurobiology of Pain* 5: 100021. doi: 10.1016/j.ynpai.2018.07.002.

Robinson, D. R. and Gebhart, G. F. (2008). Inside information: the unique features of visceral sensation. *Molecular Interventions* 8(5): 242-253.

Roelofs, S., Meijer, E., Baas, V., Dobrovolski, M., van der Staay, F. J. and Nordquist, R. E. (2019). Neurological functioning and fear responses in low and normal birth weight piglets. *Applied Animal Behaviour Science* 220: 104853. doi: 10.1016/j.applanim.2019.104853.

Rukwied, R., Dusch, M., Schley, M., Forsch, E. and Schemlz, M. (2008). Nociceptor sensititzation to mechanical and thermal stimuli in pig skin in vivo. *European Journal of Pain* 12(2): 242-250.

Rukwied, R., Schley, M., Forsch, E., Obreja, O., Dusch, M. and Schmelz, M. (2010). Nerve growth factor-evoked nociceptor sensitization in pig skin in vivo. *Journal of Neuroscience Research* 88(9): 2066-2072.

Saalfrank, A., Janssen, K. P., Ravon, M., Flisikowski, K., Eser, S., Stieger, K., Flisikowska, T., Müller-Fliedner, P., Schulze, É, Brönner, C., Gnann, A., Kappe, E., Böhm, B., Schade, B., Certa, U., Saur, D., Esposito, I., Kind, A. and Schnieke, A. (2016). A porcine model of osteosarcoma. *Oncogenesis* 5: e210. doi: 10.1038/oncsis.2016.19.

Sandercock, D. A., Gibson, I. F., Rutherford, K. M. D., Donald, R. D., Lawrence, A. B., Brash, H. M., Scott, E. M. and Nolan, A. M. (2011). The impact of prenatal stress on basal nociception and evoked responses to tail-docking and inflammatory challenge in juvenile pigs. *Physiology and Behavior* 104(5): 728-737.

Sandercock, D. A., Smith, S. H., Di Giminiani, P. and Edwards, S. A. (2016). Histopathological characterization of tail injury and traumatic neuroma development after tail docking in piglets. *Journal of Comparative Pathology* 155(1): 40-49.

Sandercock, D. A., Barnett, M. W., Coe, J. E., Downing, A. C., Nirmal, A. J., Di Giminiani, P., Edwards, S. A. and Freeman, T. C. (2019). Transcriptomics analysis of porcine caudal dorsal root ganglia in tail amputated pigs shows long-term effects on many pain-associated genes. *Frontiers in Veterinary Science* 6: 314. doi: 10.3389/fvets.2019.00314.

Schlereth, T. and Birklein, F. (2008). The sympathetic nervous system and pain. *Neuromolecular Medicine* 10(3): 141–147.

Schon, P. C., Puppe, B. and Manteuffel, G. (2001). Linear prediction coding analysis and self-organizing feature map as tools to classify stress calls of domestic pigs (Sus scrofa). *Journal of the Acoustical Society of America* 110(3 Pt 1): 1425–1431.

Schon, P. C., Puppe, B. and Manteuffel, G. (2004). Automated recording of stress vocalisations as a tool to document impaired welfare in pigs. *Animal Welfare* 13: 105–110.

Setchell, B. P., Maddocks, S. and Brooks, D. E. (1994). Anatomy, vasculature, innervation and fluids of the male reproductive tract. In: Knobil, E. and Neill, J. D. (Eds), *The Physiology of Reproduction*. New York: Raven Press, 1063–1175.

Sherrington, C. (1906). *The Integrative Action of the Nervous System*. Oxford, UK: Oxford University Press.

Simonsen, H. B., Klinken, L. and Bindseil, E. (1991). Histopathology of intact and docked pigtails. *British Veterinary Journal* 147(5): 407–412.

Sinclair, A. R. L., D'Eath, R. B., Brunton, P. J., Prunier, A. and Sandercock, D. A. (2018). Traumatic tooth shortening in commercial piglets: effects on gross tooth pathology and mRNA markers of inflammation and pain. In: International Association for the Study of Pain (IASP) 17th World Congress on Pain, 12th–15th September 2018, Boston, MA, 463236.

Sinclair, A. R. L., Tallet, C., Renouard, A., Brunton, P. J., D'Eath, R. B., Sandercock, D. A. and Prunier, A. (2019). Behaviour of piglets before and after tooth clipping, grinding or sham-grinding in the absence of social influences. In: 53rd Congress of the International Society for Applied Ethology (ISA). Bergen, Norway: OASES, 328.

Sutherland, M. A., Backus, B. L., Brooks, T. A. and McGlone, J. J. (2017). The effect of needle-free administration of local anesthetic on the behavior and physiology of castrated pigs. *Journal of Veterinary Behavior* 21: 71–76.

Sutherland, M. A., Bryer, P. J., Krebs, N. and McGlone, J. J. (2008). Tail docking in pigs: acute physiological and behavioural responses. *Animal* 2(2): 292–297.

Sutherland, M. A., Davis, B. L., Brooks, T. A. and McGlone, J. J. (2010). Physiology and behavior of pigs before and after castration: effects of two topical anesthetics. *Animal* 4(12): 2071–2079.

Sutherland, M. A., Davis, B. L., Brooks, T. A. and Coetzee, J. F. (2012). The physiological and behavioral response of pigs castrated with and without anesthesia or analgesia. *Journal of Animal Science* 90(7): 2211–2221.

Tallet, C., Rakotomahandry, M., Herlemont, S. and Prunier, A. (2019). Evidence of pain, stress and fear of human due to tail docking in suckling piglets (Sus scrofa). *Frontiers in Behavioral Neuroscience* 6: 462. doi: 10.3389/fvets.2019.00462.

Taylor, A. A. and Weary, D. M. (2000). Vocal responses of piglets to castration: identifying procedural sources of pain. *Applied Animal Behaviour Science* 70(1): 17–26.

Thodberg, K., Herskin, M. S., Jensen, T. and Jensen, K. H. (2018). The effect of docking length on the risk of tail biting, tail-directed behaviour, aggression and activity level of growing pigs kept under commercial conditions. *Animal* 12(12): 2609–2618.

Torebjörk, E., Wahren, L., Wallin, G., Hallin, R. and Koltzenburg, M. (1995). Noradrenaline-evoked pain in neuralgia. *Pain* 63(1): 11–20.

Torsney, C. and MacDermott, A. B. (2006). Disinhibition opens the gate to pathological pain signalling in superficial neurokinin 1 receptor-expressing neurons in the rat spinal cord. *Journal of Neuroscience* 26(6): 1833–1843.

Tracey, I. and Mantyh, P. W. (2007). The cerebral signature for pain perception and it's modulation. *Neuron* 55(3): 377–391.

Treede, R. D., Rief, W., Barke, A., Aziz, Q., Bennett, M. I., Benoliel, R., Cohen, M., Evers, S., Finnerup, N. B., First, M. B., Giamberardino, M. A., Kaasa, S., Kosek, E., Lavand'homme, P., Nicholas, M., Perrot, S., Scholz, J., Schug, S., Smith, B. H., Svensson, P., Vlaeyen, J. W. and Wang, S. J. (2015). A classification of chronic pain for ICD-11. *Pain* 156(6): 1003–1007.

Tsantoulas, C., Zhu, L., Saifa, Y., Grist, J., Ward, J. P. T., Raouf, R., Michael, G. J. and McMahon, S. B. (2012). Sensory neuron downregulation of the Kv9.1 potassium channel subunit mediates neuropathic pain following nerve injury. *Journal of Neuroscience* 32(48): 17502–17513.

Tucker, A. L. and Widowski, T. M. (2009). Normal profiles for deciduous dental eruption in domestic piglets: effect of sow, litter, and piglet characteristics. *Journal of Animal Science* 87(7): 2274–2281.

Turnbull, A. V. and Rivier, C. L. (1999). Regulation of the hypothalamic-pituitary-adrenal axis by cytokines: actions and mechanisms of action. *Physiological Reviews* 79(1): 1–71.

Viscardi, A. V., Hunniford, M., Lawlis, P., Leach, M. and Turner, P. V. (2017). Development of a Piglet Grimace Scale to Evaluate Piglet Pain Using Facial Expressions Following Castration and Tail Docking: A Pilot Study. *Frontiers in Veterinary Science* 4(51). doi: 10.3389/fvets.2017.00051.

Viscardi, A. V. and Turner, P. V. (2018). Use of meloxicam or ketoprofen for piglet pain control following surgical castration. *Frontiers in Veterinary Science* 5: 299. doi: 10.3389/fvets.2018.00299.

Weary, D. M., Braithwaite, L. A. and Fraser, D. (1998). Vocal response to pain in piglets. *Applied Animal Behaviour Science* 56(2–4): 161–172.

Weary, D. M. and Fraser, D. (1999). Partial tooth-clipping of suckling pigs: effects on neonatal competition and facial injuries. *Applied Animal Behaviour Science* 65(1): 21–27.

Wemelsfelder, F. and van Putten, G. (1985). *Behaviour as a Possible Indicator for Pain in Piglets*. Zeist, The Netherlands: Instituut voor Veeteeltkundig Onderzoek 'Schoonoord', 61.

White, F. A., Bhangoo, S. K. and Miller, R. J. (2005). Chemokines: integrators of pain and inflammation. *Nature Reviews. Drug Discovery* 4(10): 834–844.

White, R. G., Deshazer, J. A., Tressler, C. J., Borcher, G. M., Davey, S., Waninge, A., Parkhurst, A. M., Milanuk, M. J. and Clemens, E. T. (1995). Vocalization and physiological response of pigs during castration with or without a local anesthetic. *Journal of Animal Science* 73(2): 381–386.

Williams, A. C. and Craig, K. D. (2016). Updating the definition of pain. *Pain* 157(11): 2420–2423.

Woolf, C. J. and Mannion, R. J. (1999). Neuropathic pain: aetiology, symptoms, mechanisms, and management. *Lancet* 353(9168): 1959–1964.

Zols, S., Ritzmann, M. and Heinritzi, K. (2006). Effect of a local anaesthesia in castration of piglets. *Tieraerztliche Praxis Ausgabe Grosstiere Nutztiere* 34: 103–106.

Chapter 9

Alternatives to castration of pigs

Emma Fàbrega, IRTA, Spain

1 Introduction
2 Entire male production
3 Immunocastration
4 Sex sorting of semen
5 Case study
6 Conclusion and future trends
7 Where to look for further information
8 References

1 Introduction

Male piglets designated for pork production have traditionally been surgically castrated in many countries, to avoid the risk of boar-tainted carcasses and reduce behavioural problems. However, surgical castration is being conducted in most countries at a young age, and, in most cases, without the use of anaesthesia and analgesia. For example, under EU welfare legislation piglets can be surgically castrated before 7 days of age with no pain killers nor anaesthesia being administered (Council Directive 2008/120/EC, Anon. 2008). Surgical castration is a painful and stressful procedure (Prunier et al., 2006) and has become a major welfare issue. Behavioural alterations for several days after the procedure indicative of post-operative pain have been reported (Wemelsfelder and van Putten, 1985; Hay et al., 2003; Hansson et al., 2011), whereas results on physiological indicators have been less consistent (for a review see Rault et al., 2011). For that reason, the search for alternatives to piglet castration which could guarantee a proper meat quality and adequate behaviour, especially for those male pigs kept until heavy live weights, is crucial. These alternatives to piglet castration should be able to tackle the factors responsible for boar taint, as summarized in Fig. 1, and provide a proper welfare to non-surgically castrated pigs. Boar taint is an unpleasant odour in fat and meat that is mainly associated with uncastrated pigs. The main compounds responsible for boar taint are androstenone, skatole and to a lesser extent indole (Annor-Frempong et al., 1997; Claus et al., 1994; Rius and García-Regueiro, 2001).

http://dx.doi.org/10.19103/AS.2020.0081.09

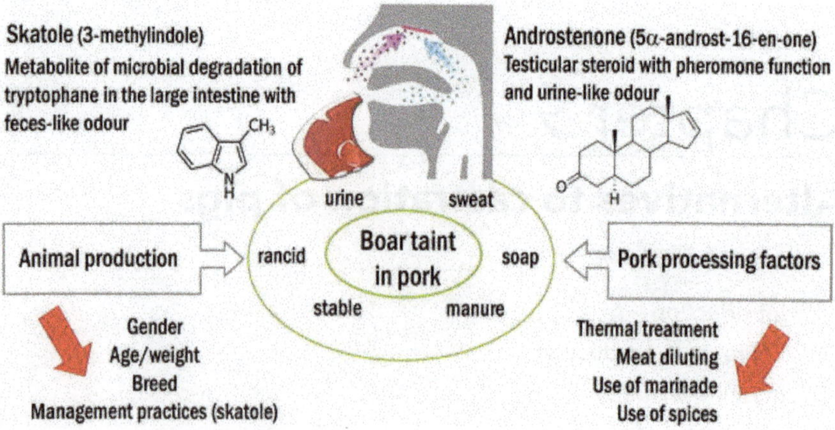

Figure 1 Main factors influencing the main two components responsible for boar taint (Androestone and Skatole). Figure from Čandek-Potokar et al., 2017.

In Europe, already in October 2010, representatives of major stakeholders signed the 'European Declaration on alternatives to surgical castration of pigs' (Anon. 2010a) and committed themselves to a roadmap which should have ended up in a voluntary end of male pig surgical castration by 1 January 2018. Despite this commitment, about 63% of all male piglets in Europe were still surgically castrated in 2017, most of them without any pain-relieving methods or anaesthesia even though alternatives exist and are already in place in some European countries (UK, Spain, Ireland) and non-European countries (Brazil, Australia) (Backus et al., 2018; de Briyne et al., 2016). The situation had not changed much in 2019 when preparing this chapter (Kress et al., 2019). The reason why some farmers are reluctant to stop castration may be found in the difficult trade-off that has to be made to implement some of the alternatives of piglet castration.

This chapter will discuss the feasibility, advantages and drawbacks of the main alternatives to piglet castration which would be the following:

- production of entire males;
- immunocastration; and
- semen sorting.

It has to be considered, however, that the strategies to substitute surgical castration without pain-reliving methods in many countries have been based so far in the provision of anaesthesia or analgesia (see Table 1), discussed in a previous chapter.

The success of entire male production as an alternative to surgical castration has been said to depend, up to a great extent, on the implementation of breeding, management or nutritional strategies, which will be discussed.

Table 1 Strategies used in different countries to substitute surgical castration without pain relief (adapted from Kress et al., 2019)

Country	Year	Alternative implemented
Spain	From 2010	80% boars
UK, Ireland	Not known	100% boars
Germany	2009	End of surgical castration without anaesthesia postponed until 2021. Meanwhile analgesia (meloxicam); anaesthesia (esp. isoflurane) required only in some organic programmes; immunocastration in some high-quality meat programmes
The Netherlands	2009	Surgical castration under anaesthesia (CO_2) for export market Increase of entire male production
Denmark	2011-2019	Analgesia (meloxicam); from 2019 on, plus local anaesthesia (procaine)
Belgium	2011	Surgical castration with analgesia (meloxicam) for export market Immunocastration for domestic retail market
France	2013	Analgesia (meloxicam)
Austria	2017	Analgesia (meloxicam)
Sweden	2016	Local anaesthesia (lidocaine) with analgesia (meloxicam); one smaller retailer prefers immunocastration
Norway	2002	Local anaesthesia (lidocaine) with analgesia (meloxicam)
Switzerland	2010	Anaesthesia (isoflurane)
Brazil, Australia	From 2010	80-100% immunocastration

Moreover, a brief mention on the consequences for meat quality of using these methods and on potential methods to detect boar taint at the abattoir will be described.

2 Entire male production

Raising entire males is the most straightforward alternative to physical castration. From a welfare perspective, entire male pig production has advantages and disadvantages (EFSA, 2004). Whereas a painful procedure is avoided, entire males present a higher incidence of aggressive and mounting behaviour, resulting in more skin lesions, leg and feet injuries and impairments also on carcass and meat quality (Prunier et al., 2006; von Borell et al., 2009; Ebschke et al., 2014, Fig. 2). From a performance and meat-quality perspective, the major issue with regards to raising entire males is the capacity to guarantee non-boar-tainted carcasses. However, entire males present certain advantages

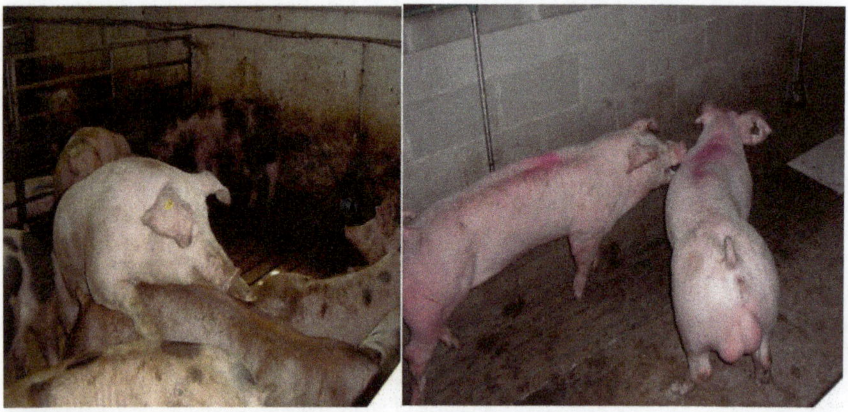

Figure 2 Higher incidence of mounting (a) and aggressive behaviour (b) is seen in entire male production systems (Pictures: IRTA).

Table 2 Advantages of entire male production compared to castrates in some performance indicators (adapted from Cost action IPEMA, 2019)

Parameter	Boar vs. barrow	Literature
Protein accretion (g/d)	+11%	Quiniou et al., 2010
Feed conversion ratio (kg/kg)	−10%	
Average daily feed intake (kg/d)	−11%	
Subcutaneous fat (%)	−16%	Pauly et al., 2009
N-excretion (kg/animal)	−20%	Dämmgen et al., 2013

like a better efficiency (i.e. feed conversion ratio) and higher protein accretion at the expense of fat (Table 2). On environmental grounds, boars have a better nitrogen retention as compared to castrates and, therefore, a lower environmental impact (review in: Pauly et al., 2012; Kress et al., 2019). In the final report of the European Union about best practices on the production, the processing and the marketing of meat from uncastrated males, the better feed conversion ratio was valued in 7.11 euros/pig.

Taking into account the advantages and disadvantages of entire male production, several projects such as the ALCASDE or CAMPIG and expert opinions have summarized that the issues to be addressed for entire male production to be a proper alternative would be:

1 To study the consumer acceptance of pork from entire male pigs;
2 To evaluate the effectiveness of the preventive measures to avoid boar taint and to guarantee proper welfare standards; and
3 To determine the accuracy and availability of on-line detection methods.

2.1 Consumer acceptance of pork from entire male pigs

The risk of boar taint is the main quality problem in pork production with entire males. For the detection and quantification of the three main compounds responsible for boar taint (androstenone, skatole and indole), a wide range of analytical methods, covering different approaches and clean-up procedures have been developed. One of the major constraints to understand consumer acceptance of pork from entire males has been the lack of a single reference method and clear cut-off thresholds. However, regardless of the selected instrumental or sensory-based sorting methods used, most of the literature reports that when pork from entire male pigs instead of females reach the market, the degree of consumers' dissatisfaction due to deviation in odour (+6.5%) and flavour (3.0%) increases (Bonneau et al., 2000; review by Bee et al., 2015). According to different studies, the degree of dislike for odour primarily originates from the underlying skatole and less from the androstenone concentration (Matthews et al., 2000; Font-i-Furnols, 2012a; Mörlein et al., 2019). By contrast, with respect to flavour, both boar-taint compounds appear as similarly important for the degree of dissatisfaction. A peculiarity related to androstenone is that whereas about one-third of the consumers are anosmic (inability to perceive odour) to androstenone, a similar proportion of consumers is highly sensitive and rejects pork with even very low (< 0.5 ppm) androstenone concentrations. Such a high variability in perception has not been reported for skatole. Skatole-tainted carcasses are rejected by most of the consumers at levels above 0.25 ppm (Font-i- Furnols, 2012a; Lunde et al., 2012). In the framework of the CAMPIG project (Backus et al., 2014), standardized patties of either castrate or boar meat all prepared at one site and containing a fat content of 15–20% were offered to consumers from Germany, Denmark, France, Italy, Poland, Russia and China. Moreover, an internet-based survey among 11 292 consumers was conducted in 10 European member state countries (Belgium, Denmark, France, Germany, Greece, Italy, Latvia, The Netherlands, Poland and Spain) and four other countries (China, Russia, South Korea and the United States of America), to assess consumer attitudes towards alternatives of surgical castration. The overall conclusion was that increasing levels of skatole and, to a lesser degree androstenone, in the meat, lower the consumer acceptance of boar meat compared to castrate meat. The effect of skatole was found over a whole range of skatole levels tested in the study whereas the effect of androstenone was found only for samples that were low in skatole and for consumers that were sensitive to androstenone. Consumers perceived 'health and quality' as the most important production aspect of pork, whereas the aspect 'animal welfare' was the second most important in a number of countries. From those conclusions, the authors suggested that pig companies have to determine their own thresholds, depending on the products and given their own risk assessment. To assist on this risk assessment, a preference map depicting the importance of skatole in the liking of boar meat patties and also the relevance of androstenone when skatole levels were low was created. From the large differences between countries in the survey on attitudes, it was suggested that pork-producing companies should use this knowledge to develop communication and selling strategies to place products from boar meat in specific market segments.

It is important to note that there are several other issues of entire male production, which are associated to the quality of meat and fat. The quality issues are negatively affecting either meat intended for fresh consumption or (even more notably) reduce the processing aptitude of such meat (Bonneau et al., 2018; Čandek-Potokar et al., 2015a), especially in the case of dry-cured products. It has also been confirmed by meta-analytic studies that entire males exhibit tougher meat compared to other sex categories (Pauly et al., 2012; Batorek et al., 2012). Increased toughness may be ascribed to factors like lower intramuscular fat content, reduced water-holding capacity and increased protein oxidation (Škrlep et al., 2019). In meat products, the reduced adiposity of entire males has been associated to excessive desiccation, and may cause lower processing yields and harder texture, in addition to higher salt uptake during processing of dry-cured hams (Škrlep et al., 2016). The fat of entire males is more polyunsaturated (Pauly et al., 2012), and this softer texture may difficult the packing process. In minced meat products like dry-fermented sausages, unsaturated fat may cause additional problems related to proper drying, texture and surface oiliness.

2.2 Effectiveness of the preventive measures to avoid boar taint and to guarantee proper welfare standards

It has been argued that the implementation of different on-farm strategies could help to reduce the risk of boar-tainted carcasses, as well as enhancing welfare in intact male production systems. Those strategies can be classified under the following different areas: nutrition, genetics and management.

2.2.1 Nutritional strategies

When producing entire males, the nutrition should be adjusted to their specific requirements for an efficient protein deposition. Moreover, nutrition can also be used to reduce the boar taint risk. In this context, it has to be pointed out that the dietary manipulations mainly work with skatole and indole levels, but androstenone is less susceptible and according to Zamaratskaia and Rasmussen (2018) there is still a lack of knowledge on some of the physiological processes involved in androstenone metabolism to find a way to reduce it through dietary means. With regard to skatole, some dietary ingredients can be used to minimize bacterial degradation of tryptophan in the large intestine, ultimately resulting in lower skatole and indole production (Claus and Raab, 1999; Bee et al., 2015).

A meta-analysis by Dunshea et al. (2013) concluded that it is essential to assure the maximum feed intake to take advantage of the inherited protein deposition potential of entire male pigs and immunocastrates (as further developed in next paragraph, Table 3). *Ad libitum* feed access may allow the maximization of energy intake and the fulfilment of lysine requirements in entire pigs. However, entire males normally show a lower feed intake compared to barrows and gilts and, therefore, a greater lysine requirement in the range of 5–8% for the body weight range of 25–50 kg and 50 kg to slaughter, respectively, may still be required (Quiniou et al., 2010; Bee et al., 2015).

On the other hand, it has been found that an increased energy input in the hindgut can alter the intestinal microbiota in such a way that less skatole is produced and this finally results in reduced skatole levels in the fat (Wesoly and Weiler, 2012). A strategy to achieve this is by feeding feedstuffs containing non-digestible fermentable polysaccharides to entire male pigs a few weeks before slaughter. The IPEMA cost action has published a list of ingredients with boar taint-reducing capacities, with a complete review of different studies and

Table 3 Performance parameters of castrated males (CM), immunocastrated pigs (IM), uncastrated males (NCM) and females (FE), before and after second vaccination

	CM	IM	NCM	FE
ADG(g/day)				
Up to V2	928.16[a] (24.75)[1]	840.76[b] (19.68)	836.46[b] (19.97)	869.26[ab] (24.14)
From V2 to slaughter	844.97[b] (38.81)	1160.04[a] (30.96)	825.32[b] (31.41)	679.71[c] (38.67)
Daily feed intake (g/day)				
Up to V2	2136.72[a] (63.73)	1761.95[b] (50.95)	1773.96[b] (51.51)	1873.37[b] (64.49)
From V2 to slaughter	2894.76[b] (93.53)	3446.65[a] (74.77)	2486.55[c] (75.72)	2417.69[c] (92.59)
Feed conversion rate				
Up to V2	2.31[a] (0.03)	2.10[b] (0.02)	2.12[b] (0.03)	2.16[b] (0.03)
From V2 to slaughter	3.43[a] (0.14)	2.98[b] (0.11)	3.20[b] (0.11)	3.78[a] (0.14)

V2: second vaccination; ADG: average daily gain; DFI: daily food intake.
Least square means with different superscripts within a row are different at a minimum value of $P < 0.05$ (adapted from Fàbrega et al., 2010).
[1] Standard errors.

proposals (https://zenodo.org/record/1460483#.XdKPvldKiUk). As a summary, some feedstuffs known to have a positive impact on skatole (and indole) levels are: raw potato starch (Lösel et al., 2006; Pauly et al., 2008), chicory in various forms (inulin, root, fructans) (Byrne et al., 2008; Øverland et al., 2011; Aluwé et al., 2017; Heyrman et al., 2018), Jerusalem artichoke (Vhile et al., 2012) and lupin seeds (Aluwé et al., 2009; Øverland et al., 2011). Other studies indicate that also secondary plant metabolites such as hydrolysable tannins, found, for example, in chestnuts, have the potential to reduce skatole (Čandek-Potokar et al., 2015b) or androstenone levels (Bee et al., 2017), although none of the results was consistent, and modify carcass and fat characteristics (Rezar et al., 2017). Increasing the levels of crude protein in the diet was also found to reduce skatole levels in serum (Lundström et al., 1994), but not all studies have reported the same effects (Quiniou et al., 2015) and it has even been suggested that boars should be fed diets with a reduction in crude protein (Backus et al., 2016).

With regards to androstenone, a number of studies suggested an effect of dietary components (Zamaratskaia and Rasmussen, 2018), but results have been less consistent. For example, Chen et al. (2007) demonstrated that plasma-free androstenone levels significantly decreased in the pigs fed with raw potato starch for 2 weeks. Fat androstenone levels were also slightly lower in these pigs, although this decrease did not reach statistical significance. Feeding male pigs with 10% dried chicory root for 16 days before slaughter reduced fat androstenone levels by up-regulation of major androstenone-metabolising enzyme 3β-hydroxysteroid dehydrogenase (Rasmussen et al., 2012). This study, however, was performed on limited number of animals and further research is needed to elucidate effect of chicory root and its active component on androstenone metabolism.

Recently, a commercial product called TAINTSTOP (Dumoulin, Belgium, patent pending EP17207866.9) has been presented as an innovative feed concept (Janssens et al., 2018), which combines a mix of different fibrous materials, specific feed additives and some rules of feed formulation in relation to the supply of fibre and protein (exact feed formulation not available yet). Heyrman et al. (2018) found that supplementing this in-feed additive for 2–3 weeks previous to slaughter reduced neck fat indole, skatole and androstenone. Another example of a boar taint-reducing diet was investigated at Wageningen University (Backus et al. (2016), consisting in the combination of: (1) reducing the crude protein content from 15.0% to 13.3%; (2) increasing the level of fermentable non-starch polysaccharides from 9.65% to 13.75% by adding 10% sugar beet pulp and 3% chicory pulp (inulin) to the diet; (3) adding 0.5% sepiolite (a clay mineral that among others binds nitrogen) to the diet; (4) adding synthetic tryptophan to the diet to decrease the level of indigestible tryptophan; and (5) adding 0.25% benzoic acid to the diet. These authors only found a tendency for lower levels of skatole, but no effect on androstenone. Overall, these approaches of combining different feedstuffs with capacity to reduce boar taint requires further investigation to establish the optimal combinations, dosages and durations of feeding, and appears as a promising possibility if the price is affordable for farmers.

When dealing with nutritional strategies, it is not only the ingredients provided which could reduce the risk of boar taint, but also the feeding regimes. Skatole level was lower in boars fed via a long trough than in boars fed by a single space feeder (Backus et al., 2016). Therefore, it has been suggested that providing adequate feeding space/eating places, as well as proper supply of water from the drinking system may have positive effects not only on animal welfare, but also on the risk of boar taint. The same authors found that on observational studies involving 70 commercial farms, feeding wet by-products and a diet with a high level of amino acids were associated, according to farmers, to less sexual and aggressive behaviour and skin lesions.

2.2.2 Management strategies

The behaviour of entire males differs from that of females and castrates, requiring a need to adapt the rearing system. Entire males engage in more aggressive encounters than castrates (Fredriksen et al., 2008; von Borell et al., 2009; Ebschke et al., 2014) and females (Rydhmer et al., 2006), and more mounting behaviour is also typical of entire males (Rydhmer et al., 2006). The investigation on the role of androstenone in aggression has yielded some controversial results: whereas Giersing et al. (2000) found a relationship between aggressive behaviour of boars and their androstenone levels in the backfat, Zamaratskaia et al. (2005) did not report the same correlation. More recently, Parois et al. (2017) found that dominant boars have a greater concentration of fat androstenone and performed a higher number of agonistic acts. However, no association between dominance status and plasma testosterone or oestradiol was found, and the authors suggested that this dichotomy could be related to the role of androstenone in controlling social behaviour, whereas it may not be involved in the control of reproductive function of males. Recent findings show that penile injuries may also be a welfare problem for boars, because when they reach puberty, they can completely extrude the penis and this may trigger other pen mates to bite it (Weiler et al., 2016).

In any case, different management strategies, most of them intending to reduce the general levels of stress, have been studied with regards to their effects on reducing the risk of aggression and mounting and of boar taint.

- *Group composition/mixing*. The three possibilities to group the entire males include sibling groups (same litter), mixed-sex groups (entire males and females, not necessarily siblings) and single-sex groups. There is evidence that sibling (or farrow-to-finish) groups may be best adapted to the needs of entire males. Fredriksen et al. (2006, 2008) found reduced levels of aggression, skin lesions and androstenone concentrations in these groups. In contrast, Backus et al. (2016) found no differences in boar taint between boars kept in litters (farrow-to-finish) and single-sex entire male groups. Fàbrega et al. (2013) and Rydhmer et al. (2013) presented results on a practice which nowadays is receiving more attention by farmers: early socialization. In both studies, piglets from different litters were allowed to mix when still at the farrowing crates. Although both studies did not report an effect on boar taint levels, reduced levels of aggression after regrouping, after split marketing or at the slaughter line (Fig. 3) were found, suggesting that welfare of entire males can be improved by socializing piglets and by keeping them in intact groups during raising and at slaughter.

Studies investigating the effect of mixed-gender versus single-sex housing are not consistent in their conclusions. Boyle and Björklund (2007) found a higher rate of agonistic interactions during feeding between single-sex males as well as

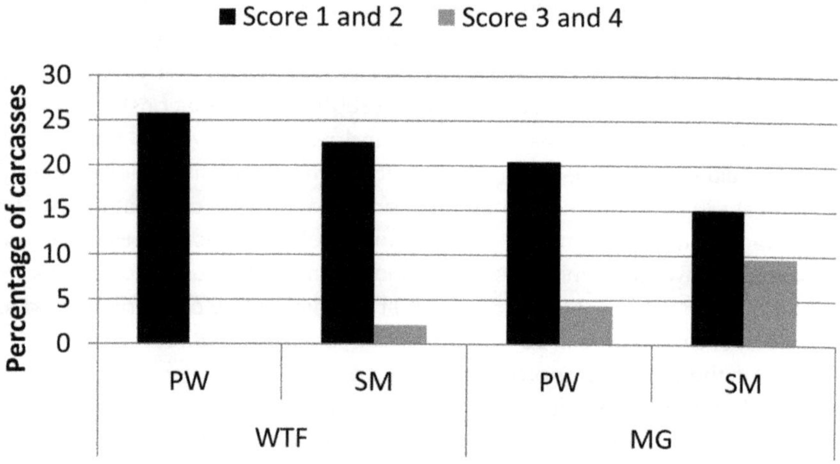

Figure 3 Percentage of carcasses presenting skins lesions with scores 1 and 2 (low/mild) or 3 and 4 (severe) in a wean-to-finish (WTF) system allowing early socialization of groups and no mixing after compared to a conventional system with mixing (MG), after slaughtering all pen together (PW) or implementing split marketing (SM). From Fàbrega et al. (2013).

a trend for more mounting compared to mixed-sex groups. This suggestion that males in single-sex groups may present a poorer welfare has been challenged by other studies (Rydhmer et al., 2006) reporting no differences between entire males in single-sex and mixed-sex groups. Another disadvantage of mixed-sex groups is the possibility of pregnancies in females, especially when they are slaughtered late in heavy pig production systems. Along that line, the onset of puberty is related to age and live weight (Einarsson et al., 1979) and depends, to a great extent, on genetics (Wilson et al., 1977). Other factors such as nutrition (Prunier and Quesnel, 2000) or social environment may play a role in puberty onset as well. It has been suggested that contact between boars and gilts enhances the development of the testes (Andersson et al., 1999) or that the introduction of a mature boar to female pigs triggers the attainment of puberty (Patterson et al., 2002). However, contradictory to this, results of a study by Salmon and Edwards (2006) showed that boars with gilt visual contact were less physiologically mature at slaughter and showed reduced levels of sexual behaviours. This could have indicated that visual contact of males and females could help to overcome the problems of mixed-group housing, while providing the advantages of male-female contact. However, studies published later on by Fàbrega et al. (2011) and Heyrman et al. (2018) have indicated that growing entire male pigs with gilts in the same compartment (but not pen) does not affect boar taint development, while the effect on performance remains inconsistent. With regards to group size and boar taint risk, few studies are available, but in a survey conducted by van Wagenberg et al. (2013) among 152 Dutch pig producers smaller group size was associated with lower farm-level boar-taint frequency. In accordance, Thomsen et al. (2016) found that entire male pigs kept in organic farms in groups of 30 as compared to smaller groups of 15 presented more skin lesions.

In line with that, Backus et al. (2016) reported that having more than 30 animals per pen were associated with a higher probability of high boar-taint prevalence levels.

- *Hygiene/Floor type/environmental conditions/lightning*. Early studies indicated that the housing conditions such as pen surface per animal (Hansen et al., 1994), floor type (Kjeldsen, 1993), hygiene conditions and the cleaning schedule (Hansen et al., 1995) could influence the prevalence of skatole in the pig carcass. Backus et al. (2016) also reported that clean pens, clean pigs and hygienic conditions were associated with lower boar-taint prevalence levels. However, the contribution of faecal contamination of the skin to skatole levels in adipose tissue is not clear and has been subject to controversial debate for decades. Hansen et al. (1994) suggested that the transdermal diffusion after soiling of ^{3}H-skatole could explain the elevated skatole concentrations in adipose tissue in dirty pigs. However, in other studies this diffusion was not proven, even with daily cleaning control pigs compared to pigs anointed twice with its own faeces (Aluwé et al., 2011a, Bekaert et al., 2012). More recently, Wesoly et al. (2016) confirmed skatole diffusion through the skin, in a study on sampling procedures, but they also remarked that the final impact of soiling on skatole concentrations was low, because of the strong local limitation of skin diffusion and the low variability of skatole concentrations along the carcass found. The provision of enrichment material, especially of straw or other roughages, has a great potential to reduce agonistic interactions and aggression and its consequences like skin lesions (reviewed in a Special Issue by Animals, 'Enrichment material of pigs', Fàbrega, 2019). Some studies have indicated that enriched housing conditions may help to reduce levels of agonistic interaction in entire males (Holinger et al., 2015). Levels of androstenone in this investigation were lower than those noted in other published studies (Rydhmer et al., 2013), but control group without enrichment was not present in the study of Holinger et al. (2015). Prunier et al. (2013) found that skin lesions were higher in a control-barren environment compared to an enriched one, but plasma testosterone, oestradiol-17*β*, fat androstenone and skatole, and weight of testes did not differ between the housing systems. Therefore, it might be concluded that enrichment certainly has a potential to improve welfare of entire males, by reducing agonistic behaviour and its consequences, but the impact of enrichment on boar taint needs further studies to be elucidated. Backus et al. (2016) also pointed out that a partly open wall, wider gaps of the slats and the provision of an increasing light schedule around delivery of the pigs can reduce aggressive behaviour in boars. In that sense, it has been found that season can affect sexual maturation, with an accelerated pubertal development (Prunier et al., 2013) and highest reproductive success (Claus and Weiler, 1985) described in autumn (i.e. with short photoperiod or day length). However, results on the effects of season on levels of androstenone have not been totally consistent in all studies. Whereas the expected higher levels of androstenone associated with short day periods have been reported by Andersson et al. (1998a) and Thomsen et al. (2015), contradictory results by Fredriksen et al. (2006) or Andersson et al. (1998b) showed the lowest levels of androstenone in the period from October to March. This aspect may be of special relevance to pigs kept in systems with natural lightning, such as organic production. In that sense, Thomsen et al. (2015) found that the effect of season interacted with weight, resulting in a greater difference in androstenone levels between seasons at heavier slaughter weights. However, Prunier et al. (2013) found a similar influence of season in conventional housing and in housing with access to an outdoor run, and Backus et al. (2016) found no effect of green light or increasing light regime on mounting behaviour.

- *Pre-slaughter conditions/slaughter weight or age.* A potential effect of stress-related hormones like cortisol on androstenone, skatole and indole formation was suggested by some early studies (Claus et al., 1994). Transport and lairage are meant to be periods with a high concentration of stressful factors for pigs. As shown by Rydhmer et al. (2013) and van Staaveren et al. (2015), mixing unfamiliar entire male pigs during transport and lairage at the abattoir will stimulate mounting behaviour and aggressive behaviour resulting in a higher risk of skin lesions, but with uncertain effects on boar taint. In that sense, Wesoly et al. (2015) concluded that androstenone and skatole could respond differently to pre-slaughter conditions. They found androstenone to be influenced by the duration of the transport, whereas skatole was affected by both transport and pre-unloading times, possibly reflecting stressful conditions and fighting encounters before slaughter. They suggested a reduction of both pre-unloading and transport times to reduce the risk of boar-tainted carcasses. The time in lairage has not yielded consistent results: whereas Heyrman et al. (2017) reported a trend for higher boar-taint scores associated to a longer time in lairage, the same authors could not confirm this effect later on when comparing from <1 h to >3 h in lairage, suggesting, thus, that boar taint does not appear to increase when pigs stay in lairage for longer than 3 h. In contrast, the incidence of skin lesions was higher for the pigs kept in lairage for a longer period. Fasting time period before slaughter was shown to influence boar taint levels in pigs (Kjeldsen, 1993), but this effect has not been deeply investigated thereafter.

Slaughtering at a younger age or weight than would be done for other pig types is done for entire males across many farms. Age is generally considered to be a more effective measure of boar taint risk than weight, but weight cut-offs appear to be a common industry practice (as an example, maximum weight for fresh meat market 105–110 kg in Spain; 115 kg in Denmark or Belgium). Countries like Ireland and the United Kingdom that have traditionally produced entire male pigs address the incidence of boar taint by slaughtering at low weight and before sexual maturity. As it will be further described, there is so far no international accepted and validated on-line method available for the measurements of boar taint in carcasses that throughout fulfils the requirement for a highly streamlined industry at the abattoirs. Therefore, slaughter weight/age is still used as a preventive measure.

2.2.3 Genetics and methods to detect boar taint

The genetic background has been found to influence the risk of boar taint development (Aluwé et al., 2011b; Robic et al., 2014; Parois et al., 2018). Mathur et al. (2012) evidenced that the maternal lines Landrace and Yorkshire had higher concentrations of androstenone and skatole in the dorsal fat compared to a paternal line Piétrain type. The Duroc breed also presents even higher levels of androstenone (Oskam et al., 2010). Developing breeding strategies against the main compounds of boar taint seems promising, and could become very effective due to the high heritabilities reported (Robic et al., 2008; Baes et al., 2013; Strathe et al., 2013; Parois et al., 2015). Moreover, it would probably receive a high consumer acceptance and result in favourable effects on relevant performance traits, like feed efficiency and carcass leanness, with an important economic impact (Bee et al., 2015). Windig et al. (2012) and Strathe et al. (2013) reported low positive genetic correlations between carcass fat depth and boar taint compounds, suggesting that selection to increase lean meat content in pigs may slowly decrease the concentration of skatole equivalents and androstenone. Since androstenone biosynthesis is controlled by the same mechanisms than other sexual steroids, Zamaratskaia and Squires (2009) suggested that there are candidate genes and genetic markers that could be used to identify animals with a decreased genetic capacity to accumulate androstenone in fat while maintaining the normal levels of testicular steroids characteristic of males. Moreover, as previously mentioned, Parois et al. (2017) also found greater concentrations of androstenone and higher percentage of agonistic behaviour in dominant boars, which were not associated with levels of plasma testosterone or oestradiol. This would suggest again that selection against boar taint compounds, if targeting the adequate candidate genes, could have a favourable effect on social behaviour, whereas not affecting reproductive traits. Caution should be taken, though, with potential interactions with other characters of interest. An early review by Robic et al. (2008) reported a mean heritability for androstenone of 0.56, with a wide variability (0.25–0.88), whereas for skatole it is lower (0.19–0.54). The genetic correlation between both compounds is positive, with a value of about 0.4 (Parois et al., 2018). Several studies have been conducted to identify the zones of the genome with a quantitative effect on boar taint. Some of the results that are listed by NAGRP (National Animal Genome Research Program) in the web site https://www.animalgenome.org, indicate that the zones detected depend on the genetic lines used in the detection protocols and also on the methods to evaluate boar taint. Some zones of the genome have been located in the chromosomes 6, 7, 10 and 14 for androstenone and/or skatole (de Campos et al., 2015; Große-Brinkhaus et al., 2015).

To include such traits into practical breeding schemes requires that their assessment is reliable. The most widely used methods of detection of boar taint, which are still under development and need to be refined, are (for a review see the Boarcheck final report, Haugen et al., 2014):

1 Instrumental methods (spectrophotometry, immunology), for the determination of androstenone or skatole levels. At present, instrumental methods are not commercially available, although some prototypes based on spectrometry are already being tested and the rapid advance of technology probably will turn new developments in affordable equipment.
2 Sensory methods, evaluating boar odour by trained persons, mainly the so-called human nose scoring system as seen in Fig. 4.

It must be said that the heritabilities of these traits seem to be lower when boar taint is assessed by sensorial methods as compared to instrumental ones (Parois et al., 2018). The sensory evaluation by the Human Nose is said to be cheap and fast, but can be performed only on carcasses of siblings of

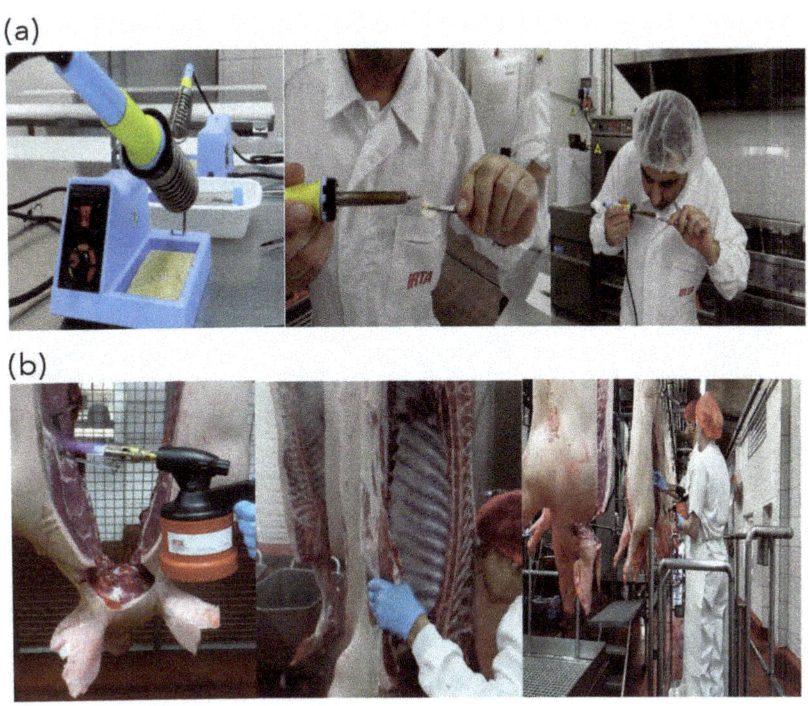

Figure 4 Use of the human nose method to assess boar taint risk (a) at-line in the laboratory, and (b) at the slaughterhouse (pictures from IRTA).

selection candidates (Mathur et al., 2012, 2013). The analytical determination of androstenone, skatole or indole can be carried out either on back fat samples collected at the abattoir from offspring of breeding boars or on biopsy fat samples from selection candidates, the later seeming more efficient and faster (Baes et al., 2013). The response to selection may be improved by including genomic information, which allows the estimation of breeding values based on genome-wide marker maps (Haberland et al., 2014; Lukić et al., 2015). Haberland et al. (2014) modelled a terminal sire-line breeding programme comparing information from the biopsy-based performance testing (Baes et al., 2013), from human nose scoring at the breeding station (Windig et al., 2012) or on genomic selection against either chemical compounds or human nose. The authors concluded that breeding against boar taint using the biopsy-based performance testing is an effective method to optimize both selection against the chemically determined boar-taint compounds and selection against human nose scores of boar taint, in terms of yearly obtained genetic gain and variable costs per selection candidate. Bee et al. (2015) reported studies in the specific Swiss pig population in which reductions of 50% in the levels of boar taint could be expected within 6, 7 and 8 years of selection. In a simulation study, Frieden et al. (2012) calculated that for the sire line four generations would be required to reduce the number of boars with high levels of androstenone, whereas to implement this breeding goal for dam lines could take up to nine generations, because an antagonistic relationship between maternal fertility and androstenone was assumed.

The major constraints with regards to the breeding strategies is that there are still some uncertainties about potential effects on fertility, sexual maturity, average daily gain and meat-quality traits such as drip loss and intramuscular fat content (Strathe et al., 2013; Haberland et al., 2014; Mercat et al., 2015). Although some studies seem to indicate that there is not a strong genetic antagonism between levels of androstenone and size of litter (Strathe et al., 2013) or other reproductive traits (Hidalgo et al., 2014), the carcass and meat-quality differences between entire males and castrates seem to be more genetically determined (Parois et al., 2018). Probably because of these uncertainties, breeding organizations seemed to be initially reluctant to include boar taint in the genetic selection. However, it appears that the increasing knowledge on which traits to breed against and the advance in the omics technologies, has encouraged breeding companies in different European countries to include boar taint as a factor in their selection methods, in some instances in cooperation with retailers or processors (for a review, see EU report 2019). Examples of farmers using low boar-taint sperm obtained after a research project involving both industry and university actors have been collected by projects like EUPIG (Anon. 2017) and are available online (https://www.eupig.eu/meat-quality/male-fatteners-without-boar-taint).

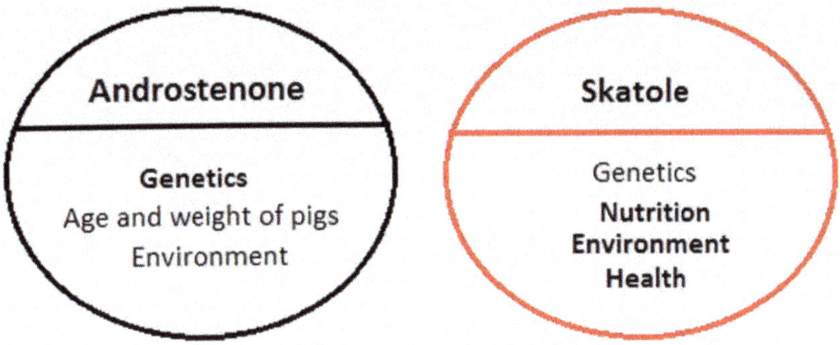

Figure 5 Main factors influencing androstenone and skatole level (adapted from Parois et al., 2018), in bold the ones having a higher impact.

As a summary, the main factors influencing each of the two principal boar-taint compounds are represented in Fig. 5.

3 Immunocastration

3.1 Mechanism of action

Immunocastration is an active immunization against gonadotropin-releasing hormone (GnRH), a key hormone of the endocrine cascade regulating reproductive functions. As a result, the pig's immune system produces antibodies against its own hypothalamic hormone GnRH, postponing the onset of the pubertal development by the suppression of the hypothalamic pituitary gonadal axis (Thompson, 2000). The antibodies bind endogenous GnRH suppressing the testicular functions for at least 10 weeks (Thompson, 2000), which may resume thereafter (Claus et al., 2008; Einarsson et al., 2009, Fig. 6). GnRH is transported by the hypophysial portal systems from the anterior hypothalamus to the anterior pituitary. The short exposure time of the GnRH to the blood system during transport from the hypothalamus to the pituitary gland is the only opportunity for exposure of GnRH to circulating antibodies. If antibodies specific to GnRH are exposed to GnRH during this stage, GnRH will bind to them and be neutralized, either by the prevention of its diffusion through capillaries or by the occupation of the binding site of GnRH to the anterior pituitary. The GnRH itself is too small to be immunogenic, and various techniques have been used to enhance the animal's system into recognizing GnRH as foreign. Such techniques typically involve conjugating GnRH to a large, non-self or foreign protein using various sites and methods of conjugation. This strategy includes the use of GnRH or a modified GnRH (truncated or repeated, with or without amino acid substitution) as antigenic target linked to a carrier substance (Ayalew, 2019). Antigens which are conjugated through its C

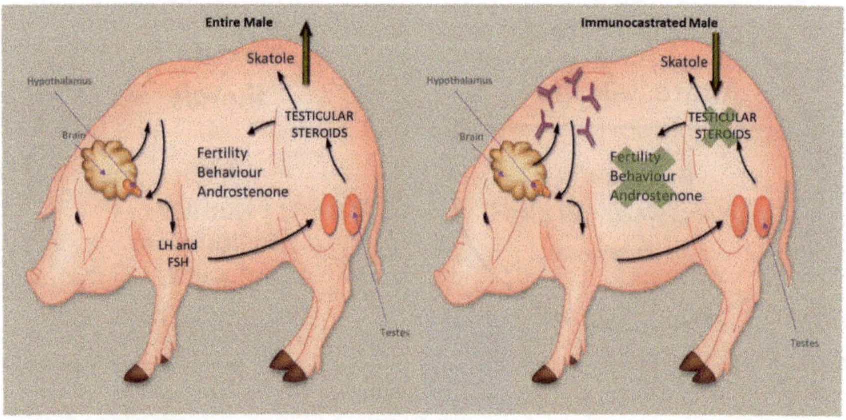

Figure 6 Physiological response to immunocastration in male pigs (adapted from Fàbrega et al., 2008, by permission of SUIS).

terminus seem to produce a higher specific antibody response than constructs in which GnRH is conjugated through its N terminus (Chang et al., 2015). The type of adjuvant used and the number of immunizations required to elicit a response are two constraining factors that hamper the use of immunovaccines. Commercial vaccines have been designed to minimize adverse reactions to the adjuvant species as well as to minimize frequency of vaccinations. However, other factors also influence the results seen from immunocastration, such as the interval between vaccinations, the timing of the second vaccination relative to slaughter, the age of the animal, the duration of the study, as well as the nutritional strategy applied (Needham et al., 2017).

In pigs, in order to be effective, a minimum of two consecutive vaccinations are required in at least in a 4 weeks interval (Fig. 7). Subcutaneous injections are given at the base of the ear with a special vaccinator designed by the vaccine producer to prevent accidental self-injection. After the first vaccination, some GnRH antibodies are already released but their concentration is not sufficient to limit gonadal axis activity (Kress et al., 2018). Within one week after the second vaccination, the production of GnRH antibodies increases markedly and suppresses testicular steroid synthesis and in consequence spermatogenesis (Claus et al., 2007; Kubale et al., 2013). The decrease in testosterone and estradiol concentrations occurs within a week, followed by a reduction in IGF-I secretion (Claus et al., 2007). The hormonal feedback system trigger an initial increase of GnRH by the hypothalamus due to the drop in testosterone, but after that the synthesis of GnRH in the hypothalamus is reduced (Han et al., 2017). Therefore, the endocrine system, growth performance or the behaviour of immunocastrates are boar-like until the second vaccination. Thereafter they become more comparable to barrows in all traits. The only way to maintain

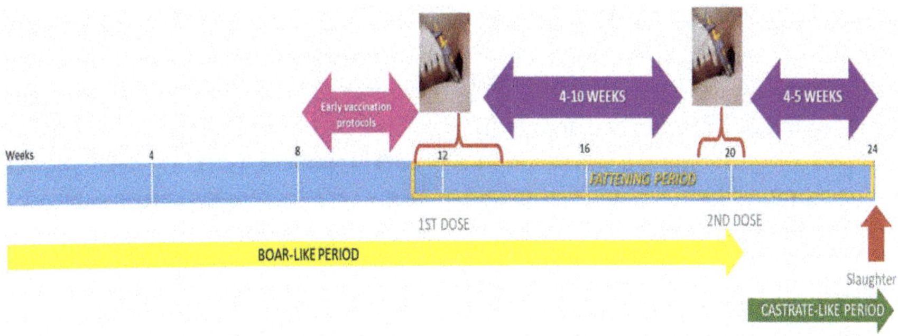

Figure 7 Possible schedules of immunocastration.* *Early vaccination protocols are not common.*

the anabolic advantage of boars is the appropriate timing of the antibody formation leading to a tailored cessation of the testicular steroid synthesis. This avoids the accumulation of boar taint till slaughter, although the anabolic effects of testicular hormones are still maintained during the main part of the fattening period.

The first vaccination could be applied at 8–9 weeks of age (Čandek-Potokar et al., 2017), although this is not recommended if piglets are not raised on the same farm because traceability for the second vaccination would be very difficult. Therefore, the first vaccination is currently being administered early in the fattening period, at an age of about 12 weeks. As previously mentioned, the recommended time between the second vaccination and slaughter is about 4–5 weeks, which allows the already accumulated androstenone and skatole from adipose tissue to be released. The longer is the time elapsed from the second vaccination to slaughter, the higher is the difference between immunocastrates and entire males and/or the similarity to surgical castrates. Long-term studies have revealed a recovery of testicular function after 10–24 weeks (Claus et al., 2008; Einarsson et al., 2009), but a third vaccination is only recommended when animals are slaughtered at heavy weights or higher ages, as common in production systems such as the Italian Parma pig production or Spanish Iberian pigs (Pinna et al., 2015; Martínez-Macipe et al., 2016). The timing of the second vaccination has been evaluated by different studies, in an attempt to properly balance conflicting aims like the desirable boar-like performance efficiency towards the welfare benefits of lower aggression in barrows and their higher fat deposition and meat quality (no boar taint). A recent meta-analysis of Nautrup et al. (2018) compared immunocastrates which had been administered the second dose before and after 4.5 weeks before slaughter. The animals vaccinated later were more boar-like in their growth and carcass characteristics than the immunocastrates vaccinated more than 4.5 weeks before slaughter. In some studies, the vaccination protocol of first vaccination/second vaccination between 10–14 and 16–20 weeks of age or 11–21 and 11–18 weeks of age have not revealed significant differences in growth performance and carcass composition (Andersson et al., 2012; Sattler et al., 2014), but welfare problems (Reiter et al., 2017) were reported: the percentage of penile injuries was found low (16.7%) if animals are vaccinated early and increased up to 41.7% if animals are vaccinated late. These last authors did not recommend however early vaccination (10–14 weeks) because of boar taint consequences.

3.2 Effects of immunocastration on boar taint, performance, carcass and meat quality and behaviour and welfare

3.2.1 Boar taint

Immunocastration blocks the synthesis of testicular steroids, including androstenone, by interfering with the hypothalamic-pituitary-gonadal axis. Androstenone production is suppressed as a consequence of suppressed testicular function. The approach with immunocastration, therefore, does not only prevent androstenone formation selectively but also reduces the synthesis of anabolic steroids. Immunocastration also reduces the level of another boar-taint compound, skatole (D'Occhio, 1993; Dunshea et al., 2001; Zamaratskaia et al., 2008a). Even though skatole is produced in the intestine by microbial degradation of amino acid tryptophan and the immunocastration has no direct effect on skatole synthesis, reduction of skatole levels in immunocastrated pigs is related to hindered production of androstenone and oestrogens. Sensory studies confirmed that immunocastration effectively and consistently reduces the risk of boar taint (reviews from Mackinnon and Pearce, 2007b; Batorek et al., 2012 and Nautrup et al., 2018; Zamaratskaia et al., 2008a, Font-i-Furnols et al., 2012b).

Although most studies suggest that effects of immunocastration in boar-taint control are similar to those of physical castration, it has also been found that some pigs may be less responsive to immunocastration (Batorek et al., 2012) and, thus, still high levels of androstenone and skatole in fat may be encountered (Batorek et al., 2012; Heyrman et al., 2019). This may be due to the influence of other factors such as breed, health problems or age at sexual maturity. The age of reaching puberty in boars, as well as the mechanisms of androstenone and skatole, differs between breeds. Prevalence of boar taint has been found to be higher for Duroc breed, compared to Hampshire, Yorkshire and Landrace (Zamaratskaia and Squires, 2009). Piétrain-sired pigs generally have a lower growth rate than Duroc-sired pigs, so slaughter age differs between fattening pigs from different sire lines (Liu et al., 2007). Therefore, some authors suggest that the state of pubertal maturity, behaviour and fat and plasma boar-taint compounds in fat and plasma may differ at second vaccination due to breed or individual differences, and this, in turn, may affect the response and pattern of diminution of androstenone/skatole.

3.2.2 Growth, performance parameters and carcass and meat characteristics

In general, immunocastrated pigs grow faster and have better feed conversion ratio than castrated pigs. Compared to entire male pigs, immunocastrated pigs consume more feed and also grow faster. The higher boar-specific anabolic potential and reduced feed intake can be exploited until the second vaccination (Fàbrega et al., 2010; Batorek et al., 2012). After the second vaccination, feed intake of immocastrates increases significantly, but compared to barrows they grow still more efficiently (Fàbrega et al., 2010; Batorek et al., 2012; Nautrup et al., 2018, Table 3). A recent meta-analysis (Nautrup et al., 2018) including 78 studies showed that, over the whole fattening period, immunocastrates have higher average daily gains than boars and barrows, whereas their feed conversion ratio is intermediate. These authors also concluded that when feeding ractopamine, which is not allowed in the EU, but it is used in other countries, the difference in average daily gain became even more favourable for immunocastrates.

The same meta-analysis by Nautrup et al. (2018) concluded that although carcasses from immuncastrates present a lower dressing percentage compared to castrated pigs, they are leaner and yield a more valuable meat. The dressing percentage of boars is inferior to that of barrows mainly due to higher weights of the genital tract. Immunocastrates present an even lower dressing percentage due to a higher volume of the gastrointestinal tract (Millet et al., 2011). In terms of carcass lean content, immunocastrates present an intermediate position between boars and barrows (Batorek et al., 2012). The meat of immunocastrates is similar to that of barrows (Font-i-Furnols et al., 2012b; Kress et al., 2019), both presenting higher levels of intramuscular fat and lower shear force compared to boars. The fatty acid profile of immunocastrates is also comparable to that of barrows, with less PUFAs than boars, which is particularly important for the production of dry-cured products (Čandek-Potokar et al., 2017). Meta-analysis of studies comparing immunocastrates and castrates raised for the production of high-quality cured products have suggested immunocastration as a suitable alternative to physical castration, since the requirements of backfat thickness and intramuscular fat for these products would still be fulfilled (Čandek-Potokar et al., 2017; Nautrup et al., 2018). The timing of the second vaccination seems to allow for a product quality tailored to the different pork markets. Given the fact that reproductive function may recover, a triple vaccination protocol has been proposed for these systems producing heavy pigs. Pinna et al. (2015) compared castrates with immunocastrates receiving a double or triple vaccination and showed higher levels of boar-taint compounds for those pigs vaccinated only twice. They concluded that a three-dose immunocastration protocol would be recommended to meet the requirements for the Italian PDO hams. In Iberian pigs (Martínez-Macipe et al., 2016), a triple vaccination protocol was found to result in no major differences on carcass or technological and sensory meat quality in females, whereas immunocastration of Iberian pig males produced somewhat leaner carcasses with less intramuscular fat and lower tenderness than surgical castration. At present new vaccination protocols are being tested for Iberian pigs to obtain better meat-quality parameters. For example, Font-i-Furnols et al. (2017) compared a pre-pubertal protocol (vaccination at 4.5, 5.5 and 9 months of age) with a pre-finishing protocol (at 10, 11 and 14 months of age), and found a higher lean percentage for the pre-finishing vaccinated pigs, whereas those vaccinated pre-pubertally presented higher fat and bone and lower lean proportion.

3.2.3 Behaviour and welfare

Immunocastration itself has been considered a relatively welfare-friendly alternative to surgical castration, since it prevents a painful procedure and the risk of wound infection in piglets. Piglet mortality was found to be higher in surgically castrated pigs in the first week after birth, compared to intact piglets (6.3% vs. 3.6%, Morales et al., 2017), especially in those piglets with low or medium live weight. When considering the possibility of pain produced by injection itself, McGlone et al. (2016) evaluated the effects of an intramuscular or subcutaneous injection of a saline solution on behaviour and stress physiology (cortisol) of piglets, and found no significant changes 1 h after injections. In early studies, some local reactions were detected by palpation in 6.25% of immunocastrates at the slaughterhouse (Einarsson, 2006) and increase in Pig-MAP after the two injections was also reported by Fàbrega et al. (2010). Plasma release of this acute phase protein is stimulated by stress-related hormones such as adrenocorticotropic hormone (ACTH) and glucocorticoids (Gruys et al., 1994) by non-specific reactions after tissue damage that can also appear as a result of infectious, inflammatory or neoplastic processes, or together with an immune response (Le Floc'h, 2003). Fàbrega et al. (2010) suggested that the plasma release in their study was probably more linked to the vaccine-like immune response triggered by immunocastration, rather than to a more general stress response, because the increase after the second dose was smaller, although a possible effect of handling could not be disentangled. In outdoor systems, the extra handling required to administer the vaccine may cause some practical difficulties that could trigger acute stress situations for pigs.

The male-like behaviour that immunocastrates show until the second vaccination may cause a higher incidence of aggression and mounting behaviour, and a higher number of skin lesions, compared to castrates (Fàbrega et al., 2010; Rydhmer et al., 2010). However, if the second vaccination is fully effective, social and sexual behaviour of immunocastrates changes relatively soon after the injection (about 1 week), and levels of aggressive encounters and mountings are reduced to the level of surgically castrated pigs (Cronin et al., 2003; Baumgartner et al., 2010; Fàbrega et al., 2010; Rydhmer et al., 2010; Puls et al., 2017). Significant behavioural differences in social, manipulating, and aggressive behaviour exist between immunocastrates and boars, even 16 or 22 weeks after an effective second immunization (Zamaratskaia et al., 2008b). A calmer behaviour is desirable not only to improve animal welfare, but also for carcass quality, since it reduces the skin lesions produced after fighting, especially if unfamiliar pigs are mixed prior to slaughter (e.g. during transport or lairage). Along that line, Guay et al. (2013) assessed human-pig interactions and handling during transport in immunocastrated pigs and barrows, and found little differences. Skin lesions are not only attributable to fighting, but also to mounting activity. Lameness and skeletal problems were reported in a 15% of all boars (Rydhmer et al., 2006); scratches in the head region were found to be much reduced in immunocastrates and absent in barrows compared to boars (Einarsson, 2006); and the higher incidence of shoulder skin lesions in immunocastrates compared to barrows before second vaccination was found to disappear after the second dose (Schmidt et al., 2011). Penile injuries found in boars may also be reduced by immunocastration, being the percentage of decrease more important if vaccination is applied early (16.7% vs. 41.7% in early vs. late vaccination (Reiter et al., 2017). Kress et al. (2018) also found a reduction in aggressive behaviour and in penile injuries in immunocastrates compared to boars regardless of the housing environment (stressful, conventional or with outdoor access).

Another important aspect worth considering for welfare in immunocastrated males is related to changes in feeding behaviour. Feed intake after second vaccination is considerably increased, up to a 25% of increase in meal size (Cronin et al., 2003; Weiler et al., 2013; Fàbrega et al., 2010; van den Broeke et al., 2016). Therefore, their feeding requirements need to be adapted to guarantee satiety, and avoid negative effects on body composition. Moreover, if fed under a restricted regime, immunocastrates showed similar levels of aggression and incidence of skin lesions than restricted fed entire males, and higher when compared to entire males fed ad lib or to castrates (Batorek et al., 2012). Therefore, restricted feeding of immunocastrates during late finishing period is not recommended.

3.3 Other practical considerations of immunocastration

3.3.1 Reliability and efficacy

Some review papers have tried to quantify and provide explanation on why some pigs are non-responders to immunocastration (Čandek-Potokar et al., 2017). No

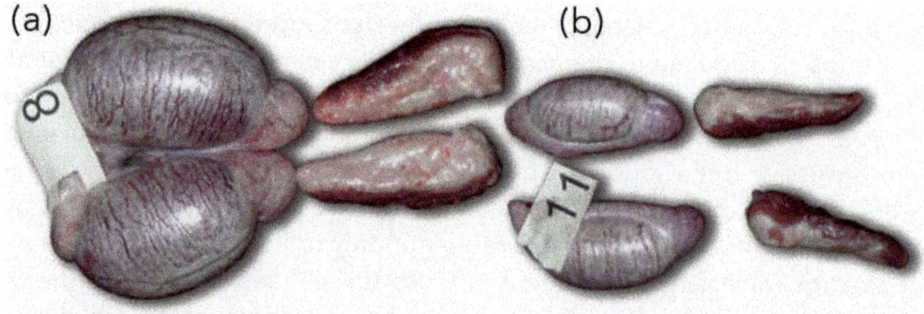

Figure 8 Reduced size of testes is one of the consequences of a sufficient immunocastration protocol; (a) entire male; (b) immunocastrated (pictures from IRTA).

single definition for non-responders exists, but the review papers established two criteria like: (1) presence of enlarged testes (similar to boars, Fig. 8) and (2) high boar-taint levels (above the threshold of 0.5–1 ppm for androstenone). At present, there are two available vaccines at commercial level, although one of them is not licensed in Europe and most of the scientific results are published for the first vaccine (Improvac©) launched in 1998 in Australia and New Zealand (Mackinnon and Pearce, 2007a). A 0–3% of non-responders for the Improvac© vaccine are assumed (Čandek-Potokar et al., 2017). Only commercial results are available for the other vaccine by Ceva (Kiss et al., 2017), with adequate results according to the company. Reasons for non-responding animals may be: (1) lack of a proper protocol of vaccination (i.e. accidentally missing one of the two-fold doses; incorrect application); (2) inadequate handling or storage of the vaccine; (3) interaction with the health or stress status of the pig, because only healthy pigs are suitable for vaccination. In that sense, a recent study from Kress et al. (2018) showed that even under additional stress before and after the vaccination, all immunocastrates showed a sufficient immune reaction with sufficient titers and low testicular steroid production. However, insufficient immunizations may occur under practical conditions and, therefore, a need for systems to properly detect boar taint on line, like when raising entire males, could be recommendable.

3.3.2 Farmers' acceptance

The stakeholders' perception of the procedure is very heterogeneous and varies between countries. When immunocastration was introduced, the first barrier to overcome with farmers was to develop a safety system to administer the vaccine (Fig. 9). But even after the danger of self-injection has been properly addressed with secure devices, still big differences in acceptance are present: whereas in Australia and Brazil more than 50% of the male pigs are immunocastrated (Mancini et al., 2017; D'Souza et al., 2018),

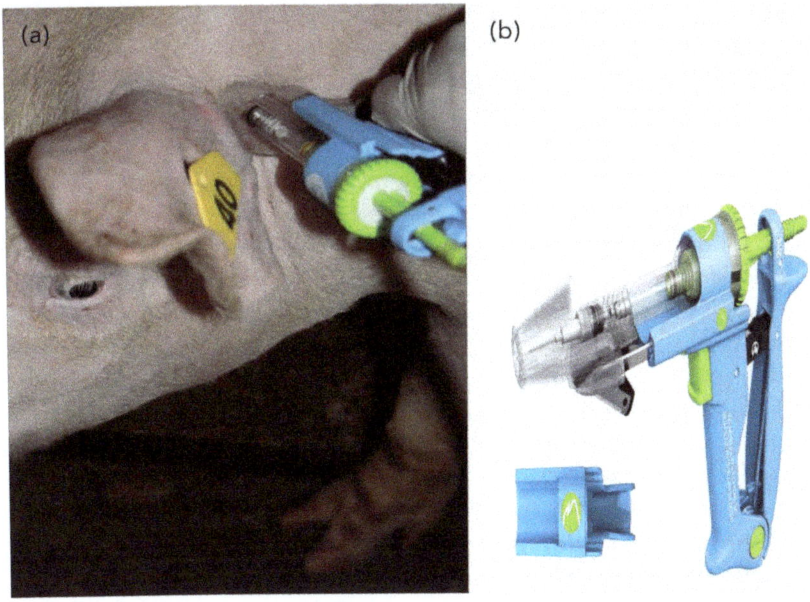

Figure 9 (a) Adequate injection site and (b) safe vaccination system, with retractile needle (original picture produced with permission from Zoetis Inc. November 2019).

Tuyttens et al. (2011) found that most European pig farmers preferred to continue with surgical castration, even though the ethical debate had started to put a lot of pressure on them. The market share of immunocastrated male pigs in Europe has kept at a low level, with only a 15% in Belgium (Backus et al., 2018). The reasons behind this reluctance in European farmers may be associated to different factors, with lack of knowledge being one of the starting points to be overcome according to Tuyttens et al. (2011) and Schübeler and Koch (2018). However, another interesting study from Aluwé et al. (2015) pointed out that practical experience with immunocastration increased Belgium farmers' concerns towards this method, being consumer acceptance, reliable prevention of boar taint, economic efficiency and the risk of accidental self-injection as the most important drawbacks. In the future, the farmers need to avoid surgical castration of pigs due to societal or industry pressure, decrease in the price of the vaccination protocol, and clear trust from consumers may change this still low engagement of farmers in some areas like Europe.

3.3.3 Societal acceptance

A review by Mancini et al. (2017) showed that the majority of consumers are unaware that male piglets are castrated surgically without anaesthesia or analgesia. Furthermore, only a few consumers knew of immunocastration. When making purchase decisions at the counter, attributes like meat quality, safety and animal welfare are weighed and result in the acceptance or rejection of pork from immunocastrates. In terms of animal welfare and product quality, consumers rate immunocastration more positively than pork production with boars or barrows, but are more skeptical about food safety and prices (Fredriksen et al., 2011; Kress et al., 2019). Di Pasquale et al. (2019) showed that Italian consumers rate meat from immunocastrates more positively than meat from surgical castrates or entire males, with a low risk perception of immunocastration and a higher willingness to pay for products from immunocastrated male pigs. Fredriksen et al. (2011) showed that consumer concerns can be minimized by information programmes from public authorities. Furthermore, some studies show that information material for target groups increases consumer acceptance of immunocastration - especially if audio-visual techniques are employed (Tuyttens et al., 2011).

3.3.4 Immunocastration as an alternative for high-quality or organic production systems

In the production of traditional high-quality pork products such as dry-cured hams and shoulders, animals are slaughtered at higher live weights. In Iberian high-quality production systems, boars and gilts are castrated surgically either to prevent boar taint or undesirable performance losses in female animals during oestrus (Martínez-Macipe et al., 2016), or in extensive free-ranging housing systems in order to prevent unwanted pregnancies during fattening (Dalmau et al., 2015). The spaying of female pigs is restricted by EU Directive 2008/120/EC (Anon. 2008), but national regulations like RD1221/2009 in Spain allow some exceptions for extensive systems, and provided that castration is performed under deep anaesthesia and analgesia by a veterinarian. Therefore, immunocastration could be a feasible alternative for those production systems, requiring three vaccinations as previously explained (Pinna et al., 2015). Consumers accepted immunocastration for the production of Parma ham if animal welfare, product quality and consumer safety are guaranteed (Di Pasquale et al., 2019).

The issue becomes a bit more controversial in organic pig production, because of possible conflicts with the legal requirements for organic production and with organic consumer attitudes. In a study from Grela et al. (2013) immunocastration in organic pig production was evaluated favourably from a productive and meat quality perspective. However, Heid and Hamm (2012) showed that German consumers were sceptical about immunocastration in organic pork production because they were worried about residues in meat and, moreover, a lower percentage considered immunocastration as a severe interference with the integrity and nature of the individual and was, therefore, deemed inappropriate for organic farming. In that sense, it has to be mentioned that as part of the European Medicines Agency approval process for Improvac® (Anon. 2010b, EMA, 2019), food safety was evaluated and several studies tested hormonal and oral efficacy of the synthetic antigen used in the vaccine.

3.3.5 Sustainability of pig production systems using immunocastration

According to Kress et al. (2019), in an exhaustive review based on the SuSi Eranet project, 'Immunocastration is a technique to improve pork quality, animal welfare, economic profitability, and environmental protection, which can contribute to a more sustainable pork production in Europe'. However, these authors point out that conflicting aims exist within each pillar of sustainability, as well as between the three pillars which have to be balanced against each other in the production process (Fig. 10). Those systems aiming at optimizing welfare through an early second vaccination, present synergistic aims with obtaining high product quality products with low boar-taint levels and higher levels of intramuscular fat. However, these advantages have to be balanced against the higher anabolic potential of boars which can create economic and environmental benefits. The later the second vaccination is applied, the better its effects for the environment and for farm profitability. The authors

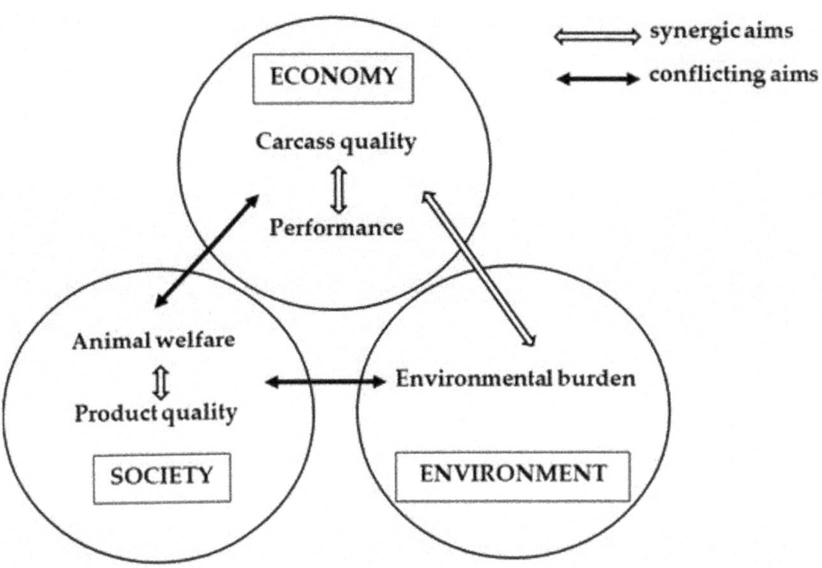

Figure 10 Sustainability aspects of the use of immunocastration according to the three pillars: economy, society and environment (from Kress et al., 2019).

suggest that synergistic aims exist between the pillar of economy and the pillar of environment. On the other hand, conflicting aims between these two pillars and the pillar of society also exist. Therefore, the timing of the second vaccination is essential in order to produce a meat quality tailored to various market segments and with different impacts on sustainability.

4 Sex sorting of semen

Whereas for other species such as cattle the use of sexed semen is a common practice, the same is not likely to become available for swine in the near future (Vazquez et al., 2009), because of price and technological difficulties. The system is based on high speed flow cytometry which sorts DNA-stained spermatozoa with difference in size between X and Y chromosome sperm. Problems for its commercial application in pigs are related to the low survival of the sorted spermatozoa, the too slow speed of the process and the prices. However, the rapid development of instruments, probes and bioinformatics tools may help in the future to overcome these problems.

5 Case study

For an interesting collection of different case studies with regards to alternatives to piglet castration, and to the way the pig industry is adapting to the challenge of stopping surgical castration, the document 'Establishing best practices on the production, the processing and the marketing of meat from uncastrated pigs or pigs vaccinated against boar taint (immunocastrated) (2019)' can be visited (https://ec.europa.eu/food/sites/food/files/animals/docs/aw_prac_farm_pigs_cast-alt_establishing-best-practices.pdf).

The selected example here refers to the adaptation of the vaccination protocol to the production in heavy pig production systems, like Parma or Iberian pigs. Iberian pigs are reared to produce high-quality products mainly because of their fat content and composition. Ham is the most valuable part of these pigs and it is important to ensure quality, mainly in terms of amount of fat content and sensory quality. To avoid the boar taint problem, surgical castration was traditionally being in male pigs. When immunocastration was available, it became an option for both males and females in the Iberian pig production system, especially for the extensive one to avoid unwanted pregnancies of females. However, the long life cycle of Iberian pigs required that the vaccination protocol was adapted with the inclusion of a third dose. Moreover, the effect of the vaccine on the final carcass composition might be modified by changing the vaccination schedule, that is, applying an early or a late vaccination or modifying the time between doses. Therefore, trials were conducted in Iberian pigs in extensive systems (Montanera) to evaluate the influence of two immunocastration protocols: pre-pubertal (P) and pre-finishing (F) on the

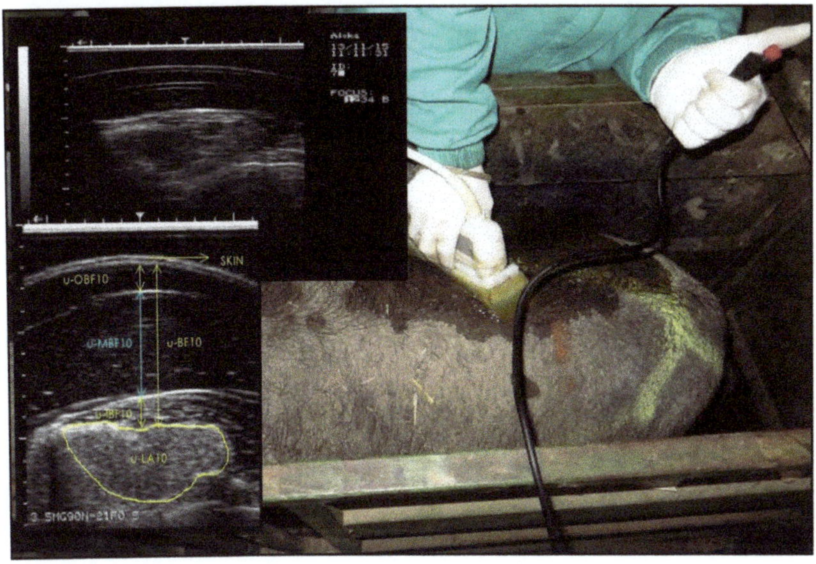

Figure 11 Ultrasound measurements of fat thickness in live pigs (a) over the loin at the 10th rib level (Image from IRTA).

composition of hams from Iberian pigs by means of computed tomography (CT). Early immunocastrated pigs received their doses at 4.5, 5.5 and 9 months and late immunocastrated (11, 12 and 14 months). Computed tomography (CT) was used to validate other non-invasive technologies like in vivo ultrasounds (Figs 11 and 12), because results are in line with those obtained *post mortem* with a ruler directly on the carcass. Early immunocastration (pre-pubertal) increased the carcass weight of the pigs and produced fattiest carcasses than late immunocastration (pre-finishing) when animals were slaughtered at the same age (García-Gudiño et al., 2019).

Thus, the Iberian pig industry has now an alternative to traditional surgical castration, and can adapt the vaccination protocol according to the requirements for the final product in terms of age and weight of slaughter and fatness.

6 Conclusion and future trends

Piglet castration is a painful procedure which has for a long time been carried out to avoid the risk of boar taint in meat, and to reduce aggressive and sexual behaviours to guarantee a proper welfare for entire males. Raising pigs in modern intensive husbandry systems without castration faces the pig industry with the need to solve these problems. At present, the main two available alternatives are: raising entire males and immunocastration. A summary of the advantages and disadvantages of both of them are summarized in Table 4.

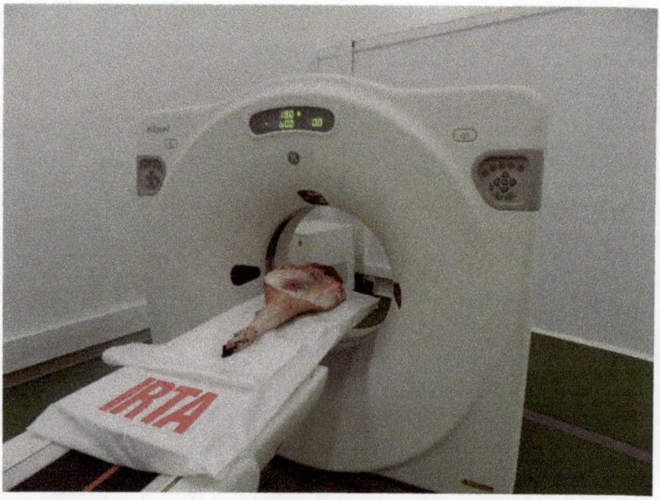

Figure 12 Computed tomography scanning of a ham (Image from IRTA).

Regardless of the alternative being used, a certain risk for boar-tainted carcass still exist, and especially when keeping males pigs intact a need to sort carcasses on line may be required.

Although boar-taint control has been a matter of study during the last three decades, there are still knowledge gaps that could be further elucidated. Moreover, raising entire males presents as well some challenges to keep proper standards of animal welfare. Main areas of future research would be:

- Development of rapid instrumental methods, and affordable, for the chemical detection of boar taint on line;
- Reference methods, harmonized and mutually recognized, for the measurement of each of the compounds responsible for boar taint;
- Better understanding of the metabolism of each of the compounds of boar taint and its interactions;
- New breeding, nutritional or management strategies to reduce boar taint risk;
- Production systems and management of entire males during rearing, transport and slaughter to reduce sexual and aggressive behaviours; and
- Alternatives to surgical castration for heavy pig production systems, that totally guarantee the meat-quality standards that these systems normally aim for.

If the future of pig production systems aims at stopping surgical castration, a fine-tuned integration between proper management strategies and on-line detection methods will be required, as well as the definition of strategies on

Table 4 Advantages and disadvantages of the commercially available alternatives to surgical piglet castration (adapted from Čandek-Potokar et al., 2017)[1]

Alternative	Advantages	Disadvantages
Entire male production (EM)	Better productive efficiency Lower N-excretion No castration pain or wounds	Risk of boar taint Higher incidence of aggression and mounting Lower technological meat quality for some parameters
EM with nutritional strategies	Better productive efficiency Reduced risk of high skatole levels No castration pain or wounds	Cost of certain feed ingredients No total guarantee of risk of boar taint (androstenone) Lower technological meat quality for some parameters Higher incidence of aggression and mounting Not a solution for heavy pig production
EM slaughtered at young age/low weight	Better productive efficiency Reduced risk of high skatole and androstenone No castration pain or wounds	No total guarantee of risk of boar taint Lower technological meat quality for some parameters Higher incidence of aggression and mounting Not a solution for heavy pig production
EM with breeding strategies	Better productive efficiency Reduced risk of high (skatole and) androstenone No castration pain or wounds	No total guarantee of risk of boar taint Lower technological meat quality for some parameters Higher incidence of aggression and mounting Risk of influence on growth and reproduction Not a solution for heavy pig production
Immunocastration	Better productive efficiency Available for males and females Reduced risk of high androstenone and skatole Applicable to heavy pig production Potential better sustainability	Cost Safety of workers needs to be guaranteed Authorisation of drug Consumer acceptance

[1] Use of surgical castration with anaesthesia and analgesia has not been included in this chapter, as it has been included in a previous chapter. Sex sorting has not been included either because it is not commercially available due to technological limitations and cost.

how to use carcasses detected with a potential risk of boar taint and deal with the higher incidence of aggressive and sexual behaviour of entire males.

7 Where to look for further information

7.1 Further reading

In the European Union, the pressure against piglet castration has led to an important amount of projects producing reports, most of them available on-line, such as:

- PIGCAS (https://ec.europa.eu/food/sites/food/files/animals/docs/aw_prac_farm_pigs_cast-alt_research_civic_pt1-synthesis_20131202.pdf) and ALCASDE (https://ec.europa.eu/food/sites/food/files/animals/docs/aw_prac_farm_pigs_cast-alt_research_alcasade_final-report.pdf) reports can be downloaded from the internet.
- The 'Study and economic analysis of the costs and benefits of ending surgical castration of pigs' (2013) provides detailed information on economic figures (https://ec.europa.eu/food/sites/food/files/animals/docs/aw_prac_farm_pigs_cast-alt_research_civic_pt1-synthesis_20131202.pdf).
- 'The 1st and 2nd progress reports from the European Declaration on alternatives to surgical castration' (Anon. 2010a), summarize the progress done by research, but also on practical implementation of alternatives (https://ec.europa.eu/food/sites/food/files/animals/docs/aw_prac_farm_pigs_cast-alt_declaration_progress-report_20141028.pdf; https://www.boarsontheway.com/wp-content/uploads/2018/08/Second-progress-report-2015-2017-final-1.pdf).
- The BOARCHECK project ended with the report 'A study on rapid methods for boar taint used or being developed at slaughter plants in the European Union' (2012). Although at present new methodologies are rapidly evolving (especially instrumental, with prototypes being tested), this report is a proper starting point (https://ec.europa.eu/food/sites/food/files/animals/docs/aw_prac_farm_pigs_cast-alt_research_boarcheck_20140901.pdf).
- The CAMPIG project (2014) produced the report 'Consumer acceptance in the European Union and in third countries of pig meat obtained from male pigs not surgically castrated', which provides a deeper inside on the consumer attitudes (https://ec.europa.eu/food/sites/food/files/animals/docs/aw_prac_farm_pigs_cast-alt_research_campig_20140706.pdf).
- The CASTRUM study (2016) specifically addressed the issue of alternatives to castration for pigs for traditional products, including the use of anaesthesia and analgesia and providing a very detailed comparison of advantages and disadvantages of each alternative. The report can be

downloaded from the internet 'Pig castration: methods of anaesthesia and analgesia for all pigs and other alternatives for pigs used in traditional products' (https://www.boarsontheway.com/wp-content/uploads/2018/09/Castrum-study.pdf).

- The report 'Establishing best practices on the production, the processing and the marketing of meat from uncastrated pigs or pigs vaccinated against boar taint (immunocastrated)' (2019) gives an overview of how the different alternatives have been implemented on practical grounds (https://ec.europa.eu/food/sites/food/files/animals/docs/aw_prac_farm_pigs_cast-alt_establishing-best-practices.pdf).
- The open access book chapter 'Immunocastration as Alternative to Surgical Castration in Pigs' (Čandek-Potokar et al., 2017) is a very comprehensive review of immunocastration (https://www.intechopen.com/books/theriogenology/immunocastration-as-alternative-to-surgical-castration-in-pigs).
- The web page https://www.boarsontheway.com/ is being updated on new information of alternatives to pig castration.

7.2 Key journals/conferences

- Meat Science for the technological, sensorial and boar taint meat-quality issues.
- Applied Animal Behaviour Science, Animal, Animals, Animal Welfare, Physiology and behaviour, PLoS ONE cover the aspects related to animal welfare and behaviour.
- The IPEMA COST Action (CA 15215, 2016–2020) which involves 25 EU countries and also third countries, organizes annual meetings, workshops and training seminars every year (http://www.ca-ipema.eu/about-ipema). Dissemination materials can also be downloaded from the webpage.
- Specific sessions on meat quality of non-castrated pigs or welfare aspects may be dealt at ICOMST, ASAS, EAAP or ISAE annual conferences.

7.3 Major international research projects

- The already mentioned in the 'further reading'.
- The SuSi project is an Era-Net-funded project involving seven European countries, and which aims at evaluating the sustainability of pig production with immunocastration (https://era-susan.eu/content/susi).

8 References

Aluwé, M., Bekaert, K. M., Tuyttens, F. A. M., Vanhaecke, L., De Smet, S., De Brabander, H. F., De Brabander, D. L. and Millet, S. (2011a). Influence of soiling on boar taint in boars, *Meat Sci.* 87(3), 175–179.

Aluwé, M., Heyrman, E., Theis, S., Sieland, C., Thurman, K. and Millet, S. (2017). Chicory fructans in pig diet reduce skatole in back fat of entire male pigs, *Res. Vet. Sci.* 115, 340–344.

Aluwé, M., Millet, S., Bekaert, K. M., Tuyttens, F. A. M., Vanhaecke, L., De Smet, S. and De Brabander, D. L. (2011b). Influence of breed and slaughter weight on boar taint prevalence in entire male pigs, *Animal* 5(8), 1283–1289.

Aluwé, M., Millet, S., Nijs, G., Tuyttens, F. A. M., Verheyden, K., De Brabander, H. F., De Brabander, D. L. and Van Oeckel, M. J. (2009). Absence of an effect of dietary fibre or clinoptilolite on boar taint in entire male pigs fed practical diets, *Meat Sci.* 82(3), 346–352.

Aluwé, M., Vanhonacker, F., Millet, S. and Tuyttens, A. M. (2015).Influence of hands-on experience on pig farmers' attitude towards alternatives for surgical castration of male piglets, *Res. Vet. Sci.* 103, 80–86.

Andersson, H., Rydhmer, L., Lundstrom, K., Wallgren, M., Andersson, K. and Forsberg, M. (1998b). Influence of artificial light regimens on sexual maturation and boar taint in entire male pigs, *Anim. Reprod. Sci.* 51(1), 31–43.

Andersson, H., Wallgren, M., Rydhmer, L., Lundstrom, K., Andersson, K. and Forsberg, M. (1998a). Photoperiodic effects on pubertal maturation of spermatogenesis, pituitary responsiveness to exogenous GnRH, and expression of boar taint in crossbred boars, *Anim. Reprod. Sci.* 54(2), 121–137.

Andersson, H. K., Hullberg, A., Malmgren, L., Lundström, K., Rydhmer, L. and Squires, J. E. (1999). Sexual maturity in entire male pigs: environmental effects, relations to skatole level and female puberty, *Acta Agric. Scan.* 49(2), 103–112.

Andersson, K., Brunius, C., Zamaratskaia, G. and Lundström, K. (2012). Early vaccination with Improvac®: effects on performance and behaviour of male pigs, *Animal* 6(1), 87–95.

Annor-Frempong, I. E., Nute, G. R., Whittington, F. W. and Wood, J. D. (1997). The problem of taint in pork–II. The influence of skatole, androstenone and indole, presented individually and in combination in a model lipid base, on odour perception, *Meat Sci.* 47(1–2), 49–61.

Anon (2004). Opinion of the Scientific Panel on Animal Health and Welfare on a request from the Commission related to welfare aspects of the castration of piglets, *EFSA J.* 91, 1–18, Available at: https://www.efsa.europa.eu/en/efsajournal/pub/91 (Accessed on 15 June 2019).

Anon. (2008). Laying down minimum standards for the protection of pigs, Council Directive 2008/120/EC, Available at: http://eur-lex.europa.eu/legal-content/EN/TXT/?uri=CELEX%3A32008L0120 (Accessed on 24 April 2018).

Anon. (2010a). European declaration on alternatives to surgical castration of pigs, Available at: https://ec.europa.eu/food/sites/food/files/animals/docs/aw:prac_farm_pigs_cast-alt_declaration_en.pdf (Accessed on 31 March 2019).

Anon. (2010b). -scientific discussion, Available at: https://www.ema.europa.eu/en/documents/scientific-discussion/improvac-epar-scientific-discussion_en.pdf (Accessed on 30 April 2019).

Anon. (2017). EUPIG Project, dissemination material, Available at: https://www.eupig.eu/meat-quality/male-fatteners-without-boar-taint (Accessed on 20 September 2019).

Anon. (2019). Establishing best practices on the production, the processing and the marketing of meat from uncastrated pigs or pigs vaccinated against boar taint (immunocastrated). Final Report, European Commission, Luxembourg.

Ayalew, G. (2019). A review on the effect of immunocastration against gonadal physiology and boar taint, *Biomed. Nurs.* 5(1), 26–40.

Backus, G., Snoek, H., Oliver, M. A., Font-Furnols, M. and Aluwé, M. (2014). Consumer acceptance in the European Union and in third countries of pig meat obtained from male pigs not surgically castrated. CAMPIG report, European Commission.

Backus, G., Higuera, M., Juul, N., Nalon, E. and de Briyne, N. (2018). European declaration on alternatives to surgical castration of pigs. Second Progress Report 2015–2017, Brussels, Available at: https://www.boarsontheway.com/wp-content/uploads/2018/08/Second-progress-report-2015-2017-final-1.pdf (accessed on 26 April 2019).

Backus, G., van den Broek, E., van der Fels, B., Heres, L., Imminka, V. M., Knol, E. F., Kornelis, M., Mathur, P. K., van der Peet-Schwering, C., van Riel, J. W., Snoek, H. M., de Smet, A., Tacken, G. M. L., Valeeva, N. I. and van Wagenberg, C. P. A. (2016). Evaluation of producing and marketing entire male pigs, *NJAS Wageningen Journ. Life Sci.* 76, 29–41.

Baes, C., Mattei, S., Luther, H., Ampuero, S., Sidler, X., Bee, G., Spring, P. and Hofer, A. (2013). A performance test for boar taint compounds in live boars, *Animal* 7(5), 714–720.

Batorek, N., Čandek-Potokar, M., Bonneau, M. and Van Milgen, J. (2012). Meta-analysis of the effect of immunocastration on production performance, reproductive organs and boar taint compounds in pigs, *Animal* 6(8), 1330–1338.

Baumgartner, J., Laister, S., Koller, M., Pfützner, A., Grodzycki, M., Andrews, S. and Schmoll, F. (2010). The behaviour of male fattening pigs following either surgical castration or vaccination with a GnRF vaccine, *Appl. Anim. Behav. Sci.* 124(1–2), 28–34.

Bee, G., Chevillon, P. and Bonneau, M. (2015).Entire male pig production in Europe, *Anim. Prod. Sci.* 55(12), 1347–1359.

Bee, G., Silacci, P., Ampuero-Kragten, S., Čandek-Potokar, M., Wealleans, A., Litten-Brown, J. and Mueller-Harvey, I. (2017). Hydrolysable tannin-based diet rich in gallotannins has a minimal impact on pig performance but significantly reduces salivary and bulbourethral gland size, *Animal* 11(9), 1617–1625.

Bekaert, K. M., Aluwé, M., Millet, S., Goethals, K., Nijs, G., Isebaert, S., De Brabander, D. L., Verheyden, K., De Brabander, H. F., Vanhaecke, L. and Tuyttens, F. A. (2012). Predicting the likelihood of developing boar taint: early physical indicators in entire male pigs, *Meat Sci.* 92(4), 382–385.

Bonneau, M., Čandek-Potokar, M., Škrlep, M., Font-i-Furnols, M., Aluwé, M. and Fontanesi, L. (2018). Potential sensitivity of pork production situations aiming at high-quality products to the use of entire male pigs as an alternative to surgical castrates, *Animal* 12(6), 1287–1295.

Bonneau, M., Walstra, P., Claudi-Magnussen, C., Kempster, A. J., Tornberg, E., Fischer, K., Diestre, A., Siret, F., Chevillon, P., Claus, R., Dijksterhuis, G., Punter, P., Matthews, K. R., Agerhem, H., Béague, M. P., Oliver, M. A., Gispert, M., Weiler, U., von Seth, G., Leask, H., Font I Furnols, M., Homer, D. B. and Cook, G. L. (2000). An international study on the importance of androstenone and skatole for boar taint: IV. Simulation studies on consumer dissatisfaction with entire male pork and the effect of sorting carcasses on the slaughter line, main conclusions and recommendations, *Meat Sci.* 54(3), 285–295.

Boyle, L. A. and Björklund, L. (2007). Effects of fattening boars in mixed or single sex groups and split marketing on pig welfare, *Anim. Welf.* 16, 259–262.

Byrne, D. V., Thamsborg, S. M. and Hansen, L. L. (2008). A sensory description of boar taint and the effects of crude and dried chicory roots (*Cichorium intybus* L.) and inulin feeding in male and female pork, *Meat Sci.* 79(2), 252-269.

Čandek-Potokar, M., Škrlep, M. and Batorek Lukač, N. (2015a). "Raising entire males or immunocastrates - outlook on meat quality", *Proced, Food Sci.* 5, 30–33.

Čandek-Potokar, M., Škrlep, M., Batorek Lukač, N., Zamaratskaia, G., Prevolnik Povše, M., Velikonja Bolta, Š, Kubale, V. and Bee, G. (2015b). Hydrolysable tannin fed to entire male pigs affects intestinal production, tissue deposition and hepatic clearance of skatole, *Vet. J.* 204(2), 162-167.

Čandek-Potokar, M., Škrlep, M. and Zamaratskaia, G. (2017). Immunocastration as alternative to surgical castration in pigs. In: Payan-Carreira, R. (Ed), *Theriogenology*, IntechOpen, 109-126, Available at: https://www.intechopen.com/books/theriogenology/immunocastration-as-alternative-to-surgical-castration-in-pigs.

Chang, C., Varamini, P., Giddam, A. K., Mansfeld, F. M., D'Occhio, M. J. and Toth, I. (2015). Investigation of structure-activity relationships of synthetic anti-gonadotropin releasing hormone vaccine candidates, *ChemMedChem* 10(5), 901-910.

Chen, G., Zamaratskaia, G., Andersson, H. K. and Lundström, K. (2007). Effects of raw potato starch and live weight on fat and plasma skatole, indole and androstenone levels measured by different methods in entire male pigs, *Food Chem.* 101(2), 439–448.

Claus, R., Lacorn, M., Danowski, K., Pearce, M. C. and Bauer, A. (2007). Short-term endocrine and metabolic reactions before and after second immunization against GnRH in boars, *Vaccine* 25(24), 4689–4696.

Claus, R. and Raab, S. (1999). Influences on skatole formation from tryptophan in the pig colon. In: Huether, G., Kochen, W., Simat, T. J. and Steinhart, H. (Eds), *Tryptophan, Serotonin, and Melatonin. Advances in Experimental Medicine and Biology* (vol. 467), Springer, Boston, MA.

Claus, R., Rottner, S. and Rueckert, C. (2008). Individual return to Leydig cell function after GnRH-immunization of boars, *Vaccine* 26(35), 4571–4578.

Claus, R. and Weiler, U. (1985). Influence of light and photoperiodicity on pig prolificacy, *J. Reprod. Fertil. Suppl.* 33, 185–197.

Claus, R., Weiler, U. and Herzog, A. (1994). Physiological aspects of androstenone and skatole formation in the boar–a review with experimental data, *Meat Sci.* 38(2), 289–305.

Cronin, G. M., Dunshea, F. R., Butler, K. L., McCauley, I., Barnett, J. L. and Hemsworth, P. H. (2003). The effects of immune and surgical-castration on the behaviour and consequently growth of group-housed, male finisher pigs, *Appl. Anim. Behav. Sci.* 81(2), 111-126.

Dalmau, A., Velarde, A., Rodríguez, P., Pedernera, C., Llonch, P., Fàbrega, E., Casal, N., Mainau, E., Gispert, M., King, V., Slootmans, N., Thomas, A. and Mombarg, M. (2015). Use of an anti-GnRF vaccine to suppress estrus in crossbred Iberian female pigs, *Theriogenology* 84(3), 342–347.

Dämmgen, U., Berk, A., Otten, C., Brade, W., Hutchings, N. J., Haenel, H.-D., Rösemann, C., Dänicke, S. and Schwerin, M. (2013). Anticipated changes in the emissions of greenhouse gases and ammonia from pork production due to shifts from fattening of barrows towards fattening of boar, *Landbauforsch Appl. Agric. For. Res.* 63(1), 47-46.

de Briyne, N., Berg, C., Blaha, T. and Temple, D. (2016). Pig castration: will the EU manage to ban pig castration by 2018? *Porc. Health Manag.* 2, 29.

de Campos, C. F., Lopes, M. S., e Silva, F. F., Veroneze, R., Knol, E. F., Sávio Lopes, P. and Guimarães, S. E. F. (2015). Genomic selection for boar taint compounds and carcass traits in a commercial pig population, *Livest. Sci.* 174, 10–17.

Di Pasquale, J., Nannoni, E., Sardi, L., Rubini, G., Salvatore, R., Bartoli, L., Adinolfi, F. and Martelli, G. (2019).Towards the abandonment of surgical castration in pigs: how is immunocastration perceived by Italian consumers?, *Animals* 9(5), 198.

D'Occhio, M. J. (1993). Immunological suppression of reproductive functions in male and female mammals, *Anim. Reprod. Sci.* 33(1–4), 345–372.

Dunshea, F. R., Allison, J. R., Bertram, M., Boler, D. D., Brossard, L., Campbell, R., Crane, J. P., Hennessy, D. P., Huber, L., de Lange, C., Ferguson, N., Matzat, P., McKeith, F., Moraes, P. J., Mullan, B. P., Noblet, J., Quiniou, N. and Tokach, M. (2013). The effect of immunization against GnRF on nutrient requirements of male pigs: a review, *Animal* 7(11), 1769–1778.

Dunshea, F. R., Colantoni, C., Howard, K., McCauley, I., Jackson, P., Long, K. A., Lopaticki, S., Nugent, E. A., Simons, J. A., Walker, J. and Hennessy, D. P. (2001). Vaccination of boars with a GnRH vaccine (Improvac®) eliminates boar taint and increases growth performance, *J. Anim. Sci.* 79(10), 2524–2535.

D'Souza, D. N., Hewitt, R. J. E. and van Barneveld, R. J. (2018). Pork production with entire males and immunocastrates in Australia, *Adv. Anim. Biosci.* 9, s58.

Ebschke, S., von Borell, E. and Weber, M. (2014). Animal welfare assessment of fattening boars with entire males and anti GnRF vaccinated male swine, *Züchtungskunde* 86(5), 342–357.

Einarsson, S. (2006). Vaccination against GnRH: pros and cons, *Acta Vet. Scand.* 48(S1), s10.

Einarsson, S., Andersson, K., Wallgren, M., Lundström, K. and Rodriguez-Martinez, H. (2009). Short- and long-term effects of immunization against gonadotropin-releasing hormone, using Improvac™, on sexual maturity, reproductive organs and sperm morphology in male pigs, *Theriogenology* 71(2), 302–310.

Einarsson, S., Holtman, M., Larsson, K., Settergren, I. and Bane, A. (1979). The effect of two different feed levels on the development of reproductive organs in boars, *Acta Vet. Scand.* 20(1), 1–9.

European Medicines Agency (2019). EPAR-scientific discussion, Available at: https://www.ema.europa.eu/en/documents/scientific-discussion/improvac-epar-scientific-discussion_en.pdf (Accessed on 30 April 2019).

Fàbrega, E. (2019). How far are we from providing pigs appropriate environmental enrichment?, *Animals* 9(10), 721.

Fàbrega, E., Gispert, M., Tibau, J., Hortós, M., Oliver, M. A. and Font-i-Furnols, M. F. (2011). Effect of housing system, slaughter weight and slaughter strategy on carcass and meat quality, sex organ development and androstenone and skatole levels in Duroc finished entire male pigs, *Meat Sci.* 89(4), 434–439.

Fàbrega, E., Puigvert, X., Soler, J., Tibau, J. and Dalmau, A. (2013). Effect of on-farm mixing and slaughter strategy on behaviour, welfare and pro-ductivity in Duroc finished entire male pigs, *Appl. Anim. Behav. Sci.* 143(1), 31–39.

Fàbrega, E., Soler, J., Cros, J., Gispert, M., Tibau, J. and Velarde, A. (2008). Resultados de diversas alternativas a la castración quirúrgica de cerdos. *Suis* 59, 26–34.

Fàbrega, E., Velarde, A., Cros, J., Gispert, M., Suárez, P., Tibau, J. and Soler, J. (2010). Effect of vaccination against gonadotrophin-releasing hormone, using Improvac®, on growth performance, body composition, behaviour and acute phase proteins, *Livest. Sci.* 132(1–3), 53–59.

Font-i-Furnols, M. (2012a). Consumer studies on sensory acceptability of boar taint: a review, *Meat Sci.* 92(4), 319–329.

Font-i-Furnols, M., García, J., Gispert, M., Brun, A., Hernández-García, F., Izquierdo, M. M. and Blanco-Penedo, I. (2017). Effect of the immunocastration protocols on ham composition evaluated with computed tomography. Book of Abstracts of the 4th Fatty Pig Science & Utilization International Conference, Spain.

Font-i-Furnols, M., Gispert, M., Soler, J., Diaz, M., García Regueiro, J. A., Diaz, I. and Pearce, M. C. (2012b). Effect of vaccination against gonadotrophin-releasing factor on growth performance, carcass, meat and fat quality of male Duroc pigs for dry-cured ham production, *Meat Sci.* 91(2), 148–154.

Fredriksen, B., Lium, B. M., Marka, C. H., Heier, B. T., Dahl, E., Choinski, J. U. and Nafstad, O. (2006). Entire male pigs in a farrow-to-finish system. Effects on androstenone and skatole, *Livest. Sci.* 102(1–2), 146–154.

Fredriksen, B., Lium, B. M., Marka, C. H., Mosveen, B. and Nafstad, O. (2008). Entire male pigs in farrow-to-finish pens–effects on animal welfare, *Appl. Anim. Behav. Sci.* 110(3–4), 258–268.

Fredriksen, B., Johnsen, A. M. S. and Skuterud, E. (2011). Consumer attitudes towards castration of piglets and alternatives to surgical castration, *Res. Vet. Sci.* 90(2), 352–357.

Frieden, L., Neuhoff, C., Große-Brinkhaus, C., Cinar, M. U., Schellander, K., Looft, C. and Tholen, E. (2012). Breeding potential of selection against boar taint, *Züchtungskunde* 84(5), 394–411.

García-Gudiño, J., Gispert, M., Brun, A., Blanco-Penedo, I. and Font-i-Furnols, M. (2019). Non-destructive technologies *in vivo* and *post mortem* to evaluate hams' characteristics from early and late immunocastrated Iberian pigs, 6th Fatty PIG, Science and Utilization International Conference, Badajoz, Spain.

Giersing, M., Lundstrom, K. and Andersson, A. (2000). Social effects and boar taint: significance for production of slaughter boars (*Sus scrofa*), *J. Anim. Sci.* 78(2), 296–305.

Grela, E. R., Kowalczuk-Vasilev, E. and Klebaniuk, R. (2013). Performance, pork quality and fatty acid composition of entire males, surgically castrated or immunocastrated males, and female pigs reared under organic system, *Pol. Journ. Vet. Sci.* 16(1), 107–114.

Große-Brinkhaus, C., Storck, L. C., Frieden, L., Neuhoff, C., Scjeññander, K., Looft, C. and Tholen, E. (2015). Genome-wide association analyses for boar taint components and testicular traits revealed gegions having pleiotropic effects, *BMC Genet.* 16, 36.

Gruys, E., Obwolo, M. J. and Toussaint, M. (1994). Diagnostic influence of the major acute phase proteins in veterinary clinical chemistry - a review, *Vet. Bullet.* 64, 1009–1018.

Guay, K., Salgado, G., Thompson, G., Backus, B., Sapkota, A., Chaya, W. and McGlone, J. J. (2013). Behavior and handling of physically and immunologically castrated market pigs on farm and going to market, *J. Anim. Sci.* 91(11), 5410–5417.

Haberland, A. M., Luther, H., Hofer, A., Tholen, E., Simianer, H., Lind, B. and Baes, C. (2014). Efficiency of different selection strategies against boar taint in pigs, *Animal* 8(1), 11–19.

Han, X., Zhou, Y., Zeng, Y., Sui, F., Liu, Y., Tan, Y., Cao, X., Du, X., Meng, F. and Zeng, X. (2017). Effects of active immunization against GnRH versus surgical castration on hypothalamic-pituitary function in boars, *Theriogenology* 97, 89–97.

Hansen, L. L., Larsen, A. E., Jensen, B. B., Hansenmoller, J. and Bartongade, P. (1994). Influence of stocking rate and feaces deposition in the pen at different temperatures on skatole concentration (boar taint) in subcutaneous fat, *Anim. Sci.* 59(1), 99–110.

Hansen, L. L., Larsen, A. E. and Hansen-Moller, J. (1995). Influence of keeping pigs heavily fouled with feces plus urine on skatole and indole concentration (boar taint) in subcutaneous fat, *Acta Agric. Scand. A* 45(3), 178–185.

Hansson, M., Lundeheim, N., Nyman, G. and Joansson, G. (2011). Effect of local anaesthesia and/or analgesia on pain responses induced by piglet castration, *Acta Vet. Scand.* 53(1), 34.

Haugen, J.-E., van Wagenberg, C., Backus, G., Nielsen, B. E., Claus, B., Bonneau, M., Panella-Riera, N. and Aluwé, M. (2014). Boarcheck: a study on rapid methods for boar taint used or being developed at slaughter plants in the European Union. D5.2 Final Report, Nofima AS, Available at: https://ec.europa.eu/food/sites/food/files/animals/docs/aw:prac_farm_pigs_cast-alt_research_boarcheck_20140901.pdf (Accessed on 15th september 2019).

Hay, M., Vulin, A., Genin, S., Sales, P. and Prunier, A. (2003). Assessment of pain induced by castration in piglets: behavioral and physiological responses over the subsequent 5 days, *Appl. Anim. Behav. Sci.* 82(3), 201–218.

Heid, A. and Hamm, U. (2012). Consumer attitudes towards alternatives to piglet castration without pain relief in organic farming: qualitative results from Germany, *J. Agr. Environ. Ethic.* 25(5), 687–706.

Heyrman, E., Kowalskia, E., Milleta, S., Tuyttens, F. A. M., Ampe, B., Janssens, S., Buys, N., Wauters, J., Vanhaecke, L. and Aluwé, M. (2019). Monitoring of behavior, sex hormones and boar taint compounds during the vaccination program for immunocastration in three sire lines, *Res. Vet. Sci.* 124, 293–302.

Heyrman, E., Millet, S., Tuyttens, F. A. M., Ampe, B., Janssens, S., Buys, N., Wauters, J., Vanhaecke, L. and Aluwé, M. (2017). Olfactory evaluation of boar taint: effect of factors measured at slaughter and link with boar taint compounds, *Animal* 11(11), 2084-2093.

Heyrman, E., Millet, S., Tuyttens, F. A. M., Ampe, B., Janssens, S., Buys, N., Wauters, J., Vanhaecke, L. and Aluwé, M. (2018). On farm intervention studies on reduction of boar taint prevalence: feeding strategies, presence of gilts and time in lairage, *Res. Vet. Sci.* 118, 508–516.

Hidalgo, A. M., Bastiaansen, J. W., Harlizius, B., Knol, E. F., Lopes, M. S., de Koning, D. J. and Groenen, M. A. (2014). Asian low-androstenone haplotype on pig chromosome 6 does not unfavorably affect production and reproduction traits, *Anim. Genet.* 45(6), 874–877.

Holinger, M., Früh, B. and Hillmann, E. (2015). Group composition for fattening entire male pigs under enriched housing conditions-influences on behaviour, injuries and boar taint compounds, *Appl. Anim. Behav. Sci.* 165, 47–56.

IPEMA COST ACTION 15215 (2019). Factsheet on pork production with entire males (produced by core group Marijke Aluwe, Ge Backus, Giuseppe Bee, Michel Bonneau, Eberhard von Borell, Meta Čandek-Potokar, Olena Doran, Maria Font-i-Furnols, Catherine Larzul, Martin Skrlep, Igor Tomasevic, Liliana Tudoreanu, Mandes Verhaagh, Ulrike Weiler), Available at: http://www.ca-ipema.eu/ (Accessed on 10 October 2019).

Janssens, G., Boulet, R., Reuter, W. and Hardy, N. (2018). Influence of Taintstop, an innovative feed concept, on boar taint. Presentation at IPEMA Meeting Oeiras

(Portugal), Available at: http://www.ca-ipema.eu/oeiras-presentations (Accessed on 25 July 2019).
Kiss, I., Palya, R. and Mazerolles, P. (2017). "Ceva Valora®" as an efficient alternative to surgical castration, *The Pig Site*, Available at: https://thepigsite.com/articles/ceva-valora-as-an-efficient-alternative-to-surgical-castration (Accessed on 25 June 2019).
Kjeldsen, N. (1993). Practical experience with production and slaughter of entire male pigs. In: Bonneau, M. (Ed.), *Measurement and Prevention of Boar Taint in Entire Male Pigs* (Vol. 60). ,Institut National de la Recherche Agronomique, Paris, France, 137–144.
Kress, K., Millet, S., Labussière, É, Weiler, U. and Stefanski, V. (2019). Sustainability of pork production with immunocastration in Europe, *Sustainability* 11(12), 3335.
Kress, K., Weiler, U. and Stefanski, V. (2018). Influence of housing conditions on antibody formation and testosterone after Improvac vaccinations, *Adv. Anim. Biosci.* 9, s19.
Kubale, V., Batorek-Lukač, N., Škrlep, M., Prunier, A., Bonneau, M., Fazarinc, G. and Candek-Potokar, M. (2013). Steroid hormones, boar taint compounds, and reproductive organs in pigs according to the delay between immunocastration and slaughter, *Theriogenology* 79(1), 69–80.
Le Floc'h (2003). Protein and amino acid metabalism during inflammation and immune system activation. Animal Welfare and Acute Phase Proteins. Fourth European Colloquium on Acute Phase Proteins, Segovia, Spain.
Zamaratskaia, G., Maribo, H., Canibe, N., Milet, S., Quiniou, N. and Bee, G. List of feed ingredients with boar taint reducing capacities. Available at: http://www.ca-ipema.eu/papers, Available at: https://zenodo.org/record/1460483#.XdKPvldKiUk.
Liu, G., Jennen, D. G. J., Tholen, E., Juengst, H., Kleinwächter, T., Hölker, M., Tesfaye, D., Un, G., Schreinemachers, H. J., Murani, E., Ponsuksili, S., Kim, J. J., Schellander, K. and Wimmers, K. (2007). A genome scan reveals QTL for growth, fatness, leanness and meat quality in a Duroc-Pietrain resource population, *Anim. Genet.* 38(3), 241-252.
Lösel, D., Lacorn, M., Buttner, D. and Claus, R. (2006). Flavor improvement in pork from barrows and gilts via inhibition of intestinal skatole formation with resistant potato starch, *J. Agric. Food Chem.* 54(16), 5990–5995.
Lukić, B., Pong-Wong, R., Rowe, S. J., Koning, D. J., Velander, I., Haley, C. S., Archibald, A. L. and Woolliams, J. A. (2015). Efficiency of genomic prediction of boar taint reduction in Danish Landrace pigs, *Anim. Genet.* 46(6), 607–616.
Lunde, K., Egelandsdal, B., Skuterud, E., Mainland, J. D., Lea, T., Hersleth, M. and Matsunami, H. (2012). Genetic variation of an odorant receptor OR7D4 and sensory perception of cooked meat containing Androstenone, *PLoS ONE* 7(5), e35259.
Lundström, K., Malmfors, B., Stern, S., Rydhymer, L., Eliasson-Selling, L., Mortensen, A. B. and Mortensen, H. P. (1994). Skatole levels in pigs selected for high lean tissue growth rate on different protein levels, *Live. Prod. Sci.* 38(2), 125–132.
Mackinnon, J. D. and Pearce, M. C. (2007a). Improvac™ (Pfizer Animal Health): an immunological product for the control of boar taint in male pigs: I. Boar taint and its control and the mode of action, safety and efficacy of Improvac™, *Pig J.* 59, 29–67.
Mackinnon, J. D. and Pearce, M. C. (2007b). Improvac™ (Pfizer Animal Health): an immunological product for the control of boar taint in entire male pigs: II. Practical applications in pig production and potential production benefits, *Pig J.* 59, 68–90.
Mancini, M. C., Menozzi, D. and Arfini, F. (2017). Immunocastration: economic implications for the pork supply chain and consumer perception. An assessment of existing research, *Livest. Sci.* 203, 10–20.

Martínez-Macipe, M., Rodríguez, P., Izquierdo, M., Gispert, M., Manteca, X., Mainau, E., Hernández, F. I., Claret, A., Guerrero, L. and Dalmau, A. (2016). Comparison of meat quality parameters in surgical castrated versus vaccinated against gonadotrophin-releasing factor male and female Iberian pigs reared in free-ranging conditions, *Meat Sci.* 111, 116–121.

Mathur, P. K., ten Napel, J., Bloemhof, S., Heres, L., Knol, E. F. and Mulder, H. A. (2012). A human nose scoring system for boar taint and its relationship with androstenone and skatole, *Meat Sci.* 91(4), 414–422.

Mathur, P. K., ten Napel, J., Crump, R. E., Mulder, H. A. and Knol, E. F. (2013). Genetic relationship between boar taint compounds, human nose scores, and reproduction traits in pigs, *J Anim Sci.* 91(9), 4080–4089. DOI: 10.2527/jas.2013-6478.

Matthews, K. R., Homer, D. B., Punter, P., Beague, M. P., Gispert, M., Kempster, A. J., Agerhem, H., Claudi-Magnussen, C., Fischer, K., Siret, F., Leask, H., Font I Furnols, M. and Bonneau, M. (2000). An international study on the importance of androstenone and skatole for boar taint: III. Consumer survey in seven European countries, *Meat Sci.* 54(3), 271–283.

McGlone, J., Guay, K. and Garcia, A. (2016). Comparison of intramuscular or subcutaneous injections vs. castration in pigs–impacts on behavior and welfare, *Animals* 6(9), 52.

Mercat, M. J., Prunier, A., Muller, N., Hassenfratz, C. and Larzul, C. (2015). Relation entre la production spermatique et le risque d'odeur de mâle entier des descendants de race pure ou croisés, *J. Rech. Porcine* 47, 241–246.

Millet, S., Gielkens, K., Brabander, D. D. and Janssens, G. P. J. (2011). Considerations on the performance of immunocastrated male pigs, *Animal* 5(7), 1119–1123.

Morales, J., Dereu, A., Manso, A., de Frutos, L., Piñeiro, C., Manzanilla, E. G. and Wuyts, N. (2017). Surgical castration with pain relief affects the health and productive performance of pigs in the suckling period, *Porc. Health Manag.* 3, 18.

Mörlein, D., Aluwé, M., Backus, G., Bonneau, M., Brockhoff, P., Chevillon, P., Christensen, R., Font-i-Furnols, M., Gertheiss, J., Meier-Dinkel, L., Mörlein, J., Oertel, E., Oliver, A., Tuyttens, F., van den Broeke, A. and Aaslyng, M. (2019). Drivers of (dis)liking: systematic pairwise reference tests to reveal the relationship between boar taint and consumer acceptance. Poster Presentation at ICOMST 2019, Germany.

Nautrup, B. P., Vlaenderen, I. V., Aldaz, A. and Mah, C. K. (2018). The effect of immunization against gonadotropin-releasing factor on growth performance, carcass characteristics and boar taint relevant to pig producers and the pork packing industry: a meta-analysis, *Res. Vet. Sci.* 119, 182–195.

Needham, T., Lambrechts, H. and Hoffman, L. C. (2017). Castration of male livestock and the potential of immunocastration to improve animal welfare and production traits: invited review, *S. Afr. J. Anim. Sci.* 47(6), 731–742.

Oskam, I. C., Lervik, S., Tajet, H., Dahl, E., Ropstad, E. and Andresen, Ø (2010). Differences in testosterone, androstenone, and skatole levels in plasma and fat between pubertal purebred Duroc and Landrace boars in response to human chorionic gonadotrophin stimulation, *Theriogenology* 74(6), 1088–1098.

Øverland, M., Kjos, N. K., Fauske, A. K., Teige, J. and Sørum, H. (2011). Easily fermentable carbohydrates reduce skatole formation in the distal intestine of entire male pigs, *Live Sci.* 140(1–3), 206–217.

Parois, S. P., Prunier, A., Mercat, M. J., Merlot, E. and Larzul, C. (2015). Genetic relationships between measures of sexual development, boar taint, health, and aggressiveness in pigs, *J. Anim. Sci.* 93(8), 3749–3758.

Parois, S., Larzul, C. and Prunier, A. (2017). Associations between the dominance status and sexual development, skin lesions or feeding behaviour of intact male pigs, *Appl. Anim. Behav. Sci.* 187, 15-22.

Parois, S., Bonneau, M., Chevillon, P., Larzul, C., Quiniou, N., Robic, A. and Prunier, A. (2018). Odeurs indésirables de la viande de porcs mâles non castrés: problèmes et solutions potentielles, *INRA Prod. Anim.* 31(1), 23-36.

Patterson, J. L., Willis, H. J., Kirkwood, R. N. and Foxcroft, G. R. (2002). Impact of boar exposure on puberty attainment and breeding outcomes in gilts, *Theriogenology* 57, 2015-2025.

Pauly, C., Spring, P., O'Doherty, J. V., Ampuero Kragten, S. and Bee, G. (2008). Performances, meat quality and boar taint of castrates and entire male pigs fed a standard and a raw potato starch-enriched diet, *Animal* 2(11), 1707-1715.

Pauly, C., Spring, P., O'Doherty, J. V., Ampuero Kragten, S. and Bee, G. (2009). Growth performance, carcass characteristics and meat quality of group-penned surgically castrated, immunocastrated (Improvac®) and entire male pigs and individually penned entire male pigs, *Animal* 3(7), 1057-1066.

Pauly, C., Luginbühl, W., Ampuero, S. and Bee, G. (2012). Expected effects on carcass and pork quality when surgical castration is omitted – results of a meta-analysis study, *Meat Sci.* 92(4), 858-862.

Pinna, A., Schivazappa, C., Virgili, R. and Parolari, G. (2015). Effect of vaccination against gonadotropin-releasing hormone (GnRH) in heavy male pigs for Italian typical dry-cured ham production, *Meat Sci.* 110, 153-159.

Prunier, A. and Quesnel, H. (2000). Nutritional influences on the hormonal control of reproduction in female pigs, *Livest. Prod. Sci.* 63(1), 1-16.

Prunier, A., Bonneau, M., Von Borell, E. H., Cinotti, S., Gunn, M., Fredriksen, B., Giersing, M., Morton, D. B., Tuyttens, F. A. M. and Velarde, A. (2006). A review of the welfare consequences of surgical castration in piglets and the evaluation of non-surgical methods, *Anim. Welf.* 15, 277-289.

Prunier, A., Brillouet, A., Merlot, E., Meunier-Salaun, M. C. and Tallet, C. (2013). Influence of housing and season on pubertal development, boar taint compounds and skin lesions of male pigs, *Animal* 7(12), 2035-2043.

Puls, C. L., Rojo, A., Matzat, P. D., Schroeder, A. L. and Ellis, M. (2017). Behavior of immunologically castrated barrows in comparison to gilts, physically castrated barrows, and intact male pigs, *J. Anim. Sci.* 95(6), 2345-2353.

Quiniou, N., Chevillon, P., Commeureuc, J-P., Peltier, B., Rocher, P., Loiseau, D., Richard, R. and Lhommeau, T. (2015). Performances de croissance et risques d'odeurs de verrat de porcs mâles entiers selon les apports alimentaires en acides aminés essentiels ou en protéines, *Journ. Rech. Porcine* 47, 69-74.

Quiniou, N., Courboulay, V., Salaün, Y. and Chevillon, P. (2010). Impact of the non castration of male pigs on growth performance and behaviour - comparison with barrows and gilts, Conference at the 61st Annual Meeting of the European Association for Animal Production, Heraklion, Crete Island, Greece, paper 8, 7.

Rasmussen, M. K., Brunius, C., Zamaratskaia, G. and Ekstrand, B. (2012). Feeding dried chicory root to pigs decrease androstenone accumulation in fat by increasing hepatic 3β hydroxysteroid dehydrogenase expression, *J. Steroid Biochem. Mol. Biol.* 130(1-2), 90-95.

Rault, J. L., Lay Jr., D. C. and Marchant-Forde, J. N. (2011). Castration induced pain in pigs and other livestock, *Appl. Anim. Behav. Sci.* 135(3), 214-225.

Reiter, S., Zöls, S., Ritzmann, M., Stefanski, V. and Weiler, U. (2017). Penile injuries in immunocastrated and entire male pigs of one fattening farm, *Animals* 7(9), 71.

Rezar, V., Salobir, J., Levart, A., Tomažin, U., Škrlep, M., Batorek Lukač, N. and Čandek-Potokar, M. (2017). Supplementing entire male pig diet with hydrolysable tannins: effect on carcass traits, meat quality and oxidative stability, *Meat Sci.* 133, 95–102.

Rius, M. A. and García-Regueiro, J. A. (2001). Skatole and indole concentrations in longissimus dorsi and fat samples of pigs, *Meat. Sci.* 59(3), 285–291.

Robic, A., Faraut, T. and Prunier, A. (2014). Pathways and genes involved in steroid hormone metabolism in male pigs: a review and update, *J. Steroid Biochem. Mol. Biol.* 140, 44–55.

Robic, A., Larzul, C. and Bonneau, M. (2008). Genetic and metabolic aspects of androstenone and skatole deposition in pig adipose tissue: a review, *Genet. Sel. Evol.* 40(1), 129–143.

Rydhmer, L., Hansson, M., Lundstrom, K., Brunius, C. and Andersson, K. (2013). Welfare of entire male pigs is improved by socializing piglets and keeping intact groups until slaughter, *Animal* 7(9), 1532–1541.

Rydhmer, L., Lundström, K. and Andersson, K. (2010). Immunocastration reduces aggressive and sexual behavior in male pigs, *Animal* 4(6), 965–972.

Rydhmer, L., Zamaratskaia, G., Andersson, H. K., Algers, B., Guillemet, R. and Lundström, K. (2006). Aggressive and sexual behaviour of growing and finishing pigs reared in groups, without castration, *Acta Agric. Scand. A* 56(2), 109–119.

Salmon, E. L. and Edwards, S. A. (2006). Effects of gender contact on the behaviour and performance of entire boars and gilts from 60–130kg, *Proc. Br. Soc. Anim. Sci.* 72.

Sattler, T., Sauer, F. and Schmoll, F. (2014). Effect of time of second GnRH vaccination on feed intake, carcass quality and fatty acid composition of male fatteners compared to entire boars and barrows, *Berl. Munch. Tierarztl. Wochenschr.* 127(7–8), 290–296.

Schmidt, T., Calabrese, J. M., Grodzycki, M., Paulick, M., Pearce, M. C., Rau, F. and von Borell, E. (2011). Impact of single-sex and mixed-sex group housing of boars vaccinated against GnRF or physically castrated on body lesions, feeding behaviour and weight gain, *Appl. Anim. Behav. Sci.* 130(1–2), 42–52.

Schübeler, A. S. and Koch, L. (2018). The search for the best way. Animal welfare vs. boar taint-immunocastration in comparison to the alternatives from the producer and master views, *Fleischwirtschaft* 98, 21–23.

Škrlep, M., Čandek-Potokar, M., Lukač, N. B., Povše, M. P., Pugliese, C., Labussière, E. and Flores, M. (2016). Comparison of entire male and immunocastrated pigs for dry-cured ham production under two salting regimes, *Meat Sci.* 111, 27–37.

Škrlep, M., Tomažin, U., Batorek Lukač, N., Poklukar, K. and Čandek-Potokar, M. (2019). Proteomic profile of longissimus dorsi muscle of entire male and castrated pigs as related to meat quality, *Animal* 9(3), 74.

Strathe, A. B., Velander, I. H., Mark, T. and Kadarmideen, H. N. (2013). Genetic parameters for androstenone and skatole as indicators of boar taint and their relationship to production and litter size traits in Danish Landrace, *J. Anim. Sci.* 91(6), 2587–2595.

Thompson, D. L. (2000). Immunization against GnRH in male species (comparative aspects), *Anim. Reprod. Sci.* 60–61, 459–469.

Thomsen, R., Edwards, S. A., Jensen, B. B., Rousing, T. and Sørensen, J. T. (2015). Weight and season affects androstenone and skatole occurrence in entire male pigs in organic pig production, *Animal* 9(9), 1577–1586.

Thomsen, R., Edwards, S. A., Rousing, T., Laboriau, R. and Sørensen, J. T. (2016). Influence of social mixing and group size on skin lesions and mounting in organic entire male pigs, *Animal* 10(7), 1225–1233.

Tuyttens, F. A. M., Vanhonacker, F., Langendries, K., Aluwé, M., Millet, S., Bekaert, K. and Verbeke, W. (2011). Effect of information provisioning on attitude toward surgical castration of male piglets and alternative strategies for avoiding boar taint, *Res. Vet. Sci.* 91(2), 327–332.

van den Broeke, A., Leen, F., Aluwé, M., Ampe, B., Van Meensel, J. and Millet, S. (2016). The effect of GnRH vaccination on performance, carcass, and meat quality and hormonal regulation in boars, barrows, and gilts, *J. Anim. Sci.* 94(7), 2811–2820.

van Staaveren, N., Teixeira, D. L., Hanlon, A. and Boyle, L. A. (2015). The effect of mixing entire male pigs prior to transport to slaughter on behaviour, welfare and carcass lesions, *PLoS ONE* 10(4), e0122841.

van Wagenberg, C. P. A., Snoek, H. M., van der Fels, J. B., van der Peet-Schwering, C. M. C., Vermeer, H. M. and Heres, L. (2013). Farm and management characteristics associated with boar taint, *Animal* 7(11), 1841–1848.

Vazquez, J. M., Parrilla, I., Roca, J., Gil, M. A., Cuello, C., Vazquez, J. L. and Martínez, E. A. (2009). Sex-sorting sperm by flow cytometry in pigs: issues and perspectives, *Theriogenology* 71(1), 80–88.

Vhile, S. G., Kjos, N. P., Sorum, H. and Overland, M. (2012). Feeding Jerusalem artichoke reduced skatole level and changed intestinal microbiota in the gut of entire male pigs, *Animal* 6(5), 807–814.

von Borell, E., Baumgartner, J., Giersing, M., Jäggin, N., Prunier, A., Tuyttens, F. A. M. and Edwards, S. A. (2009). Animal welfare implications of surgical castration and its alternatives in pigs, *Animal* 3(11), 1488–1496.

Weiler, U., Götz, M., Schmidt, A., Otto, M. and Müller, S. (2013). Influence of sex and immunocastration on feed intake behavior, skatole and indole concentrations in adipose tissue of pigs, *Animal* 7(2), 300–308.

Weiler, U., Isernhagen, M., Stefanski, V., Ritzmann, M., Kress, K., Hein, C. and Zöls, S. (2016). Penile injuries in wild and domestic pigs, *Animals* 6(4), 25.

Wemelsfelder, F. and van Putten, G. (1985). Behaviour as a possible indicator for pain in piglets. Report B-260. Instituut voor Veeteeltkundig Onderzoek 'Schoonoord", Zeist, The Netherlands.

Wesoly, R., Jungbluth, I., Stefanski, V. and Weiler, U. (2015). Pre-slaughter conditions influence skatole and androstenone in adipose tissue of boars, *Meat Sci.* 99, 60–67.

Wesoly, R., Stefanski, V. and Weiler, U. (2016). Influence of sampling procedure, sampling location and skin contamination on skatole and indole concentrations in adipose tissue of pigs, *Meat Sci.* 111, 85–91.

Wesoly, R. and Weiler, U. (2012). Nutritional influences on skatole formation and skatole metabolism in the pig, *Animals* 2(2), 221–242.

Wilson, E. R., Johnson, R. K. and Wetterman, R. P. (1977). Reproductive and testicular characteristics of purebred and crossbred boars, *J. Anim. Sci.* 44, 939–947.

Windig, J. J., Mulder, H. A., Ten Napel, J., Knol, E. F., Mathur, P. K. and Crump, R. E. (2012). Genetic parameters for androstenone, skatole, indole and human nose scores as measures of boar taint and their relationship with finishing traits, *J. Anim. Sci.* 90(7), 2120–2129.

Zamaratskaia, G., Andersson, H. K., Chen, G., Andersson, K., Madej, A. and Lundström, K. (2008a). Effect of a gonadotropin-releasing hormone vaccine (Improvac®) on steroid hormones, boar taint and performance in entire male pigs, *Reprod. Domest. Anim.* 43(4), 351–359.

Zamaratskaia, G. and Rasmussen, M. K. (2018). Is it possible to reduce androstenone by dietary means?, *Adv. Anim. Biosci.* 9 (Special Issue s1), s22.

Zamaratskaia, G., Rydhmer, L., Andersson, H. K., Chen, G., Lowagie, S., Andersson, K. and Lundström, K. (2008b). Long-term effect of vaccination against gonadotropin-releasing hormone, using Improvac, on hormonal profile and behaviour of male pigs, *Anim. Reprod. Sci.* 108(1–2), 37–48.

Zamaratskaia, G., Rydhmer, L., Chen, G., Madej, A., Andersson, H. K. and Lundstrom, K. (2005). Boar taint is related to endocrine and anatomical changes at puberty but not to aggressive behaviour in entire male pigs, *Rep. Dom. Anim.* 40(6), 500–506.

Zamaratskaia, G. and Squires, E. J. (2009). Biochemical, nutritional and genetic effects on boar taint in entire male pigs, *Animal* 3(11), 1508–1521.

Chapter 10

Understanding and preventing tail biting in pigs

Sandra Edwards, Newcastle University, UK; and Anna Valros, University of Helsinki, Finland

1 Introduction
2 The prevalence and economic importance of the problem
3 Do we have a mechanistic understanding of tail biting?
4 Risk factors for the occurrence of tail biting
5 Ethical considerations associated with tail biting
6 Managing pigs without tail biting
7 Intervention measures for a tail biting outbreak
8 Conclusions
9 Future trends in research
10 Where to look for further information
11 References

1 Introduction

Tail biting is one of the most enduring behavioural problems in pigs, with serious welfare implications for both the perpetrator and recipient. It can range in expression from gentle chewing of the tail of one pig by another, which may result visibly in only reddening and minor teeth marks on the tail of the recipient, to vigorous injurious biting which can result in serious lesions and even amputation of a portion of the victim's tail. In severe cases, persistent cannibalism can result in complete tail removal and tissue damage extending up into the spine, giving rise to infection, paralysis and even death. Whilst these outcomes for the recipient may be similar in different situations, the behaviour of tail biting may develop and manifest itself in different ways. Taylor et al. (2010) described three different types of biting, based on the observed behaviour of the perpetrator. The classical description, often called 'two stage' biting, starts with increased investigation and chewing of tails of penmates which becomes more persistent over time, until finally causing one or more bleeding lesions. A different onset of tail biting, referred to as 'sudden' or 'frustration-induced'

http://dx.doi.org/10.19103/AS.2020.0081.10

has been described in which, with no prior gentle manipulation or gradual increase, a pig suddenly makes a serious and damaging bite to a previously undamaged tail. This often occurs in situations where the perpetrator appears to be thwarted in obtaining access to a resource such as food or preferred space. After both these types of onset, the presence of blood on the victim's damaged tail attracts the attention of other pigs in the group, who also begin to chew and bite on the damaged tails and escalate the problem into a large-scale and rapidly spreading outbreak. The third form of tail biting described by Taylor et al. (2010) has been called 'obsessive' or 'fanatical'. This describes an individual pig which becomes highly focussed on tail biting activity, persistently following a victim pig around within the pen or moving from one pig to another in the group and attempting to bite a tail for extended periods of time to the exclusion of all other activities. It is not known whether these individuals begin as 'two-stage' or 'sudden' biters before the behaviour increases in prevalence and severity, but once a pig exhibits 'fanatical' biting it is easy to see and highly characteristic, showing many of the features of other obsessive/compulsive behaviours in different species. In addition to these three forms of biting, focussing on the behaviour of the individual perpetrator, Valros (2018) identified a fourth form of suddenly occurring, 'epidemic-like' behaviour, where an acute change in the environment triggers a sudden and widespread outbreak of damage. This has been reported in connection with a feed delivery problem or sudden change in temperature, and the likelihood of further bites increases for each bite observed in the pen. Whether this is a fourth distinct type or a rapidly escalating case of 'frustration-induced' biting by many individuals is not yet clear.

The welfare impacts on the victim of tail biting are obvious, ranging from disturbance and mild discomfort through to severe injury, occurrence of secondary infections and ultimately death if no intervention occurs (Schrøder-Petersen and Simonsen 2001; EFSA 2007; Heinonen et al. 2010; Munsterhjelm et al. 2013a). The welfare implications for the perpetrator are more subtle. Whilst performance of the behaviour does not, in itself, compromise welfare, there is substantial evidence that exhibition of the behaviour is a symptom of some other welfare problem for that individual. This indication comes from the range of risk factors triggering the behaviour, which covers many different intrinsic and extrinsic welfare challenges, and from the physiological changes that tail biters show (see later sections on risk factors and mechanisms). Finally, an indirect welfare consequence of tail biting is that, in order to reduce the risk of problem occurrence, the great majority of pigs around the world undergo tail docking at an early age–a procedure which causes acute and possibly long-term stress and pain (see Chapter 8 by Prunier et al. in this volume for a full discussion of this). This measure does not address the basic cause of a tail biting problem, but merely reduces the negative outcomes.

2 The prevalence and economic importance of the problem

Obtaining a true picture of the worldwide prevalence of tail biting is very difficult for a number of reasons. It is normally quantified by the proportion of pigs in a given population which show tail lesions resulting from the behaviour, since this is an easily observable outcome. However, there is no agreed definition of a tail lesion, and the severity which should be reached before a lesion is counted varies widely between studies (Honeck et al. 2019). Tail damage may be classified in different ways, including by the severity of lesion, freshness of damage and proportion of tail removed (Zonderland et al. 2008; Fig. 1).

Whilst many studies refer to clinical lesions, where bleeding wounds often require treatment or other intervention (e.g. Thodberg et al. 2018; Chou et al. 2019a), other studies also count less severe lesions such as scratches or teeth marks (e.g. Zonderland et al. 2008; Scollo et al. 2016; Larsen et al. 2018a). The prevalence on farm of these minor lesions is typically three times greater than the number of clinical cases (unpublished result from the UK study of Taylor et al. 2012), but the ratio between mild and severe lesions can be much higher in abattoir studies (Valros et al. 2004; Harley et al. 2012) and the profile of lesion types shows great variation between farms (M. Heinonen and A. Valros, unpublished data). It can also be the case that the exact definition of what has been classified as a tail injury is not given (e.g. Presto Åkerfeldt et al. 2019). Such differences in definition can give rise to very different prevalence estimates from different studies. In abattoir surveys, prevalence of tail damage can also be very different between routine meat inspection records and data recorded in more detailed scientific studies (Hunter et al. 1999; Valros et al. 2004; Keeling et al. 2012).

A second problem in determining the overall prevalence of biting arises from the fact that the behaviour is sporadic over time. Thus, a study may give large batch-to-batch variation and single point estimates may yield atypically high or low prevalence depending on whether a problem is occurring at that time (Veit et al. 2016; Pandolfi et al. 2017). For this reason, prevalence estimates obtained from regular and long-term abattoir recording or on-farm welfare outcome monitoring schemes should be most reliable, but both of these also have their problems. Such long-term abattoir records are usually associated with the meat inspection process and thus, tend to record only the most severe cases with inflamed wounds or secondary infections (Hunter et al. 1999; Keeling et al. 2012; Valros et al. 2020). Since their objective relates to food safety rather than animal welfare research or audit, they seldom note minor lesions and they generally fail to distinguish between docked tails and tails which have healed following earlier injury (Valros et al. 2004). Furthermore, they give no information on pigs which fail to reach the normal slaughter channel,

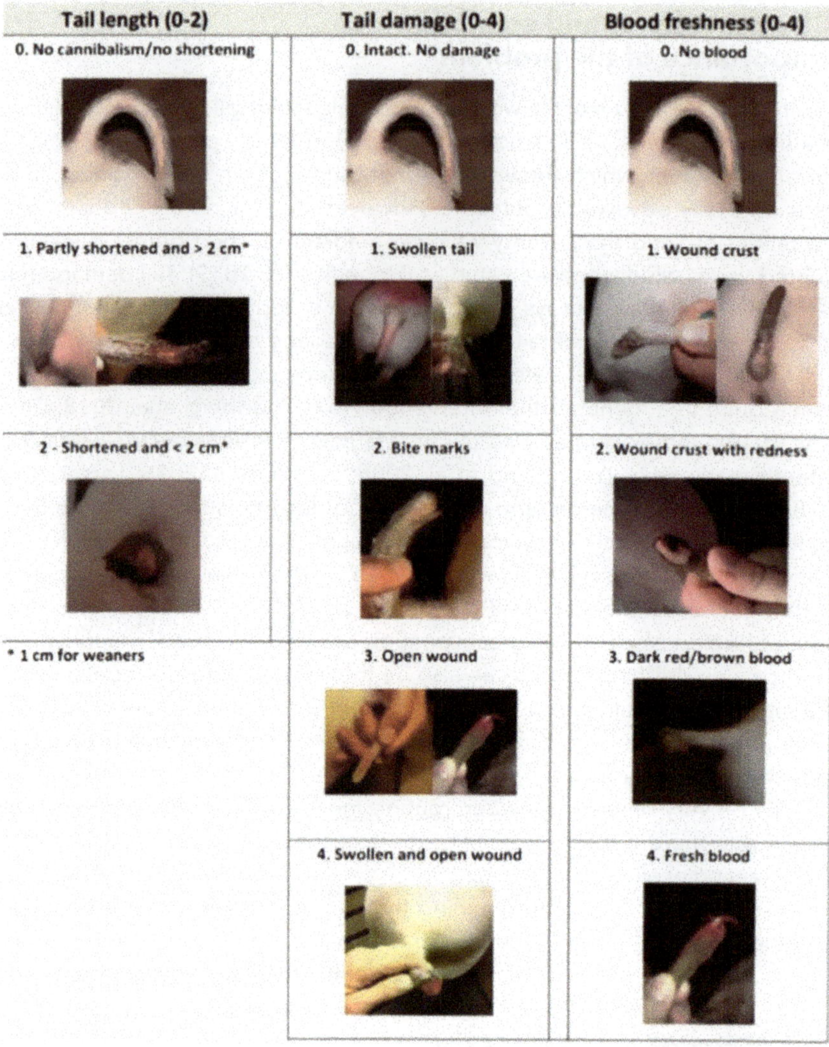

Figure 1 Method of classification for tail damage resulting from tail biting (FareWellDock project; http://farewelldock.eu. Figure courtesy of Margit Bak Jensen and Lene Juul Pedersen, Aarhus University).

either because of on-farm mortality or because of sale of casualty pigs to smaller private abattoirs. It has been reported that herd prevalence estimated by clinical examination on farm is twice that detected by carcass inspection at the abattoir (Busch et al. 2004). Schemes for on-farm assessment of welfare outcomes (e.g. WelfareQuality–Temple et al. 2011; Real Welfare–Pandolfi et al. 2017; ETT SIKAVA–Stygar et al. 2020) often record the prevalence of bitten tails as a part of their protocol, but are usually based on only a sample of pens or a quick overview of the farm and provide snapshots in time which do not always capture pigs that have previously died or been removed to hospital pens.

However, even given these difficulties in deriving a precise prevalence of tail biting, it is apparent from many reports worldwide that it represents a significant problem. From a review of available information in Europe, EFSA (2007) concluded that the prevalence of docked pigs with any signs of tail lesion at the abattoir is ~3%, with 0.5–1% showing fresh injury and infection. In undocked pigs, the abattoir prevalence of lesions is higher at 6–10%, with 2–3% showing severe damage and infection. Review of the on-farm data suggested that 30–70% of farms have experienced some degree of tail biting problem, with estimates of the prevalence of damaged tails per farm varying widely but of the order of 1–5%.

Even at these relatively low average prevalences, the economic cost of tail biting outbreaks can be substantial. The costs include both direct and indirect aspects of the management and sale of pigs which have been bitten. The more severely affected pigs require veterinary treatment and hospitalisation, incurring both medicine and labour costs, and a proportion of them which die incur lost income and further costs for carcass disposal. Even less severely affected pigs may show reduced growth and feed efficiency (Sinisalo et al. 2012) as a result of changed feeding patterns, immune activation and general behavioural disruption in the pen, whilst additional management and stockperson time are taken up in trying to resolve the outbreak. Because tail lesions pose an increased risk of local and systemic infection (Munsterhjelm et al. 2013a), there is a consequent increase in partial and total condemnation of the carcass at the abattoir, often accompanied by marketing contract penalties for pigs sold early and under-weight from problem batches. Detailed assumptions about calculation of the prevalence and cost of each of these elements can be found in various publications (Edwards 2012; D'Eath et al. 2016; Niemi et al. 2020). Zonderland et al. (2011) estimated losses in the Dutch national herd, with a 2% prevalence of tail biting in a situation where all pigs are tail-docked, to be ~ €0.36 per slaughtered pig. Edwards (2012) described a UK study where the economic impact of different levels of outbreak severity was estimated at 3.8, 5.0 and 27.5 € cent/kg carcass sold for an incidence of 11%, 17% or 35% bitten pigs within the batch. A Finnish model (cited in Munsterhjelm 2011) arrived at similar estimates €4.8 per pig when less than 50% of pens on the farm were affected, rising to €20 per pig when 80% of pens were affected, while D'Eath et al. (2016) estimated the costs of tail biting at €18.96 per victim pig. Most recently, Niemi et al. (2020), using a stochastic bio-economic model to simulate the financial return per pig space unit at different levels of occurrence of tail biting lesions, calculated a cost ranging from €16 to €35 per victim pig, depending on the farm prevalence, giving an overall cost rising to €2.3 per slaughtered pig when the prevalence was increased gradually from zero to up to 10%.

In addition to considering these costs incurred when a tail biting outbreak is experienced, there may be additional capital (e.g. greater space allowance, better ventilation systems) or variable costs (e.g. provision of additional enrichment, improved quality of feed) for changes made to buildings or management to reduce future tail biting risk. Using standard Danish production conditions as a starting point, D'Eath et al. (2016) estimated the increase in production cost of providing a lower-risk system with increased space and additional enrichment substrate to be €5.5 per pig. Niemi et al. (2020) considered a range of possible risk-reduction measures, including provision of enrichment, increased space allowance, improvement in climate control, genetic change, improvement in herd health and installation of early warning systems to detect an impending problem. The calculated costs of preventive measures varied from €0 to €6 per pig, with lower cost measures being those partially offset by other performance benefits. They demonstrated that higher cost measures would be likely to result in a net increase in profitability only when the risk of tail biting was very high. However, such extra cost might be recouped, and a change towards increased adoption of systems with undocked pigs promoted, if a higher market value for such pigs could be realised. This might be possible through financial incentives from governmental bodies (ter Beek 2015), higher market prices for welfare-labelled product or corporate social responsibility-driven initiatives from retailers (Christensen et al. 2019).

3 Do we have a mechanistic understanding of tail biting?

To understand why one individual starts and persists in tail biting behaviour whilst others in the same group do not do so requires a deeper understanding of the (neuro)physiological mechanisms which underlie the behaviour. Only with such understanding will it be possible to devise fully effective strategies to prevent or terminate tail biting outbreaks. As well as seeking differences characterising perpetrators, it has also been suggested that certain individuals may be predisposed to become victims and their characteristics are therefore also of interest. Becoming a victim of tail biting does seem to be a heritable trait (Canario and Flatres-Grall 2018), suggesting there must be some predisposing factors. However, in most of the studies discussed below, victim pigs were selected on the basis of their level of tail damage and hence it is not possible to separate cause and effect in the interpretation of differences in their (neuro) physiology. Attempts to understand tail biting at a mechanistic level are still in their infancy and are made by following two approaches: firstly by a detailed comparison of the characteristics of pigs within a group identified as tail biters, recipients or unaffected control animals, and secondly by a detailed investigation of genetic traits (both quantitative and molecular) which are associated with tail biting.

When making a comparison of biters, victims and controls it is first necessary to identify animals in these categories. Whilst identification of victims is easy because of the visible tail damage they show, identification of biters depends on behavioural observation. Since this usually occurs in pens where an outbreak has already started, it is uncertain which category of tail biting these pigs represent (two-stage, sudden or fanatical), or indeed whether they were the primary initiator of the outbreak or a subsequent recruit in the escalation. This may complicate the interpretation of results.

Early attempts to describe the characteristics of tail biting pigs focussed on their size, with many anecdotal reports that it was often 'runt' pigs which were responsible. Subsequent data suggest that this might be true for 'fanatical' biters, but not necessarily for other categories. van de Weerd et al. (2005) reported that only fanatical biters were significantly lighter than their penmates, with their lighter weight resulting not from differences at birth but from slower growth after weaning. Similarly, Beattie et al. (2005) reported that weaned piglets showing a higher prevalence of tail chewing behaviour had experienced reduced growth during the suckling period, and Zonderland et al. (2011) also found that tail-biting pigs in the post-weaning stage tended to have a lower weaning weight than non-biting contemporaries. Such data might suggest some metabolic deficiency, inducing either general or specific hunger for nutrients in these individuals, increasing their general foraging activity and making them more likely to chew and bite tails. In support of this idea, it has been shown that tail biting pigs also perform more investigatory behaviour of other types, directed to both other pigs and to objects present in the pen (Zonderland et al. 2011; Brunberg et al. 2011). Palander et al. (2013) found no signs of nutritional defects in tail biters on the basis of their intestinal morphology or blood metabolites when compared to victim or control pigs. However, their data did suggest differences in relation to the tail-biting environment, with all types of pigs in pens experiencing an outbreak differing from animals in unaffected pens in gut morphology and blood concentration of minerals and amino acids. Early studies suggested that tail biting pigs might have a specific attraction to blood which was metabolically based. Individual pigs show significant differences in their motivation to chew a blood-soaked artificial tail, and this motivation can be increased by specific nutrient deficiencies in minerals or protein (Fraser 1987; Fraser et al. 1991). Subsequently, McIntyre and Edwards (2002a) demonstrated that pigs which exhibited clinical tail biting at the onset of an outbreak also showed increased chewing of a blood-soaked artificial tail when compared with non-biting penmates.

As a result of such observations, Edwards (2006) suggested that tail biting might be associated with metabolically induced neuroendocrine changes, with possible implication of serotonin-mediated pathways in the brain controlling foraging behaviour. The serotonergic system is involved in the regulation of mood, feeding behaviour, aggressive behaviour and mental disorders in humans and other species (Carrasco and van de Kar 2003). The amino acid tryptophan is a precursor in serotonin synthesis, and dietary availability of tryptophan influences the level of serotonin in the brain. In support of the 'metabolic' theory, pigs fed diets formulated to reduce availability of tryptophan for serotonin synthesis showed reduced resting and increased foraging behaviour, and also showed increased chewing of the blood-soaked model in the artificial tail test (McIntyre and Edwards 2002b). More recently, it has been shown that biting of ears and tails can be reduced by higher levels of dietary tryptophan (Martinez-Trejo et al. 2009). Further evidence in support of a possible mechanism arising from amino acid metabolism comes from a genetic association between tail biting predisposition and lean tissue growth rate (Breuer et al. 2005; Ursinus et al. 2014a). A higher demand for tryptophan in pigs with high growth potential might mean that less is available for serotonin synthesis. Furthermore, Ursinus et al. (2014b), who classified pigs in four different phases of life (0–4 weeks, 4–8 weeks, 8–16 weeks and 16–23 weeks) according to tail biting phenotypes, showed that in phases of life during which pigs were classified as tail biters, they had lower whole blood and blood platelet serotonin (5-HT) levels and higher blood platelet 5-HT uptake velocities. These characteristics are also seen in humans with obsessive-compulsive disorders (Delorme et al. 2005). The tail biting behaviour of individual pigs was inconsistent across phases, suggesting the mechanistic basis of tail biting to be a dynamic phenomenon. Valros et al. (2015) compared the brains of 11 trios of pigs, comprising matched animals that were identified as a tail biter, a victim, or a control pig from an unaffected pen. While they found no difference between tail biters and the other phenotypic classes of pigs in the level of serotonin, the level of the main metabolite of serotonin, 5-HIAA, was increased in the prefrontal cortex of tail biters. They saw a positive correlation between the level of tryptophan in the blood, as well as the Trp:LNAA and Trp:BCAA ratios, with brain serotonin and 5-HIAA levels only in tail biters. This suggests the possibility of an enhanced serotonin metabolism in tail biting pigs, causing tryptophan to become a limiting factor for these pigs only. In a similar investigation comparing pigs with different tail biting phenotypes, Brunberg et al. (2013a,b) found differential expression of 32 transcripts in the hypothalamus and 7 transcripts in the prefrontal cortex when investigated gene expression differences in the hypothalamus and prefrontal cortex of tail biting and control pigs. These transcripts included genes associated with production traits in pigs, and sociality and novelty seeking in humans, providing a possible link to differences in exploratory behaviour.

A further reason for a growth check in 'runt' pigs could be a health disorder (Valros and Heinonen 2015), and here again changes in amino acid metabolism could be implicated in tail biting. Acute phase proteins, synthesised in response to immune activation, have a higher content of tryptophan and other amino acids such as tyrosine and phenylalanine when compared to muscle protein. Their synthesis in inflammatory states might, therefore, modify dietary requirements and induce metabolic deficiencies which stimulate exploratory oral behaviours in pigs. Depletion of tryptophan in plasma has been demonstrated in pigs following experimentally induced inflammation (Melchior et al. 2004) or a chronic immunological challenge induced by unhygienic housing conditions (Le Floc'h et al. 2009). In response to a reduction in dietary protein, pigs housed in unsanitary conditions show greater increases in activity and exploratory behaviour (Pastorelli et al. 2012) and increased damaging behaviour directed at pen mates (van der Meer et al. 2017) when compared to pigs kept in clean conditions. Such findings suggest a greater susceptibility to specific nutrient deficiency in immune-stimulated pigs. However, the relationship between poor health and tail biting has also given rise to another mechanistic explanation (Nordgreen et al. 2020). Activation of the immune system by pathogens results in the production of pro-inflammatory cytokines. Furthermore, psychological stress can also cause changes in cytokines and other immune parameters, which are similar to those seen as an infectious challenge (Zhao et al. 2019). These cytokines induce various behavioural changes that are collectively called 'sickness behaviour' and share features with mental disorders in humans such as anxiety and depression (Hart 1988). Nordgreen et al. (2020) propose, therefore, that cytokines might mediate the effect of different environmental stressors on an individual pig to increase the likelihood that it shows damaging behaviour, such as tail biting or ear biting. A systemic increase in cytokine production can influence the brain by vagal stimulation, by transport across the blood-brain barrier or by synthesis within the central nervous system. In human clinical studies, treatment with pro-inflammatory cytokines induces irritability, short temper, extreme emotional lability and cognitive impairment. Furthermore, in a link with the previously outlined metabolic mechanism, cytokines can increase synthesis of the enzyme indoleamine 2, 3 deoxygenase 1 (IDO), such that relatively more tryptophan goes into kynurenine synthesis rather than serotonin synthesis (O'Connor et al. 2009), giving the possibility for changed neurotransmitter patterns.

A further interesting relationship between cytokines, serotonin and behaviour is that anxiety can also be induced by cytokine treatment (Munshi et al. 2019). There are reports that tail biting pigs show increased fearfulness in a novel object test (Ursinus et al. 2014b), though Zupan et al. (2012) found the opposite result and attributed this to their increased exploratory motivation. However, in the latter paper, where patterns of heart rate and heart rate variability were examined in tail-biters, victims and control pigs, the patterns seen in biters were distinctive and similar to those found in humans with anxiety disorder such as panic. Fearful pigs may have a predisposition to become more stressed by environmental changes. Stressful situations can modify brain serotonin (Piekarzewska et al. 2000) and genetically stress-susceptible pigs have lower levels of serotonin in several brain regions than stress-tolerant pigs (Adeola et al. 1993). The changes in serotonin metabolism previously described in tail biters might, therefore, suggest that they were less well able to cope with stressful challenges. Tail biting pigs have been reported as having a higher serum cortisol level than control pigs (Rabhi et al. 2020), although Zupan and Zanella (2017) found no effect of tail-biting phenotype on diurnal salivary cortisol pattern or the cortisol release during different stressors. In the most detailed study, Munsterhjelm et al. (2013b) compared a variety of stress measures in quartets of age- and gender-matched fattening pigs including a tail biter, victim, unaffected control in the same pen and independent control in a pen without tail biting. Evidence for chronic stress in tail biters came only from an observed flattening of the normal circadian cortisol secretion pattern.

A final area of discussion regarding possible mechanisms underlying tail biting was explored in the review of Brunberg et al. (2016), which highlighted a possible involvement of the brain–gut–microbiota axis. Stressors can modify the microbiota composition and can increase intestinal wall permeability, allowing endotoxins from gut bacteria to enter the circulation and promote an immune response. Furthermore, the gut microbiota has been implicated in early programming of HPA-axis activity, and may act on the adrenal cortex and via neurotransmitters, including serotonin, to predispose the individual to develop abnormal behaviour (Dinan and Cryan 2012). Gut microbiota composition has also been shown to influence social and repetitive behaviours in other species. Whilst further research in this area is needed, a recent study did show differences in beta-diversity of the gut microbiota of pigs in tail-biting and tail-bitten groups in comparison with a negative control group, with the former showing reduced abundance of Lactobacillus species, which has been associated with disorders such as anxiety and depression in humans (Rabhi et al. 2020).

From all of these studies, a pattern seems to be emerging linking diet and health with changes in brain serotonin metabolism, modified by fearfulness and stress coping ability. The emerging picture is outlined in Fig. 2.

In many of the studies previously described, it is notable that the contrasts often show tail biters and victim pigs to have similar characteristics, which differ from those of unaffected control pigs (Palander et al. 2013; Munsterhjelm et al. 2013b; Ursinus et al. 2014b; Rabhi et al. 2020). Such findings led Munsterhjelm et al. (2013b) to suggest that these unaffected or neutral pigs may adopt a coping strategy in challenging situations which leads to less overall stress. A genome-wide association study identified single nucleotide polymorphism markers that were shared by tail biters and victims but differed from unaffected pigs in the same pen (Wilson et al. 2012). Furthermore, Brunberg et al. (2013a,b)

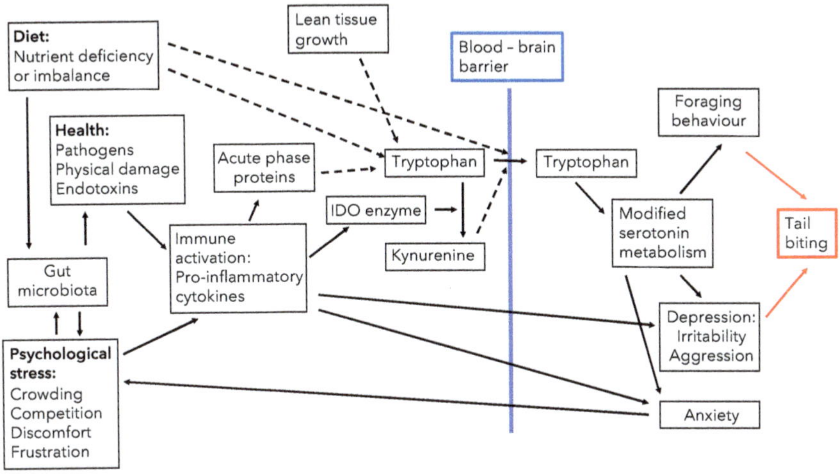

Figure 2 An overview of possible mechanistic relationships contributing to tail biting. Solid lines indicate a positive/stimulatory influence and broken lines a negative/inhibitory influence. See main text for details of each relationship.

also highlighted that neutral pigs differed most in both gene expression and behaviour from all other tail biting phenotypes. They concluded that neutral pigs must have a genetic and behavioural profile that makes them resistant to performing or receiving pig-directed abnormal behaviour, even when housed in an environment that elicits that behaviour in other pigs. Greater investigation of this possibility is therefore merited.

4 Risk factors for the occurrence of tail biting

One of the major challenges in managing tail biting successfully lies in the fact that the background of the problem behaviour is multifactorial and that risk factors and their combinations differ between farms (EFSA 2014). In practice this means that there is no generally applicable method to avoid tail biting on a specific farm. Also, a large number of risk factors for tail biting can be present on any farm, and these normally interact. Therefore, tools to identify risk factors on-farm, such as the HAT (developed in the UK, Taylor et al. 2012) and the Schwip (developed in Germany, vom Brocke et al. 2019), utilise comprehensive listings of risk factors (for more information see the later section on managing pigs without tail biting). Furthermore, the occurrence of tail biting is not only influenced by on-farm factors, such as housing and feeding-related problems, but also by intrinsic factors. Thus, not every pig will become a biter, and batches and pens will vary in their tail biting prevalence, even when housing and management do not vary. To complicate the problem even further, barn climate and air quality are important risk factors and, as these are largely influenced by outdoor conditions, risk factors will vary with weather conditions and season. However, despite the complexity of the problem, there is plenty of knowledge of which risk factors to focus on, and plenty of examples of farms where tail biting is managed to a very low level.

Studies on tail biting risk factors have either been conducted as epidemiological, on-farm assessments of risk factors (e.g. Moinard et al. 2003; Kallio et al. 2018; Grümpel et al. 2018), or as experimental studies focusing on specific, isolated factors (e.g. Telkänranta et al. 2014a,b; Larsen et al. 2018a). In addition, a few studies have asked producers for their opinion on the most efficient preventive measures (Bracke et al. 2013; Valros et al. 2016; Valros and Barber 2019; Haigh and O'Driscoll 2019). Finally, many papers refer to more sporadic anecdotal evidence. These different approaches all have their pros and cons; only experimental studies with controlled setups can prove a causal relationship, while usually concentrating on one or a few risk factors at once and thus often failing to address the multifactorial nature of the problem. On the other hand, epidemiological studies, performed either by farm visits or questionnaires, often include a large number of risk factors and have the potential to look at the interactions between these (e.g. Kallio et al. 2018). However, these studies cannot prove causalities, and often fail to some degree in separating confounding factors. One typical example of this is space allowance; even though space allowance has been indicated as a risk factor in several studies (Moinard et al. 2003; Munsterhjelm et al. 2015), it has also been indicated that this might be via a link to associated changes in for example feeder space or other resource allocation.

4.1 Extrinsic risk factors

Probably the most important factor, while also being the probably most challenging to address for reducing the risk for tail biting on any specific farm, is the experience and knowledge, as well as the motivation of the producer and the farm staff. Some of the Finnish producers taking part in the questionnaire study by Valros et al. (2016) underlined the role of the caretaker, stating that the time spent with the animals is important. One reason for this might be that the human-animal interaction itself is valuable for the pigs (Rault et al. 2020) and that they will be less stressed if they are not fearful of humans. Recently, Büttner et al. (2018) showed in an experimental study that an intensified positive human-animal interaction resulted in a lower level of tail lesions in rearing pigs. However, another mechanism for the importance of the stockperson might be the time available for observing pigs and intervening when needed. Some epidemiological studies indicate an increased risk for tail biting on larger farms (EFSA 2014; Munsterhjelm et al. 2015; Kallio et al. 2018). Probably, however, the number of staff per animal is more important than farm size *per se*. If one person is responsible for a large number of animals, early detection of tail biting outbreaks might be compromised, as well as possibilities for intervention (D'Eath et al. 2016). Moinard et al. (2003) and Valros and Barber (2019) found an increased risk of, or reported level of, tail biting damage when pens, or animals, per stockperson increased. The attitude towards tail biting and docking, however, does seem to be related to farm size, with large farms being more motivated to dock (Bracke et al. 2013), larger farms having pigs with a larger proportion of tails docked (Valros and Barber 2019) and large farms being more concerned about tail biting (Valros et al. 2016).

4.1.1 Manipulable materials

The most frequently studied risk factor for tail biting is probably the lack of proper manipulable materials for pigs (D'Eath et al. 2014), and this is commonly ranked as the most important factor by experts (EFSA 2007). This risk factor has been identified both in epidemiological studies on large sets of farms (e.g. EFSA 2014; Munsterhjelm et al. 2015; Pandolfi et al. 2017) and in experimental studies (Telkänranta et al. 2014b; Larsen et al. 2018a). It is assumed that the lack of possibility to meet the innate need for exploration in pigs causes frustration and redirection of exploration towards other pigs´ tails. The association between tail biting and use of manipulable materials is reflected in the way these two are linked in official documents within the European Union, such as the European Commission staff working document on 'best practices with a view to the prevention of routine tail-docking and the provision of enrichment materials to pigs' (EC 2016). However, when producers are surveyed, they do not rank the lack of manipulable materials or 'boredom' as an equally important risk factor (Bracke et al. 2013; Valros et al. 2016; Valros and Barber 2019; Haigh and O'Driscoll 2019). Instead, they might see manipulable materials more as an intervention strategy–something that buffers the effect of other challenges on the farm (anecdotal communication from Finnish farmers). Furthermore, the perceived efficiency of manipulable materials might be influenced by the system on the farm, and whether such materials are normally used or not. Producers in the UK differed in their perception of the efficiency of use of bedding and straw depending on how much bedding they actually use on their farm (Valros and Barber 2019).

In general, straw has been considered the most effective material to reduce tail biting risk, both by experts (EFSA 2014; Bracke et al. 2013) and by producers (Valros et al. 2016; Valros and Barber 2019). Also, large-scale epidemiological (EFSA 2014) and experimental studies (Larsen et al. 2018a) have confirmed the effectiveness of straw, and Wallgren et al. (2019a) showed that by increasing the straw use on farms already using some level of straw, tail biting severity can be reduced further without compromising pen hygiene. Straw has many of the attributes considered important to make a manipulable material functional (EFSA 2014): it is deformable, moveable and has feed-related properties. In addition, if added frequently and in appropriate amounts, it adds novelty and can be made accessible to all pigs in a pen at the same time. The way that straw is presented might, however, influence its effectiveness; Bulens et al. (2016) and Holling et al. (2017) both failed to show an effect of straw provided from dispensers on the level of tail lesions. In these studies, however, the amount of straw used by the pigs remained rather low: 8.3 g per week per pig in weaning pigs in the study by Bulens et al. (2016) and 3.5 g per day per pig in weaning pigs and 33.9 g per day per pig in finishing pigs in the study by Holling et al. (2017).

Solid objects for manipulation, such as chains, pieces of wood and plastic balls or pipes are often used in systems with constraints on the use of bedding (e.g. fully slatted floors and slurry manure systems). These are considered only moderately effective as enrichment objects (Bracke and Koene 2019; Buijs and Muns 2019) but, when well designed, can have some value for the pigs–such as the branched chain suggested by Bracke and Koene (2019). Solid objects which are more destructible than chains might be more effective, such as fresh wood (Telkänranta et al. 2014a) or jute sacks (Ursinus et al. 2014a) and their efficacy might be increased by providing a variation of objects (Chou et al. 2019b).

4.1.2 Feeding

In Finland, where tail docking has been fully banned since 2003, producers frequently refer to feeding-related issues as the most important for reducing the risk of tail biting. Feed- and water-related factors indeed came up as the main preventive factors in a farmer survey study performed in Finland (Valros et al. 2016), while the perceived importance of feed appeared lower among UK producers (Valros and Barber 2019). In both countries, however, making sure that all pigs have access to water was ranked as a very important preventive measure. Poor water availability is usually not ranked as a very important risk factor by experts (e.g. EFSA 2007), but problems with water delivery have anecdotally been reported as a big risk for tail biting outbreaks (Valros and Barber 2019).

While feed structure and composition are often referred to as very important for preventing tail biting, only a few studies have actually been able to pinpoint specific characteristics of feed. Pellet feeding (Hunter et al. 2001; Tetsuo et al. 2015) and liquid feeding (Kallio et al. 2018) have been reported as risk factors, and meal feed as a protective factor (Pandolfi et al. 2017). The importance of feed structure might, however, be confounded by the feeding system used.

Regarding feed composition, very few studies report significant effects of specific ingredients; Kallio et al. (2018) reported in an epidemiological study that feeds including wheat or whey might increase tail biting risk, while Martinez-Trejo et al. (2009) showed in an experimental study that an increased level of tryptophan in the feed could reduce tail biting prevalence. Fibre, mineral and protein levels have also been suggested as important factors (e.g. Schrøder-Petersen and Simonsen 2001; D'Eath et al. 2014; Valros et al. 2016). Some recent studies have been able to provide further proof of this; van der Meer et al. (2017) found more tail biting behaviour in pigs fed with a low as compared to a high level of protein, while Naya et al. (2019a) showed some indication of increased fibre reducing the severity of tail biting damage. However, due to a large batch effect in their study, the results were not conclusive.

The way that the feed is presented to pigs is also considered important, and there is more evidence for this link than for the composition of the feed. A large amount of tail biting has been reported to happen around the feeder (Sutherland et al. 2009; Palander et al. 2012). Palander et al. (2013) reported that, in a pen with ongoing tail biting, pigs that stayed out of trouble (non-bitten, non-biters) might have reduced feeding to avoid the risky area of the pen; these pigs showed signs of recent malnutrition, including reductions in certain minerals in the blood and changes in intestinal wall morphology. Palander et al. (2013) also showed that pigs fed ad libitum from feeders with restricted trough space had similar indications of feed restriction to those meal-fed from long troughs, suggesting an increase in competition for feed in the ad libitum fed pigs. Low feeder space when applying ad libitum feeding has, indeed, been indicated as a risk factor for tail biting in epidemiological studies (Hunter et al. 2001; Moinard et al. 2003). However, the perceived relevance of feeder space appears to be influenced by the feeding system employed. In Finland, where most finishing farms use long troughs and liquid feeding in meals, feeder space is considered much more important than by producers in the UK, where most pigs are fed ad libitum from hoppers (Valros and Barber 2019).

4.1.3 Housing and climate

When a pig production unit is in place, the housing systems, including for example pen structure, flooring and climate control, are difficult to adjust. Thus, it is important to consider management of tail biting already when planning a new unit or renovating an old one. One of the factors that have been indicated as an important risk factor in epidemiological studies is the proportion of slatted flooring in the pen (Moinard et al. 2003; Kallio et al. 2018). This is also very often anecdotally referred to as a challenge in current systems, especially as it makes it difficult to use bedding-type materials, such as straw, for manipulation (EFSA 2007). In addition, however, fully slatted floors might make it more difficult to provide a comfortable, clearly separated lying area, and might increase the risk for a high level of noxious gases (Schrøder-Petersen and Simonsen 2001).

A high stocking density is often considered to be one of the main risk factors for tail biting, also by producers (Bracke et al. 2013; Valros and Barber 2019), while the scientific proof is not fully clear. One reason for this might be the above-mentioned link between space allowance and other factors, such as feeding space. Another factor which might be linked to space allowance is the group size. However, there appears to be only little evidence confirming the role of group size. Kallio et al. (2018) found an increased risk for tail biting on farms with pigs in pens of more than 10 animals. Larsen et al. (2018a) showed that, even though animal density per se did not reduce the risk of tail biting, a combined effect of decreased density and supplying straw reduced the risk as effectively as tail docking. Regardless of possible confounding factors, several epidemiological studies do indicate a clear link between space allowance and tail biting risk, both in weaner and in finisher pigs (Moinard et al. 2003; Munsterhjelm et al. 2015; Scollo et al. 2016, 2017; Grümpel et al. 2018). In the study by Munsterhjelm et al. (2015) there was a relatively linear association between space allowance and on-farm tail biting prevalence when increasing from 0.7 m^2 to 1.5 m^2 per finishing pig.

Especially in countries with extreme climates, either due to high or low temperatures, or to a large variation in temperatures, designing a good system for climate control can be challenging. These issues are very difficult to study experimentally, but some evidence has been reviewed on the importance of temperature (Schrøder-Petersen and Simonsen 2001; D'Eath et al. 2014). Hunter et al. (2001) showed a decreased risk of tail biting on farms with artificial ventilation systems. In an experimental study, Holling et al. (2017) presented some evidence that air velocity, temperature and temperature range, as well as ammonia level (also highlighted by Scollo et al. 2017), were related to the risk of tail lesions, reinforcing earlier experiments that implicated high airspeed in inducing restlessness and injurious behaviour (Scheepens et al. 1991). Further, appropriate temperature, eliminating draughts and maintaining a good air quality were all listed within the perceived top-ten preventive measures by both UK and Finnish producers (Valros and Barber 2019). Seasonal differences in the risk of tail biting might be due to different efficacy of climate regulation systems in different weather conditions (D'Eath et al. 2014). Due to geographic differences in climate challenges, and as both heat and cold extremes can cause pigs stress, it is probable that both the on-farm risk factors related to indoor climate and the seasonal effects will vary considerably between different countries and regions.

4.1.4 Health status of the farm

It is commonly assumed that tail biting damage, in itself, causes an increased risk for secondary infections, such as lung infections, abscess formation and arthritis, with the damaged tail acting as an entrance for pathogens (Sihvo et al. 2011). On the other hand, a poor health status (Edwards 2011; Boyle et al. 2020) and low sanitary conditions (van der Meer et al. 2017) have been linked to tail biting. Some studies further report a positive effect of vaccinations and anthelmintic treatments on reducing the risk of tail biting (as reviewed by D'Eath et al. 2014). Most studies on the link between health and tail biting, however, fail to provide proof of the direction of the link–does tail biting reduce health or does poor health increase tail biting? Niemi et al. (2011) showed a temporal link between tail biting and locomotor problems; the risk of lameness was increased in pigs previously diagnosed with tail biting damage, while lameness also appeared to increase the risk for future tail biting damage. A recent study on this topic does indicate that it is not only that tail damage predisposes to secondary health problems, but also that health problems can predispose to both being bitten and to become a biter (Musterhjelm et al. 2017). Furthermore, the systemic sickness response might change the behaviour of pigs in a way that increases the risk for tail biting problems (Munsterhjelm et al. 2017, 2019; see also the previous mechanisms section).

4.2 Acute risk factors

In addition to the above listed risk factors, which are mainly of a chronic nature, that is, factors that are present on the farm during a prolonged period, or which are characteristics of the farm *per se*, tail biting outbreaks can be triggered by acute changes in the pigs' environment or feeding. There are very few actual studies reporting such effects, but there are many anecdotal reports of, for example, problems with water delivery (as mentioned previously), feeder malfunction (Paul et al. 2007; Telkänranta et al. 2014b; Scollo et al. 2017) or problems with ventilation (Holling et al. 2017) causing tail biting outbreaks. Also, health problems, such as diarrhoea in weaning pigs can predispose pigs to begin tail biting (Paul et al. 2007; J. Vugts, Personal communication). It is, of course difficult to fully manage such largely unpredictable events on a farm, but by managing the chronic risk factors to a low level, acute risk factors might be less likely to cause an outbreak. Valros (2018) described a theoretical model showing how different additive combinations of chronic and acute risk factors might increase the level of stress experienced by a pig until it exceeded the threshold necessary to trigger tail biting. Against a background of a low level of chronic risk factors, a much higher threshold for acute risk factors will occur. Producers anecdotally report, for example, that manipulable materials are most important when something else fails, such as there being a feed delay. In these challenging situations, appropriate access to manipulable materials might buffer the stress experienced by the pigs, and thus reduce the likelihood for a tail biting outbreak.

4.3 Intrinsic risk factors

Even when there are severe challenges in the housing or management of pigs, not all pigs will show tail biting behaviour. On the other hand, even on the best of farms, there might be the occasional outbreaks of tail biting. There is still a lack of evidence for the reason for this individual variation, but various predisposing characteristics have been identified.

4.3.1 Breeding and genetics

There are a few studies on the effect of breed or genetic background on the risk for tail biting behaviour but, as the identification of tail biters is rather time consuming and challenging, these studies are usually restricted by a low sample size. Sinisalo et al. (2012) studied a population including Landrace and Yorkshire pigs and found more biters to be Landrace. A higher percentage of tail biting behaviour in Landrace pigs, as compared to Yorkshire or Hampshire pigs, has also been reported in other studies (Westin 2003; Breuer et al. 2005). Some level of heritability for tail biting behaviour was further shown in Landrace pigs (Breuer et al. 2005). Finally, Breuer et al. (2003) reported more biting behaviour in Duroc pigs than in Landrace and Yorkshire pigs. On the other hand, both Westin (2003) and Sinisalo et al. (2012) found more tail biting damage in Yorkshire pigs than in Landrace and Hampshire, and Landrace pigs, respectively. Characteristics of individual genetic lines may be more important than a breed classification.

As discussed previously, problems with growth performance, and a smaller than average size of the pigs is often anecdotally mentioned as a typical characteristic of tail biters. There is some scientifically reported evidence of this link: Munsterhjelm et al. (2016) showed that pigs that were to become performers of so-called tail-in-mouth behaviour–which is suggested to sometimes escalate to actual biting behaviour–were smaller at birth than pigs that would become receivers of the behaviour after weaning. The size difference, however, disappeared after weaning, and there was no difference in growth. Recently, in an epidemiological study, Grümpel et al. (2018) showed that a poorer pen growth rate was linked to an increased risk for tail biting, although the study design did not allow for assessment of causality. Furthermore, Sinisalo et al. (2012) reported that, at arrival in the finishing unit, future performers of tail biting were smaller than other pigs. On the other hand, other studies have failed to report a difference in size between biters and other pigs (Zonderland et al. 2011; Munsterhjelm et al. 2013a). Even though the phenotypic evidence is not clear, there is some evidence from genetic studies on a link between breeding for the modern, fast growing, lean pig and an increased risk for tail biting. Breuer et al. (2005) were the first to report a positive genetic association between tail biting and lean growth. More recently, Ursinus et al. (2014a) found that growing gilts that performed tail biting grew faster and were heavier than those that did not, while Sinisalo et al. (2012) linked becoming a victim to tail biting with genetic potential for growth; pigs with tail damage had a lower genetic growth potential than pigs with no damage. An interesting approach to the question is that of looking at indirect social effects of breeding. Camerlink et al. (2015) bred pigs for a positive effect on the growth within their group and found that these pigs also exhibited less ear and tail biting behaviour.

4.3.2 Sex

Small female pigs are often said to be typical tail biters. However, while the evidence of more tail damage in barrows than females is quite clear (Valros et al. 2004; Kritas and Morrison 2004, 2007), the effect of sex on the risk of becoming a tail biter remains inconclusive. Some studies do indicate that females perform more biting and tail-in-mouth behaviour (Zonderland et al. 2010; Schrøder-Petersen et al. 2004), while others failed to find a sex difference (Zonderland et al. 2011). Group composition might further influence the outcome, with a difference in tail biting prevalence in groups with different proportions of pigs of different sexes (Kritas and Morrison 2007; Zonderland et al. 2010).

4.3.3 Age and previous experience

Certain ages or production stages are often reported as more prone to tail biting outbreaks, but there seems to be some variation between studies in when these risky periods occur. Veit et al. (2016) and Naya et al. (2019a) both reported that tail biting started in the second to third week after weaning and peaked 4–5 weeks after weaning, after which it again decreased towards the end of the grower period. On the other hand, Wallgren et al. (2019a) found that tail biting damage prevalence increased continuously during the time the pigs spent in the grower unit. In the same study, the pattern, however, differed between the four included farms for the finishing units: overall the number of intact tails decreased towards the end of the finisher period, but on one farm lesions decreased after an initial increase of damage prevalence. Larsen et al. (2018a) showed that the largest number of observations of first signs of tail damage in pens occurred already during the first week in the finishing unit. From personal experience of scoring a large number of tails at the slaughterhouse (M. Heinonen and A. Valros, unpublished data) it appears that the tail biting pattern indeed differs significantly between batches from different farms;

some batches with tail biting problems had high levels of healed tail lesions, some high levels of acute lesions, and some a combination of both. There are few studies directly comparing risk factors for tail biting at different ages, and most studies are performed with pigs in the finishing unit. However, similar risk factors, such as stocking density and low growth rate (Grümpel et al. 2018) and slatted flooring (Kallio et al. 2018), appear both in weaning pigs and finishing pigs.

Previous experiences of pigs influence their behaviour at later ages, and this is true also for tail biting. Thus, it is not only important to improve the conditions where tail biting occurs, but also consider the need to reduce risk factors earlier in the life of the animals. A few studies have indicated that it is essential to make sure that suckling piglets already get access to proper manipulable materials. Moinard et al. (2003) showed, in an epidemiological study, that farms not adding straw to the piglet creep area had a greater risk for tail biting later on than those that did this. In experimental studies, Munsterhjelm et al. (2009) showed a long-lasting effect of adding bedding material in the lactation unit; pigs that were given bedding during the first weeks of life showed less negative social behaviour during the finishing stage. Furthermore, Telkänranta et al. (2014b) found less severe tail biting damage at the age of 9 weeks in pigs that had been given extra manipulable materials in the lactation unit, in the form of ropes and newspaper. In addition, other factors of the earlier growing stages have been linked to tail biting (Prunier et al. 2020). In epidemiological studies, Smulders et al. (2008) reported that factors both in the farrowing unit (floor structure) and in the weaning unit (number of feeding places and temperature) were associated with tail biting in the finishing unit, while Grümpel et al. (2018) showed that a high level of suckling piglet losses was related to an increased tail biting risk on the farm.

Studies on alternative farrowing and lactation systems have also generated interesting results regarding the effect of early experiences on the development of damaging behaviour of piglets, although these are not specific for tail biting. van Neuwamerongen et al. (2015) showed that piglets weaned at 4 weeks of age from a multi-suckling system performed less oral manipulation of other pigs post weaning, than did piglets weaned from a conventional system with individually crated sows. Further, when comparing piglets either weaned abruptly at 4 weeks from the multi-suckling system to piglets weaned gradually (final weaning at 9 weeks) from the same system, the latter showed less negative manipulation of other piglets until at least 18 weeks of age (van Neuwamerongen et al. 2017). Naya et al. (2019b) failed to show a difference in tail lesions after weaning between piglets weaned at four or five weeks from individually housed sows, or at five weeks from group-housed sows. However, this was partly due to a strong interaction with batch, which probably overrode the effect of the treatment.

5 Ethical considerations associated with tail biting

Tail docking, namely the amputation of a large part of the tail in young piglets, is performed to reduce the risk for tail biting. Indeed, tail docking does reduce the incidence of tail biting (e.g. Lahrmann et al. 2017; Larsen et al. 2018a; Thodberg et al. 2018), but does not abolish it. Studies on slaughterhouse data have shown that a large proportion of docked pigs still have tail biting damage (Harley et al. 2012; vom Brocke et al. 2019). Docking is problematic, because it is a painful procedure, for which there is so far no fully effective and feasible pain alleviation method (Sutherland et al. 2011; Courboulay et al. 2015; Herskin et al. 2016). However, probably the most important ethical argument for questioning tail docking is that it is a means to reduce the adverse effects of keeping animals under conditions which cause them stress (D'Eath et al. 2014, 2016; Valros et al. 2015). This conflict is also reflected in current EU legislation:

> *'Neither tail-docking nor reduction of corner teeth must be carried out routinely but only where there is evidence that injuries to sows' teats or to other pigs' ears or tails have occurred. Before carrying out these procedures, other measures shall be taken to prevent tail-biting and other vices, taking into account environment and stocking densities. For this reason inadequate environmental conditions or management systems must be changed'* (Council Directive 2008/120/EC).

Even though the ban on routine tail docking has been in force for over 25 years (since 1994, Directive 91/630/EEC), and research activities have provided plenty of new information on both managing risk factors and intervening when outbreaks occur, the vast majority of EU countries still dock the majority of the pigs (De Briyne et al. 2018). Beyond the EU, the situation is very similar, with most pigs being tail docked in many countries (Nalon and De Briyne 2019). One reason for the difficulty to achieve progress might be that producers are reluctant to try non-docking (Bracke et al. 2013; Nalon and De Briyne 2019). Nalon and De Briyne (2019) suggested that this might be due to the multifactorial nature of tail biting, making it difficult to control and predict, and because producers tend to repeat habitual behaviour. A further, probably very important, reason is that raising pigs with intact tails costs more (Nalon and De Briyne 2019). When comparing different scenarios (docking in current systems versus non-docking in systems with improved housing and management to reduce risk factors), the most economical production system is the current one, namely docking and keeping production conditions as they are (D'Eath et al. 2016). However, economics is not the only force influencing how animals are kept: EU citizens are very concerned about animal welfare issues (Eurobarometer 2016). A recent NGO-campaign regarding a stricter enforcement of the tail docking legislation, as well as phasing out castration without pain relief, collected over one million signatures from different EU member states (Nalon and De Briyne 2019). Furthermore, there may be benefits to animal health and associated reductions in antimicrobial use and risks to human health from antimicrobial resistant pathogens arising from changing production systems to raise undocked pigs (Dawkins 2019).

5.1 Should tail biting be managed by other means than tail docking?

Tail biting in itself is painful to the pigs (see also Chapter 8 of this volume) and is associated with an increased risk of secondary infections (Valros and Heinonen 2015). Tail biting also reduces the economic performance of a farm substantially, through an increase in carcass condemnations, reduced growth, increased workload and medical costs, and an assumed increase in pig mortality (Niemi et al. 2012). Therefore, it is clear that, from an ethical point-of-view, it is very important to prevent the problem effectively. However, as tail biting is a behaviour associated with stress, the solution that improves the welfare of the pigs most is not only to prevent the resulting damage, but also to actually reduce the motivation for the behaviour by improving the way pigs are housed and managed. When choosing the best approach, a holistic ethical weighing, such as that suggested by Valros and Heinonen (2015), should consider both the pain experienced by pigs due to docking and to biting, and the (positive) welfare implications of adjusting production systems to allow rearing intact-tailed pigs. However, even when simply weighing pros and cons of tail docking only, one should remember that in a situation of docking, normally 100% of pigs are docked, and some are still bitten, while in a non-docking situation no pigs experience docking pain, while more might end up being bitten (Valros and Heinonen 2015). D'Eath et al. (2016) presented an ethical balance comparing a standard Danish production system in which all animals were tail docked with an improved housing system in which tails were left intact but 3.2% more pigs experienced tail biting. They concluded that, in order for docking to be the ethically more sound solution, the pain of tail biting should be at least 31.3 times worse than that of tail docking.

There are very few studies comparing the efficacy of preventing tail biting by docking tails or by using other measures. However, Larsen et al. (2018a) showed that a combination of added straw and increased space allowance decreased tail biting damage to a level comparable to that achieved by docking. Also, experience from countries where tail docking has been banned shows that it is possible to reach comparable, or at least manageable, tail biting damage levels both with and without docking of pigs (EFSA 2014; Valros and Heinonen 2015). Furthermore, the majority of producers replying to a survey in Finland, where docking was banned in 2003, stated that they would not now want to tail dock even if it was allowed, and the majority also did not see tail biting as a very serious problem on their farm (Valros et al. 2016). The extent to which a problem is perceived as serious obviously relies on subjective judgement, and will depend on influencing factors including personal circumstances, experience and social norms.

5.2 How can producers be motivated to avoid docking?

In order to achieve a situation where tail biting risks are decreased to a level where tail docking becomes unnecessary, the producer, in most cases, needs to be motivated to change both his/her production system and his/her behaviour. Knowledge about how to avoid tail biting problems is essential to achieve behavioural change (Zonderland and Zonderland-Thomassen 2016). In addition to increasing knowledge among producers on how the tail biting risk can be reduced, it should also be kept in mind that producers have different attitudes towards tail biting and docking, depending on which system they are used to, which might lead to a need for different approaches in different countries and production systems. Bracke et al. (2013) reported that Dutch producers were very reluctant to consider not tail docking, unless they were sure that no tail biting would occur. On the contrary, as stated previously, Finnish farmers had a rather negative attitude to docking (Valros et al. 2016), which is something very few of them even have experience of now. Producers in the UK who currently raise pigs with docked tails stated that 'not docking' is cruel to the animals, and is 'asking for trouble'. Some of them suggested that failing to dock is actually a welfare issue (Valros and Barber 2019). Again, the Finnish producers, who did not consider docking an option even if it was legal in Finland, often supported this view by very similar arguments to those used by UK producers to defend docking: Finnish producers referred to 'docking' as being cruel or inhumane. This is probably a typical example of defending the current practice (Spooner et al. 2014). To further motivate farmers, benefits of not docking could be underlined, and according to Nalon and De Bryine (2019), farmers do report advantages with non-docking, such as increased pride in their work, improved animal health status and a potential for better product prices. Finally, it is important to improve communication between different stakeholders, as farmers partly disagree with experts regarding how they rank risk factors for tail biting, for example by ranking the importance of using manipulable materials much lower than experts do (EFSA 2007; Bracke et al. 2013; Valros et al. 2016; Valros and Barber 2019; Haigh and O'Driscoll 2019).

6 Managing pigs without tail biting

Managing pigs without tail biting requires a high level of attention to detail, with both good overall management and day-to-day stockmanship. This is especially the case when dealing with pigs which have intact tails, where the level of risk is increased (Lahrmann et al. 2017; Thodberg et al. 2018). The key to success lies in prevention, since, once started to the extent that bleeding tails are present, a tail biting outbreak can be very difficult to control and stop (Chou et al. 2019a). Prevention of an outbreak is based on two approaches, minimising known risk factors and early identification of warning signs. Since every farm has a different profile of risk factors, decision support tools have been developed to help individual farms to assess their own situation. The first of these tools was the Husbandry Advisory Tool (HAT) (Taylor et al. 2012) which, based on scientific literature and expert opinion, provided a checklist of 83 possible risk factors and weighted their likely contribution to an outbreak. By testing the HAT on 65 commercial pig farms in England, they demonstrated that there was a significant correlation between the 'risk score' which it produced and the prevalence of tail damage on a farm. As a result of increasing awareness from using the tool, a reduction in risk factors at the end of the study was observed on 42/57 farms visited. They highlighted that, while some risk factors are structural and require substantial capital investment to change, a significant reduction in the risk of tail biting can be achieved on many farms through the systematic evaluation and modification of management practices. This HAT was subsequently made available on-line in a simplified form by the UK Agricultural and Horticultural Development Board (https://webhat.ahdb.org.uk/). Using the HAT as a starting point, a similar tool was developed for use in Germany as the 'Schwanzbeiß-Interventions-Programm' (SchwIP) (vom Brocke 2014). Dissemination of the tool through training courses for veterinarians and farm advisers showed significant variation in the evaluation of the on-farm usability of the tool, but 67% of training participants recommended training workshops for colleagues for the uptake of similar tools. SchwIP was implemented on 188 farms with tail biting problems, with farmers and external advisors deciding at each visit on appropriate measures to minimise the tail biting risk on their own farms. Across all farms, the total risk score for tail biting had decreased by a second visit in the following year, as had the prevalence of tail and ear lesions. A subsequent study showed that advising problem farms on measures to reduce tail biting risk using the SchwIP tool also helped to decrease the prevalence of tail lesions measured at the abattoir (vom Brocke et al. 2019). Whilst the SchwIP tool was initially developed for finishing pigs, it was subsequently modified and successfully applied to reduce tail biting in nursery pigs (Grümpel et al. 2019).

The collection of data at the abattoir on carcasses exhibiting damaged tails is a very important tool in risk reduction strategies. Such data can be used to monitor the farm prevalence of tail biting and provide feedback on the consequences of changes in production practice, whilst also generating large datasets which can be used in epidemiological investigation of risk factors (Hunter et al. 2001). Recent developments in sensor technology and data processing have facilitated the automation of abattoir data capture (Larsen et al. 2019a; Brünger et al. 2019; Blömke et al. 2020) and this offers important possibilities for refinement of risk reduction tools in the future.

The second key aspect in managing pigs without tail biting is early recognition of an emerging problem and prompt action before an outbreak can escalate. This currently depends on the skill of the stockperson and is something which improves with training and experience (Veit et al. 2016). The early behavioural signs that might indicate that a tail biting outbreak is imminent have been reviewed (Larsen et al. 2016). Several studies suggest that the general level of activity in a pen increases before an outbreak (Statham et al. 2009; Zonderland et al. 2011; Ursinus et al. 2014c; Larsen et al. 2019b). Changes have also been reported in non-damaging tail and ear chewing (Zonderland et al. 2011b) or chewing of enrichment objects (Ursinus et al. 2014c; Larsen et al. 2019b), though data are less consistent between studies. Changes in patterns of feeding (Wallenbeck and Keeling 2013) and drinking (Larsen et al. 2019b) have also been suggested, but show a similar lack of consistency across different studies. The most consistent sign of an impending outbreak would appear to be a change in tail posture. Zonderland et al. (2009) first highlighted that an increased prevalence of pigs with their tail held low and tucked between the legs (Fig. 3), rather than held in a natural high and curled position, could predict a tail biting outbreak at least 2–3 days before clinical damage. This observation of lowered or 'tucked' tails as a warning up to 7 days before an outbreak occurs has subsequently been reported in numerous other studies (Statham et al. 2009; Lahrmann et al. 2018a; Larsen et al. 2018b; Wedin et al. 2018; Wallgren et al. 2019b).

Figure 3 The characteristic hanging tail posture which predicts an emerging tail biting outbreak is seen in the two pigs on the left, in comparison with the normal curled tail posture in the pig on the right. (Photograph courtesy of Emma Baxter, SRUC Edinburgh).

Based on the knowledge of the behavioural changes preceding a tail biting outbreak, attempts are now being made to automate early warning systems as part of Precision Livestock Farming (PLF) systems. A 3D camera system has been developed to monitor changes in tail posture, showing an accuracy of 74% at detecting low versus not low tails (D'Eath et al. 2018). Prior to an outbreak, the prevalence of low tails occurred at an increasing rate. However, the relatively high cost of such a sophisticated detection system has led to a search for simpler and cheaper alternatives. Changes in activity level preceding a tail biting outbreak have been detected using a simpler Optical Flow system, which is less reliant on filming conditions (Li et al. 2020). However, subsequent work suggests that movement change detected by optical flow precedes, but does not predict, tail biting in pigs (Larsen et al. 2020). Not only pens developing an outbreak, but also contemporary control pens showed a similar change in movement patterns, suggesting either that the pigs were influencing each other or that both were responding to common external circumstances. Such tools may therefore raise awareness of potential risk within a room, but will not specifically identify problem pens. The generation of 'false-positive' results will be costly in investigation time and approaches to minimise this problem are needed. Given the increase in herd sizes, reduction in staffing levels and fast development of PLF technology, such approaches are likely to increase rapidly in the near future. Using sophisticated machine learning approaches to combine animal behaviour data with other routinely collected sensor data, such as on temperature and water use patterns which currently also show promise but lack specificity (Larsen et al. 2019b), may yield future progress.

7 Intervention measures for a tail biting outbreak

Once a developing problem is identified, rapid action is needed if an outbreak is to be averted. Measures to achieve this include checking for and alleviating known risk factors, such as correcting climatic deficiencies, moving animals to a larger pen or increasing enrichment. Zonderland and Zonderland-Thomassen (2016) suggested that an automated detection system for tail biting could be linked to a mechanism to automatically provide identified pens with enrichment. Early intervention, at the point when the first minor tail damage is observed, by provision of 200 g/day chopped straw on the floor, haylage at~650g/d in a spherical cage, or a hanging sisal rope with a sweet-block, has been shown to reduce the occurrence of subsequent tail biting outbreaks in nursery pigs (Lahrmann et al. 2018b).

There are few good studies on the range and efficacy of intervention measures once a serious outbreak has actually started. The unpredictability of spontaneous outbreaks, the ethical considerations precluding the deliberate induction of an outbreak under controlled conditions and the need to minimise harm as soon as a tail biting situation arises make experimental studies challenging to implement. Many suggestions therefore come from anecdotal reports or expert knowledge elicitation. Measures which have been suggested by farmers from different countries as being effective are shown in Table 1. Whilst some measures, such as removing bitten pigs and adding objects for manipulation, were deemed as being important in all surveys, the importance of other measures received less agreement. This could in many cases be linked to the predominant housing systems (fully slatted, partly slatted or straw-bedded) and cost of providing additional space in that country.

From a survey of 520 Dutch farmers, Bracke et al. (2013) identified a similar list of interventions. Removing biters or bitten animals and improving barn climate were generally viewed as being more important than dimming light and using repellents. Conventional farmers also mentioned the clipping or grinding of teeth of biting pigs, although this would not be legal in many countries, but viewed providing extra enrichment or diversion material as less useful compared to the opinions of organic farmers. Regarding addition of manipulable materials, Finnish farmers stressed the importance of adding something novel and variable to the pigs (Valros et al. 2016).

In the Finnish survey (Valros et al. 2016), feeding-related interventions were also mentioned by 14 of the 70 respondents, including addition of minerals, salts or other additional feedstuffs, mainly on the floor of the pen. Addition of salt, by increasing inclusion in the diet, sprinkling on the floor or as blocks placed in the pen, is widely reported anecdotally as a remedy to tail biting (Edwards 2011). Proprietary feed products are also marketed for tail biting remediation, but few independent tests of their efficacy have been reported. In one critical evaluation (Palander et al. 2012), a supplementary feed with wheat bran, linseed cake and additional salts (NaCl, $NaSO_2$, $NaCO_3$), a feed with wheat bran and linseed cake, and a control feed with only wheat bran were tested for efficacy in controlling an acute tail biting outbreak. None of the tested supplementary formulations reduced outbreak severity.

When asked to identify anti-biting substances used for application to the tail, UK respondents generally named proprietary products, though tar was also mentioned (Valros and Barber 2019). Bracke (2009) showed experimentally that ropes can be made less attractive to pigs by adding Stockholm tar or Dippel's oil. However, there have been few independent evaluations of the efficacy of these or of the various proprietary products in controlling tail biting outbreaks and some UK farmers reported such products as ineffective. Similarly, in Ireland 57% of respondents found sprays useful while 43% found them ineffective (Haigh and O'Driscoll 2019). A few Finnish respondents also mentioned the importance of using wound spray and medication to help bitten tail heals quickly and thus to avoid further interest in the tail (Valros et al. 2016).

Table 1 The perceived importance of different tail biting intervention measures according to farmer questionnaires

Measure	% of UK farmers mentioning measure[a]	% of Swedish farmers mentioning measure[b]	Perceived importance by Finnish farmers[c] (Median and IQ range on 1-7 scale)	Perceived importance by UK farmers[d] (Median and IQ range on 1-7 scale)	Perceived importance by Irish farmers[e] (Median and IQ range on 1-7 scale)
Removing the biter from the pen	43	54	7 (1)	7 (1)	3 (4)
Removing the bitten pig(s) from the pen	67	34	6 (1)	7 (1)	7 (0)
Adding objects for manipulation (such as rope, hosepipe, wellington boots)	51	17	5 (2)	6 (2)	7 (2)
Adding bedding material (such as straw, wood shavings/ sawdust, peat)	16	29	6(2)	6 (4)	1 (1)
Reducing stocking density			5 (2)	5 (2)	3 (6)
Making environmental interventions (e.g. adjusting lighting levels, ventilation)		20		5 (3)	
Use of anti-biting substances on the tail	25		5 (3)	4 (4)	5 (5)

[a] Hunter et al. (2001).
[b] Wallgren et al. (2016).
[c] Valros et al. (2016).
[d] Valros and Barber (2019).
[e] Haigh and O'Driscoll (2019).

There are only a few experimental studies on how to effectively intervene when a tail biting outbreak occurs. Zonderland et al. (2008) compared the effect of removing the biter to adding straw, and found these to be equally effective in controlling an outbreak in nursery pigs, although no control treatment was present. Recent studies from Denmark showed that providing straw when the first signs of biting are evident in a pen (Lahrmann et al. 2018b), or when there is already an outbreak (Lahrmann et al. 2019) can reduce the escalation of the problem in nursery pigs, although not completely abolish it. In the latter study, straw (7 g/pig/day given on the floor) prevented an escalation better (75% efficacy) than a hanging rope (65%) or a plastic toy (35%). In the most comprehensive study to date of outbreak intervention measures, Chou et al. (2019a) evaluated the effectiveness of multi-step intervention protocols in undocked finishing pigs housed on fully slatted floors. When an outbreak was identified, either the biters or the victims were removed, or enrichment (three ropes) was added. If the initial intervention failed, another intervention was randomly used until all three interventions had been deployed once. They reported that 50% of outbreaks were controlled after one intervention and 30% after 2–3 interventions, while 20% remained uncontrolled. A high initial prevalence of biting reduced intervention success, emphasising the importance of early action. When only one intervention was used, adding ropes gave the fastest amelioration. Since removing biting or bitten pigs to isolation pens has major implications for pen space needs on the farm, they also looked at methods to reintroduce these pigs to their original pens, and found that this could be successfully achieved with the aid of additional rope enrichment within 14 days.

8 Conclusions

Tail biting remains a widespread and persistent problem in pig farming, causing significant welfare and economic problems and leading to an undesirable situation of extensive adoption of tail docking as a risk reduction measure. A wide variety of risk factors for tail biting has been identified and knowledge of these has been utilised in the development of advisory tools for farm-specific risk reduction. Despite this, there is still great uncertainty on many farms about the ability to prevent or control serious sporadic outbreaks of damaging behaviour. Improved early warning systems, with potential application in future Precision Livestock Farming technologies, and a better understanding of the mechanistic basis of the problem which might facilitate genetic or pharmacological solutions to reduction or cessation of tail biting behaviour offer future progress in the rearing of pigs with an intact tail throughout their life. However, improvements to housing and management systems which are currently used in practice will also be necessary.

9 Future trends in research

At the individual pig level, further research into the mechanistic basis of tail biting is still required to determine why some pigs become tail biters when others do not. This knowledge can then be applied to aid genetic selection of lower risk populations through molecular genotyping. At present, although tail biting predisposition has some heritable component, phenotypic selection is a highly challenging process. A better mechanistic understanding will also aid the discovery of effective nutritional or pharmacological interventions to prevent a developing problem or rapidly abolish an outbreak which has started. This would give farmers greater confidence in moving to complete cessation of tail docking, which many currently consider the only effective risk management strategy.

At the group level, clarification of the necessary elements of a robust system for rearing pigs with undocked tails is still required. However, given that total freedom from risk is unlikely, improved strategies for early detection of a forthcoming outbreak are still needed. For widespread adoption, these will need to have both good sensitivity and specificity in order to be efficient in use of both cost and time. Given the increasing size of farms and reduction in labour inputs, automation of such early-warning approaches, enabling their integration in Precision Livestock Farming systems, is highly desirable.

At the farm level, more studies are needed to understand the attitudes of different categories of farmers towards tail biting and the available methods of prevention, and how behavioural change in line with the wishes of wider society can be motivated in the future.

10 Where to look for further information

Information on EU policy relating to tail biting issues can be found at:

https://ec.europa.eu/food/animals/welfare/practice/farm/pigs/tail-docking_en.

The European Reference Centre for Animal Welfare–Pigs (EURCAW-Pigs) has produced a dossier of information on tail biting and tail docking:

https://www.eurcaw.eu/en/EURCAW-Pigs/dossiers/show/tail-docking-and-tail-biting.htm.

Detailed reviews on the subject of tail biting have been published by the Animal Health and Welfare Panel of the European Food Safety Authority:

EFSA 2007. Scientific Opinion of the Panel on Animal Health and Welfare on a request from Commission on the risks associated with tail biting in pigs and possible means to reduce the need for tail docking considering the different housing and husbandry systems. *The EFSA Journal* (2007) 611, 1-13. https://efsa.onlinelibrary.wiley.com/doi/pdf/10.2903/j.efsa.2007.611.

EFSA 2014. Scientific Opinion concerning a multifactorial approach on the use of animal and non-animal-based measures to assess the welfare of pigs. *EFSA Journal* 2014; 12(5): 3702, 101 pp. doi:10.2903/j.efsa.2014.3702. https://efsa.onlinelibrary.wiley.com/doi/pdf/10.2903/j.efsa.2014.3702.

Useful websites from recent EU projects dealing with the topic of tail biting:

FareWellDock project–https://farewelldock.eu.
GroupHouseNet project–https://www.grouphousenet.eu/.
PigWatch project–https://pigwatch.net.

A web-based Decision Support Tool for reduction of tail biting risk on farm: https://webhat.ahdb.org.uk/.

11 References

Adeola, O., Ball, R. O., House, J. D. and O'Brien, P. J. (1993). Regional brain neurotransmitter concentrations in stress-susceptible pigs. *J. Anim. Sci.* 71(4), 968–974.

Beattie, V. E., Breuer, K., O'Connell, N. E., Sneddon, I. A., Mercer, J. T., Rance, K. A., Sutcliffe, M. E. M. and Edwards, S. A. (2005). Factors identifying pigs predisposed to tail biting. *Anim. Sci.* 80(3), 307–312.

Blömke, L., Volkmann, N. and Kemper, N. (2020). Evaluation of an automated assessment system for ear and tail lesions as animal welfare indicators in pigs at slaughter. *Meat Sci.* 159, 107934.

Boyle, L. A., Edwards, S. A., Bolhuis, L., Pol, F., Zupan, M., Schütze, S., Nordgreen, J., Bozakova, N., Sossidou, E. and Valros, A. (2020). The evidence for a causal link between disease and damaging behaviour in pigs.

Bracke, M. B. M. (2009). Rope test may indicate efficacy of tail-biting treatments in growing pigs. *Anim. Welf.* 18, 263–266.

Bracke, M. B. M., De Lauwere, C. C., Wind, S. M. M. and Zonderland, J. J. (2013). Attitudes of Dutch pig farmers towards tail biting and tail docking. *J. Agric. Environ. Eth.* 26(4), 847–868.

Bracke, M. B. M. and Koene, P. (2019). Expert opinion on metal chains and other indestructable objects as proper enrichment for intensively-farmed pigs. *PLOS ONE* 14(2), e0212610.

Breuer, K., Sutcliffe, M. E. M., Mercer, J. T., Rance, K. A., Beattie, V. E., Sneddon, I. A. and Edwards, S. A. (2003). The effect of breed on the development of adverse social behaviours in pigs. *Appl. Anim. Behav. Sci.* 84(1), 59–74.

Breuer, K., Sutcliffe, M. E. M., Mercer, J. T., Rance, K. A., O'Connell, N. E., Sneddon, I. A. and Edwards, S. A. (2005). Heritability of clinical tail-biting and its relation to performance traits. *Livest. Prod. Sci.* 93(1), 87–94.

Brunberg, E., Wallenbeck, A. and Keeling, L. J. (2011). Tail biting in fattening pigs: associations between frequency of tail biting and other abnormal behaviours. *Appl. Anim. Behav. Sci.* 133(1–2), 18–25.

Brunberg, E., Jensen, P., Isaksson, A. and Keeling, L. J. (2013a). Brain gene expression differences are associated with abnormal tail biting behavior in pigs. *Genes Brain Behav.* 12(2), 275–281.

Brunberg, E., Jensen, P., Isaksson, A. and Keeling, L. J. (2013b). Behavioural and brain gene expression profiling in pigs during tail biting outbreaks–evidence of a tail biting resistant phenotype. *PLoS ONE* 8(6), e66513.

Brunberg, E. I., Rodenburg, T. B., Rydhmer, L., Kjaer, J. B., Jensen, P. and Keeling, L. J. (2016). Omnivores going astray: a review and new synthesis of abnormal behavior in pigs and laying hens. *Front. Vet. Sci.* 3, 57.

Brünger, J., Dippel, S., Koch, R. and Veit, C. (2019). 'Tailception': using neural networks for assessing tail lesions on pictures of pig carcasses. *Animal* 13(5), 1030 -1036.

Buijs, S. and Muns, R. (2019). A review of the effects of non-straw enrichment on tail biting in pigs. *Animals* 9(10), 824.

Bulens, A., Biesemans, C., van Beirendonck, S., van Thielen, J. and Driessen, B. (2016). The effect of a straw dispenser on behavior and lesions in weanling pigs. *J. Vet. Behav.* 12, 49–53.

Busch, M. E., Wachmann, H., Nielsen, E. O., Petersen, H. H. and Nielsen, J. P. (2004). Tail biting–can routine meat inspection data be used for classification of herds? Proceedings of the 18th IPVS Congress, Hamburg, Germany.

Büttner, K., Czycholl, I., Basler, H. and Krieter, J. (2018). Effects of an intensified human-animal interaction on tail biting in pigs during the rearing period. *J. Agric. Sci.* 156(8), 1039–1046.

Camerlink, I., Ursinus, W. W., Bijma, P., Kemp, B. and Bolhuis, J. E. (2015). Indirect genetic effects for growth rate in domestic pigs alter aggressive and manipulative biting behaviour. *Behav. Genet.* 45(1), 117–126.

Canario, L. and Flatres-Grall, L. (2018). Genetics of tail-biting receipt in gilts from the Tai Zumu line. Proceedings of the World Congress on Genetics Applied to Livestock Production 11, 572.

Carrasco, G. A. and van de Kar, L. D. (2003). Neuroendocrine pharmacology of stress. *Eur. J. Pharmacol.* 463(1–3), 235–272.

Chou, J. Y., O'Driscoll, K., D'Eath, R. B., Sandercock, D. A. and Camerlink, I. (2019a). Multi-step tail biting outbreak intervention protocols for pigs housed on slatted floors. *Animals* 9(8), 582.

Chou, J. Y., Drique, C. M. V., Sandercock, D. A., D'Eath, R. B. and O'Driscoll, K. (2019b). Rearing undocked pigs on fully slatted floors using multiple types and variations of enrichment. *Animals* 9(4), 139.

Christensen, T., Denver, S. and Sandøe, P. (2019). How best to improve farm animal welfare? Four main approaches viewed from an economic perspective. *Anim. Welf.* 28(1), 95–106.

Council Directive 2008/120/EC of 18 December 2008 laying down minimum standards for the protection of pigs. Council of the European Union.

Courboulay, V., Gillardeau, M., Meunier-Salan, M. C. and Prunier, A. (2015). La Prise en charge de la douleur lors de la caudectomie et de la castration des porcelets. *J. Rech. Porc* 47, 235–240.

Dawkins, M. E. (2019). Animal welfare as preventative medicine. *Anim. Welf.* 28(2), 137–141.
D'Eath, R. B., Arnott, G., Turner, S. P., Jensen, T., Lahrmann, H. P., Busch, M. E., Niemi, J. K., Lawrence, A. B. and Sandøe, P. (2014). Injurious tail biting in pigs: how can it be controlled in existing systems without tail docking? *Animal* 8(9), 1479–1497.
D'Eath, R. B., Niemi, J. K., Vosough Ahmadi, B., Rutherford, K. M. D., Ison, S. H., Turner, S. P., Anker, H. T., Jensen, T., Busch, M. E., Jensen, K. K., Lawrence, A. B. and Sandøe, P. (2016). Why are most EU pigs tail docked? Economic and ethical analysis of four pig housing and management scenarios in the light of EU legislation and animal welfare outcomes. *Animal* 10(4), 687–699.
D'Eath, R. B., Jack, M., Futro, A., Talbot, D., Zhu, Q., Barclay, D. and Baxter, E. M. (2018). Automatic early warning of tail biting in pigs: 3D cameras can detect lowered tail posture before an outbreak. *PLoS ONE* 13(4), e0194524.
De Briyne, N., Berg, C., Blaha, T., Palzer, A. and Temple, D. (2018). Phasing out pig tail docking in the EU–present state, challenges and possibilities. *Porc. Health Manag.* 4, 1–9.
Delorme, R., Betancur, C., Callebert, J., Chabane, N., Laplanche, J. L., Mouren-Simeoni, M., Launay, J. and Leboyer, M. (2005). Platelet serotonergic markers as endophenotypes for obsessive-compulsive disorder. *Neuropsychopharmacol* 30(8), 1539–1547.
Dinan, T. G. and Cryan, J. F. (2012). Regulation of the stress response by the gut microbiota: implications for psychoneuroendocrinology. *Psychoneuroendocrinology* 37(9), 1369–1378.
Edwards, S. A. (2006). Tail biting in pigs: understanding the intractable problem. *Vet. J.* 171(2), 198–199.
Edwards, S. A. (2011). What do we know about tail biting today? *Pig J.* 66, 81–86.
Edwards, S. A. (2012). Tail biting in pigs–an international overview. Proceedings of the 17 Internationale Fachtagung zum Thema Tierschutz, Nurtingen 12–13 March 2012, Deutsche Veterinarmedizinische Gesellschaft e.V., pp 11–26.
EFSA. (2007). Scientific opinion of the panel on animal health and welfare on a request from commission on the risks associated with tail biting in pigs and possible means to reduce the need for tail docking considering the different housing and husbandry systems. *EFSA J.* 611, 1–13.
EFSA. (2014). Scientific opinion concerning a multifactorial approach on the use of animal and non-animal-based measures to assess the welfare of pigs. *EFSA J.* 12, 3702.
Eurobarometer. (2016). Special Eurobarometer 442. Report: Attitudes of Europeans towards Animal Welfare (vol. 8). Available at: http://ec.europa.eu/COMMFrontOffice/publicopinion/index.cfm/ResultDoc/download/DocumentKy/7134, (accessed on 1 July 2020).
European Commission. (2016). Commission staff working document on best practices with a view to the prevention of routine tail-docking and the provision of enrichment materials to pigs. Available at: https://ec.europa.eu/food/sites/food/files/animals/docs/aw:practice_farm_pigs_stfwrkdoc_en.pdf, (accessed on 1 July 2020).
Fraser, D. (1987). Mineral-deficient diets and the pig's attraction to blood, implications for tail-biting. *Can. J. Anim. Sci.* 67(4), 909–918.
Fraser, D., Bernon, D. E. and Ball, R. O. (1991). Enhanced attraction to blood by pigs with inadequate dietary protein supplementation. *Can. J. Anim. Sci.* 71, 603–606.
Grümpel, A., Krieter, J., Veit, C. and Dippel, S. (2018). Factors influencing the risk for tail lesions in weaner pigs (Sus scrofa). *Livest. Sci.* 216, 219–226.

Grümpel, A., Krieter, J. and Dippel, S. (2019). Reducing estimated tail biting risk in German weaner pigs using a management tool. *Vet. J.* 254, 105406.

Haigh, A. and O'Driscoll, K. (2019). Irish pig farmer's perceptions and experiences of tailbiting. *Porc. Health Mgmt.* 5, 30.

Harley, S., More, S. J., O'Connell, N. E., Hanlon, A., Teixeira, D. and Boyle, L. (2012). Evaluating the prevalence of tail biting and carcase condemnations in slaughter pigs in the Republic and Northern Ireland, and the potential of abattoir meat inspection as a welfare surveillance tool. *Vet. Rec.* 171(24), 621.

Hart, B. L. (1988). Biological basis of the behavior of sick animals. *Neurosci. Biobehav. Rev.* 12, 123–137.

Heinonen, M., Orro, T., Kokkonen, T., Munsterhjelm, C., Peltoni, O. and Valros, A. (2010). Tail biting induces a strong acute phase response and tail-end inflammation in finishing pigs. *Vet. J.* 184, 303–307.

Herskin, M. S., Di Giminiani, P. and Thodberg, K. (2016). Effects of administration of a local anaesthetic and/or an NSAID and of docking length on the behaviour of piglets during 5 h after tail docking. *Res. Vet. Sci.* 108, 60–67.

Holling, C., Grosse Beilage, E., Vidondo, B. and Nathues, C. (2017). Provision of straw by a foraging tower–effect on tail biting in weaners and fattening pigs. *Porc. Health Mgmt.* 3, 4.

Honeck, A., Gertz, M., grosse Beilage, E. and Krieter, J. (2019). Comparison of different scoring keys for tail-biting in pigs to evaluate the importance of one common scoring key to improve the comparability of studies–a review. *Appl. Anim. Behav. Sci.* 221, 104873.

Hunter, E. J., Jones, T. A., Guise, H. J., Penny, R. H. C. and Hoste, S. (1999). Tail biting in pigs. I: the prevalence at six UK abattoirs and the relationship of tail biting with docking, sex and other carcass damage. *Pig J.* 43, 18–32.

Hunter, E. J., Jones, T. A., Guise, H. J., Penny, R. H. C. and Hoste, S. (2001). The relationship between tail biting in pigs, docking procedure and other management practices. *Vet. J.* 161(1), 72–79.

Kallio, P. A., Janczak, A., Valros, A. E., Edwards, S. A. and Heinonen, M. (2018). Case control study on environmental, nutritional and management-based risk factors for tail-biting in long-tailed pigs. *Anim. Welf.* 27(1), 21–34.

Keeling, L. J., Wallenbeck, A., Larsen, A. and Holmgren, N. (2012). Scoring tail damage in pigs: an evaluation based on recordings at Swedish slaughterhouses. *Acta Vet. Scand.* 54, 32.

Kritas, S. K. and Morrison, R. B. (2004). An observational study on tail biting in commercial grower-finisher barns. *J. Swine Health Prod.* 12, 17–22.

Kritas, S. K. and Morrison, R. B. (2007). Relationships between tail biting in pigs and disease lesions and condemnations at slaughter. *Vet. Rec.* 160(5), 149–152.

Lahrmann, H. P., Busch, M. E., D'Eath, R. B., Forkman, B. and Hansen, C. F. (2017). More tail lesions among undocked than tail docked pigs in a conventional herd. *Animal* 11(10), 1825–1831.

Lahrmann, H. P., Hansen, C. F., D'Eath, R. B., Busch, M. E. and Forkman, B. (2018a). Tail posture predicts tail biting outbreaks at pen level in weaner pigs. *Appl. Anim. Behav. Sci.* 200, 29–35.

Lahrmann, H. P., Hansen, C. F., D'Eath, R. B., Busch, M. E., Nielsen, J. P. and Forkman, B. (2018b). Early intervention with enrichment can prevent tail biting outbreaks in weaner pigs. *Livest. Sci.* 214, 272–277.

Lahrmann, H. P., Faustrup, J. F., Hansen, C. F., D'Eath, R. B., Nielsen, J. P. and Forkman, B. (2019). The effect of straw, rope, and bite-rite treatment in weaner pens with a tail biting outbreak. *Animals* 9(6), 365.

Larsen, H. D., Nielsen, G. G. B., Black, P., Vorup, P. and Hass, M. A. (2019a). Test report. Description and functionality of a prototype system to measure and document tail length and tail lesions at the slaughterhouse. Available at: https://www.dti.dk › media › 74221_Final -Pigwatch test report D21, (accessed on 18 December 2019).

Larsen, M. L. V., Andersen, H. M.-L. and Pedersen, L. J. (2016). Can tail damage outbreaks in the pig be predicted by behavioural change? *Vet. J.* 209, 50–56.

Larsen, M. L. V., Andersen, H. M.-L. and Pedersen, L. J. (2018a). Which is the most preventive measure against tail damage in finisher pigs: tail docking, straw provision or lowered stocking density? *Animal* 12(6), 1260–1267.

Larsen, M. L. V., Andersen, H. M.-L. and Pedersen, L. J. (2018b). Tail posture as a detector of tail damage and an early detector of tail biting in finishing pigs. *Appl. Anim. Behav. Sci.* 209, 30–35.

Larsen, M. L. V., Andersen, H. M.-L and Pedersen, L. J. (2019b). Changes in activity and object manipulation before tail damage in finisher pigs as an early detector of tail biting. *Animal* 13(5), 1037–1044.

Larsen, M. L. V., Pedersen, L. J., Edwards, S., Albanie, S. and Dawkins, M. S. (2020). Movement change detected by optical flow precedes, but does not predict, tailbiting in pigs. *Livest. Sci.* 240: 104136.

Le Floc'h, N., Lebellego, L., Matte, J. J., Melchior, D. and Sève, B. (2009) The effect of sanitary status degradation and dietary tryptophan content on growth rate and tryptophan metabolism in weaning pigs. *J. Anim. Sci.* 87(5), 1686–1694.

Li, Y. Z., Johnston, L. J. and Dawkins, M. S. (2020). Utilization of optical flow algorithms to monitor development of tail biting outbreaks in pigs. *Animals* 10(2), 323.

Martinez-Trejo, G., Ortega-Cerrilla, M. E., Rodarte-Covarrubias, L. F., Herrera-Haro, J. G., Figueroa-Velasco, J. L., Figueroa-Velasco, J. L., Galindo-Maldonado, F., Sánchez-Martínez, O. and Lara-Bueno, A. (2009). Aggressiveness and productive performance of piglets supplemented with tryptophan. *J. Anim. Vet. Adv.* 8, 608–611.

McIntyre, J. and Edwards, S. A. (2002a). Preference for blood and behavioural measurements of known tail biting pigs compared to control penmates. Proceedings of the 36th International Conference (vol. P93), International Society for Applied Ethology, Egmond aan Zee, The Netherlands.

McIntyre, J. and Edwards, S. A. (2002b). An investigation into the effect of tryptophan on tail chewing behaviour of growing pigs. *Proc. Br.Soc. Anim. Sci.* 2002, 34–34.

Melchior, D., Sève, B. and Le Floc'h, N. (2004). Sève. *J. Anim. Sci.* 82(4), 1091–1099.

Moinard, C., Mendl, M., Nicol, C. J. and Green, L. E. (2003). A case control study of on-farm risk factors for tail biting in pigs. *Appl. Anim. Behav. Sci.* 81(4), 333– 355.

Munshi, S., Parrilli, V. and Rosenkrnaz, J. A. (2019). Peripheral anti-inflammatory cytokine interleukin-10 treatment mitigates interleukin-1 beta–induced anxiety and sickness behaviors in adult male rats. *Behav. Brain Res.* 372, 8.

Munsterhjelm, C. (2011). The consequence of a tail biting outbreak. Proceedings of the 3rd European Symposium of Porcine Health Management, Espoo, Finland, pp 45–47.

Munsterhjelm, C., Peltoniemi, O. A. T., Heinonen, M., Halli, O., Karhapaa, M. and Valros, A. (2009). Experience of moderate bedding affects behaviour of growing pigs. *Appl. Anim. Behav. Sci.* 118(1–2), 42–53.

Munsterhjelm, C., Simola, O., Keeling, L., Valros, A. and Heinonen, M. (2013a). Health parameters in tail biters and bitten pigs in a case-control study. *Animal* 7(5), 814–821.

Munsterhjelm, C., Brunberg, E., Heinonen, M., Keeling, L. and Valros, A. (2013b). Stress measures in tail biters and bitten pigs in a matched case-control study. *Anim. Welf.* 22(3), 331–338.

Munsterhjelm, C., Heinonen, M. and Valros, A. (2015). Application of the Welfare Quality® animal welfare assessment system in Finnish pig production, part II: associations between animal-based and environmental measures of welfare. *Anim. Welf.* 24(2), 161–172.

Munsterhjelm, C., Heinonen, M. and Valros, A. (2016). Can tail-in-mouth -behaviour in weaned piglets be predicted by behaviour and performance? *Appl. Anim. Behav. Sci.* 184, 16–24.

Munsterhjelm, C., Nordgreen, J., Aae, F., Heinonen, M., Olstad, K., Aasmundstad, T., Janczak, A. M. and Valros, A. (2017). To be blamed or pitied? The effect of illness on social behavior, cytokine levels and feed intake in undocked boars. *Physiol. Behav.* 179, 298–307.

Munsterhjelm, C., Nordgreen, J., Aae, F., Heinonen, M., Valros, A. and Janczak, A. M. (2019). Sick and grumpy: changes in social behaviour after a controlled immune stimulation in group-housed gilts. *Physiol. Behav.* 198, 76–83.

Nalon, E. and De Briyne, N. (2019). Efforts to ban the routine tail docking of pigs and to give pigs enrichment materials via EU law: where do we stand a quarter of a century on? *Animals* 9(4), 132–137.

Naya, A., Gertz, M., Hasler, M., große Beilage, E. and Krieter, J. (2019a). Does a higher content of fibre in the piglet diet have an influence on tail biting in growing pigs? *Livest. Sci.* 223, 133–137.

Naya, A., Traulsen, I., Gertz, M., Hasler, M., Burfeind, O., große Beilage, E. and Krieter, J. (2019b). Is tail biting in growing pigs reduced by a prolonged suckling period? *Appl. Anim. Behav. Sci.* 211, 41–46.

Niemi, J. K., Sinisalo, A., Valros, A. and Heinonen, M. (2011). The timing and treatment of tail biting in fattening pigs. 24th NJF Congress "Food, Feed, Fuel and Fun–Nordic Light on Future Land Use and Rural Development" NJF Report 7 (vol. 6).

Niemi, J. K., Sinisalo, A., Valros, A. and Heinonen, M. (2012). Market and policy-oriented incentives to provide animal welfare: the case of tail biting. Proceedings 126th EAAE Seminar, Capri, Italy, June 27–29, 2012.

Niemi, J. K., Edwards, S., Piette, D., Papanastasiou, D. K., Stygar, A. and Valros, A. (2020). Cost-effectiveness of strategies to reduce damaging behavior in pigs. Report of GroupHouseNet Project.

Nordgreen, J., Edwards, S. A., Boyle, L. A., Bolhuis, J. E., Veit, C., Sayyari, A., Marin, D. E., Dimitrov, I., Janczak, A. M. and Valros, A. (2020). A proposed role for pro-inflammatory cytokines in damaging behaviour in pigs. *Front. Vet. Sci.* 7, 646.

O'Connor, J. C., Lawson, M. A., Andre, C., Moreau, M., Lestage, J., Castanon, N., Kelley, K. W. and Dantzer, R. (2009). Lipopolysaccharide-induced depressive-like behavior is mediated by indoleamine 2,3 dioxygenase activation in mice. *Mol. Psychiatry* 14(5), 511–522.

Palander, P. A., Valros, A., Heinonen, M. and Edwards, S. A. (2012). Supplementary feed offered to a tail biting pen changes feeding behaviour, feed intake, growth and tail health of fattening pigs. Proceedings of the 46th Congress of the International Society for Applied Ethology, Wageningen Academic Publisher, Vienna, p 156.

Palander, P. A., Heinonen, M., Simpura, I., Edwards, S. A. and Valros, A. E. (2013). Jejunal morphology and blood metabolites in tail biting, victim and control pigs. *Animal* 7(9), 1523–1531.

Pandolfi, F., Stoddart, K., Wainwright, N., Kyriazakis, I. and Edwards, S. A. (2017). The 'real welfare' scheme: benchmarking welfare outcomes for commercially farmed pigs. *Animal* 11(10), 1816–1824.

Pastorelli, H., Le Floc'h, N., Merlot, E., Meunier-Salaün, M. C., van Milgen, J. and Montagne, L. (2012). Sanitary housing conditions modify the performance and behavioural response of weaned pigs to feed- and housing-related stressors. *Animal* 6(11), 1811–1820.

Paul, E. S., Moinard, C., Green, L. E. and Mendl, M. (2007). Farmers' attitudes to methods for controlling tail biting in pigs. *Vet. Rec.* 160(23), 803–805.

Piekarzewska, A. B., Rosochacki, S. J. and Sender, G. (2000). The effect of acute restraint stress on regional brain neurotransmitter levels in stress-susceptible Pietrain pigs. *J. Vet. Med. A* 47(5), 257–269.

Presto Åkerfeldt, M., Lindberg, J. E., Göransson, L. and Andersson, K. (2019). Effects of reducing dietary content of crude protein and indispensable amino acids on performance and carcass traits of single-phase- and 2-phase-fed growing-finishing pigs. *Livest. Sci.* 224, 96–101.

Prunier, A., Averos, X., Dimitrov, I., Edwards, S. A., Hillmann, E., Holinger, M., Ilieski, V., Leming, R., Tallet, C., Turner, S. P., Zupan, M. and Camerlink, I. (2020). Review on early life predisposing factors for biting in pigs. *Animal* 14(3), 570–587.

Rabhi, N., Thibodeau, A., Côté, J.-C., Devillers, N., Laplante, B., Fravalo, P., Larivière-Gauthier, G., Thériault, W. P., Faucitano, L., Beauchamp, G. and Quessy, S. (2020). Association between intestinal microbiota composition and tail-biting in pigs. Available at: https://www.researchsquare.com/article/rs-21643/v1.

Rault, J.-L., Waiblinger, S., Boivin, X.. and Hemsworth, P. (2020). The power of a positive human-animal relationship for animal welfare. *Front. Vet. Sci.*, doi: 10.3389/fvets.2020.590867/abstract.

Scheepens, C. J. M., Hessing, M. J. C., Laarakker, E., Schouten, W. G. P. and Tielen, M. J. M. (1991). Influences of intermittent daily draught on the behaviour of weaned pigs. *Appl. Anim. Behav. Sci.* 31(1–2), 69–82.

Schrøder-Petersen, D. L. and Simonsen, H. B. (2001). Tail biting in pigs. *Vet. J.* 162(3), 196–210.

Schrøder-Petersen, D. L., Heiskanen, T. and Ersboll, A. K. (2004). Tail-in-mouth behaviour in slaughter pigs, in relation to internal factors such as: age, size, gender, and motivational background. *Acta Agric. Scand. A* 54(3), 159–166.

Scollo, A., Contiero, B. and Gottardo, F. (2016). Frequency of tail lesions and risk factors for tail biting in heavy pig production from weaning to 170 kg live weight. *Vet. J.* 207, 92–98.

Scollo, A., Gottardo, F., Contiero, B. and Edwards, S. A. (2017). A cross-sectional study for predicting tail biting risk in pig farms using classification and regression tree analysis. *Prev. Vet. Med.* 146, 114–120.

Sihvo, H. -K., Simola, O., Munsterhjelm, C. and Syrjä, P. (2011). Systemic spread of infection in tail-bitten pigs. Proceedings of the 29th Meeting of ESVP/ECVP, Sweden, 119.

Sinisalo, A., Niemi, J. K., Heinonen, M. and Valros, A. (2012). Tail biting and production performance in fattening pigs. *Livest. Sci.* 143(2–3), 220–225.

Smulders, D., Hautekiet, V., Verbeke, G. and Geers, R. (2008). Tail and ear biting lesions in pigs: an epidemiological study. *Anim. Welf.* 17, 61–69.

Spooner, J. M., Schuppli, C. A. and Fraser, D. (2014). Attitudes of Canadian pig producers toward animal welfare. *J. Agric. Environ. Eth.* 27(4), 569–589.

Statham, P., Green, L., Bichard, M. and Mendl, M. (2009). Predicting tail-biting from behaviour of pigs prior to outbreaks. *Appl. Anim. Behav. Sci.* 121(3–4), 157–164.

Stygar, A. H., Chantziaras, I., Toppari, I., Maes, D. and Niemi, J. K. (2020). High biosecurity and welfare standards in fattening pig farms are associated with reduced antimicrobial use. *Animal* 14(10), 2178–2186. https://doi.org/10.1017/S1751731120000828.

Sutherland, M., Bryer, P. J., Krebs, N. and McGlone, J. J. (2009). The effect of method of tail docking on tail-biting behaviour and welfare of pigs. *Anim. Welf.* 18, 561–570.

Sutherland, M. A., Davis, B. L. and McGlone, J. J. (2011). The effect of local or general anesthesia on the physiology and behavior of tail docked pigs. *Animal* 5(8), 1237–1246.

Taylor, N. R., Main, D. C. J., Mendl, M. and Edwards, S. A. (2010). Tail-biting: a new perspective. *Vet. J.* 186(2), 137–147.

Taylor, N. R., Parker, R. M. A., Mendl, M., Edwards, S. A. and Main, D. C. J. (2012). Prevalence of risk factors for tail biting on commercial farms and intervention strategies. *Vet. J.* 194(1), 77–83.

Telkänranta, H., Bracke, M. B. M. and Valros, A. (2014a). Fresh wood reduces tail and ear biting and increases exploratory behaviour in finishing pigs. *Appl. Anim. Behav. Sci.* 161, 51–59.

Telkänranta, H., Swan, K., Hirvonen, H. and Valros, A. (2014b). Chewable materials before weaning reduce tail biting in growing pigs. *Appl. Anim. Behav. Sci.* 157, 14–22.

Temple, D., Dalmau, A., Ruíz de la Torre, J. L., Manteca, X. and Velarde, A. (2011). Application of the Welfare Quality® protocol on intensively kept growing pigs in Spain. *J. Vet. Behav.* 6, 138–149.

ter Beek, V. (2015). German state introduces premium for entire pig tails. Pig Progress, 26 June 2015.

Tetsuo, W., Munehiko, N., Kenichi, N., Toshiya, H. and Masato, A. (2015). Examination of managing methods to relieve tail biting in piglets–effects of chain/rope, sodium chloride and diet form. *Jap. J. Swine Sci.* 52, 123–134.

Thodberg, K., Herskin, M. S., Jensen, T. and Jensen, K. H. (2018). The effect of docking length on the risk of tail biting, tail-directed behaviour, aggression and activity level of growing pigs kept under commercial conditions. *Animal* 12(12), 2609–2618.

Ursinus, W. W., Wijnen, H. J., Bartels, A. C., Dijvesteijn, N., van Reenen, C. G. and Bolhuis, J. E. (2014a). Damaging biting behaviors in intensively kept rearing gilts: the effect of jute sacks and relations with production characteristics. *J. Anim. Sci.* 92(11), 5193–5202.

Ursinus, W. W., van Reenen, C. G., Reimert, I. and Bolhuis, J. E. (2014b). Tail biting in pigs: blood serotonin and fearfulness as pieces of the puzzle? *PLoS ONE* 9(9), e107040.

Ursinus, W. W., van Reenen, C. G., Kemp, B. and Bolhuis, J. E. (2014c). Tail biting behaviour and tail damage in pigs and the relationship with general behaviour: predicting the inevitable? *Appl. Anim. Behav. Sci.* 156, 22–36.

Valros, A., Ahlstrom, S., Rintala, H., Hakkinen, T. and Saloniemi, H. (2004). The prevalence of tail damage in slaughter pigs in Finland and associations to carcass condemnations. *Acta Agric. Scand. A* 54(4), 213–219.

Valros, A. and Heinonen, M. (2015). Save the pig tail. *Porc. Health Mgmt.* 1, 2.

Valros, A., Palander, P., Heinonen, M., Munsterhjelm, C., Brunberg, E., Keeling, L. and Piepponen, P. (2015). Evidence for a link between tail biting and central monoamine metabolism in pigs (Sus scrofa domestica). *Physiol. Behav.* 143, 151–157.

Valros, A., Munsterhjelm, C., Hanninen, L., Kauppinen, T. and Heinonen, M. (2016). Managing undocked pigs–on-farm prevention of tail biting and attitudes towards tail biting and docking. *Porc. Health Mgmt.* 2, 2.

Valros, A. (2018). Tail biting. In: *Advances in Pig Welfare* Spinka, M. (Ed.), Woodhead Publishing, Sawston, UK, pp. 137–166.

Valros, A. and Barber, C. (2019). Producer perceptions of the prevention of tail biting on UK farms: association to bedding use and tail removal proportion. *Animals* 9(9), 628.

Valros, A., Välimäki, E., Nordgren, H., Vugts, J., Fàbrega, E. and Heinonen, M. (2020). Intact tails as a welfare indicator in finishing pigs? Scoring of tail lesions and defining intact tails in undocked pigs at the abattoir. *Front. Vet. Sci.* 7, 405.

van de Weerd, H. A., Docking, C. M., Day, J. E. L. and Edwards, S. A. (2005). The development of harmful social behaviour in pigs with intact tails and different enrichment backgrounds in two housing systems. *Anim. Sci.* 80(3), 289–298.

van der Meer, Y. V., Gerrits, W. J. J., Jansman, A. J. M., Kemp, B. and Bolhuis, J. E. (2017). A link between damaging behavior in pigs, sanitary conditions, and dietary protein and amino acid supply. *PLoS ONE* 12(5), e0174688.

van Nieuwamerongen, S. E., Soede, N. M., van der Peet-Schwering, C. M. C., Kemp, B. and Bolhuis, J. E. (2015). Development of piglets raised in a new multi-litter housing system vs. conventional single-litter housing until 9 weeks of age. *J. Anim. Sci.* 93(11), 5442–5454.

van Nieuwamerongen, S. E., Soede, N. M., van der Peet-Schwering, C. M. C., Kemp, B. and Bolhuis, J. E. (2017). Gradual weaning during an extended lactation period improves performance and behavior of pigs raised in a multi-suckling system. *Appl. Anim. Behav. Sci.* 194, 24–35.

Veit, C., Traulsen, I., Hasler, M., Tölle, K.-H., Burfeind, O., grosse Beilage, Eg. and Krieter, J. (2016). Influence of raw material on the occurrence of tail-biting in undocked pigs. *Livest. Sci.* 191, 125–131.

vom Brocke, A. L. (2014). A step to reducing tail biting in finisher pigs: can a management tool help pigs and farmers? PhD thesis, Georg-August-University Göttingen, Göttingen, Germany.

vom Brocke, A. L., Karnholz, C., Madey-Rindermann, D., Gauly, M., Leeb, C., Winckler, C., Schrader, L. and Dippel, S. (2019). Tail lesions in fattening pigs: relationships with postmortem meat inspection and influence of a tail biting management tool. *Animal* 13(4), 835–844.

Wallenbeck, A. and Keeling, L. J. (2013). Using data from electronic feeders on visit frequency and feed consumption to indicate tail biting outbreaks in commercial pig production. *J. Anim. Sci.* 91(6), 2879–2884.

Wallgren, T., Westin, R. and Gunnarson, S. (2016). A survey of straw use and tail biting in Swedish pig farms rearing undocked pigs. *Acta Vet. Scand.* 58(1), 84.

Wallgren, T., Larsen, A., Lundeheim, N., Westin, R. and Gunnarsson, S. (2019a). Implication and impact of straw provision on behaviour, lesions and pen hygiene on commercial farms rearing undocked pigs. *Appl. Anim. Behav. Sci.* 210, 26–37.

Wallgren, T., Larsen, A. and Gunnarsson, S. (2019b). Tail posture as an indicator of tail biting in undocked finishing pigs. *Animals* 9(1), 18.

Wedin, M., Baxter, E. M., Jack, M., Futro, A. and D'Eath, R. B. (2018). Early indicators of tail biting outbreaks in pigs. *Appl. Anim. Behav. Sci.* 208, 7-13.

Westin, R. (2003). *Svansbitning hos gris relaterat till individuell tillväxt och ras (Tail Biting in Pigs Related to Individual Growth and Breed).* Department of Animal Environment and Health, Sveriges Lantbruksuniversitet (SLU), Veterinärmedicinska fakulteten, Veterinärprogrammet.

Wilson, K., Zanella, R., Ventura, C., Johansen, H. L., Framstad, T., Janczak, A., Zanella, A. J. and Neibergs, H. L. (2012). Identification of chromosomal locations associated with tail biting and being a victim of tail-biting behaviour in the domestic pig (Sus scrofa domesticus). *J. Appl. Genet.* 53(4), 449-456.

Zhao, X. N., Cao, F. R., Liu, Q., Li, X. S., Xu, G. Y., Liu, G., Zhang, Y., Yang, X., Yi, S., Xu, F., Fan, K. and Ma, J. (2019). Behavioral, inflammatory and neurochemical disturbances in LPS and UCMS-induced mouse models of depression. *Behav. Brain Res.* 364, 494-502.

Zonderland, J. J., Wolthuis-Fillerup, M., van Reenen, C. G., Bracke, M. B. M., Kemp, B., Den Hartog, L. Ad and Spoolder, H. A. M. (2008). Prevention and treatment of tail biting in weaned piglets. *Appl. Anim. Behav. Sci.* 110(3-4), 269- 281.

Zonderland, J. J., van Riel, J. W., Bracke, M. B. M., Kemp, B., den Hartog, L. A. and Spoolder, H. A. M. (2009). Tail posture predicts tail damage among weaned piglets. *Appl. Anim. Behav. Sci.* 121(3-4), 165-170.

Zonderland, J. J., Bracke, M. B. M., den Hartog, L. A., Kemp, B. and Spoolder, H. A. M. (2010). Gender effects on tail damage development in single- or mixed-sex groups of weaned piglets. *Livest. Sci.* 129(1-3), 151-158.

Zonderland, J. J., Schepers, F., Bracke, M. B. M., den Hartog, L. A., Kemp, B. and Spoolder, H. A. M. (2011). Characteristics of biter and victim piglets apparent before a tailbiting outbreak. *Animal* 5(5), 767-775.

Zonderland, J. J. and Zonderland-Thomassen, M. A. (2016). Behavioural change by pig producers is the key factor in raising pigs with intact tails. *Vet. J.* 211, 1-2.

Zupan, M., Janczak, A. M., Framstad, T. and Zanella, A. J. (2012). The effect of biting tails and having tails bitten in pigs. *Physiol. Behav.* 106(5), 638-644.

Zupan, M. and Zanella, A. J. (2017). Peripheral regulation of stress and fear responses in pigs from tail-biting pens. *R. Bras. Zootec.* 46(1), 33-38.

Chapter 11

The role of enrichment in optimizing pig behaviour and welfare

Sandra Düpjan and Liza R. Moscovice, Institute of Behavioural Physiology – Leibniz Institute for Farm Animal Biology Dummerstorf, Germany; and Birger Puppe, Institute of Behavioural Physiology – Leibniz Institute for Farm Animal Biology Dummerstorf and University of Rostock, Germany

1 Introduction
2 Environmental enrichment and animal welfare
3 The role of environmental enrichment in pig behaviour and welfare
4 Evaluating the biological relevance of different types of enrichment
5 Conclusions
6 Where to look for further information
7 References

1 Introduction

In modern pig farming, housing environments are primarily optimized for efficiency and productivity. This often results in environments that are easy to maintain and that meet the farmers' economy-driven demands, but are typically not designed with the animals' needs first and foremost in mind (see Fig. 1). In laying down minimum standards for the protection of pigs, the EU mandates that 'pigs must have permanent access to a sufficient quantity of material to enable proper investigation and manipulation activities, such as straw, hay, wood, sawdust, mushroom compost, peat or a mixture of such, which does not compromise the health of the animals' (European Union, 2009: Council Directive 2008/120/EC). The EU mandate focusses on providing organic materials for enrichment (European Commission, 2016), an idea that has not found broad application either in the EU or more internationally (van de Weerd and Ison, 2019). At an international level, many countries do not even mention enrichment in their national regulations for farm animals. For example, there are no enrichment requirements for pigs under US Industry Guidelines (National Pork Board, 2018), or in Australia (Prevention of Cruelty to Animals Act, 1986; Australian Livestock Management Act, 2010). There are however several

http://dx.doi.org/10.19103/AS.2020.0081.11

Figure 1 Barren standard housing for fattening pigs. Photo courtesy of M. Lechner.

optional welfare certifications that require different levels of enrichment, mostly regarding provision of enrichment items or access to the outdoors.

Across the EU, directive 2008/120/EC has been implemented inconsistently into national regulations, so that many countries provide non-organic enrichment items or no enrichment at all (Nalon and De Briyne, 2019). Even where the necessity of environmental enrichment for pigs is accepted per se, questions related to which types of enrichment are more effective and feasible to implement, and how much is 'a sufficient quantity' of enrichment remain unresolved. One also needs to take economic aspects into consideration, including how much extra work is required for the implementation of enrichment (see European Commission, 2014, page 141, Table 4.39 for the cost of implementing directive 2008/120/EC on exemplary farms) or whether an item might reduce productivity (for example due to increased energy expenditure for higher activity levels, which will reduce the energy available for growth). Also, the potential positive or negative impacts on animal health need to be considered. And finally, while the EU directive focusses primarily on encouraging investigation and manipulation of objects to meet pigs' exploratory needs, there is compelling scientific evidence that pigs also have complex social relationships and cognitive abilities. Thus, any serious attempts to provide enrichment to pigs will benefit from taking an integrative approach that considers how both the physical and social environment interact to promote or impair cognitive stimulation and emotional well-being.

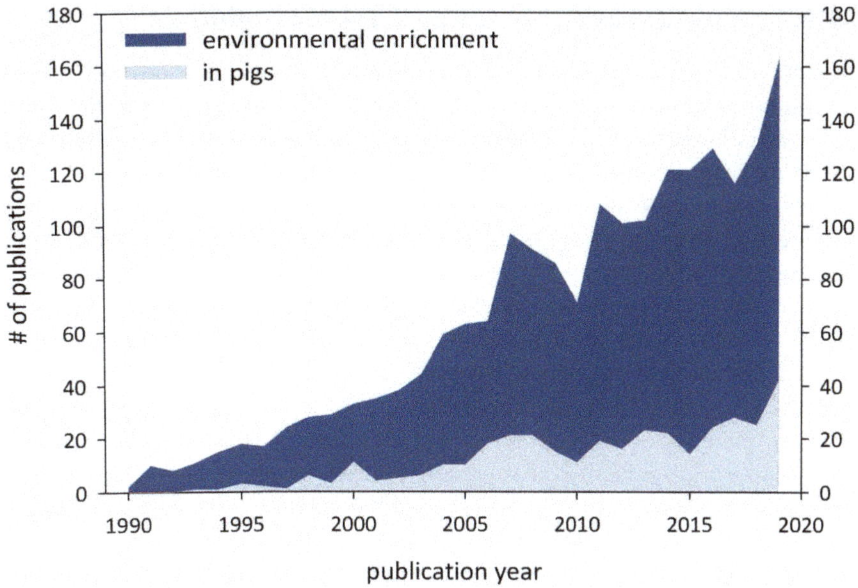

Figure 2 Trends in enrichment research. Number of publications considering environmental enrichment for animals in general, and enrichment for pigs in particular[1], over the last three decades.

This chapter reviews a full range of enrichment possibilities, ranging from artificial point-source objects and organic materials to social partners and cognitive challenges. A search in the scientific literature indicates that over the last three decades the annual number of studies mentioning 'environmental enrichment' is increasing, with research in pigs keeping pace with this general trend (see Fig. 2[1]). In the following sections, we review the existing body of evidence for welfare-enhancing effects of different types of enrichment. We highlight current as well as emerging innovative enrichment solutions that have the potential to increase the biological relevance of modern housing environments and thereby improve animal welfare. The focus is on whether a potential enrichment item promotes natural behaviours, reduces abnormal behaviours, and whether these effects are sustainable over the long-term. We conclude with a comparative evaluation of enrichment from an animal welfare point of view, taking into account the various trade-offs involved in implementing different items on a larger-scale in conventional livestock facilities and highlighting the potential of social and cognitive enrichment.

1 search on March 11th 2020 in the Web of Science for topic='environmental enrichment' AND research area=(veterinary sciences OR agriculture OR behavioural sciences) AND document type=article, timespan: 1990-2019, index: SCI-EXPANDED; second search within results for topic=((pig OR pigs OR swine OR sow OR sows OR gilt OR gilts OR boar OR boars OR piglet OR piglets OR pork OR 'Sus scrofa') NOT 'guinea pig')

2 Environmental enrichment and animal welfare

Despite approximately 8000 years of domestication, pigs are still driven by motivations and needs evolved by their wild ancestors and under semi-naturalistic conditions show similarities in their behavioural repertoire with wild boars (Stolba and Wood-Gush, 1989). They are a highly explorative and social species, prone to developing abnormal behaviours such as stereotypies (e.g. belly nosing, sham chewing) and injurious behaviours (e.g. tail-biting) when deprived of essential stimuli or resources. In order to meet the pigs' behavioural needs, researchers recommend increasing the 'biological relevance' of housing environments in order to give pigs the opportunity to execute more of their evolved behavioural repertoire (Newberry, 1995). This can be achieved by implementing effective environmental enrichment.

Environmental enrichment comprises all aspects of an animal's environment that facilitate engagement in species-specific behaviours, including exploration, manipulation, foraging and also social interactions. Thus to evaluate the effectiveness of various enrichment for pigs, it is important to have a basic understanding of their natural behavioural repertoire, which is often impaired in conventional livestock housing. Each species-specific behavioural repertoire evolved to maximize the biological fitness of the individual (e.g. Fraser et al., 1997). On a proximate level, animals evolved a motivational and affective system that 'rewards' behaviours that increase biological fitness and 'punishes' behaviours that impair fitness, via the association of such behaviours with positive and negative affective states, respectively. Animals are thus highly motivated to behave in ways that support their survival and reproduction and elicit positive affective states, while they avoid potential harm to their physical integrity and reproductive success, which would result in negative affective states. The paradox of modern farm animal husbandry systems is that while they provide the animals with many essential resources (e.g. water, feed, reproductive success via artificial insemination), they deprive them of the opportunity to use their natural behavioural repertoire to actively work for these resources (e.g. via foraging: de Jonge et al., 2008). In considering enrichment options, it is therefore important to distinguish between what an animal wants and what it likes. As Gygax (2017) described in his theoretical paper, wanting means being motivated to acquire something, while liking means enjoying having it. While at first glance these seem congruent, wanting and liking have separate underlying motivational drives, so that, for example, providing an animal with food without giving them the ability to forage for it will only fulfil the 'liking' motivational drive. Pigs are highly motivated to search for food, and in complex, semi-naturalistic environments spend more than half of their active time foraging (Stolba and Wood-Gush, 1989). Thus, while providing food in a trough is sufficient for pigs to receive adequate energy and nutrients, their motivational drive for foraging cannot be met by such a regime and the results are evident in a range of abnormal oral behaviours and impaired immune systems (e.g. Douglas et al., 1998).

Captive environments that lack in biological relevance can lead to impairments in pig behaviour and health, or welfare. In an attempt to summarize the aspects of animal welfare that are most relevant for farmed animals, Puppe and colleagues (2012) stated that 'welfare is the state of physical and mental health resulting from the process of behavioural and physiological adaptation when coping with environmental challenges, and the associated subjective experiences and emotional evaluation, all in the light of individual and/or cognitive needs and abilities.' This definition highlights several key aspects that are relevant when discussing environmental enrichment. First, it is important to note that both physical and mental health contribute to an animal's welfare state. While physical health has long been the major focus of attention in evaluating environmental enrichment (e.g. in ensuring that enrichment does not pose a threat to the animal's health by transmitting diseases), mental health is also integral to the concept of satisfying an animal's various motivational drives. Puppe and colleagues (2012) also mention adaptation in their definition. As an omnivore living across Asia, Europe and North Africa, with introduced populations in North and South America, New Zealand and Australia, wild boars have evolved facing highly versatile climates with significant seasonal changes in resource availability. While the natural habitat of wild boars seems in every aspect different from conventional pig housing, both captive and wild pigs are capable of adapting to various environments and can cope with a broad range of challenges. As such, challenges are not detrimental to welfare per se, but deviations from an optimal range that are so extreme in intensity or duration that animals are not able to adapt will cause distress and impair the welfare of the individual (e.g. Korte et al., 2007). For example, short-term social isolation can occur in both the wild and on farms, and a pig will respond to this challenge by engaging in contact calls and appetitive behaviours in order to re-establish contact with the group. However, if attempts to cope fail and pigs experience isolation for an extended period, this is likely to impact negatively on their welfare, because pigs are typically not adapted to long-term isolation (see section on social enrichment below). The definition of welfare by Puppe and colleagues (2012) also mentions an individual's subjective experiences and emotional evaluation. This leads to the last aspect of welfare in the context of enrichment, and that is individuality. Individual needs vary related to several factors, from genetics to sex, age, dominance or reproductive status. Specific enrichment items may thus be more or less effective in inducing natural behaviour depending on the specific attributes of an individual. For example, sows use nest-building material like jute sacks primarily in the period shortly before parturition. In some cases the enrichment needs of one individual may also be in conflict with the enrichment needs of others. For example, dominant individuals are likely to monopolize point-source enrichment items, which is in line with their natural behaviour but is not beneficial to subordinate individuals.

The following section summarizes the existing literature on the role of environmental enrichment on pig behaviour and welfare, taking the key aspects of welfare outlined here into account.

3 The role of environmental enrichment in pig behaviour and welfare

If we adopt the idea that environmental enrichment refers to anything that stimulates or enables the range of behaviours that evolved in the wild boar, there are various categories of enrichment items that can be evaluated in this respect. Looking at farm suppliers' portfolios, artificial objects are clearly the most common items used for enrichment, although there is limited evidence as to their effectiveness in promoting natural behaviours and reducing abnormal behaviours over the long-term. In the debate over which types of enrichment are most effective and feasible to implement, several, sometimes, conflicting issues need to be considered (e.g. Buijs and Muns, 2019). First and foremost, enrichment of the physical environment should satisfy pigs' need for exploration. Therefore, we need to know which characteristics make an item interesting enough for pigs to engage with, and for how long they will remain interested in it. Furthermore, since pigs are a highly social and intelligent species, having access to relevant social stimuli and cognitive challenges is likely to be as important as having access to rooting materials in terms of supporting an individual`s motivational drives. Referring again to the literature on environmental enrichment in pigs over the past three decades, 59% of studies (*n*= 214) focus on physical enrichment, dealing with organic materials (50%, n= 181 studies)[2] or artificial objects (28%, n= 102 studies)[3], with many of these studies mentioning both (*n* = 69). In contrast, only a handful of studies consider cognitive abilities or social behaviour when investigating enrichment options. The following sections will give an overview of these four main types of enrichment, how they are currently being implemented in conventional or other types of facilities, and what impact different items have on animal welfare.

3.1 Artificial objects as enrichment

There are a large number of artificial point-source objects marketed as 'pig toys' (Fig. 3a). Most of them are plastic objects small enough to chew on, which

2 Topic: (straw OR hay OR wood OR sawdust OR compost OR peat OR jute OR sisal OR rope) within the search results from Footnote 1

3 Topic: ((plastic OR rubber OR chain OR object OR toy OR ball) NOT 'novel object') within the search results from Footnote 1

are mounted on a chain that can be installed on the walls or hung from the ceiling of the pen. The material is designed to be long-lasting and resistant to tooth wear. Pigs mainly explore their environment, for example, when foraging, using their snout (both nose and mouth). Therefore, at first glance these types of items seem suitable to elicit investigation and manipulation by pigs, while meeting the farmers' demand for lasting items that do not require frequent replacement (to reduce costs both for the item itself and for the work to replace

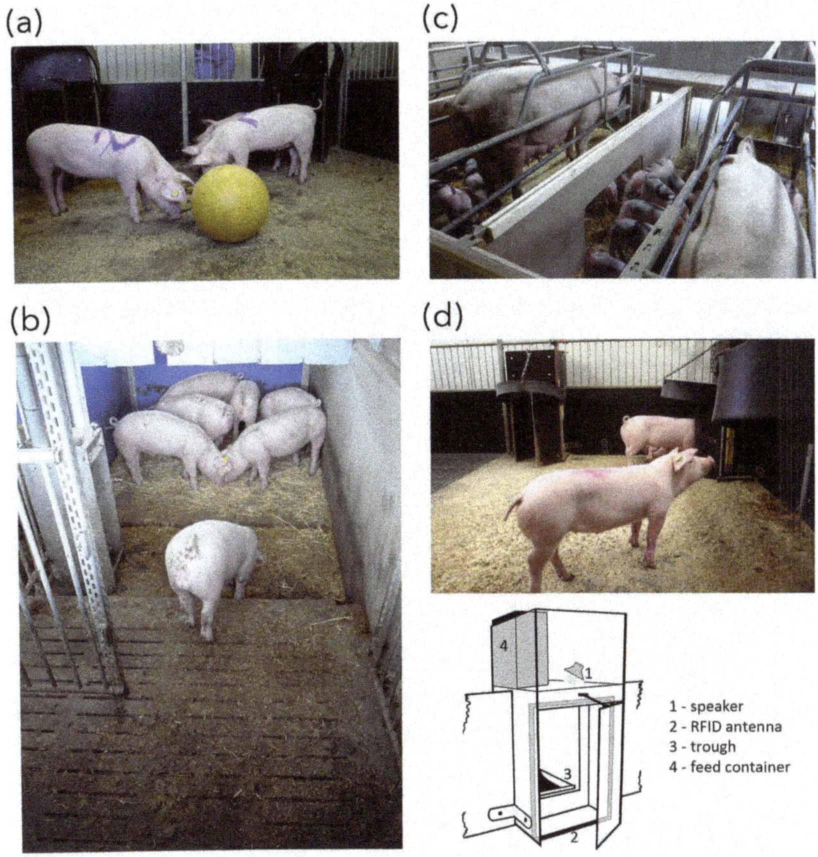

Figure 3 Exemplary pictures showing housing conditions with different levels of enrichment. (a) Housing with an artificial object (yellow plastic ball) and small amounts of organic litter (sawdust). Photo: FBN; (b) Housing with organic enrichment (straw bedding) in part of the pen. Photo © Bildungs- u. Wissenszentrum Boxberg (Landesanstalt für Schweinezucht-LSZ) 2020, all rights reserved; (c) An example of modified farrowing pens that facilitate co-mingling between litters. Photo courtesy of M. Farish, SRUC; (d) Housing with cognitive enrichment (call feeding station; details in the drawing) and small amounts of organic litter (sawdust). Photo, drawing: FBN.

it). However, their lasting quality also means that these objects are immutable, and there is a vast body of research indicating that change is very important to pigs. Immutable objects may be explored for up to two days, but after that they are not interesting to pigs (e.g. Gifford et al., 2007). According to, for example, the Australian Royal Society for the Protection of Cruelty against Animals (RSPCA), pig enrichment items should be novel, ingestible, and destructible, have a smell, and/or be chewable (see also van de Weerd et al., 2003). Pig chewing toys are, by definition, chewable, but they are not novel for long, they are not ingestible, most often do not have a smell, and while after weeks (or months) of use they can be destroyed, it is unlikely that a pig will perceive them as destructible. One item that has been discussed quite controversially is the Bite-Rite (Ikadan, Denmark), a plastic cone with four rubber rods. The benefit of this toy is that while it shares the characteristics of other chewing toys it can induce social play with its four rods, and there is evidence that pigs tend to remain interested in this object for weeks (van de Weerd et al., 2006). However, there are conflicting results on its ability to reduce tail-biting, and thus improve welfare, with some studies even finding that it increases the frequency of this abnormal oral behaviour (Lahrmann et al., 2019).

In conclusion, artificial enrichment items do not meet all criteria for effective enrichment and thus cannot be considered sufficient to meet pig welfare needs.

3.2 Organic materials as enrichment

In addition to the EU, the US Certified Humane programme and the Australian RSPCA claim that pigs should be provided with material that enables rooting/foraging behaviour. For this, organic materials are ideal (Fig. 3b). Organic materials found in pig housing comprise wood (most often small blocks mounted on a chain as an organic alternative to artificial chewing toys), straw (chopped as litter or provided long in racks) and different types of litter (e.g. wood shavings, saw dust, peat), or ropes (mostly sisal). Organic materials fulfil the requirements for enrichment items of being chewable and destructible, and they are (often) ingestible and have a smell. There is abundant evidence on the welfare-promoting effects of providing organic substrates, from increased exploration to reduced tail-biting as well as improved immune function among others (van de Weerd and Day, 2009; see Table 1 for further examples). The criterion of novelty, however, largely depends on the way the material is presented. Straw has been the most intensively investigated organic enrichment item, and there is strong evidence suggesting that it reduces tail-biting and other abnormal behaviours while motivating pigs to explore their environment when provided continuously, for example, as deep straw bedding (see Table 1 and reviews by Bracke et al., 2006; van de Weerd and Day, 2009). Providing straw from a dispenser is an effective means of reducing damaging

Table 1 Animal welfare-related behavioural and physiological changes of pigs in environments enriched with artificial objects and/or organic materials (* barren standard housing as control unless stated otherwise; ** effects for enriched vs. control unless stated otherwise; please note that we cannot provide a full review here, but chose studies that exemplify a larger body of research)

Intervention*	Evidence for enrichment effects**	Citations
Chain, rubber hose, straw rack or provision of straw on the floor twice daily	Lowest incidence of damaged tails with straw twice daily followed by straw rack	Zonderland et al. (2008)
Dispensers for straw, feed or liquid vs. straw bedding in the growing period	Highest prevalence of tail-biting with liquid dispenser and barren control housing, lowest with straw bedding	van de Weerd et al. (2006)
Fresh wood vs. artificial objects (in an enriched standard housing with straw rack, metal chain, fresh wood shavings daily)	More undamaged tails with wood than artificial objects in enriched standard housing	Telkänranta et al. (2014)
Greater space allowance with peat and straw rack from birth on	More activity and exploration, less harmful social and aggressive behaviour	Beattie et al. (1995, 2000a,b)
	Less fearful in a novel object test	Beattie et al. (2000a)
	Faster learning (operant task and maze)	Sneddon et al. (2000)
	Lower adrenal weights at slaughter	Beattie et al. (2000a)
Straw of three different lengths for growing pigs	Reduced nosing, aggression and tail-biting with straw, highest reduction of tail-biting when full- or half-chopped	Day et al. (2008)
Straw bedding from birth on	More activity, exploration and play	Bolhuis et al. (2005)
	Less oral behaviours directed at pen mates	Bolhuis et al. (2005); de Jong et al. (1998); de Jong et al. (2000b)
	Improved long-term memory (maze)	de Jong et al. (2000a)
	Higher baseline salivary cortisol	de Jong et al. (1998)
	Blunted circadian cortisol rhythm in barren housed pigs	de Jong et al. (2000a)
	No salivary cortisol increase during transport and in lairage	de Jong et al. (2000b)
	Lower humoral immune responses to an immunization in low-resisting individuals (as assessed via a backtest)	Bolhuis et al. (2003)

(*Continued*)

Table 1 (*Continued*)

Intervention*	Evidence for enrichment effects**	Citations
Straw bedding in the fattening period	Less behaviour and tail-biting directed at pen mates Reduced lameness, less severe bursitis, lower acute phase protein titres at slaughter	Scott et al. (2006)
Substrate-enriched from birth to weaning and/or post-weaning (2×2 factorial design)	Increased aggression at weaning in pre-weaning enriched pigs, but increased exploration and less oral manipulation in post-weaning enriched pigs	Melotti et al. (2011)
Three different space allowances pre-weaning, the larger two with straw	More locomotion and play behaviour, less food-related aggression in pigs from largest pen between 3 and 6 months of age	Chaloupková et al. (2007)
Switch between enriched (greater space allowance, access to straw and objects) and barren housing	More optimistic choices in a cognitive bias test	Douglas et al. (2012)
Access to mushroom compost	Fewer pigs involved in oral behaviours directed at pen mates, less tail-biting	Beattie et al. (2001)

behaviours such as tail-biting (e.g. van de Weerd et al., 2006). Zonderland and colleagues (2008) found that a small amount of straw provided twice daily was more effective as enrichment than straw provided from a rack. Adding edible parts can increase such effects and keep pigs interested in the enrichment for longer (Studnitz et al., 2007; Holm et al., 2008). Housing gestating sows on straw can also have positive trans-generational effects on the welfare of their offspring (Tatemoto et al., 2019). However, straw bedding, though beneficial both for lying comfort and as an explorable, manipulable material, is rarely found beyond organic farming due to the higher cost-effectiveness of slatted flooring, and the increased workload for the farmer associated with providing frequent and intermittent enrichment. Telkänranta and colleagues (2014) suggested fresh wood as a low-cost enrichment item for Finnish pig farmers, who often have access to wood on their properties. This is just one example highlighting the value of local solutions regarding easily available organic enrichment materials.

To summarize, organic materials meet many of the suggested criteria for appropriate enrichment for pigs, and especially deep straw bedding has been proven effective in reducing abnormal behaviours and improving welfare (Table 1). Despite the obvious benefits of providing organic materials, one has to bear in mind that such interventions mainly satisfy the animals' foraging motivation. It remains unclear whether the separation of foraging from the actual feed intake can fully satisfy the animals' drive to work for their food. Moreover, other motivations such as the motivation to engage in social interactions or to master cognitive challenges remain unanswered.

3.3 Implementing social enrichment in captive settings

Just as a better understanding of their natural exploratory and foraging behaviours can lead to more effective environmental enrichment for pigs, a better understanding of their evolved social environment can lead to more effective social enrichment, referring to aspects of the environment that help animals to obtain rewarding social interactions while avoiding unwanted interactions. In this section, we review the evolved social environment of domestic pigs and demonstrate how this knowledge can be applied to develop effective social enrichment in captive settings.

Similar to their wild relatives, domesticated pigs remain highly motivated to live in social groups (reviewed in Gonyou, 2009) and exhibit behavioural and physiological signs of distress when isolated from group members for short periods (Kanitz et al., 2009; Ruis et al., 2001), which can lead to chronic stress and health impairments if isolation persists over the long-term (Ruis et al., 2001). The natural group structure of pigs consists of stable associations of multiple adult females and their maturing offspring, who may share overlapping territories with additional social groups (Stolba and Wood-Gush, 1989). Apart from temporary periods during farrowing, wild sows are rarely separated from their social groups and form long-term social relationships with other females (Podgórski et al., 2014). In contrast, males typically leave their natal group before sexual maturity and spend much of their adult life solitary, joining social groups only during mating periods (Gonyou, 2009). Like wild boars, domestic pigs form linear dominance hierarchies based on sex, age, size and strength (Marchant-Forde and Marchant-Forde, 2005). Social enrichment for domestic pigs should provide opportunities to seek out positive social interactions with appropriate social partners, which may differ by age, sex and dominance rank, while avoiding or reducing the frequency of unwanted social interactions.

A review of the literature indicates that maintaining a subset of stable social relationships can help pigs to cope with environmental stressors, in a process known as social buffering (Hennessy et al., 2009, see also Table 2). Evidence for social buffering is widespread in animals, but the effectiveness of specific partners in reducing stress is also influenced by partner familiarity and the quality of the social relationship (Kikusui et al., 2006). The frequent re-mixing of pigs with unfamiliar individuals that begins at weaning in conventional production facilities represents a social stressor that is associated with increased release of the stress hormone cortisol (Colson et al., 2012; Lund et al., 1998), elevated rates of aggression (Puppe, 1998; Weary et al., 2008) and decreases in immunocompetence (Moore et al., 1994). The consequences of re-grouping for pigs vary based on inter-individual differences in dominance and coping strategies. For more dominant individuals who have more control over the occurrence of aggression, re-grouping may represent a minor, short-term stressor, and winning in aggressive conflicts may even boost immunocompetence (Tuchscherer et al., 1998). However, subordinate individuals are more likely to experience the unpredictable aggression following re-grouping as a chronic stressor, which may cause serious injuries, suppress the immune system and contribute to slowed growth rates and decreased productivity for sows over the long-term (Marchant-Forde and Marchant-Forde, 2005; Tuchscherer et al., 1998). Although changes in group composition are unavoidable in livestock production, being re-grouped with some familiar partners present can buffer against re-mixing stress. Familiar social partners are less likely to engage in post-mixing aggression (Colson et al., 2012; Puppe, 1998; Samarakone and Gonyou, 2009), and are more likely to provide social support in the form of affiliative social interactions such as nose-to-nose contact and relaxed nuzzling, which has been hypothesized to function in pigs similarly to social grooming in other species, in reducing physiological stress responses (Camerlink and Turner, 2013).

In one of the first studies to directly measure the impact of social buffering in pigs, Kanitz and colleagues (2014) exposed piglets to an acute period of maternal separation, which is a potent stressor, either alone, with an unfamiliar piglet or with a familiar littermate present. Piglets who went through the separation with a social partner exhibited less stress reactivity, measured by reduced release of the stress hormone cortisol and enhanced immunocompetence compared to piglets who went through the stressor alone (Kanitz et al., 2014; Tuchscherer et al., 2016). Being separated with a familiar partner present was more effective in reducing behavioural signs of distress compared to being separated with an unfamiliar partner (Kanitz et al., 2014). Interestingly, while the sex of the familiar partner did not affect social buffering in piglets, it did influence social buffering in gilts, who received more social support from same-sex partners in response to restraint stress (Reimert et al., 2014). This result is consistent with the evolved social environment of pigs reviewed above, in which adult females maintain long-term social relationships with other females and have only infrequent contact with adult males (Stolba and Wood-Gush, 1989).

Table 2 Examples of changes to the social or physical environment that promote social enrichment in domestic pigs. A greater number of citations indicate relatively more evidence for a given effect, although the list is not exhaustive. Studies finding positive effects of social enrichment over weeks or months following an intervention are indicated as 'long-term'

Intervention	Evidence for social enrichment	Citations
Co-mingling of lactating piglets with members of other litters (vs. kept only with littermates prior to weaning)	Reduced post-weaning aggression and skin lesions	Camerlink et al. (2018); Kutzer et al. (2009); Wattanakul et al. (1997); Weary et al. (1999, 2002)
	Reduced aggression in resident-intruder tests (long-term)	Camerlink et al. (2018); D'Eath (2005); Olsson et al. (1999)
	Reduced behavioural reactivity to novel environmental and social challenges	Hillmann et al. (2003)
	Greater rates of post-weaning social play	Donaldson et al. (2002); Salazar et al. (2018)
Re-mixing with some stable social partners present (vs. re-mixing without familiar partners present)	Reduced post-mixing aggression	Colson et al. (2012); Puppe (1998)
	Reduced post-mixing salivary cortisol	Colson et al. (2012); Lund et al. (1998)
	Reduced post-mixing lymphocyte and leukocyte counts	Lund et al. (1998)
	Reduced post-mixing sickness and morbidity (long-term)	Moore et al. (1994)
Having social partner present during exposure to acute stressor (vs. exposure to stressor alone)	Reduced behavioural signs of distress	Kanitz et al. (2014)
	Reduced HPA-axis reactivity; Altered expression of brain corticosteroid receptors	Kanitz et al. (2014); Kanitz et al. (2016)
	Reduced inflammatory activity (IL-6 production and TNF-alpha release)	Tuchscherer et al. (2014)
Adding hide areas to pens	Reduced post-mixing aggression	McGlone and Curtis (1985)
Housing in larger groups (while maintaining constant floor space per pig)	Reduced post-mixing aggression	Samarkone and Gonyou (2009)
	Reduced aggression in resident-intruder test (long-term)	Samarkone and Gonyou (2009); Turner et al. (2001)

Although having access to some stable social partners goes a long way towards achieving social enrichment goals, contexts such as re-grouping that involve exposure to unfamiliar individuals are likely to trigger increased aggression among at least a subset of pigs for whom dominance relationships are unclear. In addition to physical fights, pigs also use a range of visual, vocal and olfactory signals to indicate dominance or submission, which helps to reduce the frequency and intensity of physical aggression (Marchant-Forde and Marchant-Forde, 2005). There is evidence that pigs need experience to learn the appropriate social cues that indicate dominance status, and that even short-term exposure to unfamiliar partners early in life can help pigs to develop such social competencies (Camerlink et al., 2018; Salazar et al., 2018; Fig. 3c). Allowing piglets from different litters to co-mingle prior to weaning increases post-weaning play behaviour (Salazar et al., 2018) and reduces post-weaning aggression (Wattanakul et al., 1997; Weary et al., 1999), even when weaned with unfamiliar individuals (Salazar et al., 2018). Early life socialization also has long-term effects on social competence, as pigs exposed to unfamiliar individuals before weaning were better at assessing social situations and responding appropriately to social challenges later in life (Camerlink et al., 2018).

There are also aspects of the physical environment that can help pigs to avoid unwanted social interactions, by achieving more visual or spatial separation from less-preferred social partners. One fairly easy way to achieve this is by including visual barriers, such as extra corners, blinds, kennels or pop-holes, in which pigs can remain at least partially out of view (reviewed in: Marchant-Forde and Marchant-Forde, 2005). Such interventions may be especially helpful during critical re-mixing periods while hierarchies are being established (McGlone and Curtis, 1985). Counter-intuitively, increasing group size may also be an effective means of reducing aggression, provided that floor space allowance per pig remains stable (Samarakone and Gonyou, 2009; Turner et al., 2001). One reason is that increasing the pen size for larger groups also increases the amount of total free space available at any given time, giving pigs more opportunities to avoid each other (McGlone and Newby, 1994). There is also evidence that pigs in larger social groups may shift to a less-aggressive social strategy (Samarakone and Gonyou, 2009) by forming stable sub-groups and minimizing contact with other sub-groups, similar to conditions in the wild (Stolba and Wood-Gush, 1989). This process can be supported by further partitioning of the pens into multiple group-laying and feeding areas (Bünger and Kallweit, 1994). Given that modern pig production facilities increasingly favour group sizes in the hundreds (Marchant-Forde and Marchant-Forde, 2005), the initial evidence that partitioning of large spaces may promote natural sub-grouping in pigs with positive effects on social behaviour is a promising line of research that deserves more attention.

In summary, social enrichment in pigs refers to the ability of individuals to demonstrate a full range of affiliative social behaviours that help to cope with stress, while minimizing unwanted social interactions that can cause stress and injuries. The interventions described here can be implemented with relatively minor adjustments to current husbandry practices, and will go a long way towards ensuring that domestic pigs experience more enriched social lives.

3.4 Cognitive enrichment

Cognitive enrichment, first coined by Milgram (2003) to describe the effects of a prolonged learning program on cognitive plasticity in dogs, is a subset of enrichment that refers to the use of cognitive skills to solve problems and to gain control over some aspect of the environment. It should necessarily involve some form of reward for the animal, which could be either internal (such as the mental state of satisfaction) or external (such as food or another valued resource), and which has been correlated to valid measures of animal welfare (Clark, 2017). The concept of cognitive enrichment has been applied to humans in the context of fostering developmental neuroplasticity (Petrosini et al., 2009) and to laboratory, farm and zoo animals to improve welfare under captive conditions (Clark, 2017). In a stronger operational sense, cognitive enrichment occurs when a cognitively challenging task is added to the environment that causes verifiable cognitive stimulation, that is, by inducing specific, short-term problem-solving activities (Ernst et al., 2005; Puppe et al., 2007).

It is well known that pigs have advanced skills in spatial learning, memory and social cognition, and that they show at least basic levels of time perception and conscious awareness of basic emotions and sensations (Mendl et al., 2010; Nawroth et al., 2018). In this respect, it is astonishing that only in recent years has there been an increase in research to develop cognitive enrichment solutions for pigs within their housing environments. In addition to the implications for investigating mechanisms related to learning and cognition itself, successful problem-solving behaviour, especially when it is positively motivated and rewarded instrumental behaviour (Manteuffel et al., 2009a,b), can provide biologically relevant enrichment to the housing conditions of domestic pigs (Meehan and Mench, 2007), and is therefore an effective means of improving their welfare. In this sense, cognitive challenges fulfil an important need of animals to perform cognitive-related behaviour (wanting). In fact, de Jonge and colleagues (2008) showed that pigs express contra-free loading in a natural foraging task, meaning that they prefer to forage than to receive freely available food from a trough. In their study, the challenge was to find hidden food in one compartment, which was preferred over other compartments of the same structure with freely available food. This also suggests that it is important for functional cognitive enrichment to present tasks that meet the animals' species-specific needs and abilities as much as possible, and for pigs a food reward in the context of foraging seems particularly appropriate.

In order to develop an applicable cognitive enrichment solution for domestic pigs, Ernst and colleagues (2005) have designed an experimental feeding system ('call feeding stations'; Fig. 3d) using food-rewarded learning of discriminatory and instrumental cognitive tasks. Specifically, pigs first learned to link an individual acoustic signal to the delivery of food in one of several feeding stations (classical conditioning), and were then called to the feeding station by broadcasting of their individual signal, and rewarded with the delivery of food when they entered the correct feeding station (operant conditioning). To solve this task, pigs need to identify their own individual signal, and discriminate it from the signals of their group members. Zebunke and colleagues (2011) demonstrated that pigs quickly learn the task and had no behavioural or physiological response to hearing a groupmate's call, while showing positive arousal when hearing their own. In a later study, breeding sows fed in the call feeding stations during gestation could even recall their signal after a six-week break (Kirchner et al., 2012). Table 3 shows that providing pigs with this cognitive enrichment has a variety of positive effects on behaviour and physiology that can clearly be interpreted as signs of an enhanced quality of life and improved welfare. This system is now being adapted for implementation in commercial pig husbandry, for the group housing of sows with electronic feeding stations (Kirchner et al., 2012; Manteuffel et al., 2011). In particular, positive effects on management, feeding efficiency, general health and social behaviour can be expected, which should compensate for the additional costs of the system. In broad agreement with this approach, other authors have shown that cognitive tasks using an electronic feeding system are effective in reducing aggressive behaviour in suckling or weaning piglets and in improving animal welfare (Rautenberg et al., 2013; Sonoda et al., 2013).

In addition, there is evidence that an enriched physical environment (e.g. pens with more space, straw and objects to manipulate) also improves cognitive performance (Bolhuis et al., 2013; Grimberg-Henrici et al., 2016), cognitive judgement (Douglas et al., 2012) and decision-making (van der Staay et al., 2017) in pigs. This suggests that the complex interaction between environment and cognition has broader potential to promote sustainable improvements in husbandry, management and animal welfare, which requires further research. Although cognitive enrichment appears to be the least studied form of enrichment (Clark, 2017), technological advances incorporated into the housing environment that adequately challenge the cognitive abilities of intelligent species like pigs (e.g. in relation to feeding) have much potential to promote animal welfare.

4 Evaluating the biological relevance of different types of enrichment

In this section, we evaluate the types of enrichment listed in the previous sections for their biological relevance and related ability to satisfy the wanting/

Table 3 Animal welfare-related behavioural and physiological changes in pigs in a cognitively enriched housing system developed by Ernst et al. (2005) compared to conventionally housed controls

Organismic system	Measured response	Citations
Behaviour	Increased locomotor activity and exploration	Puppe et al. (2007) Zebunke et al. (2013)
	Reduced aggression	
	Reduced belly nosing and fear reactions	
Muscle physiology	Improved meat quality	Fiedler et al. (2005)
	More oxidative muscle fibres	
Immune defence	Improved wound healing	Ernst et al. (2006)
	Increased IgG blood levels	
	Enhanced *in vitro* T-cell proliferation	
Brain reward system	Downregulation of opioid receptor expression in the amygdala	Kalbe and Puppe (2010)
Autonomic nervous system	Autonomic tone indicates enhanced arousal and positive affective appraisal	Zebunke et al. (2011)

liking motivational drives, which contributes to their effectiveness by promoting natural behaviours and reducing abnormal behaviours over the long-term. This approach provides an innovative perspective on which types of enrichment are best suited to improve animal welfare from a biological point of view and may also be achievable given the economic perspectives that currently guide the implementation of enrichment in livestock facilities (see European Commission, 2014, 2018).

As elaborated above, physical enrichment is the most common type of enrichment found on farms. However, as this type of enrichment mainly comprises artificial objects, specifically small chewing toys (Nalon and De Briyne, 2019; van de Weerd and Ison, 2019), their ability to increase the biological relevance of a housing environment for pigs is doubtful. If such toys are attractive to animals, they will likely be monopolized by dominant individuals. Therefore, providing one toy to a group, as is the current practice, does not represent sufficient enrichment. But more importantly, while small chewing toys might meet the animal's drive to nose and chew objects, one may ask which functional category of behaviours is served here: is it exploration of the environment, or can it be assigned to foraging? In case of the former, the unchanging, almost indestructible nature of the objects will maintain interest for only a short time. In case of the latter, nosing and chewing are only a small part of foraging, providing mostly the anticipation but no or only a 'faux' consumption experience. This might activate the wanting-system, but most likely not the liking motivational drive. While organic materials are much better-suited substrates for rooting and exploration within the foraging context, they are not necessarily digestible, which again severs the link between wanting and liking. We have no experimental evidence on the welfare consequences of disrupting this link with its feedback loops and adaptations to the individual's current state (Gygax, 2017).

While the cognitive enrichment suggested above might seem like a highly artificial situation for the animals, it is likely to increase the biological relevance of housing. In the wild, pigs rely on their cognitive skills when foraging. The auditory system plays a pivotal role in intra-specific communication in pigs and can obviously serve other functions as well, even if it is not primarily used to locate food. This enrichment allows pigs to use their operant and instrumental learning abilities and exert more control over their environment. One additional benefit of the system is that the 'natural' dominance hierarchy within the group can be accounted for by feeding dominant individuals earlier during the day, and this system thereby has the potential to reduce the heightened levels of feeding-related aggression that are often associated with feeding stations on farms (Kirchner et al., 2012). As described above, pigs are a highly social species; but their adaptations to deal with social stress and competition for limited resources are often overstrained in captive environments. The restrictive feeding of breeding sows (which is necessary to avoid extensive weight gain and associated health issues) is likely to increase feeding motivation and therefore increase feeding competition. In addition, in a space-restricted pen without hiding places, subordinate animals cannot avoid dominant animals and may not be able to use appropriate signals or spatial distancing to avoid agonistic interactions. This combination is particularly problematic for pig welfare, but there is promising evidence that it can be solved with a more controlled feeding system like the call-feeding-stations (Kirchner et al., 2012).

Pigs are also a highly flexible species as mentioned above, and even some relatively minor changes in husbandry practices may help them to have more satisfying social lives in captivity. Allowing pigs (especially females) to maintain some stable, long-term social relationships will facilitate social buffering in order to better cope with various environmental challenges, such as re-mixing into different social groups. Having access to stable relationships allow pigs to seek contact with preferred partners (wanting) and engage in positive interactions such as social grooming (liking). Proper partitioning of the housing environment can also reduce the detrimental effects of excessive social aggression and promote the formation of large groups into more stable sub-groups. Taken together, appropriate management practices and pen designs that help pigs to preferentially associate with preferred partners and to avoid others will enable pigs to express a greater range of their evolved social behaviours and in doing so will increase the biological relevance of captive housing.

5 Conclusions

Environmental enrichment (be it physical, social or cognitive) is one of the most potent tools to promote more natural behaviour, reduce abnormal behaviour, and thereby improve the welfare of pigs. An increasing body of research highlights the beneficial effects of different types of enrichment over different time scales. However, the implementation of this knowledge on farms is currently inconsistent (van de Weerd and Ison, 2019) and requires further attention.

Real progress in providing effective enrichment under modern farming conditions requires a change in all involved parties, from farmers to consumers to policy makers, as to what qualities are required for an environment to be considered enriched and how much enrichment is 'a sufficient quantity'. For this, we need to consider the full range of evolved behaviours and motivations of pigs to properly evaluate the potential but also the limitations of specific types of enrichment. For example, when we combine enrichment items that become limited resources with inadequate social conditions that deprive animals of control over their social interactions, the implementation of enrichment will not only be insufficient, but might even reduce welfare for some group members. In addition, this review emphasizes the need to carefully assess the impact of environmental enrichment. Some artificial objects may be triggers of social play, specifically those that can be used by several individuals at the same time, and this may increase their value as enrichment items, especially for immature individuals who engage in more social play. Furthermore, while challenges are often incorrectly regarded as detrimental for welfare, challenges that animals can solve, such as cognitive challenges posed by the implementation of cognitive enrichment, can improve pig welfare. Unfortunately, this fairly novel approach has not been widely applied on farms yet. Taking into account the dimension of time, both cognitive and social enrichment appear far superior to providing point-source objects, as their impact on welfare has greater potential to be long-lasting in comparison to the often short-term effects of providing artificial objects or even organic materials.

Under current conditions, it seems that the most commonly used types of enrichment are not the most effective in promoting welfare, while the types of enrichment with the highest potential to improve welfare are the least used (Nalon and De Briyne, 2019; van de Weerd and Ison, 2019). However, we do not think this is a lost cause. Choices about implementing enrichment often depend primarily on economic viability (cf. European Commission, 2018). Therefore, implementation of effective enrichment will also require a change in the market (i.e. among all stakeholders, from farmers to consumers) that brings more income to the farmers to make such changes economically feasible. Whether this occurs via increased prices for meat products or another system of reimbursement is open to discussion. In order to provide decision-makers with scientifically sound recommendations regarding which enrichment to implement, we need to fill current gaps in research. Specifically, research investigating long-term effects of different types of enrichment on both welfare and productivity is scarce but essential. Re-evaluating the available options for environmental enrichment based on the frameworks of biological relevance and wanting/liking, together with a re-thinking of the economic structures involved, will lead to meaningful enrichment options for farmed pigs in the future.

6 Where to look for further information

6.1 Further reading

6.1.1 Reviews

- Newberry, R. C. (1995), 'Environmental Enrichment – Increasing the Biological Relevance of Captive Environments', *Appl. Anim. Behav. Sci.*, 44, 229–243.
- Rushen, J. and de Passillé, A. M. B. (1992), 'The Scientific Assessment of the Impact of Housing on Animal-Welfare – A Critical-Review', *Can. J. Anim. Sci.*, 72, 721–743.
- van de Weerd, H. and Ison, S. (2019), 'Providing Effective Environmental Enrichment to Pigs: How Far Have We Come?', *Animals,* 9, 2019.
- Godyn, D., Nowicki, J. and Herbut, P. (2019), 'Effects of Environmental Enrichment on Pig Welfare-A Review', *Animals,* 9, 2019.
- Studnitz, M., Jensen, M. B. and Pedersen, L. J. (2007), 'Why Do Pigs Root and in What Will They Root? A Review on the Exploratory Behaviour of Pigs in Relation to Environmental Enrichment', *Appl. Anim. Behav. Sci.*, 107, 183–197.
- Bracke, M. B. M., Zonderland, J. J., Lenskens, P., Schouten, W. G. P., Vermeer, H., Spoolder, H. A. M., Hendriks, H. J. M., and Hopster, H. (2006), 'Formalised Review of Environmental Enrichment for Pigs in Relation to Political Decision Making', *Appl. Anim. Behav. Sci.*, 98, 165–182.

6.1.2 Books

- Špinka, M. (Ed.) (2018), *Advances in Pig Welfare,* Woodhead Publishing.
- Appleby, M. C., Olsson, I. A. S. and Galindo, F. (Eds.) (2018), *Animal Welfare,* 3rd Edition, CABI.
- Young, R. J. (2003), *Environmental Enrichment for Captive Animals,* UFAW.

6.2 Key journals and conferences

6.2.1 Journals

- Animal Welfare.
- Applied Animal Behaviour Science.

6.2.2 Conferences

- Congress of the International Society for Applied Ethology (ISAE).
- International Conference on the Assessment of Animal Welfare at Farm and Group Level (WAFL).
- Universities Federation for Animal Welfare (UFAW) Animal Welfare Conference.

7 References

Australian Livestock Management Act (2010). Available at: https://www.legislation.vic.gov.au/in-force/acts/livestock-management-act-2010/006 (Last accessed 31.03.2020).

Beattie, V. E., Walker, N. and Sneddon, I. A. (1995). Effects of environmental enrichment on behavior and productivity of growing pigs, *Anim. Welf.* 4, 207–220.

Beattie, V. E., O'Connell, N. E., Kilpatrick, D. J. and Moss, B. W. (2000a). Influence of environmental enrichment on welfare-related behavioural and physiological parameters in growing pigs, *Anim. Sci.* 70(3), 443–450.

Beattie, V. E., O'Connell, N. E. and Moss, B. W. (2000b). Influence of environmental enrichment on the behaviour, performance and meat quality of domestic pigs, *Livest. Prod. Sci.* 65(1–2), 71–79.

Beattie, V. E., Sneddon, I. A., Walker, N. and Weatherup, R. N. (2001). Environmental enrichment of intensive pig housing using spent mushroom compost, *Anim. Sci.* 72(1), 35–42.

Bolhuis, J. E., Oostindjer, M., Hoeks, C. W. F., de Haas, E. N., Bartels, A. C., Ooms, M. and Kemp, B. (2013). Working and reference memory of pigs (*Sus scrofa domesticus*) in a holeboard spatial discrimination task: the influence of environmental enrichment, *Anim. Cogn.* 16(5), 845–850.

Bolhuis, J. E., Parmentier, H. K., Schouten, W. G. P., Schrama, J. W. and Wiegant, V. M. (2003). Effects of housing and individual coping characteristics on immune responses of pigs, *Physiol. Behav.* 79(2), 289–296.

Bolhuis, J. E., Schouten, W. G. P., Schrama, J. W. and Wiegant, V. M. (2005). Behavioural development of pigs with different coping characteristics in barren and substrate-enriched housing conditions, *Appl. Anim. Behav. Sci.* 93(3–4), 213–228.

Bracke, M. B. M., Zonderland, J. J., Lenskens, P., Schouten, W. G. P., Vermeer, H., Spoolder, H. A. M., Hendriks, H. J. M. and Hopster, H. (2006). Formalised review of environmental enrichment for pigs in relation to political decision making, *Appl. Anim. Behav. Sci.* 98(3–4), 165–182.

Buijs, S. and Muns, R. (2019). A review of the effects of non-straw enrichment on tail biting in pigs, *Animals* 9(10), 824.

Bünger, B. and Kallweit, E. (1994). Ethological studies on pregnant sows in a dynamic group-housing system with rationed individual feeding by one electronic feeder station, *Landbauforsch Völkenrode* 44, 155–166.

Camerlink, I., Farish, M., D'Eath, R. B., Arnott, G. and Turner, S. P. (2018). Long term benefits on social behaviour after early life socialization of piglets, *Animals* 8(11), 192.

Camerlink, I. and Turner, S. P. (2013). The pig's nose and its role in dominance relationships and harmful behaviour, *Appl. Anim. Behav. Sci.* 145(3–4), 84–91.

Chaloupková, H., Illmann, G., Bartoš, L. and Špinka, M. (2007). The effect of pre-weaning housing on the play and agonistic behaviour of domestic pigs, *Appl. Anim. Behav. Sci.* 103(1–2), 25–34.

Clark, F. E. (2017). Cognitive enrichment and welfare: current approaches and future directions, *Anim. Behav. Cogn.* 4(1), 52–71.

Colson, V., Martin, E., Orgeur, P. and Prunier, A. (2012). Influence of housing and social changes on growth, behaviour and cortisol in piglets at weaning, *Physiol. Behav.* 107(1), 59–64.

Day, J. E. L., van de Weerd, H. A. and Edwards, S. A. (2008). The effect of varying lengths of straw bedding on the behaviour of growing pigs, *Appl. Anim. Behav. Sci.* 109(2–4), 249–260.

D'Eath, R. B. (2005). Socialising piglets before weaning improves social hierarchy formation when pigs are mixed post-weaning, *Appl. Anim. Behav. Sci.* 93(3–4), 199–211.

de Jong, I. C., Ekkel, E. D., van de Burgwal, J. A., Lambooij, E., Korte, S. M., Ruis, M. A. W., Koolhaas, J. M. and Blokhuis, H. J. (1998). Effects of strawbedding on physiological responses to stressors and behavior in growing pigs, *Physiol. Behav.* 64(3), 303–310.

de Jong, I. C., Prelle, I. T., van de Burgwal, J. A., Lambooij, E., Korte, S. M., Blokhuis, H. J. and Koolhaas, J. M. (2000a). Effects of environmental enrichment on behavioral responses to novelty, learning, and memory, and the circadian rhythm in cortisol in growing pigs, *Physiol. Behav.* 68(4), 571–578.

de Jong, I. C., Prelle, I. T., van de Burgwal, J. A., Lambooij, E., Korte, S. M., Blokhuis, H. J. and Koolhaas, J. M. (2000b). Effects of rearing conditions on behavioural and physiological responses of pigs to preslaughter handling and mixing at transport, *Can. J. Anim. Sci.* 80(3), 451–458.

de Jonge, F. H., Tilly, S. H., Baars, A. M. and Spruijt, B. M. (2008). On the rewarding nature of appetitive feeding behaviour in pigs (*Sus scrofa*): do domesticated pigs contrafreeload?, *Appl. Anim. Behav. Sci.* 114(3–4), 359–372.

Donaldson, T. M., Newberry, R. C., Špinka, M. and Cloutier, S. (2002). Effects of early play experience on play behaviour of piglets after weaning, *Appl. Anim. Behav. Sci.* 79(3), 221–231.

Douglas, C., Bateson, M., Walsh, C., Bedue, A. and Edwards, S. A. (2012). Environmental enrichment induces optimistic cognitive biases in pigs, *Appl. Anim. Behav. Sci.* 139(1–2), 65–73.

Douglas, M. W., Cunnick, J. E., Pekas, J. C., Zimmerman, D. R. and von Borell, E. H. (1998). Impact of feeding regimen on behavioral and physiological indicators for feeding and motivation and satiety, immune function, performance of gestating sows, *J. Anim. Sci.* 76(10), 2589–2595.

Ernst, K., Puppe, B., Schön, P. C. and Manteuffel, G. (2005). A complex automatic feeding system for pigs aimed to induce successful behavioural coping by cognitive adaptation, *Appl. Anim. Behav. Sci.* 91(3–4), 205–218.

Ernst, K., Tuchscherer, M., Kanitz, E., Puppe, B. and Manteuffel, G. (2006). Effects of attention and rewarded activity on immune parameters and wound healing in pigs. *Physiol. Behav.* 89(3), 448–456.

European Commission (2014). Directorate-General for Agriculture and Rural Development AGRI-2011-EVAL-08 'Final Report Assessing farmers' cost of compliance with EU legislation in the fields of environment, animal welfare and food safety', Available at: https://ec.europa.eu/info/sites/info/files/food-farming-fisheries/key_policies/documents/ext-study-farmer-costs-fulltext_2014_en.pdf (Last accessed 24.03.2020).

European Commission (2016). Staff working document on best practices with a view to the prevention of routine tail-docking and the provision of enrichment materials to pigs, Available at: https://ec.europa.eu/food/sites/food/files/animals/docs/aw:practice_farm_pigs_stfwrkdoc_en.pdf (Last accessed 24.03.2020).

European Commission (2018). Report from the Commission to the European Parliament and the Council 'on the impact of animal welfare international activities on the competitiveness of European livestock producers in a globalized world', Available at: https://ec.europa.eu/food/sites/food/files/animals/docs/aw:international_publication-report_en.pdf#page=2. (Last accessed 24.03.2020).

European Union (2009). Council Directive 2008/120/EC of 18 December 2008 laying down minimum standards for the protection of pigs, *Off. J. Eur. Union* L 47/5, Available at: http://data.europa.eu/eli/dir/2008/120/2019-12-14. (Last accessed 24.03.2020).

Fiedler, I., Küchenmeister, U., Ender, K., Haider, W., Ernst, K., Puppe, B. and Manteuffel, G. (2005). Reaktion der Muskulatur auf eine stimulierende Haltung – Befunde am Kotelettmuskel (*M. longissimus*) von Landrasse-Schweinen, *Dtsch. Tierärztl Wschr.* 112, 363–368.

Fraser, D., Weary, D. M., Pajor, E. A. and Milligan, B. N. (1997). A scientific conception of animal welfare that reflects ethical concerns, *Anim. Welf.* 6, 187–205.

Gifford, A. K., Cloutier, S. and Newberry, R. C. (2007). Objects as enrichment: effects of object exposure time and delay interval on object recognition memory of the domestic pig, *Appl. Anim. Behav. Sci.* 107(3–4), 206–217.

Gonyou, H. W. (2009). The social behaviour of pigs. In: Keeling, L. J. and Gonyou, H. W. (Eds) *Social Behaviour in Farm Animals*, UK: CABI Publishing, pp. 147–176.

Grimberg-Henrici, C. G. E., Vermaak, P., Bolhuis, J. E., Nordquist, R. E. and van der Staay, F. J. (2016). Effects of environmental enrichment on cognitive performance of pigs in a spatial holeboard discrimination task, *Anim. Cog.* 19(2), 271–283.

Gygax, L. (2017). Wanting, liking and welfare: the role of affective states in proximate control of behaviour in vertebrates, *Ethology* 123(10), 689–704.

Hennessy, M. B., Kaiser, S. and Sachser, N. (2009). Social buffering of the stress response: diversity, mechanisms, and functions, *Front. Neuroendocrinol.* 30(4), 470–482.

Hillmann, E., Von Hollen, F., Bünger, B., Todt, D. and Schrader, L. (2003). Farrowing conditions affect the reactions of piglets towards novel environment and social confrontation at weaning, *Appl. Anim. Behav. Sci.* 81(2), 99–109.

Holm, L., Jensen, M. B., Pedersen, L. J. and Ladewig, J. (2008). The importance of a food feedback in rooting materials for pigs measured by double demand curves with and without a common scaling factor, *Appl. Anim. Behav. Sci.* 111(1-2), 68-84.

Kalbe, C. and Puppe, B. (2010). Long-term cognitive enrichment affects opioid receptor expression in the amygdala of domestic pigs, *Genes Brain Behav.* 9(1), 75-83.

Kanitz, E., Hameister, T., Tuchscherer, M., Tuchscherer, A. and Puppe, B. (2014). Social support attenuates the adverse consequences of social deprivation stress in domestic piglets, *Horm. Behav.* 65(3), 203-210.

Kanitz, E., Hameister, T., Tuchscherer, A., Tuchscherer, M. and Puppe, B. (2016). Social support modulates stress-related gene expression in various brain regions of piglets, *Front. Behav. Neurosci.* 10, 227.

Kanitz, E., Puppe, B., Tuchscherer, M., Heberer, M., Viergutz, T. and Tuchscherer, A. (2009). A single exposure to social isolation in domestic piglets activates behavioural arousal, neuroendocrine stress hormones, and stress-related gene expression in the brain, *Physiol. Behav.* 98(1-2), 176-185.

Kikusui, T., Winslow, J. T. and Mori, Y. (2006). Social buffering: relief from stress and anxiety, *Philos. Trans. R. Soc. Lond. B Biol. Sci.* 361(1476), 2215-2228.

Kirchner, J., Manteuffel, G. and Schrader, L. (2012). Individual calling to the feeding station can reduce agonistic interactions and lesions in group housed sows, *J. Anim. Sci.* 90(13), 5013-5020.

Korte, S. M., Olivier, B. and Koolhaas, J. M. (2007). A new animal welfare concept based on allostasis, *Physiol. Behav.* 92(3), 422-428.

Kutzer, T., Bünger, B., Kjaer, J. B. and Schrader, L. (2009). Effects of early contact between non-littermate piglets and of the complexity of farrowing conditions on social behaviour and weight gain, *Appl. Anim. Behav. Sci.* 121(1), 16-24.

Lahrmann, H. P., Faustrup, J. F., Hansen, C. F., D'Eath, R. B., Nielsen, J. P. and Forkman, B. (2019). The effect of straw, rope, and Bite-Rite treatment in weaner pens with a tail biting outbreak, *Animals* 9(6), 365.

Lund, A., Wallgren, P., Rundgren, M., Artursson, K., Thomke, S. and Forsum, C. (1998). Performance, behaviour and immune capacity of domestic pigs reared for slaughter as siblings or transported and reared in mixed groups, *Acta Agric. Scand. A* 48(2), 103-112.

Manteuffel, C., Schön, P. C. and Manteuffel, G. (2011). Beyond electronic feeding: the implementation of call feeding for pregnant sows, *Comput. Electron. Agr.* 79(1), 36-41.

Manteuffel, G., Langbein, J. and Puppe, B. (2009a). Increasing farm animal welfare by positively motivated instrumental behaviour, *Appl. Anim. Behav. Sci.* 118(3-4), 191-198.

Manteuffel, G., Langbein, J. and Puppe, B. (2009b). From operant learning to cognitive enrichment in farm animal housing: bases and applicability, *Anim. Welf.* 18, 87-95.

Marchant-Forde, J. and Marchant-Forde, R. (2005). Minimizing inter-pig aggression during mixing, *Pig News Inf.* 26, 63N-71N.

McGlone, J. J. and Curtis, S. E. (1985). Behavior and performance of weanling pigs in pens equipped with hide areas, *J. Anim. Sci.* 60(1), 20-24.

McGlone, J. J. and Newby, B. E. (1994). Space requirements for finishing pigs in confinement: behavior and performance while group size and space vary, *Appl. Anim. Behav. Sci.* 39(3-4), 331-338.

Meehan, C. L. and Mench, J. A. (2007). The challenge of challenge: can problem solving opportunities enhance animal welfare?, *Appl. Anim. Behav. Sci.* 102(3-4), 246-261.

Melotti, L., Oostindjer, M., Bolhuis, J. E., Held, S. and Mendl, M. (2011). Coping personality type and environmental enrichment affect aggression at weaning in pigs, *Appl. Anim. Behav. Sci.* 133(3-4), 144-153.

Mendl, M., Held, S. and Byrne, R. W. (2010). Pig cognition, *Curr. Biol.* 20(18), R796-R798.

Milgram, N. W. (2003). Cognitive experience and its effect on age-dependent cognitive decline in beagle dogs, *Neurochem. Res.* 28(11), 1677-1682.

Moore, A. S., Gonyou, H. W., Stookey, J. M. and McLaren, D. G. (1994). Effect of group composition and pen size on behavior, productivity and immune response of growing pigs, *Appl. Anim. Behav. Sci.* 40(1), 13-30.

Nalon, E. and De Briyne, N. (2019). Efforts to ban the routine tail docking of pigs and to give pigs enrichment materials via EU law: where do we stand a quarter of a century on?, *Animals* 9(4), 132.

National Pork Board (2018). Common swine industry audit, Available at: http://d3fns0a45gcg1a.cloudfront.net/sites/all/files/documents/CommonSwineIndustryAudit/CSIA-English.pdf (Last accessed 31.03.2020).

Nawroth, C., Langbein, J. and Puppe, B. (2018). Swine cognition. In: Vonk, J. and Shackelford, T. K. (Eds), *Encyclopedia of Animal Cognition and Behavior*, Cham: Springer.

Newberry, R. C. (1995). Environmental enrichment - increasing the biological relevance of captive environments, *Appl. Anim. Behav. Sci.* 44(2-4), 229-243.

Olsson, I. A. S., De Jonge, F. H., Schuurman, T. and Helmond, F. A. (1999). Poor rearing conditions and social stress in pigs: repeated social challenge and the effect on behavioural and physiological responses to stressors, *Behav. Processes* 46(3), 201-215.

Petrosini, L., De Bartolo, P., Foti, F., Gelfo, F., Cutuli, D., Leggio, M. G. and Mandolesi, L. (2009). On whether the environmental enrichment may provide cognitive and brain reserves, *Brain Res. Rev.* 61(2), 221-239.

Podgórski, T., Lusseau, D., Scandura, M., Sönnichsen, L. and Jędrzejewska, B. (2014). Long-lasting, kin-directed female interactions in a spatially structured wild boar social network, *PLoS ONE* 9(6), e99875.

Prevention of Cruelty to Animals Act (1986). Available at: https://www.legislation.vic.gov.au/in-force/acts/prevention-cruelty-animals-act-1986/095 (Last accessed 31.03.2020).

Puppe, B. (1998). Effects of familiarity and relatedness on agonistic pair relationships in newly mixed domestic pigs, *Appl. Anim. Behav. Sci.* 58(3-4), 233-239.

Puppe, B., Ernst, K., Schön, P. C. and Manteuffel, G. (2007). Cognitive enrichment affects behavioural reactivity in domestic pigs, *Appl. Anim. Behav. Sci.* 105(1-3), 75-86.

Puppe, B., Zebunke, M., Düpjan, S. and Langbein, J. (2012). Kognitiv-emotionale Umweltbewältigung beim Hausschwein - Herausforderung für Tierhaltung und Tierschutz, *Züchtungskunde* 84, 307-319.

Rautenberg, S., Sonoda, L. T., Fels, M., Viazzi, S., Ismayilova, G., Oczak, M., Bahr, C., Guarino, M., Vranken, E., Berckmans, D. and Hartung, J. (2013). Kognitive Aufgaben als Umweltanreicherung im Abferkelstall - erste Ansätze einer Nutzung früher Verhaltenskonditionierung von Saugferkeln zur Verminderung aggressiver Verhaltensweisen während der Aufzucht, *Züchtungskunde* 85, 376-387.

Reimert, I., Bolhuis, J. E., Kemp, B. and Rodenburg, T. B. (2014). Social support in pigs with different coping styles, *Physiol. Behav.* 129, 221–229.

Ruis, M. A. W., Te Brake, J. H. A., Engel, B., Buist, W. G., Blokhuis, H. J. and Koolhaas, J. M. (2001). Adaptation to social isolation acute and long-term stress responses of growing gilts with different coping characteristics, *Physiol. Behav.* 73(4), 541–551.

Salazar, L. C., Ko, H. L., Yang, C. H., Llonch, L., Manteca, X., Camerlink, I. and Llonch, L. (2018). 'Early socialisation as a strategy to increase piglets' social skills in intensive farming conditions', *Appl. Anim. Behav. Sci.* 206, 25–31.

Samarakone, T. S. and Gonyou, H. W. (2009). Domestic pigs alter their social strategy in response to social group size, *Appl. Anim. Behav. Sci.* 121(1), 8–15.

Scott, K., Chennells, D. J., Campbell, F. M., Hunt, B., Armstrong, D., Taylor, L., Gill, B. P. and Edwards, S. A. (2006). The welfare of finishing pigs in two contrasting housing systems: fully-slatted versus straw-bedded accommodation, *Livest. Sci.* 103(1–2), 104–115.

Sneddon, I. A., Beattie, V. E., Dunne, L. and Neil, W. (2000). The effect of environmental enrichment on learning in pigs, *Anim. Welf.* 9, 373–383.

Sonoda, L. T., Fels, M., Rautenberg, S., Viazzi, S., Ismayilova, G., Oczak, M., Bahr, C., Guarino, M., Vranken, E., Berckmans, D. and Hartung, J. (2013). Cognitive enrichment in piglet rearing: an approach to enhance animal welfare and to reduce aggressive behaviour, *IRSN Vet. Sci.* 2013, 389186.

Stolba, A. and Wood-Gush, D. G. M. (1989). The behaviour of pigs in a semi-natural environment, *Anim. Sci.* 48(2), 419–425.

Studnitz, M., Jensen, M. B. and Pedersen, L. J. (2007). Why do pigs root and in what will they root? A review on the exploratory behaviour of pigs in relation to environmental enrichment, *Appl. Anim. Behav. Sci.* 107(3–4), 183–197.

Tatemoto, P., Bernardino, T., Alves, L., Cristina de Oliveira Souza, A., Palme, R. and José Zanella, A. (2019). Environmental enrichment for pregnant sows modulates HPA-axis and behavior in the offspring, *Appl. Anim. Behav. Sci.* 220, 104854. https://www.sciencedirect.com/science/article/abs/pii/S0168159118305768

Telkänranta, H., Bracke, M. B. M. and Valros, A. (2014). Fresh wood reduces tail and ear biting and increases exploratory behaviour in finishing pigs, *Appl. Anim. Behav. Sci.* 161, 51–59.

Tuchscherer, M., Kanitz, E., Puppe, B., Hameister, T. and Tuchscherer, A. (2014). Social support modulates splenocyte glucocorticoid sensitivity in piglets exposed to social deprivation stress, *Physiol. Behav.* 131, 25–32.

Tuchscherer, M., Puppe, B., Tuchscherer, A. and Kanitz, E. (1998). Effects of social status after mixing on immune, metabolic, and endocrine responses in pigs, *Physiol. Behav.* 64(3), 353–360.

Tuchscherer, M., Kanitz, E., Tuchscherer, A. and Puppe, B. (2016). Effects of social support on glucocorticoid sensitivity of lymphocytes in socially deprived piglets, *Stress* 19(3), 325–332.

Turner, S. P., Horgan, G. W. and Edwards, S. A. (2001). Effect of social group size on aggressive behaviour between unacquainted domestic pigs, *Appl. Anim. Behav. Sci.* 74(3), 203–215.

van der Staay, F. J., van Zuphen, J. A., de Ridder, M. M. and Nordquist, R. E. (2017). Effect of environmental enrichment on decision-making behavior in pigs, *Appl. Anim. Behav. Sci.* 194, 14–23.

van de Weerd, H. and Ison, S. (2019). Providing effective environmental enrichment to pigs: how far have we come?, *Animals* 9(5), E254.

van de Weerd, H. A. and Day, J. E. L. (2009). A review of environmental enrichment for pigs housed in intensive housing systems, *Appl. Anim. Behav. Sci.* 116(1), 1-20.

van de Weerd, H. A., Docking, C. M., Day, J. E. L., Avery, P. J. and Edwards, S. A. (2003). A systematic approach towards developing environmental enrichment for pigs, *Appl. Anim. Behav. Sci.* 84(2), 101-118.

van de Weerd, H. A., Docking, C. M., Day, J. E. L., Breuer, K. and Edwards, S. A. (2006). Effects of species-relevant environmental enrichment on the behaviour and productivity of finishing pigs, *Appl. Anim. Behav. Sci.* 99(3-4), 230-247.

Wattanakul, W., Stewart, A. H., Edwards, S. A. and English, P. R. (1997). Effects of grouping piglets and changing sow location on suckling behaviour and performance, *Appl. Anim. Behav. Sci.* 55(1-2), 21-35.

Weary, D. M., Pajor, E. A., Bonenfant, M., Ross, S. K., Fraser, D. and Kramer, D. L. (1999). Alternative housing for sows and litters: 2. Effects of a communal piglet area on pre- and post-weaning behaviour and performance, *Appl. Anim. Behav. Sci.* 65(2), 123-135.

Weary, D. M., Pajor, E. A., Bonenfant, M., Fraser, D. and Kramer, D. L. (2002). Alternative housing for sows and litters: Part 4. Effects of sow-controlled housing combined with a communal piglet area on pre- and post-weaning behaviour and performance', *Appl. Anim. Behav. Sci.* 76(4), 279-290.

Weary, D. M., Jasper, J. and Hötzel, M. J. (2008). Understanding weaning distress, *Appl. Anim. Behav. Sci.* 110(1-2), 24-41.

Zebunke, M., Langbein, J., Manteuffel, G. and Puppe, B. (2011). Autonomic reactions indicating positive affect during acoustic reward learning in domestic pigs, *Anim. Behav.* 81(2), 481-489.

Zebunke, M., Puppe, B. and Langbein, J. (2013). Effects of cognitive enrichment on behavioural and physiological reactions in pigs, *Physiol. Behav.* 118, 70-79.

Zonderland, J. J., Wolthuis-Fillerup, M., Van Reenen, C. G., Bracke, M. B. M., Kemp, B., den Hartog, L. A. and Spoolder, H. A. M. (2008). Prevention and treatment of tail biting in weaned piglets, *Appl. Anim. Behav. Sci.* 110(3-4), 269-281.

Part 4

Assessment of welfare states

Chapter 12

Physiological and behavioral responses to disease in pigs

M. D. Pairis-Garcia and B. K. Wagner, North Carolina State University, USA

1 Introduction
2 Physiological responses to disease
3 Behavioral responses to disease
4 Strategies for disease mitigation
5 Conclusion
6 Future trends in research
7 Where to look for further information
8 References

1 Introduction

Greater interest and concern from the general public, consumers and animal product retailers specific to on-farm animal welfare has necessitated proactivity of the swine industry on an international scale. As a result, preventing poor animal welfare on-farm no longer remains only an ethical obligation, but is now required and enforced by various stakeholders via farm assessments, third-party audits and legislative initiatives. It is well understood that disease, illness and injury compromise animal welfare. Therefore, preventing, expeditiously identifying and appropriately managing the disease is critical. To accomplish this, one must first better understand the coordinated biological response that occurs when a pig experiences a disease challenge and mitigates this response through preventative care and treatment.

In 1997, Fraser and colleagues identified three primary schools defining an animal's quality of life. This includes the school of (1) biological functioning, (2) affective states and (3) natural living. These areas of study quickly became the framework for the systematic research of animal welfare within production industries, including swine. In swine systems, measuring biological functioning (e.g. health, productivity, etc.) is a common strategy utilized to quantify animal welfare at the individual and herd level.

http://dx.doi.org/10.19103/AS.2020.0081.12

The following sections will provide critical information on the physiological and behavioral responses common to pigs experiencing disease and describe animal-based outcomes directly impacted during a disease state. In addition, timely tools to prevent and monitor disease at the herd level will be discussed and disease prevention techniques described. Lastly, we will explore strategies for managing sick pigs as the critical next step when prevention is unsuccessful.

2 Physiological responses to disease

2.1 Maintaining homeostasis

Although disease affects each individual in diverse and unique ways, key physiological changes persist across all mammals regardless of etiology. These physiological changes are driven by animals experiencing stress, defined as a real or perceived threat to homeostasis (Squires 2010). In an effort to maintain homeostasis, mammals activate the hypothalamic-pituitary-adrenal (HPA) axis and sympathetic nervous system (SNS) resulting in a cascade of physiological processes. These primary responses include elevated blood glucose and increased body temperature. As a whole, the HPA axis and SNS are the primary systems associated with disease and infection and ultimately cooperate to establish and maintain homeostasis.

Briefly, activation of the HPA axis begins with recognition of a direct threat by the brain. This threat results in the release of corticotropin-releasing hormone (CRH) from the paraventricular nucleus of the hypothalamus. A signaling cascade begins, as CRH stimulates the release of adrenocorticotropic hormone (ACTH) from the anterior pituitary gland. Increases in ACTH concentrations signal the adrenal cortex to produce and release glucocorticoids (Squires 2010). Cortisol is the glucocorticoid produced in pigs and is considered one of the final effectors of the HPA axis (Tsigos and Chrousos 2002).

Conversely, SNS activation affects homeostasis through effector molecules called catecholamines. Catecholamines - epinephrine and norepinephrine - are secreted into circulation from the adrenal medulla and subsequently activate the 'fight-or-flight' response (Squires 2010).

Blood glucose concentrations are increased via activation of the HPA axis and SNS. In response to cortisol release, stored energy is broken down into intermediate metabolites, which are then converted to glucose (Cohen et al. 2012). Meanwhile, catecholamines stimulate glycogen phosphorylase activity and lipolysis while inhibiting glycogen synthase activity and lipogenesis, ultimately increasing circulating glucose and energy availability (Squires 2010).

This state, known as hyperglycemia, is useful during illness because of the increased demand for glucose by an activated immune response. In growing pigs experiencing a disease challenge, maintenance requirements increase by

23% (Huntley et al. 2017) with the majority of this requirement difference being fulfilled by increased glucose utilization (Kvidera et al. 2017).

In addition to increased glucose utilization, fever is observed consistently across species as a standard initial physiological reaction to infection. In general, fever is beneficial as it assists immune function via promoting lymphocyte trafficking, increasing resistance to viruses and producing effector molecules (Tizard 2008). Increased body temperature also cooperates with a decreased iron concentration to inhibit pathogen growth (Hart and Hart 2019). However, if body temperatures become too great or if fever becomes chronic, severe organ damage or death can result.

In research studies assessing fever as a byproduct of disease, lipopolysaccharide (LPS) challenge is commonly used. In brief, LPS is injected intravenously and stimulates the production of pro-inflammatory cytokines 1–2 h post-injection (Nordgreen et al. 2018). This well-documented laboratory strategy induces sickness and prompts an increase in body temperature within the first four hours of LPS administration (Johnson and von Borell 1994). Work conducted by Johnson (2002) suggests that this increase in body temperature is highly regulated and intentional, providing an adaptive advantage to the pig by increasing the likelihood of recovery. However, fever can only be well-regulated if the animal can conserve and produce heat simultaneously. Therefore, sick pigs seek out warmer areas, such as the heat lamp in the farrowing crate or lie in physical contact with conspecifics. If environmental facilities do not support the animal's ability to conserve heat (i.e. poorly heated barns or wet-lying areas), fever cannot be achieved.

2.2 Immune system activation

In addition to the HPA axis and SNS responses, sickness results in immune system activation and increased cytokine activity. More recently, work conducted in swine has focused on identifying and validating biomarkers specific to inflammation and immune system activation response to evaluate the severity of diseases and injuries.

Cytokines, or signaling molecules responsible for cell-to-cell communication, can be produced by various cells throughout the body (van Reeth and Nauwynch 2000). Following disease or injury, pro-inflammatory cytokines, including IL-1β, IL-6 and TNF-α, are recruited to the site of tissue damage (Zhang and An 2007). This recruitment results in clinical signs of inflammation such as redness, heat, pain, swelling at the site of injury and activation of neuropathic pain-behavior pathways via histamine release (Cockram and Hughes 2018, Rosa and Fantozzi 2013). Cytokine production results in behavioral deviations to the pig as well, including anorexia, decreased exploratory behavior and increased attention to the site of injury (Zimmerman et al. 2019, Johnson 2002). In addition to these overt consequences, cytokines also stimulate the production of functionally important proteins and enzymes.

Acute-phase proteins (APP; for example, C-reactive protein, haptoglobin, serum amyloid A) are produced by hepatocytes in response to the release of pro-inflammatory cytokines by compromised cells (Barbé et al. 2011) and increase significantly 2–7 days post-challenge (Pomorska-Mól et al. 2011). These APP assist with inhibition of microbial growth and act as a negative feedback for the inflammatory response (Jain et al. 2011). When assessing disease challenges in pigs, C-reactive protein concentration increase in response to acute inflammation (Pomorska-Mól et al. 2011, Jain et al. 2011). Additionally, metabolic enzymes serum aspartate aminotransferase (AST) and creatine kinase (CK) produced throughout the body also increase in conjunction with tissue damage and inflammation. These biomarkers have been quantified as a means to determine the severity of inflammation or tissue damage in swine experiencing either disease or injury (Ison et al. 2016).

2.3 Assessing welfare

In addition to direct physiological and immunological consequences, sickness can also negatively affect animal welfare. Pain as a byproduct of disease or injury is clinically important and can negatively impact an animal's quality of life when left uncontrolled (Hellebrekers 2000, Ison et al. 2016).

The study of pain in pigs remains challenging as there is no stand-alone biomarker for pain recognized to date and pain assessment tools lack a 'gold standard' against which to be validated (Ison et al. 2016). Substance P, a neurotransmitter that is released directly from damaged nerve fibers at the site of injury, has been explored as a potential objective biomarker for pain over the last 15 years (Guo et al. 2004, Ison et al. 2016). Unfortunately, substance P has not been demonstrated to be a reliable biomarker for pain and injury in pigs in contrast to work conducted in cattle (Ison et al. 2016).

Thus far, research has employed three approaches for assessing pain in pigs by evaluating the effect of pain on (1) general body function and maintenance (e.g. feed/water intake, weight gain and productivity), (2) physiological responses (e.g. increased cortisol, heart rate, blood pressure) and (3) behavioral changes (e.g. vocalization, activity/inactivity, play behavior and aggression). For more detail on pain assessment in pigs, please see Chapter 8.

While the search for a proven pain assessment tool continues, estimating disease and injury prevalence, both on-farm and at the abattoir, can aide in elucidating animal welfare status. On-farm animal-based measures are central to any effective animal welfare-auditing protocol and should compose the majority of any assessment tool. For example, 50% of measures assessed in the Common Swine Industry Audit tool (primary assessment audit tool used for swine in the United States) are animal-based measures (Pairis-Garcia and Moeller 2017). Similarly, animal-based measures comprise 80% of the Welfare Quality® assessment tool used in Europe (Welfare Quality® 2009).

In addition to on-farm assessments, carcass and meat inspections at the abattoir offer further insight into disease prevalence and animal welfare status. Post-mortem lesions are assessed within a herd as a disease indicator and standard scoring systems are available to evaluate disease type and severity. For example, when evaluating the severity of respiratory disease in swine, scientists can perform macroscopic assessment of lung tissue. A greater presence of firm, mottled lesions on the lungs (likely indicative of a viral disease such as porcine reproductive and respiratory syndrome (PRRS) or swine influenza) assist in quantifying the impact and chronicity of disease on the animal. For more information on lesion scoring, please see the Agriculture and Horticulture Development Board's 'Pig Health Scheme' (AHDB 2018).

3 Behavioral responses to disease

3.1 Sickness behaviors

Animals experiencing disease exhibit several distinct behaviors, known collectively as 'sickness behavior'. Generally, sickness behavior can be characterized as behaviors that prioritize immediate needs and delay activities with longer-term benefits (Fraser and Nicol 2018). This adaptive response, designed to facilitate recovery, directly impacts pig behavior (Johnson 2002, Cockram and Hughes 2018). As described previously, the LPS challenge model has been implemented to evaluate sickness behaviors in mice, rats, cattle and swine (Munsterhjelm et al. 2019, Nordgreen et al. 2018, Burdick et al. 2011, Biesmans et al. 2013, Harden et al. 2006). This model is useful as the LPS injection mirrors clinical signs of sickness identified by veterinarians and caretakers in the field, regardless of the pathogen.

It is important to recognize that sickness behaviors are not merely a response to disease, but a motivational state 'designed to enhance disease resistance and facilitate recovery' (Johnson 2002). Therefore, the motivation or needs of the animal will change when sick to encourage recovery and reduce the physical impact of the disease on the animal. Common sickness behaviors include fever, anorexia, piloerection, inactivity and decreased social interactions (Johnson 2002, Tuchscherer et al. 2006). The following sections will describe further each sickness behavior specific to the pig.

Anorexia, or severe loss of appetite, is a well-characterized clinical sign of general illness in all vertebrates (Tizard 2008, Weary et al. 2009). Decreasing feed intake can be advantageous as the animal minimizes its physiological efforts associated with digestion and diverts those resources toward immune function (Fraser and Nicol 2018). Feeding and drinking behaviors are often highlighted as the earliest indicators of sickness and are commonly monitored by producers and caretakers (Matthews et al. 2016).

Pigs infected with bacterial (e.g. *Actinobacillus pleuropneumoniae*) or viral (e.g. PRRS virus (PRRSV)) pathogens (Image 1) exhibit decreased feed intake (Sutherland et al. 2007, Swinkel et al. 1994, Cockram and Hughes 2018). The pro-inflammatory cytokine *Il-1B* is likely the main factor influencing feeding behavior as it results in a decreased motivation to eat signaled at the level of the central nervous system (Finck and Johnson 1997). In addition to decreased feed intake, some authors have also reported infection-induced reductions in water intake (Swinkel et al. 1994), however it remains unclear if this is a direct clinical sign of infection or an indirect consequence of decreased feeding behavior.

In addition to decreased feed and water intake, animals spend a greater proportion of time lying (inactive), known clinically as lethargy. This has been demonstrated in pigs infected with PRRSV and other common pathogens such as *Mycoplasma hyopneumoniae* (Escobar et al. 2007, Johnson 2002). Activity is decreased in order to minimize energy utilization (Escobar et al. 2007), as sick pigs increase maintenance requirement by 23% (Huntley et al. 2017). Given that pigs are not consuming typical feed rations, preserving energy stores via inactivity is a key factor in managing energy. Energy is also conserved, as the animal does not need to redirect its energy resources toward the breakdown and digestion of feed.

Lastly, disease and infection can affect social behaviors. Swine are generally gregarious animals and normal social behavior includes deliberate clustering

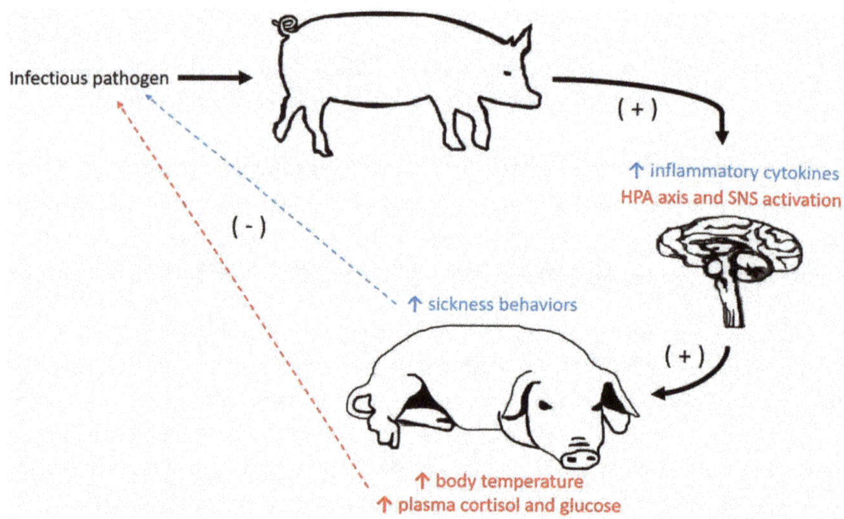

Image 1 The introduction of an infectious pathogen initiates both physiological and behavioral responses that support inhibition of the pathogen and recovery.

(Matthews et al. 2016). When sick, some animals will isolate themselves from the herd and disengage from social interaction with conspecifics. Although there is some discussion that social isolation behaviors have developed as a strategic evolutionary strategy to reduce disease transmission among a herd, decreased social behavior (e.g. play, grooming and sexual behavior) is likely the result of the animal's need to conserve energy (Deen 2010). Utilizing social behaviors as an objective tool to quantify sickness may be difficult to implement given factors such as pen size, age and temperature can alter the social behavioral dynamics of a group. In addition, given the social hierarchy of pigs is strongly established and implemented into groups, sickness behaviors may be masked to minimize and/or avoid agonistic interactions with pen mates (Hennessy et al. 2014).

The physiological and behavioral responses described above are the foundational indicators used by veterinarians and producers to detect disease in a herd. However, etiology, severity and duration of the disease process can result in deviations from the classic changes observed in the pig's physiology and behavior. In the following sections, we will briefly compare and contrast acute and chronic disease states in swine using influenza and lameness as a model.

3.1.1 Acute disease: influenza

The incidence of acute disease within a herd, even though typically short-lived, can be detrimental to animal productivity and welfare. Clinical signs and physiological changes associated with acute disease can vary dramatically, but anorexia, severe lethargy and fever are consistently observed.

The most common influenza viruses in swine are orthomyxoviruses of the influenza A group, which can result in acute respiratory disease in pigs in all stages of production (Kothalawala et al. 2006, Kaplan et al. 2015). Swine influenza can be classified as epidemic or endemic. Epidemic diseases involve an acute outbreak that moves throughout a herd rapidly, followed by rapid recovery with a low incidence of secondary bacterial infections. In contrast, endemic diseases exist permanently within a region or population and result in overt clinical signs with a high likelihood of secondary bacterial infections and morbidity. Both varieties are extremely contagious and appropriate diagnosis of either type can facilitate in a more timely treatment and management (Kothalawala et al. 2006).

The predominant IAV subtype, H1N1, is transmitted throughout a herd rapidly via nasal secretions, resulting in an acute outbreak (Neumann et al. 2009). Clinical signs of an IAV infection include coughing and sneezing, labored breathing, oculonasal discharge, fever, anorexia and prostration (Neumann et al. 2009, Detmer 2017). Though the virus is only detected within the individual pig for five to seven days post-infection, it takes two to three weeks for airways and lungs to return to normal and months for a herd to fully recover (Kothalawala et al. 2006, Janke 2014).

Though mortality is typically low in uncomplicated IAV infections (i.e., when no secondary infection is present), the economic cost of an IAV outbreak is significant. Decreased growth, feed conversion and increased medication costs all contribute to an estimated loss of US$ 3–10 per pig (Torremorell 2017). In addition to the direct economic consequences, severe acute infections can have long-term impacts on animal welfare. Secondary health issues, such as limited lung capacity, can result in chronically ill pigs that are unable to compete with healthier pen mates (Benson and Rollin 2004).

3.1.2 Chronic disease: lameness

Although chronic disease negatively affects animal productivity and welfare in the same manner as an acute disease response, its clinical presence is more subtle and its effects more prolonged. In chronic disease states, behavioral and physiological responses of the pig are often vague, indicated by a gradual decrease and loss in body condition or productivity over time. Chronic disease can be challenging for the veterinarian to diagnose and treat, as clinical signs provide limited to no specific answers.

For commercial sow herds on an international scale, lameness, resulting in deviation to an animal's gait (Karriker et al. 2013), has been identified as one of the greatest chronic disease challenges to date. The most common cause of chronic feet and leg problems in pigs are claw lesions, infectious arthritis and traumatic injuries (Benson and Rollin 2004, Heinonen et al. 2013). It is estimated that 10–20% of pigs on commercial farms are affected by lameness (Ison et al. 2016). This estimate is based on pigs demonstrating clinical indicators of lameness (i.e. non-weight bearing on a leg) and does not take into account lameness conditions, which cannot be easily observed. A study conducted in 2007 by Knauer and colleagues demonstrated that 80% of sows evaluated at slaughter had one or more foot lesions. It is likely that lameness prevalence is underestimated on-farm as we are only identifying those animals with clinical signs indicative of moderate to severe lameness.

Chronically lame sows exhibit decreased fertility, decreased litter size and poor farrowing performance compared with their healthy counterparts (Engblom et al. 2008, Anil et al. 2009). Feet and leg problems are also the most common reason for involuntary culling (Stalder et al. 2004), with an estimated 15% of sows culled due to lameness in the United States (Schenk et al. 2010). Culling females prior to the third parity results in economic inefficiency, as pig producers are neither able to pay off individual sow costs nor capitalize on the benefits of higher sow retention rates (Stalder et al. 2003, Stalder et al. 2000).

Poor reproduction, in addition to increased labor and treatment costs associated with managing and culling lame sows, results in significant economic losses to the producer and the industry as a whole (Pluym et al. 2013).

3.2 Abnormal behaviors as a byproduct of disease

In addition to behavioral adaptations deemed 'sickness behaviors', disease is also correlated with a greater prevalence of abnormal and unwanted behaviors. It has not yet been elucidated if abnormal behaviors are a byproduct of disease or if a high prevalence of abnormal behaviors predisposes animals to becoming ill, however the reality likely involves a combination of both paradigms.

In pigs, abnormal behaviors are often destructive and directed toward pen mates (e.g. chewing and biting on ears and tail). High prevalence of these unwanted behaviors results in compromised welfare for the receiver and indicates poor welfare of the performer (Diana et al. 2019a, Brunberg et al. 2016).

Such aggressive behaviors, particularly tail biting, remain a significant productivity, health and welfare issue in intensive pig production systems. Production factors, such as air quality, feeder space, feeding schedule reliability and appropriate grouping, directly affect the prevalence of tail biting and tail lesions at slaughter (Scollo et al. 2016, Sonoda et al. 2013, Moinard et al. 2003). Tail biting prompts inflammation, greater access for pathogen introduction and subsequent infection (van Staaveren et al. 2016), resulting in serious health complications.

It is well-documented that tail-bitten pigs experience more stress and are at a greater risk of infection, prompting some researchers to assess associations between tail lesions and other systemic pathologies such as lung lesions. Some authors report a moderate correlation between viscera condemnations due to pleurisy and tail lesions (Teixeira et al. 2016), while other work suggests that tail lesion scoring cannot be used as an indicator of poor lung health in pigs (van Staaveren et al. 2016). However, the presence of severe tail lesions at slaughter is strongly correlated with carcass devaluations and reduced carcass weights (Marques et al. 2012, Teixeira et al. 2016).

4 Strategies for disease mitigation

Improving animal health, productivity and welfare by reducing the severity and impact of disease remains a collective goal among producers and veterinarians in the swine industry. This has resulted in the development of several approaches to prevent and manage disease on-farm. Strategies for disease mitigation will be explored in the following sections and can be separated into two main categories: prevention and management.

4.1 Prevention

Prevention strategies aim to avoid the incidence of a disease event and include implementing protocols to both prevent and control the introduction and transmission of disease within a system. Vaccination protocols and biosecurity measures are the primary methods utilized in swine production to prevent the occurrence and spread of disease (Andres and Davies 2015). In addition, some research has suggested that providing pigs with a strategically complex environment (i.e. providing enrichment) can reduce disease risk.

4.1.1 Vaccination

Vaccination on the herd level is one proactive approach used to prevent disease. Briefly, vaccines aim to aid the animal in developing protective immunity (i.e. antibodies) against bacterial or viral pathogens. Vaccine mode of action, efficacy and duration of effect depend largely on the specific pathogen of concern and the formulated vaccine used.

Both live attenuated (i.e. viable pathogen with altered or reduced virulence) and killed (i.e. inactive pathogens) vaccines are available for use in swine. Live attenuated vaccines offer rapid, long-lasting immunity, but may be immunosuppressive and can result in disease. Killed vaccines are safer, but only offer short-term immunity and may require booster vaccinations (Ulmer et al. 2006). Producers should work closely with a veterinarian to determine a practical, farm-specialized vaccination protocol that balances overall risks and benefits.

Although vaccination protocols vary depending on farm and disease type, the swine industry as a whole relies heavily on vaccines for the prevention of common diseases (e.g. porcine parvovirus and Leptospirosis). In fact, implementing a strict vaccination protocol for Aujeszky's disease (i.e. Pseudorabies) over the last 20 years has resulted in complete eradication of the virus in US swine herds (Freuling et al. 2017).

Given the global challenge of antibiotic resistance and the risk for foreign animal disease introduction, disease prevention via vaccination is one of the main lines of defense and vital to securing pork producers access to export markets.

4.1.2 Physical biosecurity measures

Although a strong veterinary-client patient relationship enables producers to take advantage of vaccination to prevent disease, physical biosecurity measures are also a valuable alternative to prevent diseases when vaccines are not available (e.g. *Salmonella*) or have low efficacy (e.g. porcine epidemic diarrhea virus).

Isolation or quarantine of pigs entering the herd is highly encouraged. Quarantine allows isolated animals time to recover from transportation stress and to be closely monitored for clinical signs of infectious diseases. Ideally, the quarantine site is completely separate from the farm. In cases when this is not possible, isolated animals should be housed in a separate building at least 100 m away from other farm buildings (Andres and Davies 2015).

During the isolation period, veterinarians also have the opportunity to conduct appropriate diagnostic testing. Prior to being introduced into a herd, diagnostic results must rule out the possibility of specific diseases and the length of the quarantine period is dependent upon the cycle of the pathogen of interest (Zimmerman et al. 2019, Andres and Davies 2015). For example, conducting diagnostic testing to determine the presence or absence of PRRSV takes 7–14 days, therefore, new pigs should be quarantined for two weeks (Zimmerman et al. 2006).

Although quarantine systems are a useful strategy for disease prevention, only 33.6–60.9% of US farms consistently isolate animals upon arrival (Andres and Davies 2015). This is an area for great improvement to reduce disease entry and spread, which directly compromises pig welfare. Extension and outreach programs to educate and train producers on quarantine and biosecurity measures are needed, particularly in non-commercial systems with outdoor access.

In addition to vaccination and quarantine protocols, biosecurity procedures required on a daily basis contribute significantly to disease prevention. Pest control, appropriate waste management and regulating the movement of humans and animals all contribute in decreasing this risk.

Pests (e.g. rodents, birds, insects, feral animals etc.) can act as vectors for pathogen transmission to production herds either by introducing outside disease to a farm or by acting as a reservoir for resident strains to infect each new crop of animals. Pest control procedures can include a combination of a variety of tools (bait traps, poison etc.); however, the most effective way to combat pest-associated farm infection is through daily maintenance and management of the farm buildings and grounds. This includes maintaining good hygiene by removing feed spills and repairing any facility issues that increase exposure to pests (e.g. holes in exterior walls immediately) immediately (Andres and Davies 2015). Pest control is particularly important for swine as birds and rodents serve as vectors for several diseases, including *Brachyspira* and *Influenza* (Zimmerman et al. 2019).

Manure is another significant contributor to the spread of common diseases such as PRRSV and *Escherichia coli*. Several waste management strategies have been found to minimize disease transmission, including maintaining well-designed slatted floors, emptying the pit before moving in a new group of pigs and increasing the frequency of manure removal during the lactation period (Andres and Davies 2015).

Lastly, humans act as both mechanical and biological vectors for disease transmission and appropriately managing the movement of caretakers and visitors remains critical in minimizing disease risk (Andres and Davies 2015) (Image 2). Shower-in/out protocols, the use of personal protective equipment (e.g. disposable booties and Tyvek suits for visitors), enforcing suitable down times (i.e. waiting periods between farm visits) and restrictions on vehicle and visitor entry (e.g. tire wash or designated visitor parking) all contribute to supporting a healthy herd.

Regularly evaluating the quality of on-farm biosecurity can help to identify vulnerable risk areas through which a disease may enter and spread throughout a farm. Objective, science-based biosecurity scoring tools, such as Biocheck.UGent™ (https://www.biocheck.ugent.be), are available online. These tools help producers to assess potential risks by evaluating both internal and external biosecurity.

Briefly, internal biosecurity measures focus on reducing the spread of pathogens within a farm (e.g. footbaths between barns etc.), while external biosecurity measures reduce the spread of disease by focusing on contact points between the farm and the outside world (e.g. tire washing, sufficient downtimes etc.). Both internal and external biosecurity measures are critical in preventing disease and their use directly affects animal productivity, disease prevalence and antibiotic use on swine farms.

The relationship between swine farm biosecurity and pig health is a growing area of research interest. Greater biosecurity scores are positively associated with production parameters such as average daily gain, number of piglets born alive and average number weaned (Pandolfi et al. 2018, Postma et al. 2016). In addition, internal and external biosecurity scores have been negatively correlated with the frequency of individual pig treatment (Laanen et al. 2013),

Image 2 Bio-secure areas should be clearly marked.

as well as antimicrobial use (Postma et al. 2016), respectively. To date, emphasis has been placed on the development and maintenance of rigorous biosecurity measures as a main line of defense for current and future disease prevention.

4.1.3 Environment

Another, often overlooked strategy in preventing disease is to improve the environment in which production animals are housed and managed. The simplistic pen designs most commonly used in swine production (e.g. four walls, a floor, a feeder and a drinker) were developed based largely upon economical (i.e. inexpensive to build and maintain) and practical (i.e. easy to clean) considerations (Benson and Rollin 2004). However, such barren environments do not adequately occupy the pigs' attention and unwanted behaviors often emerge.

One concept to consider when discussing the necessity of providing animals with a complex environment is allostasis, or 'stability through change'. In contrast to homeostasis, which promotes the presence of a set point that the body must defend, allostasis acknowledges benefits of adaptable body systems. Facilitating a wide regulatory range, rather than making efforts to avoid change, may result in more fit pigs that are better able to handle health challenges (Korte et al. 2007).

One strategy used to support healthy, adaptable pigs is to increase environmental complexity. This can be done by providing enrichment as an outlet for pigs to express natural behaviors (e.g. rooting and chewing). Providing young pigs with a rooting substrate (e.g. straw, wood shavings, peat moss) reduces sickness behaviors and PRRSV severity (van Dixhoorn et al. 2016). In addition, providing toy enrichment can decrease the prevalence of unwanted behaviors such as aggression (Fu et al. 2018) and tail-biting (Lahrmann et al. 2018).

4.2 Management

Even with strict adherence to disease prevention protocols, it is inevitable that some pigs will become ill. In these instances, rapid detection, followed by appropriate management and animal care, is crucial in minimizing the impact of the disease at the individual and herd level.

4.2.1 Early detection

The first step in providing care to compromised pigs is identification, which can be challenging in larger production systems. Caretakers are often expected to perform regular visual assessments of the herd to identify clinical signs of disease and appropriately manage pigs with compromised health.

The development of education tools has been essential in training caretakers in appropriate evaluation, identification and timely treatment of diseased animals. Given that most caretakers have little to no formal education, interactive training tools have been designed to train caretakers using real-world scenarios. For example, in the 'Timely Euthanasia' training tool (Mullins et al. 2018), caretakers are provided with an image of a litter (Image 3a) and asked to identify the piglet requiring further evaluation or assistance. Upon selection of the correct piglet (Image 3b), the caretaker is provided with additional information, including observable signs of disease. In contrast to a classroom setting, these interactive training tools allow caretakers to learn in a way that increases understanding, retention and overall confidence.

However, human observation is subjective and time-consuming, often leading to some proportion of health problems going unnoticed. In addition, subclinical illness does not result in observable clinical signs. To combat these challenges and facilitate early disease detection, many producers

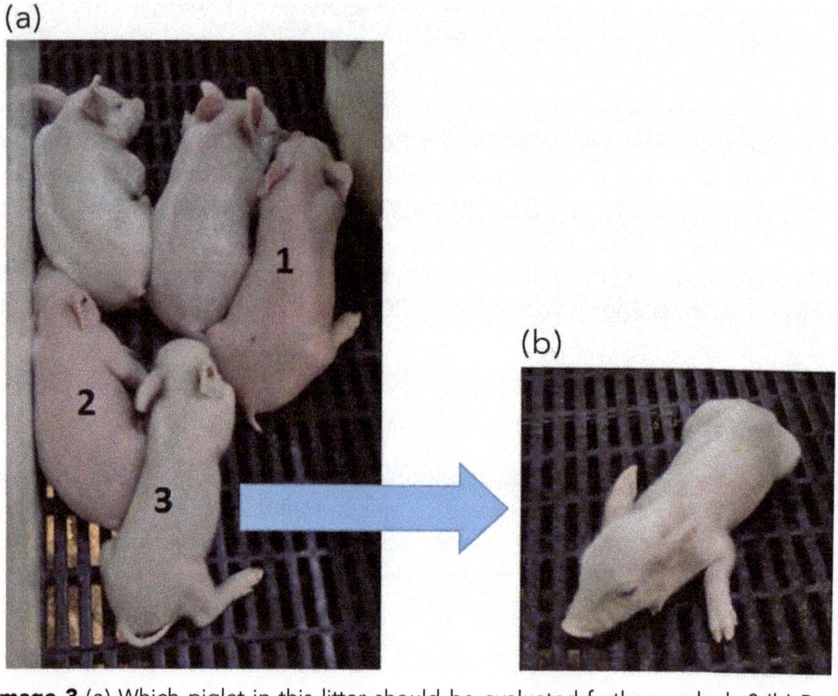

Image 3 (a) Which piglet in this litter should be evaluated further, and why? (b) Poor coloring (indicating anemia), body condition (very thin), lethargy (head down) and decreased social interaction (isolated from littermates) should alert caretakers to evaluate and monitor Piglet 3.

have implemented specific monitoring protocols that incorporate the use of technology (Matthews et al. 2016).

Monitoring changes in behavior is one effective strategy for early disease detection as behavioral changes often precede the onset of clinical signs and reflect an animal's internal state (Matthews et al. 2016, Matthews et al. 2017). As discussed previously, sick pigs will decrease feeding and drinking behaviors, socially isolate themselves from conspecifics and spend more time inactive. Precision livestock farming technologies, such as intake trackers (e.g. water flow sensors and radio frequency identification (RFID) tags), motion trackers (e.g. accelerometers) and video cameras, allow for the automated detection of these behavioral changes. Unique trade-offs exist for each detection systems, as each one varies in the type of equipment needed and amount of labor required (Matthews et al. 2016, Benjamin and Yik 2019).

Wearable accelerometers are inexpensive and allow for individualized monitoring of activity, making them among the most promising technologies for monitoring behavior in livestock. Depth cameras have been validated for detecting changes in standing and feeding durations in group-housed pigs (Matthews et al. 2017).

In addition to monitoring changes in behavior, tracking physiological changes can also enhance timely disease detection. As discussed previously, fever is the foremost initial physiological indicator of disease in mammals. Tracking changes in temperature using wearable thermistors or infrared imaging is useful for both rapid identification of fever (Benjamin and Yik 2019) and adaptive climate control (Matthews et al. 2016).

Rapid identification of illness or injury within a herd allows caretakers to remove compromised pigs and provide appropriate care in a timely manner, ultimately minimizing the impact of the disease on the individual and herd level.

4.2.2 Hospital pens

The benefits of utilizing a designated hospital pen for disease mitigation are numerous and ubiquitous. Removal of a compromised animal can positively affect the herd by reducing the risk of disease transmission to healthy herd mates. In addition, moving ill or injured animals to a designated hospital pen also increases the individual's chance of survival and recovery as it decreases the likelihood that animals will struggle to compete for resources such as feed, water and resting areas.

Although hospital pens are commonly used in swine systems across the world, significant inconsistencies in pen features (e.g. floor surface, pen cleanliness and enrichment) can result in inadequate or inappropriate space utilization (Pierozan et al. 2017). Ideally, hospital pens should be clean, well-bedded and located in an area where caretakers pass-by frequently. These features, combined with the less-crowded conditions, allow for a comfortable environment where individual pigs can be monitored more closely (Zimmerman et al. 2019).

Hospital pen requirements are highly variable between countries, leading to inconsistencies across the swine industry regarding how sick pigs are managed. Although all European Union (EU) countries have legislation regarding hospital pens, some EU countries such as Denmark, Holland and Germany have requirements that are more specific. For example, Denmark has enacted a law outlining required specifications regarding bedding, climate control and stocking density (Thomsen et al. 2016, Pierozan et al. 2017). Although utilizing a hospital pen for ill or injured animals is beneficial, further work is needed to refine standards for quality, space and pen design.

4.2.3 Pharmaceutical interventions for disease management

In addition to moving compromised animals to a designated hospital pen in order to provide comfort and improve survival, pharmacological options for disease management are also available to support active recovery. Antibiotics, or antibacterials, are the most commonly used pharmaceutical intervention available for treating infections and improving disease outcomes for pigs experiencing a health challenge.

Traditional swine production uses antibiotics to (1) treat disease therapeutically, (2) prevent disease prophylactically (i.e. individual animal) or metaphylactically (i.e. whole pen or herd) or (3) promote growth (Barton 2014). However, the emergence of antimicrobial resistance has resulted in a global health crisis, encouraging animal scientists, veterinarians and producers to reevaluate and decrease overall antibiotic use.

Today, promoting responsible and judicious use of antibiotics is not only an ethical responsibility but can also be economically beneficial. The use of prophylactic antibiotics increases input costs and may not improve herd production or health. Diana and colleagues (2019b) reported no differences between pigs provided with in-feed antibiotics and pigs that did not receive in-feed antibiotics in average daily gain, average daily feed intake and body weight in the finisher stage. In addition, consumer and retailer demand for antibiotic-free options continues to grow. In 2016, a meta-analysis by Janssen and colleagues revealed a trend among consumer research that indicated a greater willingness-to-pay for products that claim 'no antibiotic use' or 'antibiotic-free' (Janssen et al. 2016).

Antibiotic-free systems create unique challenges for producers because no animal system can completely prevent the occurrence of disease. From an animal welfare perspective, it is unethical not to provide sick pigs with necessary treatments. Therefore, antibiotic-free systems must maintain two animal flows, one that maintains the 'antibiotic-free' standards and an alternative flow to allow for the marketing of animals that required treatment. Given this trend toward antibiotic-free systems, much effort will be needed by researchers and producers to develop alternative management practices that focus on better disease prevention to minimize reliance on antibiotic treatments.

4.2.4 Timely euthanasia

Despite prevention, management and treatment efforts, it is inevitable that a subpopulation of animals will become ill to the extent that recovery is unlikely or impossible. In these instances, euthanasia, derived from the Greek terms 'eu' (good) and 'thanatos' (death), should be considered. Euthanasia is a multifactorial, decision-making process that must take into account the welfare of the animal. Euthanasia, which is commonly performed by caretakers on-farm, can be divided into three main steps: (1) identifying and providing supportive care to the compromised animal, (2) making the decision to euthanize when the animal fails to improve and (3) performing humane euthanasia in an appropriate and timely manner.

To ensure that euthanasia is being performed correctly and in a timely manner, producers should develop and enforce written euthanasia protocols that outline common timelines and appropriate euthanasia techniques. Although each euthanasia decision is case-dependent, some conditions or circumstances are unequivocal and warrant immediate euthanasia (Table 1). Training animal caretakers on each step in the euthanasia process is critical to promote uniform decision-making and safeguarding animal welfare.

Once the decision to euthanize has been made, several euthanasia techniques such as penetrating captive bolt and CO_2 are available and approved for use in swine. The age and/or size of the pig being euthanized will influence which technique is most appropriate. Caretakers should reference the written, on-farm euthanasia protocol to make this determination for each production

Table 1 Case study examples in which euthanasia is warranted

Clinical conditions	Case study descriptions
Injury	Piglet with a broken shoulder
	Second parity sow with a shoulder sore extending to the bone
Non-ambulatory	Gilt that cannot stand or walk without assistance
	Finisher pig with a torn, swollen and infected dewclaw the pig has progressed from non-weight bearing to non-ambulatory overnight
Prolapse	Multiparous sow with a rectal prolapse that has become necrotic
	A recently farrowed sow with a fresh uterine prolapse

stage. For more information on specific swine euthanasia guidelines, please see the *On-Farm Euthanasia of Swine* guidelines provided by the National Pork Board and the American Association of Swine Veterinarians (NPB 2008) and the interactive training tools available through the Common Swine Industry Audit (https://lms.pork.org/CommonIndustryAudit).

Based on prevalence and frequency, on-farm euthanasia of individual pigs, as discussed above, is an important topic to examine. However, as the risk of foreign animal disease outbreaks looms over the swine industry, mass depopulation is a critical research area. Mass depopulation, or 'welfare slaughter', is the euthanasia of healthy pigs to prevent the spread of highly contagious diseases in instances of epidemic outbreaks.

One example of a prolific enteric pathogen that necessitated mass depopulation is PEDV. Following its discovery by British veterinarians in 1971, epidemic outbreaks of PED have been confirmed throughout Asia (first confirmed epidemic outbreak: 1982), Europe (Italy: 2006; Germany and France: 2014; Belgium: 2015) and North America (US: 2013; Canada and Mexico: 2014) (Lee 2015, Kochhar 2014). In 2013, PED spread to 13 states and affected more than 200 herds throughout the United States in less than two months (Scott et al. 2016). In response to this unfortunate experience, the global swine industry has strengthened biosecurity efforts and identified the serious need for more effective means of mass depopulation.

Caretaker involvement in the euthanasia process, whether on an individual animal or a mass depopulation basis, can result in serious psychological consequences. A study by Mullins and colleagues (2017) determined that both the caring-killing paradox and compassion fatigue play important roles in the decision-making process for euthanasia and directly impact the timeliness of the procedure. Having clear, on-farm euthanasia protocols in place can help caretakers emotionally by removing some of the psychological pressure associated with making euthanasia decisions.

5 Conclusion

In summary, as animal production continues to intensify to meet the growing demand for animal protein, animal health and welfare continues to be an important topic of discussion.

Preventative measures can impede health challenges; however, some amount of disease is unavoidable. Physiological changes and behavioral responses to disease challenges can be assessed to enable early detection. Timely and appropriate disease management is critical to promote survivability and facilitate recovery. Furthermore, when animals fail to recover, euthanasia should be utilized as a tool to eliminate suffering and pain. Moving forward, swine production can be positively impacted by utilizing a comprehensive approach to assess and improve pig health and welfare on-farm.

6 Future trends in research

The future of swine production research will continue to be strongly influenced by the preeminent threat of foreign animal disease outbreaks. The devastation experienced as a result of such epidemics is undeniable and ultimately damage the swine industry on a global level. Severe financial consequences associated with international trade regulations and legislation following disease outbreaks threaten the availability of pork products worldwide, thus directly impacting the viability of the swine industry.

While researchers continue to search for vaccines for prevention and alternatives to antibiotics for disease treatment, heightened biosecurity measures can help to safeguard against foreign animal diseases. Producers should remain proactive and improve on-farm practices when possible. Overall, protecting pig health, on both the farm and global level, is everyone's responsibility.

7 Where to look for further information

In addition to the supplemental resources provided throughout this chapter, further information on specific diseases in swine can be found in *Diseases of Swine* (Zimmerman et al. 2019).

8 References

AHDB (2018). *Pig Health Scheme - Healthy Pigs for Healthy Profits*, Stoneleigh: AHDB Pork.

Andres, V. M. and Davies, R. H. (2015). Biosecurity measures to control Salmonella and other infectious agents in pig farms: a review, *Comprehensive Reviews in Food Science and Food Safety* 14(4), 317-335.

Anil, S. S., Anil, L. and Deen, J. (2009). Effect of lameness on sow longevity, *Journal of the American Veterinary Medical Association* 235(6), 734-738.

Barbé, F., Atanasova, K. and Van Reeth, K. (2011). Cytokines and acute phase proteins associated with acute swine influenza infection in pigs, *Veterinary Journal* 187(1), 48-53.

Barton, M. D. (2014). Impact of antibiotic use in the swine industry, *Current Opinion in Microbiology* 19, 9-15.

Benjamin, M. and Yik, S. (2019). Precision livestock farming in swine welfare: a review for swine practitioners, *Animals(Basel)* 9(4), 133.

Benson, G. J. and Rollin, B. E. (2004). *The Well-Being of Farm Animals: Challenges and Solutions*, Ames, IA: Blackwell Publishing.

Biesmans, S., Meert, T. F., Bouwknecht, J. A., Acton, P. D., Davoodi, N., De Haes, P., Kuijlaars, J., Langlois, X., Matthews, L. J., Ver Donck, L., Hellings, N. and Nuydens, R. (2013).

Systemic immune activation leads to neuroinflammation and sickness behavior in mice, *Mediators of Inflammation* 2013, 271359.

Brunberg, E. I., Rodenburg, T. B., Rydhmer, L., Kjaer, J. B., Jensen, P. and Keeling, L. J. (2016). Omnivores going astray: a review and new synthesis of abnormal behavior in pigs and laying hens, *Frontiers in Veterinary Science* 3(57), 1–15.

Burdick, N. C., Carroll, J. A., Hulbert, L. E., Dailey, J. W., Ballou, M. A., Randel, R. D., Willard, S. T., Vann, R. C. and Welsh, T. H., Jr. (2011). Temperament influences endotoxin-induced changes in rectal temperature, sickness behavior, and plasma epinephrine concentrations in bulls, *Innate Immunity* 17(4), 355–364.

Cockram, M. S. and Hughes, B. O. (2018). Health and disease. In: Appleby, M. C., Olsson, I. A. S. and Galindo, F. (Eds), *Animal Welfare* (3rd edn.), Cambridge: Cambridge University Press/CAB International.

Cohen, S., Janicki-Deverts, D., Doyle, W. J., Miller, G. E., Frank, E., Rabin, B. S. and Turner, R. B. (2012). Chronic stress, glucocorticoid receptor resistance, inflammation, and disease risk, *Proceedings of the National Academy of Sciences of the United States of America* 109(16), 5995–5999.

Deen, J. (2010). *Pigs: Behavior and Welfare Assessment, Encyclopedia of Animal Behaviour*, Philadelphia, PA: Elsevier Inc.

Detmer, S. E. (2017). The clinical features, pathobiology, and epidemiology of influenza infections in pigs. In: Swayne, D. E. (Ed.), *Animal Influenza* (2nd edn.), Hoboken, NJ: John Wiley & Sons, Inc.

Diana, A., Boyle, L. A., García Manzanilla, E., Leonard, F. C. and Calderón Díaz, J. A. (2019a). Ear, tail and skin lesions vary according to different production flows in a farrow-tofinish pig farm, *Porcine Health Management* 5(19), 1–9.

Diana, A., Boyle, L. A., Leonard, F. C., Carroll, C., Sheehan, E., Murphy, D. and Manzanilla, E. G. (2019b). Removing prophylactic antibiotics from pig feed: how does it affect their performance and health?, *BMC Veterinary Research* 15(1), 67.

Engblom, L., Lundeheim, N., Strandberg, E., Schneider, Mdel P., Dalin, A. M. and Andersson, K. (2008). Factors affecting length of productive life in Swedish commercial sows, *Journal of Animal Science* 86(2), 432–441.

Escobar, J., Van Alstine, W. G., Baker, D. H. and Johnson, R. W. (2007). Behaviour of pigs with viral and bacterial pneumonia, *Applied Animal Behaviour Science* 105(1–3), 42–50.

Finck, B. N. and Johnson, R. W. (1997). Anorexia, weight loss and increased plasma interleukin-6 caused by chronic intracerebroventricular infusion of interleukin-1beta in the rat, *Brain Research* 761(2), 333–337.

Fraser, D. and Nicol, C. J. (2018). Preference and motivation research. In: Appleby, M. C., Olsson, I. A. S. and Galindo, F. (Eds), *Animal Welfare* (3rd edn.), Cambridge: Cambridge University Press/CAB International.

Freuling, C. M., Müller, T. F. and Mettenleiter, T. C. (2017). Vaccines against pseudorabies virus (PrV), *Veterinary Microbiology* 206, 3–9.

Fu, L., Zhou, B., Li, H., Schinckel, A. P., Liang, T., Chu, Q., Li, Y. and Xu, F. (2018). Teeth clipping, tail docking and toy enrichment affect physiological indicators, behaviour and lesions of weaned pigs after re-location and mixing, *Livestock Science* 212, 137–142.

Guo, T. Z., Offley, S. C., Boyd, E. A., Jacobs, C. R. and Kingery, W. S. (2004). Substance P signaling contributes to the vascular and nociceptive abnormalities observed in a tibial fracture rat model of complex regional pain syndrome type I, *Pain* 108(1–2), 95–107.

Harden, L. M., du Plessis, I., Poole, S. and Laburn, H. P. (2006). Interleukin-6 and leptin mediate lipopolysaccharide-induced fever and sickness behavior, *Physiology and Behavior* 89(2), 146–155.

Hart, B. L. and Hart, L. A. (2019). Sickness behavior in animals: implications for health and wellness. In: Choe, J. C. (Ed.), *Encyclopedia of Animal Behavior* (2nd edn.), Philadelphia, PA: Elsevier Ltd.

Heinonen, M., Peltoniemi, O. and Valros, A. (2013). Impact of lameness and claw lesions in sows on welfare, health and production, *Livestock Science* 156(1-3), 2-9.

Hellebrekers, L. J. (2000). *Animal Pain: A Practice-Oriented Approach to an Effective Pain Control in Animals*, Utrecht, Netherlands: Van Der Wees Uitgeverij.

Hennessy, M. B., Deak, T. and Schiml, P. A. (2014). Sociality and sickness: have cytokines evolved to serve social functions beyond times of pathogen exposure?, *Brain, Behavior, and Immunity* 37, 15–20.

Huntley, N. F., Nyachoti, C. M. and Patience, J. F. (2017). Immune system stimulation increases nursery pig maintenance energy requirements, *Journal of Animal Science* 95(Suppl. 2), 68–69.

Ison, S. H., Lutton, R. E., Di Giminiani, P. and Rutherford, K. M. D. (2016). A review of pain assessment in pigs, *Frontiers in Veterinary Science* 3(108), 1–16.

Jain, S., Gautam, V. and Naseem, S. (2011). Acute-phase proteins: as diagnostic tool, *Journal of Pharmacy and Bioallied Sciences* 3(1), 118-127.

Janke, B. H. (2014). Influenza A virus infections in swine: pathogenesis and diagnosis, *Veterinary Pathology* 51(2), 410–426.

Janssen, M., Rödiger, M. and Hamm, U. (2016). Labels for animal husbandry systems meet consumer preferences: results from a meta-analysis of consumer studies, *Journal of Agricultural and Environmental Ethics* 29(6), 1071–1100.

Johnson, R. W. (2002). The concept of sickness behavior: a brief chronological account of four key discoveries, *Veterinary Immunology and Immunopathology* 87(3–4), 443-450.

Johnson, R. W. and von Borell, E. (1994). Lipopolysaccharide-induced sickness behavior in pigs is inhibited by pretreatment with Idomethacin, *Journal of Animal Science* 72(2), 309-314.

Kaplan, B. S., DeBeauchamp, J., Stigger-Rosser, E., Franks, J., Crumpton, J. C., Turner, J., Darnell, D., Jeevan, T., Kayali, G., Harding, A., Webby, R. J. and Lowe, J. F. (2015). Influenza virus surveillance in coordinated swine production systems, United States, *Emerging Infectious Diseases* 21(10), 1834–1836.

Karriker, L. A., Abbell, C. E., Pairis-Garcia, M. D., Holt, W. A., Sun, G., Coetzee, J. F., Johnson, A. K., Hoff, S. J. and Stalder, K. J. (2013). Validation of a lameness model in sows using physiological and mechanical measurements, *Journal of Animal Science* 91(1), 130-136.

Knauer, M., Stalder, K. J., Karriker, L., Baas, T. J., Johnson, C., Serenius, T., Layman, L. and McKean, J. D. (2007). A descriptive survey of lesions from cull sows harvested at two Midwestern U.S. facilities, *Preventive Veterinary Medicine* 82(3–4), 198–212.

Kochhar, H. (2014). *Porcine Epidemic Diarrhoea – Current Global Situation and Possible Threat for Europe*, Pairis: The World Organisation for Animal Health (OIE).

Korte, S. M., Olivier, B. and Koolhaas, J. M. (2007). A new animal welfare concept based on allostasis, *Physiology and Behavior* 92(3), 422–428.

Kothalawala, H., Toussaint, M. J. M. and Gruys, E. (2006). An overview of swine influenza, *Veterinary Quarterly* 28(2), 46–53.

Kvidera, S. K., Horst, E. A., Mayorga, E. J., Sanz-Fernandez, M. V., Abuajamieh, M. and Baumgard, L. H. (2017). Estimating glucose requirements of an activated immune system in growing pigs, *Journal of Animal Science* 95(11), 5020-5029.

Laanen, M., Persoons, D., Ribbens, S., de Jong, E., Callens, B., Strubbe, M., Maes, D. and Dewulf, J. (2013). Relationship between biosecurity and production/antimicrobial treatment characteristics in pig herds, *Veterinary Journal* 198(2), 508-512.

Lahrmann, H. P., Hansen, C. F., D'Eath, R. B., Busch, M. E., Nielsen, J. P. and Forkman, B. (2018). Early intervention with enrichment can prevent tail biting outbreaks in weaner pigs, *Livestock Science* 214, 272-277.

Lee, C. (2015). Porcine epidemic diarrhea virus: an emerging and re-emerging epizootic swine virus, *Virology Journal* 12, 193.

Marques, B. M. F. P. P., Bernardi, M. L., Coelho, C. F., Almeida, M., Morales, O. E., Mores, T. J., Borowski, S. M. and Barcellos, D. E. S. N. (2012). Inluence of tail biting on weight gain, lesions and condemnations at slaughter of inishing pigs, *Brazilian Journal of Veterinary Research* 32(10), 967-974.

Matthews, S. G., Miller, A. L., Clapp, J., Plötz, T. and Kyriazakis, I. (2016). Early detection of health and welfare compromises through automated detection of behavioural changes in pigs, *The Veterinary Journal* 217, 43-51.

Matthews, S. G., Miller, A. L., Plötz, T. and Kyriazakis, I. (2017). Automated tracking to measure behavioural changes in pigs for health and welfare monitoring, *Scientific Reports* 7(1), 17582.

Moinard, C., Mendl, M., Nicol, C. J. and Green, L. E. (2003). A case control study of on-farm risk factors for tail biting in pigs, *Applied Animal Behaviour Science* 81(4), 333-355.

Mullins, C. R., Pairis-Garcia, M. D., Campler, M. R., Anthony, R., Johnson, A. K., Coleman, G. J. and Rault, J. L. (2018). Teaching tip: the development of an interactive computer-based training program for timely and humane on-farm pig euthanasia, *Journal of Veterinary Medical Education* 45(3), 405-412.

Mullins, C. R., Pairis-Garcia, M. D., George, K. A., Anthony, R., Johnson, A. K., Coleman, G. J., Rault, J.-L. and Millman, S. T. (2017). Determination of swine euthanasia criteria and analysis of barriers to euthanasia in the United States using expert opinion, *Animal Welfare* 26(4), 449-459.

Munsterhjelm, C., Nordgreen, J., Aae, F., Heinonen, M., Valros, A. and Janczak, A. M. (2019). Sick and grumpy: changes in social behaviour after a controlled immune stimulation in group-housed gilts, *Physiology and Behavior* 198, 76-83.

Neumann, E. J., Ramirez, A. and Schwartz, K. J. (2009). *Swine Disease Manual* (4th edn.), Perry, IA: American Association of Swine Veterinarians.

Nordgreen, J., Munsterhjelm, C., Aae, F., Popova, A., Boysen, P., Ranheim, B., Heinonen, M., Raszplewicz, J., Piepponen, P., Lervik, A., Valros, A. and Janczak, A. M. (2018). The effect of lipopolysaccharide (LPS) on inflammatory markers in blood and brain and on behavior in individually-housed pigs, *Physiology and Behavior* 195, 98-111.

NPB (2008). *On-Farm Euthanasia of Swine - Recommendations for the Producer*, Perry, IA: American Association of Swine Veterinarians.

Pairis-Garcia, M. and Moeller, S. J. (2017). Animal Behavior and Well-Being Symposium: the Common Swine Industry Audit: future steps to assure positive on-farm animal welfare utilizing validated, repeatable and feasible animal-based measures, *Journal of Animal Science* 95(3), 1372-1381.

Pandolfi, F., Edwards, S. A., Maes, D. and Kyriazakis, I. (2018). Connecting different data sources to assess the interconnections between biosecurity, health, welfare, and performance in commercial pig farms in Great Britain, *Frontiers in Veterinary Science* 5, 41.

Pierozan, C. R., Dias, C. P. and da Silva, C. A. (2017). Environment, facilities, and management of hospital pens in growing and finishing pig farms: a descriptive study, *Revista Brasileira de Zootecnia* 46(11), 831–838.

Pluym, L. M., Van Nuffel, A., Van Weyenberg, S. and Maes, D. (2013). Prevalence of lameness and claw lesions during different stages in the reproductive cycle of sows and the impact on reproduction results, *Animal* 7(7), 1174–1181.

Pomorska-Mól, M., Markowska-Daniel, I., Kwit, K., Stępniewska, K. and Pejsak, Z. (2011). Kinetics of the response of four positive acute phase proteins in pigs experimentally infected with toxigenic Pasteurella multocida, *Veterinary Microbiology* 152(3–4), 429–435.

Postma, M., Backhans, A., Collineau, L., Loesken, S., Sjolund, M., Belloc, C., Emanuelson, U., Grosse Beilage, E., Nielsen, E. O., Stark, K. D. C. and Dewulf, J. (2016). Evaluation of the relationship between the biosecurity status, production parameters, herd characteristics and antimicrobial usage in farrow-to-finish pig production in four EU countries, *Porcine Health Management* 2, 9.

Rosa, A. C. and Fantozzi, R. (2013). Review: The role of histamine in neurogenic inflammation, *British Journal of Pharmacology* 170(1), 38–45.

Schenk, E., Merchant-Frode, J. and Lay, D. (2010). *Sow Lameness and Longevity*, Hoboken, NJ: Wiley-Blackwell.

Scollo, A., Contiero, B. and Gottardo, F. (2016). Frequency of tail lesions and risk factors for tail biting in heavy pig production from weaning to 170 kg live weight, *The Veterinary Journal* 207, 92–98.

Scott, A., McCluskey, B., Brown-Reid, M., Grear, D., Pitcher, P., Ramos, G., Spencer, D. and Singrey, A. (2016). Porcine epidemic diarrhea virus introduction into the United States: root cause investigation, *Preventive Veterinary Medicine* 123, 192–201.

Sonoda, L. T., Fels, M., Oczak, M., Vranken, E., Ismayilova, G., Guarino, M., Viazzi, S., Bahr, C., Berckmans, D. and Hartung, J. (2013). Tail biting in pigs--causes and management intervention strategies to reduce the behavioural disorder. A review, *Berliner und Münchener Tierärztliche Wochenschrift* 126(3–4), 104–112.

Squires, E. J. (2010). *Applied Animal Endocrinology* (2nd edn.), Cambridge: Cambridge University Press/CAB International North America.

Stalder, K. J., Knauer, M., Baas, T. J., Rothschild, M. F. and Mabry, J. W. (2004). Sow longevity, *Pig News and Information* 25(2), 53N–74N.

Stalder, K. J., Lacy, R. C., Cross, T. L. and Conatser, G. E. (2003). Financial impact of average parity of culled females in a breed-to-wean swine operation using replacement gilt net present value analysis, *Journal of Swine Health and Production* 11(2), 69–74.

Stalder, K. J., Lacy, R. C., Cross, T. L., Conatser, G. E. and Darroch, C. S. (2000). Net present value analysis of sow longevity and the economic sensitivity of net present value to changes in production, market price, feed cost, and replacement gilt costs in a farrow-to-finish operation, *The Professional Animal Scientist* 16(1), 33–40.

Sutherland, M. A., Niekamp, S. R., Johnson, R. W., Van Alstine, W. G. and Salak-Johnson, J. L. (2007). Heat and social rank impact behavior and physiology of PRRS-virus-infected pigs, *Physiology and Behavior* 90(1), 73–81.

Swinkel, J. M., Pijpers, A., Vernooy, J. C. M., Van Nes, A. V. and Verheijden, J. H. M. (1994). Effects of ketoprofen and flunixin in pigs experimentally infected with Actinobacillus pleuropneumoniae, *Journal of Veterinary Pharmacology and Therapeutics* 17(4), 299–303.

Teixeira, D. L., Harley, S., Hanlon, A., O'Connell, N. E., More, S. J., Manzanilla, E. G. and Boyle, L. A. (2016). Study on the association between tail lesion score, cold carcass weight, and viscera condemnations in slaughter pigs, *Frontiers in Veterinary Science* 3, 24.

Thomsen, P. T., Klottrup, A., Steinmetz, H. and Herskin, M. S. (2016). Attitudes of Danish pig farmers towards requirements for hospital pens, *Research in Veterinary Science* 106, 45–47.

Tizard, I. (2008). Sickness behavior, its mechanisms and significance, *Animal Health Research Reviews* 9(1), 87–99.

Torremorell, M. (2017). Immunity, diagnosis, and intervention strategies for influenza infections in pigs. In: Swayne, D. E. (Ed.), *Animal Influenza* (2nd edn.), Hoboken, NJ: John Wiley & Sons, Inc.

Tsigos, C. and Chrousos, G. P. (2002). Hypothalamic–pituitary–adrenal axis, neuroendocrine factors and stress, *Journal of Psychosomatic Research* 53(4), 865–871.

Tuchscherer, M., Kanitz, E., Puppe, B. and Tuchscherer, A. (2006). Early social isolation alters behavioral and physiological responses to an endotoxin challenge in piglets, *Hormones and Behavior* 50(5), 753–761.

Ulmer, J. B., Valley, U. and Rappuoli, R. (2006). Vaccine manufacturing: challenges and solutions, *Nature Biotechnology* 24(11), 1377–1383.

van Dixhoorn, I. D., Reimert, I., Middelkoop, J., Bolhuis, J. E., Wisselink, H. J., Groot Koerkamp, P. W., Kemp, B. and Stockhofe-Zurwieden, N. (2016). Enriched housing reduces disease susceptibility to co-infection with porcine reproductive and respiratory virus (PRRSV) and Actinobacillus pleuropneumoniae (A. pleuropneumoniae) in young pigs, *PLoS ONE* 11(9), e0161832.

van Reeth, K. and Nauwynch, H. (2000). Proinflammatory cytokines and viral respireatory disease in pigs, *Veterinary Research* 31(2), 187–213.

van Staaveren, N., Vale, A. P., Manzanilla, E. G., Teixeira, D. L., Leonard, F. C., Hanlon, A. and Boyle, L. A. (2016). Relationship between tail lesions and lung health in slaughter pigs, *Preventive Veterinary Medicine* 127, 21–26.

Weary, D. M., Huzzey, J. M. and von Keyserlingk, M. A. G. (2009). Board-Invited Review: using behavior to predict and identify ill health in animals, *Journal of Animal Science* 87(2), 770–777.

Welfare Quality® (2009). *Welfare Quality® Assessment Protocol for Pigs (Sows and Piglets, Growing and Finishing Pigs)*, Wageningen, Netherlands: Animal Science Group (ASG) Veehouderij B.V.

Zhang, J. M. and An, J. (2007). Cytokines, inflammation and pain, *International Anesthesiology Clinics* 45(2), 27–37.

Zimmerman, J., Benfield, D., Christopher-Hennings, J., Dee, S. and Stenvenson, G. (2006), *Porcine Reproductive and Respiratory Syndrom (PRRS)*, Clive, IA: Pork Information Gateway.

Zimmerman, J. J., Karriker, L. A., Ramirez, A., Schwartz, K. J., Stevenson, G. W. and Zhang, J. (2019). *Diseases of Swine* (11th edn.), Hoboken, NJ: John Wiley & Sons, Inc.

Chapter 13

Assessing emotions in pigs: determining negative and positive mental states

Eimear Murphy, Luca Melotti and Michael Mendl, Bristol Veterinary School, University of Bristol, UK

1 Introduction
2 Behavioural indicators of emotion
3 Cognitive indicators of emotion in pigs
4 Neurophysiological indicators of emotion in pigs
5 Summary and future trends
6 Where to look for further information
7 References

1 Introduction

1.1 Definitions and concepts

The study of animal emotion is a challenging area, not least because of the confusion that is generated by the terms used in this field (Anderson and Adolphs, 2014; Barrett, 2006; LeDoux, 1996; Mendl and Paul, 2020; Paul and Mendl, 2018). The concept of emotion ultimately stems from our own human experiences of mental states that we label with 'emotion words' such as 'enjoyment', 'fear', 'anger' and 'sadness' (Ekman, 1992). Because these are such potent features of our everyday experience, we are naturally interested in whether they also exist in other species and, if so, how we might assess and even measure them. The latter question is particularly relevant to understanding and monitoring the welfare of other animals, especially if one follows the widely held view that a key determinant of an animal's welfare is its mental state, in particular feelings such as pain, fear, anxiety, pleasure and other sensations and emotions (Davis and Cheeke, 1998; Fraser et al., 1997; Duncan, 1993; Paul et al., 2020; Yeates and Main, 2008). However, conscious experiences such as feelings are essentially private and inaccessible to direct measurement by third-parties, hence generating confusion about exactly what we mean when we talk about animal emotions and their assessment - are we referring to these conscious experiences, or to something else?

http://dx.doi.org/10.19103/AS.2020.0081.13

Clear definitions are essential to avoid misunderstandings. In making such definitions, we can draw a distinction between those that are descriptive and those that are prescriptive (Paul and Mendl, 2018). Descriptive definitions are dictionary-style definitions describing, as clearly as possible, how a word is used in everyday language or in a particular context such as scientific research. Descriptive definitions of emotion terms as used in scientific/psychological literature make the following points. Emotions are short-term responses to stimuli or events that comprise changes in a range of 'components' including the subject's conscious experience (usually regarded as the key feature), and its behaviour, cognitive and neurophysiological state (Paul et al., 2005; Paul and Mendl, 2018; Scherer, 1984). These changes do not always co-occur and can even dissociate in some circumstances (Winkielman et al., 2005). Importantly for our purposes, while direct measurement of conscious experience in non-human species is not currently possible (in humans we tend to rely on language as a gold-standard metric), we are able to measure the other components and can use these as indicators of animal emotions and proxies for the feelings that they may comprise. Working out whether the species of interest has the capacity for conscious experience is a separate question with a separate and expanding literature (Paul et al., 2020).

The key feature that distinguishes emotions from other mental states is that they are *valenced* – positive or negative, pleasant or unpleasant, rewarding or punishing. Valenced states are also referred to as 'affective states', an overarching term that includes both short-term 'emotions', but also longer-term and free-floating (not directly related to any one stimulus or event) 'moods'. Events inducing short-term emotions are usually relevant to an individual's well-being and survival, and longer-term moods may reflect the cumulative experience of these shorter-term emotions (Mendl et al., 2010; Mendl and Paul, 2020).

A prominent conceptual view of human emotion is the 'core affect' model (Russell, 2003) which posits that two major underlying dimensions common to all affective states are valence (positivity or negativity) and arousal (how 'activated' the organism is). Figure 1 shows the core affect model with a few examples of discrete affective states (as described by common human emotion words) located in core affect space. This model has value in animal studies by focussing on the defining characteristic of emotion (valence) in combination with the concept of arousal which is amenable to both behavioural and physiological measurement. It also defines four fundamental affective states, one in each quadrant of core affect space (high or low arousal/positive or negative), that can be identified and measured with appropriate arousal and valence metrics. Moreover, these states do not require labelling with human emotion words, something which can be a source of much confusion and interpretational difficulty, especially in species more distantly related to us (LeDoux, 1996; Mendl and Paul, 2020).

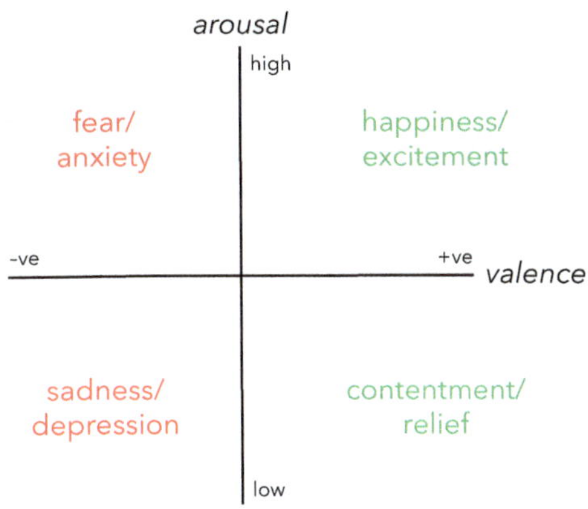

Figure 1 The core affect model with a few examples of discrete affective states (as described by common human emotion words) located in core affect space.

Most recent definitions of animal welfare stress the centrality of emotional states (e.g. Boissy et al., 2007; Dawkins, 2004; Désiré et al., 2002; Paul et al., 2005). So integral is emotion to animal welfare that Dawkins states that 'the history of animal welfare science, is, broadly speaking, the history of the search for indicators of animal emotions' (Dawkins, 2015, p. 6). However, descriptive definitions of the sort we have been considering can still leave us unsure about when an animal is in an emotional state, what that particular state might be, and even whether it is positively or negatively valenced. In seeking to develop measures of animal emotion, we thus need to have some way of providing 'ground-truth' information about what the animal's emotional state is at a particular time so that we can then identify those behavioural, neurophysiological and cognitive biomarkers that reliably co-vary with this state and hence may be valid indicators of it.

Prescriptive definitions can help here. These set out a specific set of features and conditions that precisely describe the construct that the author is referring to, 'this is what I am talking about when I use the term emotion in my research' (Izard, 2010; Paul and Mendl, 2018). A number of prescriptive definitions for animal emotion have recently been offered and three of these (Anderson and Adolphs, 2014; Dickinson and Bernard, 2010; Rolls, 2005, 2014) are reviewed in detail by Paul and Mendl (2018). Here we provide one example which builds on reinforcement learning concepts of emotion (e.g. Corr, 2008; Gray, 1975, 1987; Rolls, 2005, 2014) and can be operationalised to allow empirical investigation of the concept in animals. 'Animal affective states are elicited by rewards and punishers or their predictors. A reward is anything for which an animal will work, and a punisher is anything that it will work to escape or avoid' (Mendl and Paul, 2020; c.f. Rolls, 2005, 2014). By operationalising this prescriptive definition in terms of behaviour, it can be used to identify scenarios that, a priori, generate positively or negatively valenced affective states. 'Rewards or the absence of punishers, and associated predictions thereof, induce positive affect. Punishers or the absence of rewards, and associated predictions thereof, induce negative affect. Short-term emotion-like states follow immediately from individual rewarding or punishing events, whilst cumulative experience of events influences longer-term mood-like states' (Mendl and Paul, 2020).

Furthermore, a number of theories posit that the activity of neuro-behavioural systems involved with two fundamental processes essential to survival - Reward Acquisition (RAS) and Punishment Avoidance (PAS) - map onto core affect space as shown in Fig. 2 (Burgdorf and Panksepp, 2006; Carver, 2001; Mendl et al., 2010; Mendl and Paul, 2020; Russell and Barrett, 1999). Thus, the presence of rewards activates RAS and leads to a high-arousal positively valenced state while their absence leads to RAS deactivation and a low-arousal negative state. Likewise, punishers activate PAS resulting in high-arousal negative states while their absence leads to low-arousal positive states (Mendl and Paul, 2020). In this way, we can see how this particular prescriptive and operational definition of animal affect dovetails with the core affect concept and allows researchers to generate states lying in each quadrant of core affect space by applying/withholding behaviourally defined rewards and punishers. Behavioural, cognitive and neurophysiological biomarkers which vary reliably according to which core affect quadrant an animal is in can then be used as markers of these states. This sort of framework for inducing and 'ground-truthing' animal affective states is therefore hugely valuable for empirical research into animal emotion and, in turn, animal welfare.

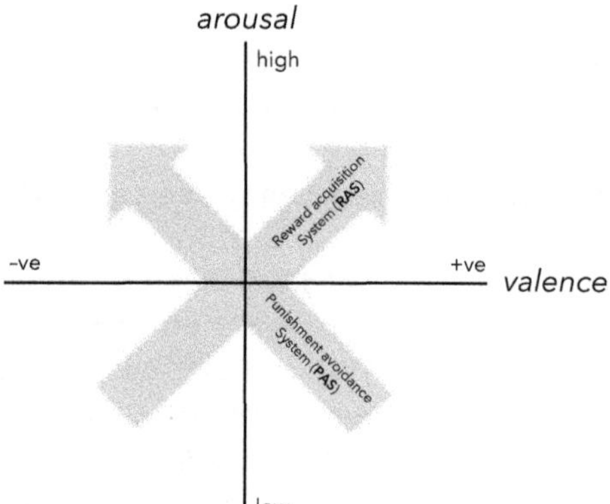

Figure 2 Reward acquisition (RAS) and punishment avoidance (PAS) processes mapped onto core affect space.

2 Behavioural indicators of emotion

2.1 Behavioural tests

Emotional responses in pigs, particularly negative emotions such as fear or anxiety, have frequently been measured using behavioural tests. Many of these tests have been adapted from research with other species, particularly rodents, and typically measure the animal's spontaneous reaction to situations involving novelty and/or social isolation. There is evidence that pigs are cautious when entering an unfamiliar space (Špinka, 2006) or demonstrate initial neophobic responses when confronted with novel stimuli. Social isolation is also considered a stressor for pigs as indicated by alterations in behaviour, physiology and the brain (Herskin and Jensen, 2000; Kanitz et al., 2005; Poletto et al., 2006; Ruis et al., 2001b; Tuchscherer et al., 2004). The advantages of such tests is that they rarely require extensive habituation or training, and require little in the way of infrastructure, making them popular on farm as quick measures of emotional responses in pigs. The disadvantage of such paradigms is that emotional responses can often be confounded by other factors, such as arousal, activity, and exploration, and thus the measures taken may lack the sensitivity to distinguish between positively and negatively valenced states. However, the inclusion of more subtle measures, specifically aimed at differentiating positive and negative states, may be a useful step forward in using such measures.

The most frequently performed of these behavioural tests with pigs is the Open Field test. Initially designed as a test for emotionality in rodents (Hall, 1934), it was later adapted for use with pigs (Beilharz and Cox, 1967). While the specific Open Field test designs used with pigs differ in a number of ways (e.g. size and shape of the test arena, the duration of the test), many features are common across tests. Typically, an animal is separated from its conspecifics and placed in an unfamiliar arena for a specified period of time, and the behavioural response is recorded. The most common behaviours recorded in studies with pigs include measures of movement/posture, vocalisation, location within the arena, explorative behaviour and elimination (Murphy et al., 2014), while less common behaviours include measures of exploration and escape attempts. Measures of location within the arena are one of the main indicators of anxiety in rodents, but there is little evidence that such measures are ethologically relevant measures of emotion in pigs (as reviewed by Forkman et al., 2007; Lind et al., 2007; Murphy et al., 2014). With respect to measures of movement/posture, there is conflicting evidence as to whether this reflects anxiety or purely activity or both. Crude measurements of physical activity by themselves are not sufficient to gauge emotional responses in the Open Field test (Rutherford et al., 2012). Instead, more subtle measures of behavioural postures and facial expressions may be better suited to tease apart emotional responses from activity and exploration, but these are rarely used. Vocalisations may be useful indicators of emotional responses (see later), but many open field studies have not distinguished between the types of vocalisations produced during the test. So while the Open Field test is likely to induce an emotional response in pigs, the measures commonly recorded are often not subtle enough to distinguish this emotional response from general arousal or activity. Few systematic approaches have been taken to validate this test and the different measures relevant to emotional responses in pigs (for an exception, see Donald et al., 2011).

Two further commonly used rodent models of anxiety have been adapted for use with pigs. The Elevated Plus Maze test involves placing the animal on an elevated four-armed platform where two arms have raised sides ('closed' arms), while two arms are more exposed ('open' arms) (Andersen et al., 2000a,b; Donald et al., 2010; Janczak et al., 2000, 2002; Rutherford et al., 2012). However, the assumption that animals should show unconditioned avoidance of the open arms does not appear to be true for pigs (Donald et al., 2010; Janczak et al., 2002), questioning the utility of this task in measuring emotional responses in pigs. Similarly the Light/Dark test, where the animal is placed in a dark compartment in a test arena and latency to enter a brighter compartment is measured (Andersen et al., 2000a,b) relies on the assumption that pigs show an unconditioned aversion to brightly lit spaces which is not the case (see Tanida et al., 1996). While both situations may provoke an emotional response due to novelty of the environment and the social isolation required to test an animal, the inherent assumptions of the tests do not appear to hold true for pigs (Murphy et al., 2014).

Another test frequently used to measure fear or anxiety responses in pigs is the Novel Object test where pigs are presented with an unfamiliar object and their latency to explore the object is measured, often along with the duration and frequency of object exploration. This test is based on the assumption that when animals encounter an unfamiliar object, they experience a conflict between their motivation to explore and a motivation to avoid novelty. Pigs are highly explorative animals, and are said to display both intrinsic exploration (i.e. motivated by curiosity) and extrinsic exploration (i.e. motivated to seek resources) (Day et al., 1995; Studnitz et al., 2007; Wood-Gush and Vestergaard, 1991). Pigs prefer novel objects to familiar items (Wood-Gush and Vestergaard, 1991), will work to gain access to exploratory material (Pedersen et al., 2005), and prefer environments where they have to forage for food rather where food is freely available (de Jonge et al., 2008). In the Novel Object test, avoidance of the stimulus (higher latencies to approach/investigate the object and lower durations of investigation) is taken to indicate higher levels of anxiety or fear.

Although the measures recorded in the Novel Object test have been relatively similar across studies, the method of presentation of the object itself has varied widely. In some studies the object was already in place in the arena before the animal/s entered, was placed by a person after the animal had entered, was slowly introduced (e.g. lowered via a rope), or was introduced suddenly into the testing environment (e.g. dropped or thrown; see Murphy et al., 2014). Further, tests have been conducted on individual animals, on groups, in the home pen, or in a novel arena with or without prior habituation. These different methods of testing and object introduction are likely to have a large effect on the emotional response of the animal. Appraisal theories pose that specific emotional responses to an event depend on characteristics including suddenness, familiarity and predictability (Veissier et al., 2009). Subtle differences in emotional responses may be masked by the suddenness with which objects are introduced into an unfamiliar arena, or by presenting items to group-housed pigs in their home pen such that they do not elicit a strong sense of neophobia. Systematic assessment of how these factors can alter the response would be a useful step towards developing a more standardised task design to facilitate comparison across studies. Further, the behavioural responses typically recorded may not be subtle enough to differentiate between differently valenced emotional responses; current research on defence cascade responses may provide a useful basis for future research.

Research into emotional responding in pigs has also looked at pigs' willingness to voluntary explore a, generally, unfamiliar human – the voluntary Human Approach test. Alternatively, studies have looked at the behavioural reaction of a pig in response to an approach by an experimenter – the Withdrawal Response test. These human interaction tests are assumed to induce an approach-avoidance conflict in the animal similar to the previous tests described above. The motivation to approach may be more dominant in the voluntary Human Approach test, while the motivation to avoid may be more dominant in the Withdrawal Response test (Waiblinger et al., 2006). While evidence suggests that pigs are motivated to approach humans (Terlouw and Porcher, 2005), the underlying motivation may be governed by exploration or aggression. Pigs with prior positive experience with humans have shorter latencies to approach (e.g. Hemsworth et al., 1994; Hemsworth et al., 1996a,b) as do pigs which have previously displayed more human-directed aggression (Marchant Forde, 2002). Further, approach latencies are less repeatable over time than withdrawal responses (Scott et al., 2009). The Withdrawal Response test therefore may provide a better measure of fear responses since it is based strongly on avoidance responses and is relatively stable over repeated testing (Murphy et al., 2014).

These behavioural tests have been, until more recently, the standard tests of emotional responses in pigs, and have focused overwhelmingly on negative emotions, such as fear or anxiety. In many cases, these tests were first developed for other species, and the inherent assumptions of the test may not be relevant for pigs (Murphy et al., 2014). While some aspects of the task designs are likely to induce an emotional response (novelty, social isolation), the common measures used do not allow for the differentiation of affective responses from levels of activity, and are not able to capture the subtle differences in behaviour which may be more indicative of emotional valence. Furthermore, there has been little systematic investigation into how aspects of task design, which can vary widely between studies, influence the responses of pigs in the test. These behavioural tests in their current form are not the best indicators of affective state in pigs without incorporating other measures to help differentiate emotion from other possible causes of the behaviour observed (Krugmann et al., 2019).

2.2 Qualitative behaviour assessment

Qualitative behaviour assessment (QBA) as proposed by Wemelsfelder et al. (2000, 2001) is an approach that sees the behaviour of the whole animal as an expression of its inner experience. This approach takes into account the behaviour of the animal as it interacts with its environment in a dynamic way, and uses it to gauge how the animal is experiencing its own situation. Human observers make qualitative assessments of the animal's behaviour and the assumption is that these qualitative judgements actually reflect the emotional experience of the animal.

QBA methodology generally follows one of two approaches. First, a free-choice profiling approach (Francoise Wemelsfelder et al., 2001) can be used, where observers rate the animal's behavioural expression according to a series of one-word descriptors (e.g. calm, distressed, agitated) they themselves generate from watching sample clips. The subsequent analysis looks for consistency across observers in the semantic meaning of the descriptors, and creates a (number of) dimensions upon which an individual animal's behavioural expression can be scored. In the second approach, often used in on-farm welfare quality protocols, instead of free-choice profiling, assessors score animals according to a fixed list of descriptors. The QBA approach has been used to study human assessment of emotional states in a variety of animal species (e.g. horses, Hintze et al., 2017; dairy cows, de Boyer des Roches et al., 2018; dogs, Arena et al., 2017) including pigs (e.g. Camerlink et al., 2016; Rutherford et al., 2012).

Aspects of reliability and validity of the QBA paradigm have been addressed in a number of studies. Studies with pigs have demonstrated that observers show a high degree of intra- and inter-observer reliability in their judgements of pigs' behavioural expression in terms of the semantic descriptors used (Czycholl et al., 2017; Francoise Wemelsfelder et al., 2001), despite potential bias due to observer background (e.g. farmer, veterinarian, animal rights activist; Francoise Wemelsfelder et al., 2012). However, assessments based on a fixed list of descriptors, as used in welfare quality protocols, were less reliable within and between observers (Czycholl et al., 2017; Temple et al., 2013) and were subject to bias due to the background/experience of observers (Duijvesteijn et al., 2014). Clarke et al. (2016) suggest that the differences in reliability may not be due to the application of free-choice profiling or fixed lists of descriptors, but how the data are analysed. While the environment in which a pig is filmed can influence observers' judgements (e.g. indoor vs. outdoor backgrounds), the effect is relatively small (Françoise Wemelsfelder et al., 2009). Observers rate pigs in extensive or enriched housing as expressing more positively valenced emotions than pigs in more intensive/barren conditions (Carreras et al., 2016b; Temple et al., 2011). Pharmacological validation of the paradigm has also been studied; observers rated pigs treated with an anxiolytic drug, Azperone, more positive on a dimension ranging from 'unsure/nervous' to 'confident/curious' than saline treated pigs (Rutherford et al., 2012). While observers did not distinguish between pigs that had either just won or just lost a fight, they did score pigs with more skin lesions (not visible on the video clips) as experiencing more negatively valenced states (Camerlink et al., 2016), indicating that observers could pick up on differences in emotional expression possibly related to pain.

The QBA approach to studying subjective states in animals has received criticism for being at risk of anthropomorphism (Würbel, 2009) and is likely to be affected by our level of similarity with the species under study. One has to accept the philosophical assumption that observers' interpretation of an emotional state in animals reflects the actual emotional state experienced by the animal. However, accepting this, the methodology is quite reliable and robust in the face of potential observer- and environmental-bias. Further, increasing evidence with different species of convergent and predictive validity of the QBA approach (e.g. Minero et al., 2018; Rutherford et al., 2012; Stockman et al., 2011) is helping to assuage critics. Application of the QBA method to assess emotion in pigs has the potential not only to discriminate between positively and negatively valenced states, but also between states of the same valence (e.g. Hintze et al., 2017). Further work aimed at specifically and systematically validating the paradigm can help establish if this is the case in research with pigs.

2.3 Vocalisations

Analysis of vocalisations has been suggested as a non-invasive method for studying the emotional state of an animal (Manteuffel et al., 2004). Vocalisations, however, can convey different messages (alarm calls, contact calls, pain) or be a reflection of arousal. They can also express an emotional state, or their nature may be altered by physical changes in the vocal apparatus due to an emotional state (Briefer, 2012). The value of specific vocalisations as measures of emotion in pigs is dependent on a number of factors - they must be honest signals (i.e. truly reflecting the animal's state), they should be specific to emotional responses, and they should allow for the distinction between negative and positive states. For example, greater hunger in piglets is signalled by a higher calling rate and calls that differ in frequency and duration (Weary and Fraser, 1995), and painful experiences, such as castration, are signalled by increased high-frequency vocalisations in piglets (Marx et al., 2003; Puppe et al., 2005; Taylor and Weary, 2000; Weary et al., 1998) suggesting some level of honest signalling.

Pigs make a wide range of calls from high-pitched 'squeals' or 'screams' to low-pitched and short 'barks', as well as low-pitched 'grunts'. High-pitched squeals and screams reflect high arousal (Held et al., 2009; Manteuffel et al., 2004) and are associated to situations of distress like social isolation, restraint, castration and food deprivation (reviewed in Held et al., 2009). Further, piglets trained to associate different tone-sequences with either a positive or a negative outcome were more likely to give high-frequency vocalisations during the anticipation period before the negative situation (Imfeld-Mueller et al., 2011). The negative valence of this type of vocalisations has been further demonstrated through pharmacological validation. Stimulation of the amygdala in pigs using acetylcholine produced high-pitched vocalisations similar to the screams associated with distress (Manteuffel et al., 2007). Conversely, pigs treated with an anxiolytic, Azaperone, vocalised less in an Open Field test than non-treated pigs (Donald et al., 2011).

Barks (low-pitched atonal calls of short duration; Chan et al., 2011) occur both during potential positive experiences, such as play, but are also used to express alarm (Newberry et al., 1988). Subtle acoustic differences in peak frequencies between barks used in contexts of play and alarm may be useful to distinguish differently valenced states (Chan, 2011; Poletto et al., 2006).

Grunts have been classified based on their duration (Marchant et al., 2001), with shorter grunts being emitted during exploration and longer grunts serving as contact calls (Leliveld et al., 2017; Marchant et al., 2001). More recent research has further investigated if emotional valence can be inferred based on the acoustic characteristics of pig grunts. Grunts emitted in contexts assumed to be positive (i.e. in an environmentally and socially enriched arena with food rewards, or in anticipation of a food reward) were significantly shorter than grunts produced in contexts assumed to be negative (i.e. during social isolation without food, or in anticipation of unpalatable food; Briefer et al., 2019; Friel et al., 2019, respectively). This finding has also been shown in wild boars (Maigrot et al., 2018). Other acoustic parameters of grunts, such as those related to the third formant (a resonance of the vocal tract that can vary based on the vocal tract length) also differ between positive and negative contexts, suggesting that pigs lower their larynx and consequently increase the length of their vocal tract when vocalising in negative situations (Briefer et al., 2019). However, acoustic parameters other than grunt duration were not found to differ between positive and negative contexts in another study, possibly due to a relatively smaller difference in emotional valence between positive and negative contexts (Friel et al., 2019). Furthermore, context (and therefore emotional valence) has been found to account for only a small part of the variation in acoustic parameters in pigs (Leliveld et al., 2016) suggesting that other factors, such as individuality (Leliveld et al., 2017) and/or motivation (Briefer, 2012), may affect pigs vocalisations.

In conclusion, looking at vocalisations within context, and using other behavioural or physiological measures which concur with the vocalisation, may allow valence to be inferred (Manteuffel et al., 2004). Further, vocalisations may even differ within the same valence allowing for more subtle differentiation between states.

2.4 Play behaviour

Play generally falls into three categories; social play between conspecifics, locomotor-rotational play and object play (Held and Špinka, 2011). Play is said to have both long-term adaptive benefits, such as providing training for future skills or unexpected situations (Pellis and Pellis, 2013; Špinka et al., 2001), as well as immediate benefits, such as conflict management, stress reduction and learning about other conspecifics (Palagi et al., 2004; Pellis and Pellis, 2013). Play has been suggested as an indicator of positive states as it generally occurs when all other needs have been met, and will reduce/cease in situations of threat (Boissy et al., 2007). However, a recent review suggests that while there is evidence that play is reduced in negative situations in normally playful animals, there is not enough research on the link between play and positive experiences to draw definite conclusions (Ahloy-Dallaire et al., 2018). In fact, play may also increase as a way of coping with challenging contexts, such as, competition (Palagi et al., 2006), social tension (Antonacci et al., 2010) or a reduction in maternal care (Bateson et al., 1990). Nevertheless, while the association between play and animal welfare is context-dependent, there is evidence that play, in itself, is rewarding, and may be accompanied by positive emotions (Held and Špinka, 2011). Indeed, during play, the endogenous opioid system is activated which modulates the hedonic properties of rewards leading to the experience of hedonic pleasure (Vanderschuren, 2010; Vanderschuren et al., 1995). Play is also contagious; thus the presence of one or few playing/playful individuals in a group can induce others to play as well (reviewed in Held and Špinka, 2011).

Pigs are highly playful animals, and as with other species, their play falls into the three categories of locomotor-rotational, object and social play. Play is more common in young pigs than adults, with the peak between two and six weeks, and can be affected by sex (reviewed in Lawrence et al., 2018). To distinguish real play from other behaviours, many authors use 'play markers' (Chalmers and Locke-Haydon, 1981), that is, indicators in a behavioural sequence which signify that the behaviour is playful. Common play markers in pigs' locomotor play include waving/tossing of the head, scampering, jumping, hopping, pawing, pivots, flops and gambolling (Bolhuis et al., 2005; Brown et al., 2018; Chan et al., 2011; Dudink et al., 2006; Newberry et al., 1988; O'Connor et al., 2010; Reimert et al., 2013). Object play in pigs entails the physical manipulation of inanimate objects using the snout or forelimbs (Horback, 2014) and often involves some of the play markers used in locomotor play such as tossing of the head. However, object manipulation can also be a form of stereotypy in pigs (Mason and Mendl, 1997), so the context in which the behaviour is occurring should be taken into account. Distinguishing social play from real conflict can be more difficult as many of the behaviours observed resemble those of antagonistic behaviour. Social play often involves one pig pushing or nudging another pig with the head or snout, mouthing parts of the other pig's body, mounting or climbing on the other pig, and chasing (Brown et al., 2018; Chaloupková et al., 2007; Dobao et al., 1985; Donaldson et al., 2002; Weller et al., 2019), and can be performed between pairs of litter/pen mates or between a piglet and a sow. The presence of play markers during these bouts, such as pivoting or head tossing, can be used as an indicator that the behaviour is playful (Newberry et al., 1988).

Research on play in pigs generally supports the conclusion that environmental conditions assumed to improve welfare are associated with increased play, especially when a comparison with a control, more barren condition is made (reviewed in Lawrence et al., 2018). Both solitary and pairs of pigs displayed locomotor and object play behaviours when given access to a compartment containing straw, digging substrate and rewards, but no play was observed when the animals were exposed to a second barren compartment (Reimert et al., 2013). Similarly, young pigs play more in enriched-housing environments compared with conventional environments (Bolhuis et al., 2005; Chaloupková et al., 2007; Martin et al., 2015), and if the ammonia concentration in their room is relatively low (<5 ppm vs. 20 ppm concentration; O'Connor et al., 2010).

Taken together, while play may not be a good indicator of past positive experiences, pigs do show more play in more stimulating environments. In understanding the link between play and affective states the interpretation of results needs to be looked at in context. Further play in itself is very likely to induce positive emotions; so while it may not reflect past experiences, environments that promote play may stimulate more experiences that are positive.

2.5 Defence cascade responses

Defence cascade responses have been proposed as a potentially promising measure of emotional valence in animals (Statham et al., 2011), and some preliminary work has been carried out with pigs. Across species, animals show an adaptive response to sudden and/or unexpected stimuli involving initial detection of the stimulus (startle response), followed by an evaluation phase (freezing response), and ending with either a defensive response or the resumption of normal behaviour (Statham et al., 2020). The startle response may represent a useful measure of emotional valence (Paul et al., 2005; Statham et al., 2020). Evidence from human and rodent research indicates that the response is increased when the subject is in a negative emotional state, but reduced when the subject is in a positive state (Statham et al., 2020). As pigs display clear startle and freeze responses to, for example, sudden and loud auditory stimuli (Blackshaw et al., 1998), followed by either fleeing or resumption of normal behaviour typical of the defence cascade response (Statham et al., 2020), interest in using these behaviours to assess emotional valence in pigs has increased.

Initial investigations revealed that, in line with predictions, pigs treated with an anxiolytic, midazolam, showed smaller startle responses and shorter freeze durations to a sudden, loud noise compared with saline-treated controls (Statham et al., 2011). Experience of prior positive or negative handling treatments, however, had no differential effect on defence cascade responses in pigs (Carreras et al., 2017). However, as the handling treatments in this study failed to influence any of the outcome measures, the possibility that the treatment itself failed to induce a difference between the groups cannot be excluded. Some factors which may modulate the response have been identified (Statham et al., 2020), such as the pigs' orientation towards the stimulus, and the baseline alertness level of the pig, and must be taken into account when using this measure to assess emotional valence in pigs. As a practical measure of emotion on farm, manual scoring of defence cascade responses are time-consuming, but efforts are underway to address this using computer vision image analysis (Statham et al., 2020).

2.6 Facial expression and body posture

In recent years there has been increased interest in the potential for facial expressions and body postures to be used as indicators of emotion and emotional valence in a wide variety of animal species, including, sheep, rats, horses, dogs and mice (Boissy et al., 2011; Caeiro et al., 2017; Dolensek et al., 2020; Finlayson et al., 2016; Hintze et al., 2016). Research into facial expressions in animals is also of interest to those wishing to investigate pain, since the first 'grimace scale' was developed for mice (Langford et al., 2010); grimace scales have been developed for a variety of species including horses, pigs, sheep and rats (e.g. Dalla Costa et al., 2014; di Giminiani et al., 2016; Häger et al., 2017; Sotocina et al., 2011; for review see McLennan et al., 2019).

Much of the research with pigs has focussed on the potential for ear and tail movements/postures to reflect positive and negative emotional valence. (Reimert et al., 2013, 2015) noted that pigs showed increased tail movement (wagging) during a rewarding event (exposure to substrate and treats in a test arena). Instead, during the aversive event (separation and short restraint), pigs were more likely to keep their ears in a backwards position, while tails remained in a more static position (curled). Similarly, Marcet Rius et al. (2018b) found that (mini) pigs showed higher tail movement duration and lower ear movement duration when presented with toys in their home pens than during a control situation. Straw provision alone however did not appear to have this effect on tail movement, but the frequency of ear movements was reduced compared to a control situation (Marcet Rius et al., 2019b). Playful interactions with enrichment and conspecifics have also been shown to be positively correlated with duration of tail movement (Marcet Rius et al., 2018a; Marcet Rius et al., 2019a). However, more intense tail wagging or tail flicking in pigs has been associated with negative experiences (such as pain or frustration; for review see, Camerlink and Ursinus, 2020). These results suggest that positive experiences in pigs (appetitive, explorative, playful) may be identified more by relaxed tail movements while negative experiences may be identified by ear position and movements.

In terms of more specific facial expressions, two approaches have been developed. First, Camerlink et al. (2018) investigated how facial expressions in pigs may change in response to different negative emotional experiences, namely aggression and fear. They identified three facial expressions involving the snout, the ears and the eyes and investigated how they changed over the course of aggressive interactions between pigs. While measures of the snout did not change throughout the interaction, pigs in retreat from the aggressive encounters, hypothesised to indicate fear, had their ears in a more backwards position and their eyes narrower when compared with before and during the aggressive encounter (Camerlink et al., 2018) Secondly, research has looked at how facial expressions may be used as indicators of pain. Di Giminiani et al., (2016) looked at how castration and tail docking affected facial expressions caused by contraction or relaxation of individual or groups of facial muscles in pigs. The facial action units that changed from pre- to post-treatment were identified and used to develop a piglet grimace scale (for sample of the Piglet Grimace Scale, see Fig. 3). This scale was then used by observers blind to the treatments to assess changes in facial expression in response to pain in neonatal piglets. It was found that piglets displayed orbital tightening, that is, narrowing of the eyes, after the tail docking procedure, indicating that, similar

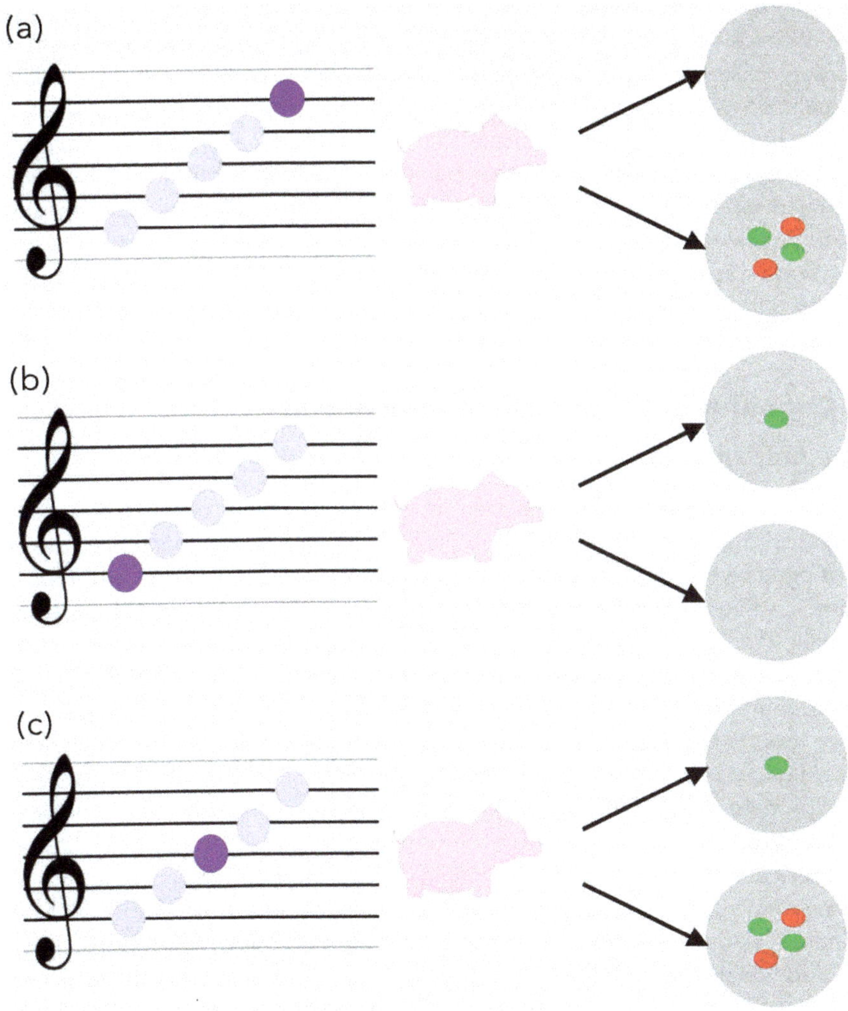

Figure 3 Example of an active choice Go-Go judgement bias task used with pigs from Murphy et al. (2013). Pigs were trained to respond to tone cues (a) predicting the presence of a large reward (positive cue) or (b) predicting the presence of a small reward (negative cue) in one of two locations. Pigs had to choose the correct location in order to obtain the reward. Once the association between cue type and reward location/quantity had been learned, (c) intermediate ambiguous tone cues were presented and the response of the pig (choice of reward location) was recorded. In response to ambiguous probe cues, a choice for the location associated with the positive cue was considered an optimistic response (i.e. expectation of a large reward), while a choice for the location associated with the negative cue was considered a pessimistic response (i.e. expectation of a small reward).

to results with other species, this may be a useful measure of painful experience in pigs (di Giminiani et al., 2016). Pigs were found to have higher scores on the grimace scale post-surgery, indicative of more pain, compared with before surgery (Vullo et al., 2020). Moreover scores were related to activity levels of piglets' post-surgery, with less-active piglets receiving higher scores (Viscardi et al., 2017). These results indicate that the pigs' grimace scales are picking up on facial expression related to pain in pigs. Such a methodology could be used to identify facial expressions related to other affective states in pigs, both positive and negative. Recent developments in automatic detection of individual pig faces using deep learning methods (Hansen et al., 2018) raise the exciting future possibility of monitoring facial expressions of identified individuals in situ on farm.

3 Cognitive indicators of emotion in pigs

3.1 *Judgement bias*

Affective states can lead to biases in various cognitive processes such as memory, attention and judgement. From human research, there is evidence that positive affective states lead to more positive judgements of ambiguous stimuli, while negative states lead to more negative judgements of the same stimuli (Blanchette and Richards, 2010). Harding et al. (2004) proposed that emotional valence in animals could be studied by measuring these differences in judgement of ambiguous stimuli, and developed the first animal-judgement bias task. Since this seminal paper, cognitive judgement biases have been investigated in a wide range of species, including birds, rodents, farm animals and primates. While these tasks have been performed in quite a variety of ways, most have, in principle, followed the original design of Harding et al. (2004). Animals are first trained on a discrimination task where the two cues along the same modality (e.g. tone frequencies; spatial location; texture gradients) offer differential outcomes (e.g. reward, punishment; reward, non-reward; large reward, small reward). Animals are trained to perform a different response depending on which cue is presented (e.g. latency to respond, Go/No-Go response; active choice, Go-Go response). Once this discrimination has been learned, animals are then presented with a series of ambiguous probe cues intermediate to the previously learned reference cues, interspersed among presentations of the reference cues. Their responses to these probe cues are scored in relation to their responses to the reference cues and interpreted as suggesting either positive or negative judgement biases. These biases are said to reflect a positive or negative underlying affective state (Mendl et al., 2009).

Two recent reviews and meta-analyses of the judgement bias research done with non-human animals have recently been published. The first has demonstrated that, based on the results from pharmacological manipulations of mood state, the paradigm demonstrates some degree of predictive validity; in general pharmacological manipulations shifted judgement bias at the probe cues in the direction predicted (Neville et al., 2020). The second demonstrated that non-pharmacological manipulations of affective state also shifted judgement bias in the direction predicted (Lagisz et al., 2020). While both of these reviews support the validity of the paradigm as a method to assess emotional valence in animals, the authors stress that some important factors need to be taken into account when using this paradigm. First, manipulations predicted to induce negative states had a larger effect than those predicted to induce positive states (Neville et al., 2020). Secondly, effects were highly variable and aspects of the task design can moderate the effects found (for details see Lagisz et al., 2020).

No systematic validation of the judgement bias paradigm has been carried out with pigs, and various task designs have been developed and implemented. Go/No-Go style tasks have been more commonly used (e.g. Carreras et al., 2015; Douglas et al., 2012; Leliveld et al., 2017) but active choice of Go/Go tasks have also been developed (e.g. Murphy et al., 2013, see also Fig. 4). Moving animals from barren to enriched housing conditions, often used to induce positive states in animals, has shown mixed effects on judgement bias in pigs; Douglas et al. (2012) found that moving pigs to enriched environment induced an optimistic bias, while Carreras et al. (2016b) found no effect. Pigs that experienced a positive-handling treatment were more likely to show an optimistic bias compared to those in negative or minimal handling treatments in one study (Brajon et al., 2015) but this result was not confirmed (Carreras et al., 2017). Social isolation was also found not to have an effect on judgement bias in piglets (Düpjan et al., 2013). Pigs classified as having a pessimistic bias also showed a longer latency to contact a novel object (Carreras et al., 2016a) demonstrating some convergent validity. Pharmacological depletion of serotonin levels in the brain was found to induce a pessimistic bias in pigs (Stracke et al., 2017a), but dietary supplementation with tryptophan to enhance brain serotonin levels also indicated a shift towards a negative bias in pigs contrary to predictions (Stracke et al., 2017b). Individual performance of pigs in repeated testing was found not to be consistent over time (Carreras et al., 2015), and pigs have shown an increasing latency to respond to unrewarded ambiguous cues with repeated testing (Murphy et al., 2013). No effects of sex on responses to ambiguous cues was found (Carreras et al., 2016a; Roelofs et al., 2017).

While the general judgement bias literature promotes the validity of the paradigm to measure emotional valence in animals (Lagisz et al., 2020; Neville et al., 2020), the results of studies with pigs are not consistent enough to promote one particular method. Much more work is needed to validate the paradigm for use with pigs and develop the optimal task design. The sensitivity of the task to pick up differences is likely affected by aspects of task design

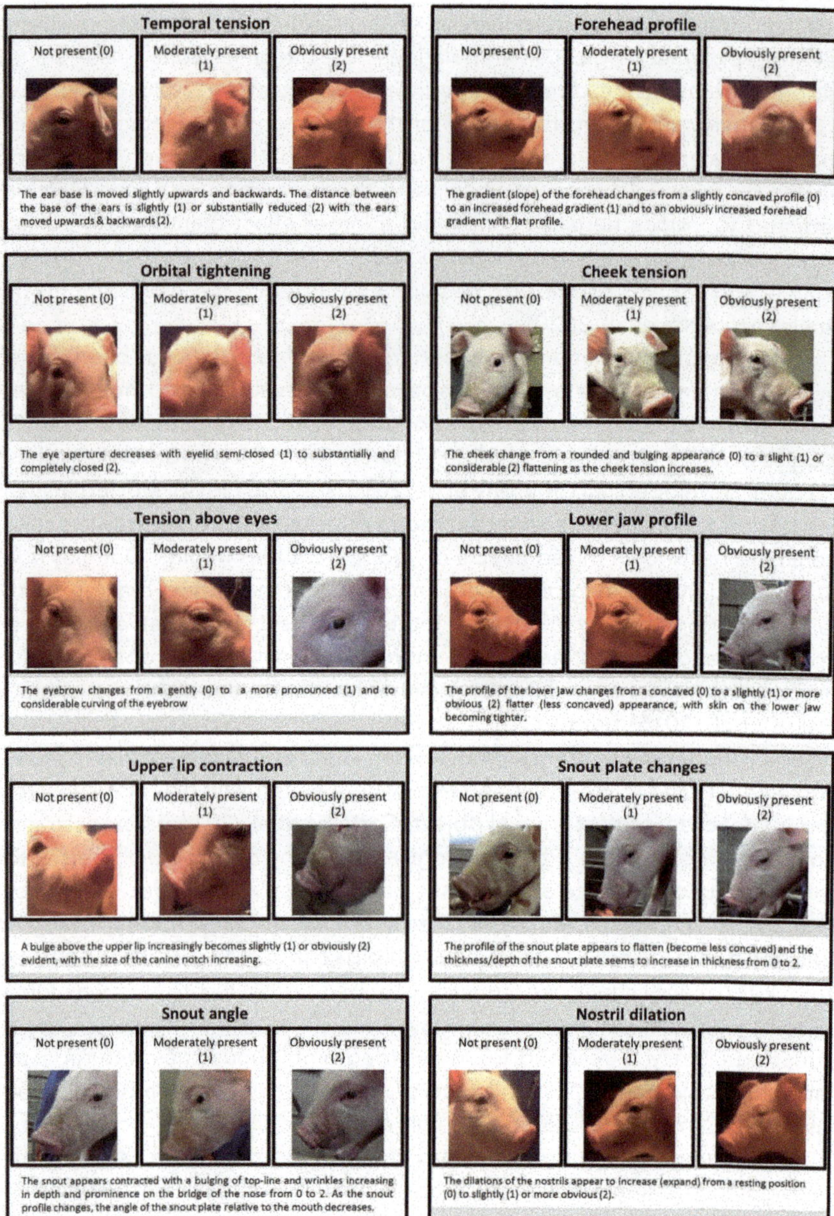

Figure 4 Sample piglet-grimace scale as developed by di Giminiani et al., 2016.

and the relative severity of the manipulations used (Lagisz et al., 2020), as well as the relatively low numbers of presentations of ambiguous cues (in order to reduce learning about the outcomes of ambiguous cues). As a further caution, many judgement bias tasks are relatively time-consuming and are impractical to run outside of a lab setting, although attempts to design tasks which are faster to run but do not compromise much on design have been developed (e.g. Hintze et al., 2018; Jones et al., 2018) but not yet tested with pigs. With a properly validated task design, the judgement bias paradigm is a promising method to differentiate between positively and negatively valenced affective states in pigs.

4 Neurophysiological indicators of emotion in pigs

4.1 Physiology of negative emotions

When it comes to looking at the interplay between physiology and negative affective state in animals it is important to include stress. Traditionally, studies of animal welfare have used the concept of stress to measure welfare, as, until recently, the study of emotion in animals was seen as too subjective. Nonetheless both the terms 'stress' and 'emotion' have many aspects in common, and have often been used synonymously in some of the animal literature (Mcmillan, 2008). However, stress is a loosely defined term often used in the literature which can take a variety of forms; an aversive stimulus is generally called a 'stressor', the 'stress response' is the physiological reaction within an animal to an aversive stimulus, and 'distress' is sometimes used to describe the psychological state of the animal (Mcmillan, 2008). Stress can be defined as an adaptive response as a consequence of the subjective appraisal of environmental challenges by an individual (Mormède et al., 2007; von Borell and Veissier, 2007). Thus stress has an adaptive value; the function of the stress response is to allow an animal to react to and potentially adapt to a challenge. In these terms, measurement of stress responses may be useful to infer about emotional state. Generally, investigations into animal stress can be viewed as the studies of responses to conditions that are likely to be aversive (Paul et al., 2005). The main focus of stress studies in animals has been on activity of the autonomic nervous system (ANS) and the hypothalamic-pituitary-adrenocortical axis (HPA axis) (Mormède et al., 2007).

In relation to measurement of emotional response, ANS activation is most commonly recorded by electrodermal (e.g. skin conductance) and cardiovascular (e.g. heart rate, blood pressure, heart rate variability) responses (Mauss and Robinson, 2009). William James (James, 1884) originally proposed that different patterns of ANS activation produced different emotions. While there is some evidence for certain patterns of ANS activation correlating with specific emotions in human research (Kreibig, 2010), such results have not been found to be consistent (Mauss and Robinson, 2009) and may in fact reflect levels of arousal (Hagemann et al., 2003). Therefore, while ANS activity may be a useful measurement of the intensity of an emotion, without other measures it may not be useful to distinguish different emotions or even simply positive from negative emotions. However, studies that used experimental treatments known to induce distress have identified a general pattern of the physiological response to negative situations. For example, subordinate pigs experiencing social defeat show a more sustained increase in heart rate compared to dominant pigs (De Jong et al., 2000; Marchant et al., 1995). Similarly, heart rate also increases after social isolation (De Jong et al., 1998) and castration (White et al., 1995). Further, both social defeat and isolation caused a more sustained cardiovascular activity in response to a novel and an aversive test (Ruis et al., 2001a, b).

Heart rate variability (HRV) in particular has been promoted as a potential measure of psycho-physiological state and emotional valence (Boissy et al., 2007; von Borell et al., 2007). Indeed, the measure of heart rate alone reflects only the net outcome of the interaction between vagal (reducing heart rate) and sympathetic (increasing heart rate) systems, and thus does not provide specific information on the activity pattern of each separate cardiac regulatory system. Instead, measures of HRV can disentangle such activity patterns (von Borell et al., 2007). HRV is a measure of the variation of the time intervals between consecutive heart beats (Malik, 1996; McCraty and Shaffer, 2015). Such variation is normally present in healthy individuals and serves to maintain the activity of the different cardiac regulatory systems within the ideal range for responding to challenges (McCraty and Shaffer, 2015). Behavioural and mental disorders, and psychological stress, are associated with a decrease in HRV (McCraty and Shaffer, 2015; von Borell et al., 2007) which suggests the experience of a negative emotional state. Research aiming to link HRV with negative emotions in pigs has been scarce. So far, both social defeat (De Jong et al., 2000) and restraint (Geverink et al., 2002) did not induce significant changes in HRV in pigs, although a decrease in HRV in the face of a stressor would be expected and has been demonstrated in other farm animals (Boissy et al., 2007). Thus, more research exploring the relationship between HRV and negative emotions in pigs is needed. Individual differences may also play an important role in how the emotional state of pigs can be affected by a stressor. Pigs with a proactive coping style show higher heart rate in feeding, resting and handling contexts compared to pigs with a reactive coping style, by means of an increased sympathetic activity and/or decreased vagal tone depending on the context considered (Krause et al., 2017).

Cortisol is the primary active hormone of the HPA axis in pigs and is often called the 'stress hormone'. Synthesis of cortisol is controlled by adrenocorticotropic hormone, ACTH, which is itself synthesised in the anterior pituitary gland. Corticotrophin-releasing hormone, CRH, and vasopressin, AVP, two neuropeptides synthesised in the hypothalamus cause the release of ACTH which is transported to the adrenal cortex where it causes the synthesis of cortisol. The HPA axis is a negative feedback system whereby cortisol in turn influences production of the neuropeptides in the hypothalamus. The HPA axis is activated in aversive situations (Mormède et al., 2007) and therefore measures of cortisol in blood or saliva are commonly used in studies of stress in animals. However, while higher levels of cortisol may be an indication of stress, there are a number of other factors which influence the amount of circulating cortisol such as the time of day/night, physical activity, health, feeding or temperature and humidity. Furthermore, there is not a linear relationship between the amount of cortisol and the apparent severity of the stress in animals (Mormède et al., 2007). So, while cortisol can provide an indication of stress in an animal, therefore an indication of negative affect, it may not give an indication of the intensity of the emotion, and could be influenced by a variety of other factors. In pigs, social isolation, restraint and social defeat increase plasma cortisol (isolation: Schrader and Ladewig, 1999; restraint: Jarvis et al., 2008; social defeat: Ruis et al., 2001a), and social isolation also induces a longer-term higher cortisol response to novel object tests (Ruis et al., 2001b). Further, mixing with unfamiliar conspecifics increases salivary cortisol (Ison et al., 2010). However less acute, longer-term treatments produced inconclusive and somewhat contradictory results. For instance, housing growing pigs at high or low density did not significantly affect salivary cortisol or alpha-amylase levels (Scollo et al., 2014). Furthermore and contrary to expectations, pigs housed in an enriched environment showed higher baseline levels of salivary cortisol than pigs kept in a barren environment (De Jong et al., 1998).

Although both elevated HPA and sympathetic activity may occur during putatively negative situations, they are also expressed in high-arousal positive situations. For example, Buwalda et al. (2012) demonstrate that male rats show almost identical corticosterone responses to the presence of an avoided aggressive-conspecific and a desired female partner. These and similar findings suggest that HPA activity and simple measures of sympathetic activation such as heart rate may primarily reflect affective arousal in preparation for action in both positively and negatively valenced situations, rather than affective valence per se. More comprehensive approaches combining multiple physiological (and behavioural) measures appear to be more promising in identifying and quantifying negative emotional states. In particular, the combined assessment of the activity of both sympathetic adrenomedullary (SAM) system and HPA axis can help quantify the stress response to unpredictable and uncontrollable situations (Koolhaas et al., 2011). For example, pigs show a rise in both cortisol and adrenaline levels after a social isolation treatment but, if such treatment is repeated several times, cortisol levels decrease over time indicating habituation to the stressor, while adrenaline levels do not consistently vary, suggesting a longer-term negative effect of social isolation (Schrader and Ladewig, 1999).

4.2 Physiology of positive emotions

Only in recent years, have physiological parameters been used to assess the positively valenced states in animals (Boissy et al., 2007). Although research in this direction is still scarce, cardiovascular activity, particularly heart rate variability (HRV) as a proxy measure of vagal tone, has been promoted as a way to assess positive affective state (Boissy et al., 2007). A good indicator of positive affective state is the relative increase in the variance between inter-beat intervals, which can be considered a proxy measure of vagal activity (reviewed in Boissy et al., 2007). Pigs that received cognitive enrichment through a cognitive paradigm involving discrimination of acoustic cues to obtain food showed both arousal, as measured by an increase in heart rate, and positive affective state as indicated by a specific pattern of the HRV parameters in response to the acoustic signals and during consumption of the food (Zebunke et al., 2011). However, pigs that were positively conditioned by receiving rewarding items (food, toy, straw) in repeated sessions mainly showed sympathetic activation (indicating arousal) in anticipation of the rewards instead of vagal activation (indicating positive valence; Leliveld et al., 2016), thus calling for further research to assess the measurability of positive affective state using HRV.

Another promising physiological indicator of positive emotional state in pigs is represented by the activity of the opioid system (Boissy et al., 2007). The endogenous opioid system acts on specific subcortical brain areas (i.e. nucleus accumbens, ventral pallidum and amygdala) and modulates the subjective hedonic experience of rewards such as play, sex and food (Vanderschuren, 2010). It differentiates from the dopamine system, which mediates the motivation towards obtaining a reward ('wanting'), and therefore can be used as a proxy measure of consummatory, hedonic pleasure ('liking': Berridge et al., 2009; Boissy et al., 2007; Vanderschuren, 2010). Using the same auditory cognitive paradigm for pigs described above (i.e. study from Zebunke et al., 2011), Kalbe and Puppe (2010) demonstrated that frequently occurring positive events in the form of cognitive enrichment modified the expression of specific brain opioid receptors in the amygdala, which is responsible for the processing of reward-sensitive stimuli (Kalbe and Puppe, 2010). This preliminary study sets the basis for further research in this area to be conducted.

The hormone oxytocin has also been proposed as a measure of improved welfare (Boissy et al., 2007; Broom and Zanella, 2004). Brain oxytocin is involved in the regulation of positive social bonding, both between mates and between mother and offspring (Winslow and Insel, 2002); thus its assessment in the context of positive social interactions could be a marker of positive emotional valence (Rault, 2016). In pigs, positive interaction with a human induced a sustained increase in oxytocin levels in the cerebrospinal fluid (Rault, 2016), supporting the above hypothesis. Another study in minipigs, which investigated peripheral levels of oxytocin in response to repeated provision of toys in the home pen, showed unclear results as whether oxytocin levels could indicate a positive emotional state or a response to challenge (i.e. blood collection; Marcet Rius et al., 2018a). However, there is mounting evidence of no consistent relationship between central and peripheral changes in oxytocin levels, since oxytocin does not pass easily through the blood-brain barrier (reviewed in Rault, 2016); thus oxytocin may be a better biomarker of positive emotional state if measured centrally. Finally, pigs which were conditioned to expect an aversive event were less likely to defecate in the presence of naïve pen mates which had received oxytocin intranasally than in the presence of control pen mates, suggesting that relatively higher levels of oxytocin in one individual can alleviate a negative state in other familiar individuals (Reimert et al., 2015).

5 Summary and future trends

The study of animal emotion is a challenging area, not only because of the difficulty in defining what we mean when we talk about terms such as emotion, affect or mental states, but also because of the variety of approaches proposed to measure these states in animals. However, in order to assess animal welfare, the development of reliable and valid measures of affective states is essential. While the key component of emotion is considered to be the conscious experience, this is not yet possible to study directly. Instead, approaches in animal research have focussed on the other components, namely, behavioural, cognitive and neurophysiological components (Paul et al., 2005; Paul and Mendl, 2018; Scherer, 1984). As with research in other species, research aimed at assessing affective states in pigs has also focussed on these three components.

In terms of behaviour, the standard approach until relatively recently was to look at unconditioned responses of pigs in a series of behaviour tests designed to provoke fear or anxiety. While some aspects of these tasks are likely to induce an emotional response in pigs, the measures used to record this are not always ethologically relevant for pigs or specific enough to differentiate between measures of emotion and activity (reviewed in Murphy et al., 2014). Further, some tasks are performed in such a variety of ways and there is little systematic study on how the different approaches may, in themselves, influence the outcome measures.

Aside from these behavioural tests, more recently, a variety of spontaneous behaviours have been proposed as potential measures of affective states in animals, including in pigs. Moving on from fear and anxiety, these approaches have also broadened the field into the study of both positive and negative states. QBA is based on the premise that people can infer an animal's emotional state from observing the behavioural expression of an animal as it interacts with its environment. Once you accept this philosophical premise, this approach is one of the few which allows for the differentiation of states of the same valence.

Pigs are quite vocal animals and make a variety of calls from squeals to barks and grunts. Squeals are assumed to be indicative of high arousal negative states, while barks can occur in positive and negative contexts. Recent work has focussed on differentiating grunts produced in positive and negative contexts based on their acoustic parameters and has identified some potential parameters for future research. Systematic study of vocalisation types and parameters could be useful to determine if the call type or acoustic structure of the call can be used to differentiate between states within the same valence. Pigs are also highly playful animals, particularly as juveniles. While the occurrence of play in itself does not necessarily indicate past positive experiences, it is likely to induce positive affect, and environments that stimulate play could potentially increase positive experiences.

Startle behaviour in response to sudden loud stimuli, part of the defence cascade, has been proposed as one method to distinguish between positive and negative affective states. Research on this behavioural response in pigs is still in the early stages. However, evidence that pharmacological manipulation of state does affect startle responses in the direction predicted makes this a promising area for future research. Furthermore, defence cascade responses could be incorporated into typical behavioural tests, such as a novel object task, as they provide a priori expectations of how affective state should influence the behavioural response.

Another growing area of study is how emotions are expressed through changes in body posture and facial expressions in pigs. Research has identified relaxed tail movements as potential indicators of positive affective states and changes in ear postures as potential indicators of negative affective states. Furthermore, some changes in facial expression (e.g. narrowing of the eyes and ears angled backwards) may even be useful for distinguishing between affective states of the same valence (e.g. fear and pain vs. aggression).

Another component of affective states is the cognitive component. Affective states can potentially be studied through their impact on cognitive processes. The most commonly studied in animal research is how biases in judgement of ambiguous stimuli can be used to distinguish between positively and negatively valenced states. There is growing evidence that, across species, both pharmacological and non-pharmacological manipulations of emotional states shifted the bias in the direction predicted (Lagisz et al., 2020; Neville et al., 2020). There is also some debate about whether the shift in bias at different probe cues may reflect differences in states of the same valence (Salmeto et al., 2011; reviewed in Roelofs et al., 2016). While the judgement bias paradigm is a promising approach to the measurement of emotion in animals, the research with pigs is patchy and a systematic approach in validating the paradigm is lacking.

There are several neurophysiological measures that can be used to infer on the affective state of an animal. Confounds in the interpretation of these measures exist, as it is sometimes difficult to disentangle emotional valence from arousal, and the assessment of one physiological parameter alone may not be sufficient to pinpoint a specific emotion or even valence. However, combining multiple physiological measures and using validated methods can provide a more comprehensive picture on emotional responses. Promising indicators of negative affective state deriving from stress physiology are the combined evaluation of the activity of both sympathetic adrenomedullary system and HPA axis (e.g. adrenalin and cortisol levels), and the assessment of cardiovascular responses, especially heart-rate variability. The latter is of particular interest as it potentially allows us to make inferences about positive affective states. Furthermore, the activity of the endogenous opioid system as a measure of hedonic pleasure, and brain oxytocin levels as a proxy measure of affective state deriving from positive social interactions, are other promising indicators of positive affect. Research in pigs, however, especially with respect to positive affective state, is still scarce and sometimes with contrasting results; thus these measures still need further validation and standardisation in methodology (e.g. for oxytocin measurement, see Rault et al., 2017).

Finally, it is worth considering that pigs, being highly social animals, can influence each other's affective state through their social and sensory interactions. Through emotional contagion, that is, a change in emotional state of one individual as a direct consequence of perceiving and sharing the emotional state of another (Hatfield et al., 1994; Preston and De Waal, 2012), the emotional response of just one or few individuals can alter the affective state of the whole social group (Held and Špinka, 2011; Špinka, 2012). In pigs, research on emotional contagion, especially on the positive spectrum, is only at an early stage, and faces the challenge of demonstrating that contagion occurs not only at the behavioural level but also internally (i.e. affecting emotional valence; Špinka, 2012). Further, some studies have failed to show contagion (e.g. through playback of distress calls; (Düpjan et al., 2011). However, this is a field calling for further investigation and several ways of assessing emotional contagion have been suggested, ranging from the use of olfactory cues, of the playback of vocalisations, the change of housing conditions, to the transfer of locomotor-rotational play (Held and Špinka, 2011; Lawrence et al., 2018; Špinka, 2012). While preliminary research suggests that particularly tail movements, defecation and play may be indicative of emotional contagion in pigs (Reimert et al., 2013, 2015, 2017), these measures first have to be properly validated as reliable indicators of affective state.

In conclusion, in this chapter we have tried to provide a critical overview of the various methods that have been proposed and applied as measures of affective states in pigs. To fully evaluate the behavioural, cognitive and neurophysiological measures proposed, it is fundamental that these measures are evaluated in a systematic way. Finding measures that vary reliably according to which core affect quadrant a pig is in is essential in establishing valid and reliable markers of affective states.

6 Where to look for further information

For some interesting discussion on the nature of emotions and their implications for animal welfare, the following articles are of relevance:

- Paul, E. S. and Mendl, M. T. (2018). Animal emotion: descriptive and prescriptive definitions and their implications for a comparative perspective. *Applied Animal Behaviour Science* 205, 202–209. https://doi.org/10.1016/j.applanim.2018.01.008.
- Paul, E. S., Sher, S., Tamietto, M., Winkielman, P. and Mendl, M. T. (2020). Towards a comparative science of emotion: affect and consciousness in humans and animals. *Neuroscience and Biobehavioral Reviews* 108, 749–770. https://doi.org/10.1016/j.neubiorev.2019.11.014.
- Boissy, A., Manteuffel, G., Jensen, M. B., Moe, R. O., Spruijt, B., Keeling, L. J., Winckler, C., Forkman, B., Dimitrov, I., Langbein, J., Bakken, M., Veissier, I. and Aubert, A. (2007). Assessment of positive emotions in animals to improve their welfare. *Physiology and Behavior* 92(3), 375–397. https://doi.org/10.1016/j.physbeh.2007.02.003.
- Lawrence, A. B., Vigors, B. and Sandøe, P. (2019). What is so positive about positive animal welfare?—a critical review of the literature. *Animals* 9(10) 783. https://doi.org/10.3390/ani9100783.

Here are relevant readings more specific to emotion in pigs:

- Held, S. D. E., Cooper, J. J. and Mendl, M. T. (2009). Advances in the study of cognition, behavioural priorities and emotions. In: Marchant-Forde, J. N. (Ed.), *The Welfare of Pigs*. Springer. https://doi.org/10.1007/978-1-4020-8909-1_3.
- Murphy, E., Nordquist, R. E. and van der Staay, F. J. (2014). A review of behavioural methods to study emotion and mood in pigs, Sus scrofa.

Applied Animal Behaviour Science 159, 9–28. https://doi.org/10.1016/j.applanim.2014.08.002.
- Lawrence, A. B., Newberry, R. C. and Špinka, M. (2018). Positive welfare: what does it add to the debate over pig welfare? *Advances in Pig Welfare*. https://doi.org/10.1016/B978-0-08-101012-9.00014-9.

Key international meetings in which the topic of emotions in animals, including pigs, is discussed:

- ISAE (International Society for Applied Ethology) annual meetings often focus on the applied assessment of animal emotions within settings involving human-animal interaction, such as farming and keeping of companion and laboratory animals.
- The biannual meeting "Behaviour", which hosts the IEC (International Ethological Conference) and ASAB (Association for the Study of Animal Behaviour) / ABS (Animal Behavior Society) conferences, also deals with the study of animal emotion from a more fundamental and theoretical perspective.
- UFAW (Universities Federation for Animal Welfare) annual meetings touch upon the assessment of animal emotions from an ethical perspective and with the aim of improving animal welfare.

7 References

Ahloy-Dallaire, J., Espinosa, J. and Mason, G. (2018). Play and optimal welfare: does play indicate the presence of positive affective states? *Behavioural Processes* 156, 3–15. https://doi.org/10.1016/j.beproc.2017.11.011.

Andersen, I. L., Bøe, K. E., Fœrevik, G., Janczak, A. M. and Bakken, M. (2000a). Behavioural evaluation of methods for assessing fear responses in weaned pigs. *Applied Animal Behaviour Science* 69(3), 227–240. https://doi.org/10.1016/S0168-1591(00)00133-7.

Andersen, I. L., Færevik, G., Bøe, K. E., Janczak, A. M. and Bakken, M. (2000b). Effects of diazepam on the behaviour of weaned pigs in three putative models of anxiety. *Applied Animal Behaviour Science* 68(2), 121–130. https://doi.org/10.1016/S0168-1591(00)00098-8.

Anderson, D. J. and Adolphs, R. (2014). A framework for studying emotions across species. *Cell Press* 157(1), 187–200. https://doi.org/10.1016/j.cell.2014.03.003.

Antonacci, D., Norscia, I. and Palagi, E. (2010). Stranger to familiar: wild strepsirhines manage xenophobia by playing. *PLoS ONE* 5(10), e13218. https://doi.org/10.1371/journal.pone.0013218.

Arena, L., Wemelsfelder, F., Messori, S., Ferri, N. and Barnard, S. (2017). Application of Free Choice Profiling to assess the emotional state of dogs housed in shelter environments. *Applied Animal Behaviour Science* 195, 72–79. https://doi.org/10.1016/j.applanim.2017.06.005.

Barrett, L. F. (2006). Are emotions natural kinds? *Perspectives on Psychological Science: A Journal of the Association for Psychological Science* 1(1), 28–58. https://doi.org/10.1111/j.1745-6916.2006.00003.x.

Bateson, P., Mendl, M. and Feaver, J. (1990). Play in the domestic cat is enhanced by rationing of the mother during lactation. *Animal Behaviour* 40(3), 514–525. https://doi.org/10.1016/S0003-3472(05)80532-9.

Beilharz, R. G. and Cox, D. F. (1967). Genetic analysis of open field behavior in swine. *Journal of Animal Science* 26(5), 988–990. https://doi.org/10.2527/jas1967.265988x.

Berridge, K. C., Robinson, T. E. and Aldridge, J. W. (2009). Dissecting components of reward: "liking", "wanting", and learning. *Current Opinion in Pharmacology* 9(1), 65–73. https://doi.org/10.1016/j.coph.2008.12.014.

Blackshaw, J. K., Blackshaw, A. W. and McGlone, J. J. (1998). Startle-freeze behaviour in weaned pigs. *International Journal of Comparative Psychology* 11(1), 30–39.

Blanchette, I. and Richards, A. (2010). The influence of affect on higher level cognition: a review of research on interpretation, judgement, decision making and reasoning. *Cognition and Emotion* 24(4), 561–595. https://doi.org/10.1080/02699930903132496.

Boissy, A., Aubert, A., Désiré, L., Greiveldinger, L., Delval, E. and Veissier, I. (2011). Cognitive sciences to relate ear postures to emotions in sheep. *Animal Welfare* 20(1), 47.

Boissy, A., Manteuffel, G., Jensen, M. B., Moe, R. O., Spruijt, B., Keeling, L. J., Winckler, C., Forkman, B., Dimitrov, I., Langbein, J., Bakken, M., Veissier, I. and Aubert, A. (2007). Assessment of positive emotions in animals to improve their welfare. *Physiology and Behavior* 92(3), 375–397. https://doi.org/10.1016/j.physbeh.2007.02.003.

Bolhuis, J. E., Schouten, W. G. P., Schrama, J. W. and Wiegant, V. M. (2005). Behavioural development of pigs with different coping characteristics in barren and substrate-enriched housing conditions. *Applied Animal Behaviour Science* 93(3–4), 213–228. https://doi.org/10.1016/j.applanim.2005.01.006.

Brajon, S., Laforest, J. P., Schmitt, O. and Devillers, N. (2015). The way humans behave modulates the emotional state of piglets. *PLoS ONE* 10(8), e0133408. https://doi.org/10.1371/journal.pone.0133408.

Briefer, E. F. (2012). Vocal expression of emotions in mammals: mechanisms of production and evidence. *Journal of Zoology* 288(1), 1–20. https://doi.org/10.1111/j.1469-7998.2012.00920.x.

Briefer, E. F., Vizier, E., Gygax, L. and Hillmann, E. (2019). Expression of emotional valence in pig closed-mouth grunts: involvement of both source- and filter-related parameters. *The Journal of the Acoustical Society of America* 145(5), 2895. https://doi.org/10.1121/1.5100612.

Broom, D. M. and Zanella, A. J. (2004). Brain measures which tell us about animal welfare. *Animal Welfare* 13, S41–S45.

Brown, S. M., Peters, R., Nevison, I. M. and Lawrence, A. B. (2018). Playful pigs: evidence of consistency and change in play depending on litter and developmental stage. *Applied Animal Behaviour Science* 198, 36–43. https://doi.org/10.1016/j.applanim.2017.09.018.

Burgdorf, J. and Panksepp, J. (2006). The neurobiology of positive emotions. *Neuroscience and Biobehavioral Reviews* 30(2), 173–187. https://doi.org/10.1016/j.neubiorev.2005.06.001.

Buwalda, B., Scholte, J., de Boer, S. F., Coppens, C. M. and Koolhaas, J. M. (2012). The acute glucocorticoid stress response does not differentiate between rewarding and aversive social stimuli in rats. *Hormones and Behavior* 61(2), 218–226. https://doi.org/10.1016/j.yhbeh.2011.12.012.

Caeiro, C., Guo, K. and Mills, D. (2017). Dogs and humans respond to emotionally competent stimuli by producing different facial actions. *Scientific Reports* 7(1), 15525. https://doi.org/10.1038/s41598-017-15091-4.

Camerlink, I., Coulange, E., Farish, M., Baxter, E. M. and Turner, S. P. (2018). Facial expression as a potential measure of both intent and emotion. *Scientific Reports* 8(1), 17602. https://doi.org/10.1038/s41598-018-35905-3.

Camerlink, I., Peijnenburg, M., Wemelsfelder, F. and Turner, S. P. (2016). Emotions after victory or defeat assessed through qualitative behavioural assessment, skin lesions and blood parameters in pigs. *Applied Animal Behaviour Science* 183, 28–34. https://doi.org/10.1016/j.applanim.2016.07.007.

Camerlink, I. and Ursinus, W. W. (2020). Tail postures and tail motion in pigs: a review. *Applied Animal Behaviour Science* 230. https://doi.org/10.1016/j.applanim.2020.105079.

Carreras, R., Arroyo, L., Mainau, E., Peña, R., Bassols, A., Dalmau, A., Faucitano, L., Manteca, X. and Velarde, A. (2016a). Effect of gender and halothane genotype on cognitive bias and its relationship with fear in pigs. *Applied Animal Behaviour Science* 177, 12–18. https://doi.org/10.1016/j.applanim.2016.01.019.

Carreras, R., Arroyo, L., Mainau, E., Valent, D., Bassols, A., Dalmau, A., Faucitano, L., Manteca, X. and Velarde, A. (2017). Can the way pigs are handled alter behavioural and physiological measures of affective state? *Behavioural Processes* 142, 91–98. https://doi.org/10.1016/j.beproc.2017.06.005.

Carreras, R., Mainau, E., Arroyo, L., Moles, X., González, J., Bassols, A., Dalmau, A., Faucitano, L., Manteca, X. and Velarde, A. (2016b). Housing conditions do not alter cognitive bias but affect serum cortisol, qualitative behaviour assessment and wounds on the carcass in pigs. *Applied Animal Behaviour Science* 185, 39–44. https://doi.org/10.1016/j.applanim.2016.09.006.

Carreras, R., Mainau, E., Rodriguez, P., Llonch, P., Dalmau, A., Manteca, X. and Velarde, A. (2015). Cognitive bias in pigs: individual classification and consistency over time. *Journal of Veterinary Behavior* 10(6), 577–581. https://doi.org/10.1016/j.jveb.2015.09.001.

Carver, C. S. (2001). Affect and the functional bases of behavior: on the dimensional structure of affective experience. *Personality and Social Psychology Review* 5(4), 345–356. https://doi.org/10.1207/S15327957PSPR0504_4.

Chalmers, N. R. and Locke-Haydon, J. (1981). Temporal patterns of play bouts in captive common marmosets (Callithrix jacchus). *Animal Behaviour* 29(4), 1229–1238. https://doi.org/10.1016/S0003-3472(81)80074-7.

Chaloupková, H., Illmann, G., Bartoš, L. and Špinka, M. (2007). The effect of pre-weaning housing on the play and agonistic behaviour of domestic pigs. *Applied Animal Behaviour Science* 103(1–2), 25–34. https://doi.org/10.1016/j.applanim.2006.04.020.

Chan, W. Y. (2011). *The Meaning of Barks: Vocal Communication of Fearful and Playful Affective States in Pigs*. Available at: https://research.libraries.wsu.edu/xmlui/bitstream/handle/2376/3517/Chan_wsu_0251E_10317.pdf?sequence=1.

Chan, W. Y., Cloutier, S. and Newberry, R. C. (2011). Barking pigs: differences in acoustic morphology predict juvenile responses to alarm calls. *Animal Behaviour* 82(4), 767-774. https://doi.org/10.1016/j.anbehav.2011.07.007.

Clarke, T., Pluske, J. R. and Fleming, P. A. (2016). Are observer ratings influenced by prescription? a comparison of Free choice profiling and fixed list methods of qualitative behavioural assessment. *Applied Animal Behaviour Science* 177, 77-83. https://doi.org/10.1016/j.applanim.2016.01.022.

Corr, P. J. (2008). *The Reinforcement Sensitivity Theory of Personality*. Cambridge University Press. https://doi.org/10.1017/CBO9780511819384.

Czycholl, I., Grosse Beilage, E. G., Henning, C. and Krieter, J. (2017). Reliability of the qualitative behavior assessment as included in the welfare quality assessment protocol for growing pigs. *Journal of Animal Science* 95(8), 3445-3454. https://doi.org/10.2527/jas.2017.1525.

Dalla Costa, E., Minero, M., Lebelt, D., Stucke, D., Canali, E. and Leach, M. C. (2014). Development of the Horse Grimace Scale (HGS) as a pain assessment tool in horses undergoing routine castration. *PLOS ONE* 9(3), e92281. https://doi.org/10.1371/journal.pone.0092281.

Davis, S. L. and Cheeke, P. R. (1998). Do domestic animals have minds and the ability to think? A provisional sample of opinions on the question. *Journal of Animal Science* 76(8), 2072-2079. https://doi.org/10.2527/1998.7682072x.

Dawkins, M. (2015). Animal welfare and the paradox of animal consciousness. In: Naguib, M., Brockmann, H. J., Mitani, J. C., Simmons, L. W., Barrett, L., Healy, S. and Slater, P. J. B. (Eds), *Advances in the Study of Behavior* (Vol. 47, pp. 5-38). Academic Press Inc. https://doi.org/10.1016/bs.asb.2014.11.001.

Dawkins, M. S. (2004). Using behaviour to assess animal welfare. *Animal Welfare* 13(5), 3-7. Available at: https://www.ingentaconnect.com/content/ufaw/aw/2004/00000013/A00101s1/art00002.

Day, J. E. L., Kyriazakis, I. and Lawrence, A. B. (1995). The effect of food deprivation on the expression of foraging and exploratory behaviour in the growing pig. *Applied Animal Behaviour Science* 42(3), 193-206. https://doi.org/10.1016/0168-1591(95)93889-9.

de Boyer des Roches, A., Lussert, A., Faure, M., Herry, V., Rainard, P., Durand, D., Wemelsfelder, F. and Foucras, G. (2018). Dairy cows under experimentally-induced Escherichia coli mastitis show negative emotional states assessed through qualitative behaviour assessment. *Applied Animal Behaviour Science* 206, 1-11. https://doi.org/10.1016/j.applanim.2018.06.004.

De Jong, I. C., Ekkel, E. D., Van De Burgwal, J. A., Lambooij, E., Korte, S. M., Ruis, M. A. W., Koolhaas, J. M. and Blokhuis, H. J. (1998). Effects of strawbedding on physiological responses to stressors and behavior in growing pigs. *Physiology and Behavior* 64(3), 303-310. https://doi.org/10.1016/S0031-9384(98)00066-3.

De Jong, I. Cd, Sgoifo, A., Lambooij, E., Korte, S. M., Blokhuis, H. J. and Koolhaas, J. M. (2000). Effects of social stress on heart rate and heart rate variability in growing pigs. *Canadian Journal of Animal Science* 80(2), 273-280. https://doi.org/10.4141/A99-085.

de Jonge, F. H., Tilly, S. L., Baars, A. M. and Spruijt, B. M. (2008). On the rewarding nature of appetitive feeding behaviour in pigs (Sus scrofa): do domesticated pigs contrafreeload? *Applied Animal Behaviour Science* 114(3-4), 359-372. https://doi.org/10.1016/j.applanim.2008.03.006.

Désiré, L., Boissy, A. and Veissier, I. (2002). Emotions in farm animals: a new approach to animal welfare in applied ethology. *Behavioural Processes* 60(2), 165–180. https://doi.org/10.1016/S0376-6357(02)00081-5.

Dickinson, A. and Bernard, B. (2010). Hedonics: the cognitive-motivational interface. In: Kringelbach, M. L. and Berridge, K. C. (Eds), *Pleasures of the Brain* (pp. 74–84). Oxford University Press. Available at: https://books.google.de/books?hl=en&lr=&id=yl2yAwAAQBAJ&oi=fnd&pg=PA74&dq=Hedonics:+the+cognitive-motivational+interface&ots=KheYsuw1ZQ&sig=5OcNGUXuGub2mDJIIH4WxUOHtTo&redir_esc=y#v=onepage&q=Hedonics%3A the cognitive-motivational interface&f=false.

di Giminiani, P., Brierley, V. L. M. H., Scollo, A., Gottardo, F., Malcolm, E. M., Edwards, S. A. and Leach, M. C. (2016). The assessment of facial expressions in piglets undergoing tail docking and castration: toward the development of the Piglet Grimace Scale. *Frontiers in Veterinary Science* 3, 100. https://doi.org/10.3389/fvets.2016.00100.

Dobao, M. T., Rodrigañez, J. and Silio, L. (1985). Choice of companions in social play in piglets. *Applied Animal Behaviour Science* 13(3), 259-266. https://doi.org/10.1016/0168-1591(85)90049-8.

Dolensek, N., Gehrlach, D. A., Klein, A. S. and Gogolla, N. (2020). Facial expressions of emotion states and their neuronal correlates in mice. *Science* 368(6486), 89–94. https://doi.org/10.1126/science.aaz9468.

Donald, R. D., Healy, S. D., Lawrence, A. B. and Rutherford, K. M. D. (2011). Emotionality in growing pigs: is the open field a valid test? *Physiology and Behavior* 104(5), 906–913. https://doi.org/10.1016/j.physbeh.2011.05.031.

Donald, R. D., Rutherford, K. and Lawrence, A. (2010). Effect of azaperone on the behaviour of pigs in an elevated-plus maize. Proceedings of the UFAW Conference, 30th June 2010.

Donaldson, T. M., Newberry, R. C., Špinka, M. and Cloutier, S. (2002). Effects of early play experience on play behaviour of piglets after weaning. *Applied Animal Behaviour Science* 79(3), 221–231. https://doi.org/10.1016/S0168-1591(02)00138-7.

Douglas, C., Bateson, M., Walsh, C., Bédué, A. and Edwards, S. A. (2012). Environmental enrichment induces optimistic cognitive biases in pigs. *Applied Animal Behaviour Science* 139(1–2), 65–73. https://doi.org/10.1016/j.applanim.2012.02.018.

Dudink, S., Simonse, H., Marks, I., de Jonge, F. H. and Spruijt, B. M. (2006). Announcing the arrival of enrichment increases play behaviour and reduces weaning-stress-induced behaviours of piglets directly after weaning. *Applied Animal Behaviour Science* 101(1–2), 86–101. https://doi.org/10.1016/j.applanim.2005.12.008.

Duijvesteijn, N., Benard, M., Reimert, I. and Camerlink, I. (2014). Same pig, different conclusions: stakeholders differ in qualitative behaviour assessment. *Journal of Agricultural and Environmental Ethics* 27(6), 1019–1047. https://doi.org/10.1007/s10806-014-9513-z.

Duncan, I. J. H.(1993). Welfare is to do with what animals feel. *Journal of Agricultural and Environmental Ethics* 6, 8–14 Available at: https://agris.fao.org/agris-search/search.do?recordID=GB19950039583.

Düpjan, S., Ramp, C., Kanitz, E., Tuchscherer, A. and Puppe, B. (2013). A design for studies on cognitive bias in the domestic pig. *Journal of Veterinary Behavior* 8(6), 485–489. https://doi.org/10.1016/j.jveb.2013.05.007.

Düpjan, S., Tuchscherer, A., Langbein, J., Schön, P. C., Manteuffel, G. and Puppe, B. (2011). Behavioural and cardiac responses towards conspecific distress calls in domestic pigs (Sus scrofa). *Physiology and Behavior* 103(5), 445–452. https://doi.org/10.1016/j.physbeh.2011.03.017.

Ekman, P. (1992). An argument for basic emotions. *Cognition and Emotion* 6(3–4), 169–200. https://doi.org/10.1080/02699939208411068.

Finlayson, K., Lampe, J. F., Hintze, S., Würbel, H. and Melotti, L. (2016). Facial indicators of positive emotions in rats. *PLoS ONE* 11(11), e0166446. https://doi.org/10.1371/journal.pone.0166446.

Forkman, B., Boissy, A., Meunier-Salaün, M. C., Canali, E. and Jones, R. B. (2007). A critical review of fear tests used on cattle, pigs, sheep, poultry and horses. *Physiology and Behavior* 92(3), 340–374. https://doi.org/10.1016/j.physbeh.2007.03.016.

Fraser, D., Weary, D., Pajor, E. and Milligan, B. (1997). A scientific conception of animal welfare that reflects ethical concerns. In: *Ethics and Animal Welfare Collection*. Available at: https://www.wellbeingintlstudiesrepository.org/ethawel/1.

Friel, M., Kunc, H. P., Griffin, K., Asher, L. and Collins, L. M. (2019). Positive and negative contexts predict duration of pig vocalisations. *Scientific Reports* 9(1), 2062. https://doi.org/10.1038/s41598-019-38514-w.

Geverink, N. A., Schouten, W. G. P., Gort, G. and Wiegant, V. M. (2002). Individual differences in behavioral and physiological responses to restraint stress in pigs. *Physiology and Behavior* 77(2–3), 451–457. https://doi.org/10.1016/S0031-9384(02)00877-6.

Gray, J. A. (1975). *Elements of a Two-Process Theory of Learning*. Available at: https://psycnet.apa.org/record/1975-24897-000.

Gray, J. A. (1987). The psychology of fear and stress. In: *Problems in the Behavioural Sciences* (2nd Ed). Cambridge University Press, Cambridge.

Hagemann, D., Waldstein, S. R. and Thayer, J. F. (2003). Central and autonomic nervous system integration in emotion. *Brain and Cognition* 52(1), 79–87. https://doi.org/10.1016/S0278-2626(03)00011-3.

Häger, C., Biernot, S., Buettner, M., Glage, S., Keubler, L. M., Held, N., Bleich, E. M., Otto, K., Müller, C. W., Decker, S., Talbot, S. R. and Bleich, A. (2017). The Sheep Grimace Scale as an indicator of post-operative distress and pain in laboratory sheep. *PLoS ONE* 12(4), e0175839. https://doi.org/10.1371/journal.pone.0175839.

Hall, C. S. (1934). Emotional behavior in the rat. I. Defecation and urination as measures of individual differences in emotionality. *Journal of Comparative Psychology* 18(3), 385–403. https://doi.org/10.1037/h0071444.

Hansen, M. F., Smith, M. L., Smith, L. N., Salter, M. G., Baxter, E. M., Farish, M. and Grieve, B. (2018). Towards on-farm pig face recognition using convolutional neural networks. *Computers in Industry* 98, 145–152. https://doi.org/10.1016/j.compind.2018.02.016.

Harding, E. J., Paul, E. S. and Mendl, M. (2004). Animal behaviour: cognitive bias and affective state. *Nature* 427(6972), 312. https://doi.org/10.1038/427312a.

Hatfield, E., Cacioppo, J. T. and Rapson, R. L. (1994). Emotional contagion. https://doi.org/10.1017/cbo9781139174138.

Held, S. D. E., Cooper, J. J. and Mendl, M. T. (2009). Advances in the study of cognition, behavioural priorities and emotions. In: Marchant-Forde, J. N. (Ed.), *The Welfare of Pigs*. Springer. https://doi.org/10.1007/978-1-4020-8909-1_3.

Held, S. D. E. and Špinka, M. (2011). Animal play and animal welfare. *Animal Behaviour* 81(5), 891–899. https://doi.org/10.1016/j.anbehav.2011.01.007.

Hemsworth, P. H., Coleman, G. J., Cox, M. and Barnett, J. L. (1994). Stimulus generalization: the inability of pigs to discriminate between humans on the basis of their previous handling experience. *Applied Animal Behaviour Science* 40(2), 129–142. https://doi.org/10.1016/0168-1591(94)90077-9.

Hemsworth, P. H., Price, E. O. and Borgwardt, R. (1996a). Behavioural responses of domestic pigs and cattle to humans and novel stimuli. *Applied Animal Behaviour Science* 50(1), 43–56. https://doi.org/10.1016/0168-1591(96)01067-2.

Hemsworth, P. H., Verge, J. and Coleman, G. J. (1996b). Conditioned approach-avoidance responses to humans: the ability of pigs to associate feeding and aversive social experiences in the presence of humans with humans. *Applied Animal Behaviour Science* 50(1), 71–82. https://doi.org/10.1016/0168-1591(96)01065-9.

Herskin, M. S. and Jensen, K. H. (2000). Effects of different degrees of social isolation on the behaviour of weaned piglets kept for experimental purposes. *Animal Welfare* 9, 237–249.

Hintze, S., Melotti, L., Colosio, S., Bailoo, J. D. J. D., Boada-Saña, M., Würbel, H. and Murphy, E. (2018). A cross-species judgement bias task: integrating active trial initiation into a spatial Go/No-go task. *Scientific Reports* 8(1), 5104. https://doi.org/10.1038/s41598-018-23459-3.

Hintze, S., Murphy, E., Bachmann, I., Wemelsfelder, F. and Würbel, H. (2017). Qualitative behaviour assessment of horses exposed to short-term emotional treatments. *Applied Animal Behaviour Science* 196, 44–51. https://doi.org/10.1016/j.applanim.2017.06.012.

Hintze, S., Smith, S., Patt, A., Bachmann, I. and Würbel, H. (2016). Are eyes a mirror of the soul? What eye wrinkles reveal about a horse's emotional state. *PLoS ONE* 11(10), e0164017. https://doi.org/10.1371/journal.pone.0164017.

Horback, K. (2014). Nosing around: play in pigs. *Animal Behavior and Cognition* 2(2), 186–196. https://doi.org/10.12966/abc.05.08.2014.

Imfeld-Mueller, S., Van Wezemael, L., Stauffacher, M., Gygax, L. and Hillmann, E. (2011). Do pigs distinguish between situations of different emotional valences during anticipation? *Applied Animal Behaviour Science* 131(3–4), 86–93. https://doi.org/10.1016/j.applanim.2011.02.009.

Ison, S. H., D'Eath, R. B., Robson, S. K., Baxter, E. M., Ormandy, E., Douglas, A. J., Russell, J. A., Lawrence, A. B. and Jarvis, S. (2010). Subordination style. *Applied Animal Behaviour Science* 124(1–2), 16–27. https://doi.org/10.1016/j.applanim.2010.02.001.

Izard, C. E. (2010). The many meanings/aspects of emotion: definitions, functions, activation, and regulation. *Emotion Review* 2(4), 363–370. https://doi.org/10.1177/1754073910374661.

James, W. (1884). What is an emotion? *Mind*, 9, 188–205. Available at: https://psycnet.apa.org/record/2006-10213-033.

Janczak, A. M., Andersen, I. L., Bøe, K. E., Færevik, G. and Bakken, M. (2002). Factor analysis of behaviour in the porcine and murine elevated plus-maze models of anxiety. *Applied Animal Behaviour Science* 77(2), 155–166. https://doi.org/10.1016/S0168-1591(02)00030-8.

Janczak, A. M., Pedersen, L. J., Jensen, K. H., Andersen, I. L., Bøe, K. E. and Bakken, M. (2000). No effect of variation in handling on behaviour in a porcine elevated plus-maze- A brief report. *Applied Animal Behaviour Science* 69(2), 169–173. https://doi.org/10.1016/S0168-1591(00)00111-8.

Jarvis, S., Moinard, C., Robson, S. K., Sumner, B. E. H., Douglas, A. J., Seckl, J. R., Russell, J. A. and Lawrence, A. B. (2008). Effects of weaning age on the behavioural and neuroendocrine development of piglets. *Applied Animal Behaviour Science* 110(1-2), 166–181. https://doi.org/10.1016/j.applanim.2007.03.018.

Jones, S., Neville, V., Higgs, L., Paul, E. S., Dayan, P., Robinson, E. S. J. and Mendl, M. (2018). Assessing animal affect: an automated and self-initiated judgement bias task based on natural investigative behaviour. *Scientific Reports* 8(1), 12400. https://doi.org/10.1038/s41598-018-30571-x.

Kalbe, C. and Puppe, B. (2010). Long-term cognitive enrichment affects opioid receptor expression in the amygdala of domestic pigs. *Genes, Brain, and Behavior* 9(1), 75–83. https://doi.org/10.1111/j.1601-183X.2009.00536.x.

Kanitz, E., Otten, W. and Tuchscherer, M. (2005). Central and peripheral effects of repeated noise stress on hypothalamic-pituitary-adrenocortical axis in pigs. *Livestock Production Science* 94(3), 213–224. https://doi.org/10.1016/j.livprodsci.2004.12.002.

Koolhaas, J. M., Bartolomucci, A., Buwalda, B., de Boer, S. F., Flügge, G., Korte, S. M., Meerlo, P., Murison, R., Olivier, B., Palanza, P., Richter-Levin, G., Sgoifo, A., Steimer, T., Stiedl, O., van Dijk, G., Wöhr, M. and Fuchs, E. (2011). Stress revisited: a critical evaluation of the stress concept. *Neuroscience and Biobehavioral Reviews* 35(5), 1291–1301. https://doi.org/10.1016/j.neubiorev.2011.02.003.

Krause, A., Puppe, B. and Langbein, J. (2017). Coping style modifies general and affective autonomic reactions of domestic pigs in different behavioral contexts. *Frontiers in Behavioral Neuroscience* 11, 103. https://doi.org/10.3389/fnbeh.2017.00103.

Kreibig, S. D. (2010). Autonomic nervous system activity in emotion: a review. *Biological Psychology* 84(3), 394–421. https://doi.org/10.1016/j.biopsycho.2010.03.010.

Krugmann, K., Warnken, F., Krieter, J. and Czycholl, I. (2019). Are behavioral tests capable of measuring positive affective states in growing pigs? *Animals: An Open Access Journal from MDPI* 9(5). https://doi.org/10.3390/ani9050274.

Lagisz, M., Zidar, J., Nakagawa, S., Neville, V., Sorato, E., Paul, E. S., Bateson, M., Mendl, M. and Løvlie, H. (2020). Optimism, pessimism and judgement bias in animals: a systematic review and meta-analysis. *Neuroscience and Biobehavioral Reviews* 118, 3–17. https://doi.org/10.1016/j.neubiorev.2020.07.012.

Langford, D. J., Bailey, A. L., Chanda, M. L., Clarke, S. E., Drummond, T. E., Echols, S., Glick, S., Ingrao, J., Klassen-Ross, T., Lacroix-Fralish, M. L., Matsumiya, L., Sorge, R. E., Sotocina, S. G., Tabaka, J. M., Wong, D., Van Den Maagdenberg, A. M. J. M., Ferrari, M. D., Craig, K. D. and Mogil, J. S. (2010). Coding of facial expressions of pain in the laboratory mouse. *Nature Methods* 7(6), 447–449. https://doi.org/10.1038/nmeth.1455.

Lawrence, A. B., Newberry, R. C. and Špinka, M. (2018). Positive welfare: what does it add to the debate over pig welfare? *Advances in Pig Welfare*. https://doi.org/10.1016/B978-0-08-101012-9.00014-9.

LeDoux, J. (1996). The emotional brain: the mysterious underpinnings of emotional life. In: *Simon & Schuster Paperbacks*. Available at: https://books.google.com/books?hl=en&lr=&id=7EJN5I8sk2wC&oi=fnd&pg=PA9&ots=085egBymYA&sig=vMT3Y5YpgJ-s-ZK_z8g8oUXUs3I.

Leliveld, L. M. C., Düpjan, S., Tuchscherer, A. and Puppe, B. (2016). Behavioural and physiological measures indicate subtle variations in the emotional valence of young pigs. *Physiology and Behavior* 157, 116–124. https://doi.org/10.1016/j.physbeh.2016.02.002.

Leliveld, L. M. C., Düpjan, S., Tuchscherer, A. and Puppe, B. (2017). Vocal correlates of emotional reactivity within and across contexts in domestic pigs (Sus scrofa). *Physiology and Behavior* 181, 117–126. https://doi.org/10.1016/j.physbeh.2017.09.010.

Lind, N. M., Moustgaard, A., Jelsing, J., Vajta, G., Cumming, P. and Hansen, A. K. (2007). The use of pigs in neuroscience: modeling brain disorders. *Neuroscience and Biobehavioral Reviews* 31(5), 728–751. https://doi.org/10.1016/j.neubiorev.2007.02.003.

Maigrot, A. L., Hillmann, E. and Briefer, E. F. (2018). Encoding of emotional valence in wild boar (Sus scrofa) calls. *Animals: An Open Access Journal from MDPI* 8(6). https://doi.org/10.3390/ani8060085.

Malik, M. (1996). Heart rate variability. Standards of measurement, physiological interpretation, and clinical use. Task Force of the European Society of Cardiology and the North American Society of Pacing and Electrophysiology. *European Heart Journal* 17(3), 354–381. https://doi.org/10.1111/j.1542-474X.1996.tb00275.x.

Manteuffel, G., Puppe, B. and Schön, P. C. (2004). Vocalization of farm animals as a measure of welfare. *Applied Animal Behaviour Science* 88(1–2), 163–182. https://doi.org/10.1016/j.applanim.2004.02.012.

Manteuffel, G., Schön, P. C., Düpjan, S., Tuchscherer, A. and Bellmann, O. (2007). Acetylcholine injection into the amygdala elicits vocalization in domestic pigs (Sus scrofa). *Behavioural Brain Research* 178(1), 177–180. https://doi.org/10.1016/j.bbr.2006.12.007.

Marcet Rius, M., Cozzi, A., Bienboire-Frosini, C., Teruel, E., Chabaud, C., Monneret, P., Leclercq, J., Lafont-Lecuelle, C. and Pageat, P. (2018a). Selection of putative indicators of positive emotions triggered by object and social play in mini-pigs. *Applied Animal Behaviour Science* 202, 13–19. https://doi.org/10.1016/j.applanim.2018.02.002.

Marcet Rius, M., Fàbrega, E., Cozzi, A., Bienboire-Frosini, C., Descout, E., Velarde, A. and Pageat, P. (2019a). Are tail and ear movements indicators of emotions in tail-docked pigs in response to environmental enrichment? *Animals: An Open Access Journal from MDPI* 9(7). https://doi.org/10.3390/ani9070449.

Marcet Rius, M., Kalonji, G., Cozzi, A., Bienboire-Frosini, C., Monneret, P., Kowalczyk, I., Teruel, E., Codecasa, E. and Pageat, P. (2019b). Effects of straw provision, as environmental enrichment, on behavioural indicators of welfare and emotions in pigs reared in an experimental system. *Livestock Science* 221, 89–94. https://doi.org/10.1016/j.livsci.2019.01.020.

Marcet Rius, M., Pageat, P., Bienboire-Frosini, C., Teruel, E., Monneret, P., Leclercq, J., Lafont-Lecuelle, C. and Cozzi, A. (2018b). Tail and ear movements as possible indicators of emotions in pigs. *Applied Animal Behaviour Science* 205, 14–18. https://doi.org/10.1016/j.applanim.2018.05.012.

Marchant, J. N., Mendl, M. T., Rudd, A. R. and Broom, D. M. (1995). The effect of agonistic interactions on the heart rate of group-housed sows. *Applied Animal Behaviour Science* 46(1–2), 49–56. https://doi.org/10.1016/0168-1591(95)00636-2.

Marchant, J. N., Whittaker, X. and Broom, D. M. (2001). Vocalisations of the adult female domestic pig during a standard human approach test and their relationships with behavioural and heart rate measures. *Applied Animal Behaviour Science* 72(1), 23–39. https://doi.org/10.1016/S0168-1591(00)00190-8.

Marchant Forde, J. N. (2002). Piglet- and stockperson-directed sow aggression after farrowing and the relationship with a pre-farrowing, human approach test. *Applied Animal Behaviour Science* 75(2), 115–132.

Martin, J. E., Ison, S. H. and Baxter, E. M. (2015). The influence of neonatal environment on piglet play behaviour and post-weaning social and cognitive development. *Applied Animal Behaviour Science* 163, 69–79. https://doi.org/10.1016/j.applanim.2014.11.022.

Marx, G., Horn, T., Thielebein, J., Knubel, B. and Von Borell, E. (2003). Analysis of pain-related vocalization in young pigs. *Journal of Sound and Vibration* 266(3), 687–698. https://doi.org/10.1016/S0022-460X(03)00594-7.

Mason, G. and Mendl, M. (1997). Do the stereotypies of pigs, chickens and mink reflect adaptive species differences in the control of foraging? *Applied Animal Behaviour Science* 53(1–2), 45–58. https://doi.org/10.1016/S0168-1591(96)01150-1.

Mauss, I. B. and Robinson, M. D. (2009). Measures of emotion: a review. *Cognition and Emotion* 23(2), 209–237. https://doi.org/10.1080/02699930802204677.

McCraty, R. and Shaffer, F. (2015). Heart rate variability: new perspectives on physiological mechanisms, assessment of self-regulatory capacity, and health risk. *Global Advances in Health and Medicine* 4(1), 46–61. https://doi.org/10.7453/gahmj.2014.073.

McLennan, K. M., Miller, A. L., Dalla Costa, E., Stucke, D., Corke, M. J., Broom, D. M. and Leach, M. C. (2019). Conceptual and methodological issues relating to pain assessment in mammals: the development and utilisation of pain facial expression scales. *Applied Animal Behaviour Science* 217, 1–15. https://doi.org/10.1016/j.applanim.2019.06.001.

McMillan, F. D. (2008). Stress, distress, and emotion: distinctions and implications for mental well-being. *Mental Health and Well-Being in Animals*. https://doi.org/10.1002/9780470384947.ch7.

Mendl, M., Burman, O. H. P., Parker, R. M. A. and Paul, E. S. (2009). Cognitive bias as an indicator of animal emotion and welfare: emerging evidence and underlying mechanisms. *Applied Animal Behaviour Science* 118(3–4), 161–181. https://doi.org/10.1016/j.applanim.2009.02.023.

Mendl, M., Burman, O. H. P. and Paul, E. S. (2010). An integrative and functional framework for the study of animal emotion and mood. *Proceedings. Biological Sciences* 277(1696), 2895–2904. https://doi.org/10.1098/rspb.2010.0303.

Mendl, M. and Paul, E. S. (2020). Animal affect and decision-making. *Neuroscience and Biobehavioral Reviews* 112, 144–163. https://doi.org/10.1016/j.neubiorev.2020.01.025.

Minero, M., Dalla Costa, E., Dai, F., Canali, E., Barbieri, S., Zanella, A., Pascuzzo, R. and Wemelsfelder, F. (2018). Using qualitative behaviour assessment (QBA) to explore the emotional state of horses and its association with human-animal relationship. *Applied Animal Behaviour Science* 204, 53–59. https://doi.org/10.1016/j.applanim.2018.04.008.

Mormède, P., Andanson, S., Aupérin, B., Beerda, B., Guémené, D., Malmkvist, J., Manteca, X., Manteuffel, G., Prunet, P., van Reenen, C. G., Richard, S. and Veissier, I. (2007). Exploration of the hypothalamic-pituitary-adrenal function as a tool to evaluate animal welfare. *Physiology and Behavior* 92(3), 317–339. https://doi.org/10.1016/j.physbeh.2006.12.003.

Murphy, E., Nordquist, R. E. and van der Staay, F. J. (2013). Responses of conventional pigs and Göttingen miniature pigs in an active choice judgement bias task. *Applied*

Animal Behaviour Science 148(1-2), 64-76. https://doi.org/10.1016/j.applanim.2013.07.011.

Murphy, E., Nordquist, R. E. and van der Staay, F. J. (2014). A review of behavioural methods to study emotion and mood in pigs, Sus scrofa. *Applied Animal Behaviour Science* 159, 9-28. https://doi.org/10.1016/j.applanim.2014.08.002.

Neville, V., Nakagawa, S., Zidar, J., Paul, E. S., Lagisz, M., Bateson, M., Løvlie, H. and Mendl, M. (2020). Pharmacological manipulations of judgement bias: a systematic review and meta-analysis. *Neuroscience and Biobehavioral Reviews* 108, 269-286. https://doi.org/10.1016/j.neubiorev.2019.11.008.

Newberry, R. C., Wood-Gush, D. G. M. and Hall, J. W. (1988). Playful behaviour of piglets. *Behavioural Processes* 17(3), 205-216. https://doi.org/10.1016/0376-6357(88)90004-6.

O'Connor, E. A., Parker, M. O., McLeman, M. A., Demmers, T. G., Lowe, J. C., Cui, L., Davey, E. L., Owen, R. C., Wathes, C. M. and Abeyesinghe, S. M. (2010). The impact of chronic environmental stressors on growing pigs, Sus scrofa (Part 1): stress physiology, production and play behaviour. *Animal: An International Journal of Animal Bioscience* 4(11), 1899-1909. https://doi.org/10.1017/S1751731110001072.

Palagi, E., Cordoni, G. and Borgognini Tarli, S. M. (2004). Immediate and delayed benefits of play behaviour: new evidence from Chimpanzees (Pan troglodytes). *Ethology* 110(12), 949-962. https://doi.org/10.1111/j.1439-0310.2004.01035.x.

Palagi, E., Paoli, T. and Tarli, S. B. (2006). Short-term benefits of play behavior and conflict prevention in Pan paniscus. *International Journal of Primatology* 27(5), 1257-1270. https://doi.org/10.1007/s10764-006-9071-y.

Paul, E. S., Harding, E. J. and Mendl, M. (2005). Measuring emotional processes in animals: the utility of a cognitive approach. *Neuroscience and Biobehavioral Reviews* 29(3), 469-491. https://doi.org/10.1016/j.neubiorev.2005.01.002.

Paul, E. S. and Mendl, M. T. (2018). Animal emotion: descriptive and prescriptive definitions and their implications for a comparative perspective. *Applied Animal Behaviour Science* 205, 202-209. https://doi.org/10.1016/j.applanim.2018.01.008.

Paul, E. S., Sher, S., Tamietto, M., Winkielman, P. and Mendl, M. T. (2020). Towards a comparative science of emotion: affect and consciousness in humans and animals. *Neuroscience and Biobehavioral Reviews* 108, 749-770 https://doi.org/10.1016/j.neubiorev.2019.11.014.

Pedersen, L. J., Holm, L., Jensen, M. B. and Jørgensen, E. (2005). The strength of pigs' preferences for different rooting materials measured using concurrent schedules of reinforcement. *Applied Animal Behaviour Science* 94(1-2), 31-48. https://doi.org/10.1016/j.applanim.2004.11.023.

Pellis, S. and Pellis, V. (2013). *The Playful Brain: Venturing to the Limits of Neuroscience*. Oneworld Publications. Available at: https://books.google.com/books?hl=en&lr=&id=khy9DwAAQBAJ&oi=fnd&pg=PT7&dq=The+Playful+Brain:+Venturing+to+the+Limits+of+Neuroscience&ots=SJU4I1prQM&sig=QswpbGuBqWrqFe39HAiBGLzEDnE.

Poletto, R., Steibel, J. P., Siegford, J. M. and Zanella, A. J. (2006). Effects of early weaning and social isolation on the expression of glucocorticoid and mineralocorticoid receptor and 11β-hydroxysteroid dehydrogenase 1 and 2 mRNAs in the frontal cortex and hippocampus of piglets. *Brain Research* 1067(1), 36-42. https://doi.org/10.1016/j.brainres.2005.10.001.

Preston, S. D. and De Waal, F. B. M. (2012). The communication of emotions and the possibility of empathy in animals. In: *Altruism and Altruistic Love: Science, Philosophy, and Religion in Dialogue*. https://doi.org/10.1093/acprof:oso/9780195143584.003.0025.

Puppe, B., Schön, P. C., Tuchscherer, A. and Manteuffel, G. (2005). Castration-induced vocalisation in domestic piglets, Sus scrofa: complex and specific alterations of the vocal quality. *Applied Animal Behaviour Science* 95(1-2), 67-78. https://doi.org/10.1016/j.applanim.2005.05.001.

Rault, J. L. (2016). Effects of positive and negative human contacts and intranasal oxytocin on cerebrospinal fluid oxytocin. *Psychoneuroendocrinology* 69, 60-66. https://doi.org/10.1016/j.psyneuen.2016.03.015.

Rault, J. L., van den Munkhof, M. and Buisman-Pijlman, F. T. A. (2017). Oxytocin as an indicator of psychological and social well-being in domesticated animals: a critical review. *Frontiers in Psychology* 8, 1521. https://doi.org/10.3389/fpsyg.2017.01521.

Reimert, I., Bolhuis, J. E., Kemp, B. and Rodenburg, T. B. (2013). Indicators of positive and negative emotions and emotional contagion in pigs. *Physiology and Behavior* 109, 42-50. https://doi.org/10.1016/j.physbeh.2012.11.002.

Reimert, I., Bolhuis, J. E., Kemp, B. and Rodenburg, T. B. (2015). Emotions on the loose: emotional contagion and the role of oxytocin in pigs. *Animal Cognition* 18(2), 517-532. https://doi.org/10.1007/s10071-014-0820-6.

Reimert, I., Fong, S., Rodenburg, T. B. and Bolhuis, J. E. (2017). Emotional states and emotional contagion in pigs after exposure to a positive and negative treatment. *Applied Animal Behaviour Science* 193, 37-42. https://doi.org/10.1016/j.applanim.2017.03.009.

Roelofs, S., Boleij, H., Nordquist, R. E. and van der Staay, F. J. (2016). Making decisions under ambiguity: judgment bias tasks for assessing emotional state in animals. *Frontiers in Behavioral Neuroscience* 10, 119. https://doi.org/10.3389/fnbeh.2016.00119.

Roelofs, S., Nordquist, R. E. and van der Staay, F. J. (2017). Female and male pigs' performance in a spatial holeboard and judgment bias task. *Applied Animal Behaviour Science* 191, 5-16. https://doi.org/10.1016/j.applanim.2017.01.016.

Rolls, E. T. (2005). *Emotion Explained*. Oxford University Press. Available at: https://doi.org/10.1093/acprof:oso/ 9780198570035.001.0001.

Rolls, E. T. (2014). *Emotion and Decision-Making Explained*. Oxford University Press. Available at: https://books.google.com/books?hl=en&lr=&id=t45nAgAAQBAJ&oi=fnd&pg=PP1&ots=OCExgmKSpC&sig=ig3E-aaZwW5W6kNbW4smagbshQs.

Ruis, M. A., de Groot, J., te Brake, J. H., Dinand Ekkel, E., van de Burgwal, J. A., Erkens, J. H., Engel, B., Buist, W. G., Blokhuis, H. J. and Koolhaas, J. M. (2001a). Behavioural and physiological consequences of acute social defeat in growing gilts: effects of the social environment. *Applied Animal Behaviour Science* 70(3), 201-225. https://doi.org/10.1016/S0168-1591(00)00150-7.

Ruis, M. A. W., Te Brake, J. H. A., Engel, B., Buist, W. G., Blokhuis, H. J. and Koolhaas, J. M. (2001b). Adaptation to social isolation acute and long-term stress responses of growing gilts with different coping characteristics. *Physiology and Behavior* 73(4), 541-551. https://doi.org/10.1016/S0031-9384(01)00548-0.

Russell, J. A. (2003). Core affect and the psychological construction of emotion. *Psychological Review* 110(1), 145-172. https://doi.org/10.1037/0033-295X.110.1.145.

Russell, J. A. and Barrett, L. F. (1999). Core affect, prototypical emotional episodes, and other things called emotion: dissecting the elephant. *Journal of Personality and Social Psychology* 76(5), 805–819. https://doi.org/10.1037//0022-3514.76.5.805.

Rutherford, K. M. D., Donald, R. D., Lawrence, A. B. and Wemelsfelder, F. (2012). Qualitative behavioural assessment of emotionality in pigs. *Applied Animal Behaviour Science* 139(3–4), 218–224. https://doi.org/10.1016/j.applanim.2012.04.004.

Salmeto, A. L., Hymel, K. A., Carpenter, E. C., Brilot, B. O., Bateson, M. and Sufka, K. J. (2011). Cognitive bias in the chick anxiety-depression model. *Brain Research* 1373, 124–130. https://doi.org/10.1016/j.brainres.2010.12.007.

Scherer, K. R. (1984). On the nature and function of emotion: a component process approach. In: Scherer, K. R. and Ekman, P. (Eds), *Approaches to Emotion* (pp. 293-317). Erlbaum, NJ.

Schrader, L. and Ladewig, J. (1999). Temporal differences in the responses of the pituitary adrenocortical axis, the sympathoadrenomedullar axis, heart rate, and behaviour to a daily repeated stressor in domestic pigs. *Physiology and Behavior* 66(5), 775–783. https://doi.org/10.1016/S0031-9384(99)00015-3.

Scollo, A., Gottardo, F., Contiero, B. and Edwards, S. A. (2014). Does stocking density modify affective state in pigs as assessed by cognitive bias, behavioural and physiological parameters? *Applied Animal Behaviour Science* 153, 26–35. https://doi.org/10.1016/j.applanim.2014.01.006.

Scott, K., Laws, D. M., Courboulay, V., Meunier-Salaün, M. C. and Edwards, S. A. (2009). Comparison of methods to assess fear of humans in sows. *Applied Animal Behaviour Science* 118(1–2), 36–41. https://doi.org/10.1016/j.applanim.2009.02.004.

Sotocina, S. G., Sorge, R. E., Zaloum, A., Tuttle, A. H., Martin, L. J., Wieskopf, J. S., Mapplebeck, J. C. S., Wei, P., Zhan, S., Zhang, S., McDougall, J. J., King, O. D. and Mogil, J. S. (2011). The Rat Grimace Scale: a partially automated method for quantifying pain in the laboratory rat via facial expressions. *Molecular Pain* 7, 55. https://doi.org/10.1186/1744-8069-7-55.

Špinka, M. (2006). How important is natural behaviour in animal farming systems? *Applied Animal Behaviour Science* 100(1–2), 117–128. https://doi.org/10.1016/j.applanim.2006.04.006.

Špinka, M. (2012). Social dimension of emotions and its implication for animal welfare. *Applied Animal Behaviour Science* 138(3–4), 170–181. https://doi.org/10.1016/j.applanim.2012.02.005.

Špinka, M., Newberry, R. C. and Bekoff, M. (2001). Mammalian play: training for the unexpected. *Quarterly Review of Biology* 76(2), 141–168. https://doi.org/10.1086/393866.

Statham, P., Campbell, N., Hannuna, S., Paul, E. S., Colborne, R., Browne, W. and Mendl, M. T. (2011). Development of a novel and automated measure of affect in pigs. Proceedings of the 5th International Conference on the Assessment of Animal Welfare at Farm and Group Level: WAFL 2011.

Statham, P., Hannuna, S., Jones, S., Campbell, N., Robert Colborne, G., Browne, W. J., Paul, E. S. and Mendl, M. (2020). Quantifying defence cascade responses as indicators of pig affect and welfare using computer vision methods. *Scientific Reports* 10(1), 8933. https://doi.org/10.1038/s41598-020-65954-6.

Stockman, C. A., Collins, T., Barnes, A. L., Miller, D., Wickham, S. L., Beatty, D. T., Blache, D., Wemelsfelder, F. and Fleming, P. A. (2011). Qualitative behavioural assessment and quantitative physiological measurement of cattle naïve and habituated to road transport. *Animal Production Science* 51(3). https://doi.org/10.1071/AN10122.

Stracke, J., Otten, W., Tuchscherer, A., Puppe, B. and Düpjan, S. (2017a). Serotonin depletion induces pessimistic-like behavior in a cognitive bias paradigm in pigs. *Physiology and Behavior* 174, 18–26. https://doi.org/10.1016/j.physbeh.2017.02.036.

Stracke, J., Otten, W., Tuchscherer, A., Witthahn, M., Metges, C. C., Puppe, B. and Düpjan, S. (2017b). Dietary tryptophan supplementation and affective state in pigs. *Journal of Veterinary Behavior* 20, 82–90. https://doi.org/10.1016/j.jveb.2017.03.009.

Studnitz, M., Jensen, M. B. and Pedersen, L. J. (2007). Why do pigs root and in what will they root? *Applied Animal Behaviour Science* 107(3–4), 183–197. https://doi.org/10.1016/j.applanim.2006.11.013.

Tanida, H., Miura, A., Tanaka, T. and Yoshimoto, T. (1996). Behavioral responses of piglets to darkness and shadows. *Applied Animal Behaviour Science* 49(2), 173–183. https://doi.org/10.1016/0168-1591(96)01039-8.

Taylor, A. A. and Weary, D. M. (2000). Vocal responses of piglets to castration: identifying procedural sources of pain. *Applied Animal Behaviour Science* 70(1), 17–26. https://doi.org/10.1016/S0168-1591(00)00143-X.

Temple, D., Manteca, X., Dalmau, A. and Velarde, A. (2013). Assessment of test-retest reliability of animal-based measures on growing pig farms. *Livestock Science* 151(1), 35–45. https://doi.org/10.1016/j.livsci.2012.10.012.

Temple, D., Manteca, X., Velarde, A. and Dalmau, A. (2011). Assessment of animal welfare through behavioural parameters in Iberian pigs in intensive and extensive conditions. *Applied Animal Behaviour Science* 131(1–2), 29–39. https://doi.org/10.1016/j.applanim.2011.01.013.

Terlouw, E. M. C. and Porcher, J. (2005). Repeated handling of pigs during rearing. I. Refusal of contact by the handler and reactivity to familiar and unfamiliar humans. *Journal of Animal Science* 83(7), 1653–1663. https://doi.org/10.2527/2005.8371653x.

Tuchscherer, M., Kanitz, E., Puppe, B., Tuchscherer, A. and Stabenow, B. (2004). Effects of postnatal social isolation on hormonal and immune responses of pigs to an acute endotoxin challenge. *Physiology and Behavior* 82(2–3), 503–511. https://doi.org/10.1016/j.physbeh.2004.04.056.

Vanderschuren, L. J. M. J. (2010). How the brain makes play fun. *American Journal of Play* 2(3), 315–337.

Vanderschuren, L. J. M. J., Stein, E. A., Wiegant, V. M. and Van Ree, J. M. (1995). Social play alters regional brain opioid receptor binding in juvenile rats. *Brain Research* 680(1–2), 148–156. https://doi.org/10.1016/0006-8993(95)00256-P.

Veissier, I., Boissy, A., Désiré, L. and Greiveldinger, L. (2009). Animals' emotions: studies in sheep using appraisal theories. *Animal Welfare* 18(4), 347–354.

Viscardi, A. V., Hunniford, M., Lawlis, P., Leach, M. and Turner, P. V. (2017). Development of a piglet grimace scale to evaluate piglet pain using facial expressions following castration and tail docking: a pilot study. *Frontiers in Veterinary Science* 4, 51. https://doi.org/10.3389/fvets.2017.00051.

von Borell, E., Langbein, J., Després, G., Hansen, S., Leterrier, C., Marchant-Forde, J., Marchant-Forde, R., Minero, M., Mohr, E., Prunier, A., Valance, D. and Veissier, I. (2007). Heart rate variability as a measure of autonomic regulation of cardiac activity for assessing stress and welfare in farm animals - a review. *Physiology and Behavior* 92(3), 293–316. https://doi.org/10.1016/j.physbeh.2007.01.007.

von Borell, E. and Veissier, I. (2007). Special section - stress and welfare in farm animals. *Physiology and Behavior* 92(3), 291–292. https://doi.org/10.1016/j.physbeh.2007.09.006.

Vullo, C., Barbieri, S., Catone, G., Graïc, J. M., Magaletti, M., Di Rosa, A., Motta, A., Tremolada, C., Canali, E. and Costa, E. D. (2020). Is the Piglet Grimace Scale (PGS) a useful welfare indicator to assess pain after cryptorchidectomy in growing pigs? *Animals* 10(3). https://doi.org/10.3390/ani10030412.

Waiblinger, S., Boivin, X., Pedersen, V., Tosi, M. V., Janczak, A. M., Visser, E. K. and Jones, R. B. (2006). Assessing the human-animal relationship in farmed species: a critical review. *Applied Animal Behaviour Science* 101(3–4), 185–242. https://doi.org/10.1016/j.applanim.2006.02.001.

Weary, D. M., Braithwaite, L. A. and Fraser, D. (1998). Vocal response to pain in piglets. *Applied Animal Behaviour Science* 56(2–4), 161–172. https://doi.org/10.1016/S0168-1591(97)00092-0.

Weary, D. M. and Fraser, D. (1995). Calling by domestic piglets: reliable signals of need? *Animal Behaviour* 50(4), 1047–1055. https://doi.org/10.1016/0003-3472(95)80105-7.

Weller, J. E., Camerlink, I., Turner, S. P., Farish, M. and Arnott, G. (2019). Socialisation and its effect on play behaviour and aggression in the domestic pig (Sus scrofa). *Scientific Reports* 9(1), 4180. https://doi.org/10.1038/s41598-019-40980-1.

Wemelsfelder, F., Hunter, T. E. A., Mendl, M. T. and Lawrence, A. B. (2001). Assessing the "whole animal": a free choice profiling approach. *Animal Behaviour* 62(2), 209–220. https://doi.org/10.1006/anbe.2001.1741.

Wemelsfelder, F., Hunter, E. A., Mendl, M. T. and Lawrence, A. B. (2000). The spontaneous qualitative assessment of behavioural expressions in pigs: first explorations of a novel methodology for integrative animal welfare measurement. *Applied Animal Behaviour Science* 67(3), 193–215. https://doi.org/10.1016/S0168-1591(99)00093-3.

Wemelsfelder, F., Hunter, A. E., Paul, E. S. and Lawrence, A. B. (2012). Assessing pig body language: agreement and consistency between pig farmers, veterinarians, and animal activists. *Journal of Animal Science* 90(10), 3652–3665. https://doi.org/10.2527/jas.2011-4691.

Wemelsfelder, F., Nevison, I. and Lawrence, A. B. (2009). The effect of perceived environmental background on qualitative assessments of pig behaviour. *Animal Behaviour* 78(2), 477–484. https://doi.org/10.1016/j.anbehav.2009.06.005.

White, R. G., DeShazer, J. A., Tressler, C. J., Borcher, G. M., Davey, S., Waninge, A., Parkhurst, A. M., Milanuk, M. J. and Clemens, E. T. (1995). Vocalization and physiological response of pigs during castration with or without a local anesthetic. *Journal of Animal Science* 73(2), 381–386. https://doi.org/10.2527/1995.732381x.

Winkielman, P., Berridge, K. C. and Wilbarger, J. L. (2005). Unconscious affective reactions to masked happy versus angry faces influence consumption behavior and judgments of value. *Personality and Social Psychology Bulletin* 31(1), 121–135. https://doi.org/10.1177/0146167204271309.

Winslow, J. T. and Insel, T. R. (2002). The social deficits of the oxytocin knockout mouse. *Neuropeptides* 36(2–3), 221–229. https://doi.org/10.1054/npep.2002.0909.

Wood-Gush, D. G. M. and Vestergaard, K. (1991). The seeking of novelty and its relation to play. *Animal Behaviour* 42(4), 599–606. https://doi.org/10.1016/S0003-3472(05)80243-X.

Würbel, H. (2009). Ethology applied to animal ethics. *Applied Animal Behaviour Science* 118(3–4), 118–127. https://doi.org/10.1016/j.applanim.2009.02.019.

Yeates, J. W. and Main, D. C. J. (2008). Assessment of positive welfare: a review. *Veterinary Journal* 175(3), 293–300. https://doi.org/10.1016/j.tvjl.2007.05.009.

Zebunke, M., Langbein, J., Manteuffel, G. and Puppe, B. (2011). Autonomic reactions indicating positive affect during acoustic reward learning in domestic pigs. *Animal Behaviour* 81(2), 481–489. https://doi.org/10.1016/j.anbehav.2010.11.023.

Chapter 14

Welfare assessment of pigs

Björn Forkman, University of Copenhagen, Denmark

1 Introduction
2 Welfare comparisons and frameworks
3 Input and output-related welfare measures
4 Required properties for welfare measures: validity, repeatability and feasibility
5 Feeding-related welfare indicators
6 Housing-related welfare indicators
7 Health- and pain-related welfare indicators
8 Behavioural welfare indicators
9 General indicators of poor and positive welfare
10 Future trends
11 Where to look for further information
12 References

1 Introduction

The purpose of a welfare protocol varies, and with it the way it is done is also varied. It may be done for different certification purposes, to assess animal welfare in scientific projects or as a tool for self-improvement (Pfeifer et al. 2020). Animal welfare relevant measures are also often included in other types of certification schemes, for example, organic protocols, or the 'Real welfare' part of the Red tractor scheme in the UK. These different aims of the assessment may lead to different approaches.

For some purposes the welfare protocol resembles legislation in that it contains a number of different criteria and that all of these need to be fulfilled for the farm to be certified. Like legislation this type of welfare assessment aims to guarantee a minimum level of animal welfare, rather than giving an assessment of the level of welfare of the animals at the farm. In this sense this approach is not a proper assessment. Most of the current certification schemes follow this approach, for example, RSPCA assured (previously Freedom food,

http://dx.doi.org/10.19103/AS.2020.0081.14

RSPCA 2016), the Dutch Beter Leven (Beter Leven 2020) and the Danish animal welfare certification scheme (FVST 2019).

Another approach to animal welfare assessment aims to assess the actual welfare state of the animals on the farm. In this approach a number of indicators are integrated to either a numerical value or alternatively to one of a number of welfare levels. In this approach there is no real minimum level for each indicator (although it is possible to use, for example, legislative minimum levels), and excellence in anyone indicator will affect the level attained. Examples of this approach are TGI 35L (Bartussek 1995) and Welfare Quality® (Botreau et al. 2009).

2 Welfare comparisons and frameworks

While the aim of most of the welfare assessment schemes is to act as certification schemes to guarantee a certain level of animal welfare to consumers and retailers, it is also possible to imagine other uses. One such is to make international comparisons to understand the processes that affect the levels of welfare in different countries with the final aim of finding strategies to improve welfare on a national level. It is difficult to find comparable information on the level of welfare for each country, but by using the national legislation as well as information from relevant certification schemes as proxys a value of the required welfare level may be found. Using this approach Sandøe et al. (2020) found that although the UK, the Netherlands and Denmark all have very similar legislation for pig production, the fact that there is much more outdoor production in the UK means that it comes out as having much better welfare than the others. Denmark and the Netherlands are similar in that they are exporting a large proportion of their pig production; however, the important market share of the animal welfare label Beter Leven gives the Netherlands a better overall score of pig welfare.

The result from this type of study has shown the important effect of animal welfare schemes to increase the overall animal welfare in a country (Sandøe et al. 2020).

Table 1 The five freedoms as defined by the Farm animal welfare council (FAWC 2009)

1. **Freedom from hunger and thirst** by ready access to fresh water and a diet to maintain full health and vigour
2. **Freedom from discomfort** by providing an appropriate environment including shelter and a comfortable resting area
3. **Freedom from pain, injury or disease** by prevention or rapid diagnosis and treatment
4. **Freedom to express (most) normal behaviour** by providing sufficient space, proper facilities and company of the animal's own kind
5. **Freedom from fear and distress** by ensuring conditions and treatment which avoid mental suffering

Table 2 The four areas of concern and 12 criteria of Welfare Quality® (Botreau et al. 2009)

Area of concern	Criteria
Good feeding	Absence of prolonged hunger
	Absence of prolonged thirst
Good housing	Comfort around resting
	Thermal comfort
	Ease of movement
Good health	Absence of injuries
	Absence of disease
	Absence of pain induced by management procedures
Appropriate behaviour	Expression of social behaviours
	Expression of other behaviours
	Good human-animal relationship
	Positive emotional state

Animal welfare is a multi-dimensional concept, and when assessing animal welfare it is therefore useful to have a framework that captures the many different aspects of it. One of the most influential animal welfare frameworks is the five freedoms (FAWC 2009). It forms the basis of the legislation in many countries as well as the animal welfare definition of the OIE (OIE 2020). The framework has been defined as four examples of 'freedom from' and one 'freedom to' (Table 1). In its first form the emphasis is therefore on the absence of bad welfare. A later addition, 'A life worth living' (FAWC 2009), has included the perception of animal welfare consisting not only of absence of bad welfare but also of positive experiences.

A more recent approach of creating a framework for animal welfare is the five domains approach (Mellor and Reid 1994; Mellor 2017). It focusses on four areas of animal welfare: feeding, housing, health and behaviour, and how they influence a fifth one, the experiences of the animal. The framework thus places an explicit and heavy emphasis on the experiences of the animal.

Both the five freedoms and the five domains frameworks are high level frameworks with broad areas of concern. Another framework that more directly leads into which indicators to apply is formulated in the Welfare Quality® project (Botreau et al. 2009). It divides animal welfare into four areas of concern which in turn are divided into 12 criteria (Table 2).

3 Input and output-related welfare measures

The measures employed in animal welfare assessments can be divided up into two groups - measures of input and measures of output. The input measures are measures of external factors that may affect the welfare of the animal. These may in turn be divided into 'resource-based measures', for example, access to rooting material or stocking density, and 'management-based measures', for example, whether pain relief is given at castration or routines for mixing pigs. Resource-based measures are very common in legislation and in many certification schemes, for example, in the ones with fixed requirements mentioned above. It is however well known that even with very similar or identical systems and resources there may still be a large variation in the welfare of the animals on the farms (e.g. Hemsworth 2018).

Output measures are measures of the effect on the animal of the input measures, for example, wounds, stereotypic behaviour or body weight. Most of these are 'animal-based measures', and the advantage with these is that they are thought to be closer to the experiences of the animal (EFSA 2015), and thus a better measure of animal welfare. A welfare assessment protocol that emphasizes the use on animal-based measures is Welfare Quality® (Botreau et al. 2007b, Table 3). It is also used in the German KBTL (Schrader et al. 2016) as well as in Assurewel (Assurewel is part of the RSPCA assured that is not included in the certification scheme at present).

If the aim of the assessment is a measure of the welfare of the animals on farm, and not only ascertaining a minimum level of welfare, then different welfare relevant indicators need to be integrated (Spoolder et al. 2003). For this to be done a common scale needs to be found on which to measure the welfare consequences associated with each indicator. Attempts have been made to make animals assign values to different resources or experiences using, for example, consumer-demand experiments (Patterson-Kane et al. 2008), for some resources this has yielded important results for, for example, different types of rooting material (Studnitz et al. 2007). However the number of studies of this kind is not very large and some of the experiences/resources are impossible to evaluate in this way, for example, the impaired welfare value of having a hernia, or being tail bitten. Another approach that has the potential of

Table 3 Illustration of the emphasis put on output (animal) measurements and input (management and resource) measurements of different animal welfare protocols, as well as the emphasis on absolute requirements and integrated assessment

Name	Output	Input	Integrated value	Abs. req.	Reference
Welfare quality®	X	(X)	X		Welfare quality® (2009)
KBTL	X	(X)			Schrader et al. (2016)
TGI 35 L		X	X		Bartussek (1995)
RSPCA assured		X		X	RSPCA (2016)
Beter Leven		X		X	Beter Leven (2020)

comparing even very different experiences, even aversive ones, is studies using the emotional bias test (see Chapter 13 of this volume). To date few studies that compare very different experiences have been done, but an example how it may be done is a study in calves comparing the effect of cow-calf separation and dehorning (Daros et al. 2014).

In practice the weights assigned to the indicators as well as their interaction is usually done through expert opinion. There are a number of different approaches to do this, which have been summarized by EFSA (2014).

Animal welfare is a multi-dimensional concept, and to capture it requires a number of measures (Fraser et al. 1997). One way of approaching this is to require that the same criteria are fulfilled thus removing the need for integration of the criteria. In many cases there is a wish for single score that reflects the measured welfare of the animal on farm. In this case the values for the welfare indicators have to be integrated. There is however no standardized way of assessing the overall welfare status of a group of animals (for a discussion of the ethical assumptions behind the different alternatives see Sandøe et al. 2019).

There are a number of considerations when attempting to integrate indicators. If, for example, the overall welfare of a farm is the average value of all measures, then this means that if a new indicator is added it will dilute the value of the existing indicators. A welfare protocol with a very large number of indicators will therefore typically give very little weight to each of them. One way of handling this may be to integrate at different levels (Botreau et al. 2007a), for example, first taking the average of all measures related to Good feeding, thereafter all the measures of Good health and so on, and subsequently taking the average of the two averages to form the final score. This means that the single measures are less devalued by the total number of measures. It also means however that if there are, for example, few measures for Good feeding compared to those for Good health then each measure in the Good feeding will come to have a very strong influence on the overall score. See Sandøe et al. (2019) for a general discussion, for a specific example in pigs see Czycholl et al. (2017) and for cattle see de Graaf et al. (2017).

4 Required properties for welfare measures: validity, repeatability and feasibility

The validity of an indicator reflects the degree to which it measures what it is supposed to measure. In animal welfare research two types of validity are often cited. The first is convergent validity, that is, if two indicators are expected to measure the same thing then they should show a positive correlation. If tail biting is a sign of stress then we would expect, for example, stress hormone levels to positively correlate with levels of tail biting (Zupan and Zanella 2017). A second weaker type of validity is face validity, it relies solely on the scientists' subjective opinion, the assumption that various diseases are indicative of bad welfare in many cases relies on face validity.

The term validity is often confused with objectivity; a measure with a high degree of objectivity is one that is not affected by the observer (see the paragraph on repeatability below). Heart rate may be considered an objective measure of fear in that it can be measured automatically with a minimum level of interference. The validity of the measure may be quite low however, since heart rate is affected by many other things than fear.

There are several forms of repeatability, with the most common being intra- and inter-observer reliability. Intra-observer reliability is the consistency of observations by the same observer, whereas inter-observer reliability is the consistency between individuals. It is important that the observers refrain from forming any prior judgement of the animal welfare on the farm since preconceived judgements have been shown to have a very strong effect on the assessment (Tuyttens et al. 2014).

It is not only the ability to assess a given condition that affects the repeatability of a measure however. A condition that is observed on an individual animal, for example, a rectal prolapse, may have a high inter-observer reliability in that everyone looking at the animal agree on the observation. If rectal prolapse is very rare it is however unlikely to have a high repeatability either between farms or over time (Friedrich et al. 2019a, 2020b). In most cases the aim of a welfare assessment is not only to give a momentary picture of the welfare state of the animals on farm, but to give an assessment that is representative of a longer time period, for example, half a year or a year. The most useful indicators for a welfare assessment is therefore not only those that relate to welfare or those with a high inter-observer reliability, but those that are common enough that they are representative of the farm over the relevant time period. In most cases it is impossible or impractical to assess all animals on a farm, and it is therefore necessary to sample the animals. Care should be taken to sample animals or pens as randomly as possible. Even with a completely random sample and with a very large sample size indicators with low prevalences are very hard to assess with any accuracy (Mullan et al. 2009).

The feasibility of an indicator is a very important aspect in choosing one. Feasibility in this context is the ease with which the indicator can be observed (does it, for example, require extensive training), the time needed to observe it and finally if any tools are needed to observe it (e.g. measuring tape). Although the time to assess an indicator may seem short, a protocol often contains many indicators and when it is to be repeated many times one visit to a large farm may take more than one day. In many cases pig farms have strict quarantine regulations and farmers may wish a bare minimum of tools to be brought into the farm, especially if they have been used in other pig farms previously.

The following sections are divided up into seven parts; the four areas of concern of Welfare Quality® (Good feeding, Good housing, Good health and Appropriate behaviour), and three additional parts, indicators of bad welfare, indicators of positive welfare, possible iceberg indicators. Each section contains one or two tables with an assessment of the current knowledge of the welfare relevance, the reproducibility and the practicality of the measures.

5 Feeding-related welfare indicators

5.1 Body condition score

A low body condition score is an effect of an intake that is lower than what is needed to maintain body weight. It is however not clear how this translates into the feeling of hunger in the animal (D'Eath et al. 2009). Charette et al. (1996) found poor inter-observer reliability, and Friedrich et al. (2019b) found poor test-retest reliability when looking at farms over a longer time period. Overall body condition score therefore seems to be both a measure with an unclear welfare relevance, and a medium reproducibility.

5.2 Roughage

Gestating sows that are kept with restricted feed need to have access to roughage, the roughage may affect some aspects of hunger, for example, giving a more natural feeding behaviour, however, the animals are still prepared to work more for feed than non-feed restricted sows (D'Eath et al. 2009). A reason for including access to roughage as an important welfare indicator not only for gestating sows but also for growing pigs, is that it decreases the risk of gastric ulcers (Herskin et al. 2016), a welfare problem that is not readily detectable. If the criterion is that the animals should have constant access to roughage then the reproducibility is probably relatively high.

5.3 Weaning age

At 21 days piglets have a less developed intestine than those weaned later (Callesen et al. 2007). The effect of weaning age on welfare relevant parameters lasts until (at least) the piglets are 6 weeks old (Miller et al. 2007; Wellock et al. 2007). This measure can only be obtained from the farmer and not observed directly. If nursing sows are used it is important to ask about the weaning age and not the time the sow spends with piglets in the farrowing unit.

5.4 Number of drinkers

It is difficult to find a good animal-based measure for thirst. A number of welfare assessment schemes recommend a maximum of 10 pigs/drinker as a resource-based measure (e.g. Welfare Quality®, RSPCA assured), but there is little scientific evidence for the threshold (e.g. Turner et al. 1999, 2000). A requirement for anyone pig to have access to at least two different drinkers can be seen as an insurance in case of one drinker not functioning properly or one drinker becoming very dirty (see Section 5.5). Number of functioning drinkers can be hard to assess for sows in the insemination or gestation unit if a system with feeding boxes is used. For growing pigs the assessment is relatively easy.

5.5 Types of drinkers

There are some indications that pigs prefer float bowl drinkers and nipple drinkers over push-lever drinkers (Torrey and Widowski 2006). Float bowl drinker are however more easily soiled than nipple drinkers (Bøe and Kjelvik 2011). There does not seem to be a very strong welfare effect of the different drinker types.

5.6 Water quality

While animals are sensitive to taste and quality of water it can be difficult to assess the criteria or level of soiling that is relevant. Unclean water is avoided (Phillips and Phillips 1999). The assessment of whether a drinker is clean or not is subjective and hard to describe accurately (Table 4).

6 Housing-related welfare indicators

6.1 Stocking density

Stocking density in itself may be an important factor. More space gives the animals the possibility of showing more behaviours (Nannoni et al. 2019), and in

Table 4 Assessment of the welfare consequence associated with the indicator, the reproducibility as well as the practicality of measuring it. Each aspect of the measure is marked from weak ('*') to strong ('***') for that aspect. For references see the relevant paragraph

Measure	Animal group	Welfare relevance	Reproduc.	Practicality
Body condition	Sows, pigs	*	*	***
Roughage	Sows/pigs	***/**	**	***
Weaning age	Piglets	**	***	**
No. func. drinkers	Sows/pigs	*	*/***	*/***
Type of drinkers	Sows, pigs, piglets	*	***	***
Water quality	Sows, pigs, piglets	**	*	**

general a lower stocking density therefore gives better welfare (Munsterhjelm et al. 2015). For an accurate evaluation of the stocking density not only the number but also the size of the pigs in the pen needs to be assessed (Table 5).

6.2 Quality of flooring

The quality of the flooring includes both the space of the solid flooring as well as possible bedding materials. A meta-analytical study showed a strong effect of the solid floor area with an increase in lying time with increasing solid floor time (growing-finishing pigs, Averós et al. 2010). A number of aspects of the quality of the floor can be assessed on the animal as prevalence of bursitis, and shoulder wounds (e.g. KilBride et al. 2009; Munsterhjelm et al. 2015); however, Friedrich et al. (2019b) found bad test-retest values for bursitis. If measures that are related to the risk of bad welfare are included in the protocol then other parts of the quality of the floor that may put the welfare of the animals at risk, such as too large distances between the slats or broken slats, can be assessed through visual inspection.

6.3 Crates

See a discussion of Baxter et al. (2012) for a discussion of the welfare consequences of different types of confinement/crates in the farrowing section. There is a consensus that crating impairs the welfare of sows, severely restricting their behavioural repertoire and, for example, preventing them from

Table 5 Assessment of the welfare consequence associated with the indicator, the reproducibility as well as the practicality of measuring it. Each aspect of the measure is marked from weak ('*') to strong ('***') for that aspect. For references see the relevant paragraph

Measure	Animal group	Welfare relevance	Reproduc.	Practicality
Stocking density	Sows, pigs, piglets	***	***	***
Flooring	Sows, pigs, piglets	***	***	***
Crates	Sows	**	**	**
Thermal comfort	Sows, pigs, piglets	**	***/*	**
Air quality	Sows, pigs	***	*	*
Enrichment	Sows, pigs, piglets	***	***	***
Manure	Sows, pigs	**	**	***
Outdoor access	Sows, pigs, piglets	**	***	***

nest building for which they are highly motivated. There is however also an ongoing discussion on the effect of crates on piglet mortality with a number of researchers arguing that crating decreases piglet mortality (e.g. Glencorse et al. 2019), whereas others have found it to have no effect (Weber et al. 2007). It is likely that factors such as the size of the pen and possibly litter size influences the results of these studies (Weber et al. 2007).

If the sows are crated in the farrowing or insemination unit the crate should give sow the possibility of lying down and getting up without hindrance. Because of the way that a sow lies down and gets up this is considerably more than the size of the sow (Moustsen and Duus 2006). Ideally the size of the crate should be dependent on the size of the sow. If it is necessary to measure her, it may be both difficult and time-consuming. Because there are now systems in which the sow is only crated for a number of days around farrowing it becomes relevant to include time in crate as well.

6.4 Thermal comfort

The thermal comfort is affected by multiple external factors, for example, draught, humidity and the temperature, as well as the temperature requirements of the animals. Because of the many factors that are involved in the thermal comfort the best way of assessing it is the behaviour of the animals. Animals that are hot will show panting behaviour, whereas animals that are cold will huddle and/or shiver. Because of the strong influence of the weather the repeatability of the measure over longer time periods may be low (Friedrich et al. 2019b). For shorter-time periods panting and huddling were found to have a good repeatability (Scott et al. 2009a). An alternative to panting is to use a resource-based measure, for example, the presence or absence of functioning sprinklers.

6.5 Air quality

Air quality, particulate matter and not least the level of ammonium, will affect the welfare of pigs (Seedorf 2013; Michiels et al. 2015). While it is possible to feel high level of ammonium as a human when entering a pig stable it may be very hard to measure the level at the level of the pigs. It is likely that there is a huge variation both between different pens and even within a pen. A large number of measurements are therefore necessary to get an accurate assessment unless the measure is to be a subjective one.

6.6 Enrichment

Apart from providing some occupation for pigs' enrichment it is also known to affect the level of injurious behaviours in pigs (Tuyttens 2005; D'Eath et al. 2014). There is good knowledge of the type of material that is best suited as enrichment (Bracke et al. 2006; Studnitz et al. 2007), with materials that are 'ingestible', 'odorous', 'chewable', 'deformable' and 'destructible' being the ones that have the best preventive effect on tail biting. Because there is good knowledge of the suitability of the materials enrichment may be recorded as a resource-based measure instead of registering the behaviour of the animals.

6.7 Manure

The level of manure on the body of the pigs is likely to change with the weather conditions. In hot weather the animals are more prone to lying on a slatted area and may start to defecate on the solid floor (Aarnink et al. 2006). Pigs with a high proportion of manure on the body are therefore a likely indication of insufficient cooling. Acceptable test-retest was found by Friedrich et al. (2019a).

6.8 Outdoor access

Outdoor access and naturalness are aspects that are important for consumers (Sørensen and Schrader 2019). Outdoor access provides a more varied environment, but at the same time may increase exposure to pathogens and adverse conditions (Wallander et al. 2016, but see Leeb et al. 2019) (Table 5).

7 Health- and pain-related welfare indicators

7.1 Lameness

Lameness is relevant for pig welfare since it may be painful in itself, but also because it prevents the animal from interacting with its environment in an unhindered way. The inter-observer reproducibility is high when trained observers are used that are looking at the same animal at the same time (Geverink et al. 2009). When observers are visiting the same pens, but at slightly different times, the repeatability is much lower however (Petersen et al. 2004). The reason given by Petersen et al. is that it may be that the lameness is intermittent. If this is the case then the repeatability for the farm is probably low. When assessing lameness the animals should preferably have walked a number of steps before being assessed. Depending on the housing system this may cause some practical problems and decrease the feasibility of the measure (Table 6).

7.2 Wounds on body

The most common wounds on the body have been caused by aggression from other pigs. Aggression is very common at mixing, so if the aim with the indicator is to have a score that is representative for a longer time period then

Table 6 Assessment of the welfare consequence associated with the indicator, the reproducibility as well as the practicality of measuring it. Each aspect of the measure is marked from weak ('*') to strong ('***') for that aspect. For references see the relevant paragraph

Measure	Animal group	Welfare relevance	Reproduc.	Practicality
Lameness	Sows, pigs, piglets	***	*	*
Wounds on body	Sows, pigs	***	**	**
Vulva lesions	Sows	***	*	*
Mortality	Sows, pigs, piglets	***	**	***
Respiratory problems	Sows, pigs	***	***	***
Prolapse	Sows, pigs	**	*	***
Enteric problems	Sows, pigs	***	*	**
Metritis and mastitis	Sows (piglets)	***	*	*
Hernias	Sows, pigs, piglets	*	***	***
Pain face	Sows, pigs, piglets	***	*	*

the pigs should not be observed immediately after mixing. If this is done the repeatability of the measure seems to be good (Friedrich et al. 2019a). (For tail and ear biting see the section on General indicators of bad welfare, below).

7.3 Vulva lesions

Vulva lesions are often related to feeding and drinking systems (Rizvi et al. 1998). Feeding systems with feeding stalls means the risk of vulva bites can be reduced. Test-retest has been reported as being poor (Friedrich et al. 2019a).

7.4 Mortality

When using mortality as an on-farm welfare indicator only animals that die on farm are relevant. Information relating to mortality on-farm has to be obtained from the farmer.

7.5 Respiratory problems – coughing, sneezing and pumping

These are well known indicators of problems. Studies have found a correlation to lung lesions at slaughter (Halloy et al. 2004) and coughing has been used as a measure of prevalence of pneumonia among pigs (e.g. Meyns et al. 2006). Friedrich et al. (2019a) found coughing to have acceptable test-retest results.

7.6 Prolapse - rectal prolapse and uterine prolapse

Prolapse is the cause of culling of sows in 1–2% of the cases (Spain: Iida et al. 2019; Sweden: Engblom et al. 2007). According to Iida et al. (2019) the risk for both rectal and uterine prolapse increases to a maximum of 16 weeks after farrowing. Additional risk factors for sows are higher parity and a small litter. Rectal prolapse in growing pigs is usually associated with an increase in intra-abdominal pressure (Scott et al. 2009b). Because of a low prevalence the repeatability is probably low.

7.7 Enteric problems - scouring and constipation

Scouring is the presence of watery faeces, diarrhoea. Especially long-term scouring may be a serious welfare problem. Using the Welfare Quality® protocol Friedrich et al. (2019a) found a low test-retest value for scouring.

7.8 Metritis and mastitis

This is an important disease complex after farrowing. The disease compromises the welfare of the sow, but also risks the milk supply to the piglets. Risk factors for the complex are constipation, soiled trough and a low flow in the drinking nipples (Jenny et al. 2015).

7.9 Hernia

Umbilical hernias are often seen as a welfare problem. They may cause pain and if very large they may hamper mobility. Although it seems that animals with umbilical hernia are more exposed when in challenging conditions, for example, at a pick-up facility before transport to slaughter, there is surprisingly little difference in the behaviour of the pigs with hernia compared to those without (Schild et al. 2015). The finding that the size of the hernias does not affect their play behaviour or performance also suggests that under normal circumstances hernias may not be an important welfare problem (Atkinson et al. 2017).

7.10 Pig grimace scale

Grimace scales, 'pain face scales', have been developed for several species. They were originally used in laboratory animals (e.g. Leach et al. 2012), but lately also in horses (Gleerup et al. 2015a) and cattle (Gleerup et al. 2015b).

A similar grimace scale has also been developed for pigs (Di Giminiani et al. 2016). While it has been used in situations in which a single animal is observed closely and shows some promise in these situations (Vullo et al. 2020) it is unlikely that at present it can be used in an on-farm setting. Experience from other grimace scales is that they are hard to observe if the animal is disturbed, for example, by the presence of a human (Table 6).

7.11 Nose ringing

Nose ringing is most commonly done in outdoor production, and it is done to prevent the pigs from rooting and thus destroying the pasture. The consequence of nose ringing is an almost total prevention of rooting behaviour, but an increase in stereotypic and sniffing behaviour (Horrell et al. 2001, but see Studnitz et al. 2003) (Table 7).

7.12 Tail docking

Tail docking is one of the most common methods for preventing tail biting. There are several studies showing that it does reduce the risk, although it does not completely eliminate it (e.g. Thodberg et al. 2018; Lahrmann et al. 2017). At the moment of docking the piglets give calls with a significantly higher energy compared to a sham-docked group, and this is usually accepted as a pain response. No subsequent difference between the two groups could be detected, neither shortly after the docking nor at 17 weeks.

7.13 Castration

In modern pig production male pigs are castrated routinely to prevent boar taint. The surgical castration is painful when it is done without pain mitigation (Taylor and Weary 2000). There are various methods for alleviating the pain, NSAIDs, local anaesthetics and general anaesthetics. The level of post-operative pain reported varies however. Hay et al. (2003) found behavioural differences between controls and castrated animals on the following five

Table 7 Assessment of the welfare consequence associated with the indicator, the reproducibility as well as the practicality of measuring it. Each aspect of the measure is marked from weak ('*') to strong ('***') for that aspect. For references see the relevant paragraph

Measure	Animal group	Welfare relevance	Reproduc.	Practicality
Nose ringing	Sows, pigs	**	***	***
Tail docking	Piglets	**	***	***
Castration	Piglets	**	***	**
Teeth clipping	Piglets	**	***	***

days, whereas using animals as their own controls Yun et al. (2019) found only minor differences between pre- and post-operative behaviours. Using a cautionary approach it would seem reasonable to grade castration into not castrated, castrated with pain alleviation and castrated without pain alleviation.

7.14 Teeth clipping or grinding

Piglets are born with 'needle teeth' (canines and third incisors) which they use to fight for access to teats. Teeth clipping is done to decrease the damage that piglets can do to each other, and piglets with clipped teeth do cause less wounds on their litter mates (Weary and Fraser 1999). While teeth clipping is faster it results in a greater incidence of unintended damages to the teeth, teeth grinding results in fewer unintended damages but takes twice as long (Sutherland 2015). The benefit of decreasing the risk of injuries in the litter has to be weighed against the pain and fear associated with the operation (Table 7).

8 Behavioural welfare indicators

8.1 Affiliative behaviours (apart from play)

Pigs are social animals and will live together in groups if they can choose. Apart from interactions between sow and piglets adult pigs do not show many affiliative behaviours. There are some indications that sows will choose to lie together with known individuals, away from unknown individuals (Camerlink et al. 2014). There is also a possibility that some of the nose-nose contact that occurs is affiliative, whether it occurs between adults (Camerlink et al. 2014), between a piglet and the sow (Portele et al. 2019) or between piglets/young pigs (Petersen et al. 1989). The best candidate for an affiliative behaviour in pigs is nose-nose contact (Table 8). (For play see Section 9.5).

8.2 Aggression

Aggression is a major welfare problem in pig production. By keeping stable groups and avoiding mixing the level of aggression can be minimized. Aggression can also occur in competition for resources and feeding system has a great influence on the level of aggression (Bench et al. 2013). Aggression typically occurs in short bouts and is therefore difficult to observe. The best measure of level of aggression are therefore wounds on the body (primarily around the head and neck, Turner et al. 2006); see in the section on Health-related indicators.

Table 8 Assessment of the welfare consequence associated with the indicator, the reproducibility as well as the practicality of measuring it. Each aspect of the measure is marked from weak ('*') to strong ('***') for that aspect. For references see the relevant paragraph

Measure	Animal group	Welfare relevance	Reproduc.	Practicality
Affiliative behaviour	Sows, pigs, piglets	*	*	**
Aggression, (behaviour)	Sows, pigs, piglets	***	*	*
Nest building	Sows	***	*	*
Fear response	Sows, pigs	***	***	**

8.3 Nest building behaviour

Sows are strongly motivated to build a nest before farrowing, and will do so if given nesting material. Provision of nesting material has been suggested to improve sow welfare (EFSA 2007). If the sows are unable to build a nest they will often show signs of frustration. Unfortunately nest building can be difficult to observe on farm since it only occurs briefly before farrowing. It is only an absence or presence of nesting material before farrowing that reveals whether the sow has the possibility to nest build.

8.4 Fear response

Fear is a natural and adaptive response to a novel stimulus. Fear of humans in production conditions is however maladaptive and may lead to stress to the extent that it even impairs reproduction (Hemsworth et al. 1989). The two most common fear tests for humans are the voluntary approach test and the forced approach test (Forkman et al. 2007). In the voluntary approach test it is the animal that is approaching the human; however, this means that an animal that is completely uninterested in humans risks itself by getting the same value as one that keeps its distance because of fear. In the forced approach test the human approaches the animal and notes the distance at which the animal starts making an avoidance response. Because of this the forced approach test is the better one of the two (Table 8).

9 General indicators of poor and positive welfare

9.1 Poor welfare: tail and ear biting

Tail biting is a behaviour that can have important consequences, both for the welfare of the animals but also economical ones. Tail and ear biting often go together and no difference will be made in this text. Tail biting outbreaks are multifactorial, among the risk factors mentioned are climatic ones (heat, drought), absence of rooting material, nutrition, mixing, feed trough length and so on. (D'Eath et al. 2014). This means that tail biting outbreaks are often hard to predict which gives a fairly low reproducibility. A possible reason for why it is so complicated to understand is that it may have different motivational backgrounds. Taylor et al. (2010) proposed that one type was a two-stage tail biting mainly caused by a lack of rooting material, a second type was 'sudden-forceful' which is an aggressive behaviour aiming at removing the attacked pig from a resource and finally a third type the 'obsessive' tail biter. Not all tail biting is severe however and sometimes it will heal spontaneously. When assessing tail damage it therefore makes sense to divide it up into more categories than just yes/no (e.g. Lahrmann et al. 2019).

9.2 Poor welfare: stereotypies

Stereotypic behaviour is repetitive, unvarying and apparently functionless (Mason and Latham 2004). The situations in which stereotypic behaviour is most often seen in modern pig production is before farrowing if the sow is not able to build a nest, and in the gestation section where stone or vacuum chewing may occur. In many cases the animal will however momentarily stop its stereotypic behaviour when something new happens, for example, when an observer arrives. An alternative to observe the behaviour is observing the 'frothy saliva' that typically results from oral stereotypies (Friedrich et al. 2020a). This makes the registration of the stereotypies much more feasible.

9.3 Poor welfare: tear staining

Tear staining (chromodacryorrhea) has been shown to be related to stress in pigs, for example, living in a barren environment (DeBoer et al. 2015) and tail- and ear-biting (Telkänranta et al. 2016). Despite it being a promising new tool for assessing animal welfare it is important to note that there is no direct relationship between level of stress and tear staining. The level of stress does affect the tear staining, but so does age (Larsen et al. 2019). Despite its promise the measure cannot be recommended until more is known about the factors affecting it.

9.4 Poor welfare: runts

Runts are characterized by being smaller than their litter mates, quite often hairy with large ears, sunken flank, visible spine and sometimes with a 'dolphin face' (Hansen et al. 2019). These piglets run a higher risk of dying during the suckling period than their litter mates (Table 9).

Table 9 Assessment of the welfare consequence associated with the indicator, the reproducibility as well as the practicality of measuring it. Each aspect of the measure is marked from weak ('*') to strong ('***') for that aspect. Unknown values are marked as '-'. For references see the relevant paragraph

Measure	Animal group	Welfare relevance	Reproduc.	Practicality
Tail and ear biting	Sows, pigs, piglets	***	*	**
Stereotypic behav.	Sows	***	***	***
Lachrymal discharge	Pig	*	-	**
Runts	Piglets	***	***	***

9.5 Positive welfare: play

Play is one of the behaviours most often suggested as an indicator of positive welfare (Boissy et al. 2007). It is a behaviour that is most commonly seen in young animals, but if stimulated with new enrichment it can also be seen in sows (Greenwood et al. 2018). A distinction is often made between social play and solitary play, and both exist in pigs. Because the social play typically takes the form of mock fighting it can in many cases be hard to say exactly when the mock fight goes over into a real fight (for an ethogram see, for example, Brown et al. 2015). Play is one of the very few indicators for positive emotions and as such may give valuable information. The results from a small experiment indicate that it may be easier to recognize play in growing pigs than in piglets (Courboulay et al. 2009)

9.6 Positive welfare: tail wagging

A recent candidate for a positive indicator is tail wagging (Marcet Rius et al. 2018). There is a correlation between time spent with a novel object and the tail wagging (Marcet Ruis et al. 2019) as well as an increase in tail wagging in a positive as opposed to a negative situation (Reimert et al 2013). For a review of the significance of tail movement in pigs, see Camerlink and Ursinus (2020). There is no data as of yet on repeatability or feasibility (Table 10).

9.7 Possible iceberg indicators

Iceberg indicators are indicators that aim to capture the welfare of animals with only one or very few measures (FAWC 2009). The attractiveness of iceberg indicators is primarily that they would require very little time and effort, and that therefore many more farms could be assessed for the same amount of money. The Welfare Quality protocol is one of the most extensive protocols, but also

Table 10 Assessment of the welfare consequence associated with the indicator, the reproducibility as well as the practicality of measuring it. Each aspect of the measure is marked from weak ('*') to strong ('***') for that aspect. Unknown values are marked as '-'. For references see the relevant paragraph

Measure	Animal group	Welfare relevance	Reproduc.	Practicality
Play	Sows, pigs, piglets	***	**	**
Tail wagging	Sows, pigs, piglets	**	-	-

one that is very time-consuming (Knierim and Winckler 2009). Attempts have therefore been done to try to find the most relevant measures and subsequently reduce the protocol (e.g. Czycholl et al. 2017; Michelsen and Hakanson subm.). One of the most drastic attempts to reduce the Welfare Quality® protocol has been done by Wadepohl et al. (2019). In this approach only five measures are included; bursa alterations, lameness, manure on the body, runts and biting wounds including tail biting. The authors state that they see this as a semi-quantitative approach that can be used to determine which farms to investigate further, rather than as a complete welfare assessment in itself.

10 Future trends

The area of on-farm animal welfare assessment is currently moving into three, not mutually exclusive, directions. There is an ever stronger emphasis on animal-based measures since these are seen as being more relevant for assessing animal welfare. Although Welfare Quality® is generally considered to be a good assessment protocol, it is also generally considered to be very long, and there is a movement towards creating smaller/cheaper version of the Welfare Quality® protocols as well as modifying some of the aggregation procedures in the protocols that have been criticized for over-emphasizing the importance of, for example, water points. Finally there is also a movement towards more automatic registrations using the knowledge gained from the current protocols (see Chapter 15 of this volume).

11 Where to look for further information

11.1 Websites

Welfare Quality - www.welfarequality.net.

Welfare Quality Network and protocols - www.welfarequalitynetwork.net/en-us/reports/assessment-protocols/.

AssureWel - www.assurewel.org/.
Real Welfare - https://pork.ahdb.org.uk/health-welfare/welfare/real-welfare/.
KTBL - www.ktbl.de/shop/produktkatalog/alle-kategorien/12617/.
International society of applied ethology (https://www.applied-ethology.org/).

11.2 Relevant conferences

ISAE annual international conference.
Welfare Assessment at Farm Level - WAFL.

12 References

Aarnink, A. J. A., Schrama, J. W., Heetkamp, M. J. W., Stefanowska, J. and Huynh, T. T. T. (2006). Temperature and body weight affect fouling of pig pens. *Journal of Animal Science* 84(8), 2224–2231. https://doi.org/10.2527/jas.2005-521.

Atkinson, M., Amezcua, R., DeLay, J., Widowski, T. and Friendship, R. (2017). Evaluation of the effect of umbilical hernias on play behaviors in growing pigs. *The Canadian Veterinary Journal* 58(10), 1065–1072.

Averós, X., Brossard, L., Dourmad, J. Y., de Greef, K. H., Edge, H. L., Edwards, S. A. and Meunier-Salaün, M. C. (2010). Quantitative assessment of the effects of space allowance, group size and floor characteristics on the lying behaviour of growing-finishing pigs. *Animals* 4(5), 777–783. https://doi.org/10.1017/S1751731109991613.

Bartussek, H. (1995). *Tiergerechtheitsindex für Mastschweine*, TGI 35/L 1995, Stand Dezember 1995, Veröffentlichungen der Bundesanstalt für alpenländische Landwirtschaft (BAL) Gumpenstein, A 8952 Irdning.

Baxter, E. M., Lawrence, A. B. and Edwards, S. A. (2012). Alternative farrowing accommodation: welfare and economic aspects of existing farrowing and lactation systems for pigs. *Animals* 6(1), 96–117. https://doi.org/10.1017/S1751731111001224.

Bench, C. J., Rioja-Lang, F. C., Hayne, S. M. and Gonyou, H. W. (2013). Group gestation sow housing with individual feeding–II: how space allowance, group size and composition, and flooring affect sow welfare. *Livestock Science* 152(2–3), 218–227. https://doi.org/10.1016/j.livsci.2012.12.020.

Beter Leven. (2020). *Varkens*. Available at: https://beterleven.dierenbescherming.nl/over-de-dieren/alle-dieren/varkens/ (Accessed 1 June 2020).

Bøe, K. E. and Kjelvik, O. (2011). Water nipples or water bowls for weaned piglets: effect on water intake, performance, and plasma osmolality. *Acta Agriculturae Scandinavica, Section A–Animal Science* 61(2), 86–91. https://doi.org/10.1080/09064702.2011.599859.

Boissy, A., Manteuffel, G., Jensen, M. B., Moe, R. O., Spruijt, B., Keeling, L. J., Winckler, C., Forkman, B., Dimitrov, I., Langbein, J., Bakken, M., Veissier, I. and Aubert, A. (2007). Assessment of positive emotions in animals to improve their welfare. *Physiology and Behavior* 92(3), 375–397. https://doi.org/10.1016/j.physbeh.2007.02.003.

Botreau, R., Bracke, M. B. M., Perny, P., Butterworth, A., Capdeville, J., Van Reenen, C. G. and Veissier, I. (2007a). Aggregation of measures to produce an overall assessment

of animal welfare. Part 2: analysis of constraints. *Animals* 1(8), 1188–1197. https://doi.org/10.1017/S1751731107000547.
Botreau, R., Veissier, I., Butterworth, A., Bracke, M. and Keeling, L. (2007b). Definition of criteria for overall assessment of animal welfare. *Animal Welfare* 16, 225–228.
Botreau, R., Veissier, I. and Perny, P. (2009). Overall assessment of animal welfare: strategy adopted in welfare quality®. *Animal Welfare* 18, 363–370.
Bracke, M. B. M., Zonderland, J. J., Lenskens, P., Schouten, W. G. P., Vermeer, H., Spoolder, H. A. M., Hendriks, H. J. M. and Hopster, H. (2006). Formalised review of environmental enrichment for pigs in relation to political decision making. *Applied Animal Behaviour Science* 98(3–4), 165–182. https://doi.org/10.1016/j.applanim.2005.08.021.
Brown, S. M., Klaffenböck, M., Nevison, I. M. and Lawrence, A. B. (2015). Evidence for litter differences in play behaviour in pre-weaned pigs. *Applied Animal Behaviour Science* 172, 17–25. https://doi.org/10.1016/j.applanim.2015.09.007.
Callesen, J., Halas, D., Thorup, F., Bach Knudsen, K. E., Kim, J. C., Mullan, B. P., Hampson, D. J., Wilson, R. H. and Pluske, J. R. (2007). The effects of weaning age, diet composition, and categorisation of creep feed intake by piglets on diarrhoea and performance after weaning. *Livestock Science* 108(1), 120–123. https://doi.org/10.1016/j.livsci.2007.01.014.
Camerlink, I., Turner, S. P., Ursinus, W. W., Reimert, I. and Bolhuis, J. E. (2014). Aggression and affiliation during social conflict in pigs. *PLoS ONE* 9(11), e113502. https://doi.org/10.1371/journal.pone.0113502.
Camerlink, I. and Ursinus, W. W. (2020). Tail postures and tail motion in pigs: a review. *Applied Animal Behaviour Science* 230, 105079. https://doi.org/10.1016/j.applanim.2020.105079.
Charette, R., Bigras-Poulin, M. and Martineau, G.-P. (1996). Body condition evaluation in sows. *Livestock Production Science* 46(2), 107–115. https://doi.org/10.1016/0301-6226(96)00022-X.
Courboulay, V., Meunier-Salaün, M. C. and Edwards, S. A. (2009). Repeatability of abnormal behaviour. In: Keeling, L. and Forkman, B. (Eds) *Welfare Quality Reports* (vol. 10, 131–140). Cardiff University, Cardiff.
Czycholl, I., Kniese, C., Schrader, L. and Krieter, J. (2017). Assessment of the multi-criteria evaluation system of the welfare quality® protocol for growing pigs. *Animals* 11(9), 1573–1580. https://doi.org/10.1017/S1751731117000210.
Daros, R. R., Costa, J. H. C., von Keyserlingk, M. A. G., Hötzel, M. J. and Weary, D. M. (2014). Separation from the dam causes negative judgement bias in dairy calves. *PLoS ONE* 9(5), e98429. https://doi.org/10.1371/journal.pone.0098429.
D'Eath, R. B., Arnott, G., Turner, S. P., Jensen, T., Lahrmann, H. P., Busch, M. E., Niemi, J. K., Lawrence, A. B. and Sandøe, P. (2014). Injurious tail biting in pigs: how can it be controlled in existing systems without tail docking? *Animals* 8(9), 1479–1497. https://doi.org/10.1017/S1751731114001359.
D'Eath, R. B., Tolkamp, B. J., Kyriazakis, I. and Lawrence, A. B. (2009). 'Freedom from hunger' and preventing obesity: the animal welfare implications of reducing food quantity or quality. *Animal Behaviour* 77(2), 275–288. https://doi.org/10.1016/j.anbehav.2008.10.028.
DeBoer, S., Garner, J., McCain, R., Lay Jr., D., Eicher, S. and Marchant-Forde, J. (2015). An initial investigation into the effects of isolation and enrichment on the welfare of laboratory pigs housed in the PigTurn® system, assessed using tear staining,

behaviour, physiology and haematology. *Animal Welfare* 24(1), 15–27. https://doi.org/10.7120/09627286.24.1.015.

de Graaf, S., Ampe, B., Winckler, C., Radeski, M., Mounier, L., Kirchner, M. K., Haskell, M. J., van Eerdenburg, F. J. C. M., des Roches, A. B., Andreasen, S. N., Bijttebier, J., Lauwers, L., Verbeke, W. and Tuyttens, F. A. M. (2017). Trained-user opinion about Welfare Quality measures and integrated scoring of dairy cattle welfare. *Journal of Dairy Science* 100(8), 6376-6388. https://doi.org/10.3168/jds.2016-12255.

Di Giminiani, P., Brierley, V. L. M. H., Scollo, A., Gottardo, F., Malcolm, E. M., Edwards, S. A. and Leach, M. C. (2016). The assessment of facial expressions in piglets undergoing tail docking and castration: toward the development of the piglet grimace scale. *Frontiers in Veterinary Science* 3, 100. https://doi.org/10.3389/fvets.2016.00100.

EFSA (European Food Safety Authority). (2007). Animal health and welfare aspects of different housing and husbandry systems for adult breeding boars, pregnant, farrowing sows and unweaned piglets–scientific opinion of the panel on animal health and welfare. *EFSA Journal* 5(10), 572. https://doi.org/10.2903/j.efsa.2007.572.

EFSA (European Food Safety Authority). (2014). Guidance on expert knowledge elicitation in food and feed safety risk assessment: guidance on expert knowledge elicitation. *EFSA Journal* 12(6), 3734. https://doi.org/10.2903/j.efsa.2014.3734.

EFSA (European Food Safety Authority). (2015). The use of animal-based measures to assess animal welfare in EU–state of the art of 10 years of activities and analysis of gaps. *EFSA Supporting Publications* 12(11). https://doi.org/10.2903/sp.efsa.2015.EN-884.

Engblom, L., Lundeheim, N., Dalin, A.-M. and Andersson, K. (2007). Sow removal in Swedish commercial herds. *Livestock Science* 106(1), 76–86. https://doi.org/10.1016/j.livsci.2006.07.002.

FAWC. (2009). *Farm Animal Welfare in Great Britain: Past, Present and Future*. Farm Animal Welfare Council, London, England.

Forkman, B., Boissy, A., Meunier-Salaün, M. C., Canali, E. and Jones, R. B. (2007). A critical review of fear tests used on cattle, pigs, sheep, poultry and horses. *Physiology and Behavior* 92(3), 340–374. https://doi.org/10.1016/j.physbeh.2007.03.016.

Fraser, D., Weary, D. M., Pajor, E. A. and Milligan, B. N. (1997). A scientific conception of animal welfare that reflects ethical concerns. *Animal Welfare* 6(3), 187–205.

Friedrich, L., Krieter, J., Kemper, N. and Czycholl, I. (2019a). Test–retest reliability of the 'welfare quality® animal welfare assessment protocol for sows and piglets'. Part 1: assessment of the welfare principle of 'appropriate behavior'. *Animals* 9(7), 398. https://doi.org/10.3390/ani9070398.

Friedrich, L., Krieter, J., Kemper, N. and Czycholl, I. (2019b). Test-retest reliability of the welfare quality assessment protocol for pigs applied to sows and piglets. Part 2: assessment of the principles good feeding, good housing, and good health1. *Journal of Animal Science* 97(3), 1143–1157. https://doi.org/10.1093/jas/skz018.

Friedrich, L., Krieter, J., Kemper, N. and Czycholl, I. (2020a). Frothy saliva–a novel indicator to assess stereotypies in sows? *Applied Animal Behaviour Science* 222, 104897. https://doi.org/10.1016/j.applanim.2019.104897.

Friedrich, L., Krieter, J., Kemper, N. and Czycholl, I. (2020b). Short communication: evaluation of the sample size of individual indicators in gestating sows concerning the welfare quality® protocol applied to sows and piglets. *Animals* 14(6), 1278–1282. https://doi.org/10.1017/S1751731119003446.

FVST. (2019). Vejledning til Fødevarestyrelsens kontrol med hjertebesætninger - svinebesætninger. Fødevarestyrelsen. *JNR*, 2018-15-19-01326, 1–9.

Geverink, N. A., Meuleman, M., van Nuffel, A., van Steenbergen, L., Hautekiet, V., Vermeulen, K., Lammens, V., van Reenen, C. G. and Tuyttens, F. A. M. (2009). Repeatability of a lameness score measured on farm. In: Keeling, L. and Forkman, B. (Eds) *Welfare Quality Reports* (vol. 10, 57–72). Cardiff University, Cardiff.

Gleerup, K. B., Forkman, B., Lindegaard, C. and Andersen, P. H. (2015a). An equine pain face. *Veterinary Anaesthesia and Analgesia* 42(1), 103–114. https://doi.org/10.1111/vaa.12212.

Gleerup, K. B., Andersen, P. H., Munksgaard, L. and Forkman, B. (2015b). Pain evaluation in dairy cattle. *Applied Animal Behaviour Science* 171, 25–32. https://doi.org/10.1016/j.applanim.2015.08.023.

Glencorse, D., Plush, K., Hazel, S., D'Souza, D. and Hebart, M. (2019). Impact of non-confinement accommodation on farrowing performance: a systematic review and meta-analysis of farrowing crates versus pens. *Animals* 9(11), 957. https://doi.org/10.3390/ani9110957

Greenwood, E., van Wettere, W., Rayner, J., Hughes, P. and Plush, K. (2018). Provision point-source materials stimulates play in sows but does not affect aggression at regrouping. *Animals* 9(1), 8. https://doi.org/10.3390/ani9010008.

Halloy, D. J., Bouhet, S., Oswald, I. P., Goret-Nicaise, M., Kobisch, M., Mainil, J. and Gustin, P. G. (2004). Pathophysiological changes occurring during *Escherichia coli* endotoxin and *Pasteurella multocida* challenge in piglets: relationship with cough and temperature and predicitive value for intensity of lesions. *Veterinary Research* 35(3), 309–324. https://doi.org/10.1051/vetres:2004012.

Hansen, C. F., Hales, J., Amdi, C. and Moustsen, V. A. (2019). Intrauterine growth-restricted piglets defined by their head shape have impaired survival and growth during the suckling period. *Animal Production Science* 59(6), 1056–1062. https://doi.org/10.1071/AN17581.

Hay, M., Vulin, A., Génin, S., Sales, P. and Prunier, A. (2003). Assessment of pain induced by castration in piglets: behavioral and physiological responses over the subsequent 5 days. *Applied Animal Behaviour Science* 82(3), 201–218. https://doi.org/10.1016/S0168-1591(03)00059-5.

Hemsworth, P. H. (2018). Key determinants of pig welfare: implications of animal management and housing design on livestock welfare. *Animal Production Science* 58(8), 1375. https://doi.org/10.1071/AN17897.

Hemsworth, P. H., Barnett, J. L., Coleman, G. J. and Hansen, C. (1989). A study of the relationships between the attitudinal and behavioural profiles of stockpersons and the level of fear of humans and reproductive performance of commercial pigs. *Applied Animal Behaviour Science* 23(4), 301–314. https://doi.org/10.1016/0168-1591(89)90099-3.

Herskin, M. S., Jensen, H. E., Jespersen, A., Forkman, B., Jensen, M. B., Canibe, N. and Pedersen, L. J. (2016). Impact of the amount of straw provided to pigs kept in intensive production conditions on the occurrence and severity of gastric ulceration at slaughter. *Research in Veterinary Science* 104, 200–206. https://doi.org/10.1016/j.rvsc.2015.12.017.

Horrell, R. I., A'Ness, P. J., Edwards, S. A. and Eddison, J. C. (2001). The use of nose-rings in pigs: consequences for rooting, other functional activities and welfare. *Animal Welfare* 10, 3–22.

Iida, R., Piñeiro, C. and Koketsu, Y. (2019). Incidences and risk factors for prolapse removal in Spanish sow herds. *Preventive Veterinary Medicine* 163, 79–86. https://doi.org/10.1016/j.prevetmed.2018.12.016.

Jenny, B., Vidondo, B., Pendl, W., Kümmerlen, D. and Sidler, X. (2015). Evaluation of risk factors for Mastitis-Metritis Agalactia in pig farms in Switzerland. *Schweizer Archiv für Tierheilkunde* 157(12), 689–696. https://doi.org/10.17236/sat00047.

KilBride, A. L., Gillman, C. E. and Green, L. E. (2009). A cross sectional study of the prevalence, risk factors and population attributable fractions for limb and body lesions in lactating sows on commercial farms in England. *BMC Veterinary Research* 5(1), 30. https://doi.org/10.1186/1746-6148-5-30.

Knierim, U. and Winckler, C. (2009). On-farm welfare assessment in cattle: validity, reliability and feasibility issues and future perspectives with special regard to the welfare quality® approach. *Animal Welfare* 18, 451–458.

Lahrmann, H. P., Busch, M. E., D'Eath, R. B., Forkman, B. and Hansen, C. F. (2017). More tail lesions among undocked than tail docked pigs in a conventional herd. *Animals* 11(10), 1825–1831. https://doi.org/10.1017/S1751731117000490.

Lahrmann, H. P., Faustrup, J. F., Hansen, C. F., D'Eath, R. B., Nielsen, J. P. and Forkman, B. (2019). The effect of straw, rope, and bite-rite treatment in Weaner pens with a tail biting outbreak. *Animals* 9(6), 365. https://doi.org/10.3390/ani9060365.

Larsen, M. L. V., Gustafsson, A., Marchant-Forde, J. N. and Valros, A. (2019). Tear staining in finisher pigs and its relation to age, growth, sex and potential pen level stressors. *Animals* 13(8), 1704–1711. https://doi.org/10.1017/S1751731118003646.

Leach, M. C., Klaus, K., Miller, A. L., Scotto di Perrotolo, M., Sotocinal, S. G. and Flecknell, P. A. (2012). The assessment of post-vasectomy pain in mice using behaviour and the mouse grimace scale. *PLoS ONE* 7(4), e35656. https://doi.org/10.1371/journal.pone.0035656.

Leeb, C., Rudolph, G., Bochicchio, D., Edwards, S., Früh, B., Holinger, M., Holmes, D., Illmann, G., Knop, D., Prunier, A., Rousing, T., Winckler, C. and Dippel, S. (2019). Effects of three husbandry systems on health, welfare and productivity of organic pigs. *Animals* 13(9), 2025–2033. https://doi.org/10.1017/S1751731119000041.

Marcet Rius, M., Cozzi, A., Bienboire-Frosini, C., Teruel, E., Chabaud, C., Monneret, P., Leclercq, J., Lafont-Lecuelle, C. and Pageat, P. (2018). Selection of putative indicators of positive emotions triggered by object and social play in mini-pigs. *Applied Animal Behaviour Science* 202, 13–19. https://doi.org/10.1016/j.applanim.2018.02.002.

Marcet Rius, M., Fabrega, E., Cozzi, A., Bienboire-Frosini, C., Descout, E., Velarde, A. and Pageat, P. (2019). Are tail and ear movements indicators of emotions in tail-docked pigs in response to environmental enrichment? *Animals* 9(7), 449. https://doi.org/10.3390/ani9070449.

Mason, G. and Latham, N. (2004). Can't stop, won't stop: is stereotypy a reliable animal welfare indicator? *Animal Welfare* 13, S57–69.

Mellor, D. (2017). Operational details of the five domains model and its key applications to the assessment and management of animal welfare. *Animals* 7(12), 60. https://doi.org/10.3390/ani7080060.

Mellor, D. and Reid, C. S. W. (1994). Concepts of animal well-being and predicting the impact of procedures on experimental animals. In: Baker, R., Jenkin, G., and Mellor, D. J. (Eds) *Improving the Well-Being of Animals in the Research Environment* (pp. 3–18). Australian and New Zealand Council for the Care of Animals in Research and Teaching, Glen Osmond, SA, Australia.

Meyns, T., Dewulf, J., de Kruif, A., Calus, D., Haesebrouck, F. and Maes, D. (2006). Comparison of transmission of Mycoplasma hyopneumoniae in vaccinated and non-vaccinated populations. *Vaccine* 24(49–50), 7081–7086. https://doi.org/10.1016/j.vaccine.2006.07.004.

Michiels, A., Piepers, S., Ulens, T., Van Ransbeeck, N., Del Pozo Sacristán, R., Sierens, A., Haesebrouck, F., Demeyer, P. and Maes, D. (2015). Impact of particulate matter and ammonia on average daily weight gain, mortality and lung lesions in pigs. *Preventive Veterinary Medicine* 121(1–2), 99–107. https://doi.org/10.1016/j.prevetmed.2015.06.011.

Miller, H. M., Carroll, S. M., Reynolds, F. H. and Slade, R. D. (2007). Effect of rearing environment and age on gut development of piglets at weaning. *Livestock Science* 108(1–3), 124–127. https://doi.org/10.1016/j.livsci.2007.01.016.

Moustsen, V. A. and Duus, L. K. (2006). Søers "rejse og lægge sig" bevægelse i forskellige farestier. Landsudvalget for Svin, Den rullende Afprøvning. Meddelelse nr. 733.

Mullan, S., Browne, W. J., Edwards, S. A., Butterworth, A., Whay, H. R. and Main, D. C. J. (2009). The effect of sampling strategy on the estimated prevalence of welfare outcome measures on finishing pig farms. *Applied Animal Behaviour Science* 119(1–2), 39–48. https://doi.org/10.1016/j.applanim.2009.03.008.

Munsterhjelm, C., Heinonen, M. and Valros, A. (2015). Application of the welfare quality® animal welfare assessment system in Finnish pig production, part II: associations between animal-based and environmental measures of welfare. *Animal Welfare* 24(2), 161–172. https://doi.org/10.7120/09627286.24.2.161.

Nannoni, E., Martelli, G., Rubini, G. and Sardi, L. (2019). Effects of increased space allowance on animal welfare, meat and ham quality of heavy pigs slaughtered at 160Kg. *PLoS ONE* 14(2), e0212417. https://doi.org/10.1371/journal.pone.0212417.

OIE. (2020). What is animal welfare? Available at: https://www.oie.int/en/animal-welfare/animal-welfare-at-a-glance/ (Accessed 1 July 2020).

Patterson-Kane, E., Pittman, M. and Pajor, E. (2008). Operant animal welfare: productive approaches and persistent difficulties. *Animal Welfare* 17, 139–148.

Petersen, H. H., Enøe, C. and Nielsen, E. O. (2004). Observer agreement on pen level prevalence of clinical signs in finishing pigs. *Preventive Veterinary Medicine* 64(2–4), 147–156. https://doi.org/10.1016/j.prevetmed.2004.05.002.

Petersen, H. V., Vestergaard, K. and Jensen, P. (1989). Integration of piglets into social groups of free-ranging domestic pigs. *Applied Animal Behaviour Science* 23(3), 223-236. https://doi.org/10.1016/0168-1591(89)90113-5.

Pfeifer, M., Koch, A., Lensches, C., Schmitt, A. O. and Hessel, E. F. (2020). Acceptance and feasibility of a guideline for the animal welfare assessment of fattening pigs from farmers' point of view. *Animals* 10(4), 711. https://doi.org/10.3390/ani10040711.

Phillips, P. A. and Phillips, M. H. (1999). Technical notes: effect of dispenser on water intake of pigs at weaning. *Transactions of the ASAE (American Society of Agricultural Engineers)* 42(5), 1471–1473.

Portele, K., Scheck, K., Siegmann, S., Feitsch, R., Maschat, K., Rault, J.-L. and Camerlink, I. (2019). Sow-piglet nose contacts in free-farrowing pens. *Animals* 9(8), 513. https://doi.org/10.3390/ani9080513.

Reimert, I., Bolhuis, J. E., Kemp, B. and Rodenburg, T. B. (2013). Indicators of positive and negative emotions and emotional contagion in pigs. *Physiology & Behavior* 109, 42-50. https://doi.org/10.1016/j.physbeh.2012.11.002.

Rizvi, S., Nicol, C. J. and Green, L. E. (1998). Risk factors for vulva biting in breeding sows in south-west England. *The Veterinary Record* 143(24), 654–658.

RSPCA. (2016). *RSPCA standards for pigs*. Available at: https://www.berspcaassured.org.uk/media/1171/rspca-standards-for-pigs-2016.pdf (Accessed 1 June 2020).

Sandøe, P., Corr, S. A., Lund, T. B. and Forkman, B. (2019). Aggregating animal welfare indicators: can it be done in a transparent and ethically robust way? *Animal Welfare* 28(1), 67–76. https://doi.org/10.7120/09627286.28.1.067.

Sandøe, P., Hansen, H. O., Rhode, H. L. H., Houe, H., Palmer, C., Forkman, B. and Christensen, T. (2020). Benchmarking farm animal welfare–A novel tool for cross-country comparison applied to pig production and pork consumption. *Animals* 10(6), 18. Available at: https://doi.org/doi:10.3390/ani10060955.

Schild, S. L., Rousing, T., Jensen, H. E., Barington, K. and Herskin, M. S. (2015). Do umbilical outpouchings affect the behaviour or clinical condition of pigs during 6h housing in a pre-transport pick-up facility? *Research in Veterinary Science* 101, 126–131. https://doi.org/10.1016/j.rvsc.2015.06.005.

Schrader, L., Czycholl, I., Krieter, J., Leeb, C., Zapf, R. and Ziron, M. (2016). *Tierschutzindikatoren: Leitfaden für die Praxis–Schwein*. KTBL, Darmstadt.

Scott, K., Guy, J. H. and Edwards, S. (2009a). The repeatability of measures of thermoregulation in pigs. In: Keeling, L. and Forkman, B. (Eds) *Welfare Quality Reports* (vol. 10, 57–72). Cardiff University, Cardiff.

Scott, K., Guy, J. H., Hautekiet, V., van Steenbergen, L., D'Eath, R. and Edwards, S. (2009b). Standardisation of health scoring in pigs. In: Keeling, L. and Forkman, B. (Eds) *Welfare Quality Reports* (vol. 10, 89–95). Cardiff University, Cardiff.

Seedorf, J. (2013). Wirkung von atmosphärischem ammo- niak auf Nutztiere - eine Kurzübersicht. *Berliner und Münchener Tierärztliche Wochenschrift* 126(3/4), 96–103.

Spoolder, H., De Rosa, G., Hörning, B., Waiblinger, S. and Wemelsfelder, F. (2003). Integrating parameters to assess on-farm welfare. *Animal Welfare* 12(4), 529–534.

Sørensen, J. T. and Schrader, L. (2019). Labelling as a tool for improving animal welfare–the pig case. *Agriculture* 9(6), 123. https://doi.org/10.3390/agriculture9060123.

Studnitz, M., Jensen, K. H. and Jørgensen, E. (2003). The effect of nose rings on the exploratory behaviour of outdoor gilts exposed to different tests. *Applied Animal Behaviour Science* 84(1), 41–57. https://doi.org/10.1016/S0168-1591(03)00144-8.

Studnitz, M., Jensen, M. B. and Pedersen, L. J. (2007). Why do pigs root and in what will they root? *Applied Animal Behaviour Science* 107(3–4), 183–197. https://doi.org/10.1016/j.applanim.2006.11.013.

Sutherland, M. A. (2015). Welfare implications of invasive piglet husbandry procedures, methods of alleviation and alternatives: a review. *New Zealand Veterinary Journal* 63(1), 52–57. https://doi.org/10.1080/00480169.2014.961990.

Taylor, A. A. and Weary, D. M. (2000). Vocal responses of piglets to castration: identifying procedural sources of pain. *Applied Animal Behaviour Science* 70(1), 17–26. https://doi.org/10.1016/S0168-1591(00)00143-X.

Taylor, N. R., Main, D. C. J., Mendl, M. and Edwards, S. A. (2010). Tail-biting: a new perspective. *The Veterinary Journal* 186(2), 137–147. https://doi.org/10.1016/j.tvjl.2009.08.028.

Telkänranta, H., Marchant-Forde, J. N. and Valros, A. (2016). Tear staining in pigs: a potential tool for welfare assessment on commercial farms. *Animals* 10(2), 318–325. https://doi.org/10.1017/S175173111500172X.

Thodberg, K., Herskin, M. S., Jensen, T. and Jensen, K. H. (2018). The effect of docking length on the risk of tail biting, tail-directed behaviour, aggression and activity level of growing pigs kept under commercial conditions. *Animals* 12(12), 2609–2618. https://doi.org/10.1017/S1751731118000563.

Torrey, S. and Widowski, T. M. (2006). A note on piglets' preferences for drinker types at two weaning ages. *Applied Animal Behaviour Science* 100(3–4), 333–341. https://doi.org/10.1016/j.applanim.2005.12.007.

Turner, S. P., Edwards, S. A. and Bland, V. C. (1999). The influence of drinker allocation and group size on the drinking behaviour, welfare and production of growing pigs. *Animal Science* 68(4), 617–624. https://doi.org/10.1017/S1357729800050645.

Turner, S. P., Farnworth, M. J., White, I. M. S., Brotherstone, S., Mendl, M., Knap, P., Penny, P. and Lawrence, A. B. (2006). The accumulation of skin lesions and their use as a predictor of individual aggressiveness in pigs. *Applied Animal Behaviour Science* 96(3–4), 245–259. https://doi.org/10.1016/j.applanim.2005.06.009.

Turner, S. P., Sinclair, A. G. and Edwards, S. A. (2000). The interaction of liveweight and the degree of competition on drinking behaviour in growing pigs at different group sizes. *Applied Animal Behaviour Science* 67(4), 321–334. https://doi.org/10.1016/S0168-1591(00)00084-8.

Tuyttens, F. A. M. (2005). The importance of straw for pig and cattle welfare: a review. *Applied Animal Behaviour Science* 92(3), 261–282. https://doi.org/10.1016/j.applanim.2005.05.007.

Tuyttens, F. A. M., de Graaf, S., Heerkens, J. L. T., Jacobs, L., Nalon, E., Ott, S., Stadig, L., Van Laer, E. and Ampe, B. (2014). Observer bias in animal behaviour research: can we believe what we score, if we score what we believe? *Animal Behaviour* 90, 273–280. https://doi.org/10.1016/j.anbehav.2014.02.007.

Vullo, C., Barbieri, S., Catone, G., Graïc, J. M., Magaletti, M., Di Rosa, A., Motta, A., Tremolada, C., Canali, E. and Dalla Costa, E. (2020). Is the Piglet Grimace Scale (PGS) a useful welfare indicator to assess pain after Cryptorchidectomy in growing pigs? *Animals* 10(3), 412. https://doi.org/10.3390/ani10030412.

Wadepohl, K., Blaha, T., Van Gompel, L., Duarte, A. S. R., Nielsen, C. L., Saatkamp, H., Wagenaar, J. A., Meemken, D. and EFFORT Group (2019). Development of a simplified on-farm animal health and welfare benchmarking tool for pig herds. *Berliner Und Munchener Tierarztliche Wochenschrift* 132(11–12), 504–512. https://doi.org/10.2376/0005-9366-18088.

Wallander, C., Frössling, J., Dórea, F. C., Uggla, A., Vågsholm, I. and Lundén, A. (2016). Pasture is a risk factor for Toxoplasma gondii infection in fattening pigs. *Veterinary Parasitology* 224, 27–32. https://doi.org/10.1016/j.vetpar.2016.05.005.

Weary, D. M. and Fraser, D. (1999). Partial tooth-clipping of suckling pigs: effects on neonatal competition and facial injuries. *Applied Animal Behaviour Science* 65(1), 21–27. https://doi.org/10.1016/S0168-1591(99)00052-0.

Weber, R., Keil, N., Fehr, M. and Horat, R. (2007). Piglet mortality on farms using farrowing systems with or without crates. *Animal Welfare* 16, 277–279.

Wellock, I. J., Fortomaris, P. D., Houdijk, J. G. M. and Kyriazakis, I. (2007). Effect of weaning age, protein nutrition and enterotoxigenic Escherichia coli challenge on the health of newly weaned piglets. *Livestock Science* 108(1–3), 102–105. https://doi.org/10.1016/j.livsci.2007.01.004.

Welfare Quality® Consortium. (2009). *Welfare Quality® Assessment Protocol for Pigs*. Welfare Quality Consortium, Lelystad, Netherlands, 40–41.

Yun, J., Ollila, A., Valros, A., Larenza-Menzies, P., Heinonen, M., Oliviero, C. and Peltoniemi, O. (2019). Behavioural alterations in piglets after surgical castration: effects of analgesia and anaesthesia. *Research in Veterinary Science* 125, 36–42. https://doi.org/10.1016/j.rvsc.2019.05.009.

Zupan, M. and Zanella, A. J. (2017). Peripheral regulation of stress and fear responses in pigs from tail-biting pens. *Revista Brasileira De Zootecnia-Brazilian Journal of Animal Science* 46(1), 33–38. https://doi.org/10.1590/S1806-92902017000100006.

Chapter 15

Advances in technologies for monitoring pig welfare

Maciej Oczak, University of Veterinary Medicine Vienna, Austria; Kristina Maschat, FFoQSI GmbH, Austria; and Johannes Baumgartner, University of Veterinary Medicine Vienna, Austria

1 Introduction
2 Good feeding
3 Good housing
4 Good health
5 Appropriate behaviour
6 Legal and ethical considerations
7 Sensor data management and modelling
8 Case study
9 Conclusion
10 Future trends in research
11 Where to look for further information
12 References

1 Introduction

Digital revolution fuelled by the exponential progress of the semiconductor technology has rapidly spread across countries, industries and socio-economic activities in the past few decades, with profound transformational effects (Savulescu 2015, Jorgenson and Vu 2016). In the agricultural industry, especially, areas of food supply chains and arable farming have been influenced by the revolution in the information and communication technology (ICT). This is in contrast to livestock farming, which has been influenced in much lesser extent (Verdouw et al. 2016).

ICT-supported management of livestock farming systems, which is also referred to as precision livestock farming (PLF), was predominantly instigated by the European research community already in the early 1990s (Banhazi et al. 2012). PLF is a tool for the management of livestock farming by means of

http://dx.doi.org/10.19103/AS.2020.0081.15

automatic, real-time monitoring/control of livestock production, reproduction, health and welfare of environmental impacts (Berckmans 2013). Technological development and progress have advanced to such an extent that accurate, powerful and affordable tools are now available. These include cameras, microphones, sensors (such as 3D accelerometers (including gyroscopes), temperature sensors, skin conductivity sensors and glucose sensors), wireless communication tools, Internet connections and cloud storage. Modern technology makes it possible to place cameras, microphones and sensors sufficiently close that they can replace the farmers' eyes and ears in monitoring individual animals (Berckmans 2014).

Apart from sensor data, there is currently an abundance of information available to livestock managers, but it is not generally structured in a way that can be applied readily. For example, a survey of producers raising beef from pastures in southern Australia showed that over 400 pieces of information could be relevant for their farms. The information comes from many sources including academic organizations, government advisors, producer magazines, media sources, technical advisers and other producers (Banhazi et al. 2012).

For monitoring and control of livestock production processes, the PLF approach makes use of modern monitoring and control theory. To achieve favourable monitoring and control of such processes, three conditions must be fulfilled (Berckmans 2006a).

The first condition to be fulfilled is that animal variables must be measured continuously and this information is analysed continuously. 'Animal variables' can be very different such as weight, activity, behaviour, drinking and feeding behaviour, feed intake, sound production, physiological variables (body temperature, respiration frequency, blood variables,). What 'continuously' means is depending on the measured variable such as 25 times a second when monitoring on-line animal activity from video images or a sample every day when monitoring animal weight.

A second condition to realise accurate animal monitoring and management is that at every moment a reliable prediction (expectation) must be available on how the animal variables will vary or how the animal will respond to environmental changes. By environment, we mean the whole of all variables that are not genetically defined. It is the continuous comparison between this prediction (in the past the experience of the farmer and now for example a mathematical model) and the actual measured values that allows to identify animal activities and to judge when something abnormal is happening.

The third condition is that this prediction together with the on-line measurements are integrated in an analysing algorithm (a number of mathematical equations implemented in a microchip) to monitor or manage the animals automatically and to achieve on-line monitoring of animal health, welfare or take control actions (climate control, feeding strategies). A schematic overview of the three conditions is shown in Fig. 1.

The PLF can greatly contribute to an objective discussion on animal welfare by providing real data to the otherwise very subjective (and sometimes emotional) discussion process. While PLF will not be able to necessarily resolve all welfare-related questions, it will allow interested parties to detect and act upon time periods when animals are kept under suboptimal conditions (Banhazi et al. 2012).

Today, early adopters are starting to use PLF products mainly in Europe, despite the PLF concept is rather new in the European pig industry. So far, there is a lack of products in the market, which would comprehensively approach monitoring the welfare of pigs. However, there are already systems available which allow monitoring individual indicators of pig welfare such as water intake or weight (Kashiha et al. 2013a, Vranken and Berckmans 2017). There are at least two major advantages of PLF technology, which will lead to its wider adoption for welfare monitoring in the coming years.

One advantage of sensor-based monitoring is its continuous character (Berckmans 2006a). This means that monitoring can be always on 7 days a week, 24 h per day. This is clearly advantageous if we compare this to the time that the farmer can directly observe animals in industrial husbandry systems. Thus, continuous monitoring can create a possibility to bring the animals closer to the farmer, especially in industrial systems, where contact between an individual

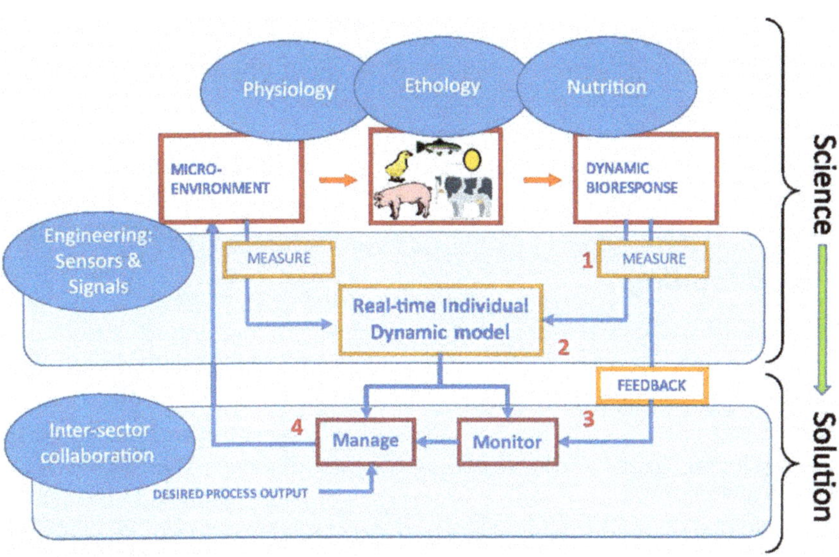

Figure 1 General scheme showing how bio-response monitoring and management of livestock animals can go from science to solutions. From Aerts, Norton, and Berckmans 2019.

animal and the farmer is very limited. The farmer can obtain insight about the animal and its welfare, thanks to sensor technology, which is not possible on the basis of short-lasting, personal audio-visual observation. Thus, the farmer can invest his time in the animals that require his attention.

The second advantage of PLF systems is that they offer real-time monitoring and this gives a possibility to respond to welfare problems immediately, when the problem occurs. This is fundamentally different from all other approaches that aim to offer a monitoring tool without improving the life of the animal under consideration. It is nice to detect a problem after an animal has arrived at the slaughterhouse. It is better to detect a problem while the animal is being reared and to take immediate management action (Berckmans 2013). However, implementation of automated monitoring technologies at the abattoir, while less immediate in terms of problem identification, may be much more cost-effective in some circumstances, for example, routine disease or welfare surveillance programmes.

In this chapter we divided advances in the development of PLF technologies for monitoring welfare in pigs into four main parts according to the grouping of welfare criteria based on four principles proposed in Welfare Quality® (WelfareQuality® 2009):

- Good feeding: Absence of prolonged hunger; absence of prolonged thirst;
- Good housing: Comfort around resting; thermal comfort; ease of movement;
- Good health: Absence of injuries; absence of disease; absence of pain induced by management procedures; and
- Appropriate behaviour: Expression of social behaviours; Expression of other behaviours; Good human-animal relationship; Positive emotional state.

2 Good feeding

Good feeding is characterized by the absence of prolonged hunger that is guaranteed by providing a suitable and appropriate diet. The evaluation method suggested by WelfareQuality® (2009) is visual inspection (growing and finishing pigs, sows) and palpation (sows) of the spine, hip and pin bones for thickness of fat tissue in order to assess body condition score (BCS). However, Maes et al. (2004) showed that back fat measurements conducted by means of ultrasonic scanning and visual scoring of the body condition of the sows are only moderately correlated (r=0.30–0.60). Although depth cameras measuring back postures have been used lately (Kuzuhara et al. 2015, Weber et al. 2014) in cows to predict body condition score and back fat, this method hasn't been adapted to pigs so far. Yet, depth image and machine learning have been applied to estimate muscle scores of in vivo pigs (Alsahaf et al. 2019).

Also, it has to be kept in mind that decreases in BCS in the context of malnutrition and diseases occur at a very late stage when the course has already become chronic. Therefore objective assessment of other indicators for feeding might be more beneficial. Automatic weight detection has been an approach to address this problem and at the same time bears the potential to increase precision of marketing pigs (Condotta et al. 2018). Recent studies have been using pigs' body dimensions assessed on the basis of top-view 2D- and 3D-camera imaging for predicting animals' mass.

Some of the authors separated pigs from their group to conduct measurements while the animal was confined (Condotta et al. 2018, Kongsro 2014) and reached very high coefficients of determination (R^2=0.99), while others have been aiming to develop non-contact methods to estimate the weight of pigs in barns with other animals (Jun et al. 2018, Pezzuolo et al. 2018, Kashiha et al. 2014) and achieved results with R^2 ranging from 0.79 to 0.96. Mobile devices have been developed (Shi et al. 2016, Wang et al. 2018) to estimate pig body sizes and commercial products for weight estimation such as Weight-Detect (PLF-Agritech Europe, Edinburgh, UK), Pigwei (Ymaging, Barcelona, Spain), eYeScan (Fig. 2) (Fancom BV, Panningen, The Netherlands), Growth Sensor (GroStat, Newport, UK), OptiSCAN (Hölscher, Emsburen, Germany), and WUGGL One (WUGGL, Lebring, Austria) have been introduced to the market (Vranken and Berckmans 2017).

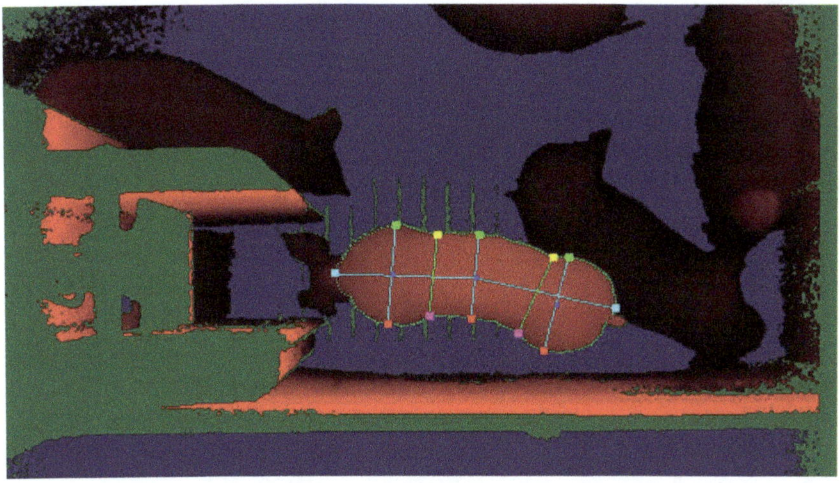

Figure 2 Image of eYeScan (3D) showing the step in the process where the body characteristics are determined to calculate the pig features such as body width, length and area. From Vranken and Berckmans 2017.

Precision feeding techniques use real-time measurements such as body weight and feed intake to provide each animal with the right amount and composition of feed at the right time (Pomar and Remus 2019, Pomar et al. 2009). They therefore can improve production strategies, meat-product quality as well as animal welfare and also have the potential to reduce environmental footprint of pig production systems. Nitrogen and phosphorous excretion (Andretta et al. 2014, Andretta et al. 2016) lysine intake (Andretta et al. 2016) and greenhouse gases emissions (Andretta et al. 2018) can be significantly reduced by increasing individual nutrient efficiency (as reviewed by Pomar and Remus (2019)).

An important feature for precision feeding is individual pig identification that is usually provided by radio-frequency identification (RFID) technology. RFID technology uses electromagnetic fields to automatically identify and track either active (with a local power source) or passive (receiving energy from an RFID reader) tags attached to objects. Its application is well known in farm animal husbandry in the context of restricted feeding, where a passive transponder is mounted to the ear of the animal and a stationary reader registers the animal's visits.

For on-farm animal management purposes low-frequency RFID systems are common as they offer high reading reliability (Brown-Brandl et al. 2017). However, they are unable to register multiple transponders simultaneously (Adrion et al. 2018) in contrast to high- and ultra-high frequency RFID systems that apply an anti-collision protocol and have higher data rates and read ranges (up to 12 m for ultra-high frequency) (Brown-Brandl et al. 2017). Recent publications in the field of animal behaviour have mainly been using high as well as ultra-high frequency RFID systems to assess feeding patterns on the basis of individual frequency and duration of feeding place visits in groups of pigs of different production stages (Maselyne et al. 2016b, Maselyne et al. 2016a, Maselyne et al. 2014, Reiners et al. 2009, Adrion et al. 2018). An online warning system for individual fattening pigs on the basis of feeding patterns determined by RFID data has been developed (Maselyne et al. 2017).

Water supply and intake can be tested and measured at group-level by a water flow-meter as a stand-alone device. Recent studies have combined the water flow-meter with RFID systems to examine drinking behaviour in individuals (Maselyne et al. 2016a, Andersen et al. 2014). Madsen et al. (2005) successfully modelled the water consumption of young growing pigs on the basis of flow-meter data by combining a growth model and a cyclic model in order to predict their drinking behaviour. However, water wastage has to be considered as a factor that might influence water intake data on the basis of water flow measurements. Kashiha et al. (2013b) applied automatic image processing based on top-view video recordings and transfer function modelling to estimate water use and reached an accuracy of 92%.

3 Good housing

Thermal comfort of pigs as one of three criteria of welfare within a principle of good housing (WelfareQuality® 2009) was traditionally monitored with sensor technology based on predetermined ambient temperature on a compartment level. Advancement in sensor technology allows placing thermometers in ear tags and measures body temperature directly on the individual animal (Andersen et al. 2008). The results of this experiment indicate that pigs adjust their behaviour to a higher ear surface temperature and they use behavioural adjustment (e.g. increased/decreased contact to pen mates) to bring their skin temperature into a preferred interval (Andersen et al. 2008).

The traditional approach of estimating thermal comfort of pigs can fall short in meeting the animals' true thermal needs because it does not integrate the effects of other contributing factors, such as drafts, humidity (particularly in hot conditions), radiation (in poorly insulated barns), floor type and/or condition (dry vs. wet, bedding vs. no bedding), nutritional plane and health status of the animal (Shao and Xin 2008).

The best indicator of the environment adequacy and thus animal comfort is animals themselves that integrate both external and internal factors, which in turn lead to distinctive resting behaviours. Huddling, resting next to one another, and spreading are the stereotypical postural patterns of group-housed animals when experiencing cold, comfortable, and warm/hot sensation, respectively (Shao and Xin 2008).

For monitoring pig behaviour as an indicator of thermal comfort 2D camera sensors were used in the research of Costa et al. (2014), Shao and Xin (2008) and Xin (1999). Costa et al. (2014) monitored behaviour of a group of fattening pigs on the basis of two variables extracted from images, the activity and occupation indexes. It was possible to automatically detect that pigs spent less time in the area of the pens with a humid floor surface and with a limited air speed. The animals occupied 'Central zone' in the barn, which was characterized by dry floor and faster air movement, more often.

In the research by Xin (1999) and Shao and Xin (2008), image analysis was used to classify thermal comfort state of group-housed pigs on the basis of their resting behaviour. The human observer first labelled images with pigs' resting behaviour as cold, comfortable and warm sensations. In the next steps images were segmented, motion detection and feature extraction was applied. Finally, the thermal comfort state was automatically classified. The test results showed over 90% of correct classification rate. This result suggests that continuous, automated classification of thermal comfort in fattening pigs measured directly on the animal is possible with modern image analysis techniques.

Image analysis was also applied in research of Nasirahmadi et al. (2017) to classify group-lying patterns of pigs (close, normal and far) in different thermal categories in commercial pig-farm conditions. An image processing technique based on Delaunay triangulation (DT) was utilized. The percentage of each defined lying pattern, obtained through calculating the perimeter of each triangle in the DT, changed significantly as the environmental temperatures increased. The authors conclude that the proposed method is an important step towards improving animal welfare in commercial farm conditions with their changeable environmental parameters.

Ease of movement was traditionally evaluated by average space available per animal in a pen. Different recommendations for appropriate space allowance were given depending on study design from which they were derived. A combination of three-dimensional cameras and image analysis allows automated measurement of space used by the animal and free space available in a pen (Fels et al. 2018). This opens up possibilities not only to assess spatial requirements for pigs in a new way but also to measure it in real-time as animals grow and take up more space in a pen (Fig. 3). Additionally, there is potential in the application of image analysis for monitoring the ease of movement of pigs by tracking the distance moved by an individual in a group (Matthews et al. 2017).

4 Good health

Animals in good health should be free of injuries (e.g. skin damage and locomotory disorders) and from disease (i.e. animal unit managers should

Figure 3 Image of a piglet group generated by the programme for image analysis. Animals identified as individuals have a green line drawn around them and are numbered 1-3. From Fels et al. 2018.

maintain high standards of hygiene and care) and should not suffer pain induced by inappropriate management, handling, slaughter or surgical procedures (WelfareQuality® 2009).

Lameness has a major impact on health and welfare of pigs and is of high economic importance, especially in sows where it is a major cause of culling (as reviewed by Heinonen et al. (2013)). Lame sows spend more time lying and less time moving and standing than healthy ones (Ala-Kurikka et al. 2017). Their stepping behaviour is increased (Conte et al. 2015), they walk more slowly, have a shorter stride length, a longer stance time and a shorter latency to lie down after feeding (Gregoire et al. 2013). So far, technologies as force plates (Conte et al. 2014, Pluym et al. 2013, Karriker et al. 2013, Mohling et al. 2014), pressure mats (Meijer et al. 2014), infrared thermography (Amezcua et al. 2014), accelerometers (Gregoire et al. 2013, Conte et al. 2015, Traulsen et al. 2016, Scheel et al. 2017), electronic feeders (Cornou et al. 2008) and image analysis (Stavrakakis et al. 2015, Pluym et al. 2013) have been applied to replace visual examination in lameness detection in pigs.

Camera-based, automated systems for the assessment of lesions have been developed for ear and tail lesions at the abattoir (Blömke et al. 2020, Brünger et al. 2019). Blömke et al. (2020) detected ear lesions with an accuracy of 95.4% and for tail lesions an accuracy of 99.5% was achieved. A future vision is automatic screening of pathological organ alterations. Initial steps towards this direction have already been taken. McKenna et al. (2018) tested a method based on auto-context segmentation on images of porcine offal in an abattoir production line. Highest dice coefficients for different organs ranged between 0.816 (heart) and 0.973 (liver).

Many studies have been focussing on automated monitoring and early detection of diseases. Munsterhjelm et al. (2015) used automatic single-space feeders in growing pigs and found a decrease of feed intake in future lame or tail-bitten individuals already 2–3 weeks before diagnosis. Electronic sow feeders have been evaluated to detect estrus, health as well as pregnancy disorders in group-housed sows (Cornou et al. 2008, Iida et al. 2017). Madsen and Kristensen (2005) applied time-series analysis on water flow-data in groups of growing pigs and were able to detect increasing water intake caused by diarrhoea 1 day before symptoms were recognised by the care-taker. Another approach to monitor disease in pig herds was developed by Martínez-Avilés et al. (2015). Pigs were infected with an attenuated strain of African swine fever and equipped with acceleration, body temperature and RFID sensors to monitor movement, body temperature and water intake. Additionally, cameras were collecting data for video image processing. This real-time monitoring system was able to identify the infection 1–3 days before qPCR detection.

Recently, infrared thermography has gained attention as a non-invasive remote sensing method that has been used to indicate thermal biometric changes in animal metabolism (McManus et al. 2016). When assessed correctly, the measured parameter has a high correlation with body temperature (as reviewed by Soerensen and Pedersen (2015)) and can be used as a feature variable, for example, for febrile responses (Cook et al. 2015) and hypothermia in newly born piglets (Kammersgaard et al. 2013). Lu et al. (2018) were able to measure ear base temperature on the basis of top-view piglet thermal images with an accuracy of 97–98%.

Most of the above-mentioned methods for disease monitoring and early detection are sensitive enough to determine a significant change in indicator values. However, most indicators are not pathognomonic and therefore allow only vague assumptions on the cause of the change. Automated monitoring of vocalizations associated with respiratory diseases permits a restriction of tentative diagnoses to those with manifestation in the respiratory tract. Audio surveillance systems that record coughing do not only allow identification of sick pig cough sounds in real time (Exadaktylos et al. 2008), but also differentiation between non-infectious and infectious coughs with further characterization of coughs caused by pathogens such as Actinobacillus pleuropneumoniae and

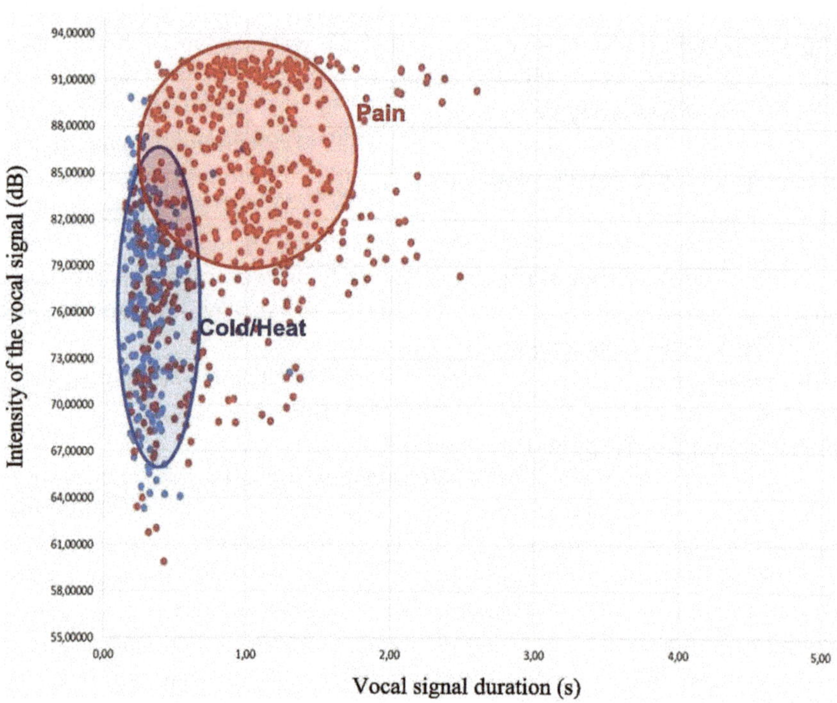

Figure 4 The output of the software using the sound wave intensity (vertical axis, dB) and the duration of the vocal signal (horizontal axis, s) during the exposition to heat and pain. From da Silva et al. (2019).

Pasteurella multocida (Ferrari et al. 2008). Chung et al. (2013) were able to make a distinction based on cough sounds between postweaning multisystemic wasting syndrome, porcine reproductive and respiratory syndrome and mycoplasma hyopneumoniae. Cough sound analysis has also been applied to assess air quality in weaner barns (Wang et al. 2019).

Moreover, sound data can also be helpful in the assessment of pain. Experiencing different stages of pain has been associated with differing vocalization patterns (Leidig et al. 2009, Taylor and Weary 2000, Weary et al. 1998, Diana et al. 2019). Pain was defined as a target variable in research from da Silva et al. (2019), who were able to predict pain on the basis of vocal intensity and duration in piglets with an accuracy of 93.0% (Fig. 4).

5 Appropriate behaviour

According to the WelfareQuality® (2009) principle, 'appropriate behaviour' animals should be able to express normal, non-harmful, social behaviours (e.g. grooming) and other normal behaviours (i.e. foraging or exploring). They should be handled well in all situations, that is, handlers should promote good human-animal relationships and negative emotions such as fear, distress, frustration or apathy should be avoided whereas positive emotions such as security or contentment should be promoted.

One of the few published attempts to automatize the detection of social behaviours was made by Šustr et al. (2001). They used the EthoVision® tracking system (Noldus Information Technology, Wageningen, the Netherlands) to provide coordinates of colour-marked body parts of two pigs and then applied an algorithm to determine whether these two pigs were in body contact, their mutual orientation and whether the pig in the 'active' position made snout contact in front of or behind the ear base of the other pig. This method might also be useful for automatic detection of play behaviour, a motivated behaviour that only occurs under currently non-life-threatening conditions (Held and Špinka 2011) and is considered as an indicator for positive welfare (Lawrence et al. 2018, Boissy et al. 2007). Other studies have been focussing on object-oriented play behaviour by equipping the object with a sensor (Zonderland et al. 2003). Another interesting approach to investigate social interactions and at the same time open new doors to a new dimension of 'social engineering' in pigs was developed by Boumans et al. (2018). They designed an in-depth agent-based model to simulate social interactions and feeding behaviour of individually and group-housed pigs and suggested the implementation of this model in research as a tool for studying the effects of social factors and group dynamics on individual variation in feeding and social interaction patterns.

Automatic detection of mounting, a behaviour that occurs during the entire growing period (Thomsen et al. 2012), but more frequently in overcrowded situations (Faucitano 2001) has been studied by Nasirahmadi et al. (2016). They achieved an accuracy of 92.7% by applying image analysis with an ellipse fitting technique.

Rooting activity as part of exploratory behaviour has been automatically detected on the basis of 3D accelerometer data with an accuracy ranging from 66.5% (Escalante et al. 2013) to 86.7% (Cornou and Lundbye-Christensen 2008). For the purpose of automatic detection of nest-building Oczak et al. (2015) observed pawing, exploratory behaviour, manipulation of pen and rack. The accuracy they found for 'nest-building activity' as combination of those four behaviours was 86%.

As the awareness and demand for assessment of positive welfare is relatively new (as reviewed by Yeates and Main (2008) and Lawrence et al. (2018)), PLF monitoring attempts have been mainly focussing on potentially harmful behaviours like aggression as welfare indicators. Although low-level aggression is considered as very important for the maintenance of a stable social hierarchy in pig groups (McGlone 1985), high occurrence (usually linked to mixing unfamiliar pigs) is associated with stress and reduced animal welfare (McGlone et al. 1980, Turner et al. 2006).

Image feature extraction analysis has been applied by Viazzi et al. (2014) to detect aggression automatically while Chen et al. (2019) combined this method with a depth sensor reaching an accuracy of 97.5%. Oczak et al. (2014) proposed a technique based on an artificial neural network that classified high and medium aggression from video recordings. Chen et al. (2018) were able to do the same classification using a kinetic energy model based on machine vision.

Abnormal behaviour can originate from unsuccessfully applied coping strategies in aversive, overstraining situations (Wechsler 1995). Tail biting often occurs in situations when environmental factors do not comply with animals' needs (Hunter et al. 2001, Moinard et al. 2003, van Putten 1969) and the animals fail to gain control over this situation. It is considered as a welfare-reducing problem (European Food Safety Authority 2007) with economic consequences for pig production.

Former studies have proved indicators such as increased pig (Statham et al. 2009, Ursinus et al. 2014, Zonderland et al. 2011) and pen (Ursinus et al. 2014) directed manipulative behaviours, object manipulation (Larsen et al. 2019a), amount of posture changes (Zonderland et al. 2011), activity (Larsen et al. 2019a, Statham et al. 2009, Ursinus et al. 2014, Zonderland et al. 2011), hanging tail posture (Lahrmann et al. 2018, Wedin et al. 2018) and low feed intake (Wallenbeck and Keeling 2013) as reliable predictors for an outbreak. Yet, only few studies have been linking these behavioural parameters to PLF technologies. Larsen et al. (2019b) collected sensor data on water usage and pen temperature to develop an algorithm for tail biting while D'Eath et al.

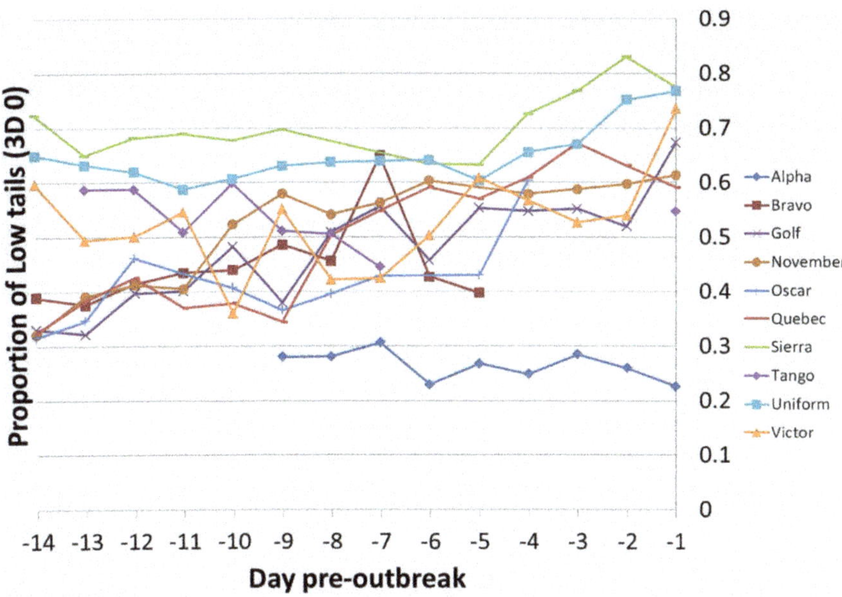

Figure 5 Proportion of 3D low tail detections (3D 0) on the days leading up to an outbreak. Data are shown for the 10 outbreak groups. Each line shows data for a different group. From D'Eath et al. (2018).

(2018) were able to detect lowered tail posture automatically on the basis of 3D-cameras before the outbreak (Fig. 5). As tail (and ear) posture and movement have been found to be indicative for emotions in pigs (Reimert et al. 2013, Marcet Rius et al. 2018), the latter method might have much potential for future applications in the context of monitoring welfare.

Another promising tool to monitor appropriate behaviour as well as emotions of pigs that has already been mentioned in the subchapter 'Good health' is audio surveillance of vocalizations (as reviewed by Manteuffel et al. (2004)). Automation techniques have been so far mainly focussing on stress responses (Schon et al. 2004, da Silva et al. 2019, Moura et al. 2008).

6 Legal and ethical considerations

In PLF systems, mainly non-invasive sensors such as cameras and microphones are used. If sensors have to be placed on (i.e. ear) or in the animal (i.e. rumen of cattle), this may under no circumstances result in significant additional pain, inflammation, functional restriction or any other risk of impaired welfare of the animal. In this regard, active sensors powered by long-life batteries can cause problems due to weight and size. In any case, the attachment of PLF sensors on or in the animal must comply with the respective animal welfare regulations

(European Union 2008b) and the animal identification regulations (European Union 2008a).

PLF-based information systems must be developed according to strict scientific criteria. The selection of suitable welfare indicators, the definition of the relevant gold standards, the assurance of reliability and validity in primary data processing and the definition of standard ranges must be based on comprehensible physiological, ethological and veterinary criteria. If inaccuracies occur at this early stage of the developmental process of PLF systems, it may have a negative impact on the health and well-being of the monitored animals due to misinformation and inadequate reaction. In order to avoid situations where poorly designed PLF products enter the market, it is recommended to consider the regulation of PLF products following the example of human medical products (European Union 2017).

One way of viewing PLF is to see it as a ground for improving the well-being of animals, because the technology highlights the care for the individual animals and their quality of life. On the other hand PLF mediates the human-animal relationship in conventional industrial livestock farming, implying that farmers and animals become subjected to new expectations. PLF redefines the notion of care, in terms of data transparency, standardization of methods for analysis, real-time collection and processing of data, remote control, and the use of digital platforms. Therefore PLF requires a redistribution of responsibilities within a wider scope of relations in the value chain (Bos et al. 2018).

PLF not only has a direct impact on animal welfare by improving animal monitoring, facilitating standardized decisions and thereby positively affecting animal care. It is also very likely that PLF technology, consisting of biometrics, Internet of Things, robotics, augmented reality and digital platforms will fundamentally change the entire animal food production chain and thereby indirectly affecting animals' quality of life.

By means of algorithms a huge amount of data is transformed in real-time to information on the health and welfare status of individual or groups of animals. Through systems like artificial intelligence, algorithms do not follow a predetermined set of rules but make use of self-learning statistical techniques. As a result, algorithm-based decisions are almost unfathomable and uncontrollable for humans. To prevent manipulation, it is therefore crucial that we understand why such algorithms make certain choices, and how to implement transparency (Royakkers et al. 2018).

From an ethical point of view, the question arises as to whether the dynamics of digitalization are changing our understanding of the subjectivity of humans and animals in conventional industrial livestock farming. According to Bos et al. (2018), the transformative effects of PLF on the human-animal relationships turn animals and human into objects. Royakkers et al. (2018) describe effects such as deanimalization, dehumanization, instrumentalization, desocialization and deskilling to those jobs that remain. However, this development has been observed in industrial livestock farming for decades.

PLF may primarily be used to enhance the viability of intensive livestock production, to make it more feasible to keep very large herds in stressful, high-density conditions with poor levels of welfare. By alerting farmers to problems at an early stage it can, to a degree, improve animal welfare and system efficiency. However, such improvements are made within a system that has inherently low potential for good welfare (Stevenson 2017). Another issue is the further consolidation of farms, as only those with the capital to invest in PLF can benefit from the 'technology treadmill' of ever-improving PLF technologies (Werkheiser 2018).

Bos et al. (2018) conclude that PLF requires a new balance of power between animal, farmer, veterinarian, industry, consumer and government in order to prevent unfair competition and exploitation. Actors with strong commercial interests are particularly likely to use PLF data to gain market power. If products become more dependent on PLF software, this strengthens PLF-manufacturers' control and how that can be utilized (Royakkers et al. 2018). This may also mean that ignoring animal welfare principles by farmers will no longer be invisible and tolerated.

7 Sensor data management and modelling

In the livestock process, the central and most complex component is the animal. Due to the time-varying behaviour of animals being living organisms, the monitoring of the PLF approach requires continuous measurements of animals' responses directly on the animal rather than in the environment surrounding the living organism (Berckmans 2017).

Since animal responses can be very fast, it is useless to carry out a survey once a year, once a month or week, or even twice a day. We need a continuous monitoring/management tool. Depending on the variable that is monitored, the word 'continuous' might mean every second (e.g. for stress monitoring) or once a day (e.g. for weight monitoring). Modern technology makes it all possible. The field data consist of a lot of numbers originating from the sensors (e.g. 240 samples per second for an accelerometer), images (e.g. 25 images per second), or sound signals (e.g. 20000 samples per second). Within the EU-PLF project (EU-PLF 2016), the PLF technology was installed in 5 broiler houses and 10 pig houses. The monitoring over 90 fattening periods for pigs in 3 years resulted in 5475 measuring days, generating more than 120 terabytes image data and 4906000 files of 5 min each with sound data (Berckmans 2017).

When the PLF technologies are installed in livestock houses, they generate a huge amount of data, and the transmission of so much data takes time, energy and money. Sending data wirelessly involves energy and costs; we should, therefore, avoid transmitting too much data and develop real-time algorithms that calculate information from the data at the lowest possible level enabling us to transmit relevant information rather than data. Therefore, we need real-time algorithms that can calculate relevant information from the data preferable on or close to the individual animal (Fig. 6) (Berckmans 2017).

Living organisms are CITD systems, which stand for complex, individually different, time-varying and dynamic systems (Berckmans 2006b). It is obvious than a living organism is much more complex than any mechanical, electronic or ICT system. The complexity of information transmission in a single cell of a living organism is much higher than most other systems that may be considered. The time-varying character of living organisms means that a living organism's

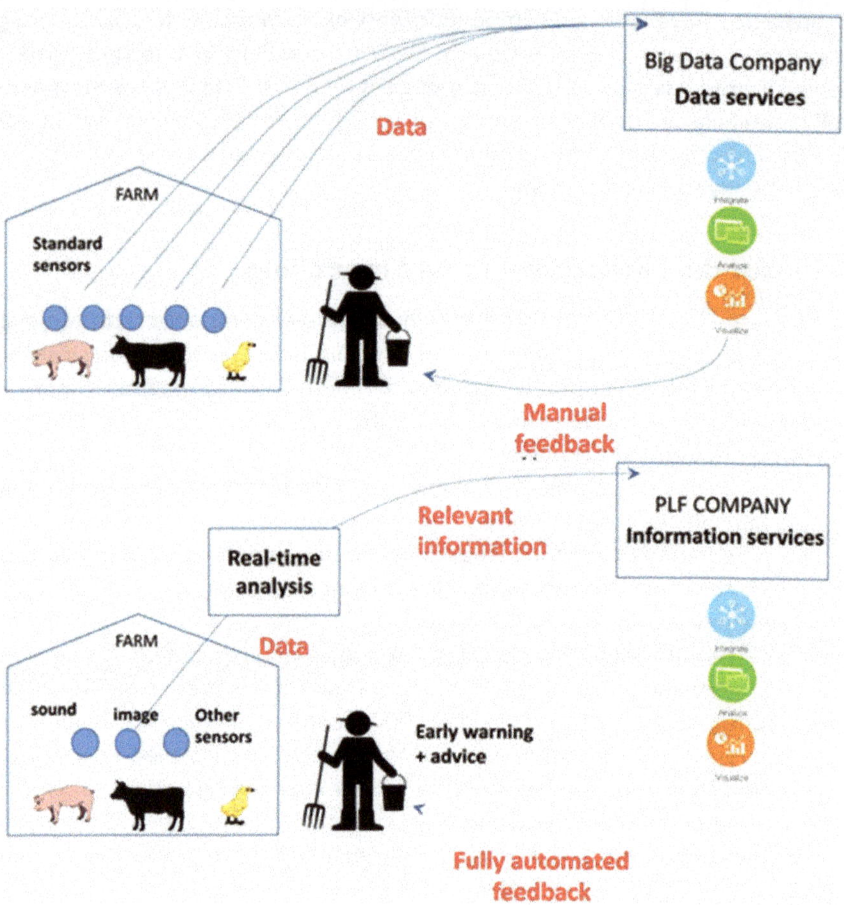

Figure 6 Scheme of a more realistic use of data versus the common idea of big data. From Berckmans 2017.

response to an environmental stimulus or stressor might be different each time it happens. A living organism is constantly looking for a good energy balance and as a consequence is continuously changing its physical condition and mental status. Of course living organisms are dynamic systems (Berckmans 2013).

The CITD nature of living organisms has an important impact on the type of algorithms that need to be developed for PLF applications. It implies that algorithms to monitor these time-varying individuals must continuously adapt to the individual and/or use principles that can be used in real time in the field application (Berckmans 2013). Algorithms from the family of dynamic models (e.g. dynamic regression, transfer function) are appropriate for modelling CITD systems (Young 2011).

The advantages of these algorithms are that they are simple, suitable for online and real-time monitoring and also are interpretable. Appropriate model order identification criteria such as Akaike (AIC) (Akaike 1974), Bayesian (BIC) (Akaike 1974) or Young Information Criterion (YIC) (Young 1984) are used together with statistical measures of how well models fit to data such as coefficient of determination, R^2, to identify models often with only few parameters.

Such simple models can be interpreted meaning that conclusions can be drawn about the underlying modelled biological process in relation to variation in model parameters (Fernández et al. 2019). Simplicity of these models also offers possibility to implement them on embedded systems, on cheap devices, close to the animal, reducing the need to transfer large amounts of sensor data to remote locations. Dynamic models can be estimated for each individual animal. Because model parameters evolve over time these models adapt accordingly to time-varying response of animals. This allows those models to achieve similar, robust performance when validated on different farms (Oczak 2018).

Algorithms from the family of machine learning (support-vector machines, artificial neural networks) that gain popularity in recent years also in the research field of animal monitoring (Nasirahmadi et al. 2017, Valletta et al. 2017, Escalante et al. 2013) are complex (1000s of parameters) and are black box models (Hansen et al. 2018). This means that these models require much more calculations than simple models. Thus, powerful hardware is needed for estimation and implementation of these models for real-time processing of sensor data. Nowadays such powerful hardware can be implemented on farms; however the cost is considerably higher in comparison to hardware cost of implementation of simple models. Black box character of machine-learning algorithms limits the possibility to gain insight on modelled biological processes as variation in 1000s of parameters cannot be related to these processes in comprehensible ways. With these disadvantages of machine-learning methods the main advantage is that very complex sensor data such as image data can be modelled with less effort for the modeller. These models possibly can be a basis for robust monitoring systems in PLF if enough variable input data is used for training (Oczak 2018).

New developments in machine-learning algorithms such as recurrent neural networks (RNN) can be suitable to model CITD systems as RNN can be individually estimated on time series data. Parameters of RNN algorithm similar to dynamic models evolve over time so that these models adapt to time-varying response of individual animals (Chen et al. 2020).

8 Case study

It is a common practice in modern intensive pig husbandry to confine sows in farrowing crates, usually for 4–5 weeks, including at least few days before the onset of farrowing. The main reason for this practice is to improve piglet survival rate by protecting new-born piglets from fatal or injurious crushing by the mother sow (King et al. 2019). However, the confinement of sows in crates has a negative impact on sows' welfare, such as limited freedom of movement, limited social interactions with new-born piglets (Melišová et al. 2011, Pedersen et al. 2013) and diminished health (Lambertz et al. 2015, Singh et al. 2017). Confinement also prevents much of the prenatal nest-building behaviour, an essential part of behavioural repertoire in sows, which starts around 24 h before parturition, is most intense 6–12 h before parturition and then decreases as parturition approaches (Wischner et al. 2009, Castrén et al. 1993). Increased physiological stress for the sow is a consequence of confinement in a crate (Jarvis et al. 2006).

A concept of temporary crating has been developed as a response to increased public concern about welfare of crated sows (Moustsen et al. 2012). According to this concept sows should be temporarily confined in farrowing crates only during the critical period of piglets' life, when piglet crushing is most probable, in the first days after farrowing (Marchant et al. 2000, BMGÖ 2012). When the crate is opened farrowing pen offers additional space for the sow, providing a compromise between the needs of the farmer, the sow and her piglets (King et al. 2019). This would allow the sow to stay unconfined during the prenatal nest-building phase from 24 h ante partum until approach of farrowing, which would have a positive impact on sows' welfare (Algers 1994). However, choosing the right moment to confine an individual sow in a farrowing crate under farm conditions in a way that makes nest-building possible and does not increase risk of piglet crushing is challenging. Due to the biological variability in gestation length, time-consuming observation of sows would be necessary. On the other hand, confinement of sows in crates based on a calculated farrowing date of the batch could either disturb adequate nest-building or the farrowing process in many sows.

Automated detection of changes in sow activity, with the use of sensor technology, was used in the research by Oczak et al. (2019) to indicate when sows should be confined in a farrowing crate. Ear tag-based acceleration data was modelled to provide two types of alarms: 'First stage' alarm indicates the beginning of nest-building behaviour while 'second stage' indicates the end of nest-building behaviour. In total, 53 sows were included in the experiment, 27 in the training data set and 26 in the validation data set. Each sow had an ear tag with an accelerometer sensor mounted on the ear. Acceleration data was modelled with Kalman filtering and fixed interval smoothing (KALMSMO) algorithm.

In the first step total physical acceleration (magnitude) was estimated from three axes of accelerometer data (xyz). In the second step total physical acceleration (magnitude) was smoothed with standard deviation calculated on a sliding window of 24 h with 15 min steps (Fig. 7).

Smoothing window of 24 h allowed the elimination of variation in the activity of animal related to diurnal rhythm. This input variable was used to estimate a trend in the activity of each sow. Changes in trends (dynamics) in the activity of sows were a basis for detection of approaching farrowing and providing first and second level alarms. To estimate the dynamics of activity of sows the KALMSMO algorithm was used that comprise as a special case – the Integrated Random Walk (Young 2011).

In the consecutive steps the fixed interval was expended recursively by 15 min steps until the trend in animal activity changed to significantly increasing. This was indicated by the input variable reaching higher value than upper confidence interval of estimated trend (Fig. 8). At this time point the 'first stage' alarm informing about approaching farrowing was raised.

In the next step KALMSMO algorithm was applied to acceleration data on a fixed interval starting from the time point of the 'first level' alarm. Next, the fixed interval was expended recursively by 15 min steps until the trend in animal activity changed to significantly decreasing. This was indicated by an input variable reaching a lower value than the lower confidence interval of the estimated trend (Fig. 9). At this time point the 'second stage' alarm was raised. This alarm could be interpreted as an indication to confine a sow in a crate.

It was possible to predict farrowing on the basis of increased activity in the validation dataset with a median of 8 h 51 min before the onset of farrowing. Alarms indicating the need for confinement of the sow in a crate were raised with a median of 2 h 3 min before the onset of farrowing (Fig. 10).

Practical application of 'first stage' alarms generated around 6–13 h before the onset of farrowing might be to warn the farmer about approaching farrowing in an automated way. This could reduce labour costs otherwise required for regular control of sows in farrowing compartments. It should also limit the need to get in close contact with sows which could interrupt their activities at a

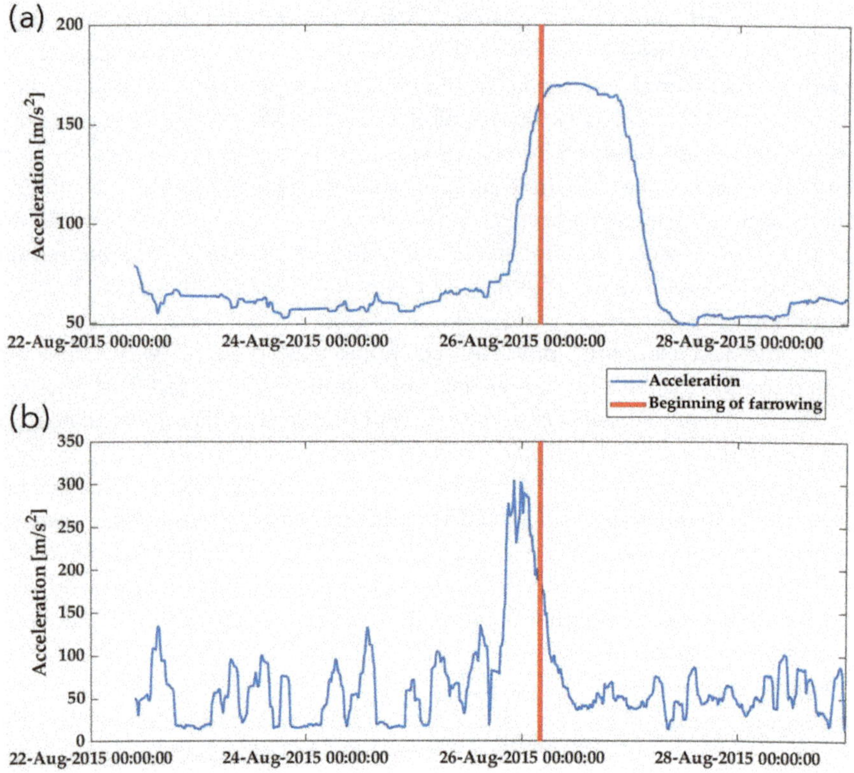

Figure 7 Standard deviation calculated on acceleration data with two sizes of sliding windows. Period depicted in the plot starts at introduction of a sow to a farrowing pen and ends around 3 days after beginning of farrowing. (a) Sliding window of 24 h with 15 min steps. Variation in activity related to diurnal rhythms was filtered out. Two rapid changes in trend in activity were at the beginning of nest-building behaviour and at the end of nest-building behaviour, when this sow started farrowing. (b) Sliding window of 2 h with 15 min steps. Peak of activity was few hours before the onset of farrowing. From Oczak et al. 2020.

time when they are sensitive to outside disturbances (Erez and Hartsock 1990, Oliviero et al. 2008).

Austrian legislation requires that in the week before farrowing, the animals should be provided with suitable nest-building material but only when the slurry system allows it (FTT 2018). Thus, when the 'first stage' alarm is raised 'nest-building' material could be provided to the sow. Providing nest-building material in higher amounts after the 'first stage' alarm is raised or only after this alarm could limit the risk of blocking the slurry system with nest-building material and improve the welfare of sows. Additionally, the procedure of preparing a farrowing pen for newborn piglets could be started when the 'first stage' alarm is raised (Traulsen et al. 2018).

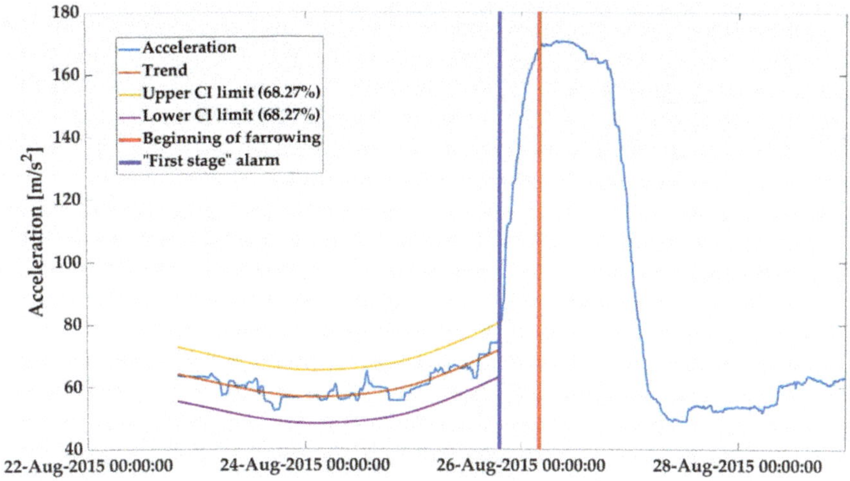

Figure 8 'First stage' alarm at 9 h before the onset of farrowing. Period depicted in the plot starts 12 h after introduction of a sow to a farrowing pen and ends around 3 days after farrowing. From Oczak et al. 2020.

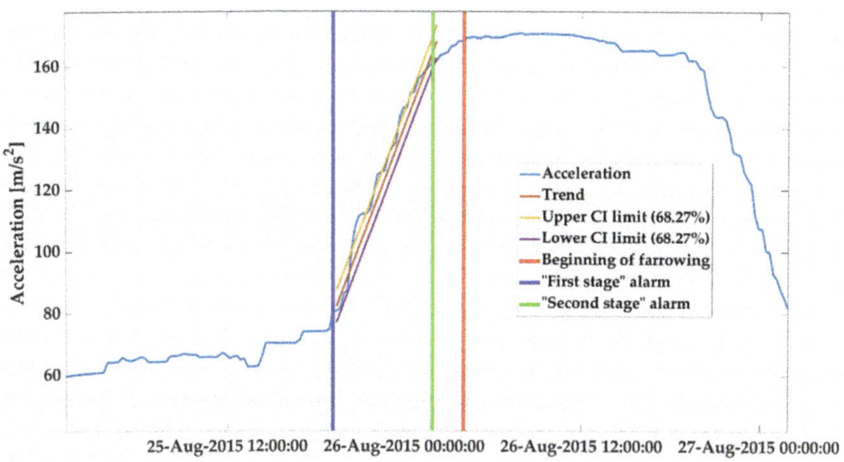

Figure 9 'Second stage' alarm at 2 h before the onset of farrowing. Period depicted in the plot starts around 24 h before the onset of farrowing and ends around 20 h after it. From Oczak et al. 2020.

According to Austrian Animal Welfare Directive, that will become mandatory in 2033, sows should only stay confined in farrowing crates during the 'critical period' of piglets' life (BMGÖ 2012). In order to reduce piglet-crushing, farmers will be permitted to confine sows in farrowing crates from the onset of farrowing to a few days after farrowing. From an animal welfare point of view, crating should start after nest-building is finished and before farrowing

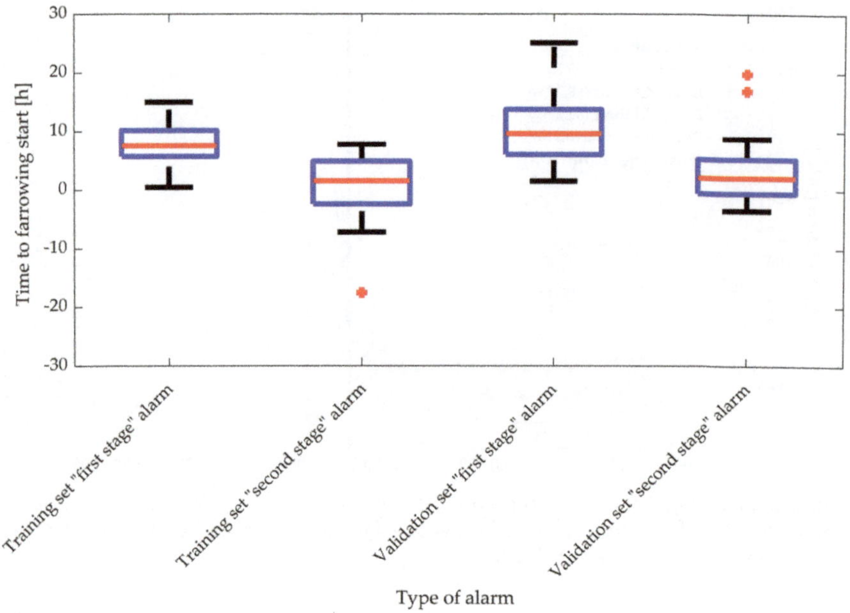

Figure 10 Distribution of duration between time of alarms and onset of farrowing in training and validation datasets. From Oczak et al. 2020.

starts. The crate has to be opened four days after farrowing (Heidinger et al. 2017). The method used in this study to generate 'second stage' alarms is based on change of trend in sow activity when nest-building behaviour ends. This change of trend is visible as 'flattening' of feature variable extracted from acceleration data (Figs 8 and 9).

The advantage of this method is that when 'second stage' alarm is raised, most of nest-building activity of a sow should be finished. Thus, confining sows after the 'second stage' alarm was raised should create little risk for those staying in crates at the time of nest-building and especially dung times of peak of nest-building behaviour. Implementation of developed model in a monitoring system of pens with the possibility of temporary crating and practical application of the tool on farms could help to achieve a compromise between the needs of the farmer, the sow and her piglets. Especially the principle of appropriate behaviour is relevant for this research as nest-building is an essential part of behavioural repertoire in sows (WelfareQuality® 2009).

9 Conclusion

With all the advantages and promises of application of PLF technology for improving possibilities of nest-building behaviour in pigs there are further scientific questions which should be answered before such system could be successfully implemented in practice. For instance, can such a system attain the expected outcome considering that many first and second stage alarms would be generated in the night when mostly there would be no staff present on farms? Assuming that practical application of developed model would be to confine sows in crates just after the 'second stage' alarm is raised, many sows that farrow in the night would stay unrestrained during farrowing.

This leads to the question of whether the application of a monitoring tool, which can be used effectively only in part of farrowing animals, is justified under practical conditions. We would hypothesize that most farmers would keep the sows confined in crates over night after they've received a 'first stage' alarm, in many cases in time of nest-building behaviour, as when 'second stage' alarm is raised during the night they would not be able to intervene in the farrowing compartment. This suggests that automated confinement of sows, without the need of human intervention is an interesting topic for further research (Fig. 11). A number of scientific questions arise from an idea of automatic confinement. Is sow in the right posture that allows automated confinement and what should that posture be? In which part of the pen a sow should be located so that a crate could be automatically closed? How a sow would react when a crate would be automatically closed?

Integration of the system for the prediction of farrowing into management procedures on pig farms should also be a topic of research (Fig. 11). It is only possible to do research on this topic when models for farrowing prediction are implemented and work in real time, at least on research farms, better than on commercial farms. Further research on changes of welfare status of sows

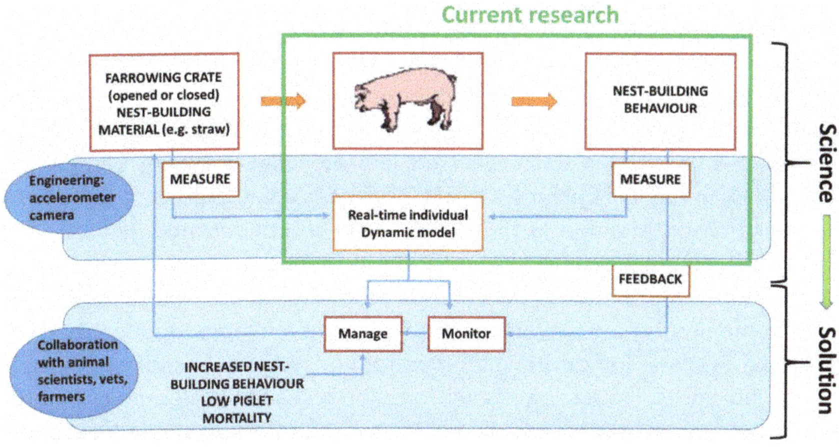

Figure 11 How farrowing prediction in pens with possibility of temporary crating can go from science to solution. Adapted from Norton et al. 2019 and Aerts, Norton and Berckmans 2019.

on farms where such system is implemented and integrated with management procedures is necessary.

Another interesting scientific objective could be to automatically detect and differentiate between behaviours that constitute nest-building behaviour, that is, rooting, pawing, manipulation of pen or crate (Oczak et al. 2015). Such a capability could provide more detailed information on sows' behaviour and possibly increase the performance of developed models. This is more likely to be successful with image analysis techniques, which have been applied for the detection of complex behaviours in livestock but not for farrowing prediction (Viazzi et al. 2014, Lee et al. 2016).

10 Future trends in research

Future development of technologies for monitoring pig welfare will be dominated by three major trends: solution-oriented research, integration of key welfare indicators and wider adoption of image analysis. These major trends will be supplemented by more validation studies on open datasets and advances in robotics.

The most important trend will be solution-oriented research, where the development of an algorithm for measuring individual welfare indicator is just a first step in developing a monitoring system that could be applied in practice (Fig. 11). In order to advance from monitoring indicators of pig welfare to the development of monitoring systems that could be integrated into farm management, it is necessary to implement developed algorithms in real-time. Real-time implementation is fundamental as farm management requires immediate action (Berckmans 2013). Once real-time implementation of an algorithm is achieved, research on the use of information provided by algorithms becomes possible. Interaction between the user, the animal and technology will be at the centre of future research on monitoring welfare and management of farms with PLF tools. To provide information to various stakeholders (e.g. farm staff, vets, retailers etc.) development of real-time dashboards in PLF will become more important in the future.

Research on monitoring pig welfare with sensor technology so far has been predominantly focussed on individual indicators such as specific diseases (Ferrari et al. 2008) or behaviours (Zonderland et al. 2003). The second future trend in monitoring pig welfare will be research on integration of monitoring individual welfare indicators into monitoring systems based on multiple welfare indicators. Thus, scientists will focus on research on algorithms that will classify the overall health of an animal instead of detecting single diseases or algorithms that will evaluate if a sow performs appropriate behaviour instead of only detecting one category of behaviour (e.g. nest-building).

Image analysis has been applied for monitoring animals in farm conditions since at least 20 years (Bloemen et al. 1997). The main advantage of this sensor technology in comparison to other sensors is that it allows monitoring in a non-invasive way, without human presence or with limited human presence. It is possible to monitor multiple animals with one sensor and image-based monitoring is most similar to the way that farmers monitor their livestock with their own eyes. Camera sensor and image analysis is also probably the most versatile monitoring technology applied so far in PLF. Thus, meaning that multiple variables (key indicators) related to welfare can be monitored with one camera sensor.

However, because of difficulties in practical application in farm conditions image analysis has been used only in a limited number of products for monitoring key indicators of welfare in pigs (e.g. OptiSCAN, WUGGL One). Difficulties of practical application in pig compartments are mainly associated to the variability of environments in which pigs are kept such as changing light conditions, ceiling heights, pen sizes and also ubiquitous dust.

Nevertheless recent advantages in camera technology and image analysis techniques led to very promising developments such as detection of multiple parts of pigs body (Psota et al. 2019), face recognition (Hansen et al. 2018) or real-time pig tracking (Labrecque et al. 2019).

Further advancement in face recognition technology might have implications for intensive livestock practices globally, allowing identification of animals without the need for RFID tags, for the purposes of welfare and growth monitoring (Hansen et al. 2018). Recent research on 'Grimace Scale' in pigs (Di Giminiani et al. 2016, Viscardi et al. 2017) opens possibilities for automated estimation of affective states in pigs, including pain, with image analysis techniques.

Advances in image analysis clearly show the potential for its wider adoption in the future for monitoring pig welfare. What might remain a challenge is the lack of public datasets with images of pigs with well-defined targets and evaluation metrics (Psota et al. 2019), such as Tencent (Shenzen, China) open dataset for general-purpose image classification with over 18 million labelled images (Tencent 2018). Most scientists work with relatively small closed datasets. We foresee that this will change in the future leading to faster progress in exciting field of image analysis for welfare monitoring.

Robots are extensively used in many agricultural sectors such as crops or vegetables. So far there has been little research done on on-farm application of robots in pig industry and especially in the context of welfare monitoring. One such example includes the development of Swine Robotics (Leola, US) which offers an entire line of robots designed to improve animal welfare, safety and production. Two examples of robots developed by Swine

Robotics are 'Boar Bot' and 'Bumper Bot'. 'Function of Boar Bot' is to control the movement of a boar when heat is checked by farm staff, which should improve he welfare of the boar. Function of 'Bumper Bot' is to help in the movement of pigs for reduction of injuries in pigs. Considering that within Europe and other developed agricultural economies, the supply of stockmen with the necessary management skills is limited (Wathes et al. 2008) and that in many fields robots are more precise, consistent and faster than human workers, we envision new developments in robotics for monitoring pig welfare on farms.

11 Where to look for further information

The following articles provide a good overview of advances in technologies for monitoring pig welfare:

- Berckmans, D. 2006. 'Automatic on-line monitoring of animals by precision livestock farming' in Geers, R. and Madec, F., eds., Livestock Production and Society. Wageningen Academic Publishers, Wageningen, 287–295.
- Berckmans, D. 2013. 'Basic principles of PLF: gold standard, labelling and field data.' In ECPLF2015, 21–29. Leuven.
- Berckmans, D. 2017. 'General introduction to precision livestock farming', Animal Frontiers, 7: 6–11.
- Norton, T., Chen, C., Larsen, M. L. V. and Berckmans, D. 2019. 'Precision livestock farming: building 'digital representations' to bring the animals closer to the farmer', Animal: 1–9.
- Wathes, C. M., Kristensen, H. H., Aerts, J. M. and Berckmans, D. 2008. 'Is precision livestock farming an engineer's daydream or nightmare, an animal's friend or foe, and a farmer's panacea or pitfall?', *Computers and Electronics in Agriculture*, 64: 2–10.

Key research in this area can be found at the following organizations, journals and conferences:

- European Association for Precision Livestock Farming (EA-PLF) (www.eaplf.eu).
- Biosystems Engineering (www.journals.elsevier.com/biosystems-engineering).
- Computers and Electronics in Agriculture (www.journals.elsevier.com/computers-and-electronics-in-agriculture).
- European Conference on Precision Livestock Farming (ECPLF).
- Asian Conference on Precision Livestock Farming (PLF Asia).

Major international research projects related to advances in technologies for monitoring pig welfare:

- BioBusiness (www.bio-business.eu).
- EU-PLF (www.eu-plf.eu).
- ALL-SMART PIGS (www.cordis.europa.eu/project/rcn/104741/factsheet/en).
- GroupHouseNet (www.grouphousenet.eu).

Top five research centres that readers can investigate:

- Measure, Model & Manage Bioresponses (M3-BIORES) (www.biw.kuleuven.be/biosyst/a2h/m3-biores).
- Precision Livestock Farming Hub (PLF-Hub) (www.vetmeduni.ac.at/PLF-Hub).
- Wageningen Livestock Research (www.wur.nl/en/Research-Results/Research-Institutes/livestock-research.htm).
- Teagasc Animal & Grassland Research and Innovation Centre (www.teagasc.ie).
- Research Institute for Agriculture, Fisheries and Food (ILVO) (www.ilvo.vlaanderen.be/language/en-US/EN/Home.aspx).

12 References

Adrion, F., Kapun, A., Eckert, F., Holland, E.-M., Staiger, M., Götz, S. and Gallmann, E. (2018). Monitoring trough visits of growing-finishing pigs with UHF-RFID, *Computers and Electronics in Agriculture* 144, 144–153.

Aerts, J. M., Norton, T. and Berckmans, D. (2019). Integration of Bioresponses in management of biological processes. Course in 1st year of Masterprogramme Biosystems Engineer, Katholieke Universiteit Leuven, pp. 360, started in 2006.

Akaike, H. (1974). A new look at the statistical model identification, *IEEE Transactions on Automatic Control* 19(6), 716–723.

Ala-Kurikka, E., Heinonen, M., Mustonen, K., Peltoniemi, O., Raekallio, M., Vainio, O. and Valros, A. (2017). Behavior changes associated with lameness in sows, *Applied Animal Behaviour Science* 193, 15–20.

Algers, B. (1994). Health, behaviour and welfare of outdoor pigs, *Pig News and Information* 15(4), 113N–115N.

Alsahaf, A., Azzopardi, G., Ducro, B., Hanenberg, E., Veerkamp, R. F. and Petkov, N. (2019). Estimation of muscle scores of live pigs using a Kinect camera, *IEEE Access* 7, 52238–52245.

Amezcua, R., Walsh, S., Luimes, P. H. and Friendship, R. M. (2014). Infrared thermography to evaluate lameness in pregnant sows, *The Canadian Veterinary Journal* 55(3), 268–272.

Andersen, H. M.-L., Dybkjær, L. and Herskin, M. S. (2014). Growing pigs' drinking behaviour: number of visits, duration, water intake and diurnal variation, *Animal: An International Journal of Animal Bioscience* 8(11), 1881–1888.

Andersen, H. M.-L., Jørgensen, E., Dybkjær, L. and Jørgensen, B. (2008). The ear skin temperature as an indicator of the thermal comfort of pigs, *Applied Animal Behaviour Science* 113(1-3), 43-56.

Andretta, I., Hauschild, L., Kipper, M., Pires, P. G. S. and Pomar, C. (2018). Environmental impacts of precision feeding programs applied in pig production, *Animal* 12(9), 1990-1998.

Andretta, I., Pomar, C., Rivest, J., Pomar, J., Lovatto, P. A. and Radünz Neto, J. (2014). The impact of feeding growing-finishing pigs with daily tailored diets using precision feeding techniques on animal performance, nutrient utilization, and body and carcass composition, *Journal of Animal Science* 92(9), 3925-3936.

Andretta, I., Pomar, C., Rivest, J., Pomar, J. and Radünz, J. (2016). Precision feeding can significantly reduce lysine intake and nitrogen excretion without compromising the performance of growing pigs, *Animal* 10(7), 1137-1147.

Banhazi, T. M., Lehr, H., Black, J., Crabtree, H., Schofield, P., Tscharke, M. and Berckmans, D. (2012). Precision livestock farming: an international review of scientific and commercial aspects, *International Journal of Agricultural and Biological Engineering* 5(3), 1-9.

Berckmans, D. (2006). Automatic on-line monitoring of animals by precision livestock farming. In: Geers, R. and Madec, F. (Eds), *Livestock Production and Society*. Wageningen: Wageningen Academic Publishers, pp. 287-295.

Berckmans, D. (2013). Basic principles of PLF: gold standard, labelling and field data. In: Berckmans, D. and Vandermeulen, J. (Eds), *6th European Conference on Precision Livestock Farming*, Leuven, Belgium, 10-12 September 2013. The Organising Committee of the 6th European Conference on Precision Livestock Farming (ECPLF), pp. 21-29.

Berckmans, D. (2014). Precision livestock farming technologies for welfare management in intensive livestock systems, *Revue Scientifique et Technique* 33(1), 189-196.

Berckmans, D. (2017). General introduction to precision livestock farming, *Animal Frontiers: The Review Magazine of Animal Agriculture* 7(1), 6-11.

Bloemen, H., Aerts, J.-M., Berckmans, D. and Goedseels, V. (1997). Image analysis to measure activity index of animals, *Equine Veterinary Journal* 29(Suppl. 23), 16-19.

Blömke, L., Volkmann, N. and Kemper, N. (2020). Evaluation of an automated assessment system for ear and tail lesions as animal welfare indicators in pigs at slaughter, *Meat Science* 159, 107934.

BMGÖ (2012). Verordnung des Bundesministers für Gesundheit, mit der die 1. Tierhaltungsverordnung geändert wird, *Bundesgesetzblatt für die Republik Österreich*, BGBl. II Nr. 61/2012.

Boissy, A., Manteuffel, G., Jensen, M. B., Moe, R. O., Spruijt, B., Keeling, L. J., Winckler, C., Forkman, B., Dimitrov, I., Langbein, J., Bakken, M., Veissier, I. and Aubert, A. (2007). Assessment of positive emotions in animals to improve their welfare, *Physiology and Behavior* 92(3), 375-397.

Bos, J. M., Bovenkerk, B., Feindt, P. H. and Van Dam, Y. K. (2018). The quantified animal: precision livestock farming and the ethical implications of objectification, *Food Ethics* 2(1), 77-92.

Boumans, I. J. M. M., de Boer, I. J. M., Hofstede, G. J. and Bokkers, E. A. M. (2018). How social factors and behavioural strategies affect feeding and social interaction patterns in pigs, *Physiology and Behavior* 194, 23-40.

Brown-Brandl, T., Maselyne, J., Adrion, F., Kapun, A., Hessel, E., Saeys, W., Van Nuffel, A. and Gallmann, E. (2017). Comparing three different passive RFID systems for behaviour monitoring in grow-finish pigs. In: *Proceedings of the* 8th European Conference on *Precision Livestock Farming*, Nantes, France, 12–14 September 2017, pp. 622–631.

Brünger, J., Dippel, S., Koch, R. and Veit, C. (2019). 'Tailception': using neural networks for assessing tail lesions on pictures of pig carcasses, *Animal* 13(5), 1030–1036.

Castrén, H., Algers, B., De Passille, A.-M., Rushen, J. and Uvnäs-Moberg, K. (1993). Preparturient variation in progesterone, prolactin, oxytocin and somatostatin in relation to nest building in sows, *Applied Animal Behaviour Science* 38(2), 91–102.

Chen, C., Zhu, W., Guo, Y., Ma, C., Huang, W. and Ruan, C. (2018). A kinetic energy model based on machine vision for recognition of aggressive behaviours among group-housed pigs, *Livestock Science* 218, 70–78.

Chen, C., Zhu, W., Liu, D., Steibel, J., Siegford, J., Wurtz, K., Han, J. and Norton, T. (2019). Detection of aggressive behaviours in pigs using a RealSence depth sensor, *Computers and Electronics in Agriculture* 166, 105003.

Chen, C., Zhu, W., Steibel, J., Siegford, J., Wurtz, K., Han, J. and Norton, T. (2020). Recognition of aggressive episodes of pigs based on convolutional neural network and long short-term memory, *Computers and Electronics in Agriculture* 169, 105166.

Chung, Y., Oh, S., Lee, J., Park, D., Chang, H.-H. and Kim, S. (2013). Automatic detection and recognition of pig wasting diseases using sound data in audio surveillance systems, *Sensors* 13(10), 12929–12942.

Condotta, I. C. F. S., Brown-Brandl, T. M., Silva-Miranda, K. O. and Stinn, J. P. (2018). Evaluation of a depth sensor for mass estimation of growing and finishing pigs, *Biosystems Engineering* 173, 11–18.

Conte, S., Bergeron, R., Gonyou, H., Brown, J., Rioja-Lang, F. C., Connor, L. and Devillers, N. (2014). Measure and characterization of lameness in gestating sows using force plate, kinematic, and accelerometer methods, *Journal of Animal Science* 92(12), 5693–5703.

Conte, S., Bergeron, R., Gonyou, H., Brown, J., Rioja-Lang, F. C., Connor, M. L. and Devillers, N. (2015). Use of an analgesic to identify pain-related indicators of lameness in sows, *Livestock Science* 180, 203–208.

Cook, N. J., Chabot, B., Lui, T., Bench, C. J. and Schaefer, A. L. (2015). Infrared thermography detects febrile and behavioural responses to vaccination of weaned piglets, *Animal* 9(2), 339–346.

Cornou, C. and Lundbye-Christensen, S. (2008). Classifying sows' activity types from acceleration patterns, *Applied Animal Behaviour Science* 111(3–4), 262–273.

Cornou, C., Vinther, J. and Kristensen, A. R. (2008). Automatic detection of oestrus and health disorders using data from electronic sow feeders, *Livestock Science* 118(3), 262–271.

Costa, A., Ismayilova, G., Borgonovo, F., Viazzi, S., Berckmans, D. and Guarino, M. (2014). Image-processing technique to measure pig activity in response to climatic variation in a pig barn, *Animal Production Science* 54(8), 1075–1083.

D'Eath, R. B., Jack, M., Futro, A., Talbot, D., Zhu, Q., Barclay, D. and Baxter, E. M. (2018). Automatic early warning of tail biting in pigs: 3D cameras can detect lowered tail posture before an outbreak, *PLoS ONE* 13(4), e0194524.

da Silva, J. P., de Alencar Nääs, I., Abe, J. M. and da Silva Cordeiro, A. F. (2019). Classification of piglet (*Sus scrofa*) stress conditions using vocalization pattern and applying paraconsistent logic Eτ, *Computers and Electronics in Agriculture* 166, 105020.

Di Giminiani, P., Brierley, V. L., Scollo, A., Gottardo, F., Malcolm, E. M., Edwards, S. A. and Leach, M. C. (2016). The assessment of facial expressions in piglets undergoing tail docking and castration: toward the development of the piglet grimace scale, *Frontiers in Veterinary Science* 3, 100.

Diana, A., Carpentier, L., Piette, D., Boyle, L. A., Berckmans, D. and Norton, T. (2019). An ethogram of biter and bitten pigs during an ear biting event: first step in the development of a Precision Livestock Farming tool, *Applied Animal Behaviour Science* 215, 26–36.

Erez, B. and Hartsock, T. G. (1990). A microcomputer-photocell system to monitor periparturient activity of sows and transfer data to remote location, *Journal of Animal Science* 68(1), 88–94.

Escalante, H. J., Rodriguez, S. V., Cordero, J., Kristensen, A. R. and Cornou, C. (2013). Sow-activity classification from acceleration patterns: a machine learning approach, *Computers and Electronics in Agriculture* 93, 17–26.

EU-PLF (2016). Project. Final Report. Bright Farm by Precision Livestock Farming. Grant agreement no: 311825.

European Food Safety Authority (2007). The risks associated with tail biting in pigs and possible means to reduce the need for tail docking considering the different housing and husbandry systems – Scientific Opinion of the Panel on Animal Health and Welfare, *EFSA Journal* 5(12), 611.

European Union (2008a). Council Directive 2008/71/EC of 15 July 2008 on the identification and registration of pigs, *Official Journal of the European Union* L 213, 31–36.

European Union (2008b). Council Directive 2008/120/EC of 18 December 2008 laying down minimum standards for the protection of pigs, *Official Journal of the European Union* L 47, 5–13

European Union (2017). Regulation (EU) 2017/745 of the European Parliament and of the Council of 5 April 2017 on medical devices, amending Directive 2001/83/EC, Regulation (EC) No 178/2002 and Regulation (EC) No 1223/2009 and repealing Council Directives 90/385/EEC and 93/42/EEC, *Official Journal of the European Union* L 117, 1–175.

Exadaktylos, V., Silva, M., Aerts, J.-M., Taylor, C. J. and Berckmans, D. (2008). Real-time recognition of sick pig cough sounds, *Computers and Electronics in Agriculture* 63(2), 207–214.

Faucitano, L. (2001). Causes of skin damage to pig carcasses, *Canadian Journal of Animal Science* 81(1), 39–45.

Fels, M., Konen, K., Hessel, E. and Kemper, N. (2018). Determination of static space occupied by individual weaner and growing pigs using an image-based monitoring system, *Journal of Agricultural Science* 156(2), 282–290.

Fernández, A. P., Norton, T., Youssef, A., Exadaktylos, V., Bahr, C., Bruininx, E., Vranken, E. and Berckmans, D. (2019). Real-time modelling of individual weight response to feed supply for fattening pigs, *Computers and Electronics in Agriculture* 162, 895–906.

Ferrari, S., Silva, M., Guarino, M. and Berckmans, D. (2008). Analysis of cough sounds for diagnosis of respiratory infections in intensive pig farming, *Transactions of the ASABE* 51(3), 1051–1055.

FTT (2018). *Handbuch Schweine*, Fachstelle für tiergerechte Tierhaltung und Tierschutz. Available at: https://www.tierschutzkonform.at/wp-content/uploads/tierschutzkonform.at-handbuch-schweine-2auflage2018.pdf.

Gregoire, J., Bergeron, R., D'Allaire, S., Meunier-Salaun, M. C. and Devillers, N. (2013). Assessment of lameness in sows using gait, footprints, postural behaviour and foot lesion analysis, *Animal* 7(7), 1163–1173.

Hansen, M. F., Smith, M. L., Smith, L. N., Salter, M. G., Baxter, E. M., Farish, M. and Grieve, B. (2018). Towards on-farm pig face recognition using convolutional neural networks, *Computers in Industry* 98, 145–152.

Heidinger, B., Stinglmayr, J. and Baumgartner, J. (2017). Projekt Pro-SAU: Ergebnisse zur Evaluierung von neuen Abferkelbuchten mit Bewegungsmöglichkeit für die Sau, *Leipziger Blaue Hefte* 469–470.

Heinonen, M., Peltoniemi, O. and Valros, A. (2013). Impact of lameness and claw lesions in sows on welfare, health and production, *Livestock Science* 156(1–3), 2–9.

Held, S. D. E. and Špinka, M. (2011). Animal play and animal welfare, *Animal Behaviour* 81(5), 891–899.

Hunter, E. J., Jones, T. A., Guise, H. J., Penny, R. H. C. and Hoste, S. (2001). The relationship between tail biting in pigs, docking procedure and other management practices, *The Veterinary Journal* 161(1), 72–79.

Iida, R., Piñeiro, C., Koketsu, Y. (2017). Behavior, displacement and pregnancy loss in pigs under an electronic sow feeder, *Journal of Agricultural Science* 9(12), 43–53.

Jarvis, S., D'Eath, R. B., Robson, S. K. and Lawrence, A. B. (2006). The effect of confinement during lactation on the hypothalamic-pituitary-adrenal axis and behaviour of primiparous sows, *Physiology and Behavior* 87(2), 345–352.

Jorgenson, D. W. and Vu, K. M. (2016). The ICT revolution, world economic growth, and policy issues, *Telecommunications Policy* 40(5), 383–397.

Jun, K., Kim, S. J. and Ji, H. W. (2018). Estimating pig weights from images without constraint on posture and illumination, *Computers and Electronics in Agriculture* 153, 169–176.

Kammersgaard, T. S., Malmkvist, J. and Pedersen, L. J. (2013). Infrared thermography – a non-invasive tool to evaluate thermal status of neonatal pigs based on surface temperature, *Animal* 7(12), 2026–2034.

Karriker, L. A., Abell, C. E., Pairis-Garcia, M. D., Holt, W. A., Sun, G., Coetzee, J. F., Johnson, A. K., Hoff, S. J. and Stalder, K. J. (2013). Validation of a lameness model in sows using physiological and mechanical measurements, *Journal of Animal Science* 91(1), 130–136.

Kashiha, M., Bahr, C., Haredasht, S. A., Ott, S., Moons, C. P. H., Niewold, T. A., Ödberg, F. O. and Berckmans, D. (2013). The automatic monitoring of pigs water use by cameras, *Computers and Electronics in Agriculture* 90, 164–169.

Kashiha, M., Bahr, C., Ott, S., Moons, C. P. H., Niewold, T. A., Ödberg, F. O. and Berckmans, D. (2014). Automatic weight estimation of individual pigs using image analysis, *Computers and Electronics in Agriculture* 107, 38–44.

King, R. L., Baxter, E. M., Matheson, S. M. and Edwards, S. A. (2019). Temporary crate opening procedure affects immediate post-opening piglet mortality and sow behaviour, *Animal* 13(1), 189–197.

Kongsro, J. (2014). Estimation of pig weight using a Microsoft Kinect prototype imaging system, *Computers and Electronics in Agriculture* 109, 32–35.

Kuzuhara, Y., Kawamura, K., Yoshitoshi, R., Tamaki, T., Sugai, S., Ikegami, M., Kurokawa, Y., Obitsu, T., Okita, M., Sugino, T. and Yasuda, T. (2015). A preliminarily study for predicting body weight and milk properties in lactating Holstein cows using a three-dimensional camera system, *Computers and Electronics in Agriculture* 111, 186–193.

Labrecque, J., Gouineau, F. and Rivest, J. (2019). Real-time individual pig tracking and behavioural metrics collection with affordable security cameras. In: O'Brien, B., Hennessy, D. and Shalloo, L. (Eds), *9th European Conference on Precision Livestock Farming*, Cork, Ireland, 26-29 August 2019. The Organising Committee of the 9th European Conference on Precision Livestock Farming (ECPLF), Teagasc, Animal & Grassland Research and Innovation Centre, Moorepark, Fermoy, Co. Cork, pp. 460-466.

Lahrmann, H. P., Hansen, C. F., D'Eath, R., Busch, M. E. and Forkman, B. (2018). Tail posture predicts tail biting outbreaks at pen level in weaner pigs, *Applied Animal Behaviour Science* 200, 29-35.

Lambertz, C., Petig, M., Elkmann, A. and Gauly, M. (2015). Confinement of sows for different periods during lactation: effects on behaviour and lesions of sows and performance of piglets, *Animal* 9(8), 1373-1378.

Larsen, M. L. V., Andersen, H. M.-L. and Pedersen, L. J. (2019a). Changes in activity and object manipulation before tail damage in finisher pigs as an early detector of tail biting, *Animal* 13(5), 1037-1044.

Larsen, M. L. V., Pedersen, L. J. and Jensen, D. B. (2019b). Prediction of tail biting events in finisher pigs from automatically recorded sensor data, *Animals* 9(7), 458.

Lawrence, A. B., Newberry, R. C. and Špinka, M. (2018) 15 - .Positive welfare: what does it add to the debate over pig welfare? In: Špinka, M. (Ed.), *Advances in Pig Welfare*. Duxford: Woodhead Publishing, 415-444.

Lee, J., Jin, L., Park, D. and Chung, Y. (2016). Automatic recognition of aggressive behavior in pigs using a Kinect depth sensor, *Sensors* 16(5), 631.

Leidig, M. S., Hertrampf, B., Failing, K., Schumann, A. and Reiner, G. (2009). Pain and discomfort in male piglets during surgical castration with and without local anaesthesia as determined by vocalisation and defence behaviour, *Applied Animal Behaviour Science* 116(2-4), 174-178.

Lu, M., He, J., Chen, C., Okinda, C., Shen, M., Liu, L., Yao, W., Norton, T. and Berckmans, D. (2018). An automatic ear base temperature extraction method for top view piglet thermal image, *Computers and Electronics in Agriculture* 155, 339-347.

Madsen, T. N., Andersen, S. and Kristensen, A. R. (2005). Modelling the drinking patterns of young pigs using a state space model, *Computers and Electronics in Agriculture* 48(1), 39-61.

Madsen, T. N. and Kristensen, A. R. (2005). A model for monitoring the condition of young pigs by their drinking behaviour, *Computers and Electronics in Agriculture* 48(2), 138-154.

Maes, D. G. D., Janssens, G. P. J., Delputte, P., Lammertyn, A. and de Kruif, A. (2004). Back fat measurements in sows from three commercial pig herds: relationship with reproductive efficiency and correlation with visual body condition scores, *Livestock Production Science* 91(1-2), 57-67.

Manteuffel, G., Puppe, B. and Schön, P. C. (2004). Vocalization of farm animals as a measure of welfare, *Applied Animal Behaviour Science* 88(1-2), 163-182.

Marcet Rius, M., Pageat, P., Bienboire-Frosini, C., Teruel, E., Monneret, P., Leclercq, J., Lafont-Lecuelle, C. and Cozzi, A. (2018). Tail and ear movements as possible indicators of emotions in pigs, *Applied Animal Behaviour Science* 205, 14-18.

Marchant, J., Rudd, A., Mendl, M., Broom, D., Meredith, M., Corning, S. and Simmins, P. (2000). Timing and causes of piglet mortality in alternative and conventional farrowing systems, *Veterinary Record* 147, 214.

Martínez-Avilés, M., Fernández-Carrión, E., López García-Baones, J. M. and Sánchez-Vizcaíno, J. M. (2015). Early detection of infection in pigs through an online monitoring system, *Transboundary and Emerging Diseases* 64(2), 364–373.
Maselyne, J., Adriaens, I., Huybrechts, T., De Ketelaere, B., Millet, S., Vangeyte, J., Van Nuffel, A. and Saeys, W. (2016a). Measuring the drinking behaviour of individual pigs housed in group using radio frequency identification (RFID), *Animal* 10(9), 1557--1566.
Maselyne, J., Saeys, W., Briene, P., Mertens, K., Vangeyte, J., De Ketelaere, B., Hessel, E. F., Sonck, B. and Van Nuffel, A. (2016b). Methods to construct feeding visits from RFID registrations of growing-finishing pigs at the feed trough, *Computers and Electronics in Agriculture* 128, 9–19.
Maselyne, J., Saeys, W., De Ketelaere, B., Mertens, K., Vangeyte, J., Hessel, E. F., Millet, S. and Van Nuffel, A. (2014). Validation of a High Frequency Radio Frequency Identification (HF RFID) system for registering feeding patterns of growing-finishing pigs, *Computers and Electronics in Agriculture* 102, 10–18.
Maselyne, J., Van Nuffel, A., Briene, P., Vangeyte, J., De Ketelaere, B., Millet, S., Van den Hof, J., Maes, D. and Saeys, W. (2017). Online warning systems for individual fattening pigs based on their feeding pattern, *Biosystems Engineering* 173, 143–156.
Matthews, S. G., Miller, A. L., PlÖtz, T. and Kyriazakis, I. (2017). Automated tracking to measure behavioural changes in pigs for health and welfare monitoring, *Scientific Reports* 7(1), 1–12.
McGlone, J. J. (1985). A quantitative ethogram of aggressive and submissive behaviors in recently regrouped pigs, *Journal of Animal Science* 61(3), 556–566.
McGlone, J. J., Kelley, K. W. and Gaskins, C. T. (1980). Lithium and porcine aggression, *Journal of Animal Science* 51(2), 447–455.
McKenna, S., Amaral, T., Plötz, T. and Kyriazakis, I. (2018). Multi-part segmentation for porcine offal inspection with auto-context and adaptive atlases, *Pattern Recognition Letters* 112, 290–296.
McManus, C., Tanure, C. B., Peripolli, V., Seixas, L., Fischer, V., Gabbi, A. M., Menegassi, S. R. O., Stumpf, M. T., Kolling, G. J., Dias, E. and Costa, J. B. G. (2016). Infrared thermography in animal production: an overview, *Computers and Electronics in Agriculture* 123, 10–16.
Meijer, E., Oosterlinck, M., van Nes, A., Back, W. and van der Staay, F. J. (2014). Pressure mat analysis of naturally occurring lameness in young pigs after weaning, *BMC Veterinary Research* 10(1), 193.
Melišová, M., Illmann, G., Andersen, I. L., Vasdal, G. and Haman, J. (2011). Can sow pre-lying communication or good piglet condition prevent piglets from getting crushed?, *Applied Animal Behaviour Science* 134(3–4), 121–129.
Mohling, C. M., Johnson, A. K., Coetzee, J. F., Karriker, L. A., Abell, C. E., Millman, S. T. and Stalder, K. J. (2014). Kinematics as objective tools to evaluate lameness phases in multiparous sows, *Livestock Science* 165, 120–128.
Moinard, C., Mendl, M., Nicol, C. J. and Green, L. E. (2003). A case control study of on-farm risk factors for tail biting in pigs, *Applied Animal Behaviour Science* 81(4), 333–355.
Moura, D. J., Silva, W. T., Naas, I. A., Tolón, Y. A., Lima, K. A. O. and Vale, M. M. (2008). Real time computer stress monitoring of piglets using vocalization analysis, *Computers and Electronics in Agriculture* 64(1), 11–18.
Moustsen, V. A., Hales, J. and Hansen, C. F. (2012). Farrowing systems with temporary crating. Report of the Free Farrowing Workshop Vienna 2011.

Munsterhjelm, C., Heinonen, M. and Valros, A. (2015). Effects of clinical lameness and tail biting lesions on voluntary feed intake in growing pigs, *Livestock Science* 181(Suppl. C), 210–219.

Nasirahmadi, A., Hensel, O., Edwards, S. A. and Sturm, B. (2016). Automatic detection of mounting behaviours among pigs using image analysis, *Computers and Electronics in Agriculture* 124(Suppl. C), 295–302.

Nasirahmadi, A., Hensel, O., Edwards, S. A. and Sturm, B. (2017). A new approach for categorizing pig lying behaviour based on a Delaunay triangulation method, *Animal* 11(1), 131–139.

Norton, T., Chen, C., Larsen, M. L. V. and Berckmans, D. (2019). Review: Precision livestock farming: building 'digital representations' to bring the animals closer to the farmer, *Animal: An International Journal of Animal Bioscience* 13(12), 3009-3017.

Oczak, M. (2018). Precision Livestock Farming for sows and weaner pigs. PhD thesis, Catholic University of Leuven.

Oczak, M., Maschat, K. and Baumgartner, J. (2019). Monitoring of approaching farrowing in pens with possibility of temporary crating on the basis of ear tag acceleration data. In: O'Brien, B., Hennessy, D. and Shalloo, L. (Eds), *European Conference on Precision Livestock Farming 2019*, Cork, Ireland, 26–29 August 2019. The Organising Committee of the 9th European Conference on Precision Livestock Farming (ECPLF), Teagasc, Animal & Grassland Research and Innovation Centre, Moorepark, Fermoy, Co. Cork, pp. 322–329.

Oczak, M., Maschat, K. and Baumgartner, J. (2020). Dynamics of sows' activity housed in farrowing pens with possibility of temporary crating might indicate the time when sows should be confined in a crate before the onset of farrowing, *Animals* 10(1), 6.

Oczak, M., Maschat, K., Berckmans, D., Vranken, E. and Baumgartner, J. (2015). Classification of nest-building behaviour in non-crated farrowing sows on the basis of accelerometer data, *Biosystems Engineering* 140(Suppl. C), 48–58.

Oczak, M., Viazzi, S., Ismayilova, G., Sonoda, L. T., Roulston, N., Fels, M., Bahr, C., Hartung, J., Guarino, M., Berckmans, D. and Vranken, E. (2014). Classification of aggressive behaviour in pigs by activity index and multilayer feed forward neural network, *Biosystems Engineering* 119, 89–97.

Oliviero, C., Pastell, M., Heinonen, M., Heikkonen, J., Valros, A., Ahokas, J., Vainio, O. and Peltoniemi, O. A. T. (2008). Using movement sensors to detect the onset of farrowing, *Biosystems Engineering* 100(2), 281–285.

Pedersen, L. J., Malmkvist, J. and Andersen, H. M.-L. (2013). Housing of sows during farrowing: a review on pen design, welfare and productivity. In: Aland, A. and Banhazi, T. (Eds), *Livestock Housing: Modern Management to Ensure Optimal Health and Welfare of Farm Animals*. Wageningen: Wageningen Academic Publishers, pp. 285–297.

Pezzuolo, A., Guarino, M., Sartori, L., González, L. A. and Marinello, F. (2018). On-barn pig weight estimation based on body measurements by a Kinect v1 depth camera, *Computers and Electronics in Agriculture* 148, 29–36.

Pluym, L. M., Maes, D., Vangeyte, J., Mertens, K., Baert, J., Van Weyenberg, S., Millet, S. and Van Nuffel, A. (2013). Development of a system for automatic measurements of force and visual stance variables for objective lameness detection in sows: SowSIS, *Biosystems Engineering* 116(1), 64–74.

Pomar, C., Hauschild, L., Zhang, G.-H., Pomar, J. and Lovatto, P. A. (2009). Applying precision feeding techniques in growing-finishing pig operations, *Revista Brasileira de Zootecnia* 38(spe), 226–237.

Pomar, C. and Remus, A. (2019). Precision pig feeding: a breakthrough toward sustainability, *Animal Frontiers: The Review Magazine of Animal Agriculture* 9(2), 52–59.
Psota, E. T., Mittek, M., Pérez, L. C., Schmidt, T. and Mote, B. (2019). Multi-pig part detection and association with a fully-convolutional network, *Sensors* 19(4), 852.
Reimert, I., Bolhuis, J. E., Kemp, B. and Rodenburg, T. B. (2013). Indicators of positive and negative emotions and emotional contagion in pigs, *Physiology and Behavior* 109(Suppl. C), 42–50.
Reiners, K., Hegger, A., Hessel, E. F., Böck, S., Wendl, G. and Van den Weghe, H. F. A. (2009). Application of RFID technology using passive HF transponders for the individual identification of weaned piglets at the feed trough, *Computers and Electronics in Agriculture* 68(2), 178–184.
Royakkers, L., Timmer, J., Kool, L. and van Est, R. (2018). Societal and ethical issues of digitization, *Ethics and Information Technology* 20(2), 127–142.
Savulescu, C. (2015). Dynamics of ICT development in the EU, *Procedia Economics and Finance* 23, 513–520.
Scheel, C., Traulsen, I., Auer, W., Muller, K., Stamer, E. and Krieter, J. (2017). Detecting lameness in sows from ear tag-sampled acceleration data using wavelets, *Animal* 11(11), 2076–2083.
Schon, P., Puppe, B. and Manteuffel, G. (2004). Automated recording of stress vocalisations as a tool to document impaired welfare in pigs, *Animal Welfare-Potters Bar Then Wheathampstead* 13(2), 105–110.
Shao, B. and Xin, H. (2008). A real-time computer vision assessment and control of thermal comfort for group-housed pigs, *Computers and Electronics in Agriculture* 62(1), 15–21.
Shi, C., Teng, G. and Li, Z. (2016). An approach of pig weight estimation using binocular stereo system based on LabVIEW, *Computers and Electronics in Agriculture* 129, 37–43.
Singh, C., Verdon, M., Cronin, G. M. and Hemsworth, P. H. (2017). The behaviour and welfare of sows and piglets in farrowing crates or lactation pens, *Animal* 11(7), 1210–1221.
Soerensen, D. D. and Pedersen, L. J. (2015). Infrared skin temperature measurements for monitoring health in pigs: a review, *Acta Veterinaria Scandinavica* 57(1), 5.
Statham, P., Green, L., Bichard, M. and Mendl, M. (2009). Predicting tail-biting from behaviour of pigs prior to outbreaks, *Applied Animal Behaviour Science* 121(3–4), 157–164.
Stavrakakis, S., Li, W., Guy, J. H., Morgan, G., Ushaw, G., Johnson, G. R. and Edwards, S. A. (2015). Validity of the Microsoft Kinect sensor for assessment of normal walking patterns in pigs, *Computers and Electronics in Agriculture* 117(Suppl. C), 1–7.
Stevenson, P. (2017). Precision livestock farming: could it drive the livestock sector in the wrong direction. Compassion in World Farming. Available at: https://www.ciwf.org.uk/media/7431928/plf-could-it-drive-the-livestock-sector-in-the-wrong-direction.pdf.
Šustr, P., Špinka, M., Cloutier, S. and Newberry, R. C. (2001). Computer-aided method for calculating animal configurations during social interactions from two-dimensional coordinates of color-marked body parts, *Behavior Research Methods, Instruments, and Computers* 33(3), 364–370.
Taylor, A. A. and Weary, D. M. (2000). Vocal responses of piglets to castration: identifying procedural sources of pain, *Applied Animal Behaviour Science* 70(1), 17–26.

Tencent (2018). Tencent ML images. Available at: https://github.com/Tencent/tencent-ml-images/tree/master/data.

Thomsen, R., Bonde, M., Kongsted, A. G. and Rousing, T. (2012). Welfare of entire males and females in organic pig production when reared in single-sex groups, *Livestock Science* 149(1-2), 118-127.

Traulsen, I., Breitenberger, S., Auer, W., Stamer, E., Muller, K. and Krieter, J. (2016). Automatic detection of lameness in gestating group-housed sows using positioning and acceleration measurements, *Animal* 10(6), 970-977.

Traulsen, I., Scheel, C., Auer, W., Burfeind, O. and Krieter, J. (2018). Using acceleration data to automatically detect the onset of farrowing in sows, *Sensors* 18(1), 170.

Turner, S. P., Farnworth, M. J., White, I. M. S., Brotherstone, S., Mendl, M., Knap, P., Penny, P. and Lawrence, A. B. (2006). The accumulation of skin lesions and their use as a predictor of individual aggressiveness in pigs, *Applied Animal Behaviour Science* 96(3-4), 245-259.

Ursinus, W. W., Van Reenen, C. G., Kemp, B. and Bolhuis, J. E. (2014). Tail biting behaviour and tail damage in pigs and the relationship with general behaviour: predicting the inevitable?, *Applied Animal Behaviour Science* 156, 22-36.

Valletta, J. J., Torney, C., Kings, M., Thornton, A. and Madden, J. (2017). Applications of machine learning in animal behaviour studies, *Animal Behaviour* 124, 203-220.

van Putten, G. (1969). An investigation into tail-biting among fattening pigs, *British Veterinary Journal* 125(10), 511-517.

Verdouw, C., Wolfert, S. and Tekinerdogan, B. (2016). Internet of Things in agriculture, *CAB Reviews: Perspectives in Agriculture, Veterinary Science, Nutrition and Natural Resources* 11(35), 1-12.

Viazzi, S., Ismayilova, G., Oczak, M., Sonoda, L. T., Fels, M., Guarino, M., Vranken, E., Hartung, J., Bahr, C. and Berckmans, D. (2014). Image feature extraction for classification of aggressive interactions among pigs, *Computers and Electronics in Agriculture* 104, 57-62.

Viscardi, A. V., Hunniford, M., Lawlis, P., Leach, M. and Turner, P. V. (2017). Development of a piglet grimace scale to evaluate piglet pain using facial expressions following castration and tail docking: a pilot study, *Frontiers in Veterinary Science* 4, 51.

Vranken, E. and Berckmans, D. (2017). Precision livestock farming for pigs, *Animal Frontiers: The Review Magazine of Animal Agriculture* 7(1), 32-37.

Wallenbeck, A. and Keeling, L. J. (2013). Using data from electronic feeders on visit frequency and feed consumption to indicate tail biting outbreaks in commercial pig production, *Journal of Animal Science* 91(6), 2879-2884.

Wang, K., Guo, H., Ma, Q., Su, W., Chen, L. and Zhu, D. (2018). A portable and automatic Xtion-based measurement system for pig body size, *Computers and Electronics in Agriculture* 148, 291-298.

Wang, X., Zhao, X., He, Y. and Wang, K. (2019). Cough sound analysis to assess air quality in commercial weaner barns, *Computers and Electronics in Agriculture* 160, 8-13.

Wathes, C. M., Kristensen, H. H., Aerts, J.-M. and Berckmans, D. (2008). Is precision livestock farming an engineer's daydream or nightmare, an animal's friend or foe, and a farmer's panacea or pitfall?, *Computers and Electronics in Agriculture* 64(1), 2-10.

Weary, D. M., Braithwaite, L. A. and Fraser, D. (1998). Vocal response to pain in piglets, *Applied Animal Behaviour Science* 56(2-4), 161-172.

Weber, A., Salau, J., Haas, J. H., Junge, W., Bauer, U., Harms, J., Suhr, O., Schönrock, K., Rothfuß, H., Bieletzki, S. and Thaller, G. (2014). Estimation of backfat thickness using extracted traits from an automatic 3D optical system in lactating Holstein-Friesian cows, *Livestock Science* 165, 129–137.

Wechsler, B. (1995). Coping and coping strategies: a behavioural view, *Applied Animal Behaviour Science* 43(2), 123–134.

Wedin, M., Baxter, E. M., Jack, M., Futro, A. and D'Eath, R. B. (2018). Early indicators of tail biting outbreaks in pigs, *Applied Animal Behaviour Science* 208, 7–13.

WelfareQuality® (2009). *Welfare Quality Assessment Protocol for Pigs (Sows and Piglets, Growing and Finishing Pigs)*. Lelystad: WelfareQuality® Consortium, 122.

Werkheiser, I. (2018). Precision livestock farming and farmers' duties to livestock, *Journal of Agricultural and Environmental Ethics* 31(2), 181–195.

Wischner, D., Kemper, N. and Krieter, J. (2009). Nest-building behaviour in sows and consequences for pig husbandry, *Livestock Science* 124(1–3), 1–8.

Xin, H. (1999). Assessing swine thermal comfort by image analysis of postural behaviors, *Journal of Animal Science* 77(Suppl. 2), 1–9.

Yeates, J. W. and Main, D. C. J. (2008). Assessment of positive welfare: a review, *The Veterinary Journal* 175(3), 293–300.

Young, P. C. (1984). *Recursive Estimation and Time-Series Analysis: An Introduction*. Chicago, IL: Springer Science & Business Media.

Young, P. C. (2011). *Recursive Estimation and Time-Series Analysis: An Introduction for the Student and Practitioner*. Berlin: Springer Science & Business Media.

Zonderland, J. J., Schepers, F., Bracke, M. B. M., den Hartog, L. A., Kemp, B. and Spoolder, H. A. M. (2011). Characteristics of biter and victim piglets apparent before a tail-biting outbreak, *Animal* 5(5), 767–775.

Zonderland, J. J., Vermeer, H., Vereijken, P. and Spoolder, H. (2003). Measuring a pig's preference for suspended toys by using an automated recording technique. *International Commission of Agricultural and Biosystems Engineering (CIGR) E-Journal* 5.

Index

95% spectral edge frequency (F95) 292

ACTH. *see* Adrenocorticotropic hormone (ACTH)
Actinobacillus pleuropneumonia (APP) 231
Acute-phase proteins (APP) 368, 434
ADFI. *see* Average daily feed intake (ADFI)
ADG. *see* Average daily gain (ADG)
Ad libitum 321
'Ad libitum' substrate 149
Adrenocorticotropic hormone (ACTH) 34, 294-296, 432
African Swine Fever (ASF) 209
Aggressive fighting 195
AI. *see* Artificial insemination (AI)
All-in-all-out (AIAO) management system 131, 215
Alternatives to castration
  advantages and disadvantages of 345
  case study 342-343
  entire male production 317-318
    advantages of 318
    boar taint detection, genetics and methods 328-331
    consumer acceptance of pork 319-320
    management strategies 323-327
    nutritional strategies 320-323
  immunocastration
    as alternative for highquality/organic production systems 340-341
    effects of, on boar taint 334-337
    farmers' acceptance 338-339
    mechanism of action 331-334
    pig production systems, sustainability of 341-342
    reliability and efficacy 337-338
    societal acceptance 339-340
  overview of 315-317
  semen, sex sorting of 342
Amino acid tryptophan 368
Androstenone
  level 331
  metabolism 321
Anecdotal experience 106
Animal welfare 106
  audit protocols 265-267
Anorexia 435
Anoxic gases exposure 264-265
ANS. *see* Autonomic nervous system (ANS)
Anti-aggressive drugs 99
Antibiotic-free systems 446
APP. *see Actinobacillus pleuropneumonia* (APP); Acute-phase proteins (APP)
Artificial insemination (AI) 104
ASF. *see* African Swine Fever (ASF)
ASF virus (ASFV) 210
Aspartate aminotransferase (AST) 434
Australian Model Code of Practice 103
Australian Royal Society for the Protection of Cruelty against Animals (RSPCA) 408
Automatic feeders 6
Autonomic nervous system (ANS) 290-291, 473, 474
Average daily feed intake (ADFI) 227
Average daily gain (ADG) 222

Barks 464
Barren standard housing 402
BCS. *see* Body condition score (BCS)
BDNF. *see* Brain-derived neurotrophic factor (BDNF)
Bedding 197
Behavioral responses to disease
  abnormal behaviors as a byproduct 439
  overview of 431-432
  sickness behaviors 435-437
    lameness, chronic disease 438-439
    swine influenza, acute disease 437-438

Best linear unbiased prediction (BLUP) 16
Blood platelet serotonin 368
BLUP. *see* Best linear unbiased prediction (BLUP)
Boar taint, immunocastration effects 315, 316, 334
  behaviour and welfare 336–337
  growth, performance parameters and carcass and meat characteristics 335–336
Body condition score (BCS) 231, 528–529
Brain-derived neurotrophic factor (BDNF) 56
Brain oxytocin 476
Breeding and gestation, pig welfare optimisation
  animal management
    animal abuse and cruelty 106
    minimising handling stress 107–108
    stockmanship 106–107
  boar 103–105
  boar exposure 88–89
  future trends 108–110
  housing gilts and sows 89–90
    fully or partially slatted, breeding pigs 102–103
    gestating sows hunger 101–102
    individual sow characteristics 103
    mixing time 99–100
    non-bedded and nonenriched environments, breeding pigs 102–103
    pen design 90–97
    specific mixing pen 97–99
    static and dynamic groups 101
  oestrus detection 88–89
  overview 87–88
  puberty stimulation 88–89

Camera-based automated systems 533
Castrated males (CM) 321
CBG. *see* Cortico-binding globulin (CBG)
Central neuropathic pain 289
Chronic/persistent pain 286
CITD systems 540–541
CK. *see* Creatine kinase (CK)
CM. *see* Castrated males (CM); Computed tomography (CM)
CO_2 exposure 263
Common Swine Industry Audit (CSIA) 232, 434
Computed tomography (CT)
  of hams 343, 344
Core affect model 456, 457
Cortico-binding globulin (CBG) 295
Corticotrophin-releasing factor (CRF) 177
Cortisol level 474, 475
Creatine kinase (CK) 434
Creep feed 60
CRF. *see* Corticotrophin-releasing factor (CRF)
Cross-fostering 130
CSIA. *see* Common Swine Industry Audit (CSIA)
Cul-de-sac arrangement 140
CXCL8 293
Cytokines 433

Delaunay triangulation (DT) 532
Developmental influences, pig behaviour
  future trends 65–66
  overview 27–28
  postnatal period environmental effects 50–51
    appetitive and ingestive behaviours 60–61
    early social environment influence 56–60
    human interaction 62–64
    pre-weaning environment complexity 51–56
    sow and piglets interaction 61–62
  prenatal environmental effects 28
    heat stress 47–48
    infectious diseases 48–49
    maternal nutrition (*see* Maternal nutrition)
    post-conception environment effects, piglet behaviour 31–34
    pre-conception environment effects, offspring behaviour 29–31
    prenatal housing conditions (*see* Prenatal housing conditions)
    sow management (*see* Sow management)
DHA. *see* Docosahexaenoic acid (DHA)
Digital revolution 525
Directive 2001/93/EC stipulates 282
Disease mitigation, strategies
  management
    early detection 443–445
    hospital pens 445–446
    pharmaceutical interventions for 446–447
    timely euthanasia 447–448

prevention
environment 443
physical biosecurity measures 440–443
vaccination 440
Docosahexaenoic acid (DHA) 43
DT. *see* Delaunay triangulation (DT)

EAAP. *see* European Federation for Animal Science (EAAP)
Early-life social effects (ELSEs) 18
EEG. *see* Electroencephalographic (EEG)
EEG median frequency (F50) 292
Electrical stunning 260, 263
Electroencephalographic (EEG) 291, 292
Electronic sow feeder (ESF) 33
Elevated Plus Maze test 460
ELSEs. *see* Early-life social effects (ELSEs)
Emotion assessing in pigs, negative and positive mental states
behavioural indicators
behavioural tests 459–462
defence cascade responses 467
facial expression and body posture 467–470
play behaviour 465–466
qualitative behaviour assessment 462–463
vocalisations 463–465
cognitive indicators, judgement bias 470–473
definitions and concepts 455–458
future trends 477–480
neurophysiological indicators
negative emotions, physiology of 473–476
positive emotions, physiology of 476–477
Enrichment role, pig behaviour and welfare
artificial objects as enrichment 406–408
cognitive enrichment 415–417
environmental enrichment and animal welfare 404–406
evaluating biological relevance of 416–418
organic materials as enrichment 408–411
overview of 401–403
social enrichment in captive settings 411–415
ESF. *see* Electronic sow feeder (ESF)
EthoVision® tracking system 535
EU directive 2008/120/EC 402
EU mandate 401
EU Regulation 264
European Federation for Animal Science (EAAP) 19
European Union (EU) 446
European Union Commission Directive 2001/93/EC 102
Euthanasia 447
EU welfare legislation 315
Extrinsic risk factors
feeding 374–375
health status, farm 377
housing and climate 375–376
manipulable materials 373–374
EYeScan 529

F95. *see* 95% spectral edge frequency (F95)
Facial action units 468
Farm animal welfare council (FAWC) 498
Farrowing and lactation, sow and piglet welfare optimisation 122–123
future trends 161–162
nutritional inputs optimisation
ensuring optimum colostrum intake, piglets 128–129
ensuring optimum milk intake 129–131
large litter management 129–131
pre-weaning solid feed intake 131–132
sow nutrition 126–128
optimal housing design, biological specifications 132–137
flooring 142–143
general system components 144–146
space 133, 138
substrate 138–140
thermal comfort 143–144
walls 140–142
overview 121–122
piglet welfare challenges 123–124
PigSAFE, zero confinement farrowing and lactation system 146–147
commercial developments 154–161
commercial trials and performance 150–154
design 147–148
initial trials 148–150
sow welfare challenges 124–126
Farrowing crates 121
Farrowing location and performance
sound-proofing and substrate 149–150
temperature and nest size 150
Fatty acid profile 335

FAWC. *see* Farm animal welfare council (FAWC)
FCI. *see* Free cortisol index (FCI)
Feeding system 94–95
Finnish survey 387
First-order nociceptive neurons synapse 286, 287
Five Freedoms 211
Flooring 95–97, 124
Floor space 90–93
Free cortisol index (FCI) 295

GABRR2 16
Gene ontology (GO) 293
Genetic selection 109
Genetics of pig behaviour
- aggressive behaviour between pigs 9–11
- characteristics, high social breeding values for growth rate 14–16
- feeding behaviour 6–7
- future trends 16–19
- growth rate and reproduction traits, direct and social effects 11–14
- maternal behaviour 4–6
- overview 3–4
- selection consequences, feed efficiency 6–7
- tail biting 8–9

Gentle handling 229
GnRH. *see* Gonadotropin-releasing hormone
GO. *see* Gene ontology (GO)
Go/Go tasks 471
Gonadotropin-releasing hormone (GnRH) 331, 332
Go/No-Go style tasks 471
Group composition/mixing 324–325
Group model 11, 12
Group size 93–94
Growing and finishing stage, pig welfare optimization
- animal health and animal care 230–231
  - culling 231–232
  - euthanasia 232
  - hospital/sick pens 231
- future trends 232–234
- group size 222–223
- living conditions, indoor systems
  - air quality 225–226
  - lighting 228–229
  - lying comfort 227–228
  - space allowance 226–227
  - stockmanship 229–230
  - temperature 223–224
- overview 209–212
- pig behavioral issues, pig welfare impact 212
  - aggression 214–215
  - belly nosing 213–214
  - oral-nasal-facial/oral behaviors 212–213
  - tail biting 213
- production systems 215–216
  - bedded indoor and outdoor systems 218–221
  - flooring 217–218
  - grow-finish pigs, alternative housing 216–217
  - hoop buildings 221–222

Grunts 464

HAT. *see* Husbandry Advisory Tool (HAT)
HCA. *see* Hydrocorticosterone acetate (HCA)
Heart rate variability (HRV) 474, 476
HFHS diets 40
Hormonal feedback system 332
Hospital pens 231
HPA. *see* Hypothalamic-pituitary-adrenal (HPA) axis
HRV. *see* Heart rate variability (HRV)
Human Approach test 461
Human nose method 329
Husbandry Advisory Tool (HAT) 384
Hydrocorticosterone acetate (HCA) 34
Hyperglycemia 432
Hypothalamic-pituitary-adrenal (HPA) axis 9, 29, 432, 433, 475

IASP. *see* International Association for the Study of Pain (IASP)
IAV infections 438
IF144. *see* Interferon-induced protein 44 (IF144)
Immunocastrated pigs (IM) 321
Indoor production systems 102
Inflammatory pain 286
Instrumental methods 329
Intensive production systems 211
Interferon-induced protein 44 (IF144) 16
International Association for the Study of Pain (IASP) 289
Intrauterine growth retardation (IUGR) 39, 124
Intrinsic risk factors
- age and previous experience 379–380
- breeding and genetics 378–379
- sex 379

Invasive management procedures, in piglets
behavioural evidence of pain 297-298
castration 298-300
tail docking 300-302
tooth resection 302-303
future trends 303-304
hormonal and metabolic evidence of pain
castration 294-296
tail docking 296-297
tooth resection 297
neural evidence of pain
castration 291-292
tooth resection 293-294
overview of 281-282
pain
anatomy and physiology of 289-291
characterisation of 287-289
definition of 284-285
in neonates 291
terminology 285-287
surgical castration 282-283
tail docking 283
tooth resection 283-284
IUGR. *see* Intrauterine growth retardation (IUGR)

Kalman filtering and fixed interval smoothing (KALMSMO) algorithm 543

Lameness 96, 109
LC-PUFA. *see* Long-chain polyunsaturated fatty acid (LC-PUFA)
Light/Dark test 460
Lipopolysaccharide (LPS) 433, 435
Locomotor play 54
Long-chain polyunsaturated fatty acid (LC-PUFA) 43
LPS. *see* Lipopolysaccharide (LPS)

Machine learning algorithms 541-542
Major histocompatibility complex two (MHCII) 48
Malaise 230
Marker-assisted selection (MAS) 16
Market access 234
MAS. *see* Marker-assisted selection (MAS)
Maternal immune activation (MIA) 48
Maternal nutrition
diet formulation and supplementation 40-44
feeding regime 38-40
flavoured diets 44-46
Metabolised energy (ME) intake 192
MHCII. *see* Major histocompatibility complex two (MHCII)
MIA. *see* Maternal immune activation (MIA)
Milk premolars 284
Monitoring technologies
case study 542-546
information and communication technology (ICT) 525-526
legal and ethical considerations 537-539
livestock production processes, condition 526-527
precision livestock farming (PLF) 525-527 (*see also* Precision livestock farming (PLF))
sensor data management and modelling 539-542
Musculoskeletal problems 95
Mycoplasma hyopneumoniae 231
Mylinated A-delta (Aδ) fibres 290

NAGRP. *see* National Animal Genome Research Program (NAGRP)
NAMI audit protocol 266-267
National Animal Genome Research Program (NAGRP) 328
NCM. *see* Uncastrated males (NCM)
NEFA. *see* Non-esterified fatty acid (NEFA)
Nest building 4, 124, 125
Neuropathic pain 289
The New Zealand Code of Welfare 104
Nociception 285
Nociceptive pain 285-286
Nociceptors 290
Non-esterified fatty acid (NEFA) 47
Nonsteroidal anti-inflammatory drug (NSAID) 127
Novel object test 460
NSAID. *see* Nonsteroidal anti-inflammatory drug (NSAID)

ONF. *see* Oral-nasal-facial (ONF) behaviors
Open field test 459, 460
Optical flow system 386
Optimising welfare
audit protocols 265-267
future trends 267-268
lairage 256
ambient control 257-258
mixing 256-257
slaughter driving 258-259
time 256
stunning and slaughter 260
exsanguination techniques 264

restrainer types 260-262
stunning methods 262-264
transport 248
animal position 250
culled sows 255-256
description 247-248
journey time 252-253
microclimate control 250-252
space allowance 253-255
vehicle design 248-250
Oral-nasal-facial (ONF) behaviors 212

PAACO. *see* Professional Animal Auditor Association (PAACO)
PAS. *see* Punishment Avoidance (PAS)
PEDv. *see* Porcine epidemic diarrhea virus (PEDv)
Pen cleanliness 186
Peripheral nervous system (PNS) 289
PHS. *see* Prenatal heat stress (PHS)
Physiological responses to disease
assessing welfare 434-435
immune system activation 433-434
maintaining homeostasis 432-433
overview of 431-432
Piétrain-sired pigs 334
Pig inventory 232
Piglet and Sow Alternative Farrowing Environment (PigSAFE) 147
Piglet castration 343
Piglet survival 151-152
Pig production 225
PigSAFE. *see* Piglet and Sow Alternative Farrowing Environment (PigSAFE)
Plasma cortisol 295, 296
PLF. *see* Precision livestock farming (PLF)
PNS. *see* Peripheral nervous system (PNS)
Porcine epidemic diarrhea virus (PEDv) 217
Porcine reproductive and respiratory syndrome (PRRS) 216
Porcine reproductive and respiratory syndrome virus (PRRSv) 48, 441, 443
Porcine respiratory disease complex (PRDC) 231
Pork Quality Assurance Plus Program 231
Posterior lesions 11
PRDC. *see* Porcine respiratory disease complex (PRDC)
Precision livestock farming (PLF) 386, 389
appropriate behaviour 535-537
good feeding 529-530
good health 532-535
good housing 531-532
legal and ethical considerations 537-539
sensor data management and modelling 539-542
Predictive adaptive responses 31
Prenatal heat stress (PHS) 47
Prenatal housing conditions
environmental enrichment 38
gestation stalls 34
group housing 34-38
Pre-slaughter conditions 327
Professional Animal Auditor Association (PAACO) 234
PRRS. *see* Porcine reproductive and respiratory syndrome (PRRS)
PRRSv. *see* Porcine reproductive and respiratory syndrome virus (PRRSv)
Punishment Avoidance (PAS) 458

QBA. *see* Qualitative behaviour assessment (QBA)
QTL. *see* Quantitative trait loci (QTL)
Qualitative behaviour assessment (QBA) 462-463, 478
Quantitative trait loci (QTL) 7
Quarantine systems 441

Radio-frequency identification (RFID) technology 530
RAS. *see* Reward Acquisition (RAS)
Recurrent neural networks (RNN) 542
Re-grouping test 15
Residual feed intake (RFI) 7
Resource monopolization 223
Respiratory disease 230
Reward Acquisition (RAS) 458
RFI. *see* Residual feed intake (RFI)
RFID. *see* Radio-frequency identification (RFID) technology
RNN. *see* Recurrent neural networks (RNN)

SAM. *see* Sympathetic adrenomedullary (SAM) system
Schwanzbeiß-Interventions-Programm (SchwIP) 384
Screams 464
Sensory methods 329
Serotonergic system 367
Sickness behaviour 369
Skatole level 331
Skin lesions 11
Slatted concrete floors 217

Slaughter weight/age 327
SNS. *see* Sympathetic nervous system (SNS)
Somatic pain 287-288
Somatosensory nervous system 289-290
Sow and piglet weight and condition 153-154
Sow management
  auditory cues 49-50
  handling 49
Sows 4
Space allowance 226
Squeals 464, 478
Startle behaviour 478
Stress hormone 474
Subcutaneous injections 332
Sufficient quantity 139
Surgical castration 295, 298, 300, 315
Sympathetic adrenomedullary (SAM) system 475
Sympathetic nervous system (SNS) 432, 433

Tail biting, understanding and prevention of
  ethical considerations associated with 381-382
    managed by tail docking 382-383
    producers motivated to avoid docking 383
  future trends 390
  intervention measures for 386-389
  managing pigs without 383-386
  mechanistic understanding of 366-371
  overview of 361-362
  prevalence and economic importance of problem 363-366
  risk factors 371-372
    acute risk factors 377-378
    extrinsic risk factors (*see* Extrinsic risk factors)
    intrinsic risk factors (*see* Intrinsic risk factors)
Tail docking 19, 292-293, 381
Tail-in-mouth behaviour 378
TAINTSTOP 322
3D camera system 386
Timely Euthanasia 444
Traditional approach 531
Trans-natal sensory continuity 50
Transportation of pigs 247-248
Tryptophan 369

Uncastrated males (NCM) 321
Unmyelinated C fibres 290
US Industry Guidelines 401

Ventilation system 224
Visceral pain 288-289
Vocalisations 460

Weaning and nursery stage, welfare optimisation
  future trends 199-200
  nursery pigs and key challenges, welfare optimisation 179-181
  optimised management, human key factor 198
  overview 177-179
  weaning impact minimisation
    elevated platform implementation, nursery unit 189-191
    environmental stress, reduce possible welfare impacts 186-189
    nutritional stress, feed and water provision optimisation 191-194
    procedures and facilities preparation 181-186
    social stress, lower welfare impacts approaches 195-198
Welfare assessment
  behavioural welfare indicators
    affiliative 511
    aggression 511
    fear response 512
    nest building 512
  comparisons and frameworks 498-499
  consequence associated 504-505, 508, 510, 512, 514-515
  description 497-498
  feeding welfare indicators
    body condition score 503
    drinkers, number of 503-504
    drinkers, types of 504
    roughage 503
    water quality 504
    weaning age 503
  future trends 515
  health and pain welfare indicators
    castration 510-511
    enteric problems 509
    grimace scale 509-510
    hernia 509
    lameness 507
    metritis and mastitis 509
    mortality 508
    nose ringing 510
    prolapse 509
    respiratory problems 508-509

tail docking 510
teeth clipping/grinding 510-511
vulva lesions 508
wounds on body 507-508
housing welfare indicators
air quality 506
crates 505-506
enrichment 506-507
manure 507
outdoor access 507
quality of flooring 505
stocking density 504-505
thermal comfort 506
illustration of 500
input and output measures 499-501
poor, general indicators of
runts 513
stereotypic behaviour 513
tail and ear biting 512-513
tear staining 513
positive, general indicators of
iceberg indicators 514-515
play 514
tail wagging 514
validity, repeatability and feasibility
measures 501-503
Welfare protocol 497
Welfare Quality criteria 499
Welfare quality protocols 463
WQ® protocol 266-267

www.ingramcontent.com/pod-product-compliance
Ingram Content Group UK Ltd.
Pitfield, Milton Keynes, MK11 3LW, UK
UKHW021008120926
8556IPUK00028B/139

9 781786 764430